MS&A

Modeling, Simulation and Applications

Volume 20

More information about this series at https://link.springer.com/bookseries/8377

Eduard Feireisl · Mária Lukáčová-Medviďová ·
Hana Mizerová · Bangwei She

Numerical Analysis of Compressible Fluid Flows

Eduard Feireisl
Institute of Mathematics
Czech Academy of Sciences
Praha, Czech Republic

Mária Lukáčová-Medvid'ová
Institute of Mathematics
Johannes Gutenberg University of Mainz
Mainz, Germany

Hana Mizerová
Department of Mathematical Analysis
and Numerical Mathematics
Comenius University in Bratislava
Bratislava, Slovakia

Bangwei She
Institute of Mathematics
Czech Academy of Sciences
Praha, Czech Republic

ISSN 2037-5255 ISSN 2037-5263 (electronic)
MS&A
ISBN 978-3-030-73787-0 ISBN 978-3-030-73788-7 (eBook)
https://doi.org/10.1007/978-3-030-73788-7

The cover picture depicts vortex structures arising in the Kelvin-Helmholtz problem. Computation was done by the second order finite volume method based on the generalized Riemann problem.

This Springer imprint is published by the registered company Springer Nature Switzerland AG
The registered company address is: Gewerbestrasse 11, 6330 Cham, Switzerland

To our children and grandchildren

Jana, Jan, Lucie, and Antonie, Benedikt

Janka

Juraj

Hanička (奕秋), 牧泽

Preface

Many real-world problems involve fluids in motion. The goal of this book is to propose a new approach to numerical analysis of the underlying *nonlinear* equations in the spirit of the celebrated *Lax equivalence theorem* stated originally by Lax [149] in the context of *linear* problems:

$$\textbf{stability} + \textbf{consistency} \quad \Leftrightarrow \quad \textbf{convergence}\,.$$

In the framework of numerical analysis, *stability* means uniform bounds on approximate solutions independent of the level of approximation (the numerical time and space step), while *consistency* means that the approximate solutions satisfy the target system of equations modulo a consistency error (truncation error) that vanishes for vanishing time and space steps of the numerical approximation.

The problem of *convergence* is quite subtle, in view of the recent results revealing the existence of *oscillations* in sequences of approximate solutions, see Chiodaroli [46], De Lellis and Székelyhidi [62, 81], Fjordholm et al. [105–107]. Accordingly, the *weak* convergence, or the convergence in terms of integral averages, seems relevant to describe the asymptotic behavior of approximate solutions at least in the context of inviscid fluids. As is well known, the weak convergence does not commute with nonlinear superposition operators which makes the asymptotic analysis rather delicate. Oscillatory sequences of approximate (numerical) solutions are not only difficult to capture by the real computational software but present mostly insurmountable difficulties even at the theoretical level. The limit objects resulting from weakly converging sequences do not coincide with distributional solutions of the target problem unless the latter is regular in which case the convergence must be strong. In fact, a similar result can be extended even to the larger class of weak admissible solutions, at least in the context of the Euler system:

$$\textbf{the limit is a weak solution of the target problem}$$
$$\Leftrightarrow \textbf{approximate solutions converge strongly},$$

see [83].

Nowadays, it is also well understood that the particular form of the equations that might seem irrelevant for the mathematical theory can be crucial for convergence of numerical methods. The use of a particular set of equations may lead to successful results, whereas alternate forms may cause oscillations or even instability.

The approach to nonlinear problems proposed in this monograph is based on the following fundamental concepts:

- **Dissipative and/or measure-valued solution** to nonlinear systems of partial differential equations that can be seen as a natural closure of the solution set with respect to the weak topology. The dissipative solutions satisfy the underlying set of differential equations modulo a perturbation related to possible oscillations in the approximate sequence. This class of solutions is large enough to accommodate all possible limits of *consistent and stable* numerical schemes. Measure-valued refers to the Young measure that is used to describe possible oscillations in the approximate sequence. Very roughly indeed, the dissipative solutions are numerical functions—expected values (or barycenters) of the associated Young measure. In particular, they reflect the observable (and computable) properties of solutions to a given problem, while the Young measure may sensitively depend on the way the problem is being approximated.
- **Weak–strong uniqueness principle** that asserts that a generalized dissipative or measure-valued solution coincides with a regular solution as long as the latter exists. This principle is quite useful in the context of problems concerning viscous fluids, where regular solutions are likely to exist under very mild assumptions on the initial data and the length of the time interval. It also gives rise to convincing convergence results in the regime, where the target system is not expected to admit spurious (weak) solutions.
- $\mathcal{K}$**–convergence** replacing weakly converging sequences of approximate solutions by their Cesàro averages that converge strongly to the corresponding asymptotic limit. The same technique is available also for the associated Young measures yielding a reliable description of possible oscillations. In such a way, weakly converging sequences can be visualized via the numerical approximations.

In accordance with the above general principles, we perform numerical analysis of the Euler and Navier–Stokes systems describing the time evolution of compressible inviscid and (linearly) viscous fluids, respectively. The numerical analysis is carried out in several steps:

- Choosing a **numerical scheme** that gives rise to a *dissipative* solution in the asymptotic limit. Typically, these are the schemes preserving some important physical properties of the underlying continuous system, such as the positivity of density and internal energy as well as the second law of thermodynamics. Such numerical schemes belong to the class of the so-called *invariant domain-preserving methods* or *structure-preserving methods*. Note that it is crucial that a suitable form of the discrete energy or entropy inequality holds at the level of numerical approximations. Correspondingly, numerical methods satisfying the latter inequalities are called energy dissipative or entropy stable methods.

- Showing **stability and consistency** of the scheme. Here stability means that there are suitable uniform bounds that render an approximating sequence precompact at least in the weak topology. Consistency means that the target system of equations is satisfied modulo a local truncation error that vanishes in the asymptotic limit.
- Performing the **limit** for vanishing numerical step obtaining a generalized solution of the problem. At this stage, we use the **weak–strong uniqueness principle** to conclude that either **(i)** the limit system admits a unique regular solution and the approximate solutions converge unconditionally (pointwise) to it, or **(ii)** the limit is a generalized (dissipative, measure-valued) solution of the target system.
- If the limit is a generalized solution, we employ the $\mathcal{K}$**–convergence**, i.e., the convergence in the sense of Cesàro averages, to identify possible oscillations (Young measure) associated to the sequence of numerical solutions.

As we shall see, at least in the context of inviscid fluids described via the Euler system, there are basically two ways how the approximate numerical solutions may approach the target system:

- Either the convergence is strong (pointwise a.a.) and then the limit is a weak solution of the limit system;
- or the convergence is weak (oscillatory) and then the limit is a generalized (measure-valued solution) of the limit system.

Both alternatives may occur and must be analyzed in a rather different manner. Rather surprisingly, the above scenarios are complementary at least in the context of compressible fluids studied in this book. By complementary, we mean that either the convergence is strong or the limit *cannot* be a weak solution of the target problem, cf. [83].

We should always keep in mind the fundamental difference between the effective *limit* of a sequence of approximate (numerical) solutions and the way how this limit is *attained*. In particular, the concept of Young measures amply used in the text below reflects the way how the solution is obtained (oscillations in the approximate sequence) rather than being an intrinsic property of the limit system. There could be different ways to attain the same limit (weak or strong) as well as different limits obtained via similar methods. Identifying a proper limit that would be physically admissible is one of the most challenging theoretical problems, in particular in light of the recent ill-posedness results provided by the method of convex integration, see Buckmaster et al. [34, 35, 38] among others.

The material treated in the book is divided into three major parts:

- **Part I [State-of-the-art]**
  Mathematical theory of the governing equations of fluid dynamics has a long history and a very active present. In spite of a concerted effort of generations of excellent mathematicians, the fundamental problems on solvability and well-posedness of the Navier–Stokes and Euler systems remain open, cf. the survey by Fefferman [79]. Recently, De Lellis and Székelyhidi launched an ambitious program to attack these issues in a new way based on the method of convex integration, cf. Buckmaster and Vicol [37] and the references cited therein. The resulting

new discoveries culminated so far in the final proof of Onsager's conjecture by Isett [131] and a remarkable nonuniqueness result for the Navier–Stokes system by Buckmaster and Vicol [38]. The message conveyed to the theory is that we must be extremely careful concerning a proper choice of the concept of generalized solutions in fluid dynamics including the underlying phase space of relevant observable variables. Our goal was to collect the up-to-date piece of information concerning the mathematical models of compressible viscous and inviscid fluid flows. We are well aware that detailed proofs of many of the mathematical statements presented would require a comprehensive preliminary material that goes beyond the scope of the present book. We therefore content ourselves with stating the relevant theorems and discussing only the highlights of the underlying ideas.

- **Part II [Generalized solutions]**
  There is a large piece of evidence and also hope that the physically relevant solutions of the (mathematically) ill-posed problems can be identified as particular limits of more complex (viscous) models, cf. Chen and Glimm [43]. Adopting the same philosophy, we identify generalized solutions with asymptotic limits of consistent approximations. These turn out to fit in a possibly larger class of generalized dissipative solutions. The dissipative solutions, however, coincide with the classical solutions if either the classical solution exists or if they enjoy certain smoothness. We discuss these properties that are fundamental for the subsequent analysis of concrete numerical schemes.
- **Part III [Numerical schemes]**
  There is a huge literature on efficient numerical methods for simulation of inviscid or viscous compressible fluids, see, e.g., Dolejší and Feistauer [71], Eymard, Gallouët, Herbin [78], Feistauer [99], Feistauer, Felcman, Straškraba [100], Kröner [143], LeVeque [151], Toro [194]. Despite enormous practical applications of such numerical methods, their convergence and error analysis remain still open in general.
  Applying general convergence results from Part II for consistent (or stable) approximations, we present the convergence analysis of some well-chosen numerical methods that are representative approximation schemes for the Euler or Navier–Stokes systems. Since the Euler system belongs to the class of hyperbolic conservation laws, we approximate them by the finite volume methods that naturally inherit the conservation property at discrete level. For the Navier–Stokes equations, we apply an upwind finite volume, a mixed finite volume – finite element of Karper [140] as well as a staggered finite difference (MAC) method. Assuming only that numerical solutions remain in a gas nondegenerate region, which reflects some boundedness of the discrete densities and temperatures, we present a detailed consistency and stability analysis of the corresponding numerical methods. Consequently, their weak convergence to generalized dissipative solutions follows. A recently developed concept of $\mathcal{K}$–convergence, cf. [90, 91] allows us to show the strong convergence of the Cesàro averages to a generalized dissipative solution. In the case that a strong solution of the underlying compressible fluid equations exists, we show the strong convergence of the numerical

methods. Moreover, if the limit of numerical solutions of the Euler system is a weak solution, then the convergence is strong.

Praha, Czech Republic — Eduard Feireisl
Mainz, Germany — Mária Lukáčová-Medvid'ová
Bratislava, Slovakia — Hana Šmitala Mizerová
Praha, Czech Republic — Bangwei She
October 2020

Acknowledgments

The research leading to the results presented in this book was supported by several grant agencies. The research of Eduard Feireisl, Hana Mizerová and Bangwei She was funded by the Czech Sciences Foundation (GAČR), Grant Agreement 18–05974S. The Institute of Mathematics of the Czech Academy of Sciences is supported by RVO:67985840. The work of Mária Lukáčová has been supported by the Deutsche Forschungsgemeinschaft (DFG, German Research Foundation)—TRR/SFB 146—Project number 233630050, TRR/SFB 165 Waves to Weather—Project A2, and by the Mainz Institute of Multiscale Modelling. She is thankful to the Gutenberg Research College of Johannes Gutenberg University Mainz for the research fellowship allowing full concentration on this project.

We are grateful to our colleagues Rémi Abgrall, Miloslav Feistauer, Song Jiang, Jiequan Li, Antonín Novotný, László Székelyhidi, Endre Süli, Eitan Tadmor and Gerald Warnecke for valuable information, advice, inspiring suggestions and fruitful discussions which helped us during the preparation of the monograph.

We wish to thank our students Simon Schneider and Andreas Schömer for careful reading of the manuscript. Simon Schneider and Yue Wang provided us with several computational results and figures.

Further, we would like to thank Alfio Quarteroni for encouraging us to publish this monograph in MS&A Series of Springer.

Most importantly, we are thankful to our families for their constant support during the work on the manuscript. They have sacrificed many evenings and weekends of family time for our research project yielding the results presented in this monograph.

Preliminary Material

This chapter contains preliminary material collected for reader's convenience. It includes the basic notation as well as mathematical tools. We tacitly suppose the reader to be familiar with this material, and, accordingly, we will refer to it throughout the whole book without further specification.

Continuous and Semicontinuous Functions

If not otherwise specified, all vector spaces are real. The symbol R^d denotes the standard d-dimensional Euclidean space, the norm on R^d being $|\cdot|$. The norm on a general (infinite-dimensional) normed linear space X is denoted $\|\cdot\|_X$. The duality pairing between a vector space X and its dual X^* is denoted as $< \cdot; \cdot >_{X^*;X}$, or simply $< \cdot; \cdot >$ in case the underlying spaces are clearly identified. If X is a Hilbert space, the symbol $< \cdot; \cdot >$ denotes the scalar product in X. The symbol $\text{span}\{M\}$ denotes the space of all finite linear combinations of vectors contained in M.

Continuous Functions

The symbol $C(Q)$ denotes the set of continuous functions defined on a topological space Q. More generally, the symbol $C(Q; X)$ denotes the space of continuous functions on Q ranging in another topological space X. If X is a Banach space, we introduce a norm

$$\|g\|_{C(Q;X)} = \sup_{y \in Q} \|g(y)\|_X.$$

If X is a Polish space, meaning a complete, separable, metrizable topological space with a metric d, we define a metric

$$d_{C(Q;X)}[f, g] = \sup_{y \in Q} d[f(y); g(y)].$$

We say that a function f is *bounded* if

$$\sup_{y \in Q} d[f(y); a] < \infty \quad \text{for some } a \in X.$$

The symbol $BC(Q; X)$ denotes the set of all bounded continuous functions on Q ranging in a Polish space X.

The symbol $C_{\text{weak}}(Q; X)$ denotes the space of functions on Q ranging in a Banach space X, continuous with respect to the weak topology. More specifically, $g \in C_{\text{weak}}(Q; X)$ if the mapping $y \mapsto \|g(y)\|_X$ is bounded and

$$y \mapsto < f; g(y) >_{X^*;X}$$

is continuous on Q for any linear form f belonging to the dual space X^*. Similarly, we define $C_{\text{weak}-(*)}(Q; X^*)$ if X^* is a dual to a Banach space X. The two concepts coincide if X is reflexive.

If X is reflexive and separable, the weak topology is metrizable on bounded sets of X by the metric

$$d_{X_{\text{weak}}}[f, g] = \sum_{n=1}^{\infty} \frac{1}{2^n} \frac{|\langle f - g; e_n \rangle|}{1 + |\langle f - g; e_n \rangle|},$$

where $\{e_n\}_{n=1}^{\infty}$ is a dense set on a unit sphere of X. Thus, if Q is compact and $B \subset X$ is bounded, we may define a metric on $C_{\text{weak}}(Q; B)$,

$$d_{C_{\text{weak}}(Q;B)}[f, g] = \sum_{n=1}^{\infty} \frac{1}{2^n} \frac{\sup_{y \in Q} |\langle f(y) - g(y); e_n \rangle|}{1 + \sup_{y \in Q} |\langle f(y) - g(y); e_n \rangle|}$$

We say that $g_n \to g$ in $C_{\text{weak}}(Q; X)$ if

- $y \mapsto \|g_n(y)\|_X$ is bounded uniformly for $n \to \infty$;
- $< f; g_n >_{X^*;X} \to < f; g >_{X^*;X}$ in $C(Q)$ for all $f \in X^*$.

Theorem 1 (Arzelà–Ascoli Theorem) Let $Q \subset R^m$ be compact and X a compact topological metric space endowed with a metric d_X. Let $\{f_n\}_{n=1}^{\infty}$ be a sequence of functions in $C(Q; X)$ that is equi-continuous, meaning, for any $\varepsilon > 0$ there is $\delta > 0$ such that

$$d_X\Big[f_n(y), f_n(z)\Big] \leq \varepsilon \text{ provided } |y - z| < \delta \text{ independently of } n = 1, 2, \ldots.$$

Then $\{f_n\}_{n=1}^{\infty}$ is precompact in $C(Q; X)$, that is, there exists a subsequence (not relabeled) and a function $f \in C(Q; X)$ such that

$$\sup_{y \in Q} d_X\Big[f_n(y), f(y)\Big] \to 0 \text{ as } n \to \infty.$$

For the proof see Kelley [141, Chapter 7, Theorem 17].

For $g : Q \to R$ we denote the *support* of g,

$$\text{supp}[g] = \text{closure}_Q\Big\{y \in Q \,\Big|\, g(y) \neq 0\Big\}.$$

The symbol $C_c(Q)$ denotes the set of functions on Q with compact support. We also define $C_c(Q; R^k)$ – the space of vector valued functions with all components in C_c. Obviously $C(Q) = C_c(Q)$ if Q is compact, in which case $C_c(Q)$ is a Banach space. The topology on $C_c(Q)$ for a general open set is more complicated and we content ourselves with the definition of *convergence* in $C_c(Q)$:

$$g_n \to g \text{ in } C_c(Q)$$
$$\Leftrightarrow \text{supp}[g_n] \subset K,\ K \subset Q \text{ compact for all } n \text{ large enough},\ g_n \to g \in C(K).$$

We denote $C_0(Q)$ the completion of $C_c(Q)$ with respect to the norm $\|\cdot\|_{C(Q)}$.

The symbol $C^k(\overline{Q})$, $Q \subset R^d$, where k is a nonnegative integer, denotes the space of functions on $\overline{Q}$ that are restrictions of k-times continuously differentiable functions on R^d. $C^{k,\nu}(\overline{Q})$, $\nu \in (0,1)$ is the subspace of $C^k(\overline{Q})$ of functions having their k-th derivatives ν-Hölder continuous in $\overline{Q}$. $C^{k,1}(\overline{Q})$ is a subspace of $C^k(\overline{Q})$ of functions whose k-th derivatives are Lipschitz on $\overline{Q}$. For a bounded domain Q, the spaces $C^k(\overline{Q})$ and $C^{k,\nu}(\overline{Q})$, $\nu \in (0,1]$ are Banach spaces with norms

$$\|u\|_{C^k(\overline{Q})} = \max_{|\alpha| \le k} \sup_{x \in Q} |\partial^\alpha u(x)|$$

and

$$\|u\|_{C^{k,\nu}(\overline{Q})} = \|u\|_{C^k(\overline{Q})} + \max_{|\alpha|=k} \sup_{(x,y)\in Q^2,\, x \neq y} \frac{|\partial^\alpha u(x) - \partial^\alpha u(y)|}{|x-y|^\nu},$$

where $\partial^\alpha u$ stands for the partial derivative $\partial_{x_1}^{\alpha_1} \dots \partial_{x_N}^{\alpha_N} u$ of order $|\alpha| = \sum_{i=1}^N \alpha_i$. The spaces $C^{k,\nu}(\overline{Q}; R^m)$ are defined in a similar way. Finally, we set $C^\infty = \cap_{k=0}^\infty C^k$.

The symbol $C_c^k(Q; R^m)$, $k \in \{0, 1, \dots, \infty\}$ denotes the vector space of functions belonging to $C^k(Q; R^m)$ and having compact support in Q. If $Q \subset R^d$ is an open set, the symbol $\mathcal{D}(Q; R^m)$ will be used alternatively for the space $C_c^\infty(Q; R^m)$ endowed with the topology induced by the convergence:

$$\varphi_n \to \varphi \in \mathcal{D}(Q)$$

if

$$\text{supp}[\varphi_n] \subset K,\ K \subset Q \text{ a compact set },\ \varphi_n \to \varphi \text{ in } C^k(K) \text{ for any } k = 0, 1, \ldots. \quad (1)$$

We write $\mathcal{D}(Q)$ instead of $\mathcal{D}(Q; R)$. The dual space $\mathcal{D}'(Q)$, $Q \subset R^m$, is the space of *distributions.*

The subscript loc refers to spaces of functions having the relevant properties on any compact subset, e.g.

$$C_{\text{loc}}(Q) = \left\{ g : Q \to R \,\middle|\, g \in C(K) \text{ for any compact } K \subset Q \right\}.$$

Lower semicontinuous functions

Let Q be a metric space. A function

$$f : Q \to R \cup \{\infty\}$$

is *lower semicontinuous (l.s.c)* if the set $f^{-1}(-\infty, a]$ is closed for any $a \in R$.

Theorem 2 (Baire Theorem)

Let Q be a metric space and

$$f : Q \to [0, \infty]$$

be a l.s.c. function.

Then there exists a sequence $\{f_n\}_{n=1}^{\infty}$, $f_n \in C(Q)$,

$$0 \le f_n \le f,\ f_n(y) \nearrow f(y) \text{ for any } y \in Q.$$

For the proof, see Baire [7] or Cobzaş et al. [57, Theorem 6.4.1].

Integrable Functions

Let Q be a topological space with the σ-algebra of Borel sets $\mathcal{B}(Q)$ and a regular Borel measure ν. We denote

$$\int_Q g \mathrm{d}\nu$$

the integral of a measurable function g. Integration with respect to the Lebesgue measure on the Euclidean space will be denoted by the symbol $\mathrm{d}y$, or more specifically, $\mathrm{d}x$ if $Q \subset R^d$, $d = 1, 2, 3$, and $\mathrm{d}t$ if $Q = (0, T)$. The Lebesgue measure of a measurable set $B \subset R^m$ will be denoted

$$|B| \equiv \int_{R^m} 1_B(y)\mathrm{d}y.$$

For a function $v = v(t, x)$, we denote

$$\left[\int_\Omega v(t,x)\,\mathrm{d}x\right]_{t=\tau_1}^{t=\tau_2} = \int_\Omega v(\tau_2, x)\,\mathrm{d}x - \int_\Omega v(\tau_1, x)\,\mathrm{d}x,$$

together with the spatial convolution

$$(u * v)(x) = \int_Q u(x-y)v(y)\,\mathrm{d}y \text{ for } Q = \mathbb{T}^d \text{ or } R^d.$$

The *Lebesgue spaces* $L^p(Q; X)$ are spaces of (Bochner) measurable functions v ranging in a Banach space X such that the norm

$$\|v\|^p_{L^p(Q;X)} = \int_Q \|v\|^p_X\,\mathrm{d}y \text{ is finite }, \ 1 \le p < \infty.$$

Similarly, $v \in L^\infty(Q; X)$ if v is (Bochner) measurable and

$$\|v\|_{L^\infty(Q;X)} = \operatorname*{ess\,sup}_{y\in Q} \|v(y)\|_X < \infty.$$

The symbol $L^p_{\mathrm{loc}}(Q; X)$ denotes the vector space of locally L^p-integrable functions, meaning

$$v \in L^p_{\mathrm{loc}}(Q; X) \text{ if } v \in L^p(K; X) \text{ for any compact set K in Q}.$$

We write $L^p(Q)$ for $L^p(Q; R)$.

Let $f \in L^1_{loc}(Q)$, where Q is an open set. A *Lebesgue point* $a \in Q$ of f in Q is characterized by the property

$$\lim_{r\to 0+} \frac{1}{|B(a,r)|}\int_{B(a,r)} f(x)\,\mathrm{d}x = f(a). \tag{2}$$

For $f \in L^1(Q)$ the set of all Lebesgue points is of full measure, meaning its complement in Q is of zero Lebesgue measure. A similar statement holds for vector valued functions $f \in L^1(Q; X)$, where X is a Banach space (see Brezis [33]). If $f \in C(Q)$, then identity (2) holds for all points a in Q.

Theorem 3 (Linear functionals on $L^p(Q; X)$)

Let $Q \subset R^m$ be a measurable set, X a Banach space that is reflexive and separable, $1 \le p < \infty$.

Then any continuous linear form $\xi \in [L^p(Q; X)]^*$ admits a unique representation $w_\xi \in L^{p'}(Q; X^*)$,

$$< \xi; v >_{L^{p'}(Q,X^*);L^p(Q;X)} = \int_Q < w_\xi(y); v(y) >_{X^*;X} \mathrm{d}y \quad \text{for all} \quad v \in L^p(Q;X),$$

where

$$\frac{1}{p} + \frac{1}{p'} = 1.$$

Moreover the norm on the dual space is given as

$$\|\xi\|_{[L^p(Q;X)]^*} = \|w_\xi\|_{L^{p'}(Q;X^*)}.$$

Accordingly, the spaces $L^p(Q;X)$ are reflexive for $1 < p < \infty$ as soon as X is reflexive and separable.

For the proof, see Gajewski, Gröger, Zacharias [109, Chapter IV, Theorem 1.14, Remark 1.9].

Identifying ξ with w_ξ, we write

$$[L^p(Q;R^k)]^* = L^{p'}(Q;R^k), \ \|\xi\|_{[L^p(Q;R^k)]^*} = \|\xi\|_{L^{p'}(Q;R^k)}, \ 1 \le p < \infty.$$

If the Banach space X in Theorem 3 is merely separable, we have

$$[L^p(Q;X)]^* = L^{p'}_{\text{weak}-(*)}(Q;X^*) \ \text{ for } \ 1 \le p < \infty,$$

where

$$L^{p'}_{\text{weak}-(*)}(Q;X^*)$$

$$\equiv \left\{ \xi : Q \to X^* \,\middle|\, y \in Q \mapsto < \xi(y); v >_{X^*;X} \ \text{ measurable for any fixed } \ v \in X, \right.$$

$$\left. y \mapsto \|\xi(y)\|_{X^*} \in L^{p'}(Q) \right\}$$

(see Edwards [72], Pedregal [177, Chapter 6, Theorem 6.14]). For simplicity we omit the subscript weak-$(*)$ in what follows if no confusion appears.

In what follows we present some fundamental inequalities that will be often applied. First, *Hölder's inequality* reads

$$\|uv\|_{L^r(Q)} \le \|u\|_{L^p(Q)}\|v\|_{L^q(Q)}, \ \frac{1}{r} = \frac{1}{p} + \frac{1}{q}, \ 1 \le p, q, r \le \infty$$

for any $u \in L^p(Q)$, $v \in L^q(Q)$, $Q \subset R^m$ (see Adams [3, Chapter 2]).

Interpolation inequality for L^p-spaces reads

$$\|v\|_{L^r(Q)} \leq \|v\|_{L^p(Q)}^{\lambda} \|v\|_{L^q(Q)}^{(1-\lambda)}, \ \frac{1}{r} = \frac{\lambda}{p} + \frac{1-\lambda}{q}, \ 1 \leq p < r < q \leq \infty, \ \lambda \in (0,1)$$

for any $v \in L^p(Q) \cap L^q(Q)$, $Q \subset R^m$ (see Adams [3, Chapter 2]).

Jensen's inequality reads

$$\Phi\left(\int_Q v \, dy\right) \leq \int_Q \Phi(v) \, dy$$

whenever Φ is convex on the range of v and $Q \subset R^m$, $|Q| = 1$, see e.g. Ziemer [200, Chapter 1, Section 1.5].

If

$$F : R^m \to (-\infty, \infty]$$

is a convex l.s.c. function, its *conjugate* F^* is defined as

$$F^*(\boldsymbol{w}) = \sup_{\boldsymbol{v} \in R^n} (\boldsymbol{v} \cdot \boldsymbol{w} - F(\boldsymbol{v})).$$

Accordingly, we have *Fenchel–Young inequality*,

$$\boldsymbol{v} \cdot \boldsymbol{w} \leq F(\boldsymbol{v}) + F^*(\boldsymbol{w}).$$

Moreover,

$$\boldsymbol{v} \cdot \boldsymbol{w} = F(\boldsymbol{v}) + F^*(\boldsymbol{w}) \ \Leftrightarrow \ \boldsymbol{w} \in \partial F(\boldsymbol{v}) \ \Leftrightarrow \ \boldsymbol{v} \in \partial F^*(\boldsymbol{w}),$$

see e.g. Ekeland and Temam [73].

Lemma 1 (Gronwall's Lemma)

Let $a \in L^1(0,T)$, $a \geq 0$, $\beta \in L^1(0,T)$, $b_0 \in R$, and

$$b(\tau) = b_0 + \int_0^\tau \beta(t) \, dt$$

be given. Let $r \in L^\infty(0,T)$ satisfy

$$r(\tau) \leq b(\tau) + \int_0^\tau a(t) r(t) \, dt \ \text{ for a.a. } \tau \in [0,T].$$

Then

$$r(\tau) \leq b_0 \exp\left(\int_0^\tau a(t) \, dt\right) + \int_0^\tau \beta(t) \exp\left(\int_t^\tau a(s) \, ds\right) dt$$

for a.a. $\tau \in [0, T]$.

For the proof, see Carroll [41].

Measures

Duals to the space of continuous functions are the spaces of measures. We start by the standard representation theorem.

Theorem 4 (Riesz Representation Theorem)
Let Q be a locally compact Hausdorff metric space. Let f be a nonnegative linear functional defined on the space $C_c(Q)$.
Then there exist a σ-algebra of measurable sets containing all Borel sets and a unique nonnegative Borel measure ν_f such that

$$\langle f; g \rangle = \int_Q g \, \mathrm{d}\nu_f \quad \text{for any} \quad g \in C_c(Q).$$

Moreover, the measure ν_f enjoys the following properties:

- $\nu_f(K) < \infty$ for any compact $K \subset Q$;
- $\nu_f(E) = \sup\{\nu_f(K) \mid K \subset E\}$ for any open set $E \subset Q$;
- $\nu_f(V) = \inf\{\nu_f(E) \mid V \subset E, \ E \ \text{open}\}$ for any Borel set V;
- If E is ν_f-measurable, $\nu_f(E) = 0$, and $A \subset E$, then A is ν_f-measurable.

For the proof see Rudin [181, Chapter 2, Theorem 2.14].

For the rest of this section, we suppose that Q is a Polish space. We denote by the symbol $\mathcal{M}^+(Q)$ the set of all nonnegative Borel measures on a topological space Q. The symbol $\mathcal{M}(Q)$ denotes the set of all signed *Radon measures* that can be identified as the space of all linear forms bounded on $C_c(Q)$. In particular, we may identify

$$[C(Q)]^* = \mathcal{M}(Q) \ \text{if} \ Q \ \text{is compact}.$$

If Q is merely locally compact, we define the space of *finite measures* $\mathcal{M}_f(Q)$ as a dual space to $C_0(Q)$.
Finally, we introduce the space of *vector valued measures* $\mathcal{M}(Q; E)$, where E is an m-dimensional space as

$$\mathcal{M}(Q; E) = \left\{ [\nu_1, \ldots, \nu_m] \,\middle|\, \nu_i \in \mathcal{M}(Q),\ i = 1, \ldots, m \right\}.$$

The symbol $\mathcal{M}^+(Q; R^{d\times d}_{\text{sym}})$ denotes the set of *positive semidefinite matrix valued measures,*

$$\mathcal{M}^+(Q; R^{d\times d}_{\text{sym}}) = \Big\{ \nu \in \mathcal{M}^+(Q; R^{d\times d}_{\text{sym}}) \Big| \int_Q \phi(\xi \otimes \xi) : \mathrm{d}\nu \geq 0 \text{ for any } \xi \in R^d,\ \phi \in C_c(Q),\ \phi \geq 0 \Big\}.$$

A measure $\nu \in \mathcal{M}^+(Q)$ is called *probability measure* if $\nu(Q) = 1$. The symbol $\mathcal{P}(Q)$ denotes the (convex) set of all Borel probability measures on Q. We say that a family $\{\nu_\alpha\}_{\alpha>0}$ of probability measures on Q is (uniformly) *tight* if for any $\varepsilon > 0$, there exists a compact $K \subset Q$ such that

$$\nu_\alpha(K) > 1 - \varepsilon \text{ for all } \alpha.$$

The set $\mathcal{P}(Q)$ can be equipped with the *narrow topology*. We say that

$$\nu_n \to \nu \text{ narrowly in } \mathcal{P}(Q) \Leftrightarrow \int_S f \mathrm{d}\nu_n \to \int_S f \mathrm{d}\nu \text{ for any } f \in BC(Q).$$

For a general sequence $\{\nu_n\}_{n=1}^\infty \subset \mathcal{M}(Q)$, we say that

$$\nu_n \to \nu \text{ weakly-(*) in } \mathcal{M}(Q) \Leftrightarrow \int_S f \mathrm{d}\nu_n \to \int_S f \mathrm{d}\nu \text{ for any } f \in C_c(Q).$$

Theorem 5 (Prokhorov Theorem)
Let Q be a Polish space.
Then a family of probability measures $\{\nu_\alpha\}_{\alpha>0}$ on S is tight if and only if its closure is sequentially compact in the space of probability measures endowed with the topology of narrow convergence.

For the proof see Prokhorov [180].

Let $\nu \in \mathcal{P}(\Pi_{i=1}^n Q_i)$ be a probability measure. We denote

$$\Pi_i \nu \in \mathcal{P}(Q_i),\ \Pi_i \nu(E) = \nu(Q_1 \times \ldots Q_{i-1} \times E \times Q_{i+1} \cdots \times Q_n),\ i = 1, \ldots, n$$

the *projection* of the measure ν. The measure Π_i is called *marginal* of ν.

Let d denote the metric on the Polish space Q. We say that a measure $\nu \in \mathcal{P}(Q)$ admits a finite p-th moment if

$$\int_Q d(q_0, y)^p \mathrm{d}\nu(y) < \infty \text{ for some } q_0 \in Q.$$

There are several concepts of *distance (metric)* on the set of probability measures:

- **Lévy–Prokhorov distance:**

$$d_P[\mu; \nu] = \inf\Big\{\varepsilon > 0 \,\Big|\, \mu(A) \le \nu(\mathcal{U}_\varepsilon(A)) + \varepsilon,\ \nu(A) \le \mu(\mathcal{U}_\varepsilon(A)) + \varepsilon,\ A \in \mathcal{B}(Q)\Big\},$$

where $\mathcal{U}_\varepsilon(A)$ denotes the ε-neighborhood of a set A and $\mathcal{B}(Q)$ the Borel sets of a set Q.
- **weak-(*) distance:**

$$d_{w-(*)}[\mu; \nu] = \sum_{n=1}^{\infty} 2^{-n} \left| \int_Q \phi_n \mathrm{d}\mu - \int_Q \phi_n \mathrm{d}\nu \right|,$$

if Q is locally compact, where $\{\phi_n\}_{n=1}^{\infty}$ is a dense set on the unit sphere in $C_0(Q)$.
- **Wasserstein p-distance:**

$$d_{W_p}[\mu, \nu] = \inf_{\lambda \in \mathcal{P}(Q \times Q), \mu = \Pi_1 \lambda, \nu = \Pi_2 \lambda} \int_{Q \times Q} d(x, y)^p \, \mathrm{d}\lambda(x, y). \tag{3}$$

Narrow convergence and convergence in the Lévy–Prokhorov distance are equivalent on a Polish space, see Villani [198, Chapter 6].

We report a duality formula for d_{W_1}.

Lemma 2 (Duality formula)

Let μ, ν be probability measures on a Polish space Q with finite first moments. Then

$$d_{W_1}[\mu, \nu] = \sup_{\phi, \|\nabla_x \phi\|_{L^\infty(Q)} \le 1} \left(\int_Q \phi \mathrm{d}\mu - \int_Q \phi \mathrm{d}\nu \right).$$

For the proof, see Villani [198, Theorem 5.10 and Remark 6.5].

Theorem 6 (Convergence in the Wasserstein distance)

Let $\{\nu_n\}_{n=1}^{\infty}$ be a sequence of probability measures on a Polish space Q with metric d. In addition, let ν_n possess finite p-th moment, $p \ge 1$.

Then the following is equivalent:

- $d_{W_p}[\nu_n, \nu] \to 0$ for a certain $\nu \in \mathcal{P}(Q)$ with finite $p-$ th moment;

- $\nu_n \to \nu$ narrowly and $\int_Q d(q_0, y)^p \mathrm{d}\nu_n \to \int_Q d(q_0, y)^p \mathrm{d}\nu$ for some $q_0 \in Q$;
- $\int_Q g(y)\mathrm{d}\nu_n \to \int_Q g(y)\mathrm{d}\nu$ for any $g \in C(Q)$ such that

$$g(y) \lesssim (1 + d(q_0; y)^p) \text{ for some } q_0 \in Q.$$

For the proof, see Villani [198, Theorem 6.9].

Finally we recall the fundamental theorem concerning parametrized (Young) measures.

Let $Q \subset R^m$ be a domain. We say that Φ is a *Caratheodory function* on $Q \times R^k$ if

$$\left\{ \begin{array}{c} \text{for a. a. } x \in Q, \text{ the function } \lambda \mapsto \Phi(x, \lambda) \text{ is continuous on } R^k; \\ \\ \text{for all } \lambda \in R^k, \text{ the function } x \mapsto \Phi(x, \lambda) \text{ is measurable on } Q. \end{array} \right\} \tag{4}$$

We say that $\{\nu_x\}_{x \in Q}$ is a *family of parametrized measures* if ν_x is a probability measure for a.a. $x \in Q$, and if

$$\left\{ \begin{array}{c} \text{the function } x \to \int_{R^k} \phi(\lambda)\, \mathrm{d}\nu_x(\lambda) \equiv < \nu_x, \phi > \text{ is measurable on } Q \\ \\ \text{for all } \phi : R^k \to R,\ \phi \in C(R^k) \cap L^\infty(R^k). \end{array} \right\} \tag{5}$$

Theorem 7 (Fundamental theorem of the theory of parametrized (Young) measures)

Let $Q \subset R^m$ be a domain. Let $\{\boldsymbol{v}_n\}_{n=1}^\infty$, $\boldsymbol{v}_n : Q \subset R^m \to R^k$ be a sequence of measurable functions such that

$$\int_Q |\boldsymbol{v}_n| \mathrm{d}y \le c$$

uniformly for $n \to \infty$.

Then there exist a subsequence (not relabeled) and a parametrized family $\{\nu_y\}_{y \in Q}$ of probability measures on R^k depending measurably on $y \in Q$ with the following property:

For any Caratheodory function $\Phi = \Phi(y, z)$, $y \in Q$, $z \in R^k$ such that

$$\Phi(\cdot, \boldsymbol{v}_n) \to \overline{\Phi} \text{ weakly in } L^1(Q),$$

we have

$$\overline{\Phi}(y) = \int_{R^k} \Phi(y, z)\, \mathrm{d}\nu_y(z) \text{ for a.a. } y \in Q.$$

For the proof, see Pedregal [177, Chapter 6, Theorem 6.2].

Sobolev Spaces

A domain $\Omega \subset R^d$ is of class $\mathcal{C}$ if for each point $x \in \partial\Omega$, there exist $r > 0$ and a mapping $\gamma : R^{d-1} \to R$ belonging to a function class$\mathcal{C}$ such that – upon rotating and relabeling the coordinate axes if necessary – we have

$$\left.\begin{aligned} \Omega \cap B(x; r) &= \{y \mid \gamma(y') < y_d\} \cap B(x, r) \\ \partial\Omega \cap B(x; r) &= \{y \mid \gamma(y') = y_d\} \cap B(x, r) \end{aligned}\right\}, \quad \text{where } y' = (y_1, \ldots, y_{d-1}),$$

where $B(x, r)$ denotes the ball centered at x of radius r. In particular, Ω is called *Lipschitz domain* if γ is Lipschitz.

If $A \subset \Gamma \equiv \partial\Omega \cap B(x; r)$, γ is Lipschitz and $f : A \to R$, then one can define the surface integral

$$\int_A f\, \mathrm{d}S_x \equiv \int_{\Phi_\gamma(A)} f(y', \gamma(y')) \sqrt{1 + \sum_{i=1}^{d-1} \left(\frac{\partial \gamma}{\partial y_i}\right)^2}\, \mathrm{d}y',$$

where $\Phi_\gamma : R^d \to R^d$, $\Phi_\gamma(y', y_N) = (y', y_N - \gamma(y'))$, whenever the (Lebesgue) integral on the right-hand side exists. If $f = 1_A$, then $S_{N-1}(A) = \int_A \mathrm{d}S_x$ is the surface measure on $\partial\Omega$ of A that can be identified with the $(d-1)$-Hausdorff measure on $\partial\Omega$ of A (cf. Evans and Gariepy [77, Chapter 4.2]EVGA). In the general case of $A \subset \partial\Omega$, one can define $\int_A f\, \mathrm{d}S_x$ using a covering $\mathcal{B} = \{B(x_i; r)\}_{i=1}^M$, $x_i \in \partial\Omega$, M finite, of $\partial\Omega$ by balls of radii r and subordinated partition of unity $\mathcal{F} = \{\varphi_i\}_{i=1}^M$, and set

$$\int_A f\, \mathrm{d}S_x = \sum_{i=1}^{M} \int_{\Gamma_i} \varphi_i f\, \mathrm{d}S_x, \quad \Gamma_i = \partial\Omega \cap B(x_i; r),$$

see Nečas [167, Section I.2] or Kufner, Fučík, John [145, Section 6.3].

A Lipschitz domain Ω admits the outer normal vector $\boldsymbol{n}(x)$ for a.a. $x \in \partial\Omega$. Here *a.a.* refers to the surface measure on $\partial\Omega$.

A differential operator ∂^α of order $|\alpha|$ can be identified with a distribution

$$\langle \partial^\alpha v; \varphi \rangle_{\mathcal{D}'(\Omega);\mathcal{D}(\Omega)} = (-1)^{|\alpha|} \int_\Omega v \partial^\alpha \varphi \, \mathrm{d}y$$

for any locally integrable function v.

The *Sobolev spaces* $W^{k,p}(\Omega; R^m)$, $1 \leq p \leq \infty$, k a positive integer, are the spaces of functions having all distributional derivatives up to order k in $L^p(\Omega; R^m)$. The norm in $W^{k,p}(\Omega; R^m)$ is defined as

$$\|\boldsymbol{v}\|_{W^{k,p}(\Omega;R^m)} = \left\{ \begin{array}{c} \left(\sum_{i=1}^m \sum_{|\alpha| \leq k} \|\partial^\alpha v_i\|_{L^p(\Omega)}^p \right)^{1/p} \ \text{if } 1 \leq p < \infty \\ \max_{1 \leq i \leq m,\, |\alpha| \leq k} \{ \|\partial^\alpha v_i\|_{L^\infty(\Omega)} \} \ \text{if } p = \infty \end{array} \right\},$$

where the symbol ∂^α stands for any partial derivative of order $|\alpha|$.

The following are basic and well-known properties of Sobolev functions, see e.g. Adams [3] or Ziemer [200].

- If $\Omega \subset R^d$ is a bounded domain with boundary of class $C^{k-1,1}$, then there exists a continuous linear operator E which maps $W^{k,p}(\Omega)$ to $W^{k,p}(R^d)$; it is called *extension operator*,

$$E(\boldsymbol{v})|_\Omega = \boldsymbol{v} \ \text{ for any } \ \boldsymbol{v} \in W^{k,p}(\Omega).$$

  For $1 < p < \infty$ the extension operator exists even for the boundary of class $C^{0,1}$.
- If $\Omega \subset R^d$ is a bounded domain with boundary of class $C^{k-1,1}$, and $1 \leq p < \infty$, then $W^{k,p}(\Omega)$ is separable and the space $C^k(\overline{\Omega})$ is its dense subspace.
- If $f : R \to R$ is a Lipschitz function and $v \in W^{1,p}(\Omega)$, then $f \circ v \in W^{1,p}(\Omega)$ and

$$\partial_{x_j}[f \circ v](x) = f'(v(x)) \partial_{x_j} v(x) \ \text{ for a.a. } x \in \Omega,$$

  see Ziemer [200, Section 2.1].
- The space $W^{1,\infty}(\Omega)$, where Ω is a bounded domain, is isometrically isomorphic to the space $C^{0,1}(\Omega)$ of Lipschitz functions on Ω.

The symbol $W_0^{k,p}(\Omega; R^m)$ denotes the completion of $C_c^\infty(\Omega; R^m)$ with respect to the norm $\|\cdot\|_{W^{k,p}(\Omega;R^m)}$. We identify $W^{0,p}(\Omega; R^m) = W_0^{0,p}(\Omega; R^m)$ with $L^p(\Omega; R^m)$.

Theorem 8 (Dual Sobolev spaces)

Let $\Omega \subset R^d$ be a domain, and let $1 \leq p < \infty$. Then the dual space $[W_0^{k,p}(\Omega)]^*$ is a proper subspace of the space of distributions $\mathcal{D}'(\Omega)$. Moreover, any linear form $f \in [W_0^{k,p}(\Omega)]^*$ admits a representation

$$\langle f; v\rangle_{[W_0^{k,p}(\Omega)]^*;W_0^{k,p}(\Omega)} = \sum_{|\alpha|\le k}\int_\Omega (-1)^{|\alpha|} w_\alpha\, \partial^\alpha v\ dx,$$
$$\text{where } w_\alpha \in L^{p'}(\Omega),\ \frac{1}{p}+\frac{1}{p'}=1. \tag{6}$$

The norm of f in the dual space is given as

$$\|f\|_{[W_0^{k,p}(\Omega)]^*} = \begin{cases} \inf\left\{\left(\sum_{|\alpha|\le k}\|w_\alpha\|_{L^{p'}(\Omega)}^{p'}\right)^{1/p'} \,\middle|\, w_\alpha \text{ satisfy } (6)\right\} \\ \qquad \text{for } 1<p<\infty; \\ \inf\left\{\max_{|\alpha|\le k}\{\|w_\alpha\|_{L^\infty(\Omega)}\} \,\middle|\, w_\alpha \text{ satisfy } (6)\right\} \\ \qquad \text{if } p=1. \end{cases}$$

The infimum is attained in both cases.

For the proof, see Adams [3, Theorem 3.8], Mazya [162, Section 1.1.14].

The dual space to the Sobolev space $W_0^{k,p}(\Omega)$ is denoted as $W^{-k,p'}(\Omega)$.

Theorem 9 (Rellich–Kondrachov Embedding Theorem)
Let $\Omega \subset R^d$ be a bounded Lipschitz domain.

(i) Then, if $kp < d$ and $p \ge 1$, the space $W^{k,p}(\Omega)$ is continuously embedded in $L^q(\Omega)$ for any

$$1 \le q \le p^* = \frac{dp}{d-kp}.$$

Moreover, the embedding is compact if $k > 0$ and $q < p^*$.

(ii) If $kp = d$, the space $W^{k,p}(\Omega)$ is compactly embedded in $L^q(\Omega)$ for any $q \in [1,\infty)$.

(iii) If $kp > d$ then $W^{k,p}(\Omega)$ is continuously embedded in $C^{k-[d/p]-1,\nu}(\overline{\Omega})$, where $[\cdot]$ denotes the integer part and

$$\nu = \begin{cases} [\frac{d}{p}]+1-\frac{d}{p} \text{ if } \frac{d}{p} \text{ is not an integer,} \\ \text{arbitrary positive number in } (0,1) \text{ if } \frac{d}{p} \text{ is an integer.} \end{cases}$$

Moreover, the embedding is compact if $0 < \nu < [\frac{d}{p}]+1-\frac{d}{p}$.

See Ziemer [200, Theorem 2.5.1, Remark 2.5.2].

The symbol $\hookrightarrow$ denotes continuous embedding, $\hookrightarrow\hookrightarrow$ indicates compact embedding. The following result may be seen as a direct consequence of Theorem 9.

Theorem 10 (Embedding Theorem for Dual Sobolev Spaces)
Let $\Omega \subset R^d$ be a bounded domain. Let $k > 0$ and $q < \infty$ satisfy

$$q > \frac{p^*}{p^* - 1}, \text{ where } p^* = \frac{dp}{d - kp} \text{ if } kp < d,$$

$$q > 1 \text{ for } kp = d,$$

or

$$q \geq 1 \text{ if } kp > d.$$

Then the space $L^q(\Omega)$ is compactly embedded into the space $W^{-k,p'}(\Omega)$, $1/p + 1/p' = 1$.

The *Sobolev–Slobodeckii spaces* $W^{k+\beta,p}(\Omega)$, $1 \leq p < \infty$, $0 < \beta < 1$, $k = 0, 1, \ldots$, where Ω is a domain in R^L, are Banach spaces of functions with finite norm

$$W^{k+\beta,p}(\Omega) = \left(\|v\|^p_{W^{k,p}(\Omega)} + \sum_{|\alpha|=k} \int_\Omega \int_\Omega \frac{|\partial^\alpha v(y) - \partial^\alpha v(z)|^p}{|y - z|^{L+\beta p}} \mathrm{d}y \, \mathrm{d}z \right)^{\frac{1}{p}},$$

see e.g. Nečas [167, Section 2.3.8].

Let $\Omega \subset R^d$ be a bounded Lipschitz domain. Referring to the notation introduced at the beginning of this section, we say that $f \in W^{k+\beta,p}(\partial\Omega)$ if $(\varphi f) \circ (\mathbb{I}', \gamma) \in W^{k+\beta,p}(R^{d-1})$ for any $\Gamma = \partial\Omega \cap B$ with B belonging to the covering $\mathcal{B}$ of $\partial\Omega$ and φ the corresponding term in the partition of unity $\mathcal{F}$ containing M components. The space $W^{k+\beta,p}(\partial\Omega)$ is a Banach space endowed with an equivalent norm $\|\cdot\|_{W^{k+\beta,p}(\partial\Omega)}$, where

$$\|v\|^p_{W^{k+\beta,p}(\partial\Omega)} = \sum_{i=1}^{M} \|(v\varphi_i) \circ (\mathbb{I}', \gamma)\|^p_{W^{k+\beta,p}(R^{d-1})}.$$

In the above formulas $(\mathbb{I}', \gamma) : R^{d-1} \to R^d$ maps y' to $(y', \gamma(y'))$. For more details see e.g. Nečas [167, Section 3.8].

If $\Omega \subset R^d$ is a bounded Lipschitz domain, the Sobolev–Slobodeckii spaces admit similar embeddings as classical Sobolev spaces. The embeddings

$$W^{k+\beta,p}(\Omega) \hookrightarrow L^q(\Omega) \text{ and } W^{k+\beta,p}(\Omega) \hookrightarrow C^s(\overline{\Omega})$$

are compact provided $(k + \beta)p < d$, $1 \leq q < \frac{dp}{d-(k+\beta)p}$, and $s = 0, 1, \ldots, k$, $(k - s + \beta)p > d$, respectively. The former embedding remains continuous (but not compact) at the border case $q = \frac{dp}{d-(k+\beta)p}$.

Theorem 11 (Trace Theorem, Gauss–Green formula)
Let $\Omega \subset R^d$ be a bounded Lipschitz domain.
Then there exists a linear operator γ_0 with the following properties:

$$[\gamma_0(v)](x) = v(x) \text{ for } x \in \partial\Omega \text{ provided } v \in C^\infty(\overline{\Omega}),$$

$$\|\gamma_0(v)\|_{W^{1-\frac{1}{p},p}(\partial\Omega)} \leq c\|v\|_{W^{1,p}(\Omega)} \text{ for all } v \in W^{1,p}(\Omega),$$

$$\ker[\gamma_0] = W_0^{1,p}(\Omega)$$

provided $1 < p < \infty$.
Conversely, there exists a continuous linear operator

$$\ell : W^{1-\frac{1}{p},p}(\partial\Omega) \to W^{1,p}(\Omega)$$

such that

$$\gamma_0(\ell(v)) = v \text{ for all } v \in W^{1-\frac{1}{p},p}(\partial\Omega)$$

provided $1 < p < \infty$.
In addition, the following formula holds:

$$\int_\Omega \partial_{x_i} uv \, \mathrm{d}x + \int_\Omega u\partial_{x_i} v \, \mathrm{d}x = \int_{\partial\Omega} \gamma_0(u)\gamma_0(v)n_i \, \mathrm{dS}_x, \; i = 1, \ldots, d,$$

for any $u \in W^{1,p}(\Omega)$, $v \in W^{1,p'}(\Omega)$, where $\boldsymbol{n}$ is the outer normal vector to the boundary $\partial\Omega$.

For the proof, see Nečas [167, Theorems 5.5, 5.7].

Fine Properties of Functions

Here we collect further results on fine properties of functions used in the book.

Poincaré inequality

Theorem 12 (Poincaré inequality)
Let $1 \leq p < \infty$, and let $\Omega \subset R^d$ be a bounded Lipschitz domain. Then the following holds:

(i) For any $A \subset \partial\Omega$ with the nonzero surface measure there exists a positive constant $c = c(p, d, A, \Omega)$ such that

$$\|v\|_{L^p(\Omega)} \leq c\left(\|\nabla v\|_{L^p(\Omega;R^d)} + \int_A |v| \, dS_x \right) \text{ for any } v \in W^{1,p}(\Omega).$$

(ii) There exists a positive constant $c = c(p, \Omega)$ such that

$$\left\| v - \frac{1}{|\Omega|} \int_\Omega v \, dx \right\|_{L^p(\Omega)} \leq c \|\nabla v\|_{L^p(\Omega;R^d)} \text{ for any } v \in W^{1,p}(\Omega).$$

This is a particular case of a more general result of Ziemer [200, Chapter 4, Theorem 4.5.1].

Functions of bounded variation

The symbol $BV(Q)$ denotes the space of functions in $L^1(Q)$ with distributional derivatives belonging to the space of measures $\mathcal{M}(Q)$. Functions belonging to $BV([0, T])$ possess well-defined right and left-hand limits and as such can be defined at *any* $t \in [0, 1]$.

We define the space $BV([0, T]; X)$ of functions of bounded variation from a real interval $[0, T]$ into a *metric space* X endowed with a metric d,

$$V_X[v] = \sup_{0 \leq t_0 < \ldots < t_m \leq T} \sum_{i=1}^{m} d(v(t_i); v(t_{i-1})).$$

Theorem 13 (Helly's Theorem)
Let $\{v_n\}_{n=1}^\infty \subset BV([0, T]; X)$ be sequence ranging in a complete metric space X that is bounded,

$$\sup_{n \geq 1} V_X[v_n] < \infty$$

and pointwise precompact,

$$\bigcup_{n=1}^{\infty} v_n(t) \text{ is precompact in } X \text{ for any } t \in [0, T].$$

Then, up to a subsequence,

$$v_n(t) \to v(t) \text{ for any } t \in [0, T], \text{ where } v \in BV([0, T]; X),$$

$$V_X[v] \le \sup_{n \ge 1} V_X[v_n].$$

For the proof, see Fleischer and Porter [108, Theorem 2.3].

Div–Curl Lemma

The celebrated Div–Curl Lemma of L. Tartar [193] (see also Murat [166]) is one of the most efficient tools in the analysis of problems with lack of compactness.

Lemma 3 Let $Q \subset R^N$ be an open set. Assume

$$\boldsymbol{U}_n \to \boldsymbol{U} \text{ weakly in } L^p(Q; R^d),$$

$$\boldsymbol{V}_n \to \boldsymbol{V} \text{ weakly in } L^q(Q; R^d),$$

where

$$\frac{1}{p} + \frac{1}{q} = \frac{1}{r} < 1.$$

In addition, let

$$\left.\begin{aligned} \operatorname{div} \boldsymbol{U}_n \equiv \nabla \cdot \boldsymbol{U}_n, \\ \mathbf{curl}\, \boldsymbol{V}_n \equiv (\nabla \boldsymbol{V}_n - \nabla^T \boldsymbol{V}_n) \end{aligned}\right\} \text{ be precompact in } \begin{cases} W^{-1,s}(Q), \\ W^{-1,s}(Q, R^{d \times d}), \end{cases}$$

for a certain $s > 1$.

Then

$$\boldsymbol{U}_n \cdot \boldsymbol{V}_n \to \boldsymbol{U} \cdot \boldsymbol{V} \text{ weakly in } L^r(Q).$$

For the proof, see [98, Lemma 11.11].

Weak compactness in the space of integrable functions

Since L^1 is neither reflexive nor dual of a Banach space, the uniformly bounded sequences in L^1 are in general not weakly relatively compact in L^1. On the other hand, the property of weak compactness is equivalent to the property of sequential weak compactness.

Theorem 14 (Weak L^1-compactness)

Let $\mathcal{U} \subset L^1(Q)$, where $Q \subset R^m$ is a bounded measurable set.

Then the following statements are equivalent:

(i) any sequence $\{v_n\}_{n=1}^{\infty} \subset \mathcal{U}$ contains a subsequence weakly converging in $L^1(Q)$;

(ii) (**Dunford–Pettis Criterion**) for any $\varepsilon > 0$ there exists $k > 0$ such that

$$\int_{\{|v| \geq k\}} |v(y)|\, \mathrm{d}y \leq \varepsilon \ \text{ for all } \ v \in \mathcal{U};$$

(iii) (**equi-integrability**) for any $\varepsilon > 0$ there exists $\delta > 0$ such for all $v \in \mathcal{U}$

$$\int_M |v(y)|\, \mathrm{d}y < \varepsilon$$

for any measurable set $M \subset Q$ such that

$$|M| < \delta;$$

(iv) (**De la Vallé–Poussin Criterion**) there exists a nonnegative function $\Phi \in C([0, \infty))$,

$$\lim_{z \to \infty} \frac{\Phi(z)}{z} = \infty,$$

such that

$$\sup_{v \in \mathcal{U}} \int_Q \Phi(|v(y)|)\, \mathrm{d}y \leq c.$$

For the proof, see Ekeland and Temam [73, Chapter 8, Theorem 1.3], Pedregal [177, Lemma 6.4].

Lemma 4 (Biting Lemma)
Let $\{v_n\}_{n=1}^{\infty} \subset L^1(Q)$ be a sequence of measurable functions, where $Q \subset R^m$ is a bounded measurable set, such that

$$\int_Q |v_n(y)|\, \mathrm{d}\, y \leq c \text{ uniformly } \text{ for } n = 1, 2, \ldots.$$

Then there exists a function $v \in L^1(Q)$, a subsequence $\{v_{n_k}\}_{k=1}^{\infty}$ of $\{v_n\}_{n=1}^{\infty}$, and nonincreasing sequence of measurable sets $\{E_\ell\}_{\ell=1}^{\infty}$, $|E_\ell| \to 0$ such that

$$v_{n_k} \to v \text{ weakly in } L^1(Q \setminus E_\ell) \text{ as } k \to \infty$$

for any fixed ℓ.

The proof can be found in Ball and Murat [9].

Finally, we record a technical lemma concerning solvability of nonlinear problems.

Theorem 15 (A fixed point Theorem)
Let M and N be positive integers. Let $C_1 > \varepsilon > 0$ and $C_2 > 0$ be real numbers. Let

$$V = \{(r, u) \in R^M \times R^N,\ r_i > 0,\ i = 1, \dots, M\},$$
$$W = \{(r, u) \in R^M \times R^N,\ \varepsilon < r_i < C_1,\ i = 1, \dots, M \text{ and } |u| \le C_2\}.$$

Let F be a continuous mapping $V \times [0, 1]$ to $R^M \times R^N$ and satisfying:

1. If $f \in V$ satisfies $F(f, \zeta) = 0$ for all $\zeta \in [0, 1]$ then $f \in W$;
2. The equation $F(f, 0) = 0$ is a linear system with respect to f and admits a solution in W.

Then there exists $f \in W$ such that $F(f, 1) = 0$.

For the proof, see Gallouët, Maltese , Novotný [115, Theorem A.1].

Space Discretization

Having introduced analytical tools we proceed with a preliminary material on discretization methods. We assume that the reader is familiar with the basic theory of finite element and finite volume methods for partial differential equations. A good source of information are the monographs Ciarlet [55], Boffi, Brezzi, Fortin [23], Dolejší, Feistauer [71], Eymard, Gallouët, Herbin [78], Feistauer, Felcman, Straškraba [100], Kröner [143], LeVeque [151], Toro [194]. To fix the notation we start by describing the computational mesh and the corresponding discrete spaces.

Let $\Omega_h \subset R^d$, $d = 2, 3$, denote a computational domain, where $h \in (0, h_0)$ for some $h_0 < 1$, is a discretization parameter. In general Ω_h denotes a polygonal approximation of the physical domain Ω and may change with h. In most cases considered in this monograph, we consider $\Omega_h = \Omega$. In particular,

$$\Omega = \mathbb{T}^d,\ \mathbb{T}^d = \Pi_{i=1}^d [a_i, b_i]|_{\{a_i, b_i\}}$$

in the case of space periodic boundary conditions, where $\mathbb{T}^d$ is the so-called flat torus. The questions arising from the approximation of Ω by a sequence of Ω_h, $h \to 0$ are discussed in Chapter 13. If not otherwise stated, we therefore drop the subscript h and identify Ω_h with Ω.

Unstructured mesh

In practical computations one can use different types of discretizations of Ω. We start by describing a general unstructured mesh.

A computational mesh $\mathcal{T}_h$ for Ω is a set of compact polygons or polyhedrons, if $d = 2$ or $d = 3$, respectively, such that

$$\overline{\Omega} = \bigcup_{K \in \mathcal{T}_h} K.$$

Here K is called an element of a computational mesh $\mathcal{T}_h$. The boundary of an element K is the union of its faces or edges. The set of all faces of an element K is denoted by $\mathcal{E}(K)$.

- The set of all faces is denoted by $\mathcal{E}$, $\mathcal{E}_{ext} = \mathcal{E} \cap \partial\Omega$ and $\mathcal{E}_{int} = \mathcal{E} \backslash \partial\Omega$ stand for the set of all exterior and interior faces, respectively. If periodic boundary conditions are applied, i.e. $\Omega = \mathbb{T}^d$, then $\mathcal{E}_{int} = \mathcal{E}$ and $\mathcal{E}_{ext} = \emptyset$.
- Each face σ is associated with a normal vector $\boldsymbol{n}_\sigma$, which we typically denote simply by $\boldsymbol{n}$. The normal vector pointing outward of K is denoted by $\boldsymbol{n}_{\sigma K}$, $\sigma \in \mathcal{E}(K)$.
- h_K stands for the diameter of an element K, and h for the (maximal) mesh size, $h = \max_{K \in \mathcal{T}_h} h_K$.

Definition 1 (UNSTRUCTURED MESH)
We speak about *unstructured mesh* $\mathcal{T}_h$ if:

- Two elements are either disjoint, or their intersection is formed by a common face, or their intersection is a common vertex. For two neighbors $K, L \in \mathcal{T}_h$ we denote $\sigma = K|L$ for $\sigma \in \mathcal{E}(K) \cap \mathcal{E}(L)$.
- $\mathcal{T}_h$ is *regular* and *quasi-uniform*, cf. [55,78], meaning that there exist positive real numbers θ_0 and c_0 which are independent of h such that

$$\inf_{K \in \mathcal{T}_h} \frac{\xi_K}{h_K} \geq \theta_0 \quad \text{and} \quad c_0 h \leq h_K. \tag{7}$$

  Here ξ_K stands for the diameter of the largest ball included in K.
- Let $|K|$ and $|\sigma|$ denote the d- and $(d-1)$-dimensional Lebesgue measure of an element K and a face σ, respectively. Then

$$|K| \approx h^d, \ |\sigma| \approx h^{d-1} \text{ for any } K \in \mathcal{T}_h, \ \sigma \in \mathcal{E}.$$

- There is a family of control points $\{x_K\}_{K \in \mathcal{T}_h}$, such that for any $\sigma = K|L$ the direction vector $\overrightarrow{x_K x_L}$ is perpendicular to σ. We denote $d_\sigma = |x_K - x_L|$.

Now we introduce a dual grid over a primary mesh $\mathcal{T}_h$.

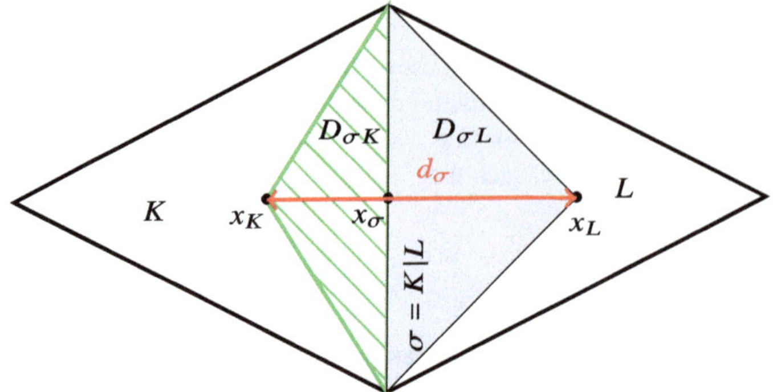

Fig. 1 Dual cell $D_\sigma = D_{\sigma K} \cup D_{\sigma L}$ for 2D unstructured mesh

Definition 2 (DUAL GRID OF AN UNSTRUCTURED MESH)
For any $\sigma = K|L \in \mathcal{E}_{int}$, a dual cell D_σ is defined as $D_\sigma \equiv D_{\sigma K} \cup D_{\sigma L}$, where $D_{\sigma K}$ (resp. $D_{\sigma L}$) is a closed area determined by the control point x_K (resp. x_L) with the common vertices of K and L, see Figure 1 for a two-dimensional illustration. Note that $D_\sigma = D_{\sigma K}$ if $\sigma \in \mathcal{E}_{ext} \cap \mathcal{E}(K)$. The dual grid of an unstructured mesh $\mathcal{D}_h$ is the union of all dual cells.

If $\mathcal{T}_h$ consists of rectangles or cuboids we speak about a structured mesh.

Structured mesh

Definition 3 (STRUCTURED MESH)
In addition to the properties of a general mesh given in Definition 1 a *structured mesh* satisfies the following conditions:

- Any $\sigma \in \mathcal{E}$ is orthogonal to one of the basis vectors of the Cartesian coordinates $\mathbf{e}_i$, $i \in \{1, \ldots, d\}$. We denote by $\mathcal{E}_i$, $i = 1, \ldots, d$ the set of all faces that are orthogonal to the unit vector $\mathbf{e}_i$ and set $\mathcal{E}_i(K) = \mathcal{E}(K) \cap \mathcal{E}_i$.
- We write $\sigma = \overrightarrow{K|L}$ if $x_L = x_K + h_i \mathbf{e}_i$ for any $\sigma \in \mathcal{E}_i$. Similarly, we write $K = \overrightarrow{[\sigma\sigma']}$ for $\sigma, \sigma' \in \mathcal{E}_i(K)$ if $x_{\sigma'} = x_\sigma + h_i \mathbf{e}_i$.
- For any $\sigma = K|L \in \mathcal{E}_i$, $i \in 1, \ldots, d$ we have $d_\sigma = h_i$. It is obvious that $|K| = h_i |\sigma|$ for any $\sigma \in \mathcal{E}_i(K)$.

Analogously to the above we will define a dual grid over a primary structured mesh.

Definition 4 (DUAL GRID OF A STRUCTURED MESH)
For any $\sigma = K|L \in \mathcal{E}_{int}$ we define a dual cell $D_\sigma \equiv D_{\sigma K} \cup D_{\sigma L}$, where $D_{\sigma K}$ (resp. $D_{\sigma L}$) is half of an element K (resp. L) adjacent to σ, see Figure 2 for a

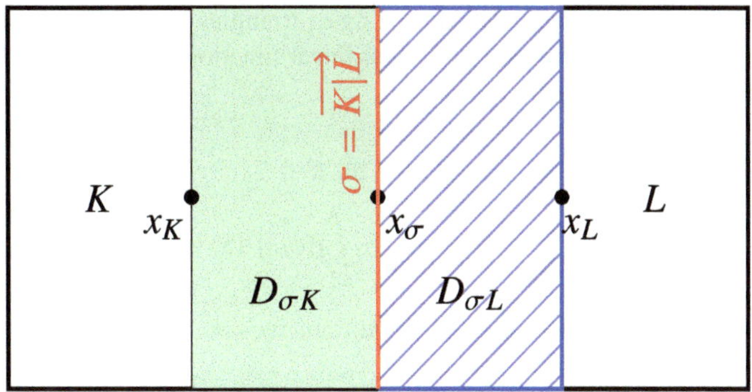

Fig. 2 Dual cell $D_\sigma = D_{\sigma K} \cup D_{\sigma L}$ for 2D structured mesh

two-dimensional example. Note that $D_\sigma = D_{\sigma K}$ if $\sigma \in \mathcal{E}_{ext} \cap \mathcal{E}(K)$. We set $\mathcal{D}_{i,h} = \{D_\sigma\}_{\sigma \in \mathcal{E}_i}$ to be the d dual grids of a structured mesh corresponding to faces perpendicular to $\mathbf{e}_i$, $i = 1, \ldots, d$.

Discrete function spaces

We continue with introducing discrete function spaces. Let f be a function defined on the computational domain Ω with a mesh $\mathcal{T}_h$, f smooth on each element $K \in \mathcal{T}_h$. For each $\sigma \in \mathcal{E}_{int}$ we define outward/inward traces in the following way

$$f^{\mathrm{out}} = \lim_{\delta \to 0+} f(x + \delta \boldsymbol{n}_\sigma), \quad f^{\mathrm{in}} = \lim_{\delta \to 0+} f(x - \delta \boldsymbol{n}_\sigma), \ x \in \sigma, \ \sigma \in \mathcal{E}_{int}. \tag{8}$$

We simply write f for f^{in} if no confusion arises. For any face $\sigma \in \mathcal{E}_{ext}$, f^{out} has to be prescribed. One possible choice is to simply set $f^{\mathrm{out}} = f^{\mathrm{in}}$ indicating the no-flux boundary condition. As we will see later in fluid dynamic models that are studied in this monograph it is more suitable to adjusted f^{out} according to the specific choice of boundary conditions. For any face $\sigma \in \mathcal{E}$ the following notation is used for a discrete jump and an average, respectively,

$$[[f]] = f^{\mathrm{out}} - f^{\mathrm{in}}, \quad \{\{f\}\} = \frac{f^{\mathrm{out}} + f^{\mathrm{in}}}{2}. \tag{9}$$

Let $\mathcal{P}_n(K)$ denote a set of polynomials of degree at most n on an element K. We introduce the *function spaces of piecewise constant* functions

$$Q_h = \left\{\phi_h \in L^1(\Omega) \,|\, \phi_h|_K = \phi_K \in \mathcal{P}_0(K) \text{ for all } K \in \mathcal{T}_h\right\}, \tag{10a}$$

$$W_h^{(i)} = \left\{\phi_h \in L^1(\Omega) \,|\, \phi_h|_{D_\sigma} = \phi_{D_\sigma} \in \mathcal{P}_0(D_\sigma) \text{ for all } D_\sigma \in \mathcal{D}_{i,h}\right\}, \tag{10b}$$

$i = 1, \ldots, d$.

Further, for unstructured mesh consisting of triangles or tetrahedrons, we introduce the space of nonconforming piecewise linear functions, the so-called *Crouzeix–Raviart* finite element space

$$V_h = \left\{ \phi_h \in L^2(\Omega) \;\middle|\; \phi_h|_K \in \mathcal{P}_1(K) \text{ for all } K \in \mathcal{T}_h, \right. \\ \left. \int_\sigma [[\phi_h]] \,\mathrm{dS}_x = 0 \text{ for } \sigma \in \mathcal{E}_{int} \right\}. \tag{10c}$$

For homogeneous Dirichlet boundary conditions we set

$$V_{0,h} = \left\{ \phi_h \in V_h \;\middle|\; \int_\sigma \phi_h \,\mathrm{dS}_x = 0 \text{ for } \sigma \in \mathcal{E}_{ext} \right\}. \tag{10d}$$

The standard projection operators associated to these spaces are defined as follows:

$$\Pi_Q : L^1(\Omega) \to Q_h, \quad \Pi_Q \phi = \sum_{K \in \mathcal{T}_h} \frac{1_K}{|K|} \int_K \phi \,\mathrm{d}x, \tag{11a}$$

$$\Pi_W^{(i)} : L^1(\Omega) \to W_h^{(i)}, \quad \Pi_W^{(i)} \phi = \sum_{\sigma \in \mathcal{E}_i} \frac{1_{D_\sigma}}{|\sigma|} \int_\sigma \phi \,\mathrm{dS}_x, \; i = 1, \dots, d, \tag{11b}$$

$$\Pi_V : W^{1,1}(\Omega) \to V_h, \quad \int_\sigma \Pi_V \phi \,\mathrm{dS}_x = \int_\sigma \phi \,\mathrm{dS}_x \text{ for any } \sigma \in \mathcal{E}. \tag{11c}$$

where 1_K and 1_{D_σ} are characteristic functions

$$1_K(x) = \begin{cases} 1 & \text{if } x \in K, \\ 0 & \text{if } x \notin K, \end{cases} \quad 1_{D_\sigma}(x) = \begin{cases} 1 & \text{if } x \in D_\sigma, \\ 0 & \text{if } x \notin D_\sigma. \end{cases} \tag{12}$$

Let $\boldsymbol{\phi}_h = (\phi_{1,h}, \dots, \phi_{d,h})$ be a vector valued function, then by $\boldsymbol{\phi}_h \in \boldsymbol{Q}_h$ (resp. $\boldsymbol{V}_h$ and $\boldsymbol{V}_{0,h}$) we mean $\phi_{i,h} \in Q_h$ (resp. V_h and $V_{0,h}$) for all $i \in \{1, \dots, d\}$. Accordingly, we extend the projection operators Π_Q and Π_V to vector valued functions componentwisely, i.e.

$$\Pi_Q \boldsymbol{\phi} = (\Pi_Q \phi_1, \dots, \Pi_Q \phi_d), \quad \Pi_V \boldsymbol{\phi} = (\Pi_V \phi_1, \dots, \Pi_V \phi_d).$$

Further, we define $\boldsymbol{\phi}_h \in \boldsymbol{W}_h \equiv W_h^{(1)} \times \dots \times W_h^{(d)}$ such that $\phi_{i,h} \in W_h^{(i)}$ for all $i = 1, \dots, d$. The corresponding projection operator for the space $\boldsymbol{W}_h$ reads

$$\Pi_W : L^1(\Omega; R^d) \to \boldsymbol{W}_h, \quad \Pi_W \boldsymbol{\phi} = \left(\Pi_W^{(1)} \phi_1, \dots, \Pi_W^{(d)} \phi_d \right). \tag{13}$$

We shall frequently write $\overline{\phi} \equiv \Pi_Q \phi$ and unify the notation for the mean value for a generic vector valued function $\boldsymbol{\phi}_h \in \boldsymbol{Q}_h \cup \boldsymbol{V}_h \cup \boldsymbol{W}_h$ at an interface $\sigma \in \mathcal{E}$:

$$\langle \boldsymbol{\phi}_h \rangle_\sigma = \begin{cases} \frac{1}{|\sigma|} \int_\sigma \boldsymbol{\phi}_h \,\mathrm{dS}_x, & \text{for } \boldsymbol{\phi}_h \in \boldsymbol{V}_h, \\ \{\{\boldsymbol{\phi}_h\}\}_\sigma, & \text{for } \boldsymbol{\phi}_h \in \boldsymbol{Q}_h, \\ \boldsymbol{\phi}_h|_\sigma, & \text{for } \boldsymbol{\phi}_h \in \boldsymbol{W}_h. \end{cases} \tag{14}$$

Discrete difference operators

Difference operators on an unstructured mesh. For piecewise linear functions $f_h \in V_h$, $\boldsymbol{g}_h \in \boldsymbol{V}_h$ it is natural to define discrete gradient and divergence as follows

$$\nabla_h f_h \equiv \nabla_x f_h \quad \text{and} \quad \mathrm{div}_h \boldsymbol{g}_h \equiv \mathrm{div}_x \boldsymbol{g}_h \quad \text{on any } K \in \mathcal{T}_h.$$

For piecewise constant functions there are several ways to define discrete difference operators. For $\phi_h \in Q_h$ and $\boldsymbol{\phi}_h = (\phi_{1,h}, \ldots, \phi_{d,h})^T \in \boldsymbol{Q}_h$ we define *discrete gradient, discrete divergence* and *discrete Laplace* operators by

$$\begin{aligned}
\nabla_h \phi_h &= \sum_{K \in \mathcal{T}_h} (\nabla_h \phi_h)_K 1_K, & (\nabla_h \phi_h)_K &= \sum_{\sigma \in \mathcal{E}(K)} \frac{|\sigma|}{|K|} \langle \phi_h \rangle_\sigma \boldsymbol{n}_{\sigma K}, \\
\nabla_{\mathcal{D}} \phi_h &= \sum_{\sigma \in \mathcal{E}} (\nabla_{\mathcal{D}} \phi_h)_\sigma 1_{D_\sigma}, & (\nabla_{\mathcal{D}} \phi_h)_\sigma &= \frac{[[\phi_h]]}{d_\sigma} \boldsymbol{n}, \\
\mathrm{div}_h \boldsymbol{\phi}_h &= \sum_{K \in \mathcal{T}_h} \left(\mathrm{div}_h \boldsymbol{\phi}_h\right)_K 1_K, & \left(\mathrm{div}_h \boldsymbol{\phi}_h\right)_K &= \sum_{\sigma \in \mathcal{E}(K)} \frac{|\sigma|}{|K|} \langle \boldsymbol{\phi}_h \rangle_\sigma \cdot \boldsymbol{n}_{\sigma K}, \\
\Delta_h \phi_h &= \sum_{K \in \mathcal{T}_h} (\Delta_h \phi_h)_K 1_K, & (\Delta_h \phi_h)_K &= \sum_{\sigma \in \mathcal{E}(K)} \frac{|\sigma|}{|K|} \frac{[[\phi_h]]}{d_\sigma}.
\end{aligned} \tag{15}$$

Note that for $\sigma \in \mathcal{E}_{ext}$ the value ϕ^{out} or $\boldsymbol{\phi}_h^{\mathrm{out}}$ is adjusted according to the numerical boundary conditions. For vector valued functions, the operators ∇_h, $\nabla_{\mathcal{D}}$, Δ_h work componentwisely, i.e.,

$$\begin{aligned}
&\nabla_h \boldsymbol{\phi}_h = \left(\nabla_h \phi_{1,h}, \ldots, \nabla_h \phi_{d,h}\right), \quad \nabla_{\mathcal{D}} \boldsymbol{\phi}_h = \left(\nabla_{\mathcal{D}} \phi_{1,h}, \ldots, \nabla_{\mathcal{D}} \phi_{d,h}\right), \\
&\Delta_h \boldsymbol{\phi}_h = \left(\Delta_h \phi_{1,h}, \ldots, \Delta_h \phi_{d,h}\right).
\end{aligned}$$

Difference operators on a structured mesh. In case of a regular structured grid, it is convenient to introduce discrete difference operators in each direction of the coordinates. For $r_h \in Q_h$ and $\mathbf{q}_h = (q_{1,h}, \ldots, q_{d,h})^T \in \boldsymbol{W}_h$ we define

$$\eth_{\mathcal{D}_i} r_h = \sum_{\sigma \in \mathcal{E}_i} 1_{D_\sigma} \left(\eth_{\mathcal{D}_i} r_h\right)_\sigma, \quad \left(\eth_{\mathcal{D}_i} r_h\right)_\sigma = \frac{r_h|_L - r_h|_K}{h}$$

for all $\sigma = \overrightarrow{K|L} \in \mathcal{E}_i$ and

$$\eth_{\mathcal{T}}^{(i)} q_{i,h} = \sum_{K\in\mathcal{T}_h} \left(\eth_{\mathcal{T}}^{(i)} q_{i,h}\right)_K 1_K, \quad \left(\eth_{\mathcal{T}}^{(i)} q_{i,h}\right)_K = \frac{q_{i,h}|_{\sigma'} - q_{i,h}|_{\sigma}}{h}$$

for all $K = \overrightarrow{[\sigma\sigma']}$, and $\sigma, \sigma' \in \mathcal{E}_i$.

With the above notations, we further define a *discrete Laplace operator*

$$\Delta_h r_h = \sum_{i=1}^d \Delta_h^{(i)} r_h, \quad \Delta_h^{(i)} r_h = \eth_{\mathcal{T}}^{(i)}(\eth_{\mathcal{D}_i} r_h), \quad i = 1, \ldots, d.$$

and the following *discrete gradient* on the dual grid and *discrete divergence* on the primary grid

$$\nabla_{\mathcal{D}} r_h = \left(\eth_{\mathcal{D}_1} r_h, \ldots, \eth_{\mathcal{D}_d} r_h\right), \quad \mathrm{div}_{\mathcal{T}} \mathbf{q}_h = \sum_{i=1}^d \eth_{\mathcal{T}}^{(i)} q_{i,h}. \tag{16}$$

It is easy to observe for $r_h \in Q_h$, $\boldsymbol{f}_h \in \boldsymbol{Q}_h$ that

$$\begin{aligned}
&\eth_{\mathcal{T}}^{(i)}\left(\Pi_W^{(i)} r_h\right) = \Pi_Q[\eth_{\mathcal{D}_i} r_h], \\
&\nabla_h r_h = \Pi_Q[\nabla_{\mathcal{D}} r_h] = \left(\eth_{\mathcal{T}}^{(1)} \Pi_W^{(1)} r_h, \ldots, \eth_{\mathcal{T}}^{(d)} \Pi_W^{(d)} r_h\right), \\
&\Delta_h r_h = \mathrm{div}_{\mathcal{T}}(\nabla_{\mathcal{D}} r_h), \\
&\mathrm{div}_h \boldsymbol{f}_h = \sum_{i=1}^d \eth_{\mathcal{T}}^{(i)} \{\{f_{i,h}\}\} = \mathrm{div}_{\mathcal{T}}\left(\Pi_{\boldsymbol{W}} \boldsymbol{f}_h\right), \\
&(\eth_{\mathcal{D}_i} r_h)_\sigma \mathbf{e}_i = (\nabla_{\mathcal{D}} r_h)_\sigma \quad \text{for } \sigma \in \mathcal{E}_i, \quad i = 1, \ldots, d.
\end{aligned} \tag{17}$$

Lemma 5 For any $\boldsymbol{\phi} \in W^{1,1}(\Omega; R^d)$ it holds

$$\int_K \mathrm{div}_{\mathcal{T}} \Pi_{\boldsymbol{W}} \boldsymbol{\phi} \, \mathrm{d}x = \int_K \mathrm{div}_x \boldsymbol{\phi} \, \mathrm{d}x \quad \text{for any } K \in \mathcal{T}_h. \tag{18}$$

Proof From the definition of $\mathrm{div}_{\mathcal{T}}$, we know that

$$\begin{aligned}
&\int_K \mathrm{div}_{\mathcal{T}} \Pi_{\boldsymbol{W}} \boldsymbol{\phi} \, \mathrm{d}x = \sum_{i=1}^d \sum_{\sigma\in\mathcal{E}_i(K)} |\sigma| \Pi_W^{(i)} \phi_i \mathbf{e}_i \cdot \boldsymbol{n}_{\sigma K} \\
&= \sum_{i=1}^d \sum_{\sigma\in\mathcal{E}_i(K)} \int_\sigma \phi_i \mathbf{e}_i \cdot \boldsymbol{n}_{\sigma K} \mathrm{dS}_x = \sum_{\sigma\in\mathcal{E}(K)} \int_\sigma \boldsymbol{\phi} \cdot \boldsymbol{n}_{\sigma K} \mathrm{dS}_x = \int_K \mathrm{div}_x \boldsymbol{\phi} \, \mathrm{d}x.
\end{aligned}$$

Interpolation errors

For simplicity, hereafter we denote by $\|\cdot\|_{L^p}$ the norm in $\|\cdot\|_{L^p(\Omega)}$, analogous notations hold for other spaces. We will also write

$$a \stackrel{<}{\sim} b \text{ if } a \leq cb,$$

where c is a positive constant independent of the discretization parameter h.

Next we list some basic interpolation inequalities, see, for instance, [56,78,115]. For $\phi \in C^1(\overline{\Omega})$, $h \in (0, h_0)$, $h_0 < 1$, we have

$$\begin{aligned}
&\left|\left[\left[\Pi_Q \phi\right]\right]\right| \stackrel{<}{\sim} h \|\phi\|_{C^1}, \left|\phi(x) - \left\{\left\{\Pi_Q \phi\right\}\right\}\right| \stackrel{<}{\sim} h \|\phi\|_{C^1}, \text{ for any } x \in \sigma \in \mathcal{E}_{int}, \\
&\left\|\phi - \Pi_Q \phi\right\|_{L^p} \stackrel{<}{\sim} h \|\phi\|_{C^1}, \left\|\Pi_Q \phi - \left\{\left\{\Pi_Q \phi\right\}\right\}\right\|_{L^p} \stackrel{<}{\sim} h \|\phi\|_{C^1}.
\end{aligned} \tag{19}$$

If additionally $\phi \in C^2(\overline{\Omega})$, $\boldsymbol{\phi} \in C^2(\overline{\Omega}; R^d)$, then we have for all $1 \leq p \leq \infty$

$$\begin{aligned}
\left\|\nabla_x \phi - \nabla_{\mathcal{D}} \Pi_Q \phi\right\|_{L^p} \stackrel{<}{\sim} h \|\phi\|_{C^2}, \quad & \left\|\nabla_x \phi - \nabla_h \left(\Pi_Q \phi\right)\right\|_{L^p} \stackrel{<}{\sim} h \|\phi\|_{C^2}, \\
\left\|\operatorname{div}_x \boldsymbol{\phi} - \operatorname{div}_h (\Pi_Q \boldsymbol{\phi})\right\|_{L^p} \stackrel{<}{\sim} h \|\boldsymbol{\phi}\|_{C^2}, \quad & \left\|\nabla_{\mathcal{D}} \Pi_Q \phi\right\|_{L^p} \stackrel{<}{\sim} \|\phi\|_{C^1}, \\
& \left\|\Delta_h \Pi_Q \phi\right\|_{L^\infty} \stackrel{<}{\sim} \|\phi\|_{C^2}.
\end{aligned} \tag{20}$$

For the Crouzeix–Raviart finite elements the following interpolation inequalities hold, see [58]. Let $\phi \in C^1(\overline{\Omega})$ then we have for all $1 \leq p \leq \infty$

$$\begin{aligned}
\left|\left[\left[\Pi_Q \Pi_V \phi\right]\right]\right| \stackrel{<}{\sim} h \|\phi\|_{C^1}, \quad & \text{for any } x \in \sigma \in \mathcal{E}_{int}, \\
\left\|\phi - \Pi_Q \Pi_V \phi\right\|_{L^p} \stackrel{<}{\sim} h \|\phi\|_{C^1}, \quad & \left\|\phi - \langle \Pi_V \phi \rangle_\sigma \right\|_{L^p} \stackrel{<}{\sim} h \|\phi\|_{C^1}.
\end{aligned} \tag{21}$$

Let $\phi \in C^2(\overline{\Omega})$, $f_h \in V_h$ then

$$\begin{aligned}
\|\phi - \Pi_V \phi\|_{L^p} + h \|\nabla_x \phi - \nabla_h \Pi_V \phi\|_{L^p} \stackrel{<}{\sim} h^2 \|\phi\|_{C^2}, \quad & 1 \leq p \leq \infty, \\
\|f_h - \langle f_h \rangle\|_{L^2(K)} \stackrel{<}{\sim} h \|\nabla_h f_h\|_{L^2(K)}, \quad & \text{for any } K \in \mathcal{T}_h.
\end{aligned} \tag{22}$$

Fundamental Discrete Inequalities

We present further important inequalities that will be used later in the analysis of numerical methods. Firstly, it is easy to check that for any $f_h, g_h \in Q_h$ the *discrete product rule*

$$[[f_h g_h]] = \{\{f_h\}\}[[g_h]] + [[f_h]]\{\{g_h\}\} = (f_h)^{\text{out}}[[g_h]] + [[f_h]](g_h)^{\text{in}} \tag{23}$$

and the algebraic identity

$$\{\{f_h g_h\}\} - \{\{f_h\}\}\,\{\{g_h\}\} = \frac{1}{4}[[f_h]][[g_h]] \tag{24}$$

hold. A direct application of the product rule (23) implies the identity

$$[[r_h \boldsymbol{f}_h]] \cdot [[\boldsymbol{f}_h]] - \frac{1}{2}[[r_h]][[|\boldsymbol{f}_h|^2]] = \{\{r_h\}\}|[[\boldsymbol{f}_h]]|^2 \tag{25}$$

for $r_h \in Q_h$, $\boldsymbol{f}_h \in \boldsymbol{Q}_h$.

Inverse estimates and trace inequality. Let $p > q \geq 1$. Then the following *inverse estimates* hold

$$\begin{aligned} &\|f_h\|_{W^{r,p}(K)} \overset{<}{\sim} h^{-r+d\left(\frac{1}{p}-\frac{1}{q}\right)} \|f_h\|_{L^q(K)}, \quad f_h \in \mathcal{P}_m(K),\ r = 0, 1,\ m = 1, 2, \ldots, \\ &\|f_h\|_{L^p(\Omega)} \overset{<}{\sim} h^{d\left(\frac{1}{p}-\frac{1}{q}\right)} \|f_h\|_{L^q(\Omega)}, \quad f_h \in Q_h \text{ or } f_h \in V_h, \end{aligned} \tag{26}$$

see, e.g., [55], and the *trace inequality*

$$\|f\|_{L^p(\partial K)} \overset{<}{\sim} h^{-\frac{1}{p}} \left(\|f\|_{L^p(K)} + h\|\nabla_x f\|_{L^p(K)}\right),\ 1 \leq p \leq \infty \ \text{ for any } \ f \in C^1(\overline{K}), \tag{27}$$

hold, see [39,111]. From the latter we readily deduce that

$$\|r_h\|_{L^p(\partial K)} \overset{<}{\sim} h^{-\frac{1}{p}} \|r_h\|_{L^p(K)}, \quad 1 \leq p \leq \infty \ \text{ for any } r_h \in \mathcal{P}_m(K), m = 0, 1, \ldots. \tag{28}$$

Let us now introduce the following $W^{1,p}$-seminorm for Q_h

$$\|r_h\|_{W^{1,p}(Q_h)} \equiv \left(\sum_{\sigma \in \mathcal{E}_{int}} \int_\sigma \frac{|[[r_h]]|^p}{d_\sigma} \mathrm{dS}_x \right)^{1/p} \qquad \text{for any } r_h \in Q_h,\ 1 \leq p < \infty.$$

For $p = 2$ we use for simplicity the notation H^1 for $W^{1,2}$-seminorm. Hereby, it is worthwhile to mention the following result on the estimate of H^1-seminorms

$$\|f_h\|_{H^1(Q_h)} \lesssim \|\nabla_h f_h\|_{L^2} \quad \text{for any } f_h \in V_h, \tag{29}$$

see [111, Lemma 2.2]. Note that the left hand side is well-defined also for $f_h \in V_h$. Moreover, if $[[r_h]] = 0$ for all $\sigma \in \mathcal{E}_{ext}$ then

$$\|r_h\|_{H^1(Q_h)} = \|\nabla_{\mathcal{D}} r_h\|_{L^2} \text{ for any } r_h \in Q_h.$$

Sobolev–Poincaré inequalities. We first recall a discrete version of Poincaré's inequality for any $f_h \in V_h$

$$\|f_h - \overline{f_h}\|_{L^2(K)} \equiv \left\|f_h - \Pi_Q f_h\right\|_{L^2(K)} \lesssim h \|\nabla_h f_h\|_{L^2(K)}, \tag{30}$$

see [176].

Theorem 16 (Discrete Sobolev–Poincaré inequalities)

Let $q \in R$ satisfying $1 \le q \le 6$ for $d = 3$, $1 \le q < \infty$ for $d = 2$. Then the following discrete versions of the Sobolev–Poincaré inequalities hold for $f_h \in V_h$

$$\|f_h\|_{L^q(\Omega)} \lesssim \|f_h\|_{L^1(\Omega)} + \|\nabla_h f_h\|_{L^2(\Omega)} \tag{31}$$

$$\|f_h - \langle f_h \rangle_\Omega\|_{L^q(\Omega)} \lesssim \|\nabla_h f_h\|_{L^2(\Omega)}, \tag{32}$$

where

$$\langle f_h \rangle_\Omega \equiv \frac{1}{|\Omega|} \int_\Omega f_h \,\mathrm{d}x.$$

If the mesh satisfies additionally the following regularity constraint:
there exists $\alpha > 0$ such that

$$\text{dist}\,(x_K, \sigma) \ge \alpha d_\sigma \quad \text{for any } K \in \mathcal{T}_h, \sigma \in \mathcal{E}(K), \tag{33}$$

then we have analogous discrete Sobolev–Poincaré inequalities for $r_h \in Q_h$

$$\|r_h\|_{L^q(\Omega)} \lesssim \|r_h\|_{L^2(\Omega)} + \|r_h\|_{H^1(Q_h)} \tag{34}$$

$$\|r_h - \langle r_h \rangle_\Omega\|_{L^q(\Omega)} \lesssim \|r_h\|_{H^1(Q_h)}, \tag{35}$$

and the discrete Poincaré inequality

$$\|r_h - \langle r_h \rangle_\Omega\|_{L^p(\Omega)} \lesssim \|r_h\|_{W^{1,p}(Q_h)}, \quad 1 \le p < \infty. \tag{36}$$

For the proof of the first inequality, see [42, Lemma 6.1], inequalities (32), (34) and (35) can be found in [148, Theorem 3.3] or [39, Theorem 7]. The discrete Poincaré inequality (36) for functions from Q_h has been proved in [22, Theorem 5].

Theorem 17 (Generalized discrete Sobolev–Poincaré inequality)
Let the hypothesis (33) hold and let $r_h \geq 0$ be such that

$$0 < c_M \leq \int_\Omega r_h \, \mathrm{d}x \quad \text{and} \quad \int_\Omega r_h^\gamma \, \mathrm{d}x \leq c_E$$

where c_M and c_E are positive constants and $\gamma > 1$.
Then there is $c = c(c_M, c_E, \gamma)$ independent of h such that

$$\begin{aligned} \|f_h\|_{L^q(\Omega)} &\leq c\left(\|f_h\|_{H^1(Q_h)} + \int_\Omega r_h |f_h| \, \mathrm{d}x \right) \\ &\leq c\|f_h\|_{H^1(Q_h)} + c\sqrt{c_M}\left(\int_\Omega r_h |f_h|^2 \, \mathrm{d}x \right)^{1/2} \end{aligned} \tag{37}$$

for any $f_h \in Q_h$ and $1 \leq q \leq 6$ for $d = 3$, $1 \leq q < \infty$ for $d = 2$.

Proof **Step 1**
By virtue of (35) we get

$$\begin{aligned} \|f_h\|_{L^q(\Omega)} &\leq \|f_h - \langle f_h \rangle_\Omega\|_{L^q(\Omega)} + \|\langle f_h \rangle_\Omega\|_{L^q(\Omega)} \stackrel{<}{\sim} \|f_h\|_{H^1(Q_h)} + |\langle f_h \rangle_\Omega| \\ &\stackrel{<}{\sim} \|f_h\|_{H^1(Q_h)} + \frac{1}{c_M} \int_\Omega r_h |\langle f_h \rangle_\Omega| \, \mathrm{d}x \\ &\stackrel{<}{\sim} \|f_h\|_{H^1(Q_h)} + \frac{1}{c_M} \int_\Omega r_h |f_h| \, \mathrm{d}x + \frac{1}{c_M} \int_\Omega r_h |f_h - \langle f_h \rangle_\Omega| \, \mathrm{d}x. \end{aligned}$$

Finally, by Hölder's inequality,

$$\int_\Omega r_h |f_h - \langle f_h \rangle_\Omega| \, \mathrm{d}x \leq \|r_h\|_{L^{q'}(\Omega)} \|f_h - \langle f_h \rangle_\Omega\|_{L^q(\Omega)} \stackrel{<}{\sim} c_E \|f_h\|_{H^1(Q_h)}.$$

This gives a desired result for $d = 2$ or

$$\gamma \geq \frac{6}{5} \text{ if } d = 3. \tag{38}$$

Step 2
To get rid of (38) we repeat the previous arguments with r_h replaced by the cut-off function

$$T_k(r_h) \equiv \min\{r_h; k\}.$$

Obviously,

$$
\begin{aligned}
c_M &\le \int_\Omega r_h \,\mathrm{d}x = \int_\Omega T_k(r_h)\,\mathrm{d}x + \int_\Omega (r_h - T_k(r_h))\,\mathrm{d}x \\
&\le \int_\Omega T_k(r_h)\,\mathrm{d}x + 2\int_{r_h \ge k} r_h \,\mathrm{d}x \le \int_\Omega T_k(r_h)\,\mathrm{d}x + 2k^{1-\gamma}\int_\Omega r_h^\gamma \,\mathrm{d}x \\
&\le \int_\Omega T_k(r_h)\,\mathrm{d}x + 2k^{1-\gamma} c_E.
\end{aligned}
$$

Now, let us choose k such that

$$
2k^{1-\gamma} c_E \le \frac{c_M}{2}.
$$

Finally, we repeat the arguments of Step 1 for $r_h \equiv T_k(r_h)$, $c_M \equiv \frac{c_M}{2}$ and $c_E \equiv |\Omega| k^{\frac{6}{5}}$.
The second inequality in (37) follows from Hölder's inequality.

Remark 1 Refining the arguments of Step 2 of the above proof we observe that the result remains true if $r_h \ge 0$ belongs to an equi-integrable set in $L^1(\Omega)$ and

$$
0 < c_M \le \int_\Omega r_h \,\mathrm{d}x.
$$

Integration by parts. Next, we show the integration by parts formulas satisfied by the discrete differential operators defined for piecewise constant functions.

Lemma 6 For any $r_h \in Q_h$, $\boldsymbol{f}_h \in \boldsymbol{Q}_h$ there holds

$$
\int_{\mathcal{E}_{int}} \left(\{\{r_h\}\}[[\boldsymbol{f}_h]] + \{\{\boldsymbol{f}_h\}\}[[r_h]]\right) \cdot \boldsymbol{n} \,\mathrm{dS}_x = 0. \tag{39}
$$

Here and hereafter we use a simplified notation $\int_{\mathcal{E}_{int}} \mathrm{dS}_x$ for $\sum_{\sigma \in \mathcal{E}_{int}} \int_\sigma \mathrm{dS}_x$.

It is worth to mention that due to the identity $\int_{\partial K} \boldsymbol{n}_{\sigma K} \mathrm{dS}_x = 0$, we have

$$
\int_{\partial K} \{\{r_h\}\} \boldsymbol{n}_{\sigma K} \mathrm{dS}_x = \frac{1}{2}\int_{\partial K} [[r_h]] \boldsymbol{n}_{\sigma K} \mathrm{dS}_x. \tag{40}
$$

Lemma 7 (Discrete integration by parts)
Let $r_h, \phi_h \in Q_h$, $\boldsymbol{f}_h, \boldsymbol{g}_h \in \boldsymbol{Q}_h$, $\mathbf{q}_h \in \boldsymbol{W}_h$ and $i \in \{1, \ldots, d\}$.
Then the following hold

$$\int_\Omega \nabla_h r_h \cdot \boldsymbol{f}_h \,\mathrm{d}x = -\int_\Omega r_h \mathrm{div}_h \boldsymbol{f}_h \,\mathrm{d}x$$

if one of the following conditions holds for all $\sigma \in \mathcal{E}_{ext}$

$$[[r_h]] = 0 = \left[\left[\boldsymbol{f}_h\right]\right] \cdot \boldsymbol{n} \quad \text{or} \quad [[r_h]] = 0 = \left\{\left\{\boldsymbol{f}_h\right\}\right\} \cdot \boldsymbol{n}$$
$$\text{or } \{\{r_h\}\} = 0 = \left[\left[\boldsymbol{f}_h\right]\right] \cdot \boldsymbol{n} \quad \text{or} \quad \{\{r_h\}\} = 0 = \left\{\left\{\boldsymbol{f}_h\right\}\right\} \cdot \boldsymbol{n}, \tag{41}$$

$$\int_\Omega \Delta_h r_h \, \phi_h \,\mathrm{d}x = -\int_\Omega \nabla_\mathcal{D} r_h \cdot \nabla_\mathcal{D} \phi_h \,\mathrm{d}x$$
$$\text{if } [[r_h]] = 0 \quad \text{or } \{\{\phi_h\}\} = 0 = [[\phi_h]] \quad \text{for all } \sigma \in \mathcal{E}_{ext}, \tag{42}$$

$$\int_\Omega q_{i,h} \eth_{\mathcal{D}_i} r_h \,\mathrm{d}x = -\int_\Omega r_h \eth^{(i)}_\mathcal{T} q_{i,h} \,\mathrm{d}x$$
$$\text{if } q_{i,h}\Big|_\sigma = 0 \quad \text{for all } \sigma \in \mathcal{E}_{ext}, \tag{43}$$

$$-\int_\Omega \mathbf{q}_h \cdot \nabla_\mathcal{D} r_h \,\mathrm{d}x = \int_\Omega r_h \mathrm{div}_\mathcal{T} \mathbf{q}_h \,\mathrm{d}x$$
$$\text{if } \mathbf{q}_h\Big|_\sigma \cdot \boldsymbol{n} = 0 \quad \text{for all } \sigma \in \mathcal{E}_{ext}. \tag{44}$$

The above equalities hold also for $\Omega = \mathbb{T}^d$ and

$$\int_{\mathbb{T}^d} \nabla_h \mathrm{div}_h \boldsymbol{g}_h \cdot \boldsymbol{f}_h \,\mathrm{d}x = -\int_{\mathbb{T}^d} \mathrm{div}_h \boldsymbol{g}_h \,\mathrm{div}_h \boldsymbol{f}_h \,\mathrm{d}x. \tag{45}$$

Proof We shall prove only the first equality as the rest are analogous. First, applying (39) and (40) and the assumption $[[r_h]] = 0 = \left[\left[\boldsymbol{f}_h\right]\right] \cdot \boldsymbol{n}$ on $\sigma \in \mathcal{E}_{ext}$ we obtain

$$\int_\Omega \nabla_h r_h \cdot \boldsymbol{f}_h \,\mathrm{d}x = \sum_{K\in\mathcal{T}_h} \boldsymbol{f}_K \cdot \int_{\partial K} \{\{r_h\}\}\boldsymbol{n}_{\sigma K} \mathrm{dS}_x$$
$$= \sum_{K\in\mathcal{T}_h} \boldsymbol{f}_K \cdot \int_{\partial K} \frac{[[r_h]]}{2} \boldsymbol{n}_{\sigma K} \mathrm{dS}_x$$
$$= \int_{\mathcal{E}_{int}} \left\{\left\{\boldsymbol{f}_h\right\}\right\} \cdot ([[r_h]]\boldsymbol{n})\mathrm{dS}_x + \frac{1}{2}\int_{\mathcal{E}_{ext}} \boldsymbol{f}_h^{\mathrm{in}} \cdot ([[r_h]]\boldsymbol{n})\mathrm{dS}_x$$
$$= -\int_{\mathcal{E}_{int}} \left[\left[\boldsymbol{f}_h\right]\right] \cdot (\{\{r_h\}\}\boldsymbol{n})\mathrm{dS}_x + \frac{1}{2}\int_{\mathcal{E}_{ext}} \left[\left[\boldsymbol{f}_h\right]\right] \cdot (r_h^{\mathrm{in}}\boldsymbol{n})\mathrm{dS}_x$$
$$= -\sum_{K\in\mathcal{T}_h} r_K \int_{\partial K} \frac{\left[\left[\boldsymbol{f}_h\right]\right]}{2} \cdot \boldsymbol{n}_{\sigma K} \,\mathrm{dS}_x$$
$$= -\sum_{K\in\mathcal{T}_h} r_K \int_{\partial K} \left\{\left\{\boldsymbol{f}_h\right\}\right\} \cdot \boldsymbol{n}_{\sigma K} \,\mathrm{dS}_x$$
$$= -\int_\Omega r_h \,\mathrm{div}_h \boldsymbol{f}_h \,\mathrm{d}x.$$

Next, we can apply the assumption $\{\{r_h\}\} = 0 = \{\{\boldsymbol{f}_h\}\} \cdot \boldsymbol{n}$ on $\sigma \in \mathcal{E}_{ext}$ and obtain analogously

$$
\begin{aligned}
\int_\Omega \nabla_h r_h \cdot \boldsymbol{f}_h \, \mathrm{d}x &= \sum_{K \in \mathcal{T}_h} \boldsymbol{f}_K \cdot \int_{\partial K} \{\{r_h\}\} \boldsymbol{n}_{\sigma K} \mathrm{dS}_x \\
&= - \int_{\mathcal{E}_{int}} [[\boldsymbol{f}_h]] \cdot \boldsymbol{n} \{\{r_h\}\} \mathrm{dS}_x + \int_{\mathcal{E}_{ext}} \{\{r_h\}\} \boldsymbol{f}_h^{\mathrm{in}} \cdot \boldsymbol{n} \, \mathrm{dS}_x \\
&= \int_{\mathcal{E}_{int}} \{\{\boldsymbol{f}_h\}\} \cdot \boldsymbol{n} [[r_h]] \mathrm{dS}_x - \int_{\mathcal{E}_{ext}} r_h^{\mathrm{in}} \, \{\{\boldsymbol{f}_h\}\} \cdot \boldsymbol{n} \, \mathrm{dS}_x \\
&= - \sum_{K \in \mathcal{T}_h} r_K \int_{\partial K} \{\{\boldsymbol{f}_h\}\} \cdot \boldsymbol{n}_{\sigma K} \, \mathrm{dS}_x \\
&= - \int_\Omega r_h \, \mathrm{div}_h \boldsymbol{f}_h \, \mathrm{d}x.
\end{aligned}
$$

Further, assuming $\{\{\boldsymbol{f}_h\}\} \cdot \boldsymbol{n} = 0 = [[r_h]]$ on the exterior faces we derive

$$
\begin{aligned}
\int_\Omega r_h \, \mathrm{div}_h \boldsymbol{f}_h \, \mathrm{d}x &= \sum_{K \in \mathcal{T}_h} r_K \int_{\partial K} \{\{\boldsymbol{f}_h\}\} \cdot \boldsymbol{n}_{\sigma K} \, \mathrm{dS}_x \\
&= - \int_{\mathcal{E}_{int}} \{\{\boldsymbol{f}_h\}\} \cdot \boldsymbol{n} [[r_h]] \mathrm{dS}_x + \int_{\mathcal{E}_{ext}} r_h^{\mathrm{in}} \, \{\{\boldsymbol{f}_h\}\} \cdot \boldsymbol{n} \, \mathrm{dS}_x \\
&= - \int_{\mathcal{E}_{int}} \{\{\boldsymbol{f}_h\}\} \cdot \boldsymbol{n} [[r_h]] \mathrm{dS}_x - \frac{1}{2} \int_{\mathcal{E}_{ext}} \boldsymbol{f}_h^{\mathrm{in}} \cdot \boldsymbol{n} \, [[r_h]] \, \mathrm{dS}_x \\
&= - \sum_{K \in \mathcal{T}_h} \boldsymbol{f}_K \cdot \int_{\partial K} \frac{[[r_h]]}{2} \boldsymbol{n}_{\sigma K} \mathrm{dS}_x = - \sum_{K \in \mathcal{T}_h} \boldsymbol{f}_K \cdot \int_{\partial K} \{\{r_h\}\} \boldsymbol{n}_{\sigma K} \mathrm{dS}_x \\
&= - \int_\Omega \nabla_h r_h \cdot \boldsymbol{f}_h \, \mathrm{d}x.
\end{aligned}
$$

The proof of (39) for the last case of $[[\boldsymbol{f}_h]] \cdot \boldsymbol{n} = 0 = \{\{r_h\}\}$ on the exterior faces is analogous.

Time Discretization

We discretize the time interval $[0, T]$ by an equidistant time grid with a time step $\Delta t > 0$ and denote

$$
f^k(x) = f(t^k, x) \; \text{ for all } \; x \in \Omega, \; t^k = k \, \Delta t \; \text{ for } k = 0, 1, \dots, N_T.
$$

In literature one can find numerous efficient ODE solvers in order to solve numerically a system of differential equations. Here we confine ourself to the first order implicit discretization in time, i.e. we apply the backward Euler method to approximate the time derivative

$$\frac{\mathrm{d}}{\mathrm{d}\,t} f(t_k, x) \approx D_t f^k(x) \equiv \frac{f^k(x) - f^{k-1}(x)}{\Delta t}, \quad k = 1, 2, \ldots, N_T, \ x \in \Omega. \tag{46}$$

We set $f^0(x)$ to be a prescribed initial condition

$$f^0(x) = f_0(x) \equiv f(0, x), \quad x \in \Omega.$$

We also introduce a piecewise constant interpolation f_Δ of discrete values f^k

$$f_\Delta(t, \cdot) = f^0 \text{ for } t < \Delta t, \& f_\Delta(t, \cdot) = f^k \text{ for } t \in [k\Delta t, (k+1)\Delta t), \ k = 1, 2, \ldots, N_T. \tag{47}$$

For a discrete function in space and time we will use the notation f_h^k, $k = 1, 2, \ldots, N_T$. For its piecewise constant interpolation in time we use a simplified notation f_h instead of more precise, but lengthy $f_{\Delta,h}$. The discrete calculus and properties introduced above for the space discretization directly apply also for the time discretization. In particular, the following result on the local truncation error will be frequently used.

Lemma 8 (Consistency error of the discrete time derivative)
Let $r_h \in Q_h$ and $\boldsymbol{m}_h \in \boldsymbol{Q}_h$. Then for any $\varphi \in C_c^2([0, T) \times \overline{\Omega})$, $\boldsymbol{\varphi} \in C_c^2([0, T) \times \overline{\Omega}; R^d)$ there hold

$$\begin{aligned} &\int_0^T \int_\Omega \Big(D_t r_h(t) \Pi_Q \varphi(t) + r_h(t) \partial_t \varphi(t)\Big) \,\mathrm{d}x \,\mathrm{d}t + \int_\Omega r_h^0 \varphi(0) \,\mathrm{d}x \\ &\lesssim \Delta t \|\varphi\|_{C^2} \|r_h\|_{L^1 L^1} + \Delta t \|\varphi\|_{C^1} \left\|r_h^0\right\|_{L^1}, \end{aligned} \tag{48a}$$

$$\begin{aligned} &\int_0^T \int_\Omega (D_t r_h(t) \Pi_V \varphi(t) + r_h(t) \partial_t \varphi(t)) \,\mathrm{d}x \,\mathrm{d}t + \int_\Omega r_h^0 \varphi(0) \,\mathrm{d}x \\ &\lesssim (\Delta t + h) \|\varphi\|_{C^2} \|r_h\|_{L^1 L^1} + (\Delta t + h) \|\varphi\|_{C^1} \left\|r_h^0\right\|_{L^1}, \end{aligned} \tag{48b}$$

$$\begin{aligned} &\int_0^T \int_\Omega \Big(D_t \boldsymbol{m}_h \cdot \overline{\Pi_{\boldsymbol{W}} \boldsymbol{\varphi}} + \boldsymbol{m}_h \cdot \partial_t \boldsymbol{\varphi}\Big) \,\mathrm{d}x \,\mathrm{d}t + \int_\Omega \boldsymbol{m}_h^0 \cdot \boldsymbol{\varphi}(0) \,\mathrm{d}x \\ &\lesssim \Delta t (\|\boldsymbol{m}_h\|_{L^1 L^1} \|\boldsymbol{\varphi}\|_{C^2} + \left\|\boldsymbol{m}_h^0\right\|_{L^1} \|\boldsymbol{\varphi}\|_{C^1}). \end{aligned} \tag{48c}$$

Proof By direct calculation we derive

$$\int_0^T \int_\Omega D_t r_h(t) \Pi_Q \varphi(t) \,\mathrm{d}x \,\mathrm{d}t = \int_0^T \int_\Omega \frac{r_h(t) - r_h(t - \Delta t)}{\Delta t} \varphi(t) \,\mathrm{d}x \,\mathrm{d}t$$

$$
\begin{aligned}
&= \frac{1}{\Delta t}\int_0^T \int_\Omega r_h(t)\varphi(t)\ \mathrm{d}x\,\mathrm{d}t - \frac{1}{\Delta t}\int_{-\Delta t}^{T-\Delta t} \int_\Omega r_h(t)\varphi(t+\Delta t)\ \mathrm{d}x\,\mathrm{d}t \\
&= -\int_0^T \int_\Omega r_h(t) D_t\varphi(t)\ \mathrm{d}x\,\mathrm{d}t + \frac{1}{\Delta t}\int_{T-\Delta t}^{T} \int_\Omega r_h(t)\varphi(t+\Delta t)\ \mathrm{d}x\,\mathrm{d}t \\
&\quad - \frac{1}{\Delta t}\int_{-\Delta t}^{0} \int_\Omega r_h(t)\varphi(t+\Delta t)\ \mathrm{d}x\,\mathrm{d}t \\
&= -\int_0^T \int_\Omega r_h(t) D_t\varphi(t)\ \mathrm{d}x\,\mathrm{d}t - \frac{1}{\Delta t}\int_0^{\Delta t} \int_\Omega r_h^0\varphi(t)\ \mathrm{d}x\,\mathrm{d}t \\
&= -\int_0^T \int_\Omega r_h(t) \partial_t\varphi(t)\ \mathrm{d}x\,\mathrm{d}t - \int_\Omega r_h^0\varphi(0)\ \mathrm{d}x + I,
\end{aligned}
$$

where thanks to Hölder's inequality and Taylor's theorem the term I is estimated by

$$
\begin{aligned}
I &= \int_0^T \int_\Omega r_h(t)(\partial_t\varphi(t) - D_t\varphi(t))\ \mathrm{d}x\,\mathrm{d}t + \int_\Omega r_h^0 \frac{1}{\Delta t}\int_0^{\Delta t} (\varphi(0)-\varphi(t))\,\mathrm{d}t\ \mathrm{d}x \\
&\lesssim \Delta t\|\varphi\|_{C^2}\|r_h\|_{L^1L^1} + \Delta t\|\varphi\|_{C^1}\left\|r_h^0\right\|_{L^1},
\end{aligned}
$$

which implies (48a). Analogously, we have

$$
\begin{aligned}
&\int_0^T \int_\Omega D_t r_h(t)\Pi_V\varphi(t)\ \mathrm{d}x\,\mathrm{d}t = \\
&= -\int_0^T \int_\Omega r_h(t) D_t\Pi_V\varphi(t)\ \mathrm{d}x\,\mathrm{d}t - \frac{1}{\Delta t}\int_0^{\Delta t} \int_\Omega r_h^0\Pi_V\varphi(t)\ \mathrm{d}x\,\mathrm{d}t \\
&= -\int_0^T \int_\Omega r_h(t) \partial_t\varphi(t)\ \mathrm{d}x\,\mathrm{d}t - \int_\Omega r_h^0\varphi(0)\ \mathrm{d}x + I,
\end{aligned}
$$

where by Hölder's inequality, Taylor's theorem, and the interpolation estimate (22) we deduce

$$
\begin{aligned}
I &= \int_0^T \int_\Omega r_h(t)(\partial_t\varphi(t) - D_t\Pi_V\varphi(t))\ \mathrm{d}x\,\mathrm{d}t \\
&\quad + \int_\Omega \frac{r_h^0}{\Delta t}\int_0^{\Delta t} (\varphi(0)-\Pi_V\varphi(t))\,\mathrm{d}t\ \mathrm{d}x \\
&= \int_0^T \int_\Omega r_h(t)(\partial_t\varphi(t) - D_t\varphi(t))\ \mathrm{d}x\,\mathrm{d}t \\
&\quad + \int_0^T \int_\Omega r_h(t) D_t\big(\varphi(t) - \Pi_V\varphi(t)\big)\ \mathrm{d}x\,\mathrm{d}t \\
&\quad + \int_\Omega r_h^0 \frac{1}{\Delta t}\int_0^{\Delta t} (\varphi(0)-\varphi(t))\,\mathrm{d}t\ \mathrm{d}x + \int_\Omega r_h^0 \frac{1}{\Delta t}\int_0^{\Delta t} (\varphi(t)-\Pi_V\varphi(t))\,\mathrm{d}t\ \mathrm{d}x \\
&\lesssim (\Delta t + h^2)\|\varphi\|_{C^2}\|r_h\|_{L^1L^1} + (\Delta t + h^2)\|\varphi\|_{C^2}\left\|r_h^0\right\|_{L^1},
\end{aligned}
$$

which completes the proof of (48b). Finally, the proof of (48c) is analogous and we omit it here.

In numerical analysis of PDEs one can often derive inequalities for discrete quantities $\{f^k\}_{k\geq 0}$ valid at each time level k. A desired uniform estimate is then typically obtained by means of the discrete counterpart of Gronwall's lemma.

Lemma 9 (Discrete Gronwall's Lemma)
Let Δt, B, a^k, b^k, c^k, w^k be nonnegative numbers for integers $k \geq 1$, and let the inequality

$$a^m + \Delta t \sum_{k=1}^{m} b^k \leq B + \Delta t \sum_{k=1}^{m} a^k w^k + \Delta t \sum_{k=1}^{m} c^k \quad \text{for } m \geq 1$$

hold. If $\Delta t w^k < 1$ for all $k = 1, \ldots, m$, then

$$a^m + \Delta t \sum_{k=1}^{m} b^k \leq \exp\left\{ \Delta t \sum_{k=1}^{m} \frac{w^k}{1 - \Delta t w^k} \right\} \left(B + \Delta t \sum_{k=1}^{m} c^k \right) \quad \text{for } m \geq 1.$$

See [128, Lemma 5.1].

Contents

About the Authors

Eduard Feireisl is a senior research worker at the Institute of Mathematics of the Czech Academy of Sciences and a full professor at the Charles University in Prague. He authored or co-authored seven research monographs and over 300 research papers registered by the database MathSciNet. His main research interest is the abstract theory of partial differential equations with application in fluid mechanics, including numerical analysis and the effect of stochastic phenomena. He is one of the leading experts in the field of mathematical fluid mechanics.

Mária Lukáčová-Medvid'ová is a professor for Applied Mathematics at the Johannes Gutenberg-University Mainz. She is a vice speaker of the Research Center Multiscale Simulation Methods for Soft Matter Systems and of the Mainz Institute for Multiscale Modeling. Her research interests lay in numerics and analysis of partial differential equations. She made important contributions to the development of structure-preserving schemes for hyperbolic conservation laws and hybrid multiscale methods for complex fluids. She received various awards, such as the Prize of the Czech Learned Society in 2002, Bronze Medal of the University of Košice in 2013 or the Gutenberg Research College Fellowship in 2020.

Hana Mizerová is an assistant professor at the Department of Mathematical Analysis and Numerical Mathematics of the Comenius University in Bratislava, Slovakia. Her research focuses on numerical analysis of partial differential equations in fluid mechanics. In 2018, she was awarded the Seal of Excellence by the European Commission.

Bangwei She is a research worker at the Institute of Mathematics of the Czech Academy of Sciences. His research interests are numerical analysis and scientific computing, particularly for partial differential equations with a focus on computational fluid dynamics.

Acronyms

List of Abbreviations

CFL	Courant–Friedrichs–Lewy
DMV	Dissipative measure-valued
DW	Dissipative weak
EOC	Experimental order of convergence
EOS	Equation of state
FE	Finite element
FV	Finite volume
GRP	Generalized Riemann problem
MAC	Marker–and–Cell
ODE	Ordinary differential equation
PDE	Partial differential equation
VFV	Viscous finite volume

List of Symbols

d	Space dimension
$\mathbf{e}_i$	Unit vector in the ith canonical direction
Ω	Physical domain
$\mathcal{T}_h$	Computational mesh
K	A generic element $K \in \mathcal{T}_h$
$\mathcal{E}$	The set of all faces of a mesh $\mathcal{T}_h$
$\mathcal{E}_{int}$	The set of all interior faces of a mesh $\mathcal{T}_h$
$\mathcal{E}_{ext}$	The set of all boundary faces of a mesh $\mathcal{T}_h$
$\mathcal{E}(K)$	The set of all faces of an element K
$\mathcal{E}_i(K)$	The set of all faces of an element K that are perpendicular to $\mathbf{e}_i$
σ	A generic face $\sigma \in \mathcal{E}$
$\mathcal{D}_{i,h}$	Dual grid of a structured mesh corresponding to faces perpendicular to $\mathbf{e}_i$

$\mathcal{D}_h$	Dual grid of an unstructured mesh
D_σ	A generic dual element $D_\sigma \in \mathcal{D}_{i,h}$ or $D_\sigma \in \mathcal{D}_h$
$\mathcal{P}_n(K)$	Polynomial of degree at most n on an element K
$[\![\cdot]\!]_\sigma$	Jump across an interface σ
$\langle\cdot\rangle_\sigma$	Mean value across an interface σ
$\{\!\{\cdot\}\!\}_\sigma$	Average of inward and outward trace at a face σ
$\overline{\cdot}$	Mean value over an element K
Q_h	Space of piecewise constant functions on $\mathcal{T}_h$
Π_Q	Projection operator associated to the space Q_h
V_h	Linear Crouzeix–Raviart finite element space on $\mathcal{T}_h$
Π_V	Projection operator associated to the space V_h
$W_h^{(i)}$	Space of piecewise constant functions on $\mathcal{D}_{i,h}$
$\Pi_W^{(i)}$	Projection operator associated to the space $W_h^{(i)}$
W_h	Space of piecewise constant functions, $W_h = \Pi_{i=1}^d W_h^{(i)}$
Π_W	Projection operator associated to the space W_h

Part I
Mathematics of Compressible Fluid Flow: The State-of-the-Art

The main objective of this introductory part is to review the key ingredients of the existing mathematical theory of compressible fluid flows. As customary in continuum mechanics, we write the basic physical principles in terms of conservation/balance laws reflecting the basic physical principles: the conservation/balance of mass, linear momentum, and energy. We introduce their weak formulation written as integral identities rather than systems of partial differential equations. As we shall see below, the weak formulation suits perfectly modern numerical methods based on the finite volume and finite element methods.

Having introduced the preliminary material, we focus on the Euler system describing the motion of a compressible inviscid fluid. We consider both the complete (full) system including the thermal effects and its barotropic simplification, where the pressure of the fluid depends solely on its mass density. In particular, we show how singularities emerge spontaneously in smooth solutions in a finite lap of time. Accordingly, we introduce the concept of weak (distributional) solution to continue the process after blowup. The property of thermodynamic stability that is crucial in the theory of dissipative solutions is also discussed.

The viscous fluids in this monograph are Newtonian, meaning the viscous tress is given by the standard Newton rheological law as a linear function of the velocity gradient. The time evolution is therefore governed by the Navier–Stokes(–Fourier) system. Unlike its inviscid (Euler) counterpart, the Navier–Stokes system admits global-in-time smooth solutions at least if the initial data are close to equilibrium and under conservative boundary conditions. Although there is no explicit example of finite time blow up, the existence of global-in-time strong solutions to the Navier–Stokes system is still an outstanding open problem. To overcome this difficulty, we content ourselves with suitable weak (distributional) solutions existing globally in time for any physically relevant data. Finally, we review some recent results on conditional regularity of weak solutions that are of great interest in numerical analysis, in particular when combined with the weak–strong uniqueness principle.

Many results are stated without proofs. The interested reader may consult the comprehensive reference material indicated in the concluding sections appended to each individual chapter.

Chapter 1
Equations Governing Fluids in Motion

Physics distinguishes four basic forms of matter: Solids, liquids, gases, and plasmas. The last three fall in the category of *fluids*.

A fluid is any body whose parts yield to any force impressed on it, and by yielding, are easily moved among themselves.

Isaac Newton, Principia Book II, 1687

There are several ways how to describe the fluid motion in the language of mathematics:

1. **Molecular dynamics** captures the fluid in its most elementary form: as a large sample of moving (rigid) particles—atoms or molecules. The equations of motion are written according to Newton's laws for each individual particle, where mutual collisions are taken account. The resulting problem consists of a large number of ordinary differential equation that are completely time reversible.
2. **Kinetic theory** replaces individual trajectories by averaged quantities. The state of the fluid is described by the density of particles having the same velocity at a given time and spatial positions. The resulting problem involves the *Boltzmann equation* that includes a collision operator. Accordingly, the problem is irreversible with respect to time.
3. **Continuum mechanics** is a phenomenological theory based on observable macroscopic variables—fields, whereas the time evolution is described in terms of *balance laws*. These are system of partial differential equations reflecting the basic physical principles: The conservation of mass, linear momentum, and energy.
4. **Turbulence theory** describes the fluids in the borderline regimes of continuum mechanics. There is no universally accepted theory of turbulence nowadays. The models consist of further averaging and/or augmenting the systems of equations provided by the existing models in continuum fluid mechanics.

This monograph develops the mathematical theory of fluids in the framework of *continuum mechanics*. A fluid in motion is described in terms of numerical values of observable macroscopic quantities—fields—depending on the time t and the spatial

E. Feireisl et al., *Numerical Analysis of Compressible Fluid Flows*, MS&A 20,
https://doi.org/10.1007/978-3-030-73788-7_1

position x. We systematically use the *Eulerian description*, where the spatial coordinate frame is attached to the physical domain $\Omega \subset R^d$, $d = 1, 2, 3$ occupied by the fluid. The reference initial time is set to be $t_0 = 0$, whereas the reference time interval is $t \in (0, T)$ with $T \leq \infty$. For obvious reasons, the final time $T < \infty$ is finite in numerical implementations. We focus on bounded spatial domains Ω avoiding problems with the far field behavior delicate to approximate in numerics. Accordingly, suitable boundary conditions must be imposed. To avoid technicalities related to the presence of the kinematic boundary, we often consider the spatially periodic boundary conditions, for which the spatial domain can be identified with the flat torus

$$\Omega = \mathbb{T}^d, \ \mathbb{T}^d = \Pi_{i=1}^d [a_i, b_i]|_{\{a_i, b_i\}}.$$

The specific length of the intervals $[a_i, b_i]$ is usually irrelevant and may be conveniently normalized. In what follows, we shall use the same symbol $\mathbb{T}^d$ to denote any spatially periodic domain. From the topological point of view, the flat torus $\mathbb{T}^d$ is a compact manifold without boundary.

1.1 Balance Laws

The cornerstone of the mathematical theory is the system of equations governing the time evolution of the fluid. They are mathematical statements of the fundamental principles of physics expressed in terms of balance laws. The time evolution of a macroscopic quantity D is described by means of its volume density $d = d(t, x)$, the flux $\boldsymbol{F} = \boldsymbol{F}(t, x)$, and the source term $s = s(t, x)$. Given a time interval $[t_1, t_2]$ and a volume element $B \subset \Omega$, the associated *balance law* can be written the form:

$$\left[\int_B d(t, x) \ \mathrm{d}x \right]_{t=t_1}^{t=t_2} = - \int_{t_1}^{t_2} \int_{\partial B} \boldsymbol{F}(t, x) \cdot \boldsymbol{n}(x) \ \mathrm{dS}_x \ \mathrm{d}t + \int_{t_1}^{t_2} \int_B s(t, x) \ \mathrm{d}x \, \mathrm{d}t, \quad (1.1)$$

where $\boldsymbol{n}$ denotes the outer normal vector to ∂B. The balance law (1.1) says that the time change of the total amount of the quantity D contained in the volume B is proportional to the amount flowing in/out through the boundary ∂B and the amount furnished by the source term s. If $s = 0$, the relation (1.1) represents a *conservation law*. Equation (1.1) holds for any choice of $t_1 \leq t_2$ and any $B \subset \Omega$. It is worth noting that a balance law written in its primitive form (1.1) requires very low regularity of the fields in question: **(i)** (local) integrability of $d, \boldsymbol{F}$, and s; **(ii)** the existence of the normal trace of the field $[d, \boldsymbol{F}] \in R^{d+1}$ on the boundary of any space-time cylinder $(t_1, t_2) \times B$.

1.1.1 Strong Versus Weak Formulation

The balance law (1.1) can be written in a more concise form as a partial differential equation on condition that all quantities involved are smooth enough. Indeed dividing (1.1) on $(t_2 - t_1)$ and performing the limit $t_2 \to t_1$ we easily obtain

$$\frac{\mathrm{d}}{\mathrm{d}t}\int_B d(t,x)\ \mathrm{d}x + \int_{\partial B} \boldsymbol{F}(t,x)\cdot\boldsymbol{n}(x)\ \mathrm{dS}_x = \int_B s(t,x)\ \mathrm{d}x$$

for any t. Next, by Gauss–Green theorem,

$$\int_{\partial B} \boldsymbol{F}(t,x)\cdot\boldsymbol{n}(x)\ \mathrm{dS}_x = \int_B \mathrm{div}_x\boldsymbol{F}(t,x)\ \mathrm{d}x;$$

whence

$$\int_B [\partial_t d(t,x) + \mathrm{div}_x\boldsymbol{F}(t,x)]\ \mathrm{d}x = \int_B s(t,x)\ \mathrm{d}x.$$

As $B \subset \Omega$ is arbitrary, we conclude that

$$\partial_t d + \mathrm{div}_x\boldsymbol{F} = s, \tag{1.2}$$

which is the differential (strong) form of the balance law (1.1).

The weak form of (1.1) is deduced from a simple observation that (1.1) can be interpreted as

$$\lim_{\delta\searrow 0}\int_0^\infty\int_{R^d}\Big[d(t,x)\partial_t\varphi_\delta(t,x) + \boldsymbol{F}(t,x)\cdot\nabla_x\varphi_\delta(t,x)\Big]\,\mathrm{d}x\,\mathrm{d}t$$
$$+\int_0^\infty\int_{R^d} s(t,x)\varphi_\delta(t,x)\ \mathrm{d}x\,\mathrm{d}t = 0,$$

where

$$\varphi_\delta \in C^1_c((t_1,t_2)\times B),\ 0 \le \varphi_\delta \le 1,\ \varphi_\delta(t,x) = 1$$

whenever

$$t_1 + \delta < t < t_2 - \delta,\ x \in B,\ \mathrm{dist}[x,\partial B] > \delta.$$

Here we have tacitly assumed that $B \subset \Omega$ is a domain with regular boundary. This motivates the following weak formulation of the balance law (1.1) in $(0,T)\times\Omega$:

$$\int_0^T \int_\Omega \Big[d(t,x)\partial_t \varphi(t,x) + \boldsymbol{F}(t,x) \cdot \nabla_x \varphi(t,x) \Big] \, dx \, dt = - \int_0^T \int_\Omega s(t,x)\varphi(t,x) \, dx \, dt \tag{1.3}$$

for any $\varphi \in C_c^1((0,T) \times \Omega)$. The function φ is usually called *test function.* Note that (1.3) is nothing other than (1.2), where the derivatives are interpreted in the sense of distributions. Obviously, any strong solution of (1.2) satisfies (1.3), and any weak solution of (1.3) with continuously differentiable d and $\boldsymbol{F}$ satisfies (1.2). This property is called *compatibility* of the weak formulation (1.3) with the equation (1.2).

Although it is customary and also more concise to formulate problems in fluid dynamics in terms of differential equations as (1.2), it is the weak formulation (1.3) that is definitely closer to the primitive form of the balance law (1.1). There is *a priori* no reason for the fields $d, \boldsymbol{F}, s$ to be continuous or even differentiable functions of t and x. As a matter of fact, the numerical approximations studied in this monograph are much closer to the weak formulation (1.3) than to its differential form (1.2).

1.1.2 Initial and Boundary Conditions

In many problems, the initial state of the quantity $d(0,\cdot)$ is known and considered as a given datum (parameter) of the problem. For $d \in C([0,T) \times \overline{\Omega})$ we simply write

$$d(0,\cdot) = d_0, \text{ where } d_0 \in C(\overline{\Omega}) \text{ is the given profile.} \tag{1.4}$$

Relation (1.4) is called *initial condition.*

The quantities appearing in the weak formulation of a balance law may be less regular therefore the meaning of their pointwise values must be clarified. The initial condition (1.4) can be easily incorporated in the weak formulation (1.3),

$$\begin{aligned} &\int_0^T \int_\Omega \Big[d(t,x)\partial_t \varphi(t,x) + \boldsymbol{F}(t,x) \cdot \nabla_x \varphi(t,x) \Big] \, dx \, dt \\ &\quad = - \int_\Omega d_0(x)\varphi(0,x) \, dx - \int_0^T \int_\Omega s(t,x)\varphi(t,x) \, dx \, dt \end{aligned} \tag{1.5}$$

by extending the class of admissible test functions to $\varphi \in C_c^1([0,T) \times \Omega)$.

As a matter of fact, any field d satisfying the weak formulation (1.3) enjoys certain continuity with respect to the time variable. To see this consider $\varphi(t,x) = \psi(t)\phi(x)$ as a test function in (1.3)

$$
\begin{aligned}
-\int_0^T \partial_t \psi(t) \left[\int_\Omega d(t,x)\phi(x) \, \mathrm{d}x \right] \mathrm{d}t \\
= \int_0^T \psi(t) \left[\int_\Omega s(t,x)\phi(x) + \boldsymbol{F}(t,x) \cdot \nabla_x \phi(x) \, \mathrm{d}x \right] \mathrm{d}t
\end{aligned}
\tag{1.6}
$$

for any $\psi \in C_c^1(0,T)$ and any $\phi \in C_c^1(\Omega)$. In addition, suppose that

$$
d \in L^\infty(0,T; L^1(\Omega)), \ \boldsymbol{F} \in L^1((0,T) \times \Omega; R^d), \ s \in L^1((0,T) \times \Omega).
$$

Relation (1.6) can be interpreted in terms of distributional derivatives as

$$
\frac{\mathrm{d}}{\mathrm{d}t} \left[t \mapsto \int_\Omega d(t,x)\phi(x) \, \mathrm{d}x \right] = G(t) \text{ in } \mathcal{D}'(0,T), \ \text{where } G \in L^1(0,T).
$$

In other words, the function of the time variable $\left[t \mapsto \int_\Omega d(t,x)\phi(x), dx\right]$ is bounded absolutely continuous in $[0,T]$ for any $\phi \in C_c^1(\Omega)$. This can be rephrased as follows: Any time $t \in [0,T)$ is a right Lebesgue point and any time $t \in (0,T]$ is a left Lebesgue point of $\left[t \mapsto \int_\Omega d(t,x)\phi(x), dx\right]$; moreover, for any $t \in (0,T)$ the value at right and left Lebesgue points of $\left[t \mapsto \int_\Omega d(t,x)\phi(x), dx\right]$ coincide for any $\phi \in C_c^1(\Omega)$. Finally, as $C_c^1(\Omega)$ is dense in $C_0(\Omega)$, we can extend the previous statement to any $\phi \in C_0(\Omega)$. Indeed let $\phi \in C_0(\Omega)$ and $\phi_n \in C_c^1(\Omega)$ such that $\phi_n \to \phi \in C_0(\Omega)$. We have

$$
\int_\Omega d(t,x)\phi(x) \, \mathrm{d}x = \int_\Omega d(t,x)\phi_n(x) \, \mathrm{d}x + \int_\Omega d(t,x)(\phi(x) - \phi_n(x)) \, \mathrm{d}x,
$$

where

$$
\left| \int_\Omega d(t,x)(\phi(x) - \phi_n(x)) \, \mathrm{d}x \right| \leq \sup_{t \in [0,T]} \|d(t,\cdot)\|_{L^1(\Omega)} \|\phi_n - \phi\|_{C_c(\Omega)} \to 0 \text{ as } n \to \infty
$$

uniformly for $t \in [0,T]$.

Let us summarize the previous discussion in the following statement.

Proposition 1.1 (Weak continuity in time) *Let $\Omega \subset R^d$ be a domain. Let*

$$
d \in L^\infty(0,T; L^1(\Omega)), \ \boldsymbol{F} \in L^1((0,T) \times \Omega; R^d), \ s \in L^1((0,T) \times \Omega)
$$

satisfy the balance law (1.3).

Then $d \in C([0,T]; \mathcal{M}(\Omega))$, *meaning any time* $t \in (0,T)$ *is a Lebesgue point of the function*

$$\left[t \mapsto \int_{\Omega} d(t,x)\phi(x)\ \mathrm{d}x \right] \ \textit{for any}\ \phi \in C_c(\Omega),$$

and

$$\left[t \mapsto \int_{\Omega} d(t,x)\phi(x)\ \mathrm{d}x \right] \in C[0,T], \quad \left| \int_{\Omega} d(t,x)\phi(x)\ \mathrm{d}x \right| \lesssim \|\phi\|_{C_c(\Omega)}$$

uniformly for $t \in (0,T)$.

Remark 1.1 The mapping $t \mapsto d(t,\cdot)$ ranges in a bounded ball of the space of finite Radon measures $\mathcal{M}(\Omega)$ that can be seen as the dual of the separable Banach space $C_0(\Omega)$. Consequently, $t \mapsto d(t,\cdot)$ may be viewed as continuous with respect to the metrics induced by the weak-(*) topology on bounded balls in $\mathcal{M}(\Omega)$.

The implementation of boundary conditions can be done in a similar way. Formally, we may prescribe the normal component of the flux $\boldsymbol{F}$ on $\partial\Omega$:

$$\boldsymbol{F} \cdot \boldsymbol{n}|_{\partial\Omega} = F_b. \tag{1.7}$$

This can be incorporated in (1.5) as

$$\begin{aligned} &\int_0^T \int_{\Omega} \Big[d(t,x)\partial_t\varphi(t,x) + \boldsymbol{F}(t,x)\cdot\nabla_x\varphi(t,x) \Big]\ \mathrm{d}x\,\mathrm{d}t \\ &= -\int_{\Omega} d_0(x)\varphi(0,x)\ \mathrm{d}x + \int_0^T \int_{\partial\Omega} F_b(t,x)\varphi(t,x)\ \mathrm{dS}_x\,\mathrm{d}t \\ &\quad - \int_0^T \int_{\Omega} s(t,x)\varphi(t,x)\ \mathrm{d}x\,\mathrm{d}t \end{aligned} \tag{1.8}$$

for any $\varphi \in C^1_c([0,T) \times \overline{\Omega})$. Finally, in view of Proposition 1.1, we may anticipate the time continuity of d and write (1.8) in a concise form:

$$\int_0^\tau \int_\Omega \left[d(t,x)\partial_t \varphi(t,x) + \boldsymbol{F}(t,x) \cdot \nabla_x \varphi(t,x) \right] \mathrm{d}x\,\mathrm{d}t$$

$$= \left[\int_\Omega d(t,x)\varphi(t,x)\ \mathrm{d}x \right]_{t=0}^{t=\tau} + \int_0^\tau \int_{\partial\Omega} F_b(t,x)\varphi(t,x)\ \mathrm{dS}_x\,\mathrm{d}t \tag{1.9}$$

$$- \int_0^\tau \int_\Omega s(t,x)\varphi(t,x)\ \mathrm{d}x\,\mathrm{d}t$$

for any $0 \leq \tau \leq T$, and any $\varphi \in C^1_c([0,T] \times \overline{\Omega})$. Note that

$$C^1_c([0,T] \times \overline{\Omega}) \neq C^1([0,T] \times \overline{\Omega})$$

if Ω is unbounded. Equation (1.9) represents the standard form of balance law used in this book.

Remark 1.2 (Vanishing normal trace) Using similar argument, we may define the *normal trace* of a differentiable function on $\partial\Omega$, where the latter may not be regular. In particular, we will often use test functions $\boldsymbol{\varphi}$ satisfying

$$\boldsymbol{\varphi} \cdot \boldsymbol{n}|_{\partial\Omega} = 0. \tag{1.10}$$

This property can be restated by means of Gauss–Green theorem as

$$\int_0^T \int_\Omega \boldsymbol{\varphi} \cdot \nabla_x w\ \mathrm{d}x\,\mathrm{d}t + \int_0^T \int_\Omega w \mathrm{div}_x \boldsymbol{\varphi}\ \mathrm{d}x\,\mathrm{d}t = 0 \tag{1.11}$$

for any $w \in C^1_c((0,T) \times R^d)$. In contrast with (1.10), formula (1.11) does not require any regularity of the boundary and of the normal vector field $\boldsymbol{n}$.

1.2 Fundamental Balance Laws in Fluid Mechanics

In obtaining the basic equations of fluid dynamics, the following strategy is applied:

- Choose the basic physical principles as conservation of mass, linear momentum, and energy.
- Identify the fields that will play the role of (unknown) phase variables as the density, the velocity/momentum, the temperature, the energy, the entropy etc. as the case may be.
- Write down the corresponding balance/conservation laws, together with the associated initial and boundary conditions.

As already pointed out we will use the *Eulerian reference system*, where the spatial reference frame is attached to the physical domain $\Omega \subset R^d$, $d = 1, 2, 3$. The macroscopic motion of the fluid is described via the (bulk) velocity field $\boldsymbol{u} = \boldsymbol{u}(t, x)$. The trajectories of hypothetical fluid particles—the so-called streamlines—are determined by a system of ordinary differential equations

$$\frac{\mathrm{d}}{\mathrm{d}t}\boldsymbol{X}(t, \boldsymbol{x}_0) = \boldsymbol{u}\left(t, \boldsymbol{X}(t, \boldsymbol{x}_0)\right), \ \boldsymbol{X}(0, \boldsymbol{x}_0) = \boldsymbol{x}_0 \in \Omega.$$

Accordingly, in the Eulerian description, the flux associated to any physical quantity d transported by the fluid must contain a *convective* component $\boldsymbol{F}_c = d\boldsymbol{u}$.

We focus on *mechanically closed* fluid systems, where the velocity is tangential to the boundary of the physical domain

$$\boldsymbol{u} \cdot \boldsymbol{n}|_{\partial\Omega} = 0, \tag{1.12}$$

where $\boldsymbol{n}$ stands for the outer normal vector to $\partial\Omega$. The boundary condition (1.12) reflects the *impermeability* of the boundary of the physical domain Ω.

Alternatively, we consider the space periodic boundary conditions, meaning the underlying physical domain is identified with the flat torus

$$\mathbb{T}^d = \Pi_{i=1}^d [a_i, b_i]|_{\{a_i, b_i\}}$$

with the period length $(b_i - a_i) > 0$ in the i-the direction. Although physically irrelevant, the case $\Omega = \mathbb{T}^d$ is often used to approximate the problems where the influence of the physical boundary can be neglected.

Of course, more complicated boundary conditions occur in numerous real world applications. In particular, the velocity as well as other relevant quantities may be prescribed on the physical boundary, together with far field conditions characterizing the behavior of the system for $|x| \to \infty$ on unbounded domains. Even the impermeability condition (1.12) sufficient to determine the boundary behavior of perfect (inviscid) fluids must be accompanied by other relevant conditions if the viscous forces are taken into account.

1.2.1 Mass Conservation

The distribution of mass in a *compressible* fluid is given by the mass density $\varrho = \varrho(t, x)$. The physical principle of *mass conservation* is encoded in the *equation of continuity*

$$\partial_t \varrho + \mathrm{div}_x(\varrho \boldsymbol{u}) = 0. \tag{1.13}$$

Applying (1.9), we obtain a weak formulation of (1.13) taking the impermeability boundary condition (1.12) into account

$$\begin{aligned}&\left[\int_{\Omega} \varrho(t,x)\varphi(t,x)\ \mathrm{d}x\right]_{t=0}^{t=\tau}\\&=\int_0^\tau\int_{\Omega}\Big[\varrho(t,x)\partial_t\varphi(t,x)+\varrho(t,x)\boldsymbol{u}(t,x)\cdot\nabla_x\varphi(t,x)\Big]\ \mathrm{d}x\,\mathrm{d}t\end{aligned} \tag{1.14}$$

for any $\varphi \in C_c^1([0,T]\times\overline{\Omega})$. Note that (1.14) remains unchanged if Ω is replaced by the flat torus $\mathbb{T}^d$.

If $\Omega \subset R^d$ is a bounded domain, or $\Omega = \mathbb{T}^d$, relation (1.14) yields the physical principle of the total mass conservation

$$\left[\int_{\Omega} \varrho(t,x)\ \mathrm{d}x\right]_{t=0}^{t=\tau} = 0$$

by considering $\varphi \equiv 1$. Similar argument on an unbounded domain is more delicate and requires certain decay of the density for $|x| \to \infty$.

Obviously, a physically relevant density must be positive or at least nonnegative if vacuum is allowed. Formally, we can compute the time evolution of the density along streamlines,

$$\begin{aligned}\frac{\mathrm{d}}{\mathrm{d}t}\varrho(t,\boldsymbol{X}(t;\boldsymbol{x}_0)) &= \partial_t\varrho(t,\boldsymbol{X}(t;\boldsymbol{x}_0))+\nabla_x\varrho(t,\boldsymbol{X}(t;\boldsymbol{x}_0))\cdot\boldsymbol{u}(t,\boldsymbol{X}(t,\boldsymbol{X}(t,\boldsymbol{x}_0))\\&= -\mathrm{div}_x\boldsymbol{u}(t,\boldsymbol{X}(t,\boldsymbol{X}(t,\boldsymbol{x}_0))\varrho(t,\boldsymbol{X}(t;\boldsymbol{x}_0)),\end{aligned}$$

obtaining

$$\begin{aligned}\varrho(0,\boldsymbol{x}_0)\int_0^\tau \exp\left(-\|\mathrm{div}_x\boldsymbol{u}(t,\cdot)\|_{L^\infty(\Omega)}\right)\ \mathrm{d}t &\le \varrho(\tau,\boldsymbol{X}(\tau,\boldsymbol{x}_0))\\&\le \varrho(0,\boldsymbol{x}_0)\int_0^\tau \exp\left(\|\mathrm{div}_x\boldsymbol{u}(t,\cdot)\|_{L^\infty(\Omega)}\right)\ \mathrm{d}t.\end{aligned} \tag{1.15}$$

Relation (1.15) indicates that the pointwise values of the density are controllable in terms of the initial data as long as

$$\int_0^T \|\mathrm{div}_x\boldsymbol{u}(t,\cdot)\|_{L^\infty(\Omega)}\ \mathrm{d}t < \infty. \tag{1.16}$$

Condition (1.16) definitely holds in the class of smooth (Lipschitz) solutions, while its satisfaction by less regular weak solutions has been so far a largely open problem. Even if the bound (1.16) is not available, relation (1.15) implies that the density remains nonnegative as soon as it is nonnegative at the initial time.

Apparently, the meaning of a velocity field in the vacuum region $\{\varrho = 0\}$ is dubious. It is more convenient to introduce the momentum $\boldsymbol{m} = \varrho \boldsymbol{u}$ that vanishes as soon as $\varrho = 0$. Accordingly, the equation of continuity (1.13) reads

$$\partial_t \varrho + \mathrm{div}_x \boldsymbol{m} = 0, \tag{1.17}$$

or, in the weak form,

$$\left[\int_\Omega \varrho(t,x)\varphi(t,x)\,\mathrm{d}x \right]_{t=0}^{t=\tau} = \int_0^\tau \int_\Omega \Big[\varrho(t,x)\partial_t \varphi(t,x) + \boldsymbol{m}(t,x)\cdot \nabla_x \varphi(t,x) \Big]\,\mathrm{d}x\,\mathrm{d}t$$

for any $\varphi \in C^1_c([0,T] \times \overline{\Omega})$. Sometimes, we may even write $1_{\varrho>0}\boldsymbol{m}$ instead of $\boldsymbol{m}$ to emphasize that the momentum $\boldsymbol{m}$ vanishes as soon as $\varrho = 0$. Unlike (1.13), however, the equation (1.17) does not provide any explicit information about positivity of ϱ even if both ϱ and $\boldsymbol{m}$ are smooth.

1.2.2 Momentum Equation – Newton's Second Law

Introducing the momentum $\boldsymbol{m} \equiv \varrho \boldsymbol{u}$ we write the physical principle of *balance of linear momentum*—Newton's Second law—in the form

$$\begin{gathered} \partial_t \boldsymbol{m} + \mathrm{div}_x \left(\frac{\boldsymbol{m} \otimes \boldsymbol{m}}{\varrho} \right) = \mathrm{div}_x \mathbb{T} + \varrho \boldsymbol{f} \\ \text{or, in terms of the velocity,} \\ \partial_t (\varrho \boldsymbol{u}) + \mathrm{div}_x (\varrho \boldsymbol{u} \otimes \boldsymbol{u}) = \mathrm{div}_x \mathbb{T} + \varrho \boldsymbol{f}. \end{gathered} \tag{1.18}$$

The tensor $\mathbb{T}$ represents the *Cauchy stress* , while $\boldsymbol{f}$ is a volume force acting on the fluid. We adopt the common mathematical definition of *fluid* as a continuum, for which $\mathbb{T}$ obeys the *Stokes' law*

$$\mathbb{T} = \mathbb{S} - p\mathbb{I}, \tag{1.19}$$

where $\mathbb{S}$ is the *viscous stress tensor* and p a scalar field called *pressure* . Accordingly, (1.18) reads

$$\partial_t (\varrho \boldsymbol{u}) + \mathrm{div}_x (\varrho \boldsymbol{u} \otimes \boldsymbol{u}) + \nabla_x p = \mathrm{div}_x \mathbb{S} + \varrho \boldsymbol{f}, \tag{1.20}$$

or, in the weak form,

$$\int_0^\tau \int_\Omega \left[\varrho \boldsymbol{u}(t,x) \cdot \partial_t \boldsymbol{\varphi} + (\varrho \boldsymbol{u} \otimes \boldsymbol{u})(t,x) : \nabla_x \boldsymbol{\varphi}(t,x) + p(t,x) \mathrm{div}_x \boldsymbol{\varphi}(t,x) \right] \mathrm{d}x \, \mathrm{d}t$$
$$= \left[\int_\Omega \varrho \boldsymbol{u}(t,x) \cdot \boldsymbol{\varphi}(t,x) \, \mathrm{d}x \right]_{t=0}^{t=\tau} + \int_0^\tau \int_\Omega \mathbb{S}(t,x) : \nabla_x \boldsymbol{\varphi} \, \mathrm{d}x \, \mathrm{d}t$$
$$- \int_0^\tau \int_\Omega \varrho \boldsymbol{f} \cdot \boldsymbol{\varphi} \, \mathrm{d}x \, \mathrm{d}t \tag{1.21}$$

for any $0 \leq \tau \leq T$, and any $\boldsymbol{\varphi} \in C^1_c([0,T] \times \Omega; R^d)$. For perfect (inviscid) fluid flows satisfying the impermeability boundary condition (1.12), the class of admissible test functions can be extended to

$$\boldsymbol{\varphi} \in C^1_c([0,T] \times \overline{\Omega}; R^d), \ \boldsymbol{\varphi} \cdot \boldsymbol{n}|_{\partial\Omega} = 0. \tag{1.22}$$

If $\mathbb{S} \neq 0$ the weak formulation (1.21), (1.22) is compatible with the *complete slip boundary condition*

$$[\mathbb{S} \cdot \boldsymbol{n}] \times \boldsymbol{n}|_{\partial\Omega} = 0. \tag{1.23}$$

In other words, the tangential component of the normal viscous stress $[\mathbb{S} \cdot \boldsymbol{n}]$ vanishes on the boundary.

Finally, we can rewrite (1.21) in terms of the momentum,

$$\int_0^\tau \int_\Omega \left[\boldsymbol{m}(t,x) \cdot \partial_t \boldsymbol{\varphi} + 1_{\varrho > 0} \frac{\boldsymbol{m} \otimes \boldsymbol{m}}{\varrho}(t,x) : \nabla_x \boldsymbol{\varphi}(t,x) + p(t,x) \mathrm{div}_x \boldsymbol{\varphi}(t,x) \right] \mathrm{d}x \, \mathrm{d}t$$
$$= \left[\int_\Omega \boldsymbol{m}(t,x) \cdot \boldsymbol{\varphi}(t,x) \, \mathrm{d}x \right]_{t=0}^{t=\tau} + \int_0^\tau \int_\Omega \mathbb{S}(t,x) : \nabla_x \boldsymbol{\varphi} \, \mathrm{d}x \, \mathrm{d}t$$
$$- \int_0^\tau \int_\Omega \varrho \boldsymbol{f} \cdot \boldsymbol{\varphi} \, \mathrm{d}x \, \mathrm{d}t \tag{1.24}$$

for any $0 \leq \tau \leq T$, and any $\boldsymbol{\varphi} \in C^1_c([0,T] \times \overline{\Omega}; R^d)$, $\boldsymbol{\varphi} \cdot \boldsymbol{n}|_{\partial\Omega} = 0$. Although condition (1.22) requires formally the existence of the outer normal vector to $\partial\Omega$, it can be also interpreted in the weak sense specified in Remark 1.2.

1.2.3 *The First Law of Thermodynamics – Energy Conservation*

The First law of thermodynamics, with the associated concept of *total energy*, is a cornerstone of the mathematical theory developed in this book. Computing the scalar product of the momentum equation (1.20) with the velocity $\boldsymbol{u}$ we obtain, after a straightforward manipulation,

$$\partial_t \left(\frac{1}{2} \varrho |\boldsymbol{u}|^2 \right) + \mathrm{div}_x \left[\left(\frac{1}{2} \varrho |\boldsymbol{u}|^2 + p \right) \boldsymbol{u} \right] - \mathrm{div}_x (\mathbb{S} \cdot \boldsymbol{u}) = -\mathbb{S} : \nabla_x \boldsymbol{u} + p \mathrm{div}_x \boldsymbol{u} + \varrho \boldsymbol{f} \cdot \boldsymbol{u}. \tag{1.25}$$

The quantity

$$e_{\mathrm{kin}} = \frac{1}{2} \varrho |\boldsymbol{u}|^2, \text{ or, in terms of the momentum, } e_{\mathrm{kin}} = \frac{1}{2} \frac{|\boldsymbol{m}|^2}{\varrho}$$

represents the *kinetic energy* of the fluid. Accordingly, equation (1.25) is a form of mechanical energy balance. It is worth noting that the kinetic energy is a *convex* function of the variables $[\varrho, \boldsymbol{m}]$. More precisely, it is convenient to define

$$e_{\mathrm{kin}}(\varrho, \boldsymbol{m}) = \begin{cases} \frac{1}{2} \frac{|\boldsymbol{m}|^2}{\varrho} \text{ for } \varrho > 0, \\ 0 \text{ if } \varrho = 0, \boldsymbol{m} = 0, \\ \infty \text{ otherwise.} \end{cases} \tag{1.26}$$

for any $[\varrho, \boldsymbol{m}] \in R^{d+1}$. A straightforward computation yields

$$\frac{\partial^2 e_{\mathrm{kin}}(\varrho, \boldsymbol{m})}{\partial \varrho^2} = \frac{|\boldsymbol{m}|^2}{\varrho^3}, \ \frac{\partial^2 e_{\mathrm{kin}}(\varrho, \boldsymbol{m})}{\partial \varrho \partial m_i} = -\frac{m_i}{\varrho^2}, \ \frac{\partial^2 e_{\mathrm{kin}}(\varrho, \boldsymbol{m})}{\partial^2 m_i} = \frac{1}{\varrho}, \ i = 1, \dots, d.$$

In particular $e_{\mathrm{kin}} = e_{\mathrm{kin}}(\varrho, \boldsymbol{m}) : R^{d+1} \to [0, \infty]$ is a convex l.s.c. function. We point out that e_{kin} *is not* strictly convex as it is linear on the lines $[\varrho, \boldsymbol{m}] = [r, \boldsymbol{b} r]$, $r > 0$.

Even if the driving force $\boldsymbol{f}$ vanishes, the right hand side of the kinetic energy balance (1.25) still contains a nonconservative source term

$$p \mathrm{div}_x \boldsymbol{u} - \mathbb{S} : \nabla_x \boldsymbol{u}.$$

As the First law of thermodynamics requires the total energy to be a conserved quantity, the latter must contain another component—the internal energy

$$e_{\mathrm{int}} = \varrho e,$$

together with the associated balance law

$$\partial_t(\varrho e) + \mathrm{div}_x(\varrho e \boldsymbol{u}) + \mathrm{div}_x \boldsymbol{q} = \mathbb{S} : \nabla_x \boldsymbol{u} - p\mathrm{div}_x \boldsymbol{u}, \tag{1.27}$$

where $\boldsymbol{q}$ is a diffusive internal energy flux. Similarly to the impermeability condition (1.12), and the complete slip boundary condition (1.23), we suppose the domain Ω is energetically (thermally) insulated, meaning

$$\boldsymbol{q} \cdot \boldsymbol{n}|_{\partial\Omega} = 0. \tag{1.28}$$

Summing up (1.25), (1.27) we obtain the desired *total energy balance*

$$\partial_t \left(\frac{1}{2}\varrho|\boldsymbol{u}|^2 + \varrho e \right) + \mathrm{div}_x \left[\left(\frac{1}{2}\varrho|\boldsymbol{u}|^2 + \varrho e \right) \boldsymbol{u} + p\boldsymbol{u} \right] + \mathrm{div}_x \left(\boldsymbol{q} - \mathbb{S} \cdot \boldsymbol{u} \right) = \varrho \boldsymbol{f} \cdot \boldsymbol{u}. \tag{1.29}$$

In accordance with the boundary conditions (1.23), (1.28), the weak form of (1.29) reads

$$\begin{aligned}
&\int_0^\tau \int_\Omega \left[\left(\frac{1}{2}\varrho|\boldsymbol{u}|^2 + \varrho e \right) \partial_t \varphi + \left(\frac{1}{2}\varrho|\boldsymbol{u}|^2 + \varrho e + p \right) \boldsymbol{u} \cdot \nabla_x \varphi \right] \, \mathrm{d}x \, \mathrm{d}t \\
&\quad = \left[\int_\Omega \left(\frac{1}{2}\varrho|\boldsymbol{u}|^2 + \varrho e \right) \varphi \, \mathrm{d}x \right]_{t=0}^{t=\tau} - \int_0^\tau \int_\Omega \left(\boldsymbol{q} - \mathbb{S} \cdot \boldsymbol{u} \right) \cdot \nabla_x \varphi \, \mathrm{d}x \, \mathrm{d}t \\
&\quad - \int_0^\tau \int_\Omega \varrho \boldsymbol{f} \cdot \boldsymbol{u} \varphi \, \mathrm{d}x \, \mathrm{d}t
\end{aligned} \tag{1.30}$$

for any $0 \leq \tau \leq T$, and any $\varphi \in C^1_c([0, T] \times \overline{\Omega})$. If Ω is bounded, the choice $\varphi \equiv 1$ in (1.30) gives rise to the total energy balance

$$\left[\int_\Omega \left(\frac{1}{2}\varrho|\boldsymbol{u}|^2 + \varrho e \right) \, \mathrm{d}x \right]_{t=0}^{t=\tau} = \int_0^\tau \int_\Omega \varrho \boldsymbol{f} \cdot \boldsymbol{u} \, \mathrm{d}x \, \mathrm{d}t. \tag{1.31}$$

We point out that validity of (1.30), (1.31) is conditioned by the impermeability of the boundary (1.12) and the energy insulation (1.23), (1.28). In other words, the fluid system is both mechanically and thermally closed.

Similarly to (1.24), the energy balance (1.30) may be written in terms of the momentum

$$\int_0^\tau \int_\Omega \left[\left(\frac{1}{2} \frac{|\boldsymbol{m}|^2}{\varrho} + \varrho e \right) \partial_t \varphi + 1_{\varrho>0} \left(\frac{1}{2} \frac{|\boldsymbol{m}|^2}{\varrho} + \varrho e + p \right) \frac{\boldsymbol{m}}{\varrho} \cdot \nabla_x \varphi \right] \mathrm{d}x \, \mathrm{d}t$$
$$= \left[\int_\Omega \left(\frac{1}{2} \frac{|\boldsymbol{m}|^2}{\varrho} + \varrho e \right) \varphi \, \mathrm{d}x \right]_{t=0}^{t=\tau} - \int_0^\tau \int_\Omega \left(\boldsymbol{q} - \mathbb{S} \cdot 1_{\varrho>0} \frac{\boldsymbol{m}}{\varrho} \right) \cdot \nabla_x \varphi \, \mathrm{d}x \, \mathrm{d}t$$
$$- \int_0^\tau \int_\Omega \boldsymbol{f} \cdot \boldsymbol{m} \varphi \, \mathrm{d}x \, \mathrm{d}t \tag{1.32}$$

for any $0 \le \tau \le T$, and any $\varphi \in C^1_c([0, T] \times \overline{\Omega})$. Here, the kinetic energy is defined via (1.26). In particular, boundedness of the total energy implies

$$\varrho \ge 0 \text{ a.a. } \in (0, T) \times \Omega, \text{ and } \varrho(t, x) = 0 \Rightarrow \boldsymbol{m}(t, x) = 0 \text{ for a.a. } (t, x) \in (0, T) \times \Omega.$$

The internal energy e may be viewed as a new unknown state variable closely related to another physical quantity—the temperature. The relation of the pressure p and the internal energy e to the (absolute) temperature ϑ is discussed in Sect. 1.3 below.

1.2.4 Basic System of Equations of Continuum Fluid Dynamics

Summing up (1.13), (1.20), (1.25), we obtain the basic system of equations of continuum fluid dynamics:

- **Mass conservation – equation of continuity**

$$\partial_t \varrho + \mathrm{div}_x(\varrho \boldsymbol{u}) = 0, \quad \text{or} \quad \partial_t \varrho + \mathrm{div}_x \boldsymbol{m} = 0; \tag{1.33}$$

- **Momentum balance – Newton's Second law**

$$\partial_t(\varrho \boldsymbol{u}) + \mathrm{div}_x(\varrho \boldsymbol{u} \otimes \boldsymbol{u}) + \nabla_x p = \mathrm{div}_x \mathbb{S} + \varrho \boldsymbol{f}, \quad \text{or} \quad \partial_t \boldsymbol{m} + \mathrm{div}_x \left(\frac{\boldsymbol{m} \otimes \boldsymbol{m}}{\varrho} \right) + \nabla_x p = \mathrm{div}_x \mathbb{S} + \varrho \boldsymbol{f}; \tag{1.34}$$

- **Energy conservation – The First law of thermodynamics**

$$\partial_t \left(\frac{1}{2} \varrho |\boldsymbol{u}|^2 + \varrho e \right) + \operatorname{div}_x \left[\left(\frac{1}{2} \varrho |\boldsymbol{u}|^2 + \varrho e \right) \boldsymbol{u} + p\boldsymbol{u} \right] + \operatorname{div}_x \left(\boldsymbol{q} - \mathbb{S} \cdot \boldsymbol{u} \right) = \varrho \boldsymbol{f} \cdot \boldsymbol{u},$$

or

$$\partial_t \left(\frac{1}{2} \frac{|\boldsymbol{m}|^2}{\varrho} + \varrho e \right) + \operatorname{div}_x \left[\left(\frac{1}{2} \frac{|\boldsymbol{m}|^2}{\varrho} + \varrho e \right) \frac{\boldsymbol{m}}{\varrho} + p \frac{\boldsymbol{m}}{\varrho} \right] + \operatorname{div}_x \left(\boldsymbol{q} - \mathbb{S} \cdot \frac{\boldsymbol{m}}{\varrho} \right) = \boldsymbol{f} \cdot \boldsymbol{m}. \tag{1.35}$$

We have also introduced conservative boundary conditions:

- **Impermeability**

$$\boldsymbol{u} \cdot \boldsymbol{n}|_{\partial\Omega} = 0, \text{ or } \boldsymbol{m} \cdot \boldsymbol{n}|_{\partial\Omega} = 0; \tag{1.36}$$

- **Complete slip**

$$[\mathbb{S} \cdot \boldsymbol{n}] \times \boldsymbol{n}|_{\partial\Omega} = 0; \tag{1.37}$$

- **Thermal insulation**

$$\boldsymbol{q} \cdot \boldsymbol{n}|_{\partial\Omega} = 0. \tag{1.38}$$

The system (1.33)–(1.35) is apparently not closed—there are more unknowns than equations. To close it, we need *constitutive relations* between specific fields.

1.3 Constitutive Relations

The *material properties* of a specific fluid are characterized by *constitutive equations*. The relation between the thermodynamic variables: the density ϱ, the pressure p, the internal energy e, and the absolute temperature ϑ is characterized through *equation of state* (EOS). *Caloric equation of state* relates the mechanical force represented by the pressure to the internal energy. *Thermal equation of state* is a relation between the internal energy and the temperature. Strictly speaking, these concepts refer to fluids in the state of *thermodynamic equilibrium*, while we use them to describe fluids *in motion* tacitly assuming the processes reestablishing the equilibrium state much faster than the observable macroscopic motion.

The well known caloric EOS in gas dynamics reads

$$p = (\gamma - 1)\varrho e, \quad \gamma > 1, \tag{1.39}$$

where γ is termed the adiabatic coefficient. The equation of state (1.39) is incomplete as it does not provide any information about the relation of the internal energy to the temperature ϑ. Introducing the total energy E, Eq. (1.39) can be rewritten in the form

$$p = \frac{1}{\gamma - 1}\left(E - \frac{1}{2}\frac{|\boldsymbol{m}|^2}{\varrho}\right).$$

We consider more general forms of EOS in Chap. 2.

1.3.1 The Second Law of Thermodynamics – Entropy

With (1.39) at hand, we return to the internal energy balance (1.27) deducing

$$\frac{1}{\gamma - 1}\varrho\Big(\partial_t \log(e) + \nabla_x \log(e) \cdot \boldsymbol{u}\Big) + \frac{1}{\gamma - 1}\mathrm{div}_x\left(\frac{\boldsymbol{q}}{e}\right)$$
$$= \frac{1}{(\gamma - 1)e}\left(\mathbb{S} : \nabla_x \boldsymbol{u} - \frac{\boldsymbol{q} \cdot \nabla_x e}{e}\right) - \varrho \mathrm{div}_x \boldsymbol{u}.$$

Moreover, it follows from the equation of continuity (1.1) that

$$\partial_t\left(\varrho \log(\varrho)\right) + \mathrm{div}_x(\varrho \log(\varrho)\boldsymbol{u}) + \varrho \mathrm{div}_x \boldsymbol{u} = 0.$$

Summing up the two equations we obtain

$$\begin{aligned} \partial_t(\varrho s(\varrho, e)) + \mathrm{div}_x\left(\varrho s(\varrho, e)\boldsymbol{u}\right) + \frac{1}{\gamma - 1}\mathrm{div}_x\left(\frac{\boldsymbol{q}}{e}\right) \\ = \frac{1}{(\gamma - 1)e}\left(\mathbb{S} : \nabla_x \boldsymbol{u} - \frac{\boldsymbol{q} \cdot \nabla_x e}{e}\right), \end{aligned} \tag{1.40}$$

which can be interpreted as a balance law for a new thermodynamic quantity

$$s(\varrho, e) = \log\left(e^{\frac{1}{\gamma - 1}}\right) - \log(\varrho)$$

called *entropy*. In accordance with the Second law of thermodynamics, the source term in (1.40) that represents the entropy production rate must be nonnegative,

$$\frac{1}{(\gamma - 1)e}\left(\mathbb{S} : \nabla_x \boldsymbol{u} - \frac{\boldsymbol{q} \cdot \nabla_x e}{e}\right) \geq 0, \tag{1.41}$$

for any physically admissible process.

Finally, we may normalize (1.40) by introducing the absolute temperature ϑ via a *thermal* EOS

$$(\gamma - 1)e = \vartheta \ \text{ or } \ e = c_v \vartheta, \quad c_v = \frac{1}{\gamma - 1}, \tag{1.42}$$

where c_v is the *specific heat at constant volume.* Thus (1.40) takes the form

$$\partial_t(\varrho s(\varrho, \vartheta)) + \mathrm{div}_x\,(\varrho s(\varrho, \vartheta)\boldsymbol{u}) + \mathrm{div}_x\left(\frac{\boldsymbol{q}}{\vartheta}\right) = \frac{1}{\vartheta}\left(\mathbb{S} : \nabla_x \boldsymbol{u} - \frac{\boldsymbol{q}\cdot\nabla_x\vartheta}{\vartheta}\right) \tag{1.43}$$

with

$$s(\varrho, \vartheta) = \log\left(\vartheta^{\frac{1}{\gamma-1}}\right) - \log(\varrho). \tag{1.44}$$

The mass density ϱ and the absolute temperature ϑ are termed *standard thermodynamic variables* and represent the *intensive* properties of the fluid—they are independent of the amount of the fluid matter. A general EOS that can be expressed in terms of ϱ and ϑ and is compatible with the Second law of thermodynamics must obey *Maxwell's equation*

$$\frac{\partial e(\varrho, \vartheta)}{\partial \varrho} = \frac{1}{\varrho^2}\left(p - \vartheta\frac{\partial p(\varrho, \vartheta)}{\partial \vartheta}\right). \tag{1.45}$$

Maxwell's equation in turn can be seen as a particular case of the general *Gibbs' relation*

$$\vartheta Ds = De + pD\left(\frac{1}{\varrho}\right). \tag{1.46}$$

Both (1.45) and (1.46) will be further discussed in Chap. 2.

1.3.2 Diffusion Transport Coefficients

The constitutive relations for the viscous stress $\mathbb{S}$ and the internal energy flux $\boldsymbol{q}$ that reflect the irreversibility of the fluid evolution must be in agreement with the Second law of thermodynamics. This means that the entropy production rate

$$\frac{1}{\vartheta}\left(\mathbb{S} : \nabla_x \boldsymbol{u} - \frac{\boldsymbol{q}\cdot\nabla_x\vartheta}{\vartheta}\right) \geq 0$$

must be nonnegative for any physically admissible process. Consequently,

$$\mathbb{S} : \nabla_x \boldsymbol{u} \geq 0, \ -\boldsymbol{q}\cdot\nabla_x\vartheta \geq 0, \tag{1.47}$$

in particular, $\mathbb{S}$ must depend on $\nabla_x \boldsymbol{u}$ and $\boldsymbol{q}$ on $\nabla_x\vartheta$.

1.3.2.1 Heat Conduction – Fourier's Law

The simplest possible choice of $\boldsymbol{q}$ leads to *Fourier's law*

$$\boldsymbol{q} = -\kappa\nabla_x\vartheta,$$

where $\kappa \geq 0$ is the heat conductivity coefficient. In this case, the internal/thermal energy balance (1.27) takes the form

$$c_v\Big[\partial_t(\varrho\vartheta) + \mathrm{div}_x(\varrho\vartheta\boldsymbol{u})\Big] - \mathrm{div}_x(\kappa\nabla_x\vartheta) = \mathbb{S} : \nabla_x\boldsymbol{u} - \varrho\vartheta\mathrm{div}_x\boldsymbol{u}. \tag{1.48}$$

It is worth noting that the Eq. (1.48), similarly to (1.15), can be used for a rigorous proof of positivity of the absolute temperature. Indeed equation (1.48) can be rewritten in the form

$$c_v\partial_t\vartheta + c_v\boldsymbol{u}\cdot\nabla_x\vartheta - \frac{1}{\varrho}\mathrm{div}_x(\kappa\nabla_x\vartheta) \geq -\vartheta\mathrm{div}_x\boldsymbol{u},$$

for which the standard parabolic comparison principle yields

$$\vartheta(\tau,x) \geq \inf_{\Omega}\vartheta(0,\cdot)\exp\left(-\frac{1}{c_v}\int_0^\tau \|\mathrm{div}_x\boldsymbol{u}(t,\cdot)\|_{L^\infty(\Omega)}\,\mathrm{d}t\right)$$

1.3.2.2 Viscous Stress – Dissipation Function

As $\mathbb{S}$ is symmetric, we can write

$$\mathbb{S} : \nabla_x\boldsymbol{u} = \mathbb{S} : \mathbb{D}\boldsymbol{u}, \text{ where } \mathbb{D}\boldsymbol{u} \equiv \frac{1}{2}\left(\nabla_x\boldsymbol{u} + \nabla_x^T\boldsymbol{u}\right).$$

We adopt a generic hypothesis that $\mathbb{S}$ depends on the symmetric velocity gradient $\mathbb{D}\boldsymbol{u}$. The simplest possible form is linear dependence that can be written as Newton's rheological law:

$$\mathbb{S} = 2\mu\left(\mathbb{D}\boldsymbol{u} - \frac{1}{d}\mathrm{trace}[\mathbb{D}\boldsymbol{u}]\mathbb{I}\right) + \lambda\mathrm{trace}[\mathbb{D}\boldsymbol{u}]\mathbb{I} = \mu\left(\nabla_x\boldsymbol{u} + \nabla_x^T\boldsymbol{u} - \frac{2}{d}\mathrm{div}_x\boldsymbol{u}\mathbb{I}\right) + \lambda\mathrm{div}_x\boldsymbol{u}\mathbb{I}, \tag{1.49}$$

where $\mu \geq 0$, $\lambda \geq 0$ are the shear and bulk viscosity coefficients, respectively.

In general, we assume that $\mathbb{S}$ and $\mathbb{D}\boldsymbol{u}$ are interrelated "implicitly" through

$$\mathbb{S} : \mathbb{D}\boldsymbol{u} = F(\mathbb{D}\boldsymbol{u}) + F^*(\mathbb{S}), \tag{1.50}$$

where we have introduced the *dissipation function* F,

$$F : R^{d\times d}_{\mathrm{sym}} \to [0,\infty] \text{ is a convex lower semicontinuous function, } F(0) = 0,$$

and

$$F^* : R^{d\times d}_{\mathrm{sym}} \to [0,\infty] \text{ its conjugate.}$$

Note that we have identified the finite dimensional Hilbert space $R^{d\times d}_{\rm sym}$ with its dual via the standard Riesz isometry. In view of Fenchel–Young inequality, relation (1.50) yields

$$\mathbb{S} \in \partial F(\mathbb{D}\boldsymbol{u}) \ \Leftrightarrow\ \mathbb{D}\boldsymbol{u} \in \partial F^*(\mathbb{S}), \tag{1.51}$$

where ∂ denotes the subdifferential. As we shall see below, the abstract formulation via (1.51) is quite convenient and gives rise to an elegant weak formulation of the problem.

1.3.3 Boundary Conditions for Diffusion Fluxes

As already stated in (1.28), the total internal energy flux through the boundary vanishes. In the present context, this gives rise to

$$\boldsymbol{q}\cdot\boldsymbol{n}|_{\partial\Omega} = -\kappa\nabla_x\vartheta\cdot\boldsymbol{n}|_{\partial\Omega} = 0. \tag{1.52}$$

For the velocity and viscous stress we recall the impermeability and complete slip conditions

$$\boldsymbol{u}\cdot\boldsymbol{n}|_{\partial\Omega} = 0,\ \ (\mathbb{S}\cdot\boldsymbol{n})\times\boldsymbol{n}|_{\partial\Omega} = 0. \tag{1.53}$$

Alternatively for viscous fluids, we may consider the *no-slip* boundary conditions for the velocity field

$$\boldsymbol{u}|_{\partial\Omega} = 0\ \Leftrightarrow \boldsymbol{u}\cdot\boldsymbol{n}|_{\partial\Omega} = \boldsymbol{u}\times\boldsymbol{n}|_{\partial\Omega} = 0. \tag{1.54}$$

If this is the case, then viscous component of the energy flux $\mathbb{S}\cdot\boldsymbol{u}\cdot\boldsymbol{n}$ vanishes on $\partial\Omega$ and no other boundary condition is needed. Moreover, if (1.54) holds, the class of test functions in the momentum equation (1.21) must be restricted to

$$\boldsymbol{\varphi} \in C^1_c([0,T]\times\Omega; R^d),$$

meaning $\boldsymbol{\varphi}$ has compact support in Ω. In contrast with (1.21), the satisfaction of the no-slip boundary condition in the context of weak solutions is enforced by means of higher regularity of $\boldsymbol{u}$ belonging to a function space of Sobolev type where the boundary traces are well defined.

1.4 Equivalent Formulation of the Total Energy Balance in the Context of Smooth Solutions

We conclude this chapter by recalling several possibilities how to express the total energy balance (1.35). In the context of strong formulation, they are completely equivalent and their particular form makes little importance. However, their use in the weak form and/or numerical implementations may lead to substantial differences discussed in the forthcoming chapters below. Adopting the EOS (1.39), (1.42), we record the following alternative formulations of the First law of thermodynamics:

- **Total energy balance**

$$\partial_t \left(\frac{1}{2}\varrho|\boldsymbol{u}|^2 + \varrho e \right) + \mathrm{div}_x \left[\left(\frac{1}{2}\varrho|\boldsymbol{u}|^2 + \varrho e \right) \boldsymbol{u} + p\boldsymbol{u} \right] + \mathrm{div}_x \left(\boldsymbol{q} - \mathbb{S} \cdot \boldsymbol{u} \right) = \varrho \boldsymbol{f} \cdot \boldsymbol{u}; \tag{1.55}$$

- **Internal/thermal energy balance**

$$c_v \Big[\partial_t(\varrho\vartheta) + \mathrm{div}_x(\varrho\vartheta\boldsymbol{u}) \Big] - \mathrm{div}_x(\kappa\nabla_x\vartheta) = \mathbb{S} : \nabla_x \boldsymbol{u} - \varrho\vartheta \mathrm{div}_x \boldsymbol{u}; \tag{1.56}$$

- **Entropy balance**

$$\partial_t(\varrho s(\varrho, \vartheta)) + \mathrm{div}_x \left(\varrho s(\varrho, \vartheta)\boldsymbol{u} \right) - \mathrm{div}_x \left(\frac{\kappa\nabla_x\vartheta}{\vartheta} \right) = \frac{1}{\vartheta} \left(\mathbb{S} : \nabla_x \boldsymbol{u} + \frac{\kappa|\nabla_x\vartheta|^2}{\vartheta} \right). \tag{1.57}$$

We point out again that the above equations are *equivalent* in the context of smooth solutions. In the *weak formulation*, however, different formulation enforce different aspects of the fluid motion and a judicious choice of the relevant field equation is needed. In the case of perfect (inviscid) fluids, where $\mathbb{S} = \boldsymbol{q} = 0$, it is only the total energy balance (1.55) that seems to be relevant for a weak formulation as the other equations contain gradient of the velocity that is not *a priori* bounded in any Lebesgue space. However, the lack of *a priori* bounds on the convective term in (1.55) forces us to use a weak formulation based on the integrated form of (1.55), combined with the *entropy inequality* :

$$\frac{\mathrm{d}}{\mathrm{d}t} \int_\Omega \left[\frac{1}{2}\varrho|\boldsymbol{u}|^2 + \varrho e \right] \,\mathrm{d}x = \int_\Omega \varrho \boldsymbol{f} \cdot \boldsymbol{u} \,\mathrm{d}x, \ \partial_t(\varrho s(\varrho, \vartheta)) + \mathrm{div}_x \left(\varrho s(\varrho, \vartheta)\boldsymbol{u} \right) \geq 0. \tag{1.58}$$

We claim that (1.58), in combination with the equation of continuity (1.33) and the momentum balance (1.34), still gives rise to any of the equations (1.55)–(1.57) (with $\mathbb{S} = \boldsymbol{q} = 0$) as soon as all quantities in question are smooth enough and the velocity field satisfies the impermeability condition (1.12). Indeed we first deduce from the estimate (1.15) that ϱ remains bounded and strictly positive as long as it is bounded and strictly positive at the initial time. Thus the second equation in (1.58), together with the equation of continuity (1.33), give rise to

$$\partial_t s + \boldsymbol{u} \cdot \nabla_x s \geq 0. \tag{1.59}$$

In particular,

$$\frac{\mathrm{d}}{\mathrm{d}t} s\,(t, \boldsymbol{X}(t, \boldsymbol{x}_0)) \geq 0 \ \text{ for any } \ \boldsymbol{x}_0 \in \Omega,$$

where $\{\boldsymbol{X}(t, \boldsymbol{x}_0)\}_{t \geq 0}$ are the streamlines introduced in Sect. 1.2. Seeing that the density is bounded above and below away from zero and s is given by the constitutive relation (1.44) we obtain a lower bound for the temperature ϑ in terms of the initial data. Consequently, the entropy inequality (1.59) divided on ϑ yields the internal energy balance

$$\partial_t(\varrho e) + \mathrm{div}_x(\varrho e \boldsymbol{u}) \geq -\varrho \vartheta \mathrm{div}_x \boldsymbol{u},$$

cf. (1.42). This relation added to the kinetic energy equation (1.25) gives rise to the total energy balance

$$\partial_t \left(\frac{1}{2} \varrho |\boldsymbol{u}|^2 + \varrho e \right) + \mathrm{div}_x \left[\left(\frac{1}{2} \varrho |\boldsymbol{u}|^2 + \varrho e + p \right) \boldsymbol{u} \right] \geq \varrho \boldsymbol{f} \cdot \boldsymbol{u}.$$

Finally, as $\boldsymbol{u} \cdot \boldsymbol{n}|_{\partial\Omega} = 0$, integration by parts yields

$$\frac{\mathrm{d}}{\mathrm{d}t} \int_\Omega \left[\frac{1}{2} \varrho |\boldsymbol{u}|^2 + \varrho e \right] \, \mathrm{d}x \geq 0 \int_\Omega \varrho \boldsymbol{f} \cdot \boldsymbol{u} \, \mathrm{d}x.$$

However, this is compatible with (1.58) only if the inequality in the entropy balance is replaced by equality. As a result, we deduce all the equivalent forms (1.55)–(1.57) of the energy equation (with $\mathbb{S} = 0$, $\boldsymbol{q} = 0$).

1.5 Conclusion, Bibliographical Remarks

The introduction of the basic equations of continuum fluid dynamics in this chapter has been quite brief and far from complete. We focused on the models including viscous as well as inviscid general compressible fluids with a particular choice of boundary conditions. The interested reader may consult the monograph by Chorin and Marsden [53] for a detailed introduction to mathematical fluid mechanics. More information can be found in Batchelor [13] or Lamb [147]. A nice introduction to the problems related to various EOS is the monograph by Eliezer, Ghatak, and Hora [74].

A more detailed exposition of the weak formulation of conservation laws in fluid dynamics can be found in [98, Chap. 1]. The problem of normal traces for general vector fields with measure divergence is treated by Chen, Torres, and Ziemer [44], [45].

Chapter 2
Inviscid Fluids: Euler System

The *Euler system* represents an iconic model of a *perfect* fluid, for which the viscous stress $\mathbb{S}$ as well as the heat flux $\boldsymbol{q}$ vanish. The relevant system of field equations reads:

- **Mass conservation or equation of continuity**

$$\partial_t \varrho + \mathrm{div}_x(\varrho \boldsymbol{u}) = 0. \tag{2.1}$$

- **Newton's Second law or momentum conservation**

$$\partial_t(\varrho \boldsymbol{u}) + \mathrm{div}_x(\varrho \boldsymbol{u} \otimes \boldsymbol{u}) + \nabla_x p = 0. \tag{2.2}$$

- **First law of thermodynamics or energy conservation**

$$\partial_t \left(\frac{1}{2} \varrho |\boldsymbol{u}|^2 + \varrho e \right) + \mathrm{div}_x \left[\left(\frac{1}{2} \varrho |\boldsymbol{u}|^2 + \varrho e + p \right) \boldsymbol{u} \right] = 0. \tag{2.3}$$

We have deliberately omitted the action of any external force for the sake of simplicity. The **Second law of thermodynamics** is enforced by the entropy balance (inequality)

$$\partial_t(\varrho s) + \mathrm{div}_x(\varrho s \boldsymbol{u}) = (\geq) 0. \tag{2.4}$$

The entropy is conserved, meaning there is equality sign in (2.4), as long as the motion is smooth. The inequality becomes relevant as soon as singularities appear. As we show below, the appearance of singularities (shock waves) is an inevitable phenomenon for the Euler system.

Apparently, the system (2.1)–(2.3) contains more unknowns than equations and constitutive equations must be imposed reflecting the material properties of the fluid. Introducing the absolute temperature ϑ we postulate **Gibbs' equation** interrelating the pressure p, the internal energy e, and the entropy s:

E. Feireisl et al., *Numerical Analysis of Compressible Fluid Flows*, MS&A 20,
https://doi.org/10.1007/978-3-030-73788-7_2

$$\vartheta Ds = De + D\left(\frac{1}{\varrho}\right)p. \tag{2.5}$$

Still there is a lot of freedom how to choose the fundamental *state variables* to describe the fluid in motion. At this stage, we point out that a judicious choice of state variables is absolutely crucial for success/failure of many numerical methods as well as their analysis.

In contrast with Chap. 1, where we have considered a very specific relation between the thermodynamic functions given by the EOS (1.39), Gibbs' equation (2.5) is fairly general including a vast class of possible EOS that are compatible with the Second law of thermodynamics. Indeed, if (2.5) holds and all quantities in question are smooth, the entropy Eq. (2.4) follows directly from the Eqs. (2.1)–(2.3). To see this, we first rewrite the energy balance (2.3) in terms of the internal energy

$$\varrho \partial_t e + \varrho \boldsymbol{u} \cdot \nabla_x e = -p \mathrm{div}_x \boldsymbol{u},$$

cf. (1.56). Dividing on ϑ we obtain

$$\varrho \frac{1}{\vartheta} \partial_t e + \varrho \boldsymbol{u} \frac{1}{\vartheta} \cdot \nabla_x e = -\frac{p}{\vartheta} \mathrm{div}_x \boldsymbol{u}.$$

Finally, we compute from (2.1)

$$\partial_t \varrho + \boldsymbol{u} \cdot \nabla_x \varrho = -\varrho \mathrm{div}_x \boldsymbol{u};$$

whence

$$\varrho \frac{1}{\vartheta} \partial_t e + \varrho \boldsymbol{u} \frac{1}{\vartheta} \cdot \nabla_x e = \varrho \frac{p}{\vartheta \varrho^2} \left(\partial_t \varrho + \boldsymbol{u} \cdot \nabla_x \varrho \right).$$

In view of Gibbs' relation (2.5) we get

$$\vartheta \partial_t s = \partial_t e - \frac{p}{\varrho^2} \partial_t \varrho, \ \vartheta \nabla_x s = \nabla_x e - \frac{p}{\varrho^2} \nabla_x \varrho$$

and the entropy balance (2.4) follows.

2.1 Euler System in Standard Variables

The Euler system so far has been considered in terms of **standard variables**:

$$\begin{aligned} \text{the mass density } \varrho &= \varrho(t, x), \\ \text{the (absolute) temperature } \vartheta &= \vartheta(t, x), \\ \text{the (bulk) velocity field } \boldsymbol{u} &= \boldsymbol{u}(t, x). \end{aligned}$$

Such a choice of reference quantities is probably most popular because these fields are also directly *observable* and *measurable* in practical experiments. Note, however, that neither the temperature nor the velocity are correctly defined on the vacuum zones, where the density ϱ vanishes. To circumvent these difficulties, the mathematical theory sometimes prefers other choices of reference quantities, such as the conservative or conservative–entropy variables discussed later in this chapter.

If the underlying physical domain is bounded, which is relevant in most numerical experiments, the system of Eqs. (2.1)–(2.3) should be supplemented with suitable boundary conditions. For the sake of simplicity, we focus on the impermeability condition

$$\boldsymbol{u} \cdot \boldsymbol{n}|_{\partial\Omega} = 0, \tag{2.6}$$

keeping in mind that many real world application my involve general in/out flow boundary conditions. In the case of unbounded physical space, the far field conditions should be prescribed, here in the form

$$\varrho \to \varrho_\infty, \ \boldsymbol{u} \to \boldsymbol{u}_\infty \ \text{as} \ |x| \to \infty. \tag{2.7}$$

Note however that unbounded domains must be replaced by finite—bounded ones—in numerical experiments. Alternatively, we may prescribe the spatially periodic conditions reducing the spatial domain to a flat torus,

$$\Omega = \Pi_{i=1}^d [a_i, b_i]|_{\{a_i, b_i\}} \equiv \mathbb{T}^d.$$

As already pointed out, such a choice eliminates all issues connected with the presence of physical boundary and may be seen as a convenient approximation of problems on "large" domains, where the boundary effect is negligible at least in a finite time lap.

As already observed in Sect. 1.4, the total energy balance may be replaced by the entropy balance

$$\partial_t(\varrho s(\varrho, \vartheta)) + \mathrm{div}_x(\varrho s(\varrho, \vartheta)\boldsymbol{u}) = 0 \tag{2.8}$$

as long as all quantities are smooth—continuously differentiable. In view of the equation of continuity (2.1), the entropy balance can be rewritten as a pure transport equation

$$\partial_t s(\varrho, \vartheta) + \boldsymbol{u} \cdot \nabla_x s(\varrho, \vartheta) = 0.$$

Multiplying the resulting equation on $Z'(s)$, where Z is a continuously differentiable function, we obtain a *renormalized version* of the entropy balance

$$\partial Z(s) + \boldsymbol{u} \cdot \nabla_x Z(s) = 0,$$

or

$$\partial_t(\varrho Z(s)) + \mathrm{div}_x\,(\varrho Z(s)\boldsymbol{u}) = 0. \tag{2.9}$$

Anticipating the inequality in (2.9) we restrict the class of test functions Z to

$$Z \in C^1(R),\ Z'(s) \geq 0 \text{ for any } s \in R, \tag{2.10}$$

and replace (2.4) by

$$\partial_t(\varrho Z(s)) + \mathrm{div}_x(\varrho Z(s)\boldsymbol{u}) \geq 0 \tag{2.11}$$

for any Z as in (2.10). Inequality (2.11) will play the role of admissibility criterion in the context of weak solutions to the Euler system. Unfortunately, as we shall see in Theorem 2.1 below, the Euler system is ill posed in the class of weak solutions even if (2.11) is satisfied.

2.1.1 Local Existence of Smooth Solutions

Consider the Euler system (2.1)–(2.3) in a bounded domain $\Omega \subset R^d$, $d = 1, 2, 3$, with the impermeability condition (2.6), and the initial data

$$\varrho(0,\cdot) = \varrho_0,\ \vartheta(0,\cdot) = \vartheta_0,\ \boldsymbol{u}(0,\cdot) = \boldsymbol{u}_0. \tag{2.12}$$

It turns out that for smooth initial data satisfying certain compatibility conditions, the initial-boundary value problem for the Euler system is well-posed *locally* in time. As we have observed, the total energy balance (2.3) is equivalent to the entropy equation (2.8), so it may be convenient to solve the system (2.1), (2.2), with (2.8) replacing (2.3).

The relevant short time existence result can be stated in the Sobolev framework of spaces $W^{k,2}(\Omega)$, $k = 1, 2, 3$. In addition to Gibbs' relation (2.5), we impose the *thermodynamic stability condition* written in the standard variables:

$$\frac{\partial p(\varrho,\vartheta)}{\partial \varrho} > 0,\ \frac{\partial e(\varrho,\vartheta)}{\partial \vartheta} > 0 \text{ for all } \varrho, \vartheta > 0. \tag{2.13}$$

We report the following result proved by Schochet [182, Theorem 1]:

Theorem 2.1 (Short time existence for complete Euler system) *Suppose that*

- *the thermodynamic functions* $p = p(\varrho,\vartheta)$, $e = e(\varrho,\vartheta)$, $s = s(\varrho,\vartheta)$ *are three times continuously differentiable for* $\varrho > 0$, $\vartheta > 0$ *and satisfy* (2.5), (2.13)*;*
- $\Omega \subset R^3$ *is a bounded domain with* C^∞*-boundary;*
- *the initial data belong to the class*

$$\varrho_0,\ \vartheta_0 \in W^{3,2}(\Omega),\ \boldsymbol{u}_0 \in W^{3,2}(\Omega; R^3),\ \varrho_0 > 0,\ \vartheta_0 > 0 \ \text{in}\ \overline{\Omega};$$

- *the compatibility conditions*

$$\partial_t^k \boldsymbol{u}_0 \cdot \boldsymbol{n}|_{\partial\Omega} = 0$$

hold for $k = 0, 1, 2$.

Then there exists $T > 0$ *such that the Euler system* (2.1), (2.2), (2.8), *with the boundary condition* (2.12), *and the initial data* $[\varrho_0, \vartheta_0, \boldsymbol{u}_0]$ *admits a classical solution in* $(0, T) \times \Omega$,

$$\varrho(t, \cdot),\ \vartheta(t, \cdot) > 0 \ \textit{in}\ \overline{\Omega}\ \textit{for}\ t \in [0, T).$$

Remark 2.1 By classical solution we mean that all functions ϱ, ϑ, and $\boldsymbol{u}$ are continuously differentiable up to the boundary and the equations hold pointwise in $(0, T) \times \Omega$. As a matter of fact, the solutions are constructed in the Sobolev classes

$$\bigcap_{k=0}^{3} C^{3-k}([0, T]; W^{k,2}(\Omega)),$$

and the desired regularity follows from the Sobolev embedding $W^{2,2}(\Omega) \hookrightarrow C(\overline{\Omega})$, $\Omega \subset R^3$.

Remark 2.2 The same result holds for $\Omega \subset R^d$, $d = 2, 3$, where one can replace $W^{3,2}$ by $W^{2,2}$ if $d = 1$. The hypothesis of C^∞-boundary can be relaxed depending on the desired regularity of solutions. However, certain regularity of the boundary is needed given the required properties of Sobolev functions.

Remark 2.3 The compatibility conditions are computed by plugging the data in the momentum Eq. (2.2). The same result holds with the periodic boundary conditions. In that case, the compatibility conditions are irrelevant.

Remark 2.4 As pointed out several times, the entropy balance (2.8) is equivalent to the total energy balance (2.3) in the framework of smooth solutions.

The proof of Theorem 2.1 is based on deriving sufficiently strong *a priori* bounds. It is nowadays well understood but still rather lengthy and technical. Our feeling is that a detailed reproduction of the proof would go beyond the scope of this monograph. We therefore omit the proof here referring the interested reader to Schochet [182].

2.1.2 Finite Time Blowup

The maximal existence time of the strong solutions to the Euler system is unfortunately finite for a fairly general class of initial data. Strong solutions develop singularities in the form of discontinuous shock waves in a finite time lap. To illuminate this phenomenon, consider the Euler system in one space dimension in the *entropy formulation*:

$$\begin{aligned}
\partial_t \varrho + \partial_x(\varrho u) &= 0,\\
\partial_t(\varrho u) + \partial_x(\varrho u^2) + \partial_x(\varrho \vartheta) &= 0,\\
\partial_t s + \partial_x s u &= 0,
\end{aligned} \tag{2.14}$$

where

$$s(\varrho, \vartheta) = c_v \log(\vartheta) - \log(\varrho).$$

Furthermore, we focus on the isentropic case setting $s(\varrho, \vartheta) = \overline{s}$, with a constant $\overline{s}$. Accordingly, problem (2.14) reduces to the isentropic Euler system

$$\begin{aligned} \partial_t \varrho + \partial_x(\varrho u) &= 0, \\ \partial_t(\varrho u) + \partial_x(\varrho u^2) + a \partial_x \varrho^\gamma &= 0 \end{aligned} \tag{2.15}$$

with

$$a = \exp\left(\frac{\overline{s}}{c_v}\right) > 0,$$

cf. also Sect. 2.3 below.

Next, we rewrite (2.15) in the form of the so-called P-system. To this end, we introduce the Lagrange mass coordinates transform

$$[t, x] \mapsto \left[t, y = y(t, x) = \int_{-\infty}^{x} \varrho(t, z) \, \mathrm{d}z \right].$$

The system (2.15) written in terms of the new independent variables $[t, y]$ reads

$$\begin{aligned} \partial_t V - \partial_y w &= 0, \\ \partial_t w + a \partial_y V^{-\gamma} &= 0, \end{aligned} \tag{2.16}$$

where $V = \varrho^{-1}$ is the specific volume and

$$w\left(t, \int_{-\infty}^{x} \varrho(t, z) \, \mathrm{d}z\right) = u(t, x)$$

is the Lagrangian velocity. The system (2.16) may be recast in a more general form as the P-system

$$\begin{aligned} \partial_t V - \partial_y w &= 0, \\ \partial_t w - \partial_y P(V) &= 0, \end{aligned} \tag{2.17}$$

where $P'(V) > 0$. Note that all of the above formulations are equivalent as soon as all quantities are smooth and the density strictly positive.

Next we show that solutions of (2.17) may develop singularities in a finite time. To this end, we first write

$$\begin{aligned} \sqrt{P'(V)} \partial_t V - \sqrt{P'(V)} \partial_y w &= 0, \\ \partial_t w - \sqrt{P'(V)} \sqrt{P'(V)} \partial_y V &= 0. \end{aligned}$$

Introducing

$$Z(V) = \int_0^V \sqrt{P'(z)}\,\mathrm{d}z \tag{2.18}$$

we obtain

$$\begin{aligned} \partial_t Z - A(Z)\partial_y w &= 0,\\ \partial_t w - A(Z)\partial_y Z &= 0 \end{aligned} \tag{2.19}$$

with

$$A(Z) = \sqrt{P'(V(Z))}, \text{ where } V = V(Z) \text{ is determined by 2.18.}$$

Problem (2.19) admits special solutions $Z = \pm w$. For $Z = -w$ we obtain

$$\partial_t Z + A(Z)\partial_y Z = 0,$$

or, for $U = A(Z)$,

$$\partial_t U + U\partial_y U = 0. \tag{2.20}$$

Now, it is easy to find initial data for which the corresponding solution of (2.20) blows up. Indeed as long as the solution U of (1.25) emanating from the initial data

$$U(0, y) = U_0(y),\ y \in R$$

remains smooth, it satisfies

$$U(t, y + tU_0(y)) = U_0(y),$$

which can be checked by differentiating this relation with respect to t. Thus if there exist $y_1 < y_2$ such that

$$U_0(y_1) > U_0(y_2),$$

then we obtain

$$U(\tau, z) = U_0(y_1) > U_0(y_2) = U(\tau, z) \text{ for } z = y_1 + \tau U_0(y_1) = y_2 + \tau U_0(y_2),$$

if

$$\tau = \frac{y_2 - y_1}{U_0(y_1) - U_0(y_2)} > 0.$$

This shows that the solution U develops singularity, specifically becomes discontinuous, at latest at the time τ.

The type of singularity we have just described is called *shock wave* in gas dynamics. Solutions remain bounded but experience discontinuities, in particular, the derivatives required for the system of equations to be satisfied do not exist in the classical sense. We point out that in the preceding example

- the singularity appeared in the finite time depending only on geometrical properties of the initial data;
- the initial data could be chosen smooth and even small in suitable sense.

Similar examples can be constructed in higher space dimension. Thus if we still prefer to retain the Euler system as a suitable mathematical model to explain phenomena in fluid dynamics, we have to enlarge the class of solutions to the weak ones introduced in Sect. 1.1.1 or even to more general objects discussed in Chap. 4.

2.1.3 Weak Solutions

Following the general strategy delineated in Sect. 1.1.1 we introduce the concept of *weak* solution to the Euler system. We start by writing the weak formulation in terms of the standard variables: the density ϱ, the temperature ϑ, and the velocity $\boldsymbol{u}$, later we consider alternative forms.

Definition 2.1 (WEAK SOLUTION TO COMPLETE EULER SYSTEM)
The trio $[\varrho, \vartheta, \boldsymbol{u}]$ is *weak (distributional) solution* of the Euler system (2.1)–(2.3) in $(0, T) \times \Omega$, with the impermeability boundary condition (2.6) and the initial data $[\varrho_0, \vartheta_0, \boldsymbol{u}_0]$ if the following is satisfied:

- (**measurability**) the quantities $\varrho = \varrho(t, x)$, $\vartheta = \vartheta(t, x)$, $\boldsymbol{u} = \boldsymbol{u}(t, x)$ are measurable functions defined for $(t, x) \in (0, T) \times \Omega$,

$$\varrho(t, x) \geq 0, \ \ \vartheta(t, x) > 0 \ \text{ for a.a. } (t, x) \in (0, T) \times \Omega; \tag{2.21}$$

- (**equation of continuity**) the integral identity

$$\int_0^T \int_\Omega \left[\varrho \partial_t \varphi + \varrho \boldsymbol{u} \cdot \nabla_x \varphi \right] \ \mathrm{d}x \, \mathrm{d}t = - \int_\Omega \varrho_0 \varphi(0, \cdot) \ \mathrm{d}x \tag{2.22}$$

  holds for any test function $\varphi \in C^1_c([0, T) \times \overline{\Omega})$;
- (**momentum equation**) the integral identity

$$\begin{aligned} &\int_0^T \int_\Omega \left[\varrho \boldsymbol{u} \cdot \partial_t \boldsymbol{\varphi} + \varrho \boldsymbol{u} \otimes \boldsymbol{u} : \nabla_x \boldsymbol{\varphi} + p(\varrho, \vartheta) \mathrm{div}_x \boldsymbol{\varphi} \right] \ \mathrm{d}x \, \mathrm{d}t \\ &\quad = - \int_\Omega \varrho_0 \boldsymbol{u}_0 \cdot \boldsymbol{\varphi}(0, \cdot) \ \mathrm{d}x \end{aligned} \tag{2.23}$$

  holds for any test function $\boldsymbol{\varphi} \in C^1_c([0, T) \times \overline{\Omega}; R^d)$, $\boldsymbol{\varphi} \cdot \boldsymbol{n}|_{\partial \Omega} = 0$;

- (**energy conservation**) the integral identity

$$\begin{aligned}
&\int_0^T \int_\Omega \left(\frac{1}{2} \varrho |\boldsymbol{u}|^2 + \varrho e(\varrho, \vartheta) \right) \partial_t \varphi \, \mathrm{d}x \, \mathrm{d}t \\
&+ \int_0^T \int_\Omega \left(\frac{1}{2} \varrho |\boldsymbol{u}|^2 + \varrho e(\varrho, \vartheta) + p(\varrho, \vartheta) \right) \boldsymbol{u} \cdot \nabla_x \varphi \, \mathrm{d}x \, \mathrm{d}t \\
&= - \int_\Omega \left(\frac{1}{2} \varrho_0 |\boldsymbol{u}_0|^2 + \varrho_0 e(\varrho_0, \vartheta_0) \right) \varphi(0, \cdot) \, \mathrm{d}x
\end{aligned} \tag{2.24}$$

holds for any test function $\varphi \in C_c^1([0, T) \times \overline{\Omega})$.

Remark 2.5 (Regularity) In Definition 2.1, we have tacitly assumed that all integrals are finite, meaning all densities and fluxes are at least integrable in $(0, T) \times \Omega$. Regularity of the spatial domain Ω has not been explicitly specified. The hypothesis $\varphi \in C^1(\overline{\Omega})$ is understood in the sense that $\varphi = \widetilde{\varphi}|_{\overline{\Omega}}$, where $\widetilde{\varphi} \in C^1(R^d)$. Similarly, the condition $\boldsymbol{\phi} \cdot \boldsymbol{n}|_\Omega$ requires the existence of the outer normal vector $\boldsymbol{n}$ at least in a suitable generalized sense. Alternatively, we may say the $\boldsymbol{\phi} \cdot \boldsymbol{n}|_{\partial\Omega} = 0$ if

$$\int_\Omega \boldsymbol{\phi} \cdot \nabla_x \psi \, \mathrm{d}x + \int_\Omega \psi \mathrm{div}_x \boldsymbol{\phi} \, \mathrm{d}x = 0 \text{ for any } \psi \in C^1(\overline{\Omega}).$$

Similar convention will be adapted in the future.

Remark 2.6 (Periodic boundary conditions) Definition 2.1 can be easily adapted to the spatially periodic boundary conditions, meaning $\Omega = \mathbb{T}^d$. Indeed, it is enough to set $\Omega = \mathbb{T}^d = \overline{\Omega}$ and omit the stipulation $\boldsymbol{\phi} \cdot \boldsymbol{n}|_{\partial\Omega} = 0$ in (2.23).

Considering a specific form of the test functions $\varphi(t, x) = \psi(t)\phi(x)$, $\boldsymbol{\varphi}(t, x) = \psi(t)\boldsymbol{\phi}(x)$ in the integral identities (2.22)–(2.24) we observe that the integral averages

$$\begin{aligned}
&t \mapsto \int_\Omega \varrho(t, x)\phi(x) \, \mathrm{d}x, \ t \mapsto \int_\Omega (\varrho \boldsymbol{u})(t, x) \cdot \boldsymbol{\phi}(x) \, \mathrm{d}x, \\
&t \mapsto \int_\Omega \left(\frac{1}{2} \varrho |\boldsymbol{u}|^2 + \varrho e(\varrho, \vartheta) \right) (t, x)\phi(x) \, \mathrm{d}x
\end{aligned} \tag{2.25}$$

are continuous functions of the time $t \in [0, T]$. Accordingly, the integral identities in Definition 2.1 can be equivalently stated in the form

$$\int_0^\tau \int_\Omega [\varrho \partial_t \varphi + \varrho \boldsymbol{u} \cdot \nabla_x \varphi] \, \mathrm{d}x \, \mathrm{d}t = \left[\int_\Omega \varrho \varphi \, \mathrm{d}x \right]_{t=0}^{t=\tau} \tag{2.26}$$

for any $0 \leq \tau \leq T$, and any $\varphi \in C^1([0,T] \times \overline{\Omega})$;

$$\int_0^\tau \int_\Omega \left[\varrho \boldsymbol{u} \cdot \partial_t \boldsymbol{\varphi} + \varrho \boldsymbol{u} \otimes \boldsymbol{u} : \nabla_x \boldsymbol{\varphi} + p(\varrho, \vartheta) \mathrm{div}_x \boldsymbol{\varphi} \right] \, \mathrm{d}x \, \mathrm{d}t = \left[\int_\Omega \varrho \boldsymbol{u} \cdot \boldsymbol{\varphi} \, \mathrm{d}x \right]_{t=0}^{t=\tau} \tag{2.27}$$

for any $0 \leq \tau \leq T$, and any$\boldsymbol{\varphi} \in C_c^1([0,T) \times \overline{\Omega}; R^d)$, $\boldsymbol{\varphi} \cdot \boldsymbol{n}|_{\partial\Omega} = 0$; and

$$\begin{aligned} &\int_0^\tau \int_\Omega \left(\frac{1}{2} \varrho |\boldsymbol{u}|^2 + \varrho e(\varrho, \vartheta) \right) \partial_t \varphi \, \mathrm{d}x \, \mathrm{d}t \\ &\quad + \int_0^\tau \int_\Omega \left(\frac{1}{2} \varrho |\boldsymbol{u}|^2 + \varrho e(\varrho, \vartheta) + p(\varrho, \vartheta) \right) \boldsymbol{u} \cdot \nabla_x \varphi \, \mathrm{d}x \, \mathrm{d}t \\ &\quad = \left[\int_\Omega \left(\frac{1}{2} \varrho |\boldsymbol{u}|^2 + \varrho e(\varrho, \vartheta) \right) \varphi \, \mathrm{d}x \right]_{t=0}^{t=\tau} \end{aligned} \tag{2.28}$$

for any $0 \leq \tau \leq T$, and any $\varphi \in C^1([0,T] \times \overline{\Omega})$.

Continuity of the integral averages stated in (2.25) can be extended to a larger class of test functions provided certain uniform bounds are available. Suppose, for instance,

$$\varrho \in L^\infty(0,T; L^\gamma(\Omega)) \ \text{ for some } \ \gamma > 1. \tag{2.29}$$

First observe that (2.26) implies that for any $t \in [0,T]$ the quantity

$$\int_\Omega \varrho(t,x) \phi(x) \, \mathrm{d}x, \ \phi \in C_c^\infty(\Omega)$$

is well defined, and, in accordance with (2.29),

$$\left| \int_\Omega \varrho(t,x) \phi(x) \, \mathrm{d}x \right| \leq c \|\phi\|_{L^{\gamma'}(\Omega)}$$

uniformly for *any* $t \in [0,T]$. As the smooth compactly supported functions are dense in $L^{\gamma'}(\Omega)$ that is a reflexive Banach space, we can identify the *instantaneous* values

$$\varrho(t,\cdot) \in L^\gamma(\Omega), \ \|\varrho(t,\cdot)\|_{L^\gamma(\Omega)} \leq c$$

uniformly for any $t \in [0,T]$. Now, given $\phi \in L^{\gamma'}(\Omega)$, $\frac{1}{\gamma} + \frac{1}{\gamma'} = 1$, we may find a sequence $\phi_n \in C_c^\infty(\Omega)$ such that

$$\phi_n \to \phi \text{ in } L^{\gamma'}(\Omega) \text{ as } n \to \infty.$$

Writing

$$t \mapsto \left[\int_{\Omega} \varrho(t,x)\phi(x) \, \mathrm{d}x = \int_{\Omega} \varrho(t,x)\phi_n(x) \, \mathrm{d}x + \int_{\Omega} \varrho(t,x)(\phi(x) - \phi_n(x)) \, \mathrm{d}x \right]$$

we observe that

$$\left[t \mapsto \int_{\Omega} \varrho(t,x)\phi_n(x) \, \mathrm{d}x \right] \to \left[t \mapsto \int_{\Omega} \varrho(t,x)\phi(x) \, \mathrm{d}x \right] \text{ uniformly in } t \in [0,T].$$

In particular,

$$t \mapsto \int_{\Omega} \varrho(t,x)\phi(x) \, \mathrm{d}x \in C[0,T] \ \text{ for any } \ \phi \in L^{\gamma'}(\Omega). \tag{2.30}$$

If (2.30) holds, we shall write

$$\varrho \in C_{\mathrm{weak}}([0,T]; L^{\gamma}(\Omega)).$$

Of course, similar treatment can be applied to the momentum $\boldsymbol{m} = \varrho \boldsymbol{u}$ and the energy $\frac{1}{2}\varrho|\boldsymbol{u}|^2 + \varrho e(\varrho, \vartheta)$ as soon as uniform bounds are available in a Lebesgue space L^p, with $p > 1$.

Remark 2.7 (Instantaneous value) Certain ambiguity appears in (2.25) as the weak solution $[\varrho, \vartheta, \boldsymbol{u}]$ is *a priori* only an integrable quantity defined for a.a. $(t,x) \in (0,T) \times \Omega$. Strictly speaking, the solutions should be modified on a set of Lebesgue measure zero to get the continuous representatives in (2.25). This set might be different for different test functions ϕ and a natural question arises if this can be done simultaneously for all ϕ's. To avoid this apparent difficulty, we may consider a regularization in terms of time averages

$$\varrho_\varepsilon(t,x) = \theta_\varepsilon * \varrho(\cdot, x) \equiv \int_0^T \theta_\varepsilon(s)\varrho(t-s,x) \, \mathrm{d}s,$$

where θ_ε is a family of regularizing kernels supported in the interval $(-\varepsilon, \varepsilon)$, with a similar definition for the other quantities in (2.26). It turns out that the integral averages

$$t \mapsto \int_{\Omega} \varrho_\varepsilon(t,x)\phi(x) \, \mathrm{d}x, \ \phi \in C_c^\infty(\Omega),$$

are well defined for $t \in (\varepsilon, T - \varepsilon)$, and, on the one hand,

• converge in $C_{\rm loc}(0,T)$ to a continuous function ϱ_ϕ for any $\phi \in C^\infty_c(\Omega)$,

and, on the other hand,

• converge to $\int_\Omega \varrho(t,\cdot)\phi \ {\rm d}x$ for a.a. $t \in (0,T)$ for any $\phi \in C^\infty_c(\Omega)$

for $\varepsilon \to 0$. Consequently, we can identify the *instantaneous value* of the density $\varrho(t,\cdot)$ as the distribution,

$$\langle \varrho(t,\cdot); \phi \rangle = \varrho_\phi(t) \text{ for } \phi \in C^\infty_c(\Omega)$$

for *any* $t \in (0,T)$. As pointed out, similar treatment can be applied to other weakly continuous quantities.

Note carefully that the weakly-in-time continuous quantities are typically the *conservative variables*: the density ϱ, the momentum $\boldsymbol{m} = \varrho \boldsymbol{u}$, and the total energy $E = \frac{1}{2}\varrho|\boldsymbol{u}|^2 + \varrho e$, namely those for which the distributional time derivative can be explicitly computed using only the weak formulation. On the other hand, the standard variables ϑ and $\boldsymbol{u}$ do not in general enjoy this property. This is one of the reasons why the conservative variables are used in the numerical methods based on the implicit/explicit time discretization of the problem.

2.1.4 Admissible Weak Solutions

As is well known, see e.g. Smoller [184], the weak solutions introduced in Definition 2.1 are not uniquely determined by the (initial) data. To save the game, several admissibility criteria have been proposed to render the problem well-posed. A natural idea is to include the Second law of thermodynamics enforced through the entropy balance equation (inequality) (2.11). Note that, unlike in the case when the solutions are smooth, the entropy balance (2.11) *does not* follow from the weak formulation (2.22)–(2.24).

Definition 2.2 (ADMISSIBLE WEAK SOLUTION TO COMPLETE EULER SYSTEM)
We say that a weak solution $[\varrho, \vartheta, \boldsymbol{u}]$ of the Euler system in the sense of Definition 2.1 is *admissible* if, in addition to (2.22)–(2.24), the entropy inequality

$$\int_0^T \int_\Omega \left[\varrho s(\varrho,\vartheta)\partial_t \varphi + \varrho s(\varrho,\vartheta)\boldsymbol{u} \cdot \nabla_x \varphi \right] \ {\rm d}x \, {\rm d}t \leq - \int_\Omega \varrho_0 s(\varrho_0,\vartheta_0)\varphi(0,\cdot) \ {\rm d}x \tag{2.31}$$

holds for any test function $\varphi \in C^1_c([0,T) \times \overline{\Omega})$, $\varphi \geq 0$.

Similarly to (2.9), we may consider the renormalized version of (2.31), namely

$$
\begin{aligned}
&\int_0^T \int_\Omega \left[\varrho Z(s(\varrho, \vartheta)) \partial_t \varphi + \varrho Z(s(\varrho, \vartheta)) \boldsymbol{u} \cdot \nabla_x \varphi \right] \, \mathrm{d}x \, \mathrm{d}t \\
&\qquad \leq - \int_\Omega \varrho_0 Z(s(\varrho_0, \vartheta_0)) \varphi(0, \cdot) \, \mathrm{d}x
\end{aligned}
\tag{2.32}
$$

for $\varphi \in C^1_c([0, T) \times \overline{\Omega})$, $\varphi \geq 0$, where Z is a nondecreasing function.

In view of the inequality sign in (2.31) the integral averages

$$
t \mapsto \int_\Omega \varrho s(\varrho, \vartheta)(t, x) \phi(x) \, \mathrm{d}x, \ \phi \in C^\infty_c(\Omega), \ \phi \geq 0
$$

are in general *not continuous* as functions of the time t. Fortunately, the function

$$
t \mapsto \left[\int_\Omega \varrho s(\varrho, \vartheta)(t, x) \phi(x) \, \mathrm{d}x - \int_0^t \left(\int_\Omega \varrho s(\varrho, \vartheta) \boldsymbol{u}(s, x) \cdot \nabla_x \phi(x) \, \mathrm{d}x \right) \mathrm{d}s \right]
$$

turns out to be *nondecreasing* in $t \in (0, T)$. As the second component

$$
t \mapsto \int_0^t \left(\int_\Omega \varrho s(\varrho, \vartheta) \boldsymbol{u}(s, x) \cdot \nabla_x \phi(x) \, \mathrm{d}x \right) \mathrm{d}s
$$

is obviously continuous, we deduce that the one-sided limits

$$
\int_\Omega \varrho s(\varrho, \vartheta)(\tau \pm, x) \phi(x) \, \mathrm{d}x = \lim_{t \to \pm \tau} \int_\Omega \varrho s(\varrho, \vartheta)(t, x) \phi(x) \, \mathrm{d}x
$$

exist for any $0 < \tau < T$, and any $\phi \in C^\infty_c(\Omega)$, $\phi \geq 0$. Similarly, we identify

$$
\int_\Omega \varrho s(\varrho, \vartheta)(0+, x) \phi(x) \, \mathrm{d}x = \lim_{t \to 0+} \int_\Omega \varrho s(\varrho, \vartheta)(t, x) \phi(x) \, \mathrm{d}x,
$$

$$
\text{and} \ \int_\Omega \varrho s(\varrho, \vartheta)(T-, x) \phi(x) \, \mathrm{d}x = \lim_{t \to T-} \int_\Omega \varrho s(\varrho, \vartheta)(t, x) \phi(x) \, \mathrm{d}x.
$$

The previous relations can be easily extended to general (sign-changing) functions, writing $\phi = \phi^+ - \phi^-$ and apply the previous result to ϕ^+ and ϕ^-, respectively.

Consequently, similarly to Remark 2.7, we may infer that the one sided instantaneous values

$$\varrho s(\varrho, \vartheta)(\tau\pm, \cdot),\ 0 < \tau < T,\ \varrho s(\varrho, \vartheta)(0+, \cdot),\ \text{and}\ \varrho s(\varrho, \vartheta)(T-, \cdot)$$

are well defined at least as distributions in Ω. Supposing, in addition,

$$\varrho s(\varrho, \vartheta) \in L^\infty(0, T; L^1(\Omega)),$$

we may deduce

$$\begin{gathered}\varrho s(\varrho, \vartheta)(\tau\pm, \cdot) \in \mathcal{M}(\overline{\Omega}),\ 0 < \tau < T,\ \varrho s(\varrho, \vartheta)(0+, \cdot) \in \mathcal{M}(\overline{\Omega}), \\ \text{and}\ \varrho s(\varrho, \vartheta)(T-, \cdot) \in \mathcal{M}(\overline{\Omega}).\end{gathered} \tag{2.33}$$

This follows from the fact that $\mathcal{M}(\overline{\Omega})$ can be identified with the dual of $C(\overline{\Omega})$, where the latter is a separable Banach space.

The renormalized entropy inequality (2.32) represents a useful tool in the analysis of weak solutions to the Euler system. In particular, it implies the *minimum entropy principle* stated below.

Proposition 2.1 (Minimum entropy principle)
Let $\varrho \geq 0$, s, $\boldsymbol{u}$,

$$\varrho, s \in L^\infty(0, T; L^1(\Omega)),\ \varrho \boldsymbol{u} \in L^\infty(0, T; L^1(\Omega; R^d))$$

satisfy the renormalized entropy inequality (2.32) *for any bounded nondecreasing function Z. Suppose that*

$$s_0(x) \geq \underline{s}\ \text{for a.a.}\ x \in \left\{ x \in \Omega \,\middle|\, \varrho_0(x) > 0 \right\}.$$

Then

$$s(\tau, x) \geq \underline{s}\ \text{for a.a.}\ x \in \left\{ x \in \Omega \,\middle|\, \varrho(\tau, x) > 0 \right\}$$

for a.a. $\tau \in (0, T)$.

Proof Using the same arguments as for the entropy, we may deduce from the renormalized entropy inequality (2.32) that the one sided limits

$$\lim_{t \to \pm\tau} \int_\Omega \varrho(t, x) Z(s(t, x))\ \mathrm{d}x$$

exist for any $\tau \in (0, T)$.

Next, we consider a spatially homogeneous test function $\varphi = \psi(t)$ in (2.32) obtaining

$$\int_0^T \int_\Omega \partial_t \psi(t) \varrho Z(s) \, dx \, dt \leq - \int_\Omega \varrho_0 Z(s_0) \psi(0) \, dx.$$

Approximating the function $1_{[0,\tau)}$ by a suitable sequence $\psi_\delta \geq 0$, $\psi_\delta(0) = 0$, we deduce

$$\int_\Omega \varrho Z(s)(\tau-, x) \, dx \geq \int_\Omega \varrho_0 Z(s_0) \, dx \text{ for any } \tau \in (0, T).$$

Finally, we consider

$$Z_n(s) = \begin{cases} 0 & \text{if } s \geq \underline{s}, \\ n(s - \underline{s}) & \text{if } s < \underline{s}. \end{cases}$$

In accordance with our hypotheses we get

$$\int_\Omega \varrho Z_n(s)(\tau-, x) \, dx \geq 0 \text{ for any } \tau \in (0, T),$$

and, letting $n \to \infty$, we obtain the desired conclusion.

Besides the entropy inequality (2.31), the minimum entropy principle is considered to be one of the fundamental properties of physically relevant weak solutions to the Euler system. Unfortunately, as we shall see in the following section, the Euler system in several dimensions is still ill posed even if the renormalized entropy balance (2.32) is imposed.

2.1.5 Ill-Posedness in the Class of Weak Solutions

The Euler system remains ill-posed even in the class of admissible weak solutions, at least in the physically relevant higher space dimensions $d = 2, 3$. Uniqueness of solutions can be restored only under more restrictive conditions that are satisfied only for certain specific initial data. We report the following negative result that can be proved by the method of convex integration, see [89, Theorem 2.6] or [47, Theorem 2.2].

Theorem 2.2 (Ill-posedness for complete Euler system) *Let $\Omega \subset R^d$, $d = 2, 3$ be a bounded domain. Suppose that the initial data ϱ_0, ϑ_0 are piecewise constant. Specifically,*

$$\Omega = \cup_{i \in I} \overline{\Omega_i}, \ \Omega_i \text{ a domain for each } i, \ |\partial \Omega_i| = 0, \ \Omega_i \cap \Omega_j = \emptyset \text{ for } i \neq j,$$

where I is at most countable set, and

$$\varrho_0|_{\Omega_i} = \varrho_{0,i}, \ \vartheta_0|_{\Omega_i} = \vartheta_{0,i},$$

where ϱ_i, ϑ_i are constants,

$$0 < \underline{\varrho} \leq \varrho_{0,i} \leq \overline{\varrho}, \ 0 < \underline{\vartheta} \leq \vartheta_{0,i} \leq \overline{\vartheta} \ \textit{for all } i \in I.$$

Then there exists $\boldsymbol{u}_0 \in L^\infty(\Omega; R^d)$ such that the Euler system (2.1)–(2.3) *with the impermeability condition* (2.5), *and the initial data*

$$\varrho(0, \cdot) = \varrho_0, \ \vartheta(0, \cdot) = \vartheta_0, \ \boldsymbol{u}(0, \cdot) = \boldsymbol{u}_0$$

admits infinitely many admissible weak solutions in $(0, T) \times \Omega$ *in the sense of Definition 2.2. The solutions belong to the class*

$$\underline{\varrho} \leq \varrho(t, x) \leq \overline{\varrho}, \ \underline{\vartheta} \leq \vartheta(t, x) \leq \overline{\vartheta} \ \textit{a.a. in } (0, T) \times \Omega,$$

$$\boldsymbol{u} \in L^\infty((0, T) \times \Omega; R^d).$$

Moreover, the entropy balance (2.31) *is satisfied as equality and the solutions are in fact* isentropic, *meaning*

the entropy density $\varrho s(\varrho, \vartheta)(t, \cdot)$ is independent of the time $t \in [0, T]$.

Remark 2.8 As a matter of fact, for given ϱ_0, ϑ_0 there is an unbounded set of the initial velocities $\boldsymbol{u}_0$ such that the conclusion of Theorem 2.2 holds.

Remark 2.9 The result does not require any structural properties to be imposed on p, e, and s. We only need these quantities to be continuous functions of $[\varrho, \vartheta]$ on the set $[\underline{\varrho}, \overline{\varrho}] \times [\underline{\vartheta}, \overline{\vartheta}]$.

Remark 2.10 The same result can be shown if the impermeability of the boundary is replaced by the periodic boundary conditions $\Omega = \mathbb{T}^d$.

A complete proof of Theorem 2.2 is lengthy and requires the complicated apparatus of convex integration developed in the context of fluid dynamics in the pioneering work of De Lellis and Székelyhidi [62]. The interested reader can find a detailed treatment of the problem in the aforementioned references [89, Theorem 2.6] [47, Theorem 2.2].

2.2 Euler System in Conservative Variables

In the context of weak solutions and also in numerous numerical experiments, it is more convenient to rewrite the Euler system in terms of **conservative variables**:

$$\text{the mass density } \varrho = \varrho(t,x),$$
$$\text{the momentum } \boldsymbol{m} = (\varrho \boldsymbol{u})(t,x),$$
$$\text{the total energy, } E = \left(\frac{1}{2}\varrho|\boldsymbol{u}|^2 + \varrho e\right)(t,x).$$

As we have observed in Sect. 2.1.3, these are exactly the quantities that possess well defined instantaneous values and are at least weakly continuous in the time variable as soon as suitable estimates on integrability of the associated fluxes are available.

The Euler system in the conservative variables reads:

$$\partial_t \varrho + \operatorname{div}_x \boldsymbol{m} = 0, \tag{2.34}$$

$$\partial_t \boldsymbol{m} + \operatorname{div}_x \left(\frac{\boldsymbol{m} \otimes \boldsymbol{m}}{\varrho}\right) + \nabla_x p = 0, \tag{2.35}$$

$$\partial_t E + \operatorname{div}_x \left[(E + p)\frac{\boldsymbol{m}}{\varrho}\right] = 0. \tag{2.36}$$

Apparently, there are certain difficulties connected with the new formulation:

(i) The equation of continuity, written in terms of momentum, does not provide any direct way how to show positivity of the mass density, not even for smooth solutions.
(ii) The density, that may in principal vanish at certain parts of the physical domain, appears in the denominator of the convective term in (2.35).
(iii) There is a rather awkward state equation relating E and e, namely

$$\varrho e = E - \frac{1}{2}\frac{|\boldsymbol{m}|^2}{\varrho}$$

where the right-hand side is *a priori* not a positive quantity as expected for e. As we shall see bellow, the entropy balance will play a crucial role to rectify these issues.

2.2.1 *The Second Law of Thermodynamics – Entropy*

In the remaining part of this section we suppose the pressure and the internal energy are interrelated through the caloric EOS introduced in Chap. 1:

$$p = (\gamma - 1)\varrho e, \ \gamma > 1, \tag{2.37}$$

giving rise to

$$p = p(\varrho, \boldsymbol{m}, E) = (\gamma - 1)\left(E - \frac{1}{2}\frac{|\boldsymbol{m}|^2}{\varrho}\right). \tag{2.38}$$

Note carefully that $p = p(\varrho, \boldsymbol{m}, E)$ is a concave function of the conservative variables. The polytropic EOS (2.37) is *incomplete* in the sense that formally closes the Euler system (2.34)–(2.36) but does not provide any piece of information concerning the temperature and entropy.

To fill the gap, we write the specific entropy $s = s(\varrho, e)$ as a function of the density and the internal energy. In view of Gibbs' relation (2.5),

$$\frac{\partial s}{\partial e}(\varrho, e) = \frac{1}{\vartheta}, \ \frac{\partial s}{\partial \varrho}(\varrho, e) = -\frac{p}{\vartheta \varrho^2} = -(\gamma - 1)\frac{e}{\vartheta \varrho} = -(\gamma - 1)\frac{e}{\varrho}\frac{\partial s}{\partial e}(\varrho, e). \tag{2.39}$$

The first relation can serve as a definition of the absolute temperature ϑ, while the second relation in (2.39) may be seen as a first order equation for $s = s(\varrho, e)$ that can be integrated yielding

$$s(\varrho, e) = \mathcal{S}\left(\frac{(\gamma - 1)e}{\varrho^{\gamma - 1}}\right) = \mathcal{S}\left(\frac{p}{\varrho^{\gamma}}\right) \tag{2.40}$$

for a certain function $\mathcal{S}$. Combining (2.37), (2.39), (2.40), and the required positivity of the absolute temperature, we get

$$\mathcal{S}' > 0.$$

Thus going back to (2.38) we obtain a formula for s being a function of the conservative variables

$$s(\varrho, \boldsymbol{m}, E) = \mathcal{S}\left(\frac{\gamma - 1}{\varrho^{\gamma}}\left(E - \frac{1}{2}\frac{|\boldsymbol{m}|^2}{\varrho}\right)\right). \tag{2.41}$$

In accordance with (2.11), the entropy is transported (increases) along the flow. In particular, choosing the initial data in such a way that the pressure (the absolute temperature) is positive,

$$E_0 - \frac{1}{2}\frac{|\boldsymbol{m}_0|^2}{\varrho_0} > 0,$$

we deduce the same property at any positive time

$$E(t, \cdot) - \frac{1}{2}\frac{|\boldsymbol{m}(t, \cdot)|^2}{\varrho(t, \cdot)} > 0, \ t > 0,$$

which is the minimum entropy principle stated in Proposition 2.1.

2.2.2 Weak Solutions

Rewriting Definition 2.1 in terms of the conservative variables $[\varrho, \boldsymbol{m}, E]$, we obtain an alternative weak formulation of the Euler system.

Definition 2.3 (WEAK FORMULATION IN CONSERVATIVE VARIABLES) Let the initial data $[\varrho_0, \boldsymbol{m}_0, E_0]$ be given satisfying

$$\varrho_0 \geq 0,\ E_0 - \frac{1}{2}\frac{|\boldsymbol{m}_0|^2}{\varrho_0} > 0 \text{ a.a. in } \Omega.$$

A trio $[\varrho, \boldsymbol{m}, E]$ is a *weak solution* of the Euler system (2.34)–(2.36), with the impermeability boundary conditions (2.6) in $(0, T) \times \Omega$ if the following holds:

- (**measurability**) the quantities $\varrho = \varrho(t, x)$, $\boldsymbol{m} = \boldsymbol{m}(t, x)$, $E = E(t, x)$ are measurable functions defined for $(t, x) \in (0, T) \times \Omega$,

$$\varrho(t, x) \geq 0,\ E(t, x) - \frac{1}{2}\frac{|\boldsymbol{m}|^2}{\varrho}(t, x) > 0 \text{ for a.a. } (t, x) \in (0, T) \times \Omega; \tag{2.42}$$

- (**equation of continuity**) the integral identity

$$\int_0^T \int_\Omega \left[\varrho \partial_t \varphi + \boldsymbol{m} \cdot \nabla_x \varphi\right] \mathrm{d}x\, \mathrm{d}t = -\int_\Omega \varrho_0 \varphi(0, \cdot)\, \mathrm{d}x \tag{2.43}$$

  holds for any test function $\varphi \in C^1_c([0, T) \times \overline{\Omega})$;
- (**momentum equation**) the integral identity

$$\begin{aligned} &\int_0^T \int_\Omega \left[\boldsymbol{m} \cdot \partial_t \boldsymbol{\varphi} + \left(1_{\varrho>0} \frac{\boldsymbol{m} \otimes \boldsymbol{m}}{\varrho}\right) : \nabla_x \boldsymbol{\varphi} + p(\varrho, \boldsymbol{m}, E) \mathrm{div}_x \boldsymbol{\varphi}\right] \mathrm{d}x\, \mathrm{d}t \\ &\quad = -\int_\Omega \boldsymbol{m}_0 \cdot \boldsymbol{\varphi}(0, \cdot)\, \mathrm{d}x \text{ with } p(\varrho, \boldsymbol{m}, E) = (\gamma - 1)\left(E - \frac{1}{2}\frac{|\boldsymbol{m}|^2}{\varrho}\right), \end{aligned} \tag{2.44}$$

  holds for any test function $\boldsymbol{\varphi} \in C^1_c([0, T) \times \overline{\Omega}; R^d)$, $\boldsymbol{\varphi} \cdot \boldsymbol{n}|_{\partial\Omega} = 0$;
- (**energy conservation**) the integral identity

$$\int_0^T \int_\Omega \left[E \partial_t \varphi + 1_{\varrho>0}\Big(E + p(\varrho, \boldsymbol{m}, E)\Big)\frac{\boldsymbol{m}}{\varrho} \cdot \nabla_x \varphi\right] \mathrm{d}x\, \mathrm{d}t = -\int_\Omega E_0 \varphi(0, \cdot)\, \mathrm{d}x \tag{2.45}$$

  holds for any test function $\varphi \in C^1_c([0, T) \times \overline{\Omega})$.

A weak solution is called *admissible* if **the entropy balance**

$$\int_0^T \int_\Omega \left[\varrho s(\varrho, \boldsymbol{m}, E) \partial_t \varphi + s(\varrho, \boldsymbol{m}, E) \boldsymbol{m} \cdot \nabla_x \varphi \right] \, \mathrm{d}x \, \mathrm{d}t \\ \leq - \int_\Omega \varrho_0 s(\varrho_0, \boldsymbol{m}_0, E_0) \varphi(0, \cdot) \, \mathrm{d}x \tag{2.46}$$

holds for any test function $\varphi \in C^1_c([0, T) \times \overline{\Omega})$, $\varphi \geq 0$, where the entropy s is given by formula (2.41).

Remark 2.11 (Kinetic energy) The kinetic energy $\frac{1}{2} \frac{|\boldsymbol{m}|^2}{\varrho}$ is defined as a function of $[\varrho, \boldsymbol{m}]$ in the following way:

$$\frac{|\boldsymbol{m}|^2}{\varrho} = \begin{cases} \frac{|\boldsymbol{m}|^2}{\varrho} \text{ if } \varrho > 0, \\ 0 \quad \text{if } \boldsymbol{m} = 0, \ \varrho \geq 0, \\ \infty \text{ otherwise.} \end{cases} \tag{2.47}$$

Accordingly, the kinetic energy defined through (2.47) is a convex, lower semicontinuous function of $[\varrho, \boldsymbol{m}] \in R^2$,

$$\text{Dom} \left[\frac{|\boldsymbol{m}|^2}{\varrho} \right] = \left\{ [\varrho, \boldsymbol{m}] \ \middle| \ \varrho > 0 \right\} \cup \{[0, 0]\}.$$

The second condition in (2.42) reflects the physical principle of positivity of the absolute temperature. As we have seen in the preceding section, it can be recovered from the minimum entropy principle even at the level of weak solutions.

2.2.3 *Thermodynamic Stability*

Let us recall the thermodynamic stability condition introduced in (2.13), namely

$$\frac{\partial p(\varrho, \vartheta)}{\partial \varrho} > 0, \quad \frac{\partial e(\varrho, \vartheta)}{\partial \vartheta} > 0. \tag{2.48}$$

In terms of the conservative variables, the same condition corresponds to the *concavity* of the total entropy

$$S(\varrho, \boldsymbol{m}, E) = \varrho s(\varrho, \boldsymbol{m}, E) = \varrho \mathcal{S} \left(\frac{\gamma - 1}{\varrho^\gamma} \left(E - \frac{1}{2} \frac{|\boldsymbol{m}|^2}{\varrho} \right) \right)$$

with respect to the variables $[\varrho, \boldsymbol{m}, E]$.

First we claim that concavity of S follows from the following structural restrictions imposed on $\mathcal{S}$

$$\mathcal{S}'(Z) > 0,\ (\gamma - 1)\mathcal{S}'(Z) + \gamma \mathcal{S}''(Z)Z < 0 \ \text{ for all } \ Z > 0. \tag{2.49}$$

To this end, we first observe that it is enough to establish concavity of the function

$$S = S(\varrho, p) = \varrho \mathcal{S}\left(\frac{p}{\varrho^\gamma}\right) \text{ with respect to the variables } [\varrho, p].$$

Next, we compute

$$\frac{\partial S(\varrho, p)}{\partial \varrho} = \mathcal{S}\left(\frac{p}{\varrho^\gamma}\right) - \gamma \mathcal{S}'\left(\frac{p}{\varrho^\gamma}\right)\frac{p}{\varrho^\gamma},\ \frac{\partial S(\varrho, p)}{\partial p} = \frac{1}{\varrho^{\gamma-1}}\mathcal{S}'\left(\frac{p}{\varrho^\gamma}\right),$$

and

$$\begin{aligned}
\frac{\partial^2 S(\varrho, p)}{\partial \varrho^2} &= -\gamma \mathcal{S}'\left(\frac{p}{\varrho^\gamma}\right)\frac{p}{\varrho^{\gamma+1}} + \gamma^2 \mathcal{S}'\left(\frac{p}{\varrho^\gamma}\right)\frac{p}{\varrho^{\gamma+1}} + \gamma^2 \mathcal{S}''\left(\frac{p}{\varrho^\gamma}\right)\frac{p^2}{\varrho^{2\gamma+1}},\\
\frac{\partial^2 S(\varrho, p)}{\partial p^2} &= \frac{1}{\varrho^{2\gamma-1}}\mathcal{S}''\left(\frac{p}{\varrho^\gamma}\right),\\
\frac{\partial^2 S(\varrho, p)}{\partial \varrho \partial p} &= \frac{1-\gamma}{\varrho^\gamma}\mathcal{S}'\left(\frac{p}{\varrho^\gamma}\right) - \gamma \mathcal{S}''\left(\frac{p}{\varrho^\gamma}\right)\frac{p}{\varrho^{2\gamma}}.
\end{aligned}$$

It follows from the hypothesis (2.49) that

$$\frac{\partial^2 S(\varrho, p)}{\partial \varrho^2} \le 0,\ \frac{\partial^2 S(\varrho, p)}{\partial p^2} \le 0.$$

Finally, we compute the determinant of the Hessian of the function $S = S(\varrho, p)$:

$$\begin{aligned}
&\det\left[\nabla^2_{\varrho,p} S(\varrho, p)\right]\\
&= \frac{1}{\varrho^{2\gamma}}\left(\gamma \mathcal{S}''(Z)\left[(\gamma-1)\mathcal{S}'(Z)Z + \gamma \mathcal{S}''(Z)Z^2\right] - \left[(1-\gamma)\mathcal{S}'(Z) - \gamma \mathcal{S}''(Z)Z\right]^2\right)\\
&= \frac{1}{\varrho^{2\gamma}}\left(-\gamma(\gamma-1)\mathcal{S}''(Z)\mathcal{S}'(Z)Z - (\gamma-1)^2 \mathcal{S}'(Z)^2\right)\\
&= -\frac{(\gamma-1)\mathcal{S}'(Z)}{\varrho^{2\gamma}}\left(\gamma \mathcal{S}''(Z)Z + (\gamma-1)\mathcal{S}'(Z)\right).
\end{aligned}$$

Thus we may infer that concavity of the function $S = S(\varrho, p)$ in $[\varrho, p]$, and, consequently of $S = S(\varrho, \boldsymbol{m}, E)$ in $[\varrho, \boldsymbol{m}, E]$, follows from (2.49).

Next, we examine the domain of $\mathcal{S}$ determined by a lower bound on the quotient $\frac{p}{\varrho^\gamma}$, or, in accordance with the relation (2.37), a lower bound on $(\gamma - 1)\frac{e}{\varrho^{\gamma-1}}$. To clarify this issue, it seems more convenient to use the standard variables $[\varrho, \vartheta]$. First observe that Gibbs' equation (2.5) yields Maxwell's equation

$$\frac{\partial e(\varrho,\vartheta)}{\partial \varrho} = \frac{1}{\varrho^2}\left(p(\varrho,\vartheta) - \vartheta \frac{\partial p(\varrho,\vartheta)}{\partial \vartheta}\right);$$

whence, in view of the caloric EOS (2.37),

$$\frac{\partial e(\varrho,\vartheta)}{\partial \varrho} = \frac{\gamma-1}{\varrho}\left(e(\varrho,\vartheta) - \vartheta \frac{\partial e(\varrho,\vartheta)}{\partial \vartheta}\right). \tag{2.50}$$

Relation (2.50) may be viewed as a first order partial differential equations that can be integrated obtaining e or p, and the entropy s is the specific form:

$$p(\varrho,\vartheta) = (\varrho\vartheta)\frac{\vartheta^{c_v}}{\varrho}P\left(\frac{\varrho}{\vartheta^{c_v}}\right) = \frac{P(Y)}{Y^\gamma}\varrho^\gamma,\ s = s(Y),\ \text{where } Y = \frac{\varrho}{\vartheta^{\frac{1}{\gamma-1}}} \tag{2.51}$$

for certain functions P and s. In addition, using $\frac{\partial e}{\partial \vartheta} > 0$, we may infer

$$Y \mapsto \frac{P(Y)}{Y^\gamma}$$

is a decreasing function of Y. We conclude that

$$\frac{p(\varrho,\vartheta)}{\varrho^\gamma} \searrow \overline{p} \geq 0 \text{ as } \vartheta \to 0+ \quad \text{for any fixed } \varrho, \tag{2.52}$$

where $\overline{p}$ is a nonnegative constant independent of ϱ. The natural domain of definition of $\mathcal{S}$ is therefore the interval $(\overline{p}, \infty)$, where, shifting $\mathcal{S}$ by an additive constant, we may assume

$$\lim_{Z\to\overline{p}+} \mathcal{S}(Z) \in \{0, -\infty\}. \tag{2.53}$$

We finish this part by stating the exact definition of the total entropy $S = \varrho s(\varrho, \boldsymbol{m}, E)$:

$$S(\varrho,\boldsymbol{m},E) = \begin{cases} \varrho\mathcal{S}\left((\gamma-1)\dfrac{E-\frac{1}{2}\frac{|\boldsymbol{m}|^2}{\varrho}}{\varrho^\gamma}\right) & \text{if } \varrho > 0,\ E \geq \frac{1}{2}\frac{|\boldsymbol{m}|^2}{\varrho} + \frac{\overline{p}}{\gamma-1}\varrho^\gamma, \\ \lim_{\varrho\to 0+}\varrho\mathcal{S}\left((\gamma-1)\frac{E}{\varrho^\gamma}\right), & \text{if } \varrho = 0,\ \boldsymbol{m} = 0,\ E > 0, \\ \lim_{E\to 0+}\left[\lim_{\varrho\to 0+}\varrho\mathcal{S}\left((\gamma-1)\frac{E}{\varrho^\gamma}\right)\right] & \text{if } \varrho = E = \boldsymbol{m} = 0, \\ -\infty \text{ otherwise.} \end{cases} \tag{2.54}$$

As we have observed, the function $S = S(\varrho, \mathbf{m}, E)$ defined via (2.54) is concave upper semicontinuous as soon as $\mathcal{S}$ satisfies (2.49) for $Z > 0$, together with (2.53).

Example The standard example in gas dynamics is the *Boyle–Mariotte law*:

$$p = \varrho\vartheta, \ e = c_v\vartheta, \ c_v = \frac{1}{\gamma - 1}, \tag{2.55}$$

for which

$$\overline{p} = 0, \ \mathcal{S}(Z) = \frac{1}{\gamma - 1}\log(Z).$$

2.2.4 Conservative-Entropy Variables

The formulation of the Euler system in terms of the conservative variables $[\varrho, \mathbf{m}, E]$ is rather awkward at the level of constitutive equations. In particular, the pressure and the internal energy—two thermostatic quantities defined *a priori* for fluid at thermodynamic equilibrium—depend on the momentum/velocity field through the total energy E. Moreover, the energy balance Eq. (2.36) as well as its weak counterpart (2.45) contain the flux

$$(E + p)\frac{\mathbf{m}}{\varrho}$$

that is not controllable in terms of available *a priori* bounds.

In meteorology it is more customary to replace the total energy E by the entropy, or its rescaled version called potential temperature, and to rewrite the Euler system in terms of the **conservative-entropy variables**:

$$\text{the density } \varrho = \varrho(t, x),$$
$$\text{the momentum } \mathbf{m} = \mathbf{m}(t, x),$$
$$\text{the total entropy } S = (\varrho s)(t, x).$$

Accordingly, the resulting system of equations reads:

$$\partial_t \varrho + \mathrm{div}_x \mathbf{m} = 0, \tag{2.56}$$

$$\partial_t \mathbf{m} + \mathrm{div}_x \left(\frac{\mathbf{m} \otimes \mathbf{m}}{\varrho} \right) + \nabla_x p(\varrho, S) = 0, \tag{2.57}$$

$$\partial_t S + \mathrm{div}_x \left(S \frac{\mathbf{m}}{\varrho} \right) = 0. \tag{2.58}$$

Similarly to (2.9), we may replace (2.58) by its renormalized version

$$\partial_t S_Z + \mathrm{div}_x \left(S_Z \frac{\boldsymbol{m}}{\varrho} \right) = 0, \tag{2.59}$$

where

$$S_Z = \varrho Z \circ S, \ Z' \geq 0.$$

Equation (1.35) may come in handy when dealing with weak solution with low integrability, where Z can be taken a suitable cut-off function.

The total energy

$$E = E(\varrho, \boldsymbol{m}, S) = \frac{1}{2} \frac{|\boldsymbol{m}|^2}{\varrho} + \varrho e(\varrho, S)$$

is now expressed as a sum of the kinetic and internal energy, where the latter is independent of the momentum. In the remaining part of this section we focus on the standard example of the perfect gas law

$$p = (\gamma - 1)\varrho e, \ e = c_v \vartheta, \ c_v = \frac{1}{\gamma - 1}, \gamma > 1.$$

After a direct manipulation, we observe that the mapping

$$p(\varrho, S) = (\gamma - 1)\varrho e(\varrho, S) = \begin{cases} \varrho^\gamma \exp\left(\frac{S}{c_v \varrho}\right) & \text{if } \varrho > 0, \ S \in R, \\ 0 & \text{if } \varrho = 0, \ S \leq 0, \\ \infty & \text{if } \varrho = 0, \ S > 0, \end{cases}$$

is convex lower semicontinuous on $[0, \infty) \times R$. Indeed it is a routine matter to compute the Hessian matrix:

$$\frac{\partial p(\varrho, S)}{\partial \varrho} = \gamma \varrho^{\gamma - 1} \exp\left(\frac{S}{c_v \varrho}\right) - \frac{S}{c_v} \varrho^{\gamma - 2} \exp\left(\frac{S}{c_v \varrho}\right),$$
$$\frac{\partial p(\varrho, S)}{\partial S} = \frac{1}{c_v} \varrho^{\gamma - 1} \exp\left(\frac{S}{c_v \varrho}\right),$$

and

$$\frac{\partial^2 p(\varrho, S)}{\partial \varrho^2} = \left[(\gamma - 1)\varrho^2 + \left((\gamma - 1)\varrho - \frac{S}{c_v} \right)^2 \right] \varrho^{\gamma-4} \exp\left(\frac{S}{c_v \varrho} \right),$$

$$\frac{\partial^2 p(\varrho, S)}{\partial S^2} = \frac{1}{c_v^2} \varrho^{\gamma-2} \exp\left(\frac{S}{c_v \varrho} \right) = \frac{1}{c_v^2} \varrho^2 \varrho^{\gamma-4} \exp\left(\frac{S}{c_v \varrho} \right),$$

$$\frac{\partial^2 p(\varrho, S)}{\partial \varrho \partial S} = \frac{1}{c_v} (\gamma - 1) \varrho^{\gamma-2} \exp\left(\frac{S}{c_v \varrho} \right) - \frac{S}{c_v^2} \varrho^{\gamma-3} \exp\left(\frac{S}{c_v \varrho} \right)$$

$$= \left[\frac{1}{c_v} (\gamma - 1)\varrho^2 - \frac{S}{c_v^2} \varrho \right] \varrho^{\gamma-4} \exp\left(\frac{S}{c_v \varrho} \right).$$

Obviously, the Hessian has positive trace, while its determinant reads

$$\varrho^{\gamma-4} \exp\left(\frac{S}{c_v \varrho} \right) \left\{ \frac{\varrho^2}{c_v^2} \left[(\gamma - 1)\varrho^2 + \left((\gamma - 1)\varrho - \frac{S}{c_v} \right)^2 \right] - \left(\frac{\gamma - 1}{c_v} \varrho^2 - \frac{S}{c_v^2} \varrho \right)^2 \right\}$$
$$= \frac{\varrho^\gamma}{c_v^2} \exp\left(\frac{S}{c_v \varrho} \right).$$

Thus for the kinetic energy

$$\frac{|\boldsymbol{m}|^2}{\varrho} = \begin{cases} \frac{|\boldsymbol{m}|^2}{\varrho} \text{ if } \varrho > 0, \ \boldsymbol{m} \in R^d, \\ 0 \text{ if } \boldsymbol{m} = 0, \\ \infty \text{ if } \varrho = 0, \boldsymbol{m} \neq 0, \end{cases} \tag{2.60}$$

which is a convex lower semicontinuous function on $[0, \infty) \times R^d$, the total energy

$$E(\varrho, \boldsymbol{m}, S) = \frac{1}{2} \frac{|\boldsymbol{m}|^2}{\varrho} + \varrho e(\varrho, S)$$

is also convex lower semicontinuous. Convexity plays an important role in the study of weak solutions as they are known to be only weakly continuous with respect to the time variable; whence their composition with a convex function is weakly lower semicontinuous. In particular, the total energy enjoys this property.

A bit at odds with basic physical principles, the system (2.47), (2.57), (2.58), (2.59) can be supplemented by the total energy balance (inequality)

$$\frac{\mathrm{d}}{\mathrm{d}t} \int_\Omega E(t) \ \mathrm{d}x = (\leq) 0, \tag{2.61}$$

where we have tacitly assumed the impermeability or spatially periodic boundary conditions. To avoid the problem of a rather unphysical phenomenon of energy dissipation in (2.61), we may return to the original Euler system and reformulate it in the conservative-entropy variables:

$$\partial_t \varrho + \mathrm{div}_x \boldsymbol{m} = 0, \tag{2.62}$$

$$\partial_t \boldsymbol{m} + \mathrm{div}_x \left(\frac{\boldsymbol{m} \otimes \boldsymbol{m}}{\varrho} \right) + \nabla_x p(\varrho, S) = 0, \tag{2.63}$$

$$\partial_t \left(\frac{1}{2} \frac{|\boldsymbol{m}|^2}{\varrho} + \varrho e(\varrho, S) \right) + \mathrm{div}_x \left[\left(\frac{1}{2} \frac{|\boldsymbol{m}|^2}{\varrho} + \varrho e(\varrho, S) + p(\varrho, S) \right) \frac{\boldsymbol{m}}{\varrho} \right] = 0, \tag{2.64}$$

supplemented with the renormalized entropy balance

$$\partial_t \left(\varrho Z \left(\frac{S}{\varrho} \right) \right) + \mathrm{div}_x \left[Z \left(\frac{S}{\varrho} \right) \boldsymbol{m} \right] \geq 0 \tag{2.65}$$

for Z nondecreasing.

Definition 2.4 (**WEAK FORMULATION IN CONSERVATIVE-ENTROPY VARIABLES**) Let the initial data $[\varrho_0, \boldsymbol{m}_0, S_0]$ be given satisfying

$$\varrho_0 \geq 0,\ E_0 = \frac{1}{2} \frac{|\boldsymbol{m}_0|^2}{\varrho_0} + \varrho_0 e(\varrho_0, S_0) < \infty \text{ a.a. in } \Omega,\ S_0 = 0 \text{ a.a. in } \{\varrho_0 = 0\}.$$

A trio $[\varrho, \boldsymbol{m}, S]$ is a *weak solution* of the Euler system (2.62)–(2.64) with the impermeability boundary conditions (2.6) in $(0, T) \times \Omega$ if the following holds:

- (**measurability, compatibility**) the quantities $\varrho = \varrho(t, x)$, $\boldsymbol{m} = \boldsymbol{m}(t, x)$, $S = E(t, x)$ are measurable functions defined for $(t, x) \in (0, T) \times \Omega$,

$$\begin{aligned} &\varrho(t, x) \geq 0, \\ &E(t, x) \equiv \left(\frac{1}{2} \frac{|\boldsymbol{m}|^2}{\varrho} + \varrho e(\varrho, S) \right)(t, x) < \infty \text{ for a.a. } (t, x) \in (0, T) \times \Omega, \\ &S(t, x) = 0 \text{ a.a. } (t, x) \text{ in } \{\varrho = 0\}\,; \end{aligned} \tag{2.66}$$

- (**equation of continuity**) the integral identity

$$\int_0^T \int_\Omega \left[\varrho \partial_t \varphi + \boldsymbol{m} \cdot \nabla_x \varphi \right] \mathrm{d}x \, \mathrm{d}t = - \int_\Omega \varrho_0 \varphi(0, \cdot) \, \mathrm{d}x \tag{2.67}$$

  holds for any test function $\varphi \in C^1_c([0, T) \times \overline{\Omega})$;

- (**momentum equation**) the integral identity

$$\int_0^T \int_\Omega \left[\boldsymbol{m} \cdot \partial_t \boldsymbol{\varphi} + \left(1_{\varrho>0} \frac{\boldsymbol{m} \otimes \boldsymbol{m}}{\varrho} \right) : \nabla_x \boldsymbol{\varphi} + 1_{\varrho>0} p(\varrho, S) \mathrm{div}_x \boldsymbol{\varphi} \right] \mathrm{d}x \, \mathrm{d}t$$
$$= - \int_\Omega \boldsymbol{m}_0 \cdot \boldsymbol{\varphi}(0, \cdot) \, \mathrm{d}x \tag{2.68}$$

holds for any test function $\boldsymbol{\varphi} \in C^1_c([0, T) \times \overline{\Omega}; R^d)$, $\boldsymbol{\varphi} \cdot \boldsymbol{n}|_{\partial\Omega} = 0$;
- (**energy conservation**) the integral identity

$$\int_0^T \int_\Omega \left(\frac{1}{2} \frac{|\boldsymbol{m}|^2}{\varrho} + \varrho e(\varrho, S) \right) \partial_t \varphi \, \mathrm{d}x \, \mathrm{d}t$$
$$+ \int_0^T \int_\Omega \left[1_{\varrho>0} \left(\frac{1}{2} \frac{|\boldsymbol{m}|^2}{\varrho} + \varrho e(\varrho, S) + p(\varrho, S) \right) \frac{\boldsymbol{m}}{\varrho} \right] \cdot \nabla_x \varphi \, \mathrm{d}x \, \mathrm{d}t \tag{2.69}$$
$$= - \int_\Omega \left(\frac{1}{2} \frac{|\boldsymbol{m}_0|^2}{\varrho_0} + \varrho_0 e(\varrho_0, S_0) \right) \varphi(0, \cdot) \, \mathrm{d}x$$

holds for any test function $\varphi \in C^1_c([0, T) \times \overline{\Omega})$.

A weak solution is called *admissible* if **the entropy balance**

$$\int_0^T \int_\Omega \left[\varrho Z \left(\frac{S}{\varrho} \right) \partial_t \varphi + \left(Z \left(\frac{S}{\varrho} \right) \boldsymbol{m} \right) \cdot \nabla_x \varphi \right] \mathrm{d}x \, \mathrm{d}t \leq - \int_\Omega \varrho_0 Z \left(\frac{S_0}{\varrho_0} \right) \varphi(0, \cdot) \, \mathrm{d}x \tag{2.70}$$

holds for any test function $\varphi \in C^1_c([0, T) \times \overline{\Omega})$, $\varphi \geq 0$, and any $Z \in C^1(R)$, $Z' \geq 0$.

Remark 2.12 As the total energy is supposed to be finite for a.a. (t, x) we deduce a natural compatibility condition

$$\boldsymbol{m}(t, x) = 0 \text{ a.a. in the vacuum set } \{\varrho = 0\}.$$

A similar condition

$$S(t, x) = 0 \text{ a.a. in the vacuum set } \{\varrho = 0\}.$$

postulated in (2.66) is merely a convention. Indeed, all other nonlinearities in the equations are set to be zero on the vacuum set, and, in view of the renormalized equation (2.70), it is convenient to *define*

$$S(t,x) = \lim_{\delta \to 0} \varrho(t,x) Z\left(\frac{S(t,x)}{\varrho(t,x)}\right), \ Z \in BC(R), \ Z_\delta(Y) \to Y \text{ as } \delta \to 0.$$

The conservative-entropy variables $[\varrho, \boldsymbol{m}, S]$ are in fact the most convenient for the analysis of numerical schemes studied in this book.

2.3 Barotropic (Isentropic) Euler System

The isentropic Euler system is formally obtained from the complete system by requiring the entropy $s = \overline{s}$ to be constant in (2.8). Consequently, the entropy balance and the equation of continuity coincide and the complete Euler system (2.1)–(2.3) reduces to the first two equations. Note that, in accordance with Gibbs' equation (2.5), the entropy is an increasing function of the total energy; whence the latter can be seen as a function of the density ϱ only. In the context of the Boyle–Mariotte law (2.55), we easily deduce the isentropic pressure-density EOS

$$p(\varrho) = a\varrho^\gamma, \ e(\varrho) = \frac{a}{\gamma - 1}\varrho^{\gamma-1}, \ \gamma > 1. \tag{2.71}$$

In general, we say that a compressible fluid is *barotropic* if the pressure $p = p(\varrho)$ depends only on the density ϱ. Barotropic models ignore the thermal dissipative effects in the fluid and reduce the motion to a purely mechanical process. An example of a barotropic model was studied in Sect. 2.1.2 to demonstrate the blow up phenomena (shock waves) in compressible fluid flows.

2.3.1 *Energy Balance*

The total energy of a barotropic fluid can be expressed in terms of the kinetic energy and the *pressure potential*

$$P(\varrho) \text{ satisfying } P'(\varrho)\varrho - P(\varrho) = p(\varrho). \tag{2.72}$$

Indeed multiplying the momentum Eq. (2.2) on $\boldsymbol{u} = \frac{\boldsymbol{m}}{\varrho}$ we recover the kinetic energy balance

$$\partial_t \left(\frac{|\boldsymbol{m}|^2}{\varrho}\right) + \text{div}_x \left(\frac{|\boldsymbol{m}|^2}{\varrho}\boldsymbol{u}\right) + \text{div}_x \left(p(\varrho)\boldsymbol{u}\right) - p(\varrho)\text{div}_x\boldsymbol{u} = 0.$$

Next, we easily check that

$$\partial_t P(\varrho) + \text{div}_x \left(P(\varrho)\boldsymbol{u}\right) + p(\varrho)\text{div}_x\boldsymbol{u} = 0;$$

whence

$$\partial_t \left(\frac{|\boldsymbol{m}|^2}{\varrho} + P(\varrho) \right) + \mathrm{div}_x \left[\left(\frac{|\boldsymbol{m}|^2}{\varrho} + P(\varrho) \right) \boldsymbol{u} \right] + \mathrm{div}_x \left(p(\varrho) \boldsymbol{u} \right) = 0. \tag{2.73}$$

Note that $\varrho \mapsto P(\varrho)$ is convex as soon as $p'(\varrho) \geq 0$; where the latter is nothing other than a thermodynamic stability condition in the context of barotropic fluids.

2.3.2 *Weak Solutions*

We introduce the concept of *weak solution* to the barotropic Euler system.

Definition 2.5 (Weak formulation of the barotropic Euler system)
Let the initial data $[\varrho_0, \boldsymbol{m}_0]$ be given satisfying

$$\varrho_0 \geq 0, \quad \frac{|\boldsymbol{m}_0|^2}{\varrho_0} < \infty \text{ a.a. in } \Omega.$$

A pair $[\varrho, \boldsymbol{m}]$ is a *weak solution* of the barotropic Euler system with the impermeability boundary conditions (2.6) in $(0, T) \times \Omega$ if the following holds:

- (**measurability**) the quantities $\varrho = \varrho(t, x)$, $\boldsymbol{m} = \boldsymbol{m}(t, x)$ are measurable functions defined for $(t, x) \in (0, T) \times \Omega$,

$$\varrho(t, x) \geq 0 \text{ for a.a. } (t, x) \in (0, T) \times \Omega; \tag{2.74}$$

- (**equation of continuity**) the integral identity

$$\int_0^T \int_\Omega \left[\varrho \partial_t \varphi + \boldsymbol{m} \cdot \nabla_x \varphi \right] \, \mathrm{d}x \, \mathrm{d}t = - \int_\Omega \varrho_0 \varphi(0, \cdot) \, \mathrm{d}x \tag{2.75}$$

  holds for any test function $\varphi \in C^1_c([0, T) \times \overline{\Omega})$;
- (**momentum equation**) the integral identity

$$\begin{aligned} &\int_0^T \int_\Omega \left[\boldsymbol{m} \cdot \partial_t \boldsymbol{\varphi} + \left(1_{\varrho > 0} \frac{\boldsymbol{m} \otimes \boldsymbol{m}}{\varrho} \right) : \nabla_x \boldsymbol{\varphi} + p(\varrho) \mathrm{div}_x \boldsymbol{\varphi} \right] \, \mathrm{d}x \, \mathrm{d}t \\ &= - \int_\Omega \boldsymbol{m}_0 \cdot \boldsymbol{\varphi}(0, \cdot) \, \mathrm{d}x \end{aligned} \tag{2.76}$$

  holds for any test function $\boldsymbol{\varphi} \in C^1_c([0, T) \times \overline{\Omega}; R^d)$, $\boldsymbol{\varphi} \cdot \boldsymbol{n}|_{\partial\Omega} = 0$.

A weak solution is called *admissible* if **the energy inequality**

$$\int_0^T \int_\Omega \left(\frac{1}{2} \frac{|\boldsymbol{m}|^2}{\varrho} + P(\varrho) \right) \partial_t \psi \ \mathrm{d}x \, \mathrm{d}t \leq - \int_\Omega \left(\frac{1}{2} \frac{|\boldsymbol{m}_0|^2}{\varrho_0} + P(\varrho_0) \right) \psi(0) \ \mathrm{d}x \quad (2.77)$$

for any test function $\psi \in C_c^1[0,T)$, $\psi \geq 0$.

Remark 2.13 The integrated energy inequality (2.77) may be strengthened to its "differential" form

$$\begin{aligned} &\int_0^T \int_\Omega \left(\frac{1}{2} \frac{|\boldsymbol{m}|^2}{\varrho} + P(\varrho) \right) \partial_t \varphi \ \mathrm{d}x \, \mathrm{d}t \\ &+ \int_0^T \int_\Omega \left[\left(\frac{1}{2} \frac{|\boldsymbol{m}|^2}{\varrho} + P(\varrho) + p(\varrho) \right) \frac{\boldsymbol{m}}{\varrho} \cdot \nabla_x \varphi \right] \ \mathrm{d}x \, \mathrm{d}t \\ &\leq - \int_\Omega \left(\frac{1}{2} \frac{|\boldsymbol{m}_0|^2}{\varrho_0} + P(\varrho_0) \right) \varphi(0,\cdot) \ \mathrm{d}x \end{aligned} \quad (2.78)$$

for any test function $\varphi \in C_c^1([0,T) \times \overline{\Omega})$, $\varphi \geq 0$.

Note that the example of formation of shock waves in Sect. 2.1.2 concerns a barotropic fluid and thus solutions of the barotropic Euler system may become singular in a finite time lap independently of smoothness and size of the initial data. By the same token, Theorem 2.2 on the ill-posedness of the Euler system applies to the barotropic case as well. Summarizing, we may infer that the barotropic Euler system shares the essential properties of the complete Euler system. It is locally well posed in the framework of smooth solutions and globally ill posed in the framework of admissible weak solutions. The existence of global in time admissible weak solutions for *any* finite energy initial data is still an open problem in the multidimensional case.

2.4 Conclusion, Bibliographical Remarks

The motion of an inviscid perfect fluid in the framework of continuum mechanics is described by the (complete) Euler system (2.1)–(2.3), or, alternatively, by the simplified isentropic (barotropic) model introduced in Sect. 2.3. The absence of the regularizing effect of viscosity results in severe mathematical difficulties and render the initial and/or boundary value problem essentially *ill-posed*. In accordance with the recent state-of-the-art, we may summarize the material of this chapter as follows:

- **Local-in-time well-posedness**. The Euler system admits local in time smooth solutions as soon as the initial data as well as the underlying physical domain are smooth. The solutions exists on a maximal time interval $[0, T_{\max})$, see e.g. the monographs of Benzoni–Gavage, Serre [18], Dafermos [60], Majda [156], Smoller [184] among others.
- **Development of singularities**. Singularities in the form of shock waves develop in a finite time for a fairly generic class of the initial data. There are standard examples of this phenomena discussed by Dafermos [60], Smoller [184], for more recent treatment we refer to Buckmaster et al. [36].
- **Effect of physical boundaries**. The absence of physical boundaries or its large distance from the initial data perturbation may extend the life span of classical solutions considerably due to dispersive phenomena, see Christodoulou and Miao [54].
- **Existence of weak solutions**. The admissible weak solutions exist globally in time for any finite energy initial data if $d = 1$, see DiPerna [64], Lions, Perthame, Souganidis, [155]. In the multidimensional case, the weak solutions are known to exist for any initial data enjoying certain regularity, see De Lellis and Székelyhidi [62], Chiodaroli et al. [46, 49, 51]. The problem is ill posed in the class of weak solutions, there are infinitely many for any initial data. They may not be admissible, however, the entropy and/or energy balance may be violated. Still, there is a vast class of (nonsmooth) initial data for which the problem is ill posed, see [81].
- **Global ill-posedness for smooth initial data**. There exist C^∞ initial data and $T > 0$ such that the barotropic Euler system in R^d, $d = 2, 3$ admits infinitely many admissible weak solutions in $(0, T)$, see Chiodaroli et al. [50]. Similar examples exist for the complete Euler system in the class of Lipschitz initial data, see Markfelder [158].

Despite the existence of infinitely many (weak) solutions for rather vast class of data, it is still not known if a global in time *admissible weak* solution exists for *given* data, at least if $d > 1$. Convergence of a sequence of *approximate solutions*, for instance those resulting from a numerical scheme, to a weak solution remains an outstanding open problem. What is more, certain numerical experiments suggest that the limit may not be a weak solution but rather a more complex object that can be described only in a statistical manner, cf. Fjordholm et al. [103, 104, 107]. This motivates the introduction of a more general concept of solution discussed in large in Part II of this monograph.

Chapter 3
Viscous Fluids: Navier–Stokes–(Fourier) System

The perfect fluids described by the Euler system represent a mathematical idealization of the model of motion of *real fluids*, for which the effect of viscosity and/or heat conductivity can not be neglected. In particular, in many real world applications more complex models must be considered. We adopt the common *mathematical definition of fluid* as a continuum for which the Cauchy stress $\mathbb{T}$ is characterized by **Stokes' law**,

$$\mathbb{T} = \mathbb{S} - p\mathbb{I}$$

where p is the pressure and $\mathbb{S}$ is the viscous stress tensor. Leaving apart for a moment the specific form of $\mathbb{S}$, we may write the field equations in the form:

- **Mass conservation or equation of continuity**

$$\partial_t \varrho + \mathrm{div}_x(\varrho \boldsymbol{u}) = 0 \tag{3.1}$$

- **Newton's Second law or momentum conservation**

$$\partial_t(\varrho \boldsymbol{u}) + \mathrm{div}_x(\varrho \boldsymbol{u} \otimes \boldsymbol{u}) + \nabla_x p = \mathrm{div}_x \mathbb{S} \tag{3.2}$$

The energy balance of a viscous fluid is usually expressed in terms of the internal energy e

- **Internal energy balance**

$$\partial_t(\varrho e) + \mathrm{div}_x(\varrho e \boldsymbol{u}) + \mathrm{div}_x \boldsymbol{q} = \mathbb{S} : \nabla_x \boldsymbol{u} - p \mathrm{div}_x \boldsymbol{u}, \tag{3.3}$$

where $\boldsymbol{q}$ denotes the internal energy flux. Note that we have deliberately omitted the influence of both mechanical and thermal sources for the sake of simplicity.

While the pressure p and the internal energy obey basically the same thermodynamic principles as perfect fluids, in particular Gibbs' relation (2.5), the specific form

E. Feireisl et al., *Numerical Analysis of Compressible Fluid Flows*, MS&A 20,
https://doi.org/10.1007/978-3-030-73788-7_3

of $\mathbb{S}$ and $\boldsymbol{q}$ is given by the material properties of a particular real fluid. *Newtonian fluids* are characterized by the linear dependence of $\mathbb{S}$ on the velocity gradient $\nabla_x^T \boldsymbol{u}$ or rather on its symmetric part

$$\mathbb{D}\boldsymbol{u} = \frac{1}{2}\left(\nabla_x \boldsymbol{u} + \nabla_x^T \boldsymbol{u}\right), \text{ or its traceless component } \frac{1}{2}\left(\nabla_x \boldsymbol{u} + \nabla_x^T \boldsymbol{u}\right) - \frac{1}{d}\mathrm{div}_x \boldsymbol{u}\mathbb{I}.$$

If the fluid is isotropic, meaning its material properties are independent of the orientation of the reference frame, the most general form of the viscous stress is

- **Newton's rheological law**

$$\mathbb{S} = \mathbb{S}(\nabla_x \boldsymbol{u}) = \mu\left(\nabla_x \boldsymbol{u} + \nabla_x^T \boldsymbol{u} - \frac{2}{d}\mathrm{div}_x \boldsymbol{u}\mathbb{I}\right) + \lambda \mathrm{div}_x \boldsymbol{u}\mathbb{I}, \tag{3.4}$$

where μ and λ are nonnegative scalars that may depend on ϱ and e as the case may be. The first deviatoric component of $\mathbb{S}$ corresponds to the shear stress while the second is called bulk stress. Accordingly, μ is called the shear viscosity coefficient and λ the bulk viscosity coefficient. We speak about a viscous fluid if $\mu > 0, \lambda \geq 0$; there are fluid models where $\lambda = 0$, for example this holds for some gases.

The internal energy flux $\boldsymbol{q}$ is usually intimately related to the diffusive transport of heat causing temperature changes. We consider $\boldsymbol{q}$ given by

- **Fourier's law**

$$\boldsymbol{q} = \boldsymbol{q}(\nabla_x \vartheta) = -\kappa \nabla_x \vartheta, \tag{3.5}$$

where ϑ is the absolute temperature and κ the heat conductivity coefficient.

The system of Eqs. (3.1)–(3.3), where $\mathbb{S}$ and $\boldsymbol{q}$ are given by (3.4) and (3.5), respectively, is called the *Navier–Stokes–Fourier system.*

3.1 Classical Solutions

To fix ideas, we consider the Navier–Stokes–Fourier system on a bounded domain $\Omega \subset R^d$ with an impermeable boundary. In terms of the velocity field $\boldsymbol{u}$, this amounts to the boundary condition

$$\boldsymbol{u} \cdot \boldsymbol{n}|_{\partial\Omega} = 0. \tag{3.6}$$

Similarly, we consider the no-flux conditions for the internal energy,

$$\boldsymbol{q} \cdot \boldsymbol{n}|_{\partial\Omega} = 0. \tag{3.7}$$

The fluid is therefore both mechanically and thermally isolated. In addition, a second boundary condition is required in view of the elliptic character of the momentum equation. We consider either the no-slip condition for the velocity

$$\boldsymbol{u} \times \boldsymbol{n}|_{\partial\Omega} = 0, \quad \text{which, together with (3.6), yields } \boldsymbol{u}|_{\partial\Omega} = 0, \tag{3.8}$$

or the complete slip (Navier) boundary condition

$$(\mathbb{S} \cdot \boldsymbol{n}) \times \boldsymbol{n}|_{\partial\Omega} = 0. \tag{3.9}$$

Of course, the space periodic boundary conditions $\Omega = \mathbb{T}^d$ can be alternatively used to avoid the treatment of physical boundary.

3.1.1 Local Existence of Smooth Solutions

The Navier–Stokes–Fourier system is a model of a nondilute fluid, where the density ϱ is expected to be bounded below away from zero. In particular, this property should be enforced through the choice of the initial data

$$\varrho(0,\cdot) = \varrho_0 > 0, \ \boldsymbol{m}(0,\cdot) = \varrho_0 \boldsymbol{u}_0, \ e(0,\cdot) = e_0. \tag{3.10}$$

In contrast with the Euler system discussed in Chapter 2, there is no explicit example of solutions to the Navier–Stokes–Fourier system emanating from smooth initial data and blowing up at a finite time. Although there is no rigorous proof, formation of shock waves is not expected. Possible singularities, if any, would be related to formation of a "gravitational collapse" when the density becomes infinite, cf. Sect. 3.4. Still the problem of existence of global in time classical solutions remains largely open, at least in the physically relevant space dimensions $d = 2, 3$.

Similarly to the Euler system, the solutions of the Navier–Stokes–Fourier system with smooth initial data are known to exist locally in time. To formulate the relevant result, we rewrite the problem in the standard (primitive) variables $[\varrho, \boldsymbol{u}, \vartheta]$:

$$\partial_t \varrho + \mathrm{div}_x(\varrho \boldsymbol{u}) = 0, \tag{3.11}$$

$$\begin{aligned} &\partial_t(\varrho \boldsymbol{u}) + \mathrm{div}_x(\varrho \boldsymbol{u} \otimes \boldsymbol{u}) + \nabla_x p(\varrho, \vartheta) \\ &\quad = \mathrm{div}_x \left(\mu \left[\nabla_x \boldsymbol{u} + \nabla_x^T \boldsymbol{u} - \frac{2}{d} \mathrm{div}_x \boldsymbol{u} \mathbb{I} \right] + \lambda \mathrm{div}_x \boldsymbol{u} \mathbb{I} \right), \end{aligned} \tag{3.12}$$

$$\begin{aligned} &\varrho c_v(\varrho, \vartheta) \Big(\partial_t \vartheta + \boldsymbol{u} \cdot \nabla_x \vartheta \Big) - \mathrm{div}_x \left(\kappa \nabla_x \vartheta \right) \\ &\quad = \left(\mu \left[\nabla_x \boldsymbol{u} + \nabla_x^T \boldsymbol{u} - \frac{2}{3} \mathrm{div}_x \boldsymbol{u} \mathbb{I} \right] + \lambda \mathrm{div}_x \boldsymbol{u} \mathbb{I} \right) : \nabla_x \boldsymbol{u} - \varrho \frac{\partial p(\varrho, \vartheta)}{\partial \vartheta} \mathrm{div}_x \boldsymbol{u}, \end{aligned} \tag{3.13}$$

where

$$c_v(\varrho, \vartheta) = \frac{\partial e(\varrho, \vartheta)}{\partial \vartheta} > 0 \text{ is the specific heat at constant volume.}$$

Note that the specific form of the last term in (3.13) follows from Gibbs' relation (1.46).

The following result is due to Valli [196, Theorem A and Remark 3.3]. It is stated for $d = 3$ and the no-slip boundary condition (3.8), however, extension for $d = 2$ and other types of boundary conditions is straightforward.

Theorem 3.1 (Local existence of smooth solutions)
Let $\Omega \subset R^3$ be a bounded domain of class $C^{2+\nu}$, $\nu > 0$. Let the initial data

$$\varrho_0, \vartheta_0 \in W^{3,2}(\Omega), \ \boldsymbol{u}_0 \in W^{3,2}(\Omega; R^3)$$

be given such that (ϱ_0, ϑ_0) belong to a compact subset of an open set $\mathcal{U} \subset (0, \infty)^2$, and satisfy the compatibility conditions

$$\boldsymbol{u}_0|_{\partial\Omega} = 0, \ \nabla_x \vartheta_0 \cdot \boldsymbol{n}|_{\partial\Omega} = 0,$$

$$\nabla_x p(\varrho_0, \vartheta_0)\Big|_{\partial\Omega} = \mathrm{div}_x \left(\mu(\varrho_0, \vartheta_0) \left[\nabla_x \boldsymbol{u}_0 + \nabla_x^T \boldsymbol{u}_0 - \frac{2}{3} \mathrm{div}_x \boldsymbol{u}_0 \mathbb{I} \right] + \lambda(\varrho_0, \vartheta_0) \mathrm{div}_x \boldsymbol{u}_0 \mathbb{I} \right) \Big|_{\partial\Omega}.$$

Suppose that the pressure $p = p(\varrho, \vartheta)$, the specific heat at constant volume $c_v = c_v(\varrho, \vartheta)$, as well as the transport coefficients $\mu = \mu(\varrho, \vartheta)$, $\lambda = \lambda(\varrho, \vartheta)$, and $\kappa = \kappa(\varrho, \vartheta)$ are three-times continuously differentiable in $\mathcal{U}$ and satisfy

$$\frac{\partial p(\varrho, \vartheta)}{\partial \varrho} > 0, \ c_v(\varrho, \vartheta) > 0, \ \mu(\varrho, \vartheta) > 0, \ \lambda(\varrho, \vartheta) \geq 0, \ \kappa(\varrho, \vartheta) > 0$$

for all $(\varrho, \vartheta) \in \mathcal{U}$.

Then there exists $T > 0$ such that the Navier–Stokes–Fourier system (3.11)–(3.13), *supplemented with the boundary conditions* (3.7), (3.8) *admits a classical solution unique in the class*

$$\varrho, \ \vartheta \in C([0, T]; W^{3,2}(\Omega)) \cap C^1([0, T]; W^{2,2}(\Omega)),$$

$$\boldsymbol{u} \in C([0, T]; W^{3,2}(\Omega; R^3)) \cap C^1([0, T]; W^{2,2}(\Omega; R^3)).$$

Remark 3.1 It can be shown that any solution belonging to the class specified in Theorem 3.1 possess all the necessary derivatives and is therefore a classical solution in the *open set* $(0, T) \times \Omega$.

3.1.2 Global Existence of Smooth Solutions

The dissipative mechanism, encoded in the Navier–Stokes–Fourier system, provides certain stability of equilibrium states that gives rise to the existence of global-in-time smooth solutions for "small" initial data. The following result is attributed to Matsumura and Nishida [159, 160], for the accommodation of various boundary conditions see Valli and Zajaczkowski [197]. To avoid technical problems connected with the boundary conditions, we state it for the periodic boundary conditions.

Theorem 3.2 (Global existence for small data) *Let $\Omega = \mathbb{T}^d$, $d = 2, 3$. Suppose there are two positive constants $\overline{\varrho} > 0$, $\overline{\vartheta} > 0$ such that*

$$\|\varrho_0 - \overline{\varrho}\|_{W^{3,2}(\mathbb{T}^d)} + \|\vartheta_0 - \overline{\vartheta}\|_{W^{3,2}(\mathbb{T}^d)} + \|\boldsymbol{u}_0\|_{W^{3,2}(\mathbb{T}^d;R^d)} < \varepsilon.$$

Let the pressure $p = p(\varrho, \vartheta)$, the specific heat at constant volume $c_v = c_v(\varrho, \vartheta)$, as well as the transport coefficients $\mu = \mu(\varrho, \vartheta)$, $\lambda = \lambda(\varrho, \vartheta)$, and $\kappa = \kappa(\varrho, \vartheta)$ be three-times continuously differentiable in some open neighborhood $\mathcal{U} \subset (0, \infty)^2$ of $[\overline{\varrho}, \overline{\vartheta}]$ and satisfy

$$\frac{\partial p(\varrho, \vartheta)}{\partial \varrho} > 0,\ c_v(\varrho, \vartheta) > 0,\ \mu(\varrho, \vartheta) > 0,\ \lambda(\varrho, \vartheta) \geq 0,\ \kappa(\varrho, \vartheta) > 0$$

for all $(\varrho, \vartheta) \in \mathcal{U}$.

Then there exists $\varepsilon_0 > 0$ such that for any $0 \leq \varepsilon < \varepsilon_0$, the Navier–Stokes–Fourier system (3.11)–(3.13) *admits a unique global-in-time solution in the class*

$$\varrho,\ \vartheta \in BC([0, \infty); W^{3,2}(\mathbb{T}^d)) \cap BC^1([0, \infty); W^{2,2}(\mathbb{T}^d)),$$

$$\boldsymbol{u} \in BC([0, \infty); W^{3,2}(\mathbb{T}^d; R^d)) \cap BC^1([0, \infty); W^{2,2}(\mathbb{T}^d; R^d)).$$

3.2 Weak Solutions, Navier–Stokes System

The problem of the existence of global-in-time solutions to the Navier–Stokes–Fourier system in the multidimensional setting for arbitrary (large) data remains open. However, there is a well developed global theory in the framework of *weak* (distributional) solutions we are about to discuss now. To explain the main ideas, we restrict ourselves to the barotropic case, where, similarly to its inviscid counterpart studied in Sect. 2.3, the thermal effects are neglected. The relevant system of equations reads:

- **Mass conservation or equation of continuity**

$$\partial_t \varrho + \operatorname{div}_x(\varrho \boldsymbol{u}) = 0 \tag{3.14}$$

- **Newton's Second law or momentum conservation**

$$\partial_t(\varrho \boldsymbol{u}) + \mathrm{div}_x(\varrho \boldsymbol{u} \otimes \boldsymbol{u}) + \nabla_x p(\varrho) = \mathrm{div}_x \mathbb{S}(\nabla_x \boldsymbol{u}) \tag{3.15}$$

- **Newton's rheological law**

$$\mathbb{S}(\nabla_x \boldsymbol{u}) = \mu \left(\nabla_x \boldsymbol{u} + \nabla_x^T \boldsymbol{u} - \frac{2}{d} \mathrm{div}_x \boldsymbol{u} \mathbb{I} \right) + \lambda \mathrm{div}_x \boldsymbol{u} \mathbb{I}. \tag{3.16}$$

The boundary conditions are stated in terms of the velocity. For the sake of simplicity, we focus on the **no-slip condition**

$$\boldsymbol{u}|_{\partial\Omega} = 0, \tag{3.17}$$

or, alternatively, the periodic boundary conditions $\Omega = \mathbb{T}^d$.

In the context of weak solutions, it is convenient to augment the system by **energy inequality**

$$\frac{\mathrm{d}}{\mathrm{d}t} \int_\Omega E \, \mathrm{d}x + \int_\Omega \mathbb{S} : \nabla_x \boldsymbol{u} \, \mathrm{d}x \leq 0, \ E \equiv \frac{1}{2} \varrho |\boldsymbol{u}|^2 + P(\varrho), \ P'(\varrho)\varrho - P(\varrho) = p(\varrho). \tag{3.18}$$

3.2.1 *Approximation Scheme*

The weak solutions to the Navier–Stokes system can be obtained via a multilevel approximation scheme that is strongly reminiscent of certain numerical schemes discussed in Part III of this book. To avoid problems with boundary conditions, we consider $\Omega = \mathbb{T}^d$.

The equation of continuity is regularized by adding *artificial viscosity*

$$\partial_t \varrho + \mathrm{div}_x(\varrho \boldsymbol{u}) = \varepsilon \Delta_x \varrho. \tag{3.19}$$

The momentum equation is solved by means of a Galerkin approximation. Specifically, we look for a velocity field

$$\boldsymbol{u} \in C([0, T]; X_n),$$

where X_n is a finite dimensional space spanned by trigonometric polynomials on $\mathbb{T}^d$. Accordingly, we solve

$$
\begin{aligned}
&\frac{\mathrm{d}}{\mathrm{d}t}\int_{\mathbb{T}^d} \varrho \boldsymbol{u}\cdot \boldsymbol{w}\ \mathrm{d}x \\
&= \int_{\mathbb{T}^d} \Big[\varrho \boldsymbol{u}\otimes \boldsymbol{u} : \nabla_x \boldsymbol{w} + p(\varrho)\mathrm{div}_x \boldsymbol{w} + \delta \varrho^\Gamma \mathrm{div}_x \boldsymbol{w}\Big]\ \mathrm{d}x \\
&- \int_{\mathbb{T}^d} \Big[\mathbb{S} : \nabla_x \boldsymbol{w} - \varepsilon(\varrho \boldsymbol{u})\cdot \Delta_x \boldsymbol{w}\Big]\ \mathrm{d}x
\end{aligned}
\tag{3.20}
$$

for any $\boldsymbol{w} \in X_n$. Here, we have added an "artificial pressure" to compensate possible lack of coercivity of p and the artificial viscosity term $\varepsilon\Delta_x(\varrho\boldsymbol{u})$ to preserve the energy balance. Writing

$$
\frac{\mathrm{d}}{\mathrm{d}t}\int_{\mathbb{T}^d} \varrho \boldsymbol{u}\cdot \boldsymbol{w}\ \mathrm{d}x = \int_{\mathbb{T}^d} \partial_t(\varrho \boldsymbol{u})\cdot \boldsymbol{w}\ \mathrm{d}x
$$

we are allowed to consider $\boldsymbol{w} = \boldsymbol{u}$ as a test function in (3.20). We have

$$
\int_{\mathbb{T}^d} \partial_t(\varrho \boldsymbol{u})\cdot \boldsymbol{u}\ \mathrm{d}x = \int_{\mathbb{T}^d} \left[\partial_t \varrho |\boldsymbol{u}|^2 + \varrho\frac{1}{2}\partial_t|\boldsymbol{u}|^2\right]\ \mathrm{d}x.
$$

Next, multiplying (3.19) on $b'(\varrho)$ we recover the renormalized variant

$$
\partial_t b(\varrho) + \mathrm{div}_x(b(\varrho)\boldsymbol{u}) + \Big(b'(\varrho)\varrho - b(\varrho)\Big)\mathrm{div}_x \boldsymbol{u} = \varepsilon \mathrm{div}_x\left(b'(\varrho)\nabla_x\varrho\right) - \varepsilon b''(\varrho)|\nabla_x \varrho|^2.
\tag{3.21}
$$

Finally, using (3.21), we deduce

$$
\int_{\mathbb{T}^d} \varrho \boldsymbol{u}\otimes\boldsymbol{u} : \nabla_x \boldsymbol{u}\ \mathrm{d}x = -\int_{\mathbb{T}^d} \mathrm{div}_x(\varrho\boldsymbol{u})\frac{1}{2}|\boldsymbol{u}|^2\ \mathrm{d}x,
$$

$$
\int_{\mathbb{T}^d} p(\varrho)\mathrm{div}_x \boldsymbol{u}\ \mathrm{d}x = -\int_{\mathbb{T}^d} \partial_t P(\varrho)\ \mathrm{d}x - \varepsilon\int_{\mathbb{T}^d} P''(\varrho)|\nabla_x\varrho|^2\ \mathrm{d}x,
$$

$$
\delta\int_{\mathbb{T}^d} \varrho^\Gamma \mathrm{div}_x \boldsymbol{u}\ \mathrm{d}x = -\delta\int_{\mathbb{T}^d} \partial_t \varrho^\Gamma\ \mathrm{d}x - \varepsilon\delta\int_{\mathbb{T}^d} \Gamma\varrho^{\Gamma-2}|\nabla_x\varrho|^2\ \mathrm{d}x,
$$

and

$$\varepsilon \int_{\mathbb{T}^d} (\varrho \boldsymbol{u}) \Delta_x \boldsymbol{u} \, \mathrm{d}x = -\varepsilon \int_{\mathbb{T}^d} \varrho |\nabla_x \boldsymbol{u}|^2 \, \mathrm{d}x - \varepsilon \int_{\mathbb{T}^d} \nabla_x \varrho \cdot \boldsymbol{u} \cdot \nabla_x \boldsymbol{u} \, \mathrm{d}x$$

$$= -\varepsilon \int_{\mathbb{T}^d} \varrho |\nabla_x \boldsymbol{u}|^2 \, \mathrm{d}x + \frac{\varepsilon}{2} \int_{\mathbb{T}^d} \Delta_x \varrho |\boldsymbol{u}|^2 \, \mathrm{d}x.$$

Summing up the previous computations, we obtain an approximate version of the energy balance

$$\frac{\mathrm{d}}{\mathrm{d}t} \int_{\mathbb{T}^d} \left[\frac{1}{2} \varrho |\boldsymbol{u}|^2 + P(\varrho) + \delta \varrho^{\Gamma} \right] \mathrm{d}x + \int_{\mathbb{T}^d} \mathbb{S}(\nabla_x \boldsymbol{u}) : \nabla_x \boldsymbol{u} \, \mathrm{d}x$$
$$+ \varepsilon \int_{\mathbb{T}^d} \left[\varrho |\nabla_x \boldsymbol{u}|^2 + P''(\varrho) |\nabla_x \varrho|^2 + \delta \varrho^{\Gamma - 2} |\nabla_x \varrho|^2 \right] \mathrm{d}x = 0. \tag{3.22}$$

As we shall see in Part III, similar relation can be deduced for suitable numerical schemes, where the ε-dependent terms represent numerical viscosity.

The energy balance (3.22) gives rise to uniform bounds called stability estimates in the numerical context. For the isentropic pressure $p(\varrho) = a\varrho^{\gamma}$, we obtain

$$\sup_{t \in [0,T]} \left\| \varrho |\boldsymbol{u}|^2 (t, \cdot) \right\|_{L^1(\mathbb{T}^d)} + \sup_{t \in [0,T]} \| \varrho(t, \cdot) \|_{L^{\gamma}(\mathbb{T}^d)} \leq c(\text{data}), \tag{3.23}$$

where $c(\text{data})$ denotes a generic positive constant depending only on the initial data. In addition, we also have

$$\int_0^T \int_{\mathbb{T}^d} \mathbb{S}(\nabla_x \boldsymbol{u}) : \nabla_x \boldsymbol{u} \, \mathrm{d}x \, \mathrm{d}t \leq c(\text{data}),$$

which, together with (3.23) and the generalized version of Korn–Poincaré inequality, yields cf. a continuous version of the Sobolev–Poincaré inequality, see [97, Theorem 10.17].

$$\| \boldsymbol{u} \|_{L^2(0,T; W^{1,2}(\mathbb{T}^d; R^d))} \leq c(\text{data}). \tag{3.24}$$

There are, of course, other estimates that depend on the approximation parameters ε and δ.

The bounds (3.23), (3.24) control the quantities ϱ, $\varrho \boldsymbol{u}$, $\varrho \boldsymbol{u} \otimes \boldsymbol{u}$ appearing in the weak formulation of the Navier–Stokes system in a reflexive space L^q for some $q > 1$. Indeed, on one hand, in view of (3.23),

$$\varrho \boldsymbol{u} \otimes \boldsymbol{u} \text{ is bounded in } L^{\infty}(0, T; L^1(\mathbb{T}^d; R^{d \times d})). \tag{3.25}$$

On the other hand, in view of the embedding relation $W^{1,2}(\mathbb{T}^d) \hookrightarrow L^q(\mathbb{T}^d)$, where $q = 6$ if $d = 3$, $q < \infty$ arbitrary of $d = 2$, we have

$$\varrho \boldsymbol{u} \otimes \boldsymbol{u} \text{ is bounded in } L^1(0,T; L^r(\mathbb{T}^d; R^{d\times d})) \text{ for some } r > 1 \tag{3.26}$$

as long as

$$\gamma > \frac{d}{2}. \tag{3.27}$$

Interpolating (3.25), (3.26) we get the desired result.

Note that similar estimates for the pressure cannot be deduced from the energy estimates. Before we discuss this issue, however, we make a small detour and inspect shortly an interesting model in fluid mechanics proposed by H. Brenner.

3.2.2 Brenner's Two-Velocity Fluid Mechanics

There is a striking similarity between the approximation scheme (3.19), (3.20) and the two velocity hydrodynamics advocated by H. Brenner. Motivated by certain experiments, Brenner argues that there are two velocities to be considered in fluid dynamics: the mass velocity $\boldsymbol{u}_m$ derived from the classical notion of mass transport, and the volume velocity $\boldsymbol{u}$ associated to the motion of individual fluid particles (molecules). The relevant field equations now read

$$\partial_t \varrho + \mathrm{div}_x(\varrho \boldsymbol{u}_m) = 0, \tag{3.28}$$

$$\partial_t(\varrho \boldsymbol{u}) + \mathrm{div}_x\,(\varrho \boldsymbol{u} \otimes \boldsymbol{u}_m) + \nabla_x p(\varrho) = \mathrm{div}_x \mathbb{S}(\nabla_x \boldsymbol{u}), \tag{3.29}$$

with

$$\mathbb{S}(\nabla_x \boldsymbol{u}) = \mu \left(\nabla_x \boldsymbol{u} + \nabla_x^T \boldsymbol{u} - \frac{2}{d} \mathrm{div}_x \boldsymbol{u} \mathbb{I} \right) + \lambda \mathrm{div}_x \boldsymbol{u} \mathbb{I}. \tag{3.30}$$

Brenner also suggests a phenomenological relation between the two velocities,

$$\boldsymbol{u} - \boldsymbol{u}_m = K \nabla_x \log(\varrho). \tag{3.31}$$

Setting $K = \varepsilon$, the Eq. (3.28) coincides with (3.19), while (3.29) differs from (3.28) by an additive term $\varepsilon \mathrm{div}_x(\varrho \nabla_x \boldsymbol{u})$ that produces the extra dissipation in (3.22). Although Brenner's model has been subjected to substantial criticism concerning its physical relevance, it is very close to certain numerical approximations discussed later in this monograph.

3.2.3 Pressure Estimates, Compactness

The uniform bounds that can be deduced from the approximate energy balance (3.22) are not strong enough to render the pressure $p(\varrho)$ integrable in L^q, $q > 1$. There are refined estimates to solve this problem, however, they are applicable only after having performed the limit $n \to \infty$ in the Galerkin approximation of the momentum Eq. (3.20). Although such a step is relatively easy to carry over in a purely theoretical existence proof, its adaptation in the context of numerical schemes is quite delicate. In numerics, we are not allowed to perform successively the limits $n \to \infty$, $\varepsilon \to 0$, and $\delta \to 0$ but must invent a way to carry over this process *simultaneously*. This may lead to technical difficulties in the analysis and unnecessary extra terms introduced in the scheme to prevent (hypothetical) oscillations. We discuss this matter later in this book.

In what follows, we tacitly assume having performed the limit $n \to \infty$ in (3.20) so that this identity holds for any sufficiently smooth test functions $\boldsymbol{w}$. Accordingly, we may rewrite (3.20) as

$$
\begin{aligned}
&\left[\int_{\mathbb{T}^d} \varrho \boldsymbol{u} \cdot \boldsymbol{\varphi} \,\mathrm{d}x\right]_{t=0}^{t=\tau} \\
&= \int_0^\tau \int_{\mathbb{T}^d} \left[\varrho \boldsymbol{u} \cdot \partial_t \boldsymbol{\varphi} + \varrho \boldsymbol{u} \otimes \boldsymbol{u} : \nabla_x \boldsymbol{\varphi} + p(\varrho) \mathrm{div}_x \boldsymbol{\varphi} + \delta \varrho^\Gamma \mathrm{div}_x \boldsymbol{\varphi}\right] \mathrm{d}x \,\mathrm{d}t \qquad (3.32) \\
&- \int_0^\tau \int_{\mathbb{T}^d} \left[\mathbb{S}(\nabla_x \boldsymbol{u}) : \nabla_x \boldsymbol{\varphi} - \varepsilon(\varrho \boldsymbol{u}) \cdot \Delta_x \boldsymbol{\varphi}\right] \mathrm{d}x \,\mathrm{d}t
\end{aligned}
$$

for any $0 \leq \tau \leq T$, and any $\boldsymbol{\varphi} \in C^1([0, T]; C^2(\mathbb{T}^d; R^d))$. This is nothing other than a weak formulation of the momentum equation (3.2) modulo the extra ε, δ-dependent terms.

Very roughly indeed, we can say the pressure estimate are obtained by "computing" the pressure from (3.32). To this end, we use the quantities

$$
\boldsymbol{\varphi} \equiv \nabla_x \Delta_x^{-1} \left[\varrho - \frac{1}{|\mathbb{T}^d|} \int_{\mathbb{T}^d} \varrho \,\mathrm{d}x\right],
$$

where Δ_x^{-1} denotes the inverse of the Laplace operator Δ_x on the space of periodic functions with zero mean, as test functions in (3.32). First note that

$$\int_{\mathbb{T}^d} \left(p(\varrho) + \delta\varrho^\Gamma\right) \mathrm{div}_x \left(\nabla_x \Delta_x^{-1} \left[\varrho - \frac{1}{|\mathbb{T}^d|} \int_{\mathbb{T}^d} \varrho \ \mathrm{d}x \right] \right) \ \mathrm{d}x$$

$$= \int_{\mathbb{T}^d} \left(p(\varrho)\varrho + \delta\varrho^{\Gamma+1} \right) \ \mathrm{d}x - \frac{1}{|\mathbb{T}^d|} \left(\int_{\mathbb{T}^d} \varrho \ \mathrm{d}x \right) \left(\int_{\mathbb{T}^d} p(\varrho) \ \mathrm{d}x + \delta \int_{\mathbb{T}^d} \varrho^\Gamma \ \mathrm{d}x \right),$$

where the second integral on the right-hand side is bounded in terms of the data.

The remaining terms can be handled as follows:

$$\int_{\mathbb{T}^d} \varrho \boldsymbol{u} \cdot \partial_t \nabla_x \Delta_x^{-1} \left[\varrho - \frac{1}{|\mathbb{T}^d|} \int_{\mathbb{T}^d} \varrho \ \mathrm{d}x \right] \ \mathrm{d}x = \int_{\mathbb{T}^d} \varrho \boldsymbol{u} \cdot \nabla_x \Delta_x^{-1} [\partial_t \varrho] \ \mathrm{d}x$$

$$= - \int_{\mathbb{T}^d} \varrho \boldsymbol{u} \cdot \nabla_x \Delta_x^{-1} \mathrm{div}_x (\varrho \boldsymbol{u}) \ \mathrm{d}x + \varepsilon \int_{\mathbb{T}^d} \varrho \boldsymbol{u} \cdot \nabla_x \varrho \ \mathrm{d}x,$$

$$\int_{\mathbb{T}^d} \varrho \boldsymbol{u} \otimes \boldsymbol{u} : \nabla_x^2 \Delta_x^{-1} \left[\varrho - \frac{1}{|\mathbb{T}^d|} \int_{\mathbb{T}^d} \varrho \ \mathrm{d}x \right] \ \mathrm{d}x = \int_{\mathbb{T}^d} \varrho \boldsymbol{u} \otimes \boldsymbol{u} : \nabla_x \Delta_x^{-1} \nabla_x [\varrho] \ \mathrm{d}x,$$

$$\int_{\mathbb{T}^d} \mathbb{S}(\nabla_x \boldsymbol{u}) : \nabla_x^2 \Delta_x^{-1} \left[\varrho - \frac{1}{|\mathbb{T}^d|} \int_{\mathbb{T}^d} \varrho \ \mathrm{d}x \right] \ \mathrm{d}x = \int_{\mathbb{T}^d} \mathbb{S}(\nabla_x \boldsymbol{u}) : \nabla_x \Delta_x^{-1} \nabla_x [\varrho] \ \mathrm{d}x$$

$$= \left(2\mu \frac{d-1}{d} + \lambda \right) \int_{\mathbb{T}^d} \varrho \mathrm{div}_x \boldsymbol{u} \ \mathrm{d}x,$$

and

$$\varepsilon \int_{\mathbb{T}^d} \varrho \boldsymbol{u} \cdot \Delta_x \nabla_x \Delta_x^{-1} \left[\varrho - \frac{1}{|\mathbb{T}^d|} \int_{\mathbb{T}^d} \varrho \ \mathrm{d}x \right] \ \mathrm{d}x = \varepsilon \int_{\mathbb{T}^d} \varrho \boldsymbol{u} \cdot \nabla_x \varrho \ \mathrm{d}x.$$

Here we have used the fact that the operator ∇_x commutes with Δ_x^{-1} thanks to the periodic boundary conditions.

Regrouping several terms in (3.32) we get a remarkable identity

$$
\begin{aligned}
&\int_0^\tau \int_{\mathbb{T}^d} \varrho \left[p(\varrho) - \left(2\mu \frac{d-1}{d} + \lambda \right) \mathrm{div}_x \boldsymbol{u} \right] \, \mathrm{d}x \, \mathrm{d}t \\
&\quad = \int_0^\tau \int_{\mathbb{T}^d} \boldsymbol{u} \cdot \left(\varrho \boldsymbol{u} \cdot \nabla_x \Delta_x^{-1} \nabla_x [\varrho] - \varrho \nabla_x \Delta_x^{-1} \mathrm{div}_x [\varrho \boldsymbol{u}] \right) \, \mathrm{d}x \, \mathrm{d}t \\
&\quad + \frac{1}{|\mathbb{T}^d|} \left(\int_{\mathbb{T}^d} \varrho \, \mathrm{d}x \right) \int_0^\tau \left(\int_{\mathbb{T}^d} p(\varrho) \, \mathrm{d}x + \delta \int_{\mathbb{T}^d} \varrho^\Gamma \, \mathrm{d}x \right) \mathrm{d}t \\
&\quad + \left[\int_{\mathbb{T}^d} \varrho \boldsymbol{u} \cdot \Delta_x^{-1} [\nabla_x \varrho] \, \mathrm{d}x \right]_{t=0}^{t=\tau}
\end{aligned}
\tag{3.33}
$$

for any $0 \leq \tau \leq T$. The quantity

$$
p(\varrho) - \left(2\mu \frac{d-1}{d} + \lambda \right) \mathrm{div}_x \boldsymbol{u}
$$

is called *effective viscous flux* and enjoys certain sequential compactness with respect to weakly converging families $\{\varrho_n, \boldsymbol{u}_n\}_{n=1}^\infty$. The first integral on the right-hand side contains the bilinear form

$$
\begin{aligned}
&\mapsto \boldsymbol{w} \cdot \nabla_x \Delta_x^{-1} \nabla_x \cdot \boldsymbol{v} - \boldsymbol{v} \cdot \nabla_x \Delta_x^{-1} \nabla_x \cdot \boldsymbol{w} \\
&= \left(\boldsymbol{w} - \nabla_x \Delta_x^{-1} \nabla_x \cdot \boldsymbol{w} \right) \cdot \nabla_x \Delta_x^{-1} \nabla_x \cdot \boldsymbol{v} - \left(\boldsymbol{v} - \nabla_x \Delta_x^{-1} \nabla_x \cdot \boldsymbol{v} \right) \cdot \nabla_x \Delta_x^{-1} \nabla_x \cdot \boldsymbol{w},
\end{aligned}
$$

where

$$
\begin{aligned}
\mathrm{div}_x \left(\boldsymbol{w} - \nabla_x \Delta_x^{-1} \nabla_x \cdot \boldsymbol{w} \right) = \mathrm{div}_x \left(\boldsymbol{v} - \nabla_x \Delta_x^{-1} \nabla_x \cdot \boldsymbol{v} \right) = 0 \\
\mathbf{curl} \nabla_x \Delta_x^{-1} \nabla_x \cdot \boldsymbol{v} = \mathbf{curl} \nabla_x \Delta_x^{-1} \nabla_x \cdot \boldsymbol{w} = 0.
\end{aligned}
$$

Thus, in view of the celebrated Div–Curl Lemma, this quantity is weakly compact $L^p \times L^q(\mathbb{T}^d)$, $\frac{1}{p} + \frac{1}{q} < 1$. The above observations play a crucial role in the theory of weak solutions to the compressible Navier–Stokes system. We refer the reader to the monographs [80, 88], Lions [154], or Novotný and Straškraba [175] for a self-contained exposition on the existence of weak solutions and related problems. Here, we content ourselves by claiming that:

- Relation (3.33), together with the already available energy bounds (3.23)–(3.26), give rise to the desired pressure estimate

$$
\int_0^T \int_{\mathbb{T}^d} \varrho p(\varrho) \, \mathrm{d}x \, \mathrm{d}t \approx \int_0^T \int_{\mathbb{T}^d} \varrho^{\gamma+1} \, \mathrm{d}x \, \mathrm{d}t \approx \int_0^T \int_{\mathbb{T}^d} p(\varrho)^{\frac{\gamma+1}{\gamma}} \, \mathrm{d}x \, \mathrm{d}t \leq cD
$$

as long as $\Gamma > 3$.

- It is possible to perform the limit $\varepsilon \to 0$ in a family of approximate solutions to obtain a weak solution of the Navier–Stokes system with the extra pressure term $\delta \varrho^\Gamma$.

As already pointed out, the second step in the delineated process is highly non-trivial and forms the heart of the mathematical theory proposed by Lions [154].

3.2.4 Global-in-Time Existence

The final step of the existence proof consists in removing the artificial pressure $\delta \varrho^\Gamma$. This is rather delicate for the following reason. The desired pointwise convergence of the sequence of approximate densities $\{\varrho_\delta\}_{\delta>0}$ is intimately related to the satisfaction of the renormalized equation of continuity

$$\partial_t b(\varrho) + \mathrm{div}_x (b(\varrho)\boldsymbol{u}) + \Big(b'(\varrho)\varrho - b(\varrho)\Big)\mathrm{div}_x \boldsymbol{u} = 0 \tag{3.34}$$

for both $\{\varrho_\delta\}_{\delta>0}$ and its weak limit ϱ. In view of the bound (3.24) on the velocity gradient in L^2, the DiPerna–Lions theory [68] requires ϱ_δ to be bounded at least in the Lebesgue space $L^2((0,T)\times \mathbb{T}^d)$ uniformly for $\delta \to 0$. Unfortunately, such a bound is not available for the pressure-density EOS with $\gamma > \frac{d}{2}$. The way to circumvent this difficulty is careful analysis of density oscillations performed in [80, Chapters 6,7].

Leaving the interested reader to consult some of the monographs [80, 88] or [174], we state the main result concerning the existence of global-in-time weak solutions for the Navier–Stokes system.

Definition 3.1 (WEAK SOLUTION) Let $\Omega \subset R^d$, $d = 2, 3$ be a bounded Lipschitz domain. We say that $[\varrho, \boldsymbol{u}]$ is *weak solution* of the Navier–Stokes system (3.14)–(3.16) in $(0,T)\times\Omega$, with the no-slip boundary condition (3.17), if the following is satisfied:

- (**weak continuity, weak differentiability**)

$$\varrho \in C_{\mathrm{weak}}([0,T]; L^\gamma(\Omega)),\ \varrho \geq 0,$$
$$(\varrho\boldsymbol{u}) \in C_{\mathrm{weak}}([0,T]; L^{\frac{2\gamma}{\gamma+1}}(\Omega; R^d)),\ \boldsymbol{u} \in L^2(0,T; W_0^{1,2}(\Omega; R^d));$$

- (**equation of continuity**) the integral identity

$$\left[\int_\Omega \varrho\varphi \,\mathrm{d}x\right]_{t=0}^{t=\tau} = \int_0^\tau \int_\Omega \left[\varrho\partial_t\varphi + \varrho\boldsymbol{u}\cdot\nabla_x\varphi\right] \,\mathrm{d}x\,\mathrm{d}t \tag{3.35}$$

holds for any $0 \leq \tau \leq T$, and any test functions $\varphi \in C^1([0,T]\times\overline{\Omega})$;

- (**renormalized equation of continuity**) the integral identity

$$\left[\int_\Omega b(\varrho)\varphi \,\mathrm{d}x\right]_{t=0}^{t=\tau} = \int_0^\tau \int_\Omega \left[b(\varrho)\partial_t\varphi + b(\varrho)\boldsymbol{u}\cdot\nabla_x\varphi + \Big(b(\varrho) - b'(\varrho)\varrho\Big)\mathrm{div}_x\boldsymbol{u}\right] \,\mathrm{d}x\,\mathrm{d}t \tag{3.36}$$

  holds for any

$$b \in C^1[0,\infty),\ b' \in C_c^\infty[0,\infty),$$

  any $0 \le \tau \le T$, and any test functions $\varphi \in C^1([0,T]\times\overline{\Omega})$;
- (**momentum equation**) the integral identity

$$\left[\int_\Omega \varrho\boldsymbol{u}\cdot\boldsymbol{\varphi} \,\mathrm{d}x\right]_{t=0}^{t=\tau} = \int_0^\tau \int_\Omega \left[\varrho\boldsymbol{u}\cdot\partial_t\boldsymbol{\varphi} + \varrho\boldsymbol{u}\otimes\boldsymbol{u} : \nabla_x\boldsymbol{\varphi} + p(\varrho)\mathrm{div}_x\boldsymbol{\varphi} - \mathbb{S}(\nabla_x\boldsymbol{u}) : \nabla_x\boldsymbol{\varphi}\right] \mathrm{d}x\,\mathrm{d}t \tag{3.37}$$

  holds for any $0 \le \tau \le T$, and any test functions $\boldsymbol{\varphi} \in C_c^1([0,T]\times\Omega; R^d)$;
- (**energy inequality**) the inequality

$$\left[\int_\Omega \left[\frac{1}{2}\varrho|\boldsymbol{u}|^2 + P(\varrho)\right](t,\cdot)\,\mathrm{d}x\right]_{t=0}^{t=\tau} + \int_0^\tau \int_\Omega \mathbb{S}(\nabla_x\boldsymbol{u}) : \nabla_x\boldsymbol{u}\,\mathrm{d}x\,\mathrm{d}t \le 0 \tag{3.38}$$

  for a.a. $0 \le \tau \le T$.

Remark 3.2 The definition can be easily modified to the space periodic boundary conditions replacing simply Ω by $\mathbb{T}^d$ everywhere, and $C_c^1([0,T]\times\Omega; R^d)$ by $C^1([0,T]\times\mathbb{T}^d; R^d)$ in (3.37).

Theorem 3.3 (Global-in-time weak solution) *Let $\Omega \subset R^d$, $d = 2, 3$ be a bounded Lipschitz domain. Suppose that the pressure p belongs to the class*

$$p \in C[0,\infty)\cap C^1(0,\infty),\ p(0) = 0,\ a_1\varrho^{\gamma-1} - a_2 \le p'(\varrho) \le a_3\varrho^{\gamma-1} + a_4,\ a_1 > 0, \tag{3.39}$$

where

$$\gamma > \frac{d}{2}. \tag{3.40}$$

Let the initial data $\varrho(0,\cdot) = \varrho_0$, $\varrho \boldsymbol{u}(0,\cdot) = \boldsymbol{m}_0$ *be given such that*

$$\varrho_0 \geq 0,\ E_0 = \int_\Omega \left[\frac{1}{2} \frac{|\boldsymbol{m}_0|^2}{\varrho_0} + P(\varrho_0) \right] \mathrm{d}x < \infty.$$

Then the Navier–Stokes system (3.14)–(3.16), *with the no-slip boundary conditions* (3.17) *admit a weak solution* $[\varrho, \boldsymbol{u}]$ *in* $(0,T) \times \Omega$ *for any* $T > 0$ *in the sense of Definition 3.1.*

Remark 3.3 (Periodic boundary conditions) The same result can be shown for the periodic boundary conditions, meaning $\Omega = \mathbb{T}^d$ as well as other types of domains. Regularity of the boundary can be relaxed as well. The reader may consult Sect. 3.4 for other related results.

Remark 3.4 (Navier–Stokes–Fourier system) The existence theory can be extended to the full Navier–Stokes–Fourier system (3.11)–(3.13). The technical difficulties arise, however, at the level of *a priori* bounds. As a result, rather severe but still physically relevant restrictions must be imposed on the constitutive relations. The interested reader may consult the monographs [80], and [97].

As a matter of fact Theorem 3.3 was proved in [80, Theorem 7.1] under additional regularity of the domain Ω. General (even non-Lipschitz) domains were treated by Kukučka [146].

3.3 Strong Solutions, Conditional Regularity, Navier-Stokes System

The local existence result stated in Theorem 3.1 extends in a straightforward manner to the barotropic problem (3.14), (3.15). Our goal is to discuss suitable *conditional regularity criteria* under which the solution remains smooth on a given time interval.

3.3.1 Local Strong Solutions

Probably optimal, with respect to regularity of the data, is the following result of Cho et al. [52, Proposition 5].

Theorem 3.4 (Local strong solution) *Let* $\Omega \subset R^3$ *be a bounded domain of class* $C^{2+\nu}$, *let* $p \in C^1[0,\infty)$. *Suppose the initial data* ϱ_0, $\boldsymbol{u}_0$ *belong to the class*

$$\varrho_0 \in W^{1,q}(\Omega) \text{ for some } q > 3,\ \varrho \geq \underline{\varrho} > 0 \text{ in } \overline{\Omega},\ \boldsymbol{u}_0 \in W_0^{1,2}(\Omega; R^3) \cap W^{2,2}(\Omega; R^3).$$

Then there exists $T > 0$ *such that the Navier–Stokes system* (3.14)–(3.16), *with the no-slip condition* (3.17) *admits a strong solution* $[\varrho, \boldsymbol{u}]$, *unique in the class*

$$\varrho \in C([0,T]; W^{1,r}(\Omega)),\ r = \min\{6, q\},$$
$$\boldsymbol{u} \in C([0,T]; W_0^{1,2} \cap W^{2,2}(\Omega; R^3)) \cap L^2_{\rm loc}([0,T); W^{2,r}(\Omega; R^3)),$$

$$\partial_t \varrho \in C([0,T); L^r(\Omega)),\ \partial_t \boldsymbol{u} \in L^2_{\rm loc}([0,T); W^{1,2}(\Omega; R^3)).$$

Remark 3.5 If the initial data enjoy higher regularity and satisfy the relevant compatibility conditions as in Theorem 3.1, then the solution belongs to the same regularity class as in Theorem 3.1.

3.3.2 Conditional Regularity

Possible blow up of the local smooth solutions cannot occur provided considerable weaker norms are controlled. This is the heart of the conditional regularity results. Here, we quote a simple criterion established by Sun et al. [186, Theorem 1.3] that comes in handy in the subsequent numerical analysis.

Theorem 3.5 (Conditional regularity) *In addition to the hypotheses of Theorem 3.4, suppose that* $p = a\varrho^\gamma$, $a > 0$, $\gamma > 1$, *and that the bulk viscosity* λ *vanishes,* $\lambda = 0$. *Let* $T > 0$ *be the maximal time interval on which the strong solution* $[\varrho, \boldsymbol{u}]$, *the existence of which is guaranteed by Theorem 3.4, exists.*

Then

$$\limsup_{t \to T+} \left(\sup_\Omega \varrho(t, \cdot) \right) = \infty. \tag{3.41}$$

Remark 3.6 The technical condition $\lambda = 0$ is not necessary, if (3.41) is replaced by

$$\limsup_{t \to T+} \left[\left(\sup_\Omega \varrho(t, \cdot) \right) + \|\boldsymbol{u}(t, \cdot)\|_{L^\infty(\Omega; R^3)} \right] = \infty \tag{3.42}$$

see Sun et al. [187, Remark 6].

Theorem 3.5 is a remarkable result and in combination with the weak-strong uniqueness principle establish for generalized solutions to the Navier–Stokes system gives rise to rather strong *convergence* statement. Namely any *bounded* sequence of consistent numerical approximations converges strongly to a smooth solution of the continuous system. We shall discuss this and similar issues later in Part II.

3.4 Conclusion, Bibliographical Remarks

The motion of linearly viscous (Newtonian) compressible fluids in the framework of continuum mechanics is described by the Navier–Stokes–Fourier system (3.1)–(3.3) or by the simplified Navier–Stokes system (3.14)–(3.16). The regularizing effect of viscosity guarantees the existence of global-in-time smooth solutions provided the initial data are smooth and sufficiently close to equilibrium. Still global existence of smooth solutions for arbitrary smooth initial data remains an outstanding open problem in the multidimensional case. The monodimensional case is quite well understood, see e.g. the monograph by Antontsev et al. [6] and the references cited therein. The existence of global-in-time smooth solutions under the additional assumption that the viscosity coefficients μ and λ depend on ϱ in a very specific way was established by Vaigant and Kazhikhov [195] in the two-dimensional physical domain.

There is a well developed theory of weak solutions described in Sect. 3.2. The technical assumption $\gamma > \frac{d}{2}$ has been relaxed to the case $\gamma \geq 1, d = 2$, by Plotnikov and Vaigant [179]. For more recent development of the theory, we refer to Bresch and Desjardins [30, 31] and Bresch and Jabin [32], among others. Despite its unquestionable theoretical impact, the theory of weak solutions is difficult to implement in the numerical analysis which is the main objective of the present monograph. The main stumbling block we have observed in Sect. 3.2 is a different character of the transport part of the underlying equations, which is the same as for the hyperbolic Euler system, and the viscous stress rending the momentum equation parabolic. The energy estimates studied in Sect. 3.2.1 are compatible with the standard finite element discretization while the pressure estimates examined in Sect. 3.2.3 require rather a finite volume approach. Carrying over the compactness proof in the discrete setting faces many technical difficulties as well. As a result, we need a mixed finite volume—finite element numerical scheme to handle the above mentioned difficulties. Such a scheme was proposed and analyzed by Karper [139, 140] (cf. also Gallouet et al. [111]), where convergence to a weak solution of the isentropic Navier–Stokes system was established under a technical assumption $\gamma > 3$. Similar approach has been adapted to the Navier–Stokes–Fourier system in a series of papers [86, 87]. To mimic the principal steps of the existence theory, the numerical scheme has to be modified by a number of artificial viscosity terms that may slow down convergence and provide an incorrect picture of the limit solution.

In the numerical part of this monograph, we frequently anticipate that models of viscous fluids are likely to possess regular solutions, at least for a "generic" class of data. We first identify all possible limits of a numerical scheme with a general object—dissipative solution. This concept goes beyond the standard class of weak solutions but shares with them a fundamental property: the *weak-strong uniqueness* principle. In particular, convergence to a dissipative solution turns out to be equivalent with (pointwise) convergence to a strong solution as long as the latter exists. If this is the case, qualitative error estimates can be also derived.

Part II
Generalized Solutions to Equations and Systems Describing Compressible Fluids

We develop a unifying framework of generalized solutions applicable to models of both inviscid (perfect) and viscous fluids. Our primer motivation is to capture all possible limits arising from various *numerical schemes*. Being aware of the facts revealed in Part I, we do not expect to obtain strong stability estimates—uniform bounds in topologies strong enough to guarantee at least pointwise convergence of numerical approximations. As we learned above, such bounds are definitely not expected in the context of inviscid fluids modeled by the Euler system and difficult to obtain for models of viscous fluids as the Navier–Stokes system. To facilitate the analysis and capture the largest class of possible limits, we consider the *weak* topology on the spaces L^p of Lebesgue integrable functions.

As is well known, the weak topology can be perfectly adapted to *linear* problems as soon as the derivatives are interpreted in the sense of distributions. This can be rephrased in terms of the celebrated *Lax equivalence principle*: Stability (*a priori* bounds) and consistency (smallness of the approximation error) for linear problems is equivalent to *convergence*. Weak convergence in the L^p-sense is essentially nothing else than convergence in the sense of integral averages. Consequently, there are two basic stumbling blocks when applying this approach to *nonlinear problems*:

• **Oscillations**.

Bounded sequences of functions may get practically out of control developing rapid oscillations varying in space and time. Consider an a-periodic continuous function $g : R \to R$,

$g(x+a) = g(x)$ for all $x \in R$, $\int_0^a g(x)dx = 0$,and a sequence

$g_n(x) = g(nx)$, $n = 1, 2, \dots$

Our goal is to describe the limit $\lim_{n\to\infty} g_n$. Apparently, the sequence $g(nx)$ does not converge pointwise, not even a.a. pointwise and not even for a subsequence. To capture its asymptotic behavior, we have to consider its averaged values,

$\int_R g_n(x)\varphi(x)\,\mathrm{d}x$, where $\varphi \in C_c^\infty(R)$.

Introducing the primitive function G,

$G(x) = \int_0^x g(z)\ \mathrm{d}z$

we easily observe that G is also continuous and periodic. Consequently, by means of the by-parts-integration formula,

$\int_R g_n(x)\varphi(x)\ \mathrm{d}x = \int_R g(nx)\varphi(x)\ \mathrm{d}x = -\frac{1}{n}\int_R G(nx)\partial_x\varphi(x)\ \mathrm{d}x \to 0$
as $n \to \infty$ for any smooth function φ. We say that the sequence $\{g_n\}_{n=1}^{\infty}$ *converges weakly* to 0, $g_n \rightharpoonup 0$. As a straightforward consequence, we deduce that a sequence

$h_n(x) = h(nx), \quad \text{where} \quad h \quad \text{is} \quad a- \quad \text{periodic},$

converges weakly to the integral average $\int_0^a h(x)\ \mathrm{d}x,\ h_n \rightharpoonup \int_0^a h(x)\ \mathrm{d}x.$

Thus, the weak convergence does not, in general, commute with nonlinear compositions, specifically

$g_n \rightharpoonup g \quad \text{does not imply} \quad H(g_n) \rightharpoonup H(g) \quad \text{if} \quad H \quad \text{is not linear}.$

Convex compositions are *weakly lower semicontinuous*,

$g_n \rightharpoonup g \quad \text{implies} \quad \int_R H(g)\varphi\,\mathrm{d}x \leq \liminf_{n\to\infty} \int_R H(g_n)\varphi \quad \mathrm{d}x \quad \text{for any} \quad \varphi \geq 0$

whenever H is convex. This can be easily seen from the subdifferential inequality

$H(g_n) \geq H(g) + \partial H(g)(g_n - g),$

or, if the function is twice continuously differentiable, from he Taylor decomposition

$H(g_n) = H(g) + H'(g)(g_n - g) + \frac{1}{2}H''(\xi)(g_n - g)^2.$

The above formula reveals the difference between the weak limit of $H(g_n)$ and $H(g)$ due to the quadratic term. Indeed one can deduce that g_n converges strongly to g only if there exists a *strictly convex* function H such that $H(g_n) \rightharpoonup H(g)$. Another, more rigorous explanation follows from the fact the convex functions are characterized by suprema of their affine minorants:

$H(z) = \sup\{h(z) \mid h \quad \text{affine} \quad h(z) \leq H(z) \quad \text{for any} \quad z\}.$

Consequently

$H(g_n) \geq h(g_n) \rightharpoonup h(g) \quad \text{for any} \quad \text{affine} \quad h \leq H.$

Convexity of certain thermodynamic functions, among which the total energy, will play a significant role in the analysis of convergence of numerical schemes.

- **Concentrations.**

Consider a sequence

$g_n(x) = ng(nx), \quad \text{where}\ g \in C_c^{\infty}(-1,1),\ g(-x) = g(x),\ g \geq 0,\ \int_R g(x)\ \mathrm{d}x = 1.$

It is easy to check that

$g_n(x) \to 0\ as\ n \to \infty$ for any $x \neq 0$, in particular $g_n \to 0\ a.a.$ in R;

$\|g_n\|_{L^1(R)} = \int_R g_n(x)\ \mathrm{d}x = \int_R g(x)\ \mathrm{d}x = 1 \quad \text{for any} \quad n = 1, 2, \ldots.$

Next, we observe that g_n *does not* converge weakly to 0. Indeed

$$\int_R g_n(x)\varphi(x)\ \mathrm{d}x$$

$$= \int_{-1/n}^{1/n} g_n(x)\varphi(x)\ \mathrm{d}x \in \left[\min_{x\in[-1/n,1/n]}\varphi(x), \quad \max_{x\in[-1/n,1/n]}\varphi(x)\right] \to \varphi(0)$$

as soon as φ is continuous. As a matter of fact, the limit object cannot be identified with an integral average of a function, it is a *measure*—the Dirac mass δ_0 concentrated at 0.

We conclude this introduction by an example of $\mathcal{K}$*–convergence* of an oscillatory sequence. Going back to the first example, we consider an oscillatory sequence

$g_n(x) = \cos(nx), \quad n = 1, 2, \ldots$

Next, we introduce the *Cesàro averages*

$$\tilde{g}_N(x) = \frac{1}{N}\sum_{n=1}^{N} g_n(x) = \frac{1}{N}\sum_{n=1}^{N}\cos(nx) = \frac{1}{N}\left(\frac{1}{2}\frac{\sin((N+\frac{1}{2})x)}{\sin(\frac{x}{2})} - \frac{1}{2}\right),$$

where the right-hand side tends to 0 as $N \to \infty$ for any $x \neq 2k\pi$, k an integer. Consequently, the *Cesàro averages* converge to the weak limit of $\{g_n\}_{n=1}^{\infty}$*pointwise* a.a. in R. In particular, as $\tilde{g}_N$ are uniformly bounded, Lebesgue dominance convergence theorem yields

$$\tilde{g}_N \to 0 \ \ in \ \ L^p_{\rm loc}(R) \ \ as \ \ N \to \infty, \ \ 1 \leq p < \infty.$$

This is a standard example of the statistical *Strong Law of Large Numbers*. Averaging may convert weakly converging sequences to strongly converging, which may be used to visualize effectively the weak limits in numerical experiments. We examine the properties of *Cesàro averages* of weakly convergent sequences in detail in Part III.

This part of the monograph is devoted to generalized solutions to the Euler and Navier–Stokes systems introduced in Part I. We go even beyond the framework of standard weak (distributional) solutions that are suitable for linear problems but do not capture the complex behavior of nonlinear terms under weak convergence. Our approach is highly inspired by the work of DiPerna [65], and DiPerna, Majda [69], [70], [66] on the measure-valued solutions to the Euler system. Our goal is to identify a class of generalized solutions large enough to accommodate limits of suitable numerical schemes but still complying with the following *condiciones sine quibus non*:

- **(E) Existence**. Generalized solutions exist globally in time for any physically admissible data. In the context of Euler/Navier Stokes system, physically admissible means the density ρ_0 should be nonnegative, the absolute temperature ϑ_0 strictly positive, the total energy finite, etc.
- **(C) Compatibility**. Generalized solutions possessing all necessary derivatives required by the corresponding system of equations are classical solutions—they satisfy the problem in the classical sense.
- **(WS) Weak–strong uniqueness**. A generalized solution coincides with the classical solution originating from the same initial data on the life span of the latter.
- **(S) Sequential stability**. The class of generalized solutions is closed with respect to the (weak) topology imposed by the available *a priori* bounds. Limits of sequences of generalized solutions are generalized solutions of the same problem.

Note carefully the subtle difference between "compatibility" and "weak–strong uniqueness" principles. Compatibility asserts that generalized solutions, similarly to distributional solutions to linear problems, represent a true extension of the concept of classical solution to the case when differentiability is not available. Weak–strong uniqueness means stability of strong solutions in the class of generalized solutions. All generalized solutions coincide with the classical one as long as the latter exists. The above-mentioned properties are absolutely crucial to carry out our program concerning numerical treatment of potentially ill-posed problems as the Euler system.

Chapter 4
Classical and Weak Solutions, Relative Energy

The concept of weak (distributional) solution for general systems of conservation/balance laws was introduced in Sect. 1.1.1 and discussed in Chaps. 2, 3. Here, we revisit the topic in detail for the Euler and Navier–Stokes systems. In the former case, we show the class is possibly not closed with respect to the available weak topologies induced by *a priori* bounds; whence extension of this concept to a larger class of objects is of interest. A similar problem for the Navier–Stokes system is more subtle. Although, as we have seen in Chap. 3, the set of weak solutions is closed for certain models, the necessary estimates may not be easily available at the level of a numerical scheme. This is the main reason why to extend the class of weak solutions to some models of viscous fluids as well.

Given a time interval $(0, T)$ and a spatial domain $\Omega \subset R^d$, we say that a solution is *classical* if it is continuous in the closed set $[0, T] \times \overline{\Omega}$, and if all relevant partial derivatives exist and are continuous in the open set $(0, T) \times \Omega$. If the boundary conditions involve derivatives, then those must be continuously extendable from $(0, T) \times \Omega$ to $(0, T) \times \overline{\Omega}$. Thus this issue is inseparable from the geometric properties and, in particular, the smoothness of the boundary $\partial\Omega$. In particular, if the boundary conditions involve the outer normal vector, then the latter must exist at any point of $\partial\Omega$. The equations as well as the boundary and initial conditions are satisfied pointwise.

In the literature, the notion of *smooth* solution and/or *smooth* domain is frequently used. Very often, "smooth" in this context does not mean of class C^∞ or even analytic but should be interpreted as "sufficiently smooth" or "as smooth as necessary". We try to avoid this a bit dubious and misleading terminology. The term *strong* solution will be used in the situation when all required derivatives, expressed in terms of the theory of distributions, can be interpreted as (locally) integrable functions.

In this chapter, we focus on the relation between weak and strong solutions. In particular, we introduce the concept of *relative energy* functional. This quantity is derived from the total energy E of the system, and, if the latter is a convex function of the state variables, represents a *Bregman distance* with respect to the (convex)

E. Feireisl et al., *Numerical Analysis of Compressible Fluid Flows*, MS&A 20,
https://doi.org/10.1007/978-3-030-73788-7_4

total energy E, see e.g. Sprung [185]. In particular, the relative energy can be used to measure the distance between a weak and strong solution starting from different (or identical) initial data. The relative energy vanishes if the data are the same for both solutions which is the desired weak-strong **(WS)** property. The relative energy may be seen as an alternative to the relative *entropy* functional introduced in the context of nonlinear conservation laws by Dafermos [59]. The approach based on the relative entropy requires integrability of the (total) energy flux. Therefore the method based on relative energy rather than entropy works efficiently in the context of weak and even more general dissipative solutions for the Euler and Navier–Stokes system, for which the available *a priori* bounds are not strong enough to render the total energy flux integrable.

Convexity of the total energy with respect to a suitable set of state variables plays a crucial role here that is intimately related to the property of *thermodynamic stability* of the fluid system. The interpretation of the notion of *thermodynamic stability* can be twofold: (1) certain material constants as compressibility and specific heat at constant volume are nonnegative, (2) equilibrium states are (linearly) stable, cf. Bechtel etal. [14]. We provide a unified approach identifying the relative energy with the Bregman distance associated to total energy.

4.1 Weak and Strong Solutions to the Euler System

The Euler system has been introduced in Chap. 2. As we have seen, there are several possible choices of basic field variables and, accordingly, the formulation of the field equations:

- **Standard (or primitive) variables**. The mass density ϱ, the (absolute) temperature ϑ, the velocity $\boldsymbol{u}$.
- **Conservative variables**. The mass density ϱ, the (total) energy E, the momentum $\boldsymbol{m}$.
- **Conservative-entropy variables**. The mass density ϱ, the (total) entropy S, the momentum $\boldsymbol{m}$.

Of course, there are other (infinitely many) possibilities how to choose the set of independent field variables.

Recall that

$$\boldsymbol{m} = \varrho \boldsymbol{u},\ E = \frac{1}{2}\frac{|\boldsymbol{m}|^2}{\varrho} + \varrho e,\ S = \varrho s,$$

where the internal energy e, the pressure p, and the entropy s satisfy Gibbs equation

$$\vartheta Ds = De + pD\left(\frac{1}{\varrho}\right). \tag{4.1}$$

In (4.1), the symbol D denotes a differential (gradient) with respect to either $[\varrho, \vartheta]$, $[\varrho, S]$ or any other choice of the independent parameters.

The field equations written in the standard variables read:

$$\partial_t \varrho + \mathrm{div}_x(\varrho \boldsymbol{u}) = 0 \text{ in } (0, T) \times \Omega; \tag{4.2}$$

$$\partial_t(\varrho \boldsymbol{u}) + \mathrm{div}_x(\varrho \boldsymbol{u} \otimes \boldsymbol{u}) + \nabla_x p(\varrho, \vartheta) = 0 \text{ in } (0, T) \times \Omega; \tag{4.3}$$

$$\partial_t \left(\frac{1}{2}\varrho |\boldsymbol{u}|^2 + \varrho e(\varrho, \vartheta) \right) + \mathrm{div}_x \left[\left(\frac{1}{2}\varrho |\boldsymbol{u}|^2 + \varrho e(\varrho, \vartheta) + p(\varrho, \vartheta) \right) \boldsymbol{u} \right] = 0 \text{ in } (0, T) \times \Omega. \tag{4.4}$$

Furthermore we consider the impermeability boundary conditions

$$\boldsymbol{u} \cdot \boldsymbol{n}|_{\partial\Omega} = 0, \tag{4.5}$$

or, alternatively, the periodic boundary conditions $\Omega = \mathbb{T}^d$.

In some real world applications, for instance in meteorology, it is more convenient to express the pressure $p = p(\varrho, s)$ as a function of the density and the entropy and to rewrite the Euler system in the form

$$\partial_t \varrho + \mathrm{div}_x(\varrho \boldsymbol{u}) = 0 \text{ in } (0, T) \times \Omega; \tag{4.6}$$

$$\partial_t(\varrho \boldsymbol{u}) + \mathrm{div}_x(\varrho \boldsymbol{u} \otimes \boldsymbol{u}) + \nabla_x p(\varrho, s) = 0 \text{ in } (0, T) \times \Omega; \tag{4.7}$$

$$\partial_t s + \nabla_x s \cdot \boldsymbol{u} = 0 \text{ in } (0, T) \times \Omega. \tag{4.8}$$

The entropy s satisfies the transport equation (4.8) and it is easy to see that

$$\partial_t Z(s) + \nabla_x Z(s) \cdot \boldsymbol{u} = 0 \text{ in } (0, T) \times \Omega \tag{4.9}$$

as soon Z is a continuously differentiable function. In particular, if $Z' > 0$, we may replace s by Z and write $p = p(\varrho, Z)$ obtaining a new problem in terms of $[\varrho, \boldsymbol{u}, Z]$. The limit case $Z \to$ const gives rise to the isentropic (barotropic) Euler system with $p = p(\varrho)$,

$$\partial_t \varrho + \mathrm{div}_x(\varrho \boldsymbol{u}) = 0 \text{ in } (0, T) \times \Omega; \tag{4.10}$$

$$\partial_t(\varrho \boldsymbol{u}) + \mathrm{div}_x(\varrho \boldsymbol{u} \otimes \boldsymbol{u}) + \nabla_x p(\varrho) = 0 \text{ in } (0, T) \times \Omega. \tag{4.11}$$

Although the systems (4.2)–(4.4) and (4.6)–(4.9) are completely equivalent in the framework of classical solutions, for which $\varrho > 0$, $\vartheta > 0$, they give rise to qualitatively different *weak* formulations of the Euler system. We will discuss this issue later in this chapter.

4.1.1 Classical Solutions to the Euler System

In this section, we suppose that $\Omega \subset R^d$ is a bounded domain with a boundary of class C^1. In particular, the outer normal vector exists at any point $x \in \partial\Omega$.

Definition 4.1 (Classical solution to the Euler system) We say that a trio $[\varrho, \vartheta, \boldsymbol{u}]$ is a *classical solution* of the Euler system (4.2)–(4.4), with the impermeability boundary condition (4.5), if

$$
\begin{aligned}
\varrho &\in C([0,T] \times \overline{\Omega}) \cap C^1((0,T) \times \Omega) \cap W^{1,\infty}((0,T) \times \Omega),\\
\vartheta &\in C([0,T] \times \overline{\Omega}) \cap C^1((0,T) \times \Omega) \cap W^{1,\infty}((0,T) \times \Omega),\\
\boldsymbol{u} &\in C([0,T] \times \overline{\Omega}; R^d) \cap C^1((0,T) \times \Omega; R^d) \cap W^{1,\infty}((0,T) \times \Omega; R^d);
\end{aligned}
$$

$$
0 < \underline{\varrho} \le \varrho(t,x),\ 0 < \underline{\vartheta} \le \vartheta(t,x) \text{ for any } t \in [0,T],\ x \in \overline{\Omega},
$$
$$
\boldsymbol{u}(t,x) \cdot \boldsymbol{n}(x) = 0 \text{ for any } t \in [0,T],\ x \in \partial\Omega;
$$

and the equations (4.2)–(4.4) hold.

Remark 4.1 Here we have tacitly assumed that both $p = p(\varrho, \vartheta)$ and $e = e(\varrho, \vartheta)$ are continuously differentiable for ϱ and ϑ bounded below away from zero.

Remark 4.2 (Strong solutions) We speak about strong solutions if $[\varrho, \vartheta, \boldsymbol{u}]$ are required to be only (globally) Lipschitz continuous.

Remark 4.3 (Periodic boundary conditions, Lipschitz domain) The above definition can be modified in an obvious way to accommodate the periodic boundary conditions $\Omega = \mathbb{T}^d$, where, of course the issue of boundary regularity is irrelevant. We recall that $\mathbb{T}^d$ can be viewed as a smooth manifold over R^d without boundary.

The requirement of the existence of the outer normal vector at *any* boundary point of Ω is rather restrictive. Recall that Lipschitz domains admit an normal vector for a.a. boundary point $x \in \Omega$, where the latter is endowed with the standard $(d-1)$-Hausdorff measure. The above definition extends to this case in a direct manner. The impermeability or zero normal trace condition (4.5) can be also reformulated in a weak form

$$
\int_\Omega \boldsymbol{u}(t,\cdot) \cdot \nabla_x \phi \, \mathrm{d}x + \int_\Omega \phi \mathrm{div}_x \boldsymbol{u}(t,\cdot) \, \mathrm{d}x = 0 \ \text{ for any } \ \phi \in C^1_c(R^d),\ t \in [0,T]. \tag{4.12}
$$

Obviously, any classical solution of the Euler system written in the form (4.2)–(4.4) is also a classical solution of the entropy formulation (4.6)–(4.8) as long as the thermodynamic functions are interrelated through Gibbs' equation (4.1).

As we observed in Sect. 2.1.2, classical solutions develop shock singularities in a finite time for a fairly general class of initial data. They obviously violate the existence condition **(E)** postulated in the introductory section of this part. Their application in the numerical analysis is therefore limited to the regime, where their life span exceeds the time desired for prediction. A relevant example is the Euler system in the low Mach number regime, considered often in meteorology, where the fluid velocity is largely dominated by the speed of sound and the occurrence of shocks is not expected. Still we are very far from a rigorous proof of such a statement, in particular in the physically relevant $3D$-case.

4.1.2 Weak (Distributional) Solutions to the Barotropic Euler System

We revisit our discussion concerning weak (distributional) solutions starting with the barotropic Euler system (4.10), (4.11), where the theory is quite simple, elegant, and self-contained. Formally, the definition is obtained via multiplying the field equations (4.10), (4.11) on a suitable *test* function and declaring the resulting identity to be a proper definition of the weak solution. Following these lines, we obtain, exactly as in Sect. 2.1.3,

$$\left[\int_{\Omega} \varrho \varphi \, \mathrm{d}x \right]_{t=0}^{t=\tau} = \int_0^{\tau} \int_{\Omega} \left[\varrho \partial_t \varphi + \varrho \boldsymbol{u} \cdot \nabla_x \varphi \right] \mathrm{d}x \, \mathrm{d}t \tag{4.13}$$

for any $0 \leq \tau \leq T$, and any $\varphi \in C^1([0, T] \times \overline{\Omega})$;

$$\left[\int_{\Omega} \varrho \boldsymbol{u} \cdot \boldsymbol{\varphi} \, \mathrm{d}x \right]_{t=0}^{t=\tau} = \int_0^{\tau} \int_{\Omega} \left[\varrho \boldsymbol{u} \cdot \partial_t \boldsymbol{\varphi} + \varrho \boldsymbol{u} \otimes \boldsymbol{u} : \nabla_x \boldsymbol{\varphi} + p(\varrho) \mathrm{div}_x \boldsymbol{\varphi} \right] \mathrm{d}x \, \mathrm{d}t \tag{4.14}$$

for any $0 \leq \tau \leq T$, and any $\boldsymbol{\varphi} \in C^1([0, T] \times \overline{\Omega}; R^d)$, $\boldsymbol{\varphi} \cdot \boldsymbol{n}|_{\partial\Omega} = 0$.

Given the recent state of the art discussed in Chap. 2, such a definition complies with the existence principle **(E)** as well as with the compatibility principle **(C)**. Specifically, the weak solutions satisfying "*only*" (4.13), (4.14) exist globally in time for any (continuous) initial data, cf. [1]. However, the problem is desperately ill-posed even in the class of smooth initial data (the reader may consult the literature collected in Sect. 2.4). In particular, the weak-strong uniqueness principle **(WS)** is violated unless suitable *admissibility conditions* are imposed. To identify the class of suitable admissible solutions, we append (4.13), (4.14) by the *energy inequality*

$$\left[\int_{\Omega} \left(\frac{1}{2} \varrho |\boldsymbol{u}|^2 + P(\varrho) \right) \mathrm{d}x \right]_{t=0}^{t=\tau} \leq 0 \tag{4.15}$$

for any $0 \leq \tau \leq T$, where P is the pressure potential satisfying

$$P'(\varrho)\varrho - P(\varrho) = p(\varrho).$$

It is easy to check the P is uniquely determined by p modulo a linear function of ϱ. On the other hand, as the boundary is impermeable, the total mass

$$M = \int_{\Omega} \varrho(t, x) \, \mathrm{d}x$$

is a constant of motion; whence (4.15) remains unchanged for any affine perturbation of P.

Boundedness of the total energy/mass in terms of the initial data is basically the only available source of *a priori* (stability) estimates for the Euler system. The weak point of the formulation (4.13)–(4.15) in terms of the standard variables is that the velocity $\boldsymbol{u}$ is not controlled by the energy on the (hypothetical) vacuum region where $\varrho = 0$. It is therefore more convenient to consider the conservative variables $[\varrho, \boldsymbol{m} = \varrho \boldsymbol{u}]$ which give rise to the concept of weak solution for the barotropic Euler system introduced in Definition 2.5 that we now reproduce for convenience:

Definition 4.2 (Weak solution of barotropic Euler system) A pair $[\varrho, \boldsymbol{m}]$ is an *admissible weak solution* of the barotropic Euler system (4.10), (4.11) with the impermeability boundary condition (4.5) in $(0, T) \times \Omega$ if the following holds:

- (**weak continuity**) the quantities $\varrho, \boldsymbol{m}$ belong to the class

$$\begin{aligned} &\varrho \in C_{\mathrm{weak}}([0, T]; L^{\gamma}(\Omega)), \ \varrho \geq 0 \text{ a.a. in } (0, T) \times \Omega \\ &\boldsymbol{m} \in C_{\mathrm{weak}}([0, T]; L^{\frac{2\gamma}{\gamma+1}}(\Omega; R^d)) \end{aligned} \tag{4.16}$$

  for some $\gamma > 1$;
- (**equation of continuity**) the integral identity

$$\left[\int_{\Omega} \varrho \varphi \, \mathrm{d}x \right]_{t=0}^{t=\tau} = \int_0^{\tau} \int_{\Omega} \left[\varrho \partial_t \varphi + \boldsymbol{m} \cdot \nabla_x \varphi \right] \, \mathrm{d}x \, \mathrm{d}t \tag{4.17}$$

  holds for any $0 \leq \tau \leq T$, and any test function $\varphi \in C^1([0, T] \times \overline{\Omega})$;

- **(momentum equation)** the integral identity

$$
\begin{aligned}
&\left[\int_\Omega \boldsymbol{m} \cdot \boldsymbol{\varphi} \, \mathrm{d}x \right]_{t=0}^{t=\tau} \\
&= \int_0^\tau \int_\Omega \left[\boldsymbol{m} \cdot \partial_t \boldsymbol{\varphi} + \left(1_{\varrho>0} \frac{\boldsymbol{m} \otimes \boldsymbol{m}}{\varrho} \right) : \nabla_x \boldsymbol{\varphi} + p(\varrho) \mathrm{div}_x \boldsymbol{\varphi} \right] \mathrm{d}x \, \mathrm{d}t
\end{aligned}
\tag{4.18}
$$

holds for any $0 \le \tau \le T$, and any test function $\boldsymbol{\varphi} \in C^1([0,T] \times \overline{\Omega}; R^d)$, $\boldsymbol{\varphi} \cdot \boldsymbol{n}|_{\partial\Omega} = 0$;
- **(energy inequality)** the integral inequality

$$
\left[\int_\Omega \left(\frac{1}{2} \frac{|\boldsymbol{m}|^2}{\varrho} + P(\varrho) \right) \psi \, \mathrm{d}x \right]_{t=0}^{t=\tau} \le \int_0^\tau \int_\Omega \left(\frac{1}{2} \frac{|\boldsymbol{m}|^2}{\varrho} + P(\varrho) \right) \partial_t \psi \, \mathrm{d}x \, \mathrm{d}t \tag{4.19}
$$

holds for any test function $\psi \in C^1[0,T]$, $\psi \ge 0$.

Remark 4.4 (Initial data) In the above, we have assumed

$$
\varrho(0,\cdot) = \varrho_0, \ \boldsymbol{m}(0,\cdot) = \boldsymbol{m}_0,
$$

where ϱ_0, $\boldsymbol{m}_0$ are the initial data, with the initial energy

$$
\int_\Omega \left[\frac{1}{2} \frac{|\boldsymbol{m}_0|^2}{\varrho_0} + P(\varrho_0) \right] \mathrm{d}x < \infty.
$$

Here, the kinetic energy $\frac{1}{2}\frac{|\boldsymbol{m}|^2}{\varrho}$ is interpreted as a convex l.s.c. function via formula (2.60).

Keeping in mind the canonical example of the isentropic pressure $p(\varrho) = a\varrho^\gamma$, we suppose

$$
\begin{aligned}
&p \in C^1[0,\infty) \cap C^2(0,\infty), \ p(0) = 0, \ p'(\varrho) > 0 \text{ for } \varrho > 0, \\
&0 < \liminf_{\varrho\to\infty} \frac{p'(\varrho)}{\varrho^{\gamma-1}} \le \limsup_{\varrho\to\infty} \frac{p'(\varrho)}{\varrho^{\gamma-1}} < \infty
\end{aligned}
\tag{4.20}
$$

for some $\gamma > 1$. Seeing that

$$
P''(\varrho) = \frac{p'(\varrho)}{\varrho} \text{ for } \varrho > 0
$$

we easily deduce that the pressure potential P is a strictly convex function of ϱ, and

$$|p(\varrho)| \stackrel{<}{\sim} \Big(c + P(\varrho) \Big) \text{ for a suitable } c > 0. \tag{4.21}$$

This means that all nonlinearities appearing in the momentum equation (4.19) are controlled in the L^1 norm by the total energy. More precisely, we have the following result.

Proposition 4.1 (Uniform stability estimates for barotropic Euler system) *Let $\Omega \subset R^d$ be a bounded domain. Let the pressure $p = p(\varrho)$ satisfy* (4.20) *with $\gamma > 1$. Suppose that $[\varrho, \boldsymbol{m}]$ is an admissible weak solution of the Euler system in the sense of Definition 4.2, with the initial data $[\varrho_0, \boldsymbol{m}_0]$ such that*

$$\int_{\Omega} \left[\frac{1}{2} \frac{|\boldsymbol{m}_0|^2}{\varrho_0} + P(\varrho_0) \right] \mathrm{d}x = E_0 < \infty. \tag{4.22}$$

Then

$$\begin{aligned} \sup_{t \in [0,T]} \left\| 1_{\varrho > 0} \frac{\boldsymbol{m} \otimes \boldsymbol{m}}{\varrho}(t, \cdot) \right\|_{L^1(\Omega; R^{d \times d})} &\leq c(E_0), \\ \sup_{t \in [0,T]} \|\varrho(t, \cdot)\|_{L^\gamma(\Omega)} \leq c(E_0), \quad & \sup_{t \in [0,T]} \|p(\varrho)(t, \cdot)\|_{L^1(\Omega)} \leq c(E_0), \\ \sup_{t \in [0,T]} \|\boldsymbol{m}(t, \cdot)\|_{L^{\frac{2\gamma}{\gamma+1}}(\Omega; R^d)} &\leq c(E_0). \end{aligned} \tag{4.23}$$

Remark 4.5 Recall that the total energy is defined as a convex function of $[\varrho, \boldsymbol{m}]$ via (2.60). In particular, hypothesis (4.22) entails nonnegativity of the initial density as well as a compatibility condition of the initial momentum on the vacuum zone,

$$\varrho_0 \geq 0, \text{ and } \varrho_0 = 0 \Rightarrow \boldsymbol{m}_0 = 0 \text{ a.a. in } \Omega.$$

Proof Taking $\psi \equiv 1$ in the energy inequality (4.19) we obtain

$$\int_{\Omega} \left[\frac{1}{2} \frac{|\boldsymbol{m}|^2}{\varrho} + P(\varrho) \right] (\tau, \cdot) \, \mathrm{d}x \leq E_0 \text{ for any } \tau \in [0, T]. \tag{4.24}$$

Note that the above inequality is indeed valid for *any* $\tau \in [0, T]$ as the energy is convex and both ϱ and $\boldsymbol{m}$ are weakly continuous as functions of the time.

In view of the hypothesis (4.20), the bound (4.24) gives rise to all estimates claimed in (4.23). Indeed to see the last bound in (4.23) write

$$\boldsymbol{m} = 1_{\varrho > 0} \frac{\boldsymbol{m}}{\sqrt{\varrho}} \sqrt{\varrho};$$

whence, by Hölder's inequality,

$$\|\boldsymbol{m}\|_{L^{\frac{2\gamma}{\gamma+1}}(\Omega;R^d)} \leq \left\| 1_{\varrho>0} \frac{\boldsymbol{m}}{\sqrt{\varrho}} \right\|_{L^2(\Omega;R^d)} \|\sqrt{\varrho}\|_{L^{2\gamma}(\Omega)} \leq c(E_0).$$

As pointed out in Remark 4.5, boundedness of the kinetic energy given by (2.60) implies that $\boldsymbol{m} = 0$ a.a. on the vacuum set $\{\varrho = 0\}$. □

Proposition 4.1 may be seen as the first step towards the stability property **(S)**. The set of all weak solutions emanating from a bounded set of initial data remains bounded uniformly in time. This definitely implies stability in suitable L^p-topologies, however, oscillations and/or concentrations may still appear and the set of weak solutions in the sense of Definition 4.2 is not likely to be closed (sequentially stable). We discuss this issue in detail in the concluding part of this chapter.

4.1.3 Relative Energy for the Barotropic Euler System

Relative energy is a simple but extremely useful tool in the analysis of nonlinear systems. It may be seen as a variant of the concept of *Bregman distance (divergence)* known in convex analysis, where the underlying convex potential is the total energy. Closely related is the concept of *relative entropy* frequently used in the analysis of conservation laws. For technical reasons that become evident in the forthcoming part of this chapter, relative energy seems better adapted to the rather poor stability bounds available for the Euler system. The possibility to compute effectively the *time evolution* of the relative energy on the sole basis of the weak formulation of the Euler system is the key ingredient of the proof of weak-strong uniqueness **(WS)** property. The relative energy for the barotropic Euler system is formally defined as

$$E\left(\varrho, \boldsymbol{m} \,\middle|\, \widetilde{\varrho}, \widetilde{\boldsymbol{u}}\right) = \frac{1}{2}\left(\frac{|\boldsymbol{m}|^2}{\varrho} - 2\boldsymbol{m}\cdot\widetilde{\boldsymbol{u}} + \varrho|\widetilde{\boldsymbol{u}}|^2\right) + P(\varrho) - P'(\widetilde{\varrho})(\varrho - \widetilde{\varrho}) - P(\widetilde{\varrho}). \tag{4.25}$$

Here $[\varrho, \boldsymbol{m}]$ represents an admissible weak solution of the barotropic Euler system, while $\widetilde{\varrho}, \widetilde{\boldsymbol{u}}$ play a role of "test" functions. The first expression is rather awkward and should be interpreted as follows: As the energy of the weak solution $[\varrho, \boldsymbol{m}]$ is finite, we have

$$\frac{1}{2}\left(\frac{|\boldsymbol{m}|^2}{\varrho} - 2\boldsymbol{m}\cdot\widetilde{\boldsymbol{u}} + \varrho|\widetilde{\boldsymbol{u}}|^2\right) = \frac{1}{2}1_{\varrho>0}\left(\frac{|\boldsymbol{m}|^2}{\varrho} - 2\boldsymbol{m}\cdot\widetilde{\boldsymbol{u}} + \varrho|\widetilde{\boldsymbol{u}}|^2\right) = 1_{\varrho>0}\frac{1}{2}\varrho|\boldsymbol{u} - \widetilde{\boldsymbol{u}}|^2, \tag{4.26}$$

where we have set $\boldsymbol{u} = \frac{\boldsymbol{m}}{\varrho}$ on the set where $\varrho > 0$. Thus if P is convex, meaning $p' \geq 0$, and $[\varrho, \boldsymbol{m}]$ represent an admissible weak solution, then $E(\varrho, \boldsymbol{m}|\widetilde{\varrho}, \widetilde{\boldsymbol{u}}) \geq 0$ a.a. in $(0, T) \times \Omega$.

Denoting

$$E(\varrho, \boldsymbol{m}) = \frac{1}{2}\frac{|\boldsymbol{m}|^2}{\varrho} + P(\varrho)$$

we can be interpret the relative energy as the *Bregman distance*

$$E\left(\varrho, \boldsymbol{m} \middle| \widetilde{\varrho}, \widetilde{\boldsymbol{u}}\right) = \mathfrak{B}_E^{\xi}\left(\varrho, \boldsymbol{m} \middle| \widetilde{\varrho}, \widetilde{\boldsymbol{m}}\right) = E(\varrho, \boldsymbol{m}) - \xi \cdot (\varrho - \widetilde{\varrho}, \boldsymbol{m} - \widetilde{\boldsymbol{m}}) - E(\widetilde{\varrho}, \widetilde{\boldsymbol{m}}),$$
$$\widetilde{\boldsymbol{m}} = \widetilde{\varrho}\widetilde{\boldsymbol{u}},\ \xi \in \partial E(\widetilde{\varrho}, \widetilde{\boldsymbol{m}}),$$

associated to the convex potential $E = E[\varrho, \boldsymbol{m}]$, see e.g. Sprung [185]. Bregman distance is not symmetric, therefore not a proper metric. However, the following holds:

$$E\left(\varrho, \boldsymbol{m} \middle| \widetilde{\varrho}, \widetilde{\boldsymbol{u}}\right) \geq 0 \text{ and } E\left(\varrho, \boldsymbol{m} \middle| \widetilde{\varrho}, \widetilde{\boldsymbol{u}}\right) = 0 \ \Leftrightarrow\ \varrho = \widetilde{\varrho},\ \boldsymbol{m} = \widetilde{\varrho}\widetilde{\boldsymbol{u}}.$$

Remark 4.6 Strictly speaking, the relative energy should be written entirely in conservative variables,

$$E\left(\varrho, \boldsymbol{m} \middle| \widetilde{\varrho}, \widetilde{\boldsymbol{m}}\right) = \frac{1}{2}\left(\frac{|\boldsymbol{m}|^2}{\varrho} - 2\boldsymbol{m} \cdot \frac{\widetilde{\boldsymbol{m}}}{\widetilde{\varrho}} + \frac{\varrho}{\widetilde{\varrho}^2}|\widetilde{\boldsymbol{m}}|^2\right) + P(\varrho) - P'(\widetilde{\varrho})(\varrho - \widetilde{\varrho}) - P(\widetilde{\varrho}),$$

which is consistent with its interpretation as Bregman distance. Considering the test functions in the standard variables, however, is more suitable in the applications, notably in the proof of weak-strong uniqueness.

Our goal is to compute

$$\left[\int_\Omega E\left(\varrho, \boldsymbol{m} \middle| \widetilde{\varrho}, \widetilde{\boldsymbol{u}}\right) \,\mathrm{d}x\right]_{t=0}^{t=\tau} = \left[\int_\Omega \left[\frac{1}{2}\frac{|\boldsymbol{m}|^2}{\varrho} + P(\varrho)\right] \,\mathrm{d}x\right]_{t=0}^{t=\tau} - \left[\int_\Omega \boldsymbol{m} \cdot \widetilde{\boldsymbol{u}} \,\mathrm{d}x\right]_{t=0}^{t=\tau}$$
$$+ \left[\int_\Omega \varrho\left(\frac{1}{2}|\widetilde{\boldsymbol{u}}|^2 - P'(\widetilde{\varrho})\right) \,\mathrm{d}x\right]_{t=0}^{t=\tau} + \left[\int_\Omega p(\widetilde{\varrho}) \,\mathrm{d}x\right]_{t=0}^{t=\tau}.$$

Remarkably, all integrals on the right-hand side can be evaluated by means of the weak formulation as soon as $\varphi = \left(\frac{1}{2}|\widetilde{\boldsymbol{u}}|^2 - P'(\widetilde{\varrho})\right)$ can be taken as a test function in (4.17), and $\boldsymbol{\varphi} = \widetilde{\boldsymbol{u}}$ as a test function in (4.18). To this end, we impose an extra hypothesis

$$\widetilde{\boldsymbol{u}} \in C^1([0, T] \times \overline{\Omega}; R^d),\ \widetilde{\boldsymbol{u}} \cdot \boldsymbol{n}|_{\partial\Omega} = 0,\ \widetilde{\varrho} \in C^1([0, T] \times \overline{\Omega}),\ \widetilde{\varrho} > 0 \text{ in } [0, T] \times \overline{\Omega}. \tag{4.27}$$

Accordingly, we have

$$\left[\int_{\Omega} \left[\frac{1}{2} \frac{|\boldsymbol{m}|^2}{\varrho} + P(\varrho) \right] \, \mathrm{d}x \right]_{t=0}^{t=\tau} \leq 0, \tag{4.28}$$

$$\left[\int_{\Omega} \boldsymbol{m} \cdot \widetilde{\boldsymbol{u}} \, \mathrm{d}x \right]_{t=0}^{t=\tau} = \int_0^\tau \int_{\Omega} \left[\boldsymbol{m} \cdot \partial_t \widetilde{\boldsymbol{u}} + \left(1_{\varrho>0} \frac{\boldsymbol{m} \otimes \boldsymbol{m}}{\varrho} \right) : \nabla_x \widetilde{\boldsymbol{u}} + p(\varrho) \mathrm{div}_x \widetilde{\boldsymbol{u}} \right] \, \mathrm{d}x \, \mathrm{d}t, \tag{4.29}$$

and

$$\begin{aligned} \left[\int_{\Omega} \varrho \left(\frac{1}{2} |\widetilde{\boldsymbol{u}}|^2 - P'(\widetilde{\varrho}) \right) \, \mathrm{d}x \right]_{t=0}^{t=\tau} &= \int_0^\tau \int_{\Omega} \left[\varrho \widetilde{\boldsymbol{u}} \cdot \partial_t \widetilde{\boldsymbol{u}} + \boldsymbol{m} \cdot \widetilde{\boldsymbol{u}} \cdot \nabla_x \widetilde{\boldsymbol{u}} \right] \, \mathrm{d}x \, \mathrm{d}t \\ &\quad - \int_0^\tau \int_{\Omega} \left[\varrho P''(\widetilde{\varrho}) \partial_t \widetilde{\varrho} + P''(\widetilde{\varrho}) \boldsymbol{m} \cdot \nabla_x \widetilde{\varrho} \right] \, \mathrm{d}x \, \mathrm{d}t \\ &= \int_0^\tau \int_{\Omega} \left[\varrho \widetilde{\boldsymbol{u}} \cdot \partial_t \widetilde{\boldsymbol{u}} + 1_{\varrho>0} \frac{\boldsymbol{m}}{\varrho} \cdot \varrho \widetilde{\boldsymbol{u}} \cdot \nabla_x \widetilde{\boldsymbol{u}} \right] \, \mathrm{d}x \, \mathrm{d}t \\ &\quad - \int_0^\tau \int_{\Omega} \left[\varrho P''(\widetilde{\varrho}) \partial_t \widetilde{\varrho} + P''(\widetilde{\varrho}) \boldsymbol{m} \cdot \nabla_x \widetilde{\varrho} \right] \, \mathrm{d}x \, \mathrm{d}t. \end{aligned}$$

Regrouping several terms we obtain, after a straightforward manipulation, the *relative energy inequality* for the barotropic Euler system,

$$\begin{aligned} \left[\int_{\Omega} E \left(\varrho, \boldsymbol{m} \,\middle|\, \widetilde{\varrho}, \widetilde{\boldsymbol{u}} \right) \, \mathrm{d}x \right]_{t=0}^{t=\tau} &\leq - \int_0^\tau \int_{\Omega} 1_{\varrho>0} \varrho \nabla_x \widetilde{\boldsymbol{u}} \cdot \left(\frac{\boldsymbol{m}}{\varrho} - \widetilde{\boldsymbol{u}} \right) \cdot \left(\frac{\boldsymbol{m}}{\varrho} - \widetilde{\boldsymbol{u}} \right) \, \mathrm{d}x \, \mathrm{d}t \\ &- \int_0^\tau \int_{\Omega} \left[p(\varrho) - p'(\widetilde{\varrho})(\varrho - \widetilde{\varrho}) - p(\widetilde{\varrho}) \right] \mathrm{div}_x \widetilde{\boldsymbol{u}} \, \mathrm{d}x \, \mathrm{d}t \\ &+ \int_0^\tau \int_{\Omega} \frac{1}{\widetilde{\varrho}} (\varrho \widetilde{\boldsymbol{u}} - \boldsymbol{m}) \cdot \left[\partial_t (\widetilde{\varrho} \widetilde{\boldsymbol{u}}) + \mathrm{div}_x \left(\widetilde{\varrho} \widetilde{\boldsymbol{u}} \otimes \widetilde{\boldsymbol{u}} \right) + \nabla_x p(\widetilde{\varrho}) \right] \, \mathrm{d}x \, \mathrm{d}t \\ &+ \int_0^\tau \int_{\Omega} \left[\left(1 - \frac{\varrho}{\widetilde{\varrho}} \right) p'(\widetilde{\varrho}) + \frac{1}{\widetilde{\varrho}} \widetilde{\boldsymbol{u}} \cdot (\boldsymbol{m} - \varrho \widetilde{\boldsymbol{u}}) \right] \left[\partial_t \widetilde{\varrho} + \mathrm{div}_x (\widetilde{\varrho} \widetilde{\boldsymbol{u}}) \right] \, \mathrm{d}x \, \mathrm{d}t. \end{aligned} \tag{4.30}$$

Now, it is clear how the weak-strong uniqueness property **(WS)** can be deduced from (4.30). On condition that the problem admits a strong solution $\widetilde{\varrho}$, $\widetilde{\varrho} > 0$, and $\widetilde{\boldsymbol{m}} = \widetilde{\varrho}\widetilde{\boldsymbol{u}}$ belonging to the class (4.27), the quantities $\widetilde{\varrho}, \widetilde{\boldsymbol{u}}$ can be used as test functions in (4.30). As $\widetilde{\varrho}$ and $\widetilde{\boldsymbol{u}}$ represent a strong solution, the last two integrals vanish while the first two can be "absorbed" by the left-hand side via a Gronwall type argument. We postpone the proof to the end of this section. In Chap. 5 we actually show a more general result.

Going back to Definition 4.1, we observe that the class of strong solutions is slightly larger than (4.27). It is therefore desirable to extend validity of the relative energy inequality (4.30) to a larger class of functions. Given the rather limited available integrability of the weak solutions, the optimal result in this direction is to allow $\widetilde{\varrho}$ and $\widetilde{\boldsymbol{u}}$ to be only Lipschitz continuous. We need the following auxiliary result.

Lemma 4.1 *Let $\Omega \subset R^d$ be a bounded domain of class C^2. Suppose that*

$$\widetilde{\varrho} \in W^{1,\infty}((0,T)\times\Omega), \quad \inf_{t\in(0,T), x\in\Omega} \widetilde{\varrho}(t,x) > 0,$$
$$\widetilde{\boldsymbol{u}} \in W^{1,\infty}((0,T)\times\Omega; R^d), \ \widetilde{\boldsymbol{u}}\cdot\boldsymbol{n}|_{\partial\Omega} = 0.$$

Then there exist sequences

$$\{\varrho_n\}_{n=1}^\infty, \ \varrho_n \in C^1([0,T]\times\overline{\Omega}), \ \inf_{x\in\Omega} \varrho_n(x) > 0 \text{ uniformly for } n = 1,2,\dots,$$
$$\{\boldsymbol{u}_n\}_{n=1}^\infty, \ \boldsymbol{u}_n \in C^1([0,T]\times\overline{\Omega}; R^d), \ \boldsymbol{u}_n\cdot\boldsymbol{n}|_{\partial\Omega} = 0 \text{ for all } n = 1,2,\dots$$

such that

$$\varrho_n \to \widetilde{\varrho} \text{ in } W^{1,p}((0,T)\times\Omega) \text{ for any } 1 \le p < \infty$$
$$\text{and weakly-(*) in } W^{1,\infty}((0,T)\times\Omega);$$
$$\boldsymbol{u}_n \to \widetilde{\boldsymbol{u}} \text{ in } W^{1,p}((0,T)\times\Omega; R^d) \text{ for any } 1 \le p < \infty,$$
$$\text{and weakly-(*) in } W^{1,\infty}((0,T)\times\Omega; R^d).$$

Moreover,

$$\begin{aligned} \|\nabla_x\varrho_n\|_{L^\infty((0,T)\times\Omega;R^d)} &\le c\left(\Omega; \|\nabla_x\widetilde{\varrho}\|_{L^\infty((0,T)\times\Omega;R^d)}\right), \\ \|\nabla_x\boldsymbol{u}_n\|_{L^\infty((0,T)\times\Omega;R^{d\times d})} &\le c\left(\Omega; \|\nabla_x\widetilde{\boldsymbol{u}}\|_{L^\infty((0,T)\times\Omega;R^{d\times d})}\right) \end{aligned} \tag{4.31}$$

uniformly for $n \to \infty$.

Proof **Step 1:**

As $\widetilde{\varrho}$ and $\widetilde{\boldsymbol{u}}$ are globally Lipschitz on $[0,T]\times\overline{\Omega}$, they can be extended to the whole space R^{d+1} in such a way that

$$\widetilde{\varrho} \in W^{1,\infty}(R^{d+1}), \ \widetilde{\boldsymbol{u}} \in W^{1,\infty}(R^{d+1}; R^d).$$

Using the standard regularization procedure, we may construct a sequence $\varrho_n \in C^\infty(R^{d+1})$ such that

$$\|\varrho_n\|_{W^{1,\infty}(K)} \leq c(K) \text{ uniformly for } n \to \infty$$

$$\varrho_n \to \widetilde{\varrho} \text{ in } C(K),\ \partial_t \varrho_n \to \partial_t \widetilde{\varrho},\ \nabla_x \varrho_n \to \nabla_x \widetilde{\varrho} \text{ a.a. in } K$$

for any compact $K \subset R^d$, which completes the proof for $\widetilde{\varrho}$. As a matter of fact, this procedure can carried over on arbitrary bounded domain as the extension theorem holds.

Step 2:

Obviously, the same treatment can be applied to $\widetilde{\boldsymbol{u}}$, however, we have to preserve the property of zero normal trace for the approximate sequence. First, we observe that regularization in time can be done in the same way as above. To simplify the proof, we shall therefore suppose that $\widetilde{\boldsymbol{u}} = \widetilde{\boldsymbol{u}}(x)$ is a function of x only.

As the boundary is of class C^2, there is an open neighborhood $\mathcal{U}$ of $\partial\Omega$ such that the distance function

$$d(x) = \text{dist}[x, \partial\Omega] \text{ is of class } C^2(\overline{\mathcal{U}}).$$

Next, write

$$\widetilde{\boldsymbol{u}} = \boldsymbol{v} + \boldsymbol{w}, \text{ where } \boldsymbol{v} \in W^{1,\infty} \cap C_c(\Omega; R^d),\ \boldsymbol{w} \in W^{1,\infty} \cap C_c(\overline{\Omega} \cap \mathcal{U}; R^d).$$

The function $\boldsymbol{v}$ can be approximated in the same way as in **Step 1**.

Finally, we write $\boldsymbol{w}$ as

$$\boldsymbol{w} = [\boldsymbol{w} \cdot \nabla_x d]\nabla_x d - \nabla_x d \times (\nabla_x d \times \boldsymbol{w}).$$

Note that $\nabla_x d$ is a unit vector of class $C^1(\mathcal{U}; R^d)$,

$$\nabla_x d(x) = -\boldsymbol{n}(x) \text{ for } x \in \partial\Omega. \tag{4.32}$$

Thus applying the approximation procedure of **Step 1** to $\boldsymbol{w}$, we obtain a sequence $\{\boldsymbol{w}_n\}_{n=1}^\infty$ of class C^1,

$$\nabla_x d \times (\nabla_x d \times \boldsymbol{w}_n) \to \nabla_x d \times (\nabla_x d \times \boldsymbol{w}) \text{ in } W^{1,p}(\Omega; R^d) \text{ for any } 1 \leq p < \infty,$$
$$\text{and weakly-(*) in } W^{1,\infty}(\Omega; R^d).$$

Obviously, $\nabla_x d \times (\nabla_x d \times \boldsymbol{w}_n) \cdot \boldsymbol{n}|_{\partial\Omega} = 0$ for any $n = 1, 2, \ldots$.

As the normal component of $\boldsymbol{w}$ vanishes on $\partial\Omega$, we have

$$w^N = (\boldsymbol{w} \cdot \nabla_x d) \text{ in } W_0^{1,\infty}(\mathcal{U} \cap \Omega).$$

Consequently, there is a sequence of smooth functions $W_n^N \in C_c^\infty(\mathcal{U} \cap \Omega)$ such that

$$w_n^N \to W^N \text{ in } W_0^{1,p}(\mathcal{U} \cap \Omega) \text{ and weakly-(*) in} W_0^{1,\infty}(\mathcal{U} \cap \Omega)$$

We therefore conclude

$$w_n^N \nabla_x d \to [\mathbf{w} \cdot \nabla_x d] \nabla_x d \text{ in } W_0^{1,p}(\mathcal{U} \cap \Omega; R^d) \text{ and weakly-(*) in} W_0^{1,\infty}(\mathcal{U} \cap \Omega; R^d).$$

□

4.1.3.1 Relative Energy Inequality

In view of Lemma 4.1, we may summarize the previous discussion in the following statement.

Theorem 4.1 (Relative energy inequality for barotropic Euler system) *Let $\Omega \subset R^d$, $d = 2, 3$, be a bounded domain of class C^2. Let ϱ, $\mathbf{m}$ be an admissible weak solution of the barotropic Euler system* (4.10), (4.11) *in* $(0, T) \times \Omega$, *with the impermeability boundary condition* (4.5), *in the sense of Definition 4.1, where $p \in C^1[0, \infty) \cap C^2(0, \infty)$.*

Let $\widetilde{\varrho}$, $\widetilde{\mathbf{u}}$ be (test) functions belonging to the class

$$\begin{aligned} &\widetilde{\varrho} \in W^{1,\infty}((0, T) \times \Omega), \quad \inf_{(t,x) \in (0,T) \times \Omega} \widetilde{\varrho}(t, x) > 0, \\ &\widetilde{\mathbf{u}} \in W^{1,\infty}((0, T) \times \Omega; R^d), \ \widetilde{\mathbf{u}} \cdot \mathbf{n}|_{\partial\Omega} = 0. \end{aligned} \tag{4.33}$$

Then the relative energy inequality

$$\begin{aligned} \left[\int_\Omega E\left(\varrho, \mathbf{m} \middle| \widetilde{\varrho}, \widetilde{\mathbf{u}}\right) \, \mathrm{d}x \right]_{t=0}^{t=\tau} &\leq - \int_0^\tau \int_\Omega 1_{\varrho > 0} \varrho \nabla_x \widetilde{\mathbf{u}} \cdot \left(\frac{\mathbf{m}}{\varrho} - \widetilde{\mathbf{u}} \right) \cdot \left(\frac{\mathbf{m}}{\varrho} - \widetilde{\mathbf{u}} \right) \, \mathrm{d}x \, \mathrm{d}t \\ &- \int_0^\tau \int_\Omega \left[p(\varrho) - p'(\widetilde{\varrho})(\varrho - \widetilde{\varrho}) - p(\widetilde{\varrho}) \right] \mathrm{div}_x \widetilde{\mathbf{u}} \, \mathrm{d}x \, \mathrm{d}t \\ &+ \int_0^\tau \int_\Omega \frac{1}{\widetilde{\varrho}} (\varrho \widetilde{\mathbf{u}} - \mathbf{m}) \cdot \left[\partial_t (\widetilde{\varrho} \widetilde{\mathbf{u}}) + \mathrm{div}_x \left(\widetilde{\varrho} \widetilde{\mathbf{u}} \otimes \widetilde{\mathbf{u}} \right) + \nabla_x p(\widetilde{\varrho}) \right] \, \mathrm{d}x \, \mathrm{d}t \\ &+ \int_0^\tau \int_\Omega \left[\left(1 - \frac{\varrho}{\widetilde{\varrho}} \right) p'(\widetilde{\varrho}) + \frac{1}{\widetilde{\varrho}} \widetilde{\mathbf{u}} \cdot (\mathbf{m} - \varrho \widetilde{\mathbf{u}}) \right] \left[\partial_t \widetilde{\varrho} + \mathrm{div}_x (\widetilde{\varrho} \widetilde{\mathbf{u}}) \right] \, \mathrm{d}x \, \mathrm{d}t. \end{aligned} \tag{4.34}$$

holds for any $0 \leq \tau \leq T$.

Remark 4.7 The previous result can be extended to unbounded domains, and, obviously, to the case of periodic boundary conditions $\Omega = \mathbb{T}^d$.

Remark 4.8 As a matter of fact, the total energy need not be a nonincreasing function as required in (4.19). For the relative energy inequality to hold, it is enough that

$$\int_{\Omega} \left[\frac{1}{2} \frac{|\boldsymbol{m}|^2}{\varrho} + P(\varrho) \right] (\tau, \cdot) \, \mathrm{d}x \leq \int_{\Omega} \left[\frac{1}{2} \frac{|\boldsymbol{m}_0|^2}{\varrho_0} + P(\varrho_0) \right] \, \mathrm{d}x$$

for any $0 \leq \tau \leq T$.

4.1.3.2 Weak-Strong Uniqueness

The weak-strong uniqueness property **(WS)** is a straightforward corollary of the relative energy inequality established in Theorem 4.1. As a matter of fact, we establish a more general result later in Chap. 5.

Theorem 4.2 (Weak-strong uniqueness) *Let $\Omega \subset R^d$, $d = 2, 3$, be a bounded domain of class C^2. Let ϱ, $\boldsymbol{m}$ be an admissible weak solution of the barotropic Euler system* (4.10), (4.11) *in $(0, T) \times \Omega$, with the impermeability boundary condition* (4.5), *in the sense of Definition 4.2. Let the pressure p satisfy the growth condition* (4.20). *Let $\widetilde{\varrho}$, $\widetilde{\boldsymbol{u}}$ be a strong solution of the same problem belonging to the class*

$$\widetilde{\varrho} \in W^{1,\infty}((0, T) \times \Omega), \quad \inf_{(t,x) \in (0,T) \times \Omega} \widetilde{\varrho}(t, x) > 0,$$

$$\widetilde{\boldsymbol{u}} \in W^{1,\infty}((0, T) \times \Omega; R^d), \ \widetilde{\boldsymbol{u}} \cdot \boldsymbol{n}|_{\partial\Omega} = 0,$$

and such that

$$\varrho(0, \cdot) = \widetilde{\varrho}(0, \cdot), \ \boldsymbol{m}(0, \cdot) = \widetilde{\varrho}(0, \cdot)\widetilde{\boldsymbol{u}}(0, \cdot).$$

Then

$$\varrho = \widetilde{\varrho}, \ \boldsymbol{m} = \widetilde{\varrho}\widetilde{\boldsymbol{u}} \ \textit{in} \ (0, T) \times \Omega.$$

Proof The proof is an easy application of the relative energy inequality (4.34) in combination with the standard Gronwall type argument. Indeed plugging the strong solution in (4.34) we have only to observe that

$$\begin{aligned} &\left| \int_{\Omega} 1_{\varrho > 0} \varrho \nabla_x \widetilde{\boldsymbol{u}} \cdot \left(\frac{\boldsymbol{m}}{\varrho} - \widetilde{\boldsymbol{u}} \right) \cdot \left(\frac{\boldsymbol{m}}{\varrho} - \widetilde{\boldsymbol{u}} \right) \, \mathrm{d}x \right| \\ &+ \left| \int_{\Omega} \left[p(\varrho) - p'(\widetilde{\varrho})(\varrho - \widetilde{\varrho}) - p(\widetilde{\varrho}) \right] \mathrm{div}_x \widetilde{\boldsymbol{u}} \, \mathrm{d}x \right| \\ &\stackrel{<}{\sim} c \left(\|\nabla_x \widetilde{\boldsymbol{u}}\|_{L^\infty}, \|\widetilde{\varrho}\|_{L^\infty} \right) \int_{\Omega} E \left(\varrho, \boldsymbol{m} \,\middle|\, \widetilde{\varrho}, \widetilde{\boldsymbol{u}} \right) \, \mathrm{d}x. \end{aligned} \tag{4.35}$$

As the pressure satisfies (4.20), the pressure potential P is strictly convex; whence

$$|p(\varrho) - p'(\widetilde{\varrho})(\varrho - \widetilde{\varrho}) - p(\widetilde{\varrho})| \leq c\,(\min \widetilde{\varrho}, \max \widetilde{\varrho}) \Big(P(\varrho) - P'(\widetilde{\varrho})(\varrho - \widetilde{\varrho}) - P(\widetilde{\varrho}) \Big) \tag{4.36}$$

whenever

$$\varrho \in \left[\frac{1}{2} \min_{(0,T)\times\Omega} \widetilde{\varrho}, 2 \max_{(0,T)\times\Omega} \widetilde{\varrho} \right].$$

Next, by the same token,

$$\inf_{0 \leq \varrho \leq \frac{1}{2} \min_{(0,T)\times\Omega} \widetilde{\varrho}} \Big(P(\varrho) - P'(\widetilde{\varrho})(\varrho - \widetilde{\varrho}) - P(\widetilde{\varrho}) \Big) > 0$$

and, by virtue of (4.21), the inequality (4.36) holds also in the regime $\varrho \geq 2 \max_{(0,T)\times\Omega} \widetilde{\varrho}$. Clearly (4.36) implies (4.35). □

Remark 4.9 (Domain regularity) The assumption on regularity of the spatial domain Ω may seem rather restrictive, in particular in numerical applications, where the underlying domain is typically a polygon. Note, however, that *existence* of a smooth solution requires similar restrictions. Fortunately, the problem is irrelevant in the case of periodic boundary conditions.

4.1.3.3 Relative Energy—Summary

With future applications of the method in mind, we conclude by collecting the principal steps of the proof of the weak-strong **(WS)** uniqueness principle:

- **Conservative state variables**. Identify a suitable set *conservative* state variables, the time evolution of which can be expressed in terms of the weak formulation. These quantities enjoy certain kind of continuity in the time variable.
- **Convex energy**. Express the total energy as a *convex* function of these state variables.
- **Relative energy**. Relative energy is the (integrated) *Bregman distance* associated to the energy between a weak solutions and suitable test functions. Its time evolution can be identified by means of the energy balance and the weak formulation of the field equations. The energy balance is *indispensable* at this stage; whence only admissible weak solutions are eligible.
- **Weak-strong uniqueness principle**. Use the strong solution as a test function in the relative energy inequality.

4.1.4 Weak (Distributional) Solutions to the Complete Euler System

The weak solutions to the complete Euler system have been introduced in Sect. 2.1.3. Unfortunately, however, the bounds provided by the energy balance Eq. (4.4), are not strong enough to control the energy flux

$$\left(\frac{1}{2}\varrho|\boldsymbol{u}|^2 + \varrho e\right)\boldsymbol{u} + p\boldsymbol{u}$$

in the space of integrable functions. This is a serious obstacle when dealing with sequences of approximate solutions since the limit in the energy equation is problematic. From this point of view, it seems more convenient to work with *entropy* rather than the *energy* balance. Indeed the renormalized entropy equation (4.9), written in the form

$$\partial_t\left(\varrho Z(s)\right) + \mathrm{div}_x\left(\varrho Z(s)\boldsymbol{u}\right) = 0, \tag{4.37}$$

requires only ϱ and $\boldsymbol{m} = \varrho\boldsymbol{u}$ to be integrable as soon as Z is a bounded function. We can therefore consider the system consisting of the equation of continuity (4.6), the momentum equation (4.7), and the entropy balance (4.37). Similarly to the barotropic case, the system can be appended by the total energy inequality

$$\frac{\mathrm{d}}{\mathrm{d}t}\int_\Omega \left[\frac{1}{2}\varrho|\boldsymbol{u}|^2 + \varrho e\right] \mathrm{d}x \leq 0, \ \text{with } e = e(\varrho, s),$$

as an *admissibility condition*. This approach is frequently used in meteorological models that describe fluids in the low Mach number regime, where occurrence of shock waves or other singularities is not expected; whence all formulations are equivalent.

In order to accommodate more general regimes of fluid motion, in particular those when the entropy is not conserved, we propose the following problem as a basis for the weak formulation

$$\partial_t\varrho + \mathrm{div}_x\boldsymbol{m} = 0, \tag{4.38}$$

$$\partial_t\boldsymbol{m} + \mathrm{div}_x\left(\frac{\boldsymbol{m}\otimes\boldsymbol{m}}{\varrho}\right) + \nabla_x p = 0, \tag{4.39}$$

$$\partial_t(\varrho Z(s)) + \mathrm{div}_x\left(Z(s)\boldsymbol{m}\right) \geq 0, \ Z'(s) \geq 0, \tag{4.40}$$

together with the admissibility condition

$$\frac{\mathrm{d}}{\mathrm{d}t}\int_\Omega \left[\frac{1}{2}\frac{|\boldsymbol{m}|^2}{\varrho} + \varrho e\right] \mathrm{d}x \leq 0, \tag{4.41}$$

where $p = p(\varrho, s)$, $e = e(\varrho, s)$ are determined by appropriate EOS. Of course, the energy balance (4.41) is conditioned by the impermeability of the boundary

$$\boldsymbol{m} \cdot \boldsymbol{n}|_{\partial\Omega} = 0. \tag{4.42}$$

The resulting system may be seen underdetermined as we have replaced the entropy *equation* (4.37) by *inequality* (4.40). Moreover, the total energy is no longer conserved which is at odds with the First law of thermodynamics encoded in the complete Euler system. However, as we shall see below, the compatibility property **(C)** as well as the weak-strong uniqueness principle **(WS)** remain in force.

The exact definition of *generalized weak solution* to the complete Euler system presented below can be seen as the first attempt of relaxation of the concept of "standard" weak solution. We formulate the problems in terms of ϱ, $\boldsymbol{m}$, and s for definiteness. Of course, other settings are possible and the modification of the definition is straightforward.

Definition 4.3 (GENERALIZED WEAK SOLUTION TO COMPLETE EULER SYSTEM) A trio $[\varrho, \boldsymbol{m}, s]$ is called *generalized weak solution* of the complete Euler system (4.2)–(4.4) with the impermeability boundary condition (4.5) in $(0, T) \times \Omega$ if the following holds:

- (**weak continuity**) the quantities ϱ, $\boldsymbol{m}$ belong to the class

$$\begin{aligned}
&\varrho \in C_{\rm weak}([0,T]; L^{\gamma}(\Omega)),\ \varrho \geq 0 \text{ a.a. in } (0,T) \times \Omega,\\
&\boldsymbol{m} \in C_{\rm weak}([0,T]; L^{\frac{2\gamma}{\gamma+1}}(\Omega; R^d)),\\
&\varrho Z(s) = S_1^Z + S_2^Z, \text{ where } S_1^Z \in C_{\rm weak}([0,T]; L^{\gamma}(\Omega)),\\
&\tau \mapsto \int_{\Omega} S_2^Z(\tau, \cdot)\phi \ {\rm d}x \text{ nondecreasing for any } \phi \in C^1(\overline{\Omega}),\ \phi \geq 0,
\end{aligned} \tag{4.43}$$

  for some $\gamma > 1$, and for any $Z \in C^1(R)$, $Z' \geq 0$, Z concave, $Z(s) \leq \overline{Z}$ for any s;
- (**constitutive equations**) the pressure $p = p(\varrho, s)$ and the internal energy $e = e(\varrho, s)$ are determined by a given EOS,

$$p \in L^1((0,T) \times \Omega),\ \varrho e \in L^1((0,T) \times \Omega);$$

- (**equation of continuity**) the integral identity

$$\left[\int_{\Omega} \varrho \varphi \ {\rm d}x \right]_{t=0}^{t=\tau} = \int_0^{\tau} \int_{\Omega} \left[\varrho \partial_t \varphi + \boldsymbol{m} \cdot \nabla_x \varphi \right] \ {\rm d}x \, {\rm d}t \tag{4.44}$$

  holds for any $0 \leq \tau \leq T$, and any test function $\varphi \in C^1([0,T] \times \overline{\Omega})$;

- (**momentum equation**) the integral identity

$$\left[\int_{\Omega} \boldsymbol{m} \cdot \boldsymbol{\varphi} \, \mathrm{d}x\right]_{t=0}^{t=\tau} = \int_0^{\tau} \int_{\Omega} \left[\boldsymbol{m} \cdot \partial_t \boldsymbol{\varphi} + \left(1_{\varrho>0} \frac{\boldsymbol{m} \otimes \boldsymbol{m}}{\varrho}\right) : \nabla_x \boldsymbol{\varphi} + p \mathrm{div}_x \boldsymbol{\varphi}\right] \mathrm{d}x \, \mathrm{d}t \tag{4.45}$$

holds for any $0 \leq \tau \leq T$, and any test function $\boldsymbol{\varphi} \in C^1([0, T] \times \overline{\Omega}; R^d)$, $\boldsymbol{\varphi} \cdot \boldsymbol{n}|_{\partial\Omega} = 0$;
- (**entropy inequality**) the integral inequality

$$\left[\int_{\Omega} \varrho Z(s) \varphi \, \mathrm{d}x\right]_{t=0}^{t=\tau+} \geq \int_0^{\tau} \int_{\Omega} \left[\varrho Z(s) \partial_t \varphi + Z(s) \boldsymbol{m} \cdot \nabla_x \varphi\right] \mathrm{d}x \, \mathrm{d}t \tag{4.46}$$

holds for any $0 \leq \tau < T$, any test function $\varphi \in C^1([0, T] \times \overline{\Omega})$, $\varphi \geq 0$, and any $Z \in C^1(R)$, $Z' \geq 0$, Z concave, $Z(s) \leq \overline{Z}$ for any s;
- (**energy inequality**) the integral inequality

$$\left[\int_{\Omega} \left(\frac{1}{2} \frac{|\boldsymbol{m}|^2}{\varrho} + \varrho e\right) \mathrm{d}x\right]_{t=0}^{t=\tau} \leq 0 \tag{4.47}$$

holds for a.a. $0 < \tau < T$.

Remark 4.10 The upper bound $\tau+$ in (4.46) can be replaced by $\tau-$ for any $0 < \tau \leq T$. Note that, in view of (4.43) the one sided limits $\int_{\Omega} \varrho Z(s) \varphi(\tau\pm) \, \mathrm{d}x$ exist for any test function $\varphi \in C^1([0, T] \times \overline{\Omega})$.

We are ready to show the compatibility property (**C**).

Proposition 4.2 (Compatibility) *Suppose that $\Omega \subset R^d$, $d = 2, 3$ is a bounded domain with C^2 boundary. Let $[\varrho, \boldsymbol{m}, s]$ be an admissible weak solution of the Euler system in the sense of Definition 4.3. Let $\vartheta = \vartheta(\varrho, s)$ be given by the (implicit) constitutive relation $s(\varrho, \vartheta) = s$. Suppose that*

$$\mathrm{ess} \inf_{(0,T)\times\Omega} \varrho > 0, \ \mathrm{ess} \inf_{(0,T)\times\Omega} \vartheta > 0$$

and that ϱ, $\boldsymbol{u} = \frac{\boldsymbol{m}}{\varrho}$, ϑ belong to the regularity class of classical solutions specified in Definition 4.1.

Then $[\varrho, \vartheta, \boldsymbol{u}]$ is a classical solution in the sense of Definition 4.1.

Remark 4.11 Given $\varrho > 0$ and $s = s(\varrho, \vartheta)$, the temperature is uniquely determined by the value of the entropy s.

Proof As $\varrho > 0$ we may introduce the velocity $\boldsymbol{u}$ as well as the temperature ϑ (cf. Remark 4.11). Now it is standard to observe that the equation of continuity as well as the momentum equation are satisfied in the classical way:

$$\partial_t \varrho + \mathrm{div}_x(\varrho \boldsymbol{u}) = 0, \tag{4.48}$$

$$\partial_t(\varrho \boldsymbol{u}) + \mathrm{div}_x(\varrho \boldsymbol{u} \otimes \boldsymbol{u}) + \nabla_x p = 0. \tag{4.49}$$

Next, integrating (4.44) by parts and using (4.48), we obtain

$$\int_0^T \int_{\partial\Omega} \varphi \varrho \boldsymbol{u} \cdot \boldsymbol{n} \, \mathrm{dS}_x \, \mathrm{d}t = 0$$

for any $\varphi \in C_c^1((0, T) \times \overline{\Omega})$. This yields

$$\boldsymbol{u} \cdot \boldsymbol{n}|_{\partial\Omega} = 0. \tag{4.50}$$

Multiplying (4.49) on $\boldsymbol{u}$ and using (4.48) we derive the kinetic energy balance

$$\partial_t \left(\frac{1}{2} \varrho |\boldsymbol{u}|^2 \right) + \mathrm{div}_x \left(\frac{1}{2} \varrho |\boldsymbol{u}|^2 \boldsymbol{u} \right) + \mathrm{div}_x(p\boldsymbol{u}) = p\mathrm{div}_x \boldsymbol{u}. \tag{4.51}$$

Furthermore, by virtue of (4.48) and Gibbs' equation (4.1),

$$\frac{1}{\varrho} \partial_t \varrho + \frac{1}{\varrho} \boldsymbol{u} \nabla_x \varrho = -\mathrm{div}_x \boldsymbol{u}, \ \vartheta \frac{\partial s}{\partial \varrho} = \frac{\partial e}{\partial \varrho} - \frac{p}{\varrho^2};$$

whence

$$p\mathrm{div}_x \boldsymbol{u} = -\frac{p}{\varrho} \partial_t \varrho - \frac{p}{\varrho} \boldsymbol{u} \nabla_x \varrho = \varrho \vartheta \frac{\partial s}{\partial \varrho} \partial_t \varrho - \varrho \frac{\partial e}{\partial \varrho} \partial_t \varrho + \varrho \vartheta \frac{\partial s}{\partial \varrho} \boldsymbol{u} \cdot \nabla_x \varrho - \varrho \frac{\partial e}{\partial \varrho} \boldsymbol{u} \cdot \nabla_x \varrho. \tag{4.52}$$

Finally, we deduce from the equation of continuity (4.48) and the entropy inequality (4.46) that

$$\varrho \vartheta \left(\frac{\partial s}{\partial \varrho} \partial_t \varrho + \frac{\partial s}{\partial \vartheta} \partial_t \vartheta \right) + \varrho \vartheta \boldsymbol{u} \cdot \left(\frac{\partial s}{\partial \varrho} \nabla_x \varrho + \frac{\partial s}{\partial \vartheta} \nabla_x \vartheta \right) = \varrho \vartheta \, \partial_t s + \varrho \vartheta \boldsymbol{u} \cdot \nabla_x s \geq 0. \tag{4.53}$$

Combining (4.51)–(4.53) we may infer that

$$\partial_t \left(\frac{1}{2} \varrho |\boldsymbol{u}|^2 + \varrho e \right) + \mathrm{div}_x \left(\frac{1}{2} \varrho |\boldsymbol{u}|^2 \boldsymbol{u} + \varrho e \boldsymbol{u} \right) + \mathrm{div}_x(p\boldsymbol{u}) \geq 0.$$

This inequality, integrated over Ω and the compared with the energy inequality (4.47), yields the total energy balance (4.4). Of course, this step requires the impermeability boundary condition (4.50). □

4.1.5 Lower Bound on the Entropy

As shown in Proposition 2.1, an interesting consequence of the renormalized entropy inequality is a version of *minimum principle* for the entropy. The result transfer directly to the present setting.

Theorem 4.3 (Minimum entropy principle) *Let* $[\varrho, \mathbf{m}, s]$ *be a generalized weak solution to the complete Euler system in* $(0, T) \times \Omega$ *in the sense of Definition 4.3. Suppose that*

$$\varrho Z(s)(0, \cdot) = \varrho_0 Z(s_0), \ \textit{where} \int_\Omega \varrho_0 \, \mathrm{d}x > 0, \ s_0(x) \geq \underline{s} \textit{ for a.a. } x \in \Omega.$$

Then

$$s(t, x) \geq \underline{s} \textit{ a.a. in the set } \left\{ (t, x) \in (0, T) \times \Omega \ \middle| \ \varrho(t, x) > 0 \right\}.$$

4.1.6 Relative Energy for the Complete Euler System

Using the general principles introduced in Sect. 4.1.3, we identify the relative energy for the complete Euler system. We emphasize once more that the quantity we obtain is indeed a relative *energy* and not *entropy*. We give several definitions in terms of the standard, conservative, and entropy-conservative variables. We will finally identify the *entropy-conservative variables* as the only choice of phase variables that renders the total energy convex and agrees with all general principles stated at the end of Sect. 4.1.3. The key property to besides Gibbs' relation to be satisfied by the equation of state is the **hypothesis of thermodynamic stability** formulated in the *standard* variables as follows

$$\frac{\partial p(\varrho, \vartheta)}{\partial \varrho} > 0, \ \frac{\partial e(\varrho, \vartheta)}{\partial \vartheta} > 0. \tag{4.54}$$

The physical meaning of the former condition is positive *compressibility*, while the latter expresses *positivity of the specific heat at constant volume* of the fluid in question.

4.1.6.1 Relative Energy in the Standard Variables

For given $\widetilde{\vartheta} > 0$, we introduce the *ballistic energy* functional,

$$H_{\widetilde{\vartheta}}(\varrho, \vartheta) = \varrho \left(e(\varrho, \vartheta) - \widetilde{\vartheta} s(\varrho, \vartheta) \right).$$

Expressed in terms of the standard variables, the relative energy reads

$$E \left(\varrho, \vartheta, \boldsymbol{m} \middle| \widetilde{\varrho}, \widetilde{\vartheta}, \widetilde{\boldsymbol{u}} \right) = \frac{1}{2} \varrho \left| \frac{\boldsymbol{m}}{\varrho} - \widetilde{\boldsymbol{u}} \right|^2 + H_{\widetilde{\vartheta}}(\varrho, \vartheta) - \frac{\partial H_{\widetilde{\vartheta}}(\widetilde{\varrho}, \widetilde{\vartheta})}{\partial \varrho} (\varrho - \widetilde{\varrho}) - H_{\widetilde{\vartheta}}(\widetilde{\varrho}, \widetilde{\vartheta}). \tag{4.55}$$

Furthermore, we write

$$\begin{aligned} H_{\widetilde{\vartheta}}(\varrho, \vartheta) &- \frac{\partial H_{\widetilde{\vartheta}}(\widetilde{\varrho}, \widetilde{\vartheta})}{\partial \varrho} (\varrho - \widetilde{\varrho}) - H_{\widetilde{\vartheta}}(\widetilde{\varrho}, \widetilde{\vartheta}) \\ &= H_{\widetilde{\vartheta}}(\varrho, \vartheta) - H_{\widetilde{\vartheta}}(\varrho, \widetilde{\vartheta}) + H_{\widetilde{\vartheta}}(\varrho, \widetilde{\vartheta}) - \frac{\partial H_{\widetilde{\vartheta}}(\widetilde{\varrho}, \widetilde{\vartheta})}{\partial \varrho} (\varrho - \widetilde{\varrho}) - H_{\widetilde{\vartheta}}(\widetilde{\varrho}, \widetilde{\vartheta}). \end{aligned}$$

As e and s are interrelated through Gibbs' equation (2.5), a direct manipulation yields

$$\frac{\partial H_{\widetilde{\vartheta}}(\varrho, \vartheta)}{\partial \vartheta} = \frac{\varrho}{\vartheta} (\vartheta - \widetilde{\vartheta}) \frac{\partial e(\varrho, \vartheta)}{\partial \vartheta}$$

and

$$\frac{\partial^2 H_{\widetilde{\vartheta}}(\varrho, \widetilde{\vartheta})}{\partial \varrho^2} = \frac{1}{\varrho} \frac{\partial p(\varrho, \widetilde{\vartheta})}{\partial \varrho}.$$

Consequently,

$$\vartheta \mapsto H_{\widetilde{\vartheta}}(\varrho, \vartheta) - H_{\widetilde{\vartheta}}(\varrho, \widetilde{\vartheta}) + H_{\widetilde{\vartheta}}(\varrho, \widetilde{\vartheta})$$

is a nonnegative function attaining strict minimum at $\vartheta = \widetilde{\vartheta}$ for any fixed ϱ, and

$$\varrho \mapsto H_{\widetilde{\vartheta}}(\varrho, \widetilde{\vartheta}) - \frac{\partial H_{\widetilde{\vartheta}}(\widetilde{\varrho}, \widetilde{\vartheta})}{\partial \varrho} (\varrho - \widetilde{\varrho}) - H_{\widetilde{\vartheta}}(\widetilde{\varrho}, \widetilde{\vartheta}) \text{ is strictly convex.}$$

In particular,

$$E \left(\varrho, \vartheta, \boldsymbol{m} \middle| \widetilde{\varrho}, \widetilde{\vartheta}, \widetilde{\boldsymbol{u}} \right) = 0 \text{ only if } \varrho = \widetilde{\varrho}, \ \vartheta = \widetilde{\vartheta}, \ \boldsymbol{m} = \widetilde{\varrho} \widetilde{\boldsymbol{u}}$$

whenever $\widetilde{\varrho} > 0$. Thus, similarly to its counterpart introduced in Sect. 4.1.3 for the barotropic Euler system, the relative energy represents a "distance" between $[\varrho, \vartheta, \boldsymbol{m}]$ and $[\widetilde{\varrho}, \widetilde{\vartheta}, \widetilde{\varrho} \widetilde{\boldsymbol{u}}]$. Note, however, that the relative energy is definitely not convex with respect to the standard variables $[\varrho, \vartheta, \boldsymbol{m}]$.

4.1.6.2 Relative Energy in the Conservative Variables

Now, we pass to the conservative variables. Note that formula (4.55) is rather awkward, containing the derivatives of the ballistic energy. Seeing that

$$\widetilde{\varrho}\frac{\partial(e-\widetilde{\vartheta}s)(\widetilde{\varrho},\widetilde{\vartheta})}{\partial\varrho}=\frac{p(\widetilde{\varrho},\widetilde{\vartheta})}{\widetilde{\varrho}}$$

we may rewrite the relative energy in the form

$$\begin{aligned} E\left(\varrho,\vartheta,\boldsymbol{m}\Big|\widetilde{\varrho},\widetilde{\vartheta},\widetilde{\boldsymbol{u}}\right) &= \left(\frac{1}{2}\frac{|\boldsymbol{m}|^2}{\varrho}+\varrho e(\varrho,\vartheta)\right)-\left(\frac{1}{2}\widetilde{\varrho}|\widetilde{\boldsymbol{u}}|^2+\widetilde{\varrho}e(\widetilde{\varrho},\widetilde{\vartheta})\right)\\ &\quad+\frac{1}{2}\widetilde{\varrho}|\widetilde{\boldsymbol{u}}|^2-\boldsymbol{m}\cdot\widetilde{\boldsymbol{u}}+\frac{1}{2}\varrho|\widetilde{\boldsymbol{u}}|^2-\widetilde{\vartheta}\left(\varrho s(\varrho,\vartheta)-\widetilde{\varrho}s(\widetilde{\varrho},\widetilde{\vartheta})\right)\\ &\quad-\left(e(\widetilde{\varrho},\widetilde{\vartheta})-\widetilde{\vartheta}s(\widetilde{\varrho},\widetilde{\vartheta})+\frac{p(\widetilde{\varrho},\widetilde{\vartheta})}{\widetilde{\varrho}}\right)(\varrho-\widetilde{\varrho}). \end{aligned} \tag{4.56}$$

Next, we recall the definition of the conservative variables,

$$\varrho,\boldsymbol{m},E=\frac{1}{2}\frac{|\boldsymbol{m}|^2}{\varrho}+\varrho e.$$

Writing $p=p(\varrho,e)$, $s=s(\varrho,e)$ we may use Gibbs' equation (2.5) to compute

$$\frac{\partial s}{\partial e}(\varrho,e)=\frac{1}{\vartheta},\ \frac{\partial s}{\partial\varrho}(\varrho,e)=-\frac{p}{\vartheta\varrho^2}. \tag{4.57}$$

Let

$$S=S(\varrho,\boldsymbol{m},E)=\varrho s\left(\varrho,\frac{1}{\varrho}\left(E-\frac{1}{2}\frac{|\boldsymbol{m}|^2}{\varrho}\right)\right)$$

be the total entropy. With help of (4.57) we compute

$$\begin{aligned} \frac{\partial S(\varrho,\boldsymbol{m},E)}{\partial\varrho} &= s-\frac{p}{\vartheta\varrho}-\frac{E}{\vartheta\varrho}+\frac{1}{\vartheta}\frac{|\boldsymbol{m}|^2}{\varrho^2}=\frac{1}{\vartheta}\left(\vartheta s-\frac{p}{\varrho}-e+\frac{1}{2}\frac{|\boldsymbol{m}|^2}{\varrho^2}\right),\\ \nabla_{\boldsymbol{m}}S(\varrho,\boldsymbol{m},E) &= -\frac{1}{\varrho\vartheta}\boldsymbol{m},\\ \frac{\partial S(\varrho,\boldsymbol{m},E)}{\partial E} &= \frac{1}{\vartheta}. \end{aligned}$$

Setting

$$\tilde{E}=\frac{1}{2}\widetilde{\varrho}|\widetilde{\boldsymbol{u}}|^2+\widetilde{\varrho}e(\widetilde{\varrho},\widetilde{\vartheta}),\ \ \widetilde{\boldsymbol{m}}=\widetilde{\varrho}\widetilde{\boldsymbol{u}},$$

we rewrite the relative energy in the conservative variables

$$
\begin{aligned}
& E\left(\varrho, \boldsymbol{m}, E \middle| \widetilde{\varrho}, \widetilde{\vartheta}, \widetilde{\boldsymbol{u}}\right) \\
&= E - \widetilde{\vartheta} S(\varrho, \boldsymbol{m}, E) - \boldsymbol{m} \cdot \widetilde{\boldsymbol{u}} + \frac{1}{2}\varrho |\widetilde{\boldsymbol{u}}|^2 + p(\widetilde{\varrho}, \widetilde{\vartheta}) \\
&- \left(e(\widetilde{\varrho}, \widetilde{\vartheta}) - \widetilde{\vartheta} s(\widetilde{\varrho}, \widetilde{\vartheta}) + \frac{p(\widetilde{\varrho}, \widetilde{\vartheta})}{\widetilde{\varrho}} \right) \varrho \\
&= -\widetilde{\vartheta} \Big[S(\varrho, \boldsymbol{m}, E) - S(\widetilde{\varrho}, \widetilde{\boldsymbol{m}}, \widetilde{E}) \\
&- \frac{\partial S(\widetilde{\varrho}, \widetilde{\boldsymbol{m}}, \widetilde{E})}{\partial \varrho} (\varrho - \widetilde{\varrho}) - \nabla_{\boldsymbol{m}} S(\widetilde{\varrho}, \widetilde{\boldsymbol{m}}, \widetilde{E}) \cdot (\boldsymbol{m} - \widetilde{\boldsymbol{m}}) - \frac{\partial S(\widetilde{\varrho}, \widetilde{\boldsymbol{m}}, \widetilde{E})}{\partial E} (E - \widetilde{E}) \Big] .
\end{aligned}
\tag{4.58}
$$

Identity (4.58) reveals the intimate relation between the relative energy and relative entropy that differ by a multiplicative factor $\widetilde{\vartheta}$. It also shows that the thermodynamic stability hypothesis (4.54) may be expressed in term of *concavity* of the total entropy S with respect to the conservative variables $[\varrho, \boldsymbol{m}, E]$.

4.1.6.3 Relative Energy in the Conservative-Entropy Variables

The relative energy, expressed in terms of the conservative-entropy variables—the density ϱ, the momentum $\boldsymbol{m}$, and the total entropy $S = \varrho s$—fits in the general framework introduced in the preceding section and may be seen as the Bregman distance associated to the total energy. Indeed returning to (4.58) we obtain

$$
\begin{aligned}
& E\left(\varrho, \boldsymbol{m}, S \middle| \widetilde{\varrho}, \widetilde{\vartheta}, \widetilde{\boldsymbol{u}}\right) \\
&= E - \widetilde{\vartheta} S - \boldsymbol{m} \cdot \widetilde{\boldsymbol{u}} + \frac{1}{2}\varrho |\widetilde{\boldsymbol{u}}|^2 + p(\widetilde{\varrho}, \widetilde{\vartheta}) - \left(e(\widetilde{\varrho}, \widetilde{\vartheta}) - \widetilde{\vartheta} s(\widetilde{\varrho}, \widetilde{\vartheta}) + \frac{p(\widetilde{\varrho}, \widetilde{\vartheta})}{\widetilde{\varrho}} \right) \varrho \\
&= \frac{1}{2}\varrho \left| \frac{\boldsymbol{m}}{\varrho} - \widetilde{\boldsymbol{u}} \right|^2 + \varrho e(\varrho, S) - \left(e(\widetilde{\varrho}, \widetilde{\vartheta}) - \widetilde{\vartheta} s(\widetilde{\varrho}, \widetilde{\vartheta}) + \frac{p(\widetilde{\varrho}, \widetilde{\vartheta})}{\widetilde{\varrho}} \right) (\varrho - \widetilde{\varrho}) \\
&- \widetilde{\vartheta}(S - \widetilde{S}) - \widetilde{\varrho} e(\widetilde{\varrho}, \widetilde{\vartheta}),
\end{aligned}
$$

where we have denoted $\widetilde{S} = \widetilde{\varrho} s(\widetilde{\varrho}, \widetilde{\vartheta})$.

Using Gibbs' relation (2.5) we check easily that

$$
\frac{\partial \left(\varrho e(\varrho, S) \right)}{\partial \varrho} = e(\varrho, S) - \vartheta \frac{S}{\varrho} + \frac{p(\varrho, S)}{\varrho},
$$

and

$$
\frac{\partial \left(\varrho e(\varrho, S) \right)}{\partial S} = \vartheta(\varrho, S).
$$

Consequently, we may infer that

$$
\begin{aligned}
E & \left(\varrho, \boldsymbol{m}, S \middle| \widetilde{\varrho}, \widetilde{\boldsymbol{u}}, \widetilde{S}\right) \\
& = \frac{1}{2}\varrho\left|\frac{\boldsymbol{m}}{\varrho} - \widetilde{\boldsymbol{u}}\right|^2 + \varrho e(\varrho, S) - \frac{\partial(\widetilde{\varrho}e(\widetilde{\varrho}, \widetilde{S}))}{\partial \varrho}(\varrho - \widetilde{\varrho}) \\
& - \frac{\partial(\widetilde{\varrho}e(\widetilde{\varrho}, \widetilde{S}))}{\partial S}(\widetilde{\varrho}, \widetilde{S})(S - \widetilde{S}) - \widetilde{\varrho}e(\widetilde{\varrho}, \widetilde{S}),
\end{aligned}
\tag{4.59}
$$

where we have replaced $\widetilde{\vartheta}$ by $\widetilde{S}$. We may infer that, similarly to the barotropic case, the relative energy expressed in terms of the conservative-entropy variables is the Bregman distance associated to the total energy

$$
E(\varrho, \boldsymbol{m}, S) = \frac{1}{2}\frac{|\boldsymbol{m}|^2}{\varrho} + \varrho e(\varrho, S).
$$

Of course, to see that the relative energy is Bregman distance, we have to rewrite it in terms of the conservative-entropy variables $[\widetilde{\varrho}, \widetilde{\boldsymbol{m}}, \widetilde{S}]$.

4.1.6.4 Thermodynamic Stability

A direct comparison of (4.55), (4.58), and (4.59) reveals equivalent formulation of the **hypothesis of thermodynamic stability**, namely:

- **Standard variables.**

$$
\begin{aligned}
& \frac{p(\varrho, \vartheta)}{\partial \varrho} > 0 \text{ (positive compressibility)}, \\
& \frac{e(\varrho, \vartheta)}{\partial \vartheta} > 0 \text{ (positive specific heat at constant volume)}
\end{aligned}
$$

- **Conservative variables.**

$$
(\varrho, \boldsymbol{m}, E) \mapsto S(\varrho, \boldsymbol{m}, E) \text{ concave}
$$

- **Conservative-entropy variables.**

$$
(\varrho, \boldsymbol{m}, S) \mapsto E(\varrho, \boldsymbol{m}, S) \equiv \frac{1}{2}\frac{|\boldsymbol{m}|^2}{\varrho} + \varrho e(\varrho, S) \text{ convex}
$$

4.1.7 Relative Energy Inequality for the Complete Euler System

Finally, we derive a relative energy inequality for the complete Euler system. Similarly to Theorem 4.1, we use only the weak formulation of the problem specified in Definition 4.3. We start by rewriting (4.56) in the form

$$
\begin{aligned}
&E\left(\varrho, \boldsymbol{m}, s \middle| \widetilde{\varrho}, \widetilde{\vartheta}, \widetilde{\boldsymbol{u}}\right) \\
&= \left[\frac{1}{2}\frac{|\boldsymbol{m}|^2}{\varrho} + \varrho e(\varrho, s)\right] - \widetilde{\vartheta}\varrho\Big[s - \widetilde{s}(\widetilde{\varrho}, \widetilde{\vartheta})\Big] - \boldsymbol{m}\cdot\widetilde{\boldsymbol{u}} + \frac{1}{2}\varrho|\widetilde{\boldsymbol{u}}|^2 \\
&+ p(\widetilde{\varrho}, \widetilde{\vartheta}) - \left(\widetilde{e}(\widetilde{\varrho}, \widetilde{\vartheta}) + \frac{\widetilde{p}(\widetilde{\varrho}, \widetilde{\vartheta})}{\widetilde{\varrho}}\right)\varrho.
\end{aligned}
$$

As we have seen above, there are several choices of phase variables, all of them being essentially equivalent. Here we have opted for $[\varrho, \boldsymbol{m}, s]$ for the weak solutions, while the test functions $[\widetilde{\varrho}, \widetilde{\vartheta}, \widetilde{\boldsymbol{u}}]$ correspond to the standard variables. Note that the above formula formally coincides with its isentropic (barotropic) counterpart (4.25), with $s = \text{const}$, and

$$
p = p(\varrho),\ \varrho e = P(\varrho),\ P'(\varrho)\varrho - P(\varrho) = p(\varrho).
$$

We have deliberately used the different symbols p and $\widetilde{p}$, e and $\widetilde{e}$, s and $\widetilde{s}$ to distinguish between the thermodynamic functions related to the weak solution expressed in terms of ϱ and s and those written in terms of the standard variables $\widetilde{\varrho}$ and $\widetilde{\vartheta}$.

Pursuing step by step the arguments of Sect 4.1.3, we may calculate the time increments

$$
\left[\int_\Omega E\left(\varrho, s, \boldsymbol{m} \middle| \widetilde{\varrho}, \widetilde{\vartheta}, \widetilde{\boldsymbol{u}}\right)(t, \cdot)\ \mathrm{d}x\right]_{t=0}^{t=\tau}
$$

in terms of the weak formulation (4.44)–(4.46) as long as the quantities $[\widetilde{\varrho}, \widetilde{\vartheta}, \widetilde{\boldsymbol{u}}]$ are sufficiently regular to be used as test functions. Similarly to (4.27) we therefore require

$$
\begin{aligned}
&\widetilde{\boldsymbol{u}} \in C^1([0, T] \times \overline{\Omega}; R^d),\ \widetilde{\boldsymbol{u}}\cdot\boldsymbol{n}|_{\partial\Omega} = 0, \\
&\widetilde{\varrho},\ \widetilde{\vartheta} \in C^1([0, T] \times \overline{\Omega}),\ \widetilde{\varrho} > 0,\ \widetilde{\vartheta} > 0 \text{ in } [0, T] \times \overline{\Omega}.
\end{aligned} \tag{4.60}
$$

In virtue of (4.47),

$$
\left[\int_\Omega \left[\frac{1}{2}\frac{|\boldsymbol{m}|^2}{\varrho} + \varrho e\right]\ \mathrm{d}x\right]_{t=0}^{t=\tau} \leq 0
$$

for a.a. $\tau \in (0, T)$. Next, considering $\widetilde{\boldsymbol{u}}$ as a test function in the momentum equation (4.45), we obtain

$$\left[\int_{\Omega} \boldsymbol{m} \cdot \widetilde{\boldsymbol{u}} \, \mathrm{d}x \right]_{t=0}^{t=\tau} = \int_0^{\tau} \int_{\Omega} \left[\boldsymbol{m} \cdot \partial_t \widetilde{\boldsymbol{u}} + \left(1_{\varrho>0} \frac{\boldsymbol{m} \otimes \boldsymbol{m}}{\varrho} \right) : \nabla_x \widetilde{\boldsymbol{u}} + p \mathrm{div}_x \widetilde{\boldsymbol{u}} \right] \, \mathrm{d}x \, \mathrm{d}t.$$

Similarly, it follows from the weak formulation of the equation of continuity (4.44) that

$$\left[\int_{\Omega} \varrho \left(\frac{1}{2} |\widetilde{\boldsymbol{u}}|^2 - \widetilde{e}(\widetilde{\varrho}, \widetilde{\vartheta}) - \frac{\widetilde{p}(\widetilde{\varrho}, \widetilde{\vartheta})}{\widetilde{\varrho}} \right) \, \mathrm{d}x \right]_{t=0}^{t=\tau}$$
$$= \int_0^{\tau} \int_{\Omega} \left[\varrho \widetilde{\boldsymbol{u}} \cdot \partial_t \widetilde{\boldsymbol{u}} + 1_{\varrho>0} \frac{\boldsymbol{m}}{\varrho} \cdot \varrho \widetilde{\boldsymbol{u}} \cdot \nabla_x \widetilde{\boldsymbol{u}} \right] \, \mathrm{d}x \, \mathrm{d}t$$
$$- \int_0^{\tau} \int_{\Omega} \left[\varrho \partial_t \left(\widetilde{e}(\widetilde{\varrho}, \widetilde{\vartheta}) + \frac{\widetilde{p}(\widetilde{\varrho}, \widetilde{\vartheta})}{\widetilde{\varrho}} \right) + \boldsymbol{m} \cdot \nabla_x \left(\widetilde{e}(\widetilde{\varrho}, \widetilde{\vartheta}) + \frac{\widetilde{p}(\widetilde{\varrho}, \widetilde{\vartheta})}{\widetilde{\varrho}} \right) \right] \, \mathrm{d}x \, \mathrm{d}t.$$

Finally, we use the entropy inequality (4.46) with $\widetilde{\vartheta} > 0$ as test function to deduce

$$\left[\int_{\Omega} \varrho Z(s) \widetilde{\vartheta} \, \mathrm{d}x \right]_{t=0}^{t=\tau} \geq \int_0^{\tau} \int_{\Omega} \left[\varrho Z(s) \partial_t \widetilde{\vartheta} + Z(s) \boldsymbol{m} \cdot \nabla_x \widetilde{\vartheta} \right] \, \mathrm{d}x \, \mathrm{d}t$$

for any Z as in (4.46). Moreover, in view of (4.44),

$$\left[\int_{\Omega} \varrho \widetilde{\vartheta} \widetilde{s}(\widetilde{\varrho}, \widetilde{\vartheta}) \, \mathrm{d}x \right]_{t=0}^{t=\tau}$$
$$= \int_0^{\tau} \int_{\Omega} \widetilde{s}(\widetilde{\varrho}, \widetilde{\vartheta}) \left[\varrho \partial_t \widetilde{\vartheta} + \boldsymbol{m} \cdot \nabla_x \widetilde{\vartheta} \right] \, \mathrm{d}x \, \mathrm{d}t$$
$$+ \int_0^{\tau} \int_{\Omega} \widetilde{\vartheta} \left[\varrho \partial_t \widetilde{s}(\widetilde{\varrho}, \widetilde{\vartheta}) + \boldsymbol{m} \cdot \nabla_x \widetilde{s}(\widetilde{\varrho}, \widetilde{\vartheta}) \right] \, \mathrm{d}x \, \mathrm{d}t.$$

Thus, introducing a modified relative energy

$$E_Z\left(\varrho, s, \boldsymbol{m} \middle| \widetilde{\varrho}, \widetilde{\vartheta}, \widetilde{\boldsymbol{u}}\right) = \left[\frac{1}{2}\frac{|\boldsymbol{m}|^2}{\varrho} + \varrho e(\varrho, s)\right] - \widetilde{\vartheta}\varrho\Big[Z(s) - \widetilde{s}(\widetilde{\varrho}, \widetilde{\vartheta})\Big] - \boldsymbol{m}\cdot\widetilde{\boldsymbol{u}}$$
$$+ \frac{1}{2}\varrho|\widetilde{\boldsymbol{u}}|^2 + \widetilde{p}(\widetilde{\varrho}, \widetilde{\vartheta}) - \left(\widetilde{e}(\widetilde{\varrho}, \widetilde{\vartheta}) + \frac{\widetilde{p}(\widetilde{\varrho}, \widetilde{\vartheta})}{\widetilde{\varrho}}\right)\varrho, \tag{4.61}$$

and summing up the preceding calculations, we arrive at a *relative energy inequality* in the form

$$\left[\int_\Omega E_Z\left(\varrho, s, \boldsymbol{m} \middle| \widetilde{\varrho}, \widetilde{\vartheta}, \widetilde{\boldsymbol{u}}\right)\,\mathrm{d}x\right]_{t=0}^{t=\tau}$$
$$\leq -\int_0^\tau\int_\Omega 1_{\varrho>0}\frac{(\varrho\widetilde{\boldsymbol{u}} - \boldsymbol{m})\otimes(\varrho\widetilde{\boldsymbol{u}} - \boldsymbol{m})}{\varrho} : \nabla_x\widetilde{\boldsymbol{u}}\,\mathrm{d}x\,\mathrm{d}t$$
$$-\int_0^\tau\int_\Omega \Big[p(\varrho, s) - \widetilde{p}(\widetilde{\varrho}, \widetilde{\vartheta})\Big]\mathrm{div}_x\widetilde{\boldsymbol{u}}\,\mathrm{d}x\,\mathrm{d}t$$
$$+\int_0^\tau\int_\Omega (\varrho\widetilde{\boldsymbol{u}} - \boldsymbol{m})\cdot\left[\partial_t\widetilde{\boldsymbol{u}} + \widetilde{\boldsymbol{u}}\cdot\nabla_x\widetilde{\boldsymbol{u}} + \frac{1}{\widetilde{\varrho}}\nabla_x\widetilde{p}(\widetilde{\varrho}, \widetilde{\vartheta})\right]\mathrm{d}x\,\mathrm{d}t \tag{4.62}$$
$$-\int_0^\tau\int_\Omega \left[\varrho Z(s)\partial_t\widetilde{\vartheta} + Z(s)\boldsymbol{m}\cdot\nabla_x\widetilde{\vartheta}\right]\mathrm{d}x\,\mathrm{d}t$$
$$+\int_0^\tau\int_\Omega \left[\varrho\widetilde{s}(\widetilde{\varrho}, \widetilde{\vartheta})\partial_t\widetilde{\vartheta} + \widetilde{s}(\widetilde{\varrho}, \widetilde{\vartheta})\boldsymbol{m}\cdot\nabla_x\widetilde{\vartheta}\right]\mathrm{d}x\,\mathrm{d}t$$
$$+\int_0^\tau\int_\Omega \left[(\widetilde{\varrho} - \varrho)\frac{1}{\widetilde{\varrho}}\partial_t\widetilde{p}(\widetilde{\varrho}, \widetilde{\vartheta}) + (\widetilde{\varrho} - \varrho)\frac{1}{\widetilde{\varrho}}\widetilde{\boldsymbol{u}}\cdot\nabla_x\widetilde{p}(\widetilde{\varrho}, \widetilde{\vartheta})\right]\mathrm{d}x\,\mathrm{d}t,$$

where we have used Gibbs' equation (4.1) to handle the terms $p(\widetilde{\varrho}, \widetilde{\vartheta})$, $e(\widetilde{\varrho}, \widetilde{\vartheta})$ and $s(\widetilde{\varrho}, \widetilde{\vartheta})$.

Finally, using Lemma 4.1 we may extend the class of admissible test functions similarly to Theorem 4.1. We are ready to formulate the relative energy inequality for the complete Euler system.

Theorem 4.4 (Relative energy inequality for complete Euler system) *Let $\Omega \subset R^d$, $d = 2, 3$, be a bounded domain of class C^2. Let $[\varrho, \boldsymbol{m}, s]$ be an admissible weak solution of the complete Euler system (4.2)–(4.4), with the impermeability boundary condition (4.5) in $(0, T) \times \Omega$ in the sense of Definition 4.3. Suppose that the thermodynamic functions $p = p(\varrho, \vartheta)$, $e = e(\varrho, \vartheta)$, $s = s(\varrho, \vartheta)$ are of class $C^2(0, \infty)^2$. Let $\widetilde{\varrho}$, $\widetilde{\vartheta}$, $\widetilde{\boldsymbol{u}}$ be (test) functions belonging to the class*

$$
\begin{aligned}
&\widetilde{\varrho},\ \widetilde{\vartheta} \in W^{1,\infty}((0,T)\times\Omega), \quad \inf_{(t,x)\in(0,T)\times\Omega} \widetilde{\varrho}(t,x) > 0, \quad \inf_{(t,x)\in(0,T)\times\Omega} \widetilde{\vartheta}(t,x) > 0, \\
&\widetilde{\boldsymbol{u}} \in W^{1,\infty}((0,T)\times\Omega; R^d),\ \widetilde{\boldsymbol{u}}\cdot\boldsymbol{n}|_{\partial\Omega} = 0.
\end{aligned}
\tag{4.63}
$$

Let Z be as in (4.46), and let E_Z be the relative energy defined through (4.61).

Then the relative energy inequality

$$
\begin{aligned}
&\left[\int_\Omega E_Z\left(\varrho, s, \boldsymbol{m} \middle| \widetilde{\varrho}, \widetilde{\vartheta}, \widetilde{\boldsymbol{u}}\right) \,\mathrm{d}x \right]_{t=0}^{t=\tau} \\
&\leq - \int_0^\tau \int_\Omega 1_{\varrho>0} \frac{(\varrho\widetilde{\boldsymbol{u}} - \boldsymbol{m}) \otimes (\varrho\widetilde{\boldsymbol{u}} - \boldsymbol{m})}{\varrho} : \nabla_x \widetilde{\boldsymbol{u}} \,\mathrm{d}x\,\mathrm{d}t \\
&- \int_0^\tau \int_\Omega \left[p - \widetilde{p}\right] \mathrm{div}_x \widetilde{\boldsymbol{u}} \,\mathrm{d}x\,\mathrm{d}t \\
&+ \int_0^\tau \int_\Omega (\varrho\widetilde{\boldsymbol{u}} - \boldsymbol{m}) \cdot \left[\partial_t \widetilde{\boldsymbol{u}} + \widetilde{\boldsymbol{u}} \cdot \nabla_x \widetilde{\boldsymbol{u}} + \frac{1}{\widetilde{\varrho}} \nabla_x \widetilde{p} \right] \mathrm{d}x\,\mathrm{d}t \\
&- \int_0^\tau \int_\Omega \left[\varrho Z(s) \partial_t \widetilde{\vartheta} + Z(s) \boldsymbol{m} \cdot \nabla_x \widetilde{\vartheta} \right] \mathrm{d}x\,\mathrm{d}t + \int_0^\tau \int_\Omega \left[\varrho \widetilde{s} \partial_t \widetilde{\vartheta} + \widetilde{s} \boldsymbol{m} \cdot \nabla_x \widetilde{\vartheta} \right] \mathrm{d}x\,\mathrm{d}t \\
&+ \int_0^\tau \int_\Omega \left[(\widetilde{\varrho} - \varrho) \frac{1}{\widetilde{\varrho}} \partial_t \widetilde{p} + (\widetilde{\varrho} - \varrho) \frac{1}{\widetilde{\varrho}} \widetilde{\boldsymbol{u}} \cdot \nabla_x \widetilde{p} \right] \mathrm{d}x\,\mathrm{d}t
\end{aligned}
\tag{4.64}
$$

holds for a.a. $\tau \in (0, T)$. *Here we have denoted* $p = p(\varrho, s)$ *the pressure related to the weak solution, while* $\widetilde{p} = p(\widetilde{\varrho}, \widetilde{\vartheta})$, $\widetilde{s} = s(\widetilde{\varrho}, \widetilde{\vartheta})$ *denote thermodynamic functions written in terms of* $\widetilde{\varrho}$ *and* $\widetilde{\vartheta}$.

Relation (4.64) simplifies considerably if written in the conservative-entropy variables $[\varrho, \boldsymbol{m}, S]$, and, accordingly, $[\widetilde{\varrho}, \widetilde{\boldsymbol{u}}, \widetilde{S}]$. Indeed we first observe that

$$\int_0^\tau \int_\Omega \left[(\widetilde{\varrho} - \varrho) \frac{1}{\widetilde{\varrho}} \partial_t \widetilde{p} + (\widetilde{\varrho} - \varrho) \frac{1}{\widetilde{\varrho}} \widetilde{\boldsymbol{u}} \cdot \nabla_x \widetilde{p} \right] \, \mathrm{d}x \, \mathrm{d}t$$
$$= \int_0^\tau \int_\Omega \left[(\widetilde{\varrho} - \varrho) \frac{1}{\widetilde{\varrho}} \frac{\partial p(\widetilde{\varrho}, \widetilde{S})}{\partial \varrho} (\partial_t \widetilde{\varrho} + \widetilde{\boldsymbol{u}} \cdot \nabla_x \widetilde{\varrho}) \right] \, \mathrm{d}x \, \mathrm{d}t$$
$$+ \int_0^\tau \int_\Omega \left[(\widetilde{\varrho} - \varrho) \frac{1}{\widetilde{\varrho}} \frac{\partial p(\widetilde{\varrho}, \widetilde{S})}{\partial S} \left(\partial_t \widetilde{S} + \widetilde{\boldsymbol{u}} \cdot \nabla_x \widetilde{S}\right) \right] \, \mathrm{d}x \, \mathrm{d}t$$
$$= \int_0^\tau \int_\Omega \left[(\varrho - \widetilde{\varrho}) \frac{\partial p(\widetilde{\varrho}, \widetilde{S})}{\partial \varrho} \mathrm{div}_x \widetilde{\boldsymbol{u}} \right] \, \mathrm{d}x \, \mathrm{d}t$$
$$+ \int_0^\tau \int_\Omega \left[(\widetilde{\varrho} - \varrho) \frac{1}{\widetilde{\varrho}} \frac{\partial p(\widetilde{\varrho}, \widetilde{S})}{\partial \varrho} (\partial_t \widetilde{\varrho} + \mathrm{div}_x (\widetilde{\varrho} \widetilde{\boldsymbol{u}})) \right] \, \mathrm{d}x \, \mathrm{d}t$$
$$+ \int_0^\tau \int_\Omega \left[(\widetilde{\varrho} - \varrho) \frac{1}{\widetilde{\varrho}} \frac{\partial p(\widetilde{\varrho}, \widetilde{S})}{\partial S} \left(\partial_t \widetilde{S} + \mathrm{div}_x (\widetilde{\boldsymbol{u}} \widetilde{S})\right) \right] \, \mathrm{d}x \, \mathrm{d}t$$
$$- \int_0^\tau \int_\Omega \left[(\widetilde{\varrho} - \varrho) \frac{\widetilde{S}}{\widetilde{\varrho}} \frac{\partial p(\widetilde{\varrho}, \widetilde{S})}{\partial S} \mathrm{div}_x \widetilde{\boldsymbol{u}} \right] \, \mathrm{d}x \, \mathrm{d}t.$$

Consequently, the inequality (4.64) can be written as

$$
\begin{aligned}
&\left[\int_\Omega E_Z\left(\varrho, S, \boldsymbol{m} \Big| \widetilde{\varrho}, \widetilde{S}, \widetilde{\boldsymbol{u}}\right) \ \mathrm{d}x \right]_{t=0}^{t=\tau} \\
&\leq - \int_0^\tau \int_\Omega 1_{\varrho>0} \frac{(\varrho\widetilde{\boldsymbol{u}} - \boldsymbol{m}) \otimes (\varrho\widetilde{\boldsymbol{u}} - \boldsymbol{m})}{\varrho} : \nabla_x \widetilde{\boldsymbol{u}} \ \mathrm{d}x \, \mathrm{d}t \\
&- \int_0^\tau \int_\Omega \left[p(\varrho, S) - (\varrho - \widetilde{\varrho}) \frac{\partial p(\widetilde{\varrho}, \widetilde{S})}{\partial \varrho} - (S - \widetilde{S}) \frac{\partial p(\widetilde{\varrho}, \widetilde{S})}{\partial S} - p(\widetilde{\varrho}, \widetilde{S}) \right] \mathrm{div}_x \widetilde{\boldsymbol{u}} \ \mathrm{d}x \, \mathrm{d}t \\
&+ \int_0^\tau \int_\Omega (\varrho\widetilde{\boldsymbol{u}} - \boldsymbol{m}) \cdot \left[\partial_t \widetilde{\boldsymbol{u}} + \widetilde{\boldsymbol{u}} \cdot \nabla_x \widetilde{\boldsymbol{u}} + \frac{1}{\widetilde{\varrho}} \nabla_x \widetilde{p} \right] \ \mathrm{d}x \, \mathrm{d}t \\
&+ \int_0^\tau \int_\Omega \left[(\widetilde{\varrho} - \varrho) \frac{1}{\widetilde{\varrho}} \frac{\partial p(\widetilde{\varrho}, \widetilde{S})}{\partial \varrho} \left(\partial_t \widetilde{\varrho} + \mathrm{div}_x (\widetilde{\varrho} \widetilde{\boldsymbol{u}}) \right) \right] \ \mathrm{d}x \, \mathrm{d}t \\
&+ \int_0^\tau \int_\Omega \left[(\widetilde{\varrho} - \varrho) \frac{1}{\widetilde{\varrho}} \frac{\partial p(\widetilde{\varrho}, \widetilde{S})}{\partial S} \left(\partial_t \widetilde{S} + \mathrm{div}_x (\widetilde{\boldsymbol{u}} \widetilde{S}) \right) \right] \ \mathrm{d}x \, \mathrm{d}t \\
&- \int_0^\tau \int_\Omega \left[\varrho Z\left(\frac{S}{\varrho}\right) \partial_t \widetilde{\vartheta} + Z\left(\frac{S}{\varrho}\right) \boldsymbol{m} \cdot \nabla_x \widetilde{\vartheta} \right] \ \mathrm{d}x \, \mathrm{d}t \\
&+ \int_0^\tau \int_\Omega \left[\frac{\varrho}{\widetilde{\varrho}} \widetilde{S} \left(\partial_t \widetilde{\vartheta} + \frac{\boldsymbol{m}}{\varrho} \cdot \nabla_x \widetilde{\vartheta} \right) \right] \ \mathrm{d}x \, \mathrm{d}t \\
&+ \int_0^\tau \int_\Omega \left[\left(\frac{\varrho}{\widetilde{\varrho}} \widetilde{S} - S \right) \frac{\partial p(\widetilde{\varrho}, \widetilde{S})}{\partial S} \mathrm{div}_x \widetilde{\boldsymbol{u}} \right] \ \mathrm{d}x \, \mathrm{d}t,
\end{aligned}
\tag{4.65}
$$

with $\widetilde{S} = \widetilde{\varrho} s(\widetilde{\varrho}, \widetilde{\vartheta})$.

Finally, we let $Z(s) \nearrow s$ obtaining

$$
\begin{aligned}
&\int_0^\tau \int_\Omega \left[\varrho Z\left(\frac{S}{\varrho}\right) \partial_t \widetilde{\vartheta} + Z\left(\frac{S}{\varrho}\right) \boldsymbol{m} \cdot \nabla_x \widetilde{\vartheta} \right] \ \mathrm{d}x \, \mathrm{d}t \\
&\to \int_0^\tau \int_\Omega \left[S \partial_t \widetilde{\vartheta} + S \frac{\boldsymbol{m}}{\varrho} \cdot \nabla_x \widetilde{\vartheta} \right] \ \mathrm{d}x \, \mathrm{d}t,
\end{aligned}
$$

and

$$
E_Z\left(\varrho, S, \boldsymbol{m} \ \Big| \ \widetilde{\varrho}, \widetilde{S}, \widetilde{\boldsymbol{u}}\right) \to E\left(\varrho, S, \boldsymbol{m} \ \Big| \ \widetilde{\varrho}, \widetilde{S}, \widetilde{\boldsymbol{u}}\right).
$$

Consequently,

$$
\begin{aligned}
&-\int_0^\tau \int_\Omega \left[\varrho Z\left(\frac{S}{\varrho}\right) \partial_t \widetilde{\vartheta} + Z\left(\frac{S}{\varrho}\right) \boldsymbol{m} \cdot \nabla_x \widetilde{\vartheta} \right] \mathrm{d}x\, \mathrm{d}t \\
&+\int_0^\tau \int_\Omega \left[\frac{\varrho}{\widetilde{\varrho}} \widetilde{S} \left(\partial_t \widetilde{\vartheta} + \frac{\boldsymbol{m}}{\varrho} \cdot \nabla_x \widetilde{\vartheta} \right) \right] \mathrm{d}x\, \mathrm{d}t \\
&+\int_0^\tau \int_\Omega \left[\left(\frac{\varrho}{\widetilde{\varrho}} \widetilde{S} - S \right) \frac{\partial p(\widetilde{\varrho}, \widetilde{S})}{\partial S} \mathrm{div}_x \widetilde{\boldsymbol{u}} \right] \mathrm{d}x\, \mathrm{d}t \\
&\to \int_0^\tau \int_\Omega \left[\left(\frac{\varrho}{\widetilde{\varrho}} \widetilde{S} - S \right) \left(\partial_t \widetilde{\vartheta} + \frac{\boldsymbol{m}}{\varrho} \cdot \nabla_x \widetilde{\vartheta} \right) \right] \mathrm{d}x\, \mathrm{d}t \\
&+\int_0^\tau \int_\Omega \left[\left(\frac{\varrho}{\widetilde{\varrho}} \widetilde{S} - S \right) \frac{\partial p(\widetilde{\varrho}, \widetilde{S})}{\partial S} \mathrm{div}_x \widetilde{\boldsymbol{u}} \right] \mathrm{d}x\, \mathrm{d}t,
\end{aligned}
$$

where, furthermore,

$$
\begin{aligned}
&\int_0^\tau \int_\Omega \left[\left(\frac{\varrho}{\widetilde{\varrho}} \widetilde{S} - S \right) \left(\partial_t \widetilde{\vartheta} + \frac{\boldsymbol{m}}{\varrho} \cdot \nabla_x \widetilde{\vartheta} \right) \right] \mathrm{d}x\, \mathrm{d}t \\
&\quad + \int_0^\tau \int_\Omega \left[\left(\frac{\varrho}{\widetilde{\varrho}} \widetilde{S} - S \right) \frac{\partial p(\widetilde{\varrho}, \widetilde{S})}{\partial S} \mathrm{div}_x \widetilde{\boldsymbol{u}} \right] \mathrm{d}x\, \mathrm{d}t \\
&\quad = \int_0^\tau \int_\Omega \left[\left(\frac{\varrho}{\widetilde{\varrho}} \widetilde{S} - S \right) \left(\partial_t \widetilde{\vartheta} + \widetilde{\boldsymbol{u}} \cdot \nabla_x \widetilde{\vartheta} + \frac{\partial p(\widetilde{\varrho}, \widetilde{S})}{\partial S} \mathrm{div}_x \widetilde{\boldsymbol{u}} \right) \right] \mathrm{d}x\, \mathrm{d}t \\
&\quad + \int_0^\tau \int_\Omega \left(\frac{\varrho}{\widetilde{\varrho}} \widetilde{S} - S \right) \left(\frac{\boldsymbol{m}}{\varrho} - \widetilde{\boldsymbol{u}} \right) \cdot \nabla_x \widetilde{\vartheta} \, \mathrm{d}x\, \mathrm{d}t
\end{aligned}
$$

Summing up the previous computation we may rewrite (4.65) in the form

$$
\begin{aligned}
&\left[\int_{\Omega} E\left(\varrho, S, \boldsymbol{m} \middle| \widetilde{\varrho}, \widetilde{S}, \widetilde{\boldsymbol{u}}\right) \,\mathrm{d}x\right]_{t=0}^{t=\tau} \\
&\leq -\int_0^\tau \int_\Omega 1_{\varrho>0} \frac{(\varrho\widetilde{\boldsymbol{u}} - \boldsymbol{m}) \otimes (\varrho\widetilde{\boldsymbol{u}} - \boldsymbol{m})}{\varrho} : \nabla_x \widetilde{\boldsymbol{u}} \,\mathrm{d}x\,\mathrm{d}t \\
&- \int_0^\tau \int_\Omega \left[p(\varrho, S) - (\varrho - \widetilde{\varrho}) \frac{\partial p(\widetilde{\varrho}, \widetilde{S})}{\partial \varrho} - (S - \widetilde{S}) \frac{\partial p(\widetilde{\varrho}, \widetilde{S})}{\partial S} - p(\widetilde{\varrho}, \widetilde{S})\right] \mathrm{div}_x \widetilde{\boldsymbol{u}} \,\mathrm{d}x\,\mathrm{d}t \\
&+ \int_0^\tau \int_\Omega (\varrho\widetilde{\boldsymbol{u}} - \boldsymbol{m}) \cdot \left[\partial_t \widetilde{\boldsymbol{u}} + \widetilde{\boldsymbol{u}} \cdot \nabla_x \widetilde{\boldsymbol{u}} + \frac{1}{\widetilde{\varrho}} \nabla_x \widetilde{p}\right] \,\mathrm{d}x\,\mathrm{d}t \\
&+ \int_0^\tau \int_\Omega \left[(\widetilde{\varrho} - \varrho) \frac{1}{\widetilde{\varrho}} \frac{\partial p(\widetilde{\varrho}, \widetilde{S})}{\partial \varrho} \left(\partial_t \widetilde{\varrho} + \mathrm{div}_x (\widetilde{\varrho}\widetilde{\boldsymbol{u}})\right)\right] \,\mathrm{d}x\,\mathrm{d}t \\
&+ \int_0^\tau \int_\Omega \left[(\widetilde{\varrho} - \varrho) \frac{1}{\widetilde{\varrho}} \frac{\partial p(\widetilde{\varrho}, \widetilde{S})}{\partial S} \left(\partial_t \widetilde{S} + \mathrm{div}_x (\widetilde{\boldsymbol{u}}\widetilde{S})\right)\right] \,\mathrm{d}x\,\mathrm{d}t \\
&+ \int_0^\tau \int_\Omega \left[\left(\frac{\varrho}{\widetilde{\varrho}} \widetilde{S} - S\right) \left(\partial_t \widetilde{\vartheta} + \widetilde{\boldsymbol{u}} \cdot \nabla_x \widetilde{\vartheta} + \frac{\partial p(\widetilde{\varrho}, \widetilde{S})}{\partial S} \mathrm{div}_x \widetilde{\boldsymbol{u}}\right)\right] \,\mathrm{d}x\,\mathrm{d}t \\
&+ \int_0^\tau \int_\Omega \left(\frac{\varrho}{\widetilde{\varrho}} \widetilde{S} - S\right) \left(\frac{\boldsymbol{m}}{\varrho} - \widetilde{\boldsymbol{u}}\right) \cdot \nabla_x \widetilde{\vartheta} \,\mathrm{d}x\,\mathrm{d}t.
\end{aligned}
\tag{4.66}
$$

Similarly to Sect. 4.1.3, the relative energy inequality (4.65), or in the form (4.66), can be used to show the weak-strong uniqueness principle. Indeed the relative energy

$$
E\left(\varrho, \boldsymbol{m}, S \;\middle|\; \widetilde{\varrho}, \widetilde{\boldsymbol{u}}, \widetilde{S}\right) \approx E\left(\varrho, \boldsymbol{m}, S \;\middle|\; \widetilde{\varrho}, \widetilde{\boldsymbol{m}} = \widetilde{\varrho}\widetilde{\boldsymbol{u}}, \widetilde{S}\right)
$$

corresponds to the *Bregman distance* associated to the total energy $E(\varrho, \boldsymbol{m}, S)$ – a convex function of the conservative-entropy variables $[\varrho, \boldsymbol{m}, S]$. In comparison with the barotropic case, the proof of the weak-strong uniqueness is more involved, and we postpone it to Chap. 5, where a more general class of weak solutions is treated.

4.2 Weak and Strong Solutions to the Navier–Stokes System, Relative Energy

The concepts of weak and strong solution to the Navier–Stokes(–Fourier) system have been introduced and discussed in Sect. 3. Here, we focus on deriving the relative energy inequality similarly to the Euler system. As a matter of fact, the relative energy functional remains the same as for the inviscid fluid models.

4.2.1 Relative Energy for the Navier–Stokes System

We restrict ourselves to the *barotropic* Navier–Stokes system :

$$\partial_t \varrho + \mathrm{div}_x(\varrho \boldsymbol{u}) = 0, \tag{4.67}$$

$$\partial_t(\varrho \boldsymbol{u}) + \mathrm{div}_x(\varrho \boldsymbol{u} \otimes \boldsymbol{u}) + \nabla_x p(\varrho) = \mathrm{div}_x \mathbb{S}. \tag{4.68}$$

For definiteness, we consider the non-slip boundary condition

$$\boldsymbol{u}|_{\partial\Omega} = 0. \tag{4.69}$$

As the viscous stress is in many cases a function of the velocity gradient, it is more convenient to work in the frame of standard variables. The energy balance (inequality) takes the form

$$\int_\Omega \left[\frac{1}{2} \varrho |\boldsymbol{u}|^2 + P(\varrho) \right] (\tau, \cdot) \,\mathrm{d}x + \int_0^\tau \int_\Omega \mathbb{S} : \nabla_x \boldsymbol{u} \,\mathrm{d}x \,\mathrm{d}t \leq \int_\Omega \left[\frac{1}{2} \varrho_0 |\boldsymbol{u}_0|^2 + P(\varrho_0) \right] \mathrm{d}x \tag{4.70}$$

where $\varrho_0, \boldsymbol{u}_0$ are the initial data and P the pressure potential, $P'(\varrho)\varrho - P(\varrho) = p(\varrho)$. For the moment, we deliberately leave open the specific choice of the rheological relation for the viscous stress $\mathbb{S}$, keeping in mind only the Second law of thermodynamics requiring

$$\mathbb{S} : \nabla_x \boldsymbol{u} \geq 0.$$

The relative energy, written in terms of standard variables ϱ and $\boldsymbol{u}$ reads

$$E\left(\varrho, \boldsymbol{u} \,\middle|\, \widetilde{\varrho}, \widetilde{\boldsymbol{u}}\right) = \frac{1}{2} \varrho |\boldsymbol{u} - \widetilde{\boldsymbol{u}}|^2 + P(\varrho) - P'(\widetilde{\varrho})(\varrho - \widetilde{\varrho}) - P(\widetilde{\varrho}) \tag{4.71}$$

cf. (4.25). Note that E *is not* a convex function of $[\varrho, \boldsymbol{u}]$.

The weak solutions to the Navier–Stokes system were introduced in Definition 3.1. The relative entropy inequality can be now derived mimicking the procedure applied to the Euler system in Sect. 4.1.3. Indeed the only two steps to be modified are:

(**i**) the energy inequality (4.28) that should read

$$\left[\int_{\Omega} \left[\frac{1}{2} \varrho |\boldsymbol{u}|^2 + P(\varrho) \right] \, \mathrm{d}x \right]_{t=0}^{t=\tau} + \int_0^{\tau} \int_{\Omega} \mathbb{S} : \nabla_x \boldsymbol{u} \, \mathrm{d}x \, \mathrm{d}t \leq 0,$$

(**ii**) the relation (4.29) that should be replaced by

$$\left[\int_{\Omega} \varrho \boldsymbol{u} \cdot \widetilde{\boldsymbol{u}} \, \mathrm{d}x \right]_{t=0}^{t=\tau}$$

$$= \int_0^{\tau} \int_{\Omega} \left[\varrho \boldsymbol{u} \cdot \partial_t \widetilde{\boldsymbol{u}} + (\varrho \boldsymbol{u} \otimes \boldsymbol{u}) : \nabla_x \widetilde{\boldsymbol{u}} + p(\varrho) \mathrm{div}_x \widetilde{\boldsymbol{u}} - \mathbb{S} : \nabla_x \widetilde{\boldsymbol{u}} \right] \, \mathrm{d}x \, \mathrm{d}t,$$

where

$$\widetilde{\boldsymbol{u}} \in C^1([0,T] \times \overline{\Omega}; R^d), \ \widetilde{\boldsymbol{u}}|_{\partial \Omega} = 0$$

is a "test function". Putting together all the remaining integrals exactly as in Sect. 4.1.3, we deduce the *relative energy inequality* in the form

$$\begin{aligned}
&\left[\int_{\Omega} E\left(\varrho, \boldsymbol{u} \,\middle|\, \widetilde{\varrho}, \widetilde{\boldsymbol{u}} \right) \, \mathrm{d}x \right]_{t=0}^{t=\tau} + \int_0^{\tau} \int_{\Omega} (\mathbb{S} - \widetilde{\mathbb{S}}) : (\nabla_x \boldsymbol{u} - \nabla_x \widetilde{\boldsymbol{u}}) \, \mathrm{d}x \, \mathrm{d}t \\
&\leq - \int_0^{\tau} \int_{\Omega} \varrho \nabla_x \widetilde{\boldsymbol{u}} \cdot (\boldsymbol{u} - \widetilde{\boldsymbol{u}}) \cdot (\boldsymbol{u} - \widetilde{\boldsymbol{u}}) \, \mathrm{d}x \, \mathrm{d}t \\
&- \int_0^{\tau} \int_{\Omega} \left[p(\varrho) - p'(\widetilde{\varrho})(\varrho - \widetilde{\varrho}) - p(\widetilde{\varrho}) \right] \mathrm{div}_x \widetilde{\boldsymbol{u}} \, \mathrm{d}x \, \mathrm{d}t \\
&+ \int_0^{\tau} \int_{\Omega} \frac{\varrho}{\widetilde{\varrho}} (\widetilde{\boldsymbol{u}} - \boldsymbol{u}) \cdot \left[\partial_t (\widetilde{\varrho} \widetilde{\boldsymbol{u}}) + \mathrm{div}_x \left(\widetilde{\varrho} \widetilde{\boldsymbol{u}} \otimes \widetilde{\boldsymbol{u}} \right) + \nabla_x p(\widetilde{\varrho}) - \mathrm{div}_x \widetilde{\mathbb{S}} \right] \mathrm{d}x \, \mathrm{d}t \\
&+ \int_0^{\tau} \int_{\Omega} \left[\left(1 - \frac{\varrho}{\widetilde{\varrho}} \right) p'(\widetilde{\varrho}) + \frac{\varrho}{\widetilde{\varrho}} \widetilde{\boldsymbol{u}} \cdot (\boldsymbol{u} - \widetilde{\boldsymbol{u}}) \right] \left[\partial_t \widetilde{\varrho} + \mathrm{div}_x (\widetilde{\varrho} \widetilde{\boldsymbol{u}}) \right] \mathrm{d}x \, \mathrm{d}t \\
&+ \int_0^{\tau} \int_{\Omega} \left(\frac{\varrho}{\widetilde{\varrho}} - 1 \right) (\widetilde{\boldsymbol{u}} - \boldsymbol{u}) \cdot \mathrm{div}_x \widetilde{\mathbb{S}} \, \mathrm{d}x \, \mathrm{d}t
\end{aligned} \tag{4.72}$$

for any pair of test functions $\widetilde{\varrho}, \widetilde{\boldsymbol{u}}$ in the class

$$\begin{aligned}&\widetilde{\varrho} \in C^1([0,T] \times \overline{\Omega}), \widetilde{\varrho} > 0,\ \widetilde{\boldsymbol{u}} \in C^1([0,T] \times \overline{\Omega}; R^d),\ \widetilde{\boldsymbol{u}}|_{\partial\Omega} = 0,\\ &\quad \text{and } \widetilde{\mathbb{S}} \in C^1([0,T] \times \overline{\Omega}; R^{d\times d}_{\rm sym}),\end{aligned} \tag{4.73}$$

cf. (4.30).

Remark 4.12 The tensor $\widetilde{\mathbb{S}}$ has been added "artificially" to relative energy inequality. Of course, we consider $\widetilde{S} = \widetilde{\mathbb{S}}(\nabla_x \widetilde{\boldsymbol{u}})$ in future applications to the weak-strong uniqueness problem.

In comparison with the relative energy inequality (4.34) for the barotropic Euler system, the relation (4.72) contains an extra term

$$\int_0^\tau \int_\Omega \left(\frac{\varrho}{\widetilde{\varrho}} - 1 \right) (\widetilde{\boldsymbol{u}} - \boldsymbol{u}) \cdot {\rm div}_x \widetilde{\mathbb{S}} \ {\rm d}x\, {\rm d}t.$$

Note carefully that this integral *cannot* be controlled by the relative energy on the vacuum zone $\{\varrho = 0\}$ as

$$\left(\frac{\varrho}{\widetilde{\varrho}} - 1 \right) (\widetilde{\boldsymbol{u}} - \boldsymbol{u}) \cdot {\rm div}_x \widetilde{\mathbb{S}} \approx (\widetilde{\boldsymbol{u}} - \boldsymbol{u}) \text{ for } \varrho \to 0.$$

Although vacuum is not expected to appear *spontaneously* in viscous fluid flows, a rigorous proof is not yet available. In order to show the weak-strong uniqueness principle, the term

$$\int_0^\tau \int_\Omega (\mathbb{S} - \widetilde{\mathbb{S}}) : (\nabla_x \boldsymbol{u} - \nabla_x \widetilde{\boldsymbol{u}}) \ {\rm d}x\, {\rm d}t$$

must be used. Details can be found in Chap. 5, where a large class of generalized solutions is considered.

4.3 Conclusion, Bibliographical Remarks

We have introduced the concept of weak (distributional) solution and the relative energy to both the Euler and the Navier–Stokes system. These are objects solving the problem in the sense of generalized derivatives (distributions); satisfying automatically the compatibility principle **(C)**. If supplemented by a suitable form of the energy balance, they also satisfy the weak-strong uniqueness principle **(WS)**. These results are based on the concept of relative energy and the associated relative energy inequality for the weak solutions. In particular, the relative energy can be interpreted as the *Bregman distance* associated to the total energy of the system, cf. e.g. Sprung [185]. Results of this type go back to the pioneering paper by Dafermos [59] in the context of general conservation laws with involutions and are intimately related to the thermodynamic stability of the fluid system, see e.g. Bechtel et al. [14]. Since then the method of relative energy/entropy has found numerical applications, in particular in problems of weak-strong uniqueness for models of viscous and inviscid fluids, see e.g. Brenier et al. Székelyhidi [26], Germain [117], Gwiazda et al. [126], Mellet and Vasseur [164], the survey paper of Wiedemann [199], and the references therein. Here, we have postponed the rigorous proofs to the next chapter, where even more general class of *dissipative* solutions will be introduced.

There are two main issues that make analysis of numerical schemes in the framework of *weak solutions* quite delicate:

- Despite the numerous examples provided by the method of convex integration, see e.g., Chiodaroli et al. [49, 51], the global-in-time existence of *admissible* weak solutions for the Euler system is still an open problem. Although the weak solutions are known to exist for the Navier–Stokes(–Fourier) system, the existence proof is rather delicate, limited to severe restrictions imposed on constitutive relations, and difficult to implement in the proof of convergence of a numerical scheme, unless the latter is of very special type, cf. Karper [139, 140].
- The set of all weak solutions emanating from given initial data for the Euler system is not sequentially closed, meaning the property **(S)** listed in the introduction to Part II.

To demonstrate that the class of weak solutions to the barotropic Euler system fails to comply with the stability property **(S)**, we report the following result proved in [25, Proposition 2.1].

Theorem 4.5 *Let $\Omega \subset R^d$, $d = 2, 3$ be a bounded domain. Let $\varrho_0 \in L^\infty(\Omega)$, $\varrho_0 > 0$ be given.*

Then there exists a sequence of weak solutions $[\varrho_n, \boldsymbol{m}_n]$ to the Euler system (4.10), (4.11) *in $(0, T) \times \Omega$, with the impermeability condition* (4.5)*, such that $\varrho_n = \varrho_n(x)$ depends only on the spatial variable, and*

$$\varrho_n \to \varrho_0 \text{ weakly-(*) in } L^\infty(\Omega),\ \boldsymbol{m}_n \to 0 \text{ weakly-(*) in } L^\infty((0,T) \times \Omega; R^N),$$

$$\liminf_{n \to \infty} \int_\Omega |\varrho_n - \varrho_0| \, \mathrm{d}x > 0.$$

Note that the limit ϱ_0 is arbitrary, while the momentum limit $\boldsymbol{m} \equiv 0$. Thus the limit is a solution of the Euler system only if $\varrho_0 = \overline{\varrho}$ is a (positive) constant. In particular, a weak limit of a sequence of weak solutions to the barotropic Euler system may not be a weak solution.

In the following chapter, we extend the class of weak solutions to more general objects commonly known as *measure-valued* or *dissipative* solutions. We show that the dissipative solutions comply with all the requirements introduced above. In particular, they exist globally in time for any physically relevant initial data, they are compatible with classical solutions, they obey the weak-strong uniqueness principle, and the solution set associated to given initial data is compact.

Chapter 5
Generalized Weak Solutions

We identify a class of generalized solutions to the Euler and Navier–Stokes system that comply with the four basic requirements stated at the beginning of Part II. They exist globally in time and represent the asymptotic limits of sequences of approximate solutions (**property (E)**); they are compatible with strong solutions, meaning they solve the problem in the classical way as long as they are smooth (**property (C)**); they comply with the weak-strong uniqueness principle (**property (WS)**); the set of all solutions emanating from fixed initial data is compact (**property (S)**). Our aim is to identify the largest class possible of objects that would meet the requirements listed above. Motivated by future applications in numerical analysis, the generalized solutions will be identified as asymptotic limits of suitable approximations. We distinguish between **stable approximations** satisfying uniform (energy) bounds and approximating exactly the given data, and **consistent approximations** that satisfy, in addition, the underlying system of equations modulo an approximate error vanishing in the asymptotic limit.

Although our aim is to put solutions of the Euler and the Navier–Stokes system in a unified framework, our general philosophy is different concerning viscous and inviscid fluid models:

- We tacitly anticipate that solutions of the *Navier–Stokes system* are regular. This is indeed the case if the initial data are smooth and solutions remain uniformly bounded in the space L^∞ on the desired time interval $(0, T)$. Let us assume that we are given a family of approximate solutions resulting from a numerical scheme. The line of arguments we follow in the viscous case can be described as follows: (i) Suppose that a family of numerical solutions is bounded uniformly with respect to the time step and the space discretization parameter. (ii) Identify the set of possible limits of the approximate solutions – the generalized weak solutions of the problem. (iii) As the family of numerical solutions is bounded, we use the weak-strong uniqueness principle to show that the limit is in fact the (unique) strong solution of the problem. (iv) As the limit is a strong solution, we may use

E. Feireisl et al., *Numerical Analysis of Compressible Fluid Flows*, MS&A 20,
https://doi.org/10.1007/978-3-030-73788-7_5

estimates derived from the relative energy inequality to evaluate the numerical error.

- As for the *Euler system*, the exact solution remains smooth only on a certain maximal time interval. If this is the case, we may deduce that convergence of approximate solutions is strong and even evaluate the error. As soon as the solution develops singularities, the asymptotic limit belongs to the class of generalized solutions. In such a situation, the approximate sequence of numerical solutions necessarily develops oscillations that may be described in terms of the Young measure. In such a case, we propose a method how to visualize (compute) effectively the associated Young measure.

As we have observed in the preceding chapters, the distribution of the *total energy* plays a crucial role in the analysis of compressible fluid flows. Generalized solutions satisfying some form of the energy balance will be called *dissipative solutions*. The dissipative solutions will be defined/identified as limits of *consistent approximation schemes*. Very roughly indeed, a consistent approximation is sequences of functions satisfying the underlying system of equations modulo a certain error that vanishes in the asymptotic limit. For the purpose of this monograph, the consistent approximations will be represented by numerical schemes. The concept is, however, more general and includes many other kinds of approximations, in particular, the vanishing viscosity limit passage from the Navier–Stokes to the Euler system. We also consider more general *stable approximations* admitting certain uniform bounds and approaching the given data in the asymptotic limit.

In many cases, the *energy estimates* are the only uniform bounds available for an approximate sequence. This fact is reflected in consistent approximations that can be controlled only in the rather weak L^p-norms. Accordingly, the dissipative solutions are identified with *weak* limits of approximate sequences in the L^p-topology. Here and hereafter, whenever we speak about *weak* convergence, we mean convergence of integral averages. Specifically, we say that a sequence $\{U_n\}_{n=1}^{\infty}$ converges weakly to a limit U if the functions $U, U_n, n = 1, 2, \ldots$ are integrable in a domain Q, and

$$\int_B U_n \,\mathrm{d}y \to \int_B U \,\mathrm{d}y \text{ as } n \to \infty \text{ for any Borel set } B \subset Q$$

or, equivalently,

$$\int_Q \varphi U_n \,\mathrm{d}y \to \int_Q \varphi U \,\mathrm{d}y$$

for any smooth function φ.

The notoriously known stumbling block when applying weak convergence to *nonlinear* problems is its incompatibility with nonlinear superpositions. In general,

$$U_n \to U \text{ weakly } \not\Rightarrow b(U_n) \to b(U) \text{ weakly}$$

unless b is an affine function. As a matter of fact,

$$U_n \to U, \;\; b(U_n) \to b(U) \;\Rightarrow\; U_n \to U \text{ strongly (a.a. for a subsequence)}$$

as soon as b is strictly convex. This can be formally seen by applying Taylor's formula,

$$b(U_n) = b(U) + b'(U)(U_n - U) + \frac{1}{2} b''(\xi)|U_n - U|^2.$$

Thus if b'' is positive bounded below, we conclude

$$\int_B |U_n - U|^2 \, dy \to 0 \text{ for any Borel set } B,$$

which implies, at least for a suitable subsequence, the desired pointwise convergence.

The knowledge of the weak limit of an approximate sequence is therefore insufficient to identify the limit problem as long as the underlying equations are nonlinear. A piece of information on possible oscillations and/or concentrations of the approximations must be retained as an integral part of the definition of generalized solutions. This idea leads to various concepts of measure-valued solutions discussed at the beginning of this chapter. In this approach, the exact numerical value of solution is replaced by a probability measure (Young measure) defined on the associated phase space.

The *dissipative solutions*, introduced later in this chapter, can be interpreted as *expected* values of the underlying Young measure. The definition is, however, *intrinsic* without explicit reference to any approximating sequence. The piece of information pertinent to "turbulent" character of the fluid motion is encoded in a single quantity termed *Reynolds stress* that forms an integral part of the definition of dissipative solution. The Reynolds stress can be seen as a contribution to the viscous forces counterbalanced by the energy dissipation defect in the energy balance equation. As a result, the dissipative solutions enjoy the compatibility as well as the weak-strong uniqueness property. The Reynolds stress can be identified as a positively semidefinite tensor valued measure resulting from possible oscillations and/or concentrations in the approximate sequence. Accordingly, the Reynolds stress vanishes as soon as the approximations converge strongly in some L^p-topology. If this is the case, the resulting dissipative solution is also a weak (distributional) solution of the problem. In Part III, we disclose a striking fact that these properties are equivalent at least for the Euler system. More specifically, the Reynolds stress vanishes and the convergence is strong if and only if the limit is a weak solution. In other words, the limits of *oscillatory* consistent approximations *are not* weak (distributional) solutions of the Euler system.

5.1 Measure-Valued Solutions to the Euler System

Consider a sequence $\{U_n\}_{n=1}^{\infty}$ of measurable functions,

$$U_n : Q \to R^k, \text{ where } Q \subset R^m \text{ is a domain.}$$

Identifying

$$U_n(y) \approx \delta_{U_n(y)},$$

where $\delta_U \in \mathcal{P}(R^k)$ denotes the Dirac measure in R^k supported at U, we may interpret U_n as a mapping

$$U_n : y \in Q \mapsto \delta_{U_n(y)} \in \mathcal{P}(R^k) \text{ that is weakly-(*) measurable.}$$

Recall that weak-(*) measurability here means that for any $b \in BC(R^k)$, the mapping

$$y \in Q \mapsto \left\langle \delta_{U_n(y)}; b \right\rangle = b(U_n(y)) \text{ is measurable.} \tag{5.1}$$

Our goal is to identify the limit object generated by the sequence of parametrized measures $\delta_{U_n(y)}$, $y \in Q$.

In view of (5.1), the sequence $\{U_n\}_{n=1}^{\infty}$ can be identified with a bounded sequence in the Banach space

$$L^{\infty}_{\text{weak}-(*)}(Q; \mathcal{M}(R^k)) = \left[L^1(0,T; C_0(R^k))\right]^*.$$

As the space $L^1(0,T; C_0(R^k))$ is separable, there exists a subsequence (not relabeled here) such that

$$U_n \to \mathcal{V} \text{ weakly-(*) in } L^{\infty}_{\text{weak}-(*)}(Q; \mathcal{M}(R^k)). \tag{5.2}$$

The limit quantity $\mathcal{V}$ is called the *Young measure* associated or generated by the sequence U_n. The Young measure $\mathcal{V}$ is often interpreted as a parameterized family of Borel measures $\{\mathcal{V}_y\}_{y \in Q}$. Recall that the weak-(*) convergence claimed in (5.2) means

$$\int_Q \varphi(y) b\left(U_n(y)\right) \, dy \to \int_Q \varphi(y) \left\langle \mathcal{V}_y; b(\widetilde{U}) \right\rangle \, dy \text{ for any } \varphi \in L^1(Q),\ b \in C_0(R^k) \tag{5.3}$$

where

$$\left\langle \mathcal{V}_y; b(\widetilde{U}) \right\rangle \equiv \int_{R^k} b(\widetilde{U}) \, d\mathcal{V}_y(\widetilde{U}).$$

As $\mathcal{V}$ is a (weak-(*)) limit of probability measures (Dirac masses),

$$\mathcal{V}_y \in \mathcal{M}^+(R^k) \text{ and } \|\mathcal{V}_y\|_{\mathcal{M}(R^k)} \leq 1 \tag{5.4}$$

for a.a. $y \in Q$.

Remark 5.1 One should always keep in mind that a given sequence $\{\boldsymbol{U}_n\}_{n=1}^{\infty}$ may generate *different* Young measures for different subsequences.

Frequently, we omit the subscript y and use the symbol $\mathcal{V}$ to denote the measure-valued mapping

$$\mathcal{V} : y \in Q \to \mathcal{M}^+(R^k), \ \mathcal{V} \in L^{\infty}_{\text{weak}-(*)}(Q; \mathcal{M}^+(R^k)).$$

We also systematically use the symbol $\widetilde{\boldsymbol{U}}$ to denote the "dummy" variables and write

$$\left\langle \mathcal{V}; b(\widetilde{\boldsymbol{U}}) \right\rangle \text{ to denote the measurable function } y \in Q \mapsto \int_{R^k} b(\widetilde{\boldsymbol{U}}) \, \mathrm{d}\mathcal{V}_y(\widetilde{\boldsymbol{U}}).$$

5.1.1 Integrability, Concentrations

The Young measure as introduced in the preceding part may not be a probability measure unless certain restrictions are imposed on the generating sequence $\{\boldsymbol{U}_n\}_{n=1}^{\infty}$. For $b \in C_0(R^k)$ and a bounded Borel set $B \subset R^m$, $|B| > 0$, we have

$$\frac{1}{|B|} \int_B b(\boldsymbol{U}_n) \, \mathrm{d}y \to \frac{1}{|B|} \int_B \left\langle \mathcal{V}_y; b(\widetilde{\boldsymbol{U}}) \right\rangle \, \mathrm{d}y,$$

which can be understood as

$$\frac{1}{|B|} \int_B \delta_{\boldsymbol{U}_n(y)} \, \mathrm{d}y \to \frac{1}{|B|} \int_B \mathcal{V}_y \mathrm{d}y \text{ weakly-(*) in } \mathcal{M}(R^k).$$

We desire to identify the necessary condition for the sequence of probability measures $\{\frac{1}{|B|} \int_B \delta_{\boldsymbol{U}_n(y)} \, \mathrm{d}y\}_{n=1}^{\infty}$ to be tight; whence narrowly precompact. To this end, we estimate

$$\frac{1}{|B|} \int_B 1_{|\boldsymbol{U}_n| \geq k} \, \mathrm{d}y = \frac{1}{|B|} \int_B 1_{|\boldsymbol{U}_n| \geq k} \frac{h(|\boldsymbol{U}_n|)}{h(|\boldsymbol{U}_n|)} \, \mathrm{d}y \leq \frac{1}{h(k)} \frac{1}{|B|} \int_B h(|\boldsymbol{U}_n|) \, \mathrm{d}y \, \mathrm{d}y,$$

where $h : [0, \infty) \to [0, \infty)$, $h(0) = 0$, h is strictly increasing, $h(Z) \to \infty$ as $Z \to \infty$. Consequently, the sequence of probability measures

$$\frac{1}{|B|}\int_B \delta_{\boldsymbol{U}_n(y)}\,\mathrm{d}y,\ n = 1, 2, \ldots$$

is (uniformly) tight if there is $h \in C^1[0, \infty)$, $h(0) = 0$, $h'(Z) > 0$, $\lim_{Z\to\infty} h(Z) = \infty$ so that

$$\int_Q h(|\boldsymbol{U}_n|)\,\mathrm{d}y \le c \text{ uniformly for } n \to \infty. \tag{5.5}$$

If (5.5) holds we may infer that

$$\frac{1}{|B|}\int_B \mathcal{V}_y \mathrm{d}y \in \mathcal{P}(R^k) \text{ for any Borel set } B \subset Q,\ |B| > 0$$

which, combined with (5.4), gives rise to the desired conclusion

$$\mathcal{V}_y \in \mathcal{P}(R^k) \text{ for a.a. } y \in Q. \tag{5.6}$$

Obviously, the condition (5.5) is satisfied with $h(Z) = Z$ if the sequence $\{\boldsymbol{U}_n\}_{n=1}^{\infty}$ is bounded in $L^1(Q)$,

$$\int_Q |\boldsymbol{U}_n|\,\mathrm{d}y \le c \text{ uniformly for } n \to \infty. \tag{5.7}$$

The previous discussion can be summarized in the following statement.

Proposition 5.1 (Young measure)
Let $Q \subset R^m$ be a domain and let $\{\boldsymbol{U}_n\}_{n=1}^{\infty}$ be a sequence of measurable vectorial functions such that

$$\boldsymbol{U}_n \in L^1(Q; R^k),\ \|\boldsymbol{U}_n\|_{L^1(Q;R^k)} \le c \textit{ uniformly for } n \to \infty. \tag{5.8}$$

Then there exists a subsequence $\{\boldsymbol{U}_{n_k}\}_{k=1}^{\infty}$, and a parametrized family of probability measures $\{\mathcal{V}_y\}_{y\in Q}$,

$$\mathcal{V}_y \in \mathcal{P}(R^k) \textit{ for a.a. } y \in Q,\ y \in Q \mapsto \mathcal{V}_y \in \mathcal{P}(R^k) \textit{ weakly-(*) measurable,}$$

such that

$$\int_Q \varphi(y) b(\boldsymbol{U}_{n_k}(y))\,\mathrm{d}y \to \int_Q \varphi(y) \left\langle \mathcal{V}_y; b(\widetilde{\boldsymbol{U}})\right\rangle \mathrm{d}y \textit{ as } k \to \infty$$

for any $\varphi \in L^1(Q)$ and any $b \in BC(R^k)$.

Remark 5.2 Note that (5.8) can be replaced by the considerably weaker assumption

$$\boldsymbol{U}_n \text{ measurable}, \int_Q h(|\boldsymbol{U}_n|)\,\mathrm{d}y \leq c \text{ uniformly for } n \to \infty,$$

where $h(Z) \to \infty$ for $Z \to \infty$.

5.1.1.1 Convergence of Nonlinear Superpositions, Concentrations

Proposition 5.1 does not identify the limits of superpositions $b(\boldsymbol{U}_n)$ for b having certain growth for large values of its argument. Consider the situation

$$\boldsymbol{U}_n \in L^1(Q; R^k),\ \|\boldsymbol{U}_n\|_{L^1(Q;R^k)} \leq c,\ \|b(\boldsymbol{U}_n)\|_{L^1(Q)} \leq c,\ b \in C(R^k) \tag{5.9}$$

uniformly for $n = 1, 2, \dots$ In addition, suppose we already know that $\{\boldsymbol{U}_n\}_{n=1}^{\infty}$ generates a Young measure $\{\mathcal{V}_y\}_{y \in Q}$. We claim that b is integrable with respect to $\mathcal{V}_y$, more specifically,

$$\left\langle \mathcal{V}_y; b(\widetilde{\boldsymbol{U}}) \right\rangle \text{ is finite for a.a. } y \in Q, \text{ and } y \in Q \mapsto \left\langle \mathcal{V}_y; b(\widetilde{\boldsymbol{U}}) \right\rangle \in L^1(Q).$$

Decomposing $b = b^+ - b^-$ we may suppose that $b \geq 0$. Now, consider a sequence $b_n \in C_c(R^k)$,

$$0 \leq b_n \nearrow b \text{ in } R^k.$$

On one hand, by the Lévy theorem,

$$\left\langle \mathcal{V}_y; b_n(\widetilde{\boldsymbol{U}}) \right\rangle \nearrow \left\langle \mathcal{V}_y; b(\widetilde{\boldsymbol{U}}) \right\rangle \in [0, \infty] \text{ for a.a } y \in Q.$$

On the other hand,

$$\begin{aligned} &\sup_j \left\| b(\boldsymbol{U}_j) \right\|_{L^1(Q)} \\ &\quad \geq \lim_{j \to \infty} \int_B b_n(\boldsymbol{U}_j)\,\mathrm{d}y = \int_B \left\langle \mathcal{V}_y; b_n(\widetilde{\boldsymbol{U}}) \right\rangle \mathrm{d}y \nearrow \int_B \left\langle \mathcal{V}_y; b(\widetilde{\boldsymbol{U}}) \right\rangle \mathrm{d}y \end{aligned}$$

as $n \to \infty$ for any Borel set $B \subset Q$, which proves the claim. We have shown the following result.

Proposition 5.2 (Young measure of composition)
Let $Q \subset R^m$ be a given domain, let $\{\boldsymbol{U}_n\}_{n=1}^{\infty}$ be a sequence of measurable vectorial functions ranging in R^k, and let $b \in C(R^k)$ such that

$$\boldsymbol{U}_n \in L^1(Q; R^k),\ \|\boldsymbol{U}_n\|_{L^1(Q;R^k)} \leq c,\ \|b(\boldsymbol{U}_n)\|_{L^1(Q)} \leq c \text{ uniformly for } n \to \infty.$$

Suppose, in addition, that $\{\boldsymbol{U}_n\}_{n=1}^{\infty}$ *generates a Young measure* $\{\mathcal{V}_y\}_{y\in Q}$.
Then

$$\left\langle \mathcal{V}_y; b(\widetilde{\boldsymbol{U}})\right\rangle \text{ is finite for a.a. } y \in Q, \text{ and } y \in Q \mapsto \left\langle \mathcal{V}_y; b(\widetilde{\boldsymbol{U}})\right\rangle \in L^1(Q).$$

Remark 5.3 The statements of Propositions 5.1, 5.2 can be localized, meaning, the space $L^1(Q)$ may be replaced by $L^1_{\text{loc}}(Q)$.

Finally, consider a sequence of functions $\{\boldsymbol{U}_n\}_{n=1}^{\infty}$ satisfying the hypotheses of Proposition 5.2, along with $\{b(\boldsymbol{U}_n)\}_{n=1}^{\infty}$. In addition, suppose that $Q \subset R^m$ is a bounded domain. As $\{b(\boldsymbol{U}_n)\}_{n=1}^{\infty}$ is only uniformly integrable, we may assert that

$$b(\boldsymbol{U}_n) \to \overline{b(\boldsymbol{U})} \text{ weakly-(*) in } \mathcal{M}(\overline{Q}),$$

up to a suitable subsequence, where the space of (finite) measures $\mathcal{M}(\overline{Q})$ is identified with the dual space of $C(\overline{Q})$.

Furthermore, suppose that

$$\boldsymbol{U}_n \to \boldsymbol{U} \text{ weakly in } L^1(Q; R^k)$$

up to a suitable subsequence, meaning $\{\boldsymbol{U}_n\}_{n=1}^{\infty}$ is equi-integrable in Q. Thus, extracting a subsequence several times as the case may be, we end up with three quantities:

$$\overline{b(\boldsymbol{U})} \in \mathcal{M}(\overline{Q}),\ y \mapsto \left\langle \mathcal{V}_y; b(\widetilde{\boldsymbol{U}})\right\rangle \in L^1(Q), \text{ and } b(\boldsymbol{U}) - \text{a measurable function on } Q.$$

Definition 5.1 (DEFECT MEASURES)
Let $Q \subset R^m$ be a bounded domain. Let

$$\{\boldsymbol{U}_n\}_{n=1}^{\infty},\quad \|\boldsymbol{U}_n\|_{L^1(Q;R^k)} \le c$$

be a sequence generating a Young measure $\{\mathcal{V}_y\}_{y\in Q}$. Let $b \in C(R^k)$,

$$|b(\boldsymbol{V})| \le c(1 + |\boldsymbol{V}|),$$

satisfy

$$b(\boldsymbol{U}_n) \to \overline{b(\boldsymbol{U})} \text{ weakly-(*) in } \mathcal{M}(\overline{Q}).$$

We define the following quantities:

- **Concentration defect**

$$\overline{b(\boldsymbol{U})} - \left\{ y \mapsto \left\langle \mathcal{V}_y; b(\widetilde{\boldsymbol{U}})\right\rangle \right\} \in \mathcal{M}(\overline{Q})$$

- **Oscillation defect**

$$\left\{ y \mapsto \left\langle \mathcal{V}_y; b(\widetilde{\boldsymbol{U}}) \right\rangle \right\} - b(\boldsymbol{U}) - \text{ a measurable function on } Q.$$

Remark 5.4 If b is convex l.s.c, it follows from Jensen's inequality that

$$b(\boldsymbol{U}(y)) = b\left(\left\langle \mathcal{V}_y; \widetilde{\boldsymbol{U}} \right\rangle \right) \leq \left\langle \mathcal{V}_y; b(\widetilde{\boldsymbol{U}}) \right\rangle.$$

In particular, the oscillation defect is nonnegative, and

$$0 \leq \left\{ y \mapsto \left\langle \mathcal{V}_y; b(\widetilde{\boldsymbol{U}}) \right\rangle \right\} - b(\boldsymbol{U}) \in L^1(Q).$$

Intuitively, if the concentration defect of a sequence $\{b(\boldsymbol{U}_n)\}_{n=1}^{\infty}$ vanishes, then the sequence converges weakly in L^1. Indeed we report the following result:

Lemma 5.1 *Let $Q \subset R^m$ be a bounded domain, and let $\{\boldsymbol{U}_n\}_{n=1}^{\infty}$ be sequence of vector valued functions,*

$$\boldsymbol{U}_n : Q \to R^k, \quad \int_Q |\boldsymbol{U}_n| \, \mathrm{d}y \leq c \text{ uniformly for } n \to \infty,$$

generating a Young measure $\mathcal{V}_y \in \mathcal{P}(R^k)$, $y \in Q$. Let $E : R^k \to [0, \infty]$ be a l.s.c. function such that $E(\boldsymbol{U}_n) \in L^1(Q)$, and

$$E(\boldsymbol{U}_n) \to \left\langle \mathcal{V}; E(\widetilde{\boldsymbol{U}}) \right\rangle \text{ weakly-(*) in } \mathcal{M}(\overline{Q}), \ \left\langle \nu; E(\widetilde{\boldsymbol{U}}) \right\rangle \in L^1(Q). \tag{5.10}$$

Then

$$E(\boldsymbol{U}_n) \to \left\langle \mathcal{V}; E(\widetilde{\boldsymbol{U}}) \right\rangle \text{ weakly in } L^1(Q).$$

Remark 5.5 In (5.10), we have identified

$$\left\langle \mathcal{V}; E(\widetilde{\boldsymbol{U}}) \right\rangle \text{ with a function on } Q, \ y \mapsto \left\langle \mathcal{V}_y; E(\widetilde{\boldsymbol{U}}) \right\rangle, \ y \in Q.$$

Proof As E is l.s.c bounded below, there is an increasing sequence of bounded continuous functions,

$$E_j \in BC(R^k), 0 \leq E_j \leq j, \ E_j(\boldsymbol{U}) \nearrow E(\boldsymbol{U}) \text{ for any } \boldsymbol{U} \in R^k.$$

In accordance with (5.10),

$$\int_Q \left[E_j(\boldsymbol{U}_n) + (E - E_j)(\boldsymbol{U}_n)\right] \varphi \,\mathrm{d}y = \int_Q E(\boldsymbol{U}_n)\varphi \,\mathrm{d}y$$
$$\to \int_Q \left\langle \mathcal{V}_y; E(\widetilde{\boldsymbol{U}})\right\rangle \varphi \,\mathrm{d}y = \int_Q \left\langle \mathcal{V}_y; E_j(\widetilde{\boldsymbol{U}}) + (E - E_j)(\widetilde{\boldsymbol{U}})\right\rangle \varphi \,\mathrm{d}y$$

as $n \to \infty$ for any $\varphi \in C(\overline{Q})$. Since $\mathcal{V}$ is a Young measure generated by the sequence $\{\boldsymbol{U}_n\}_{n=1}^{\infty}$, we have

$$\int_Q E_j(\boldsymbol{U}_n)\varphi \,\mathrm{d}y \to \int_Q \left\langle \mathcal{V}_y; E_j(\widetilde{\boldsymbol{U}})\right\rangle \varphi \,\mathrm{d}y;$$

whence

$$\int_Q (E - E_j)(\boldsymbol{U}_n)\varphi \,\mathrm{d}y \to \int_Q \left\langle \mathcal{V}_y; (E - E_j)(\widetilde{\boldsymbol{U}})\right\rangle \varphi \,\mathrm{d}y \text{ as } n \to \infty. \tag{5.11}$$

As

$$E \geq E - E_j \searrow 0 \text{ as } j \to \infty,$$

we may use the Lebesgue Dominant Convergence Theorem to conclude

$$\int_Q \left\langle \mathcal{V}_y; (E - E_j)(\widetilde{\boldsymbol{U}})\right\rangle \,\mathrm{d}y \to 0 \text{ as } j \to \infty. \tag{5.12}$$

Our goal is to show equi-integrability of the sequence $\{E(\boldsymbol{U}_n)\}_{n=1}^{\infty}$. To this end, write

$$\int_{\{E(\boldsymbol{U}_n)\geq M\}} E(\boldsymbol{U}_n) \,\mathrm{d}y \leq \int_{\{E(\boldsymbol{U}_n)\geq M\}} E_j(\boldsymbol{U}_n) \,\mathrm{d}y + \int_Q (E - E_j)(\boldsymbol{U}_n) \,\mathrm{d}y$$
$$\leq j \left|\{E(\boldsymbol{U}_n) \geq M\}\right| + \int_Q (E - E_j)(\boldsymbol{U}_n) \,\mathrm{d}y$$
$$\leq \frac{j}{M} \|E(\boldsymbol{U}_n)\|_{L^1(Q)} + \int_Q (E - E_j)(\boldsymbol{U}_n) \,\mathrm{d}y.$$

Now, by virtue of (5.11), (5.12), we my fix $j = j(\varepsilon)$ large enough so that there exists n_j such that

$$\int_Q (E - E_j)(\boldsymbol{U}_n) \,\mathrm{d}y \leq \frac{\varepsilon}{2} \text{ for all } n \geq n_j.$$

Finally, we take $M = M(\varepsilon, j(\varepsilon))$ so large that

$$\frac{j(\varepsilon)}{M}\|E(\boldsymbol{U}_n)\|_{L^1(Q)} \leq \frac{\varepsilon}{2} \text{ for all } n \geq j,$$

which yields the desired conclusion

$$\int_{\{E(\boldsymbol{U}_n)\geq M(\varepsilon)\}} E(\boldsymbol{U}_n)\,\mathrm{d}y \leq \varepsilon \text{ for } n \geq n_j(\varepsilon).$$

□

Remark 5.6 As a byproduct of the construction of the functions E_j in the proof, we also obtain

$$\overline{E(\boldsymbol{U})} \geq \left\langle \mathcal{V}; E(\widetilde{\boldsymbol{U}})\right\rangle \text{ in } \mathcal{M}(\overline{Q}),$$

meaning the concentration defect is nonnegative whenever $E : R^k \to [0, \infty]$ is a l.s.c. function. Obviously the same holds for any l.s.c. function that is bounded below.

5.1.1.2 Comparing Concentration Defects

We finish this part by comparing the concentration defect associated to two different nonlinearities.

Proposition 5.3 (Comparison of concentration defects)
Let $Q \subset R^m$ be a bounded domain. Let $F : R^k \to [0, \infty]$, $G : R^k \to R$ be two Borel measurable functions. Let $\{\boldsymbol{U}_n\}_{n=1}^\infty \subset L^1(Q; R^k)$ be a sequence generating a Young measure $\{\mathcal{V}_y\}_{y\in Q}$,

$$\|F(\boldsymbol{U}_n)\|_{L^1(Q)} \leq c,\ \|G(\boldsymbol{U}_n)\|_{L^1(Q)} \leq c \text{ uniformly for } n \to \infty.$$

Let $\{F_j\}_{j=1}^\infty \subset C_c(R^k)$, $\{G_j\}_{j=1}^\infty \subset C_c(R^k)$ such that

$$\begin{aligned}
&|F_j| \stackrel{<}{\sim} (1 + F + |G|),\ |G_j| \stackrel{<}{\sim} (1 + F + |G|) \text{ uniformly for } j \to \infty,\\
&F_j(\boldsymbol{U}) \to F(\boldsymbol{U}),\ G_j(\boldsymbol{U}) \to G(\boldsymbol{U}) \text{ for any } \boldsymbol{U} \in R^k \text{ as } j \to \infty,\\
&\limsup_{j\to\infty}|G(\boldsymbol{U}) - G_j(\boldsymbol{U})| \leq \liminf_{j\to\infty}\left(F(\boldsymbol{U}) - F_j(\boldsymbol{U})\right) \text{ uniformly for } \boldsymbol{U} \in R^k.
\end{aligned} \tag{5.13}$$

Then

$$\left|\overline{G(\boldsymbol{U})} - \left\langle\mathcal{V}; G(\widetilde{\boldsymbol{U}})\right\rangle\right| \leq \overline{F(\boldsymbol{U})} - \left\langle\mathcal{V}; F(\widetilde{\boldsymbol{U}})\right\rangle.$$

Remark 5.7 The exact meaning of the inequality

$$\left|\overline{G(\boldsymbol{U})} - \langle \mathcal{V}; G(\widetilde{\boldsymbol{U}})\rangle\right| \le \overline{F(\boldsymbol{U})} - \langle \mathcal{V}; F(\widetilde{\boldsymbol{U}})\rangle.$$

is

$$\overline{F(\boldsymbol{U})} - \langle \mathcal{V}; F(\widetilde{\boldsymbol{U}})\rangle \pm \overline{G(\boldsymbol{U})} - \langle \mathcal{V}; G(\widetilde{\boldsymbol{U}})\rangle \ge 0.$$

Proof Applying the Lebesgue Dominant Convergence Theorem, we obtain

$$\langle \mathcal{V}; F_j(\widetilde{\boldsymbol{U}})\rangle \to \langle \mathcal{V}; F(\widetilde{\boldsymbol{U}})\rangle, \ \langle \mathcal{V}; G_j(\widetilde{\boldsymbol{U}})\rangle \to \langle \mathcal{V}; G(\widetilde{\boldsymbol{U}})\rangle \text{ in } L^1(Q) \text{ as } j \to \infty.$$

Consequently, it is enough to compare

$$\left|\overline{G(\boldsymbol{U})} - \langle \mathcal{V}; G_j(\widetilde{\boldsymbol{U}})\rangle\right| \text{ with } \overline{F(\boldsymbol{U})} - \langle \mathcal{V}; F_j(\widetilde{\boldsymbol{U}})\rangle.$$

We have

$$\begin{aligned}
&\int_{\overline{Q}} \varphi \, \mathrm{d}\overline{G(\boldsymbol{U})} - \int_Q \varphi(y) \langle \mathcal{V}_y; G_j(\widetilde{\boldsymbol{U}})\rangle \, \mathrm{d}y \\
&\quad = \lim_{n\to\infty} \int_Q \Big[G(\boldsymbol{U}_n) - G_j(\boldsymbol{U}_n)\Big]\varphi \, \mathrm{d}y, \ \varphi \in C(\overline{Q}),
\end{aligned}$$

and, similarly,

$$\begin{aligned}
&\int_{\overline{Q}} \varphi \, \mathrm{d}\overline{F(\boldsymbol{U})} - \int_Q \varphi(y) \langle \mathcal{V}_y; F_j(\widetilde{\boldsymbol{U}})\rangle \, \mathrm{d}y \\
&\quad = \lim_{n\to\infty} \int_Q \Big[F(\boldsymbol{U}_n) - F_j(\boldsymbol{U}_n)\Big]\varphi \, \mathrm{d}y, \ \varphi \in C(\overline{Q}).
\end{aligned}$$

It follows from the hypothesis (5.13) that for any $\varepsilon > 0$ there exists $j_0 = j_0(\varepsilon)$ such that for all $j \ge j_0$

$$\left|\overline{G(\boldsymbol{U})} - \langle \mathcal{V}; G_j(\widetilde{\boldsymbol{U}})\rangle\right| \le \overline{F(\boldsymbol{U})} - \langle \mathcal{V}; F_j(\widetilde{\boldsymbol{U}})\rangle + \varepsilon \text{ for all } j \ge j_0(\varepsilon).$$

As $\varepsilon > 0$ can be taken arbitrarily small, the desired conclusion follows. □

Corollary 5.1 *Let $Q \subset R^m$ be a bounded domain. Let $F \in C(R^k)$, $G \in C(R^k)$ such that*

$$\limsup_{|\boldsymbol{U}|\to\infty} |G(\boldsymbol{U})| \le \liminf_{|\boldsymbol{U}|\to\infty} F(\boldsymbol{U}).$$

Let $\{U_n\}_{n=1}^{\infty} \subset L^1(Q; R^k)$ *be a sequence generating a Young measure* $\{\mathcal{V}_y\}_{y \in Q}$*,*

$$\|F(U_n)\|_{L^1(Q)} \leq c \text{ uniformly for } n \to \infty.$$

Then

$$\left| \overline{G(U)} - \langle \mathcal{V}; G(\widetilde{U}) \rangle \right| \leq \overline{F(U)} - \langle \mathcal{V}; F(\widetilde{U}) \rangle .$$

Proof Consider a family of functions

$$T_j \in C_c(R^k),\ 0 \leq T_j \leq 1,\ T_j(U) = 1 \text{ for all } |U| \leq j.$$

Set

$$F_j(U) = T_j(U)F(U),\ G_j(U) = T_j(U)G(U),\ j = 1, 2, \ldots$$

It is easy to check that all hypotheses of Proposition 5.3 are satisfied; whence the desired conclusion follows. □

Corollary 5.2 *Let* $Q \subset R^M$ *be a bounded domain. Let* $F : R^k \to [0, \infty]$ *be a l.s.c. function. Let* $\{U_n\}_{n=1}^{\infty} \subset L^1(Q; R^k)$ *be a sequence generating a Young measure* $\{\mathcal{V}_y\}_{y \in Q}$*,*

$$\|F(U_n)\|_{L^1(Q)} \leq c \text{ uniformly for } n \to \infty.$$

Let $G : R^k \to R^n$*,* $G \in C(R^k; R^n)$ *be continuous functions satisfying*

$$\limsup_{|U| \to \infty} |G(U)| < \liminf_{|U| \to \infty} F(U). \tag{5.14}$$

Then

$$\overline{F(U)} - \langle \mathcal{V}; F(\widetilde{U}) \rangle \geq \left| \overline{G(U)} - \langle \mathcal{V}; G(\widetilde{U}) \rangle \right| . \tag{5.15}$$

Remark 5.8 Here $\overline{F(U)} \in \mathcal{M}^+(\overline{Q})$ and $\overline{G(U)} \in \mathcal{M}(\overline{Q}; R^n)$. The inequality (5.15) is understood as

$$\overline{F(U)} - \langle \mathcal{V}; F(\widetilde{U}) \rangle - \xi \cdot \left(\overline{G(U)} + \langle \mathcal{V}; G(\widetilde{U}) \rangle \right) \geq 0$$

for any $\xi \in R^n$, $|\xi| = 1$, cf. Remark 5.7.

Proof Note that the desired result for *continuous* functions F, G follows from Corollary 5.1 applied for any fixed ξ.

To extend it to the class of l.s.c. functions like F, we first observe that there is a sequence of continuous functions $f_n \in C(R^k)$ such that

$$0 \leq f_n \leq F,\ f_n \nearrow F.$$

In view of (5.14), there exists $R > 0$ such that

$$|\boldsymbol{G}(\boldsymbol{U})| < F(\boldsymbol{U}) \text{ whenever } |\boldsymbol{U}| > R.$$

Consider a function

$$T : C^\infty(R^k),\ 0 \le T \le 1,\ T(\boldsymbol{U}) = 0 \text{ for } |\boldsymbol{U}| \le R,\ T(\boldsymbol{U}) = 1 \text{ for } |\boldsymbol{U}| \ge R+1,$$

and construct a sequence

$$F_n(\boldsymbol{U}) = T(\boldsymbol{U}) \max\{|\boldsymbol{G}(\boldsymbol{U})|; f_n(\boldsymbol{U})\}.$$

We have

$$0 \le F_n(\boldsymbol{U}) \le F(\boldsymbol{U}) \text{ and } F_n(\boldsymbol{U}) \ge |\boldsymbol{G}(\boldsymbol{U})| \text{ for all } |\boldsymbol{U}| \ge R+1.$$

Applying Corollary 5.1 componentwise, we get

$$\overline{F_n(\boldsymbol{U})} - \left\langle \mathcal{V}; F_n(\widetilde{\boldsymbol{U}}) \right\rangle \ge \left| \overline{\boldsymbol{G}(\boldsymbol{U})} - \left\langle \mathcal{V}; \boldsymbol{G}(\widetilde{\boldsymbol{U}}) \right\rangle \right|$$

for any n. Thus the proof reduces to showing

$$\overline{F_n(\boldsymbol{U})} - \left\langle \mathcal{V}; F_n(\widetilde{\boldsymbol{U}}) \right\rangle \le \overline{F(\boldsymbol{U})} - \left\langle \mathcal{V}; F(\widetilde{\boldsymbol{U}}) \right\rangle,$$

or

$$\overline{H(\boldsymbol{U})} - \left\langle \mathcal{V}; H(\widetilde{\boldsymbol{U}}) \right\rangle \ge 0 \text{ whenever } H : R^k \to [0, \infty] \text{ is a l.s.c function.}$$

However, the latter was observed in Remark 5.6. □

5.1.2 *Consistent Approximation of the Euler System*

We consider the complete Euler system introduced in Chap. 2 written in terms of the conservative-entropy variables $[\varrho, \boldsymbol{m}, S]$

$$\partial_t \varrho + \mathrm{div}_x \boldsymbol{m} = 0, \tag{5.16}$$

$$\partial_t \boldsymbol{m} + \mathrm{div}_x \left(\frac{\boldsymbol{m} \otimes \boldsymbol{m}}{\varrho} \right) + \nabla_x p(\varrho, S) = 0, \tag{5.17}$$

$$\partial_t \left[\frac{1}{2} \frac{|\boldsymbol{m}|^2}{\varrho} + \varrho e(\varrho, S) \right] + \mathrm{div}_x \left(\left[\frac{1}{2} \frac{|\boldsymbol{m}|^2}{\varrho} + \varrho e(\varrho, S) \right] \frac{\boldsymbol{m}}{\varrho} \right) + \mathrm{div}_x \left(p(\varrho, S) \frac{\boldsymbol{m}}{\varrho} \right) = 0. \tag{5.18}$$

Furthermore, we suppose the fluid is confined to a bounded physical domain $\Omega \subset R^d$ with impermeable boundary,

$$\boldsymbol{m} \cdot \boldsymbol{n}|_{\partial\Omega} = 0. \tag{5.19}$$

Finally, we recall the entropy balance satisfied by any strong solution of (5.16)–(5.18):

$$\partial_t S + \mathrm{div}_x \left(S \frac{\boldsymbol{m}}{\varrho} \right) = 0. \tag{5.20}$$

Remark 5.9 We point out that the entropy balance (5.20) and the energy balance (5.18) are entirely *equivalent* in the context of smooth solutions. Moreover, at least in the context of gas dynamics, the energy balance reflecting the First law of thermodynamics should be retained in some form even for weak solutions, whereas (5.20) may be replaced by the inequality

$$\partial_t S + \mathrm{div}_x \left(S \frac{\boldsymbol{m}}{\varrho} \right) \geq 0. \tag{5.21}$$

The concept of dissipative solution developed in this chapter is even more general relaxing both (5.18) and (5.20) to inequalities, see Definition 5.2 below.

For definiteness, we consider the Boyle–Mariotte EOS,

$$p = (\gamma - 1)\varrho e, \ e = c_v \vartheta, \ c_v = \frac{1}{\gamma - 1}, \ \gamma > 1. \tag{5.22}$$

As observed in Sect. 2.2.4, relation (5.18) yields

$$p(\varrho, S) = \varrho^\gamma \exp\left((\gamma - 1) \frac{S}{\varrho} \right), \ \frac{\partial p(\varrho, S)}{\partial S} = \frac{1}{c_v} \frac{p(\varrho, S)}{\varrho}.$$

Consequently, the balance of internal energy reads:

$$0 = \partial_t e + \boldsymbol{u} \cdot \nabla_x e + \frac{p}{\varrho} \mathrm{div}_x \boldsymbol{u} = c_v \left[\left(\partial_t \vartheta + \boldsymbol{u} \cdot \nabla_x \vartheta \right) + \frac{\partial p(\varrho, S)}{\partial S} \mathrm{div}_x \boldsymbol{u} \right], \text{ with } \boldsymbol{u} \equiv \frac{\boldsymbol{m}}{\varrho}. \tag{5.23}$$

We recall from Sect. 2.2.4 that the extended functions

$$[\varrho, \boldsymbol{m}] \in R^{d+1} \mapsto \frac{1}{2} \frac{|\boldsymbol{m}|^2}{\varrho} = \begin{cases} \frac{1}{2} \frac{|\boldsymbol{m}|^2}{\varrho} \text{ if } \varrho > 0, \ \boldsymbol{m} \in R^d, \\ 0 \text{ if } \varrho = 0, \ \boldsymbol{m} = 0, \\ \infty \text{ otherwise,} \end{cases} \tag{5.24}$$

and

$$[\varrho, S] \in R^2 \mapsto (\gamma - 1)\varrho e(\varrho, S) = \begin{cases} \varrho^\gamma \exp\left(\frac{S}{c_v \varrho}\right) \text{ if } \varrho > 0, \ S \in R, \\ 0 \text{ if } \varrho = 0, \ S \leq 0, \\ \infty \text{ otherwise} \end{cases} \tag{5.25}$$

are convex l.s.c. ranging in $[0, \infty]$. In particular, the total energy

$$E(\varrho, \boldsymbol{m}, S) = \frac{1}{2} \frac{|\boldsymbol{m}|^2}{\varrho} + \varrho e(\varrho, S),$$

extended as in (5.24), (5.25) enjoys the same property.

Definition 5.2 **(Consistent approximation for Euler systems)**
A sequence $\{\varrho_n, \boldsymbol{m}_n, S_n\}_{n=1}^{\infty}$ is a *consistent approximation* of the Euler system (5.16)–(5.18) in $(0, T) \times \Omega$, with the boundary condition (5.19), and the initial data $[\varrho_0, \boldsymbol{m}_0, S_0]$ if:

- **(energy inequality)** there is a sequence $\{\varrho_{0,n}, \boldsymbol{m}_{0,n}, S_{0,n}\}_{n=1}^{\infty}$,

$$\begin{aligned} &\varrho_{0,n} \to \varrho_0 \text{ weakly in } L^1(\Omega), \ \boldsymbol{m}_{0,n} \to \boldsymbol{m}_0 \text{ weakly in } L^1(\Omega; R^d), \\ &S_{0,n} \to S_0 \text{ weakly in } L^1(\Omega), \end{aligned} \tag{5.26}$$

  and

$$\int_\Omega \left[\frac{1}{2} \frac{|\boldsymbol{m}_{0,n}|^2}{\varrho_{0,n}} + \varrho_{0,n} e(\varrho_{0,n}, S_{0,n})\right] \mathrm{d}x \to \int_\Omega \left[\frac{1}{2} \frac{|\boldsymbol{m}_0|^2}{\varrho_0} + \varrho_0 e(\varrho_0, S_0)\right] \mathrm{d}x < \infty \tag{5.27}$$

  satisfying

$$\begin{aligned} &\int_\Omega \left[\frac{1}{2} \frac{|\boldsymbol{m}_n|^2}{\varrho_n} + \varrho_n e(\varrho_n, S_n)\right] (\tau, \cdot) \, \mathrm{d}x \\ &\leq \int_\Omega \left[\frac{1}{2} \frac{|\boldsymbol{m}_{0,n}|^2}{\varrho_{0,n}} + \varrho_{0,n} e(\varrho_{0,n}, S_{0,n})\right] \mathrm{d}x + e_n^1 \end{aligned} \tag{5.28}$$

  for a.a. $0 \leq \tau \leq T$, where

$$e_n^1 \to 0 \text{ as } n \to \infty;$$

- **(minimum entropy principle)** there exists $\underline{s} \in R$ such that

$$S_n \geq \varrho_n \underline{s} \text{ a.a. in } (0, T) \times \Omega; \tag{5.29}$$

- (**equation of continuity**) the integral identity

$$\int_0^T \int_\Omega \left[\varrho_n \partial_t \varphi + \boldsymbol{m}_n \cdot \nabla_x \varphi \right] \, \mathrm{d}x \, \mathrm{d}t = - \int_\Omega \varrho_{0,n} \varphi(0, \cdot) \, \mathrm{d}x + e_n^2[\varphi] \tag{5.30}$$

holds for any $\varphi \in C_c^1([0,T) \times \overline{\Omega})$, where

$$e_n^2[\varphi] \to 0 \text{ as } n \to \infty \text{ for any } \varphi \in C_c^M([0,T) \times \overline{\Omega}),$$

where $M \geq 1$ is a positive integer;
- (**momentum equation**) the integral identity

$$\begin{aligned} &\int_0^T \int_\Omega \left[\boldsymbol{m}_n \partial_t \boldsymbol{\varphi} + 1_{\varrho_n > 0} \frac{\boldsymbol{m}_n \otimes \boldsymbol{m}_n}{\varrho_n} : \nabla_x \boldsymbol{\varphi} + 1_{\varrho_n > 0} p(\varrho_n, S_n) \mathrm{div}_x \boldsymbol{\varphi} \right] \mathrm{d}x \, \mathrm{d}t \\ &\quad = - \int_\Omega \boldsymbol{m}_{0,n} \boldsymbol{\varphi}(0, \cdot) \, \mathrm{d}x + e_n^3[\boldsymbol{\varphi}] \end{aligned} \tag{5.31}$$

holds for any $\boldsymbol{\varphi} \in C_c^1([0,T) \times \overline{\Omega}; R^d)$, $\boldsymbol{\varphi} \cdot \boldsymbol{n}|_{\partial\Omega} = 0$, where

$$e_n^3[\boldsymbol{\varphi}] \to 0 \text{ as } n \to \infty \text{ for any } \boldsymbol{\varphi} \in C_c^M([0,T) \times \overline{\Omega}; R^d);$$

- (**entropy inequality**) the integral inequality

$$\int_0^T \int_\Omega \left[S_n \partial_t \varphi + 1_{\varrho_n > 0} \left(S_n \frac{\boldsymbol{m}_n}{\varrho_n} \right) \cdot \nabla_x \varphi \right] \mathrm{d}x \, \mathrm{d}t \leq - \int_\Omega S_{0,n} \varphi(0, \cdot) \, \mathrm{d}x + e_n^4[\varphi] \tag{5.32}$$

holds for any $\varphi \in C_c^1([0,T) \times \overline{\Omega})$, $\varphi \geq 0$, where

$$e_n^4[\varphi] \to 0 \text{ as } n \to \infty \text{ for any } \varphi \in C_c^M([0,T) \times \overline{\Omega}).$$

Remark 5.10 The hypotheses (5.26), (5.27), together with the minimum entropy principle (5.29), are equivalent to the strong convergence of the initial data in a suitable Lebesgue space.

Remark 5.11 There are several ways how to obtain a consistent approximation. In this monograph, the consistent approximations will be provided by suitable *numerical schemes*. The value of the parameter M that characterizes the required smoothness of test functions then depends on the scheme. In the case of periodic boundary conditions $\Omega = \mathbb{T}^d$, we may consider $M = \infty$. In the case of general domain, one should be aware of possible problems related to regularity of the boundary $\partial\Omega$.

It follows from (5.27), (5.28) that the total energy

$$\mathcal{E}_n = \int_{\Omega} \left[\frac{1}{2} \frac{|\boldsymbol{m}_n|^2}{\varrho_n} + \varrho_n e(\varrho_n, \boldsymbol{m}_n) \right] (\tau, \cdot) \, \mathrm{d}x \leq \mathcal{E}_0 \tag{5.33}$$

remains bounded for a.a. τ uniformly for $n \to \infty$. In particular $\varrho_n(t, x) \geq 0$ a.a. in $(0, T) \times \Omega$, and

$$\boldsymbol{m}_n = 0 \text{ a.a. on the vacuum set } \{\varrho_n = 0\} \tag{5.34}$$

cf. (5.24). Similarly, comparing (5.25) with (5.29) we deduce

$$S_n = 0 \text{ a.a. on the vacuum set } \{\varrho_n = 0\}. \tag{5.35}$$

Next, we claim that a consistent approximation of the Euler system admits the following uniform bounds.

Lemma 5.2 (Energy bounds)
Let $\Omega \subset R^d$ be a bounded domain. Suppose that

$$\{\varrho_n, \boldsymbol{m}_n, S_n\}_{n=1}^{\infty}$$

is a consistent approximation of the complete Euler system in $Q = (0, T) \times \Omega$ in the sense of Definition 5.2.
Then

$$\operatorname*{ess\,sup}_{\tau \in (0,T)} \|\varrho_n(\tau, \cdot)\|_{L^\gamma(\Omega)} \leq c(E_0, \underline{s}); \tag{5.36}$$

$$\operatorname*{ess\,sup}_{\tau \in (0,T)} \|S_n(\tau, \cdot)\|_{L^\gamma(\Omega)} \leq c(E_0, \underline{s}); \tag{5.37}$$

$$\operatorname*{ess\,sup}_{\tau \in (0,T)} \|\boldsymbol{m}_n(\tau, \cdot)\|_{L^{\frac{2\gamma}{\gamma+1}}(\Omega; R^d)} \leq c(E_0, \underline{s}); \tag{5.38}$$

uniformly for $n \to \infty$, where the bounds depend only on the initial energy upper bound in (5.33) *and the entropy lower bound* (5.29).

Proof As $S_n \geq \underline{s}\varrho_n$, we get

$$\exp\left(\frac{S_n}{c_v \varrho_n}\right) \geq \exp\left(\frac{\underline{s}}{c_v}\right) > 0;$$

whence (5.36).

Next, by the Hölder inequality,

$$\|\boldsymbol{m}_n\|_{L^{\frac{2\gamma}{\gamma+1}}(\Omega; R^d)} \leq \|\sqrt{\varrho_n}\|_{L^{2\gamma}(\Omega)} \left\| \frac{|\boldsymbol{m}_n|}{\sqrt{\varrho_n}} \right\|_{L^2(\Omega; R^d)},$$

where the right hand side integrals are bounded in view of (5.33), (5.36).

Finally,

$$|S_n| \leq |\underline{s}|\varrho_n \text{ if } S_n \leq 0,$$

while

$$\varrho_n^\gamma \exp\left(\frac{S_n}{c_v \varrho_n}\right) = c_v^{-\gamma} \exp\left(\frac{S_n}{c_v \varrho_n}\right) \left(\frac{S}{c_v \varrho_n}\right)^{-\gamma} S_n^\gamma \geq c(\gamma) S_n^\gamma \text{ whenever } S_n \geq 0.$$

Thus (5.37) follows from (5.33), (5.36). □

Finally, we claim the bound

$$\operatorname*{ess\,sup}_{\tau \in (0,T)} \left\| \frac{S_n}{\sqrt{\varrho_n}}(\tau, \cdot) \right\|_{L^{2\gamma}(\Omega)} \leq c(E_0, \underline{s}). \tag{5.39}$$

Indeed one has

$$\left| \frac{S_n}{\sqrt{\varrho_n}} \right| \leq -\underline{s}\sqrt{\varrho_n} \text{ if } S_n \leq 0. \tag{5.40}$$

If $S_n > 0$, we get

$$\begin{aligned} \varrho_n^\gamma \exp\left(\frac{S_n}{c_v \varrho_n}\right) &= \varrho_n^\gamma \exp\left(\frac{S_n}{\sqrt{\varrho_n}} \frac{1}{c_v \sqrt{\varrho_n}}\right) \\ &= c_v^{-2\gamma} \frac{\exp\left(\frac{S_n}{\sqrt{\varrho_n}} \frac{1}{c_v \sqrt{\varrho_n}}\right)}{\left(\frac{S_n}{\sqrt{\varrho_n}} \frac{1}{c_v \sqrt{\varrho_n}}\right)^{2\gamma}} \left(\frac{S_n}{\sqrt{\varrho_n}}\right)^{2\gamma} \gtrsim \left(\frac{S_n}{\sqrt{\varrho_n}}\right)^{2\gamma}. \end{aligned} \tag{5.41}$$

Consequently, (5.39) follows from (5.40), (5.41), and the energy bounds established in Lemma 5.2.

5.1.3 Asymptotic Limit of Consistent Approximation of the Euler System

Our goal is to identify the limit of a consistent approximation $\{\varrho_n, \boldsymbol{m}_n, S_n\}_{n=1}^\infty$ for $n \to \infty$. The resulting object will represent a generalized solution of the Euler system. Noticing that the only uniform bounds available are the L^p-bounds established in Lemma 5.2, we may anticipate that this procedure takes us *beyond* the class of weak (distributional) solutions introduced in Chap. 4. In the course of the limit process, we will systematically extract subsequences of the original sequence that will be *not* relabeled unless explicitly specified.

In accordance with the uniform bounds obtained in Lemma 5.2 we may assume that

$$\begin{aligned} \varrho_n &\to \varrho \text{ weakly-(*) in } L^\infty(0,T;L^\gamma(\Omega)),\\ S_n &\to S \text{ weakly-(*) in } L^\infty(0,T;L^\gamma(\Omega)),\\ \boldsymbol{m}_n &\to \boldsymbol{m} \text{ weakly-(*) in } L^\infty(0,T;L^{\frac{2\gamma}{\gamma+1}}(\Omega;R^d)) \end{aligned} \tag{5.42}$$

extracting suitable subsequences as the case may be. In addition, by virtue of (5.28), (5.29),

$$\varrho \geq 0,\ S \geq \varrho \underline{s} \text{ a.a. in } (0,T)\times\Omega. \tag{5.43}$$

5.1.3.1 Equation of Continuity

Letting $n \to \infty$ in the approximate equation of continuity (5.30) we obtain

$$\int_0^T \int_\Omega \left[\varrho \partial_t \varphi + \boldsymbol{m} \cdot \nabla_x \varphi \right] \mathrm{d}x\,\mathrm{d}t = - \int_\Omega \varrho_0 \varphi \ \mathrm{d}x \tag{5.44}$$

for any test function $\varphi \in C_c^M([0,T) \times \overline{\Omega})$. Here, the class of test functions can be extended to $\varphi \in C_c([0,T);\overline{\Omega}) \cap W^{1,\infty}((0,T)\times\Omega)$ by a simple density argument. Moreover, exactly as in Sect. 2.1.3, we deduce

$$\varrho \in C_{\text{weak}}([0,T];L^\gamma(\Omega)).$$

Thus (5.44) can be equivalently written in the form

$$\left[\int_\Omega \varrho\varphi \ \mathrm{d}x \right]_{t=0}^{t=\tau} = \int_0^\tau \int_\Omega \left[\varrho \partial_t \varphi + \boldsymbol{m} \cdot \nabla_x \varphi \right] \mathrm{d}x\,\mathrm{d}t \tag{5.45}$$

for any $0 \leq \tau \leq T$, and any $\varphi \in W^{1,\infty}((0,T)\times\Omega)$.

5.1.3.2 Energy Balance

In accordance with (5.28), the total energy

$$E_n = E(\varrho_n, \boldsymbol{m}_n, S_n) = \frac{1}{2}\frac{|\boldsymbol{m}_n|^2}{\varrho_n} + \varrho_n e(\varrho_n, S_n)$$

is bounded in $L^\infty(0,T;L^1(\Omega)) \hookrightarrow L^\infty(0,T;\mathcal{M}^+(\overline{\Omega}))$. Consequently, again for a suitable subsequence, we may assume that

$$E(\varrho_n, \boldsymbol{m}_n, S_n) \to \overline{E(\varrho, \boldsymbol{m}, S)} \text{ weakly-(*) in } L^\infty(0, T; \mathcal{M}^+(\overline{\Omega})).$$

Besides, in accordance with Proposition 5.1, the vector valued sequence

$$(t, x) \in (0, T) \times \Omega \mapsto \left[\varrho_n(t, x), \boldsymbol{m}_n(t, x), S_n(t, x)\right] \in R^{d+2}$$

generates a Young measure $\left\{\mathcal{V}_{t,x}\right\}_{(t,x)\in(0,T)\times\Omega}$. In terms of Definition 5.1, we introduce:

- **energy concentration defect**

$$\mathfrak{E}_{cd} \equiv \overline{E(\varrho, \boldsymbol{m}, S)} - \left\langle \mathcal{V}; E(\widetilde{\varrho}, \widetilde{\boldsymbol{m}}, \widetilde{S}) \right\rangle;$$

- **energy oscillation defect**

$$\mathfrak{E}_{od} \equiv \left\langle \mathcal{V}; E(\widetilde{\varrho}, \widetilde{\boldsymbol{m}}, \widetilde{S}) \right\rangle - E(\varrho, \boldsymbol{m}, S);$$

- **energy defect**

$$\mathfrak{E} = \mathfrak{E}_{cd} + \mathfrak{E}_{od} = \overline{E(\varrho, \boldsymbol{m}, S)} - E(\varrho, \boldsymbol{m}, S).$$

As the total energy $E = E(\varrho, \boldsymbol{m}, S)$ is a convex l.s.c. function of its arguments, cf. (5.24), (5.25), we may apply Corollary 5.2 to obtain

$$\mathfrak{E}_{cd} \in L^\infty(0, T; \mathcal{M}^+(\overline{\Omega})), \tag{5.46}$$

and, by Jensen's inequality,

$$\begin{gathered} \mathfrak{E}_{od} \in L^\infty(0, T; L^1(\Omega)), \ \mathfrak{E}_{od} \geq 0 \text{ a.a. in } (0, T) \times \Omega, \\ \text{in particular, } \ \mathfrak{E} \in L^\infty(0, T; \mathcal{M}^+(\overline{\Omega})). \end{gathered} \tag{5.47}$$

Finally, it follows from (5.28), (5.27) that

$$\begin{aligned} \int_{\overline{\Omega}} \mathrm{d}\overline{E(\varrho, \boldsymbol{m}, S)}(\tau) &= \int_{\Omega} \left\langle \mathcal{V}_{\tau,x}; E(\widetilde{\varrho}, \widetilde{\boldsymbol{m}}, \widetilde{S}) \right\rangle \, \mathrm{d}x + \int_{\overline{\Omega}} \mathrm{d}\mathfrak{E}_{cd}(\tau) \\ &= \int_{\Omega} \left[\frac{1}{2} \frac{|\boldsymbol{m}|^2}{\varrho} + \varrho e(\varrho, S) \right] (\tau, \cdot) \, \mathrm{d}x + \int_{\overline{\Omega}} \mathrm{d}\mathfrak{E}(\tau) \\ &\leq \int_{\Omega} \left[\frac{1}{2} \frac{|\boldsymbol{m}_0|^2}{\varrho_0} + \varrho_0 e(\varrho_0, S_0) \right] \, \mathrm{d}x \end{aligned} \tag{5.48}$$

for a.a. $\tau \in (0, T)$. Relation (5.48) can be seen as the asymptotic limit of the total energy balance.

Remark 5.12 If, in addition to (5.28), the approximate total energy

$$\tau \mapsto \int_{\Omega} E_n(\tau, \cdot) \, dx$$

is a nonincreasing function of $\tau \in (0, T)$, the same is true for the limit

$$\tau \mapsto \int_{\Omega} \overline{E(\varrho, \boldsymbol{m}, S)}(\tau) \, dx.$$

5.1.3.3 Momentum Equation

We have everything at hand to perform the limit in the momentum equation (5.31). As the energy of the approximate solutions is finite, we deduce

$$1_{\varrho_n > 0} p(\varrho_n, S_n) = (\gamma - 1)\varrho_n e(\varrho_n, S_n) \text{ a.a. in } (0, T) \times \Omega.$$

Consequently, repeating the above arguments based on convexity of the pressure/internal energy EOS, we obtain

$$1_{\varrho_n > 0} p(\varrho_n, S_n) \to \overline{1_{\varrho > 0} p(\varrho, S)} \text{ weakly-(*) in } L^\infty(0, T; \mathcal{M}^+(\overline{\Omega})),$$

with the corresponding pressure concentration/oscillation defects:

$$\begin{aligned} \mathfrak{P}_{cd} &= \overline{1_{\varrho > 0} p(\varrho, S)} - \left\langle \mathcal{V}; 1_{\widetilde{\varrho} > 0} p(\widetilde{\varrho}, \widetilde{S}) \right\rangle \\ &= (\gamma - 1) \left[\overline{\varrho e(\varrho, S)} - \left\langle \mathcal{V}; \widetilde{\varrho} e(\widetilde{\varrho}, \widetilde{S}) \right\rangle \right] \in L^\infty(0, T; \mathcal{M}^+(\overline{\Omega})), \end{aligned}$$

and

$$\mathfrak{P}_{od} = \left\langle \mathcal{V}; 1_{\widetilde{\varrho} > 0} p(\widetilde{\varrho}, \widetilde{S}) \right\rangle - 1_{\varrho > 0} p(\varrho, S) = (\gamma - 1) \left[\left\langle \mathcal{V}; \widetilde{\varrho} e(\widetilde{\varrho}, \widetilde{S}) \right\rangle - \varrho e(\varrho, S) \right] \geq 0.$$

Here, it is important to notice that the functions

$$[\widetilde{\varrho}, \widetilde{S}] \in R^2 \mapsto 1_{\widetilde{\varrho} > 0} p(\widetilde{\varrho}, \widetilde{S}) \text{ and } [\widetilde{\varrho}, \widetilde{S}] \in R^2 \mapsto (\gamma - 1)\widetilde{\varrho} e(\widetilde{\varrho}, \widetilde{S}),$$

where the latter is defined through (5.25), *are not* the same. However, they coincide on the set

$$\mathcal{F}_1 \equiv \left\{ [\widetilde{\varrho}, \widetilde{S}] \,\middle|\, \widetilde{\varrho} > 0, \ \widetilde{S} > 0 \right\} \cup \left\{ [\widetilde{\varrho}, \widetilde{S}] \,\middle|\, \widetilde{\varrho} \geq 0, \ \widetilde{S} \geq \varrho \underline{s} \right\}.$$

Moreover, thanks to the energy inequality (5.28) and the lower bound on the entropy (5.29),

$$[\varrho_n(t,x), S_n(t,x)] \in \mathcal{F}_1 \ \text{ for a.a. } (t,x) \in (0,T) \times \Omega.$$

Similarly, by virtue of the limit energy inequality (5.48), we have

$$\mathcal{V}_{t,x}[\mathcal{F}_1] = 1 \ \text{ for a.a. } (t,x) \in (0,T) \times \Omega,$$

meaning

$$\widetilde{\varrho} e(\widetilde{\varrho}, \widetilde{S}) = 1_{\widetilde{\varrho}>0} p(\widetilde{\varrho}, \widetilde{S}) \ \mathcal{V}_{t,x} - \text{ a.s.}$$

for a.a. $(t,x) \in (0,T) \times \Omega$.

Using similar arguments, we handle the convective term. First, we introduce the set

$$\mathcal{F}_2 = \left\{ [\widetilde{\varrho}, \widetilde{\boldsymbol{m}}] \,\middle|\, \widetilde{\varrho} > 0, \ \widetilde{\boldsymbol{m}} \in R \right\} \cup \left\{ [\widetilde{\varrho}, \widetilde{\boldsymbol{m}}] \,\middle|\, \widetilde{\varrho} = 0, \ \widetilde{\boldsymbol{m}} = 0 \right\}$$

and observe, similarly to the above, that

$$[\varrho_n(t,x), \boldsymbol{m}_n(t,x)] \in \mathcal{F}_2 \text{ for a.a. } (t,x) \in (0,T),$$

and

$$\mathcal{V}_{t,x}[\mathcal{F}_2] = 1 \text{ for a.a. } (t,x) \in (0,T). \tag{5.49}$$

Next, we have

$$1_{\varrho_n>0} \frac{\boldsymbol{m}_n \otimes \boldsymbol{m}_n}{\varrho_n} \text{ bounded in } L^\infty(0,T; \mathcal{M}^+(\overline{\Omega}; R^{d\times d}_{\rm sym}));$$

whence, up to a subsequence,

$$1_{\varrho_n>0} \frac{\boldsymbol{m}_n \otimes \boldsymbol{m}_n}{\varrho_n} \to \overline{1_{\varrho>0} \frac{\boldsymbol{m} \otimes \boldsymbol{m}}{\varrho}} \text{ weakly-(*) in } L^\infty(0,T; \mathcal{M}^+(\overline{\Omega}; R^{d\times d}_{\rm sym})),$$

in particular,

$$\begin{aligned} \operatorname{tr}\left[1_{\varrho_n>0} \frac{\boldsymbol{m}_n \otimes \boldsymbol{m}_n}{\varrho} \right] &= \frac{|\boldsymbol{m}_n|^2}{\varrho_n} \\ \to \overline{\frac{|\boldsymbol{m}|^2}{\varrho}} &= \operatorname{tr}\left[\overline{1_{\varrho>0} \frac{\boldsymbol{m} \otimes \boldsymbol{m}}{\varrho}} \right] \text{ weakly-(*) in } L^\infty(0,T; \mathcal{M}^+(\overline{\Omega})). \end{aligned}$$

We introduce the defects associated to the convective term,

$$\mathfrak{C}_{cd} = \overline{1_{\varrho>0} \frac{\boldsymbol{m} \otimes \boldsymbol{m}}{\varrho}} - \left\langle \mathcal{V}; 1_{\widetilde{\varrho}>0} \frac{\widetilde{\boldsymbol{m}} \otimes \widetilde{\boldsymbol{m}}}{\widetilde{\varrho}} \right\rangle, \ \mathfrak{C}_{od} = \left\langle \mathcal{V}; 1_{\widetilde{\varrho}>0} \frac{\widetilde{\boldsymbol{m}} \otimes \widetilde{\boldsymbol{m}}}{\widetilde{\varrho}} \right\rangle - 1_{\varrho>0} \frac{\boldsymbol{m} \otimes \boldsymbol{m}}{\varrho},$$

and

$$\mathfrak{C}_d = \mathfrak{C}_{cd} + \mathfrak{C}_{od} = \overline{1_{\varrho>0} \frac{\boldsymbol{m} \otimes \boldsymbol{m}}{\varrho}} - 1_{\varrho>0} \frac{\boldsymbol{m} \otimes \boldsymbol{m}}{\varrho}.$$

An important observation is that the defects are positively semidefinite symmetric matrices, more precisely,

$$\mathfrak{C}_{cd} \in L^\infty(0, T; \mathcal{M}^+(\overline{\Omega}; R^{d \times d}_{\rm sym})),$$
$$\mathfrak{C}_{od} \in L^\infty(0, T; L^1(\Omega; R^{d \times d}_{\rm sym})), \ \mathfrak{C}_{od}(t, x) \geq 0 \text{ for a.a. } (t, x).$$

Indeed, for any $\xi \in R^d$, the function

$$[\varrho, \boldsymbol{m}] \mapsto \begin{cases} \frac{|\boldsymbol{m} \cdot \xi|^2}{\varrho} \text{ if } \varrho > 0, \\ 0 \text{ if } \varrho = 0, \ \boldsymbol{m} \cdot \xi = 0, \\ \infty \text{ otherwise} \end{cases}$$

is convex l.s.c. Consequently, in view of Corollary 5.2,

$$\mathfrak{C}_{cd} : (\xi \otimes \xi) = \overline{\frac{|\boldsymbol{m} \cdot \xi|^2}{\varrho}} - \left\langle \mathcal{V}; 1_{\widetilde{\varrho}>0} \frac{|\widetilde{\boldsymbol{m}} \cdot \xi|^2}{\widetilde{\varrho}} \right\rangle \geq 0, \tag{5.50}$$

while, by Jensen's inequality and (5.49),

$$\mathfrak{C}_{od} : (\xi \otimes \xi) = \left\langle \mathcal{V}; 1_{\widetilde{\varrho}>0} \frac{|\widetilde{\boldsymbol{m}} \cdot \xi|^2}{\widetilde{\varrho}} \right\rangle - 1_{\varrho>0} \frac{|\boldsymbol{m} \cdot \xi|^2}{\varrho} \geq 0 \tag{5.51}$$

for any $\xi \in R^d$.

Summing up the previous discussion, we can let $n \to \infty$ in the momentum equation (5.31) obtaining

$$\int_0^T \int_\Omega \left[\boldsymbol{m} \cdot \partial_t \boldsymbol{\varphi} + \left\langle \mathcal{V}; 1_{\varrho>0} \frac{\boldsymbol{m} \otimes \boldsymbol{m}}{\varrho} \right\rangle : \nabla_x \boldsymbol{\varphi} + \left\langle \mathcal{V}; 1_{\varrho>0} p(\varrho, S) \right\rangle {\rm div}_x \boldsymbol{\varphi} \right] \, {\rm d}x \, {\rm d}t$$
$$= - \int_\Omega \boldsymbol{m}_0 \cdot \boldsymbol{\varphi}(0, \cdot) \, {\rm d}x - \int_0^T \int_{\overline{\Omega}} \nabla_x \boldsymbol{\varphi} : {\rm d}\mathfrak{C}_{cd}(t) \, {\rm d}t - \int_0^T \int_{\overline{\Omega}} {\rm div}_x \boldsymbol{\varphi} \, {\rm d}\mathfrak{P}_{cd}(t) \, {\rm d}t, \tag{5.52}$$

for any test function $\boldsymbol{\varphi} \in C^M_c([0, T) \times \overline{\Omega}; R^d)$, $\boldsymbol{\varphi} \cdot \boldsymbol{n}|_{\partial\Omega} = 0$, $M \geq 1$ is a positive integer. The concentration defects

$$\mathfrak{C}_{cd} \in L^\infty(0, T; \mathcal{M}^+(\overline{\Omega}; R^{d \times d}_{\rm sym})), \ \mathfrak{P}_{cd} \in L^\infty(0, T; \mathcal{M}^+(\overline{\Omega}))$$

are controlled by the energy defect,

$$\underline{d}\ \mathfrak{C}_{cd} \leq \left(\mathrm{tr}[\mathfrak{C}_{cd}] + \mathfrak{P}_{cd}\right) \leq \overline{d}\ \mathfrak{C}_{cd} \ \text{ for certain constants } \ 0 < \underline{d} \leq \overline{d}. \tag{5.53}$$

Relation (5.53) is crucial for the weak-strong uniqueness principle we show later in this chapter. It follows from (5.52) that

$$\boldsymbol{m} \in C_{\mathrm{weak}}([0,T]; L^{\frac{2\gamma}{\gamma+1}}(\Omega; R^d));$$

whence (5.52) may be rewritten as

$$\begin{aligned}
&\left[\int_\Omega \boldsymbol{m}\cdot\boldsymbol{\varphi}\ \mathrm{d}x\right]_{t=0}^{t=\tau} \\
&= \int_0^\tau\int_\Omega \left[\boldsymbol{m}\cdot\partial_t\boldsymbol{\varphi} + \left\langle \mathcal{V}; 1_{\varrho>0}\frac{\boldsymbol{m}\otimes\boldsymbol{m}}{\varrho}\right\rangle : \nabla_x\boldsymbol{\varphi} + \left\langle\mathcal{V}; 1_{\varrho>0}p(\varrho,S)\right\rangle \mathrm{div}_x\boldsymbol{\varphi}\right]\ \mathrm{d}x\,\mathrm{d}t \\
&+ \int_0^\tau\int_{\overline{\Omega}} \nabla_x\boldsymbol{\varphi} : \mathrm{d}\mathfrak{C}_{cd}(t)\,\mathrm{d}t + \int_0^\tau\int_{\overline{\Omega}} \mathrm{div}_x\boldsymbol{\varphi}\ \mathrm{d}\mathfrak{P}_{cd}(t)\,\mathrm{d}t
\end{aligned} \tag{5.54}$$

for any $0 \leq \tau \leq T$, and any $\boldsymbol{\varphi} \in C^M([0,T]\times\overline{\Omega}; R^d)$, $\boldsymbol{\varphi}\cdot\boldsymbol{n}|_{\partial\Omega} = 0$.

Furthermore, we may rewrite (5.54) using the oscillation defects as

$$\begin{aligned}
&\left[\int_\Omega \boldsymbol{m}\cdot\boldsymbol{\varphi}\ \mathrm{d}x\right]_{t=0}^{t=\tau} \\
&= \int_0^\tau\int_\Omega \left[\boldsymbol{m}\cdot\partial_t\boldsymbol{\varphi} + 1_{\varrho>0}\frac{\boldsymbol{m}\otimes\boldsymbol{m}}{\varrho} : \nabla_x\boldsymbol{\varphi} + 1_{\varrho>0}p(\varrho,S)\mathrm{div}_x\boldsymbol{\varphi}\right]\ \mathrm{d}x\,\mathrm{d}t \\
&+ \int_0^\tau\int_\Omega \nabla_x\boldsymbol{\varphi} : \mathfrak{C}_{od}\ \mathrm{d}x\,\mathrm{d}t + \int_0^\tau\int_{\overline{\Omega}} \nabla_x\boldsymbol{\varphi} : \mathrm{d}\mathfrak{C}_{cd}(t)\,\mathrm{d}t \\
&+ \int_0^\tau\int_\Omega \mathrm{div}_x\boldsymbol{\varphi}\mathfrak{P}_{od}\ \mathrm{d}x\,\mathrm{d}t + \int_0^\tau\int_{\overline{\Omega}} \mathrm{div}_x\boldsymbol{\varphi}\ \mathrm{d}\mathfrak{P}_{cd}(t)\,\mathrm{d}t
\end{aligned} \tag{5.55}$$

for any $0 \leq \tau \leq T$, and any $\boldsymbol{\varphi} \in C^M([0,T]\times\overline{\Omega}; R^d)$, $\boldsymbol{\varphi}\cdot\boldsymbol{n}|_{\partial\Omega} = 0$. Finally, introducing the *Reynolds defect*

$$\mathfrak{R} \equiv \mathfrak{C}_{cd} + \mathfrak{C}_{od} + \left(\mathfrak{P}_{cd} + \mathfrak{P}_{od}\right)\mathbb{I} \in L^\infty(0,T; \mathcal{M}^+(\overline{\Omega}; R^{d\times d}_{\mathrm{sym}})),$$

we may rewrite (5.55) in a concise form

$$\left[\int_{\Omega} \boldsymbol{m} \cdot \boldsymbol{\varphi} \, \mathrm{d}x\right]_{t=0}^{t=\tau}$$

$$= \int_0^\tau \int_\Omega \left[\boldsymbol{m} \cdot \partial_t \boldsymbol{\varphi} + 1_{\varrho>0} \frac{\boldsymbol{m} \otimes \boldsymbol{m}}{\varrho} : \nabla_x \boldsymbol{\varphi} + 1_{\varrho>0} p(\varrho, S) \mathrm{div}_x \boldsymbol{\varphi} \right] \mathrm{d}x \, \mathrm{d}t \tag{5.56}$$

$$+ \int_0^\tau \int_{\overline{\Omega}} \nabla_x \boldsymbol{\varphi} : \mathrm{d}\mathfrak{R}(t) \, \mathrm{d}t$$

for any $0 \leq \tau \leq T$, and any $\boldsymbol{\varphi} \in C^M([0,T] \times \overline{\Omega}; R^d)$, $\boldsymbol{\varphi} \cdot \boldsymbol{n}|_{\partial\Omega} = 0$, $M \geq 1$ is a positive integer. Moreover, the trace of Reynolds defect $\mathfrak{R}$ is proportional to the energy defect $\mathfrak{E}$, more specifically,

$$\underline{d} \, \mathfrak{E} \leq \mathrm{tr}[\mathfrak{R}] \leq \overline{d} \, \mathfrak{E} \text{ for certain constants } 0 < \underline{d} \leq \overline{d}. \tag{5.57}$$

Remark 5.13 Similarly to the limit form of the equation of continuity (5.44), the weak formulation of the momentum equation (5.56) does not contain explicitly the Young measure $\{\mathcal{V}_{t,x}\}_{(t,x)\in(0,T)\times\Omega}$.

5.1.3.4 Entropy Balance

Our ultimate goal is to perform the limit in the entropy inequality (5.32). Using the uniform bound (5.39) we may write

$$1_{\varrho_n>0} S_n \frac{\boldsymbol{m}_n}{\varrho_n} = 1_{\varrho_n<0} \frac{S_n}{\sqrt{\varrho_n}} \frac{\boldsymbol{m}_n}{\sqrt{\varrho_n}},$$

$$1_{\varrho_n>0} \frac{S_n}{\sqrt{\varrho_n}} \text{ bounded in } L^\infty(0,T; L^{2\gamma}(\Omega)),$$

$$1_{\varrho_n>0} \frac{\boldsymbol{m}_n}{\sqrt{\varrho_n}} \text{ bounded in } L^\infty(0,T; L^2(\Omega; R^d)).$$

Consequently, again up to a suitable subsequence,

$$1_{\varrho_n>0} S_n \frac{\boldsymbol{m}_n}{\varrho_n} \to \overline{1_{\varrho>0} S \frac{\boldsymbol{m}}{\varrho}} \text{ weakly-(*) in } L^\infty(0,T; L^{\frac{2\gamma}{\gamma+1}}(\Omega; R^d)).$$

One is tempted to say that

$$\overline{1_{\varrho>0}S\frac{\boldsymbol{m}}{\varrho}} = \left\langle \mathcal{V}; 1_{\widetilde{\varrho}>0}\widetilde{S}\frac{\widetilde{\boldsymbol{m}}}{\widetilde{\varrho}} \right\rangle, \tag{5.58}$$

which is indeed true, however, the argument is not straightforward as we deal with a composition with a *discontinuous* Borel function. To see (5.58), consider a sequence of functions

$$\chi_k(\varrho) \in C^\infty(R),\ \chi_k(z) = \begin{cases} 0 \text{ if } \varrho \leq \frac{1}{2k}, \\ \in [0,1] \text{ if } \frac{1}{2k} < \varrho < \frac{1}{k}, \\ 1 \text{ if } \varrho \geq \frac{1}{k}. \end{cases}$$

Writing

$$1_{\varrho_n>0}S_n\frac{\boldsymbol{m}_n}{\varrho_n} = 1_{\varrho_n>0}(1-\chi_k(\varrho_n))S_n\frac{\boldsymbol{m}_n}{\varrho_n} + \chi_k(\varrho_n)S_n\frac{\boldsymbol{m}_n}{\varrho_n}$$

we deduce

$$\chi_k(\varrho_n)S_n\frac{\boldsymbol{m}_n}{\varrho_n} \to \left\langle \mathcal{V}; \chi_k(\widetilde{\varrho})\widetilde{S}\frac{\widetilde{\boldsymbol{m}}}{\widetilde{\varrho}} \right\rangle \text{ weakly-(*) in } L^\infty(0,T;L^{\frac{2\gamma}{\gamma+1}}(\Omega;R^d)),$$

where, by the Lebesgue Dominant Convergence Theorem,

$$\left\langle \mathcal{V}; \chi_k(\widetilde{\varrho})\widetilde{S}\frac{\widetilde{\boldsymbol{m}}}{\widetilde{\varrho}} \right\rangle \to \left\langle \mathcal{V}; 1_{\widetilde{\varrho}>0}\widetilde{S}\frac{\widetilde{\boldsymbol{m}}}{\widetilde{\varrho}} \right\rangle \text{ as } k \to \infty.$$

Consequently, we have to show that

$$\begin{aligned} &\int_0^T \int_\Omega 1_{\varrho_n>0}(1-\chi_k(\varrho_n)) \left| S_n\frac{\boldsymbol{m}_n}{\varrho_n} \right| \,\mathrm{d}x\,\mathrm{d}t \\ &\leq \int_0^T \int_\Omega 1_{0<\varrho_n\leq\frac{1}{k}} \left| S_n\frac{\boldsymbol{m}_n}{\varrho_n} \right| \,\mathrm{d}x\,\mathrm{d}t \to 0 \text{ as } k \to \infty \end{aligned} \tag{5.59}$$

uniformly in $n = 1, 2 \ldots$

To see (5.59) we evoke the energy estimates, specifically (5.28), (5.29), obtaining

$$\operatorname*{ess\,sup}_{t\in(0,T)} \int_\Omega 1_{\varrho_n>0}\varrho_n^\gamma \exp\left((\gamma-1)\frac{S_n}{\varrho_n}\right) \,\mathrm{d}x \leq c,\ S_n \geq \varrho_n \underline{s} \text{ uniformly for } n \to \infty.$$

Consequently, for any $\beta \geq 1$, there is a constant $c = c(\beta, \underline{s})$ such that

$$\int_\Omega 1_{\varrho_n>0}\varrho_n^{\gamma-\beta}|S_n|^\beta \ \mathrm{d}x \leq c(\beta, \underline{s}) \int_\Omega 1_{\varrho_n>0}\varrho_n^\gamma \exp\left((\gamma-1)\frac{S_n}{\varrho_n}\right) \ \mathrm{d}x.$$

In particular, for $\beta = 3\gamma$ we obtain

$$\operatorname*{ess\,sup}_{t\in(0,T)} \int_\Omega 1_{\varrho_n>0}\left(\frac{S_n}{\varrho_n^{\frac{2}{3}}}\right)^{3\gamma} \ \mathrm{d}x \leq c.$$

Going back to (5.59) we conclude

$$\begin{aligned} \int_\Omega 1_{0<\varrho_n\leq\frac{1}{k}}\left|S_n\frac{\boldsymbol{m}_n}{\varrho_n}\right| \ \mathrm{d}x &= \int_\Omega 1_{0<\varrho_n\leq\frac{1}{k}}\varrho_n^{\frac{1}{6}}\left|\frac{S_n}{\varrho_n^{\frac{2}{3}}}\frac{\boldsymbol{m}_n}{\sqrt{\varrho_n}}\right| \ \mathrm{d}x \\ &\leq k^{-\frac{1}{6}}\int_\Omega 1_{\varrho_n>0}\left|\frac{S_n}{\varrho_n^{\frac{2}{3}}}\frac{\boldsymbol{m}_n}{\sqrt{\varrho_n}}\right| \ \mathrm{d}x \to 0 \end{aligned} \tag{5.60}$$

for $k \to \infty$ uniformly in $t \in [0, T]$ and $n = 1, 2, \ldots$. We have shown (5.59); whence (5.58).

Summing up the previous discussion, we may write the limit entropy balance letting $n \to \infty$ in (5.32):

$$\int_0^T\int_\Omega \left[S\partial_t\varphi + \left\langle\mathcal{V}; 1_{\widetilde{\varrho}>0}\left(\widetilde{S}\frac{\widetilde{\boldsymbol{m}}}{\widetilde{\varrho}}\right)\right\rangle\cdot\nabla_x\varphi\right] \ \mathrm{d}x\,\mathrm{d}t \leq -\int_\Omega S_0\varphi(0,\cdot) \ \mathrm{d}x \tag{5.61}$$

for any $\varphi \in C_c^M([0, T) \times \overline{\Omega})$, $\varphi \geq 0$, $M \geq 1$ being a positive integer. Moreover, the validity of (5.61) can be extended to the class of test functions $\varphi \in C_c([0, T) \times \overline{\Omega}) \cap W^{1,\infty}((0, T) \times \Omega)$, $\varphi \geq 0$, via a density argument.

The total entropy, in general, is not weakly continuous in time due to the inequality in (5.61). However, we may deduce that the function

$$\tau \mapsto \int_\Omega S(\tau, x)\phi(x) \ \mathrm{d}x + \int_0^\tau\int_\Omega \left\langle\mathcal{V}; 1_{\widetilde{\varrho}>0}\left(\widetilde{S}\frac{\widetilde{\boldsymbol{m}}}{\widetilde{\varrho}}\right)\right\rangle\cdot\nabla_x\phi(x) \ \mathrm{d}x\,\mathrm{d}t$$

is nondecreasing for any $\phi \in C(\overline{\Omega})$, $\phi \geq 0$. In particular, we may correctly define the one-sided limits

$$\int_\Omega S(\tau\pm, x)\phi(x) \ \mathrm{d}x,$$

see Sect. 2.1.4 for details. Accordingly, the entropy balance (5.61) can be written in the form

$$\left[\int_{\Omega} S\varphi \,\mathrm{d}x\right]_{t=\tau_1-}^{t=\tau_2+} \geq \int_{\tau_1}^{\tau_2}\int_{\Omega}\left[S\partial_t\varphi + \left\langle \mathcal{V};\, 1_{\widetilde{\varrho}>0}\left(\widetilde{S}\frac{\widetilde{\boldsymbol{m}}}{\widetilde{\varrho}}\right)\right\rangle \cdot \nabla_x\varphi\right]\mathrm{d}x\,\mathrm{d}t \qquad (5.62)$$

for any $0 \leq \tau_1 \leq \tau_2 < T$, and any $\varphi \in W^{1,\infty}((0,T)\times\Omega)$, $\varphi \geq 0$. Here, we have set

$$S(0-,\cdot) = S_0.$$

5.1.3.5 Conclusion

The equation of continuity (5.45), the momentum balance (5.54), the entropy inequality (5.62), together with the energy inequality (5.48) characterize the target problem satisfied by the asymptotic limit of a consistent approximation of the complete Euler system (5.16)–(5.19). Note that the same result holds for the space periodic boundary conditions $\Omega = \mathbb{T}^d$. In this case, the test functions $\boldsymbol{\varphi}$ in the momentum balance (5.54) do not have to satisfy any additional boundary conditions, and validity of (5.54) can be extended to $\boldsymbol{\varphi} \in C^1([0,T]\times\Omega; R^d)$ by density argument. The Young measure $\{\mathcal{V}_{t,x}\}_{(t,x)\in(0,T)\times\Omega}$ represents a measure-valued solution of the system introduced in detail in the next section.

5.1.4 Measure-Valued Solutions to the Complete Euler System

Motivated by the discussion in the preceding section, we introduce the concept of dissipative measure-valued solution to the complete Euler system.

Definition 5.3 (DISSIPATIVE MEASURE-VALUED (DMV) SOLUTION TO EULER SYSTEM)

Let $\Omega \subset R^d$, $d = 1, 2, 3$ be a bounded domain. A parametrized probability measure $\{\mathcal{V}_{t,x}\}_{(t,x)\in(0,T)\times\Omega}$,

$$\mathcal{V} \in L^\infty((0,T)\times\Omega; \mathcal{P}(R^{d+2})),\ R^{d+2} = \left\{[\widetilde{\varrho}, \widetilde{\boldsymbol{m}}, \widetilde{S}] \,\middle|\, \widetilde{\varrho} \in R,\ \widetilde{\boldsymbol{m}} \in R^d,\ \widetilde{S} \in R\right\},$$

is called *dissipative measure-valued (DMV) solution* of the Euler system (5.16)–(5.18), with the boundary conditions (5.19), and the initial conditions $[\varrho_0, \boldsymbol{m}_0, S_0]$ if the following holds:

- (**lower bound on density and entropy**)

$$\mathcal{V}_{t,x}\Big[\{\widetilde{\varrho} \geq 0,\ \widetilde{S} \geq \underline{s}\widetilde{\varrho}\}\Big] = 1 \text{ for a.a. } (t,x); \tag{5.63}$$

- (**energy inequality**) the integral inequality

$$\begin{aligned} &\int_{\Omega} \left\langle \mathcal{V}_{\tau,x}; \frac{1}{2}\frac{|\widetilde{\boldsymbol{m}}|^2}{\widetilde{\varrho}} + \widetilde{\varrho} e(\widetilde{\varrho}, \widetilde{S}) \right\rangle \,\mathrm{d}x + \int_{\overline{\Omega}} \mathrm{d}\mathfrak{E}_{cd}(\tau) \\ &\leq \int_{\Omega} \left[\frac{1}{2}\frac{|\boldsymbol{m}_0|^2}{\varrho_0} + \varrho_0 e(\varrho_0, S_0) \right] \,\mathrm{d}x \end{aligned} \tag{5.64}$$

holds for a.a. $0 \leq \tau \leq T$, with the energy concentration defect

$$\mathfrak{E}_{cd} \in L^{\infty}(0,T; \mathcal{M}^{+}(\overline{\Omega}));$$

- (**equation of continuity**)

$$\langle \mathcal{V}; \widetilde{\varrho} \rangle \in C_{\text{weak}}([0,T]; L^{\gamma}(\Omega)),\ \left\langle \mathcal{V}_{0,x}; \widetilde{\varrho} \right\rangle = \varrho_0(x) \text{ for a.a. } x \in \Omega,$$

and the integral identity

$$\left[\int_{\Omega} \langle \mathcal{V}; \widetilde{\varrho} \rangle \,\varphi \,\mathrm{d}x \right]_{t=0}^{t=\tau} = \int_0^{\tau} \int_{\Omega} \Big[\langle \mathcal{V}; \widetilde{\varrho} \rangle \,\partial_t \varphi + \langle \mathcal{V}; \widetilde{\boldsymbol{m}} \rangle \cdot \nabla_x \varphi \Big] \,\mathrm{d}x \,\mathrm{d}t \tag{5.65}$$

for any $0 \leq \tau \leq T$, and any $\varphi \in W^{1,\infty}((0,T) \times \Omega)$;

- (**momentum equation**)

$$\langle \mathcal{V}; \widetilde{\boldsymbol{m}} \rangle \in C_{\text{weak}}([0,T]; L^{\frac{2\gamma}{\gamma+1}}(\Omega; R^d)),\ \left\langle \mathcal{V}_{0,x}; \widetilde{\boldsymbol{m}} \right\rangle = \boldsymbol{m}_0(x) \text{ for a.a. } x \in \Omega,$$

and the integral identity

$$
\begin{aligned}
&\left[\int_{\Omega} \langle \mathcal{V}; \widetilde{\boldsymbol{m}} \rangle \cdot \boldsymbol{\varphi} \, \mathrm{d}x \right]_{t=0}^{t=\tau} \\
&= \int_0^\tau \int_\Omega \left[\langle \mathcal{V}; \widetilde{\boldsymbol{m}} \rangle \cdot \partial_t \boldsymbol{\varphi} + \left\langle \mathcal{V}; 1_{\widetilde{\varrho}>0} \frac{\widetilde{\boldsymbol{m}} \otimes \widetilde{\boldsymbol{m}}}{\widetilde{\varrho}} \right\rangle : \nabla_x \boldsymbol{\varphi} \right] \mathrm{d}x \, \mathrm{d}t \\
&+ \int_0^\tau \int_\Omega \left\langle \mathcal{V}; 1_{\widetilde{\varrho}>0} p(\widetilde{\varrho}, \widetilde{S}) \right\rangle \mathrm{div}_x \boldsymbol{\varphi} \, \mathrm{d}x \, \mathrm{d}t \\
&+ \int_0^\tau \int_{\overline{\Omega}} \nabla_x \boldsymbol{\varphi} : \mathrm{d}\mathfrak{R}_{cd}(t) \, \mathrm{d}t
\end{aligned}
\tag{5.66}
$$

for any $0 \leq \tau \leq T$, and any $\boldsymbol{\varphi} \in C^M([0, T] \times \overline{\Omega}; R^d)$, $\boldsymbol{\varphi} \cdot \boldsymbol{n}|_{\partial\Omega} = 0$, $M \geq 1$, with Reynolds concentration defect

$$
\mathfrak{R}_{cd} \in L^\infty(0, T; \mathcal{M}^+(\overline{\Omega}; R^{d \times d}_{\mathrm{sym}}))
$$

satisfying

$$
\underline{d} \, \mathfrak{E}_{cd} \leq \mathrm{tr}[\mathfrak{R}_{cd}] \leq \overline{d} \, \mathfrak{E}_{cd} \text{ for some constants } 0 < \underline{d} \leq \overline{d}; \tag{5.67}
$$

- **(entropy balance)**

$$
\int_\Omega \left\langle \mathcal{V}_{\tau\pm,x}; \widetilde{S} \right\rangle \phi(x) \, \mathrm{d}x \equiv \lim_{t \to \tau\pm} \int_\Omega \left\langle \mathcal{V}_{t,x}; \widetilde{S} \right\rangle \phi(x) \, \mathrm{d}x \text{ exists for any } 0 \leq \tau < T,
$$

$$
\int_\Omega \left\langle \mathcal{V}_{0-,x}; \widetilde{S} \right\rangle \phi(x) \, \mathrm{d}x \equiv \int_\Omega S_0 \phi \, \mathrm{d}x \text{ for any } \phi \in C(\overline{\Omega}),
$$

and the integral inequality

$$
\begin{aligned}
&\left[\int_\Omega \left\langle \mathcal{V}; \widetilde{S} \right\rangle \varphi \, \mathrm{d}x \right]_{t=\tau_1-}^{t=\tau_2+} \\
&\geq \int_{\tau_1}^{\tau_2} \int_\Omega \left[\left\langle \mathcal{V}; \widetilde{S} \right\rangle \partial_t \varphi + \left\langle \mathcal{V}; 1_{\widetilde{\varrho}>0} \left(\widetilde{S} \frac{\widetilde{\boldsymbol{m}}}{\widetilde{\varrho}} \right) \right\rangle \cdot \nabla_x \varphi \right] \mathrm{d}x \, \mathrm{d}t
\end{aligned}
\tag{5.68}
$$

for any $0 \leq \tau_1 \leq \tau_2 < T$, and any $\varphi \in W^{1,\infty}((0, T) \times \Omega)$, $\varphi \geq 0$.

Remark 5.14 Similar definition can be used in the case of periodic boundary conditions $\Omega = \mathbb{T}^d$. The only modification concerns the class of test functions $\boldsymbol{\varphi}$ in the momentum equation (5.66), namely,

$$\boldsymbol{\varphi} \in C^1([0,T] \times \mathbb{T}^d; R^d).$$

Note that (5.66) contains the concentration defect measure $\mathfrak{R}_{cd}$ acting on $\nabla_x \boldsymbol{\varphi}$, which requires the latter to be continuous in $\overline{\Omega}$ for a.a. $t \in (0,T)$.

The measure $\mathfrak{E}_{cd}$ reflects the energy *concentration defect*, while $\mathfrak{R}_{cd}$ is *Reynolds concentration defect*. As we have seen in the preceding section, they describe possible concentrations in the sequence of approximate solutions. The important property of DMV solutions is the energy balance (5.64) that is crucial for proving the weak DMV-strong uniqueness principle later in this chapter. Summarizing the discussion in Sect. 5.1.3 we obtain the following result.

Theorem 5.1 (Asymptotic limit of consistent approximation)
Let $\Omega \subset R^d$, $d = 1, 2, 3$ be a bounded domain. Let $\{\varrho_n, \boldsymbol{m}_n, S_n\}_{n=1}^\infty$ be a consistent approximation of the complete Euler system specified in Definition 5.2.
Then, up to a subsequence,

- $\{\varrho_n, \boldsymbol{m}_n, S_n\}_{n=1}^\infty$ *generates a Young measure* $\left\{\mathcal{V}_{t,x}\right\}_{(t,x)\in(0,T)\times\Omega}$

$$\mathcal{V}_{t,x} \in \mathcal{P}(R^{d+2}),\ R^{d+2} = \left\{\widetilde{\varrho} \in R, \widetilde{\boldsymbol{m}} \in R^d,\ \widetilde{S} \in R\right\};$$

-
$$E_n(\varrho_n, \boldsymbol{m}_n, S_n) = \left[\frac{1}{2}\frac{|\boldsymbol{m}_n|^2}{\varrho_n} + \varrho_n e(\varrho_n, \boldsymbol{m}_n)\right] \to \overline{E(\varrho, \boldsymbol{m}, S)}$$

  weakly-(*) in $L^\infty(0,T; \mathcal{M}(\overline{\Omega}))$;

-
$$1_{\varrho_n>0}\left(\frac{\boldsymbol{m}_n \otimes \boldsymbol{m}_n}{\varrho_n} + p(\varrho_n, S_n)\mathbb{I}\right) \to \overline{1_{\varrho>0}\left(\frac{\boldsymbol{m} \otimes \boldsymbol{m}}{\varrho} + p(\varrho, S)\mathbb{I}\right)}$$

  weakly-(*) in $L^\infty(0,T; \mathcal{M}(\overline{\Omega}; R^{d\times d}_{\rm sym}))$.

The Young measure $\{\mathcal{V}_{t,x}\}_{(t,x)\in(0,T)\times\Omega}$ is a dissipative measure-valued solution of the Euler system in the sense of Definition 5.3, with the defect measures

$$\mathfrak{E}_{cd} = \overline{E(\varrho, \boldsymbol{m}, S)} - \left\langle \mathcal{V}; E(\widetilde{\varrho}, \widetilde{\boldsymbol{m}}, \widetilde{S})\right\rangle \in L^\infty(0,T; \mathcal{M}^+(\overline{\Omega})),$$

$$\mathfrak{R}_{cd} = \overline{1_{\varrho>0}\left(\frac{\boldsymbol{m} \otimes \boldsymbol{m}}{\varrho} + p(\varrho, S)\mathbb{I}\right)} - \left\langle \mathcal{V}; 1_{\widetilde{\varrho}>0}\left(\frac{\widetilde{\boldsymbol{m}} \otimes \widetilde{\boldsymbol{m}}}{\widetilde{\varrho}} + p(\widetilde{\varrho}, \widetilde{S})\mathbb{I}\right)\right\rangle$$
$$\in L^\infty(0,T; \mathcal{M}^+(\overline{\Omega}; R^{d\times d}_{\rm sym})).$$

The class of DMV solutions is the smallest one to capture all possible limits of consistent approximations, at least in view of the available estimates. Note, however, that the definition of the DMV solution is intrinsic, meaning it does require any kind of generating sequence. The following theorem reveals some basic properties of DMV solutions.

Theorem 5.2 (Properties of DMV solutions)
Let $\Omega \subset R^d$, $d = 1, 2, 3$ be a bounded domain.

- *Let $\{\mathcal{V}^n\}_{n=1}^{\infty}$ be a sequence of DMV solutions in the sense of Definition 5.3 starting from the initial data $[\varrho_{0,n}, \boldsymbol{m}_{0,n}, S_{0,n}]$ satisfying*

$$\varrho_{0,n} \to \varrho_0 \text{ weakly in } L^1(\Omega),\ \boldsymbol{m}_{0,n} \to \boldsymbol{m}_0 \text{ weakly in } L^1(\Omega; R^d),$$
$$S_{0,n} \to S_0 \text{ weakly in } L^1(\Omega),$$

$$\int_{\Omega} \left[\frac{1}{2} \frac{|\boldsymbol{m}_{0,n}|^2}{\varrho_{0,n}} + \varrho_{0,n} e(\varrho_{0,n}, S_{0,n}) \right] \,\mathrm{d}x \to \int_{\Omega} \left[\frac{1}{2} \frac{|\boldsymbol{m}_0|^2}{\varrho_0} + \varrho_0 e(\varrho_0, S_0) \right] \,\mathrm{d}x < \infty.$$

Then, up to a subsequence,

$$\mathcal{V}^n \to \mathcal{V} \text{ weakly-(*) in } L^{\infty}(0, T; \mathcal{P}(R^{d+2})),$$

where $\mathcal{V}$ is a DMV solution starting from the initial data $[\varrho_0, \boldsymbol{m}_0, S_0]$.
- *The set of all DMV solutions starting from the* same *initial data is convex.*

Remark 5.15 Note that a convex combination of two DMV solutions starting from *different* initial data may not be a DMV solution as the convex combination of the initial data may violate the energy inequality (5.64).

Proof The arguments are very similar to those in Sect. 5.1.3. First observe that the energy balance (5.64), together with the lower bound for the entropy (5.63), imply that a family of measures

$$\left\{ \frac{1}{|B|} \int_B \mathcal{V}^n_{t,x} \,\mathrm{d}x \,\mathrm{d}t \right\}_{n=1}^{\infty},\ B \subset (0, T) \times \Omega \text{ an open set}$$

is uniformly tight. Consequently, up to a subsequence,

$$\mathcal{V}^n \to \mathcal{V} \text{ weakly-(*) in } L^{\infty}(0, T; \mathcal{P}(R^{d+2})),$$

where

$$\mathcal{V}_{t,x} \left\{ \widetilde{\varrho} \geq 0,\ \widetilde{S} \geq \underline{s} \widetilde{\varrho} \right\} = 1.$$

Next, as the energy $E = E(\widetilde{\varrho}, \widetilde{\boldsymbol{m}}, \widetilde{S})$ is a convex function, we apply Jensen's inequality to the energy balance (5.64) to obtain uniform bounds on the expected values similar to (5.42). In particular,

$$\begin{aligned}
\left\langle \mathcal{V}^n; \widetilde{\varrho} \right\rangle &\to \langle \mathcal{V}; \widetilde{\varrho} \rangle \text{ weakly-(*) in } L^\infty(0, T; L^\gamma(\Omega)),\\
\left\langle \mathcal{V}^n; \widetilde{\boldsymbol{m}} \right\rangle &\to \langle \mathcal{V}; \widetilde{\boldsymbol{m}} \rangle \text{ weakly-(*) in } L^\infty(0, T; L^{\frac{2\gamma}{\gamma+1}}(\Omega; R^d)),\\
\left\langle \mathcal{V}^n; \widetilde{S} \right\rangle &\to \left\langle \mathcal{V}; \widetilde{S} \right\rangle \text{ weakly-(*) in } L^\infty(0, T; L^\gamma(\Omega)).
\end{aligned}$$

Obviously, the limit satisfies the equation of continuity (5.65).

Consider the cut-off functions

$$\chi_k(\varrho) \in C^\infty(R),\ \chi_k(z) = \begin{cases} 0 \text{ if } \varrho \leq \frac{1}{2k}, \\ \in [0, 1] \text{ if } \frac{1}{2k} < \varrho < \frac{1}{k}, \\ 1 \text{ if } \varrho \geq \frac{1}{k}, \end{cases}$$

and

$$T_k(Y) \in C^\infty(R),\ T_k(Y) = \begin{cases} 0 \text{ if } Y \leq \frac{1}{k}, \\ T_k'(Y) \geq 0,\ T_k(Y) \leq k \text{ otherwise.} \end{cases},$$
$$T_k(Y) \nearrow Y \text{ as } k \to \infty \text{ for any } Y \geq 0.$$

Now, we rewrite the kinetic energy in the form

$$\frac{1}{2}\frac{|\widetilde{\boldsymbol{m}}|^2}{\widetilde{\varrho}} = \chi_k(\widetilde{\varrho})\frac{1}{2}\frac{T_k(|\widetilde{\boldsymbol{m}}|)^2}{\widetilde{\varrho}} + (1 - \chi_k(\widetilde{\varrho}))\frac{1}{2}\frac{|\widetilde{\boldsymbol{m}}|^2}{\widetilde{\varrho}} + \chi_k(\widetilde{\varrho})\frac{1}{2}\frac{|\widetilde{\boldsymbol{m}}|^2 - T_k(|\widetilde{\boldsymbol{m}}|)^2}{\widetilde{\varrho}}, \tag{5.69}$$

and, similarly the convective term,

$$\begin{aligned}
1_{\widetilde{\varrho}>0}\frac{\widetilde{\boldsymbol{m}} \otimes \widetilde{\boldsymbol{m}}}{\widetilde{\varrho}} &= \chi_k(\widetilde{\varrho})\frac{\widetilde{\boldsymbol{m}} \otimes \widetilde{\boldsymbol{m}}}{\widetilde{\varrho}}\frac{T_k(|\widetilde{\boldsymbol{m}}|)^2}{|\widetilde{\boldsymbol{m}}|^2} + 1_{\widetilde{\varrho}>0}(1 - \chi_k(\widetilde{\varrho}))\frac{\widetilde{\boldsymbol{m}} \otimes \widetilde{\boldsymbol{m}}}{\widetilde{\varrho}} \\
&\quad + \chi_k(\widetilde{\varrho})\frac{\widetilde{\boldsymbol{m}} \otimes \widetilde{\boldsymbol{m}}}{\widetilde{\varrho}}\left(1 - \frac{T_k(|\widetilde{\boldsymbol{m}}|)^2}{|\widetilde{\boldsymbol{m}}|^2}\right).
\end{aligned} \tag{5.70}$$

Thanks to the energy balance (5.64)

$$\mathcal{V}^n_{t,x}\Big[\{\widetilde{\varrho} > 0\} \cup \{\widetilde{\varrho} = 0,\ \widetilde{\boldsymbol{m}} = 0\} \Big] = \mathcal{V}_{t,x}\Big[\{\widetilde{\varrho} > 0\} \cup \{\widetilde{\varrho} = 0,\ \widetilde{\boldsymbol{m}} = 0\} \Big] = 1$$

for a.a. (t, x). Consequently,

$$\left\langle \mathcal{V}^n; \chi_k(\widetilde{\varrho})\frac{1}{2}\frac{T_k(|\widetilde{\boldsymbol{m}}|)^2}{\widetilde{\varrho}} \right\rangle \to \left\langle \mathcal{V}; \chi_k(\widetilde{\varrho})\frac{1}{2}\frac{T_k(|\widetilde{\boldsymbol{m}}|)^2}{\widetilde{\varrho}} \right\rangle \text{ weakly-(*) in } L^\infty((0, T) \times \Omega)$$

as $n \to \infty$, and, by the Lebesgue Dominant Convergence Theorem,

$$\left\langle \mathcal{V}_{t,x}; \chi_k(\widetilde{\varrho}) \frac{1}{2} \frac{T_k(|\widetilde{\boldsymbol{m}}|)^2}{\widetilde{\varrho}} \right\rangle \to \left\langle \mathcal{V}_{t,x}; \frac{1}{2} \frac{|\widetilde{\boldsymbol{m}}|^2}{\widetilde{\varrho}} \right\rangle \text{ as } k \to \infty \text{ for a.a. } t, x.$$

Similarly,

$$\left\langle \mathcal{V}^n; \chi_k(\widetilde{\varrho}) \frac{\widetilde{\boldsymbol{m}} \otimes \widetilde{\boldsymbol{m}}}{\widetilde{\varrho}} \frac{T_k(|\widetilde{\boldsymbol{m}}|)^2}{|\widetilde{\boldsymbol{m}}|^2} \right\rangle \to \left\langle \mathcal{V}; \chi_k(\widetilde{\varrho}) \frac{\widetilde{\boldsymbol{m}} \otimes \widetilde{\boldsymbol{m}}}{\widetilde{\varrho}} \frac{T_k(|\widetilde{\boldsymbol{m}}|)^2}{|\widetilde{\boldsymbol{m}}|^2} \right\rangle$$
$$\text{weakly-(*) in } L^\infty((0,T) \times \Omega; R^{d\times d}_{\rm sym}) \text{ as } n \to \infty,$$

and

$$\left\langle \mathcal{V}_{t,x}; \chi_k(\widetilde{\varrho}) \frac{\widetilde{\boldsymbol{m}} \otimes \widetilde{\boldsymbol{m}}}{\widetilde{\varrho}} \frac{T_k(|\widetilde{\boldsymbol{m}}|)^2}{|\widetilde{\boldsymbol{m}}|^2} \right\rangle \to \left\langle \mathcal{V}_{t,x}; 1_{\widetilde{\varrho}>0} \frac{\widetilde{\boldsymbol{m}} \otimes \widetilde{\boldsymbol{m}}}{\widetilde{\varrho}} \right\rangle \text{ as } k \to \infty \text{ for a.a. } t, x.$$

The remaining terms in (5.69), (5.70) will contribute to the concentration defect. We have

$$\left\langle \mathcal{V}^n; (1 - \chi_k(\widetilde{\varrho})) \frac{1}{2} \frac{|\widetilde{\boldsymbol{m}}|^2}{\widetilde{\varrho}} \right\rangle \to \overline{(1 - \chi_k(\varrho)) \frac{1}{2} \frac{|\boldsymbol{m}|^2}{\varrho}}$$
$$\text{weakly-(*) in } L^\infty(0,T; \mathcal{M}^+(\overline{\Omega})) \text{ as } n \to \infty,$$

$$\overline{(1 - \chi_k(\varrho)) \frac{1}{2} \frac{|\boldsymbol{m}|^2}{\varrho}} \to \mathfrak{E}^1_{cd} \text{ weakly-(*) in } L^\infty(0,T; \mathcal{M}^+(\overline{\Omega})) \text{ as } k \to \infty.$$

Similarly,

$$\left\langle \mathcal{V}^n; 1_{\widetilde{\varrho}>0} (1 - \chi_k(\widetilde{\varrho})) \frac{\widetilde{\boldsymbol{m}} \otimes \widetilde{\boldsymbol{m}}}{\widetilde{\varrho}} \right\rangle \to \overline{1_{\varrho>0} (1 - \chi_k(\varrho)) \frac{\boldsymbol{m} \otimes \boldsymbol{m}}{\varrho}}$$
$$\text{weakly-(*) in } L^\infty(0,T; \mathcal{M}^+(\overline{\Omega}; R^{d\times d}_{\rm sym})) \text{ as } n \to \infty,$$

and

$$\overline{1_{\varrho>0} (1 - \chi_k(\varrho)) \frac{\boldsymbol{m} \otimes \boldsymbol{m}}{\varrho}} \to \mathfrak{R}^1_{cd} \text{ weakly-(*) in } L^\infty(0,T; \mathcal{M}^+(\overline{\Omega}; R^{d\times d}_{\rm sym})) \text{ as } k \to \infty,$$

where

$$\mathfrak{E}^1_{cd} = \frac{1}{2} {\rm tr}[\mathfrak{R}^1_d].$$

Applying the same treatment to the remaining terms in (5.69), (5.70), and also to the internal energy/pressure we may infer that the limit quantities satisfy the energy balance (5.64), (5.65), with the concentration defect measure satisfying (5.67).

It remains to perform the limit in the entropy balance (5.68). We use the same argument as in Sect. 5.1.3. We write the entropy flux as

$$\left\langle \mathcal{V}^n; 1_{\widetilde{\varrho}} \widetilde{S} \frac{\widetilde{\boldsymbol{m}}}{\widetilde{\varrho}} \right\rangle = \left\langle \mathcal{V}^n; \chi_k(\widetilde{\varrho}) \widetilde{S} \frac{\widetilde{\boldsymbol{m}}}{\widetilde{\varrho}} \right\rangle + \left\langle \mathcal{V}^n; (1 - \chi_k(\widetilde{\varrho})) \widetilde{S} \frac{\widetilde{\boldsymbol{m}}}{\widetilde{\varrho}} \right\rangle,$$

where

$$\left\langle \mathcal{V}^n; \chi_k(\widetilde{\varrho}) \widetilde{S} \frac{\widetilde{\boldsymbol{m}}}{\widetilde{\varrho}} \right\rangle \to \left\langle \mathcal{V}; \chi_k(\widetilde{\varrho}) \widetilde{S} \frac{\widetilde{\boldsymbol{m}}}{\widetilde{\varrho}} \right\rangle \text{ weakly-(*) in } L^\infty(0, T; L^{\frac{2\gamma}{\gamma+1}}(\Omega; R^d)),$$

and

$$\left\langle \mathcal{V}_{t,x}; \chi_k(\widetilde{\varrho}) \widetilde{S} \frac{\widetilde{\boldsymbol{m}}}{\widetilde{\varrho}} \right\rangle \to \left\langle \mathcal{V}_{t,x}; 1_{\widetilde{\varrho}>0} \widetilde{S} \frac{\widetilde{\boldsymbol{m}}}{\widetilde{\varrho}} \right\rangle \text{ as } k \to \infty \text{ for a.a. } t, x.$$

The remaining term

$$\left\langle \mathcal{V}^n; (1 - \chi_k(\widetilde{\varrho})) \widetilde{S} \frac{\widetilde{\boldsymbol{m}}}{\widetilde{\varrho}} \right\rangle$$

can be handled exactly as in (5.60), namely

$$\left| \left\langle \mathcal{V}^n; (1 - \chi_k(\widetilde{\varrho})) \widetilde{S} \frac{\widetilde{\boldsymbol{m}}}{\widetilde{\varrho}} \right\rangle \right| \leq c k^{-\frac{1}{6}} \left\langle \mathcal{V}^n; 1_{\widetilde{\varrho}>0} \left| \frac{\widetilde{S}}{\widetilde{\varrho}^{\frac{2}{3}}} \frac{\widetilde{\boldsymbol{m}}}{\sqrt{\widetilde{\varrho}}} \right| \right\rangle,$$

where, thanks to the energy and entropy estimates

$$\left\langle \mathcal{V}^n; 1_{\widetilde{\varrho}>0} \left| \frac{\widetilde{S}}{\widetilde{\varrho}^{\frac{2}{3}}} \frac{\widetilde{\boldsymbol{m}}}{\sqrt{\widetilde{\varrho}}} \right| \right\rangle \text{ is bounded in } L^\infty(0, T; L^1(\Omega)) \text{ uniformly for } n \to \infty.$$

As convexity of the set of all DMV solutions emanating from the same initial data is obvious, the proof is complete. □

5.1.5 *Measure-Valued Solutions to the Barotropic (Isentropic) Euler System*

The isentropic Euler system may be seen as a particular case of the complete system, in which the entropy is supposed to be constant. As a result, the pressure p,

$$p = p(\varrho, S) = \varrho^\gamma \exp\left((\gamma - 1) \frac{S}{\varrho} \right) = a\varrho^\gamma, \ a > 0, \ \gamma > 1,$$

depends only on the density. Accordingly, the entropy balance (5.20) become superfluous, and the complete system reduces to

$$\partial_t \varrho + \mathrm{div}_x \boldsymbol{m} = 0,$$
$$\partial_t \boldsymbol{m} + \mathrm{div}_x \left(\frac{\boldsymbol{m} \otimes \boldsymbol{m}}{\varrho} \right) + \nabla_x p(\varrho) = 0, \ p(\varrho) = a\varrho^\gamma, \tag{5.71}$$

together with the impermeability boundary condition

$$\boldsymbol{m} \cdot \boldsymbol{n}|_{\partial\Omega} = 0. \tag{5.72}$$

We occasionally speak about *barotropic* Euler system if the pressure $p = p(\varrho)$ depends only on the mass density.

Obviously, all the results obtained for the complete system transfer without any essential modification to the barotropic Euler system. We summarize the relevant material in the two sections below.

5.1.5.1 Consistent Approximation

Similarly to the complete Euler system, we introduce consistent approximation to its isentropic (or barotropic) version.

Definition 5.4 (CONSISTENT APPROXIMATION OF BAROTROPIC EULER SYSTEM)

A sequence $\{\varrho_n, \boldsymbol{m}_n\}_{n=1}^\infty$ is a *consistent approximation* of the barotropic Euler system (5.71) in $(0, T) \times \Omega$, with the boundary condition (5.72), and the initial data $[\varrho_0, \boldsymbol{m}_0]$ if:

- (**energy inequality**) there is a sequence $\{\varrho_{0,n}, \boldsymbol{m}_{0,n}\}_{n=1}^\infty$,

$$\varrho_{0,n} \to \varrho_0 \text{ weakly in } L^1(\Omega), \ \boldsymbol{m}_{0,n} \to \boldsymbol{m}_0 \text{ weakly in } L^1(\Omega; R^d), \tag{5.73}$$

and

$$\int_\Omega \left[\frac{1}{2} \frac{|\boldsymbol{m}_{0,n}|^2}{\varrho_{0,n}} + P(\varrho_{0,n}) \right] \mathrm{d}x \to \int_\Omega \left[\frac{1}{2} \frac{|\boldsymbol{m}_0|^2}{\varrho_0} + P(\varrho_0) \right] \mathrm{d}x < \infty,$$
$$P'(\varrho)\varrho - P(\varrho) = p(\varrho), \tag{5.74}$$

satisfying

$$\int_\Omega \left[\frac{1}{2} \frac{|\boldsymbol{m}_n|^2}{\varrho_n} + P(\varrho_n) \right] (\tau, \cdot) \, \mathrm{d}x \le \int_\Omega \left[\frac{1}{2} \frac{|\boldsymbol{m}_{0,n}|^2}{\varrho_{0,n}} + P(\varrho_{0,n}) \right] \mathrm{d}x + e_n^1 \tag{5.75}$$

for a.a. $0 \leq \tau \leq T$, where

$$e_n^1 \to 0 \text{ as } n \to \infty;$$

- (**equation of continuity**) the integral identity

$$\int_0^T \int_\Omega \left[\varrho_n \partial_t \varphi + \boldsymbol{m}_n \cdot \nabla_x \varphi \right] \, \mathrm{d}x \, \mathrm{d}t = - \int_\Omega \varrho_{0,n} \varphi(0, \cdot) \, \mathrm{d}x + e_n^2[\varphi] \tag{5.76}$$

holds for any $\varphi \in C_c^1([0, T) \times \overline{\Omega})$, where

$$e_n^2[\varphi] \to 0 \text{ as } n \to \infty \text{ for any } \varphi \in C_c^M([0, T) \times \overline{\Omega}),$$

where $M \geq 1$ is a positive integer;
- (**momentum equation**) the integral identity

$$\begin{aligned} \int_0^T \int_\Omega & \left[\boldsymbol{m}_n \cdot \partial_t \boldsymbol{\varphi} + 1_{\varrho_n > 0} \frac{\boldsymbol{m}_n \otimes \boldsymbol{m}_n}{\varrho_n} : \nabla_x \boldsymbol{\varphi} + p(\varrho_n) \mathrm{div}_x \boldsymbol{\varphi} \right] \mathrm{d}x \, \mathrm{d}t \\ & = - \int_\Omega \boldsymbol{m}_{0,n} \cdot \boldsymbol{\varphi}(0, \cdot) \, \mathrm{d}x + e_n^3[\boldsymbol{\varphi}] \end{aligned} \tag{5.77}$$

holds for any $\boldsymbol{\varphi} \in C_c^1([0, T) \times \overline{\Omega}; R^d)$, $\boldsymbol{\varphi} \cdot \boldsymbol{n}|_{\partial\Omega} = 0$, where

$$e_n^3[\boldsymbol{\varphi}] \to 0 \text{ as } n \to \infty \text{ for any } \boldsymbol{\varphi} \in C_c^M([0, T) \times \overline{\Omega}; R^d)$$

with a positive integer $M \geq 1$.

5.1.5.2 Measure-Valued Solutions

Exactly as in Sect. 5.1.3, the asymptotic limit of a consistent approximation generates a DMV solution. To be consistent with the definition for the complete system, we still assume

$$p(\varrho) = (\gamma - 1)\varrho e(\varrho), \text{ meaning } p(\varrho) = a\varrho^\gamma, \ P(\varrho) = \frac{a}{\gamma - 1}\varrho^\gamma. \tag{5.78}$$

Definition 5.5 (**Dissipative measure-valued (DMV) solution to barotropic Euler system**)
Let $\Omega \subset R^d$, $d = 1, 2, 3$ be a bounded domain. Suppose the pressure is related to the internal energy through (5.78). A parametrized probability measure $\{\mathcal{V}_{t,x}\}_{(t,x) \in (0,T) \times \Omega}$,

$$\mathcal{V} \in L^{\infty}((0,T)\times\Omega; \mathcal{P}(R^{d+1})),\ R^{d+1} = \left\{ [\widetilde{\varrho}, \widetilde{\boldsymbol{m}}] \;\middle|\; \widetilde{\varrho} \in R,\ \widetilde{\boldsymbol{m}} \in R^d \right\},$$

is called *dissipative measure-valued (DMV) solution* of the Euler system (5.71), (5.72), with the initial conditions $[\varrho_0, \boldsymbol{m}_0]$ if the following holds:

- (**energy inequality**) the integral inequality

$$\int_{\Omega} \left\langle \mathcal{V}_{\tau,x}; \frac{1}{2}\frac{|\widetilde{\boldsymbol{m}}|^2}{\widetilde{\varrho}} + P(\widetilde{\varrho}) \right\rangle \mathrm{d}x + \int_{\overline{\Omega}} \mathrm{d}\mathfrak{E}_{cd}(\tau) \leq \int_{\Omega} \left[\frac{1}{2}\frac{|\boldsymbol{m}_0|^2}{\varrho_0} + P(\varrho_0) \right] \mathrm{d}x \tag{5.79}$$

holds for a.a. $0 \leq \tau \leq T$, with the energy concentration defect

$$\mathfrak{E}_{cd} \in L^{\infty}(0,T; \mathcal{M}^{+}(\overline{\Omega}));$$

- (**equation of continuity**)

$$\langle \mathcal{V}; \widetilde{\varrho} \rangle \in C_{\mathrm{weak}}([0,T]; L^{\gamma}(\Omega)),\ \left\langle \mathcal{V}_{0,x}; \widetilde{\varrho} \right\rangle = \varrho_0(x) \text{ for a.a. } x \in \Omega,$$

and the integral identity

$$\left[\int_{\Omega} \langle \mathcal{V}; \widetilde{\varrho} \rangle \varphi \, \mathrm{d}x \right]_{t=0}^{t=\tau} = \int_0^{\tau} \int_{\Omega} \left[\langle \mathcal{V}; \widetilde{\varrho} \rangle \, \partial_t \varphi + \langle \mathcal{V}; \widetilde{\boldsymbol{m}} \rangle \cdot \nabla_x \varphi \right] \mathrm{d}x \, \mathrm{d}t \tag{5.80}$$

for any $0 \leq \tau \leq T$, and any $\varphi \in W^{1,\infty}((0,T)\times\Omega)$;
- (**momentum equation**)

$$\langle \mathcal{V}; \widetilde{\boldsymbol{m}} \rangle \in C_{\mathrm{weak}}([0,T]; L^{\frac{2\gamma}{\gamma+1}}(\Omega; R^d)),\ \left\langle \mathcal{V}_{0,x}; \widetilde{\boldsymbol{m}} \right\rangle = \boldsymbol{m}_0(x) \text{ for a.a. } x \in \Omega,$$

and the integral identity

$$\begin{aligned} &\left[\int_{\Omega} \langle \mathcal{V}; \widetilde{\boldsymbol{m}} \rangle \cdot \boldsymbol{\varphi} \, \mathrm{d}x \right]_{t=0}^{t=\tau} \\ &= \int_0^{\tau} \int_{\Omega} \left[\langle \mathcal{V}; \widetilde{\boldsymbol{m}} \rangle \cdot \partial_t \boldsymbol{\varphi} + \left\langle \mathcal{V}; 1_{\widetilde{\varrho}>0} \frac{\widetilde{\boldsymbol{m}} \otimes \widetilde{\boldsymbol{m}}}{\widetilde{\varrho}} \right\rangle : \nabla_x \boldsymbol{\varphi} + \langle \mathcal{V}; p(\widetilde{\varrho}) \rangle \, \mathrm{div}_x \boldsymbol{\varphi} \right] \mathrm{d}x \, \mathrm{d}t \\ &+ \int_0^{\tau} \int_{\overline{\Omega}} \nabla_x \boldsymbol{\varphi} : \mathrm{d}\mathfrak{R}_{cd}(t) \, \mathrm{d}t \end{aligned} \tag{5.81}$$

for any $0 \leq \tau \leq T$, and any $\boldsymbol{\varphi} \in C^M([0,T]\times\overline{\Omega}; R^d)$, $\boldsymbol{\varphi}\cdot\boldsymbol{n}|_{\partial\Omega} = 0$, $M \geq 1$, with the Reynolds concentration defect

$$\mathfrak{R}_{cd} \in L^\infty(0,T; \mathcal{M}^+(\overline{\Omega}; R^{d\times d}_{\text{sym}}))$$

satisfying

$$\underline{d}\ \mathfrak{E}_{cd} \leq \text{tr}[\mathfrak{R}_{cd}] \leq \overline{d}\ \mathfrak{E}_{cd} \text{ for some constants } 0 < \underline{d} \leq \overline{d}. \tag{5.82}$$

As a direct consequence of the energy inequality (5.79), we get

$$\mathcal{V}_{t,x}\Big[\{\widetilde{\varrho} > 0\} \cup \{\widetilde{\varrho} = 0,\ \widetilde{\boldsymbol{m}} = 0\} \Big] = 1 \text{ for a.a. } (t,x), \tag{5.83}$$

meaning $\varrho \geq 0$ and $\boldsymbol{m}$ vanishes on the vacuum zone $\mathcal{V}_{t,x}$-a.s.

Finally, we reformulate Theorem 5.1 in the context of the barotropic Euler system.

Theorem 5.3 (Asymptotic limit of consistent approximation – barotropic case)

Let $\Omega \subset R^d$, $d = 1, 2, 3$ be a bounded domain. Let the pressure $p = p(\varrho)$ be related to the internal energy through (5.78). Let $\{\varrho_n, \boldsymbol{m}_n\}_{n=1}^\infty$ be a consistent approximation of the barotropic Euler system specified in Definition 5.4.

Then, up to a subsequence,

- *$\{\varrho_n, \boldsymbol{m}_n\}_{n=1}^\infty$ generates a Young measure $\{\mathcal{V}_{t,x}\}_{(t,x)\in(0,T)\times\Omega}$,*

$$\mathcal{V}_{t,x} \in \mathcal{P}(R^{d+1}),\ R^{d+1} = \left\{\widetilde{\varrho} \in R, \widetilde{\boldsymbol{m}} \in R^d\right\};$$

- $$E_n(\varrho_n, \boldsymbol{m}_n) = \left[\frac{1}{2}\frac{|\boldsymbol{m}_n|^2}{\varrho_n} + P(\varrho_n)\right] \to \overline{E(\varrho, \boldsymbol{m})} \text{ weakly-(*) in } L^\infty(0,T; \mathcal{M}(\overline{\Omega}));$$

- $$1_{\varrho_n>0}\frac{\boldsymbol{m}_n \otimes \boldsymbol{m}_n}{\varrho_n} + p(\varrho_n)\mathbb{I} \to \overline{1_{\varrho>0}\frac{\boldsymbol{m}\otimes\boldsymbol{m}}{\varrho} + p(\varrho)\mathbb{I}}$$
$$\text{weakly-(*) in } L^\infty(0,T; \mathcal{M}(\overline{\Omega}; R^{d\times d}_{\text{sym}})).$$

The Young measure $\{\mathcal{V}_{t,x}\}_{(t,x)\in(0,T)\times\Omega}$ is a dissipative measure-valued solution of the barotropic Euler system in the sense of Definition 5.5, with the defect measures

$$\mathfrak{E}_{cd} = \overline{E(\varrho, \boldsymbol{m})} - \langle \mathcal{V}; E(\widetilde{\varrho}, \widetilde{\boldsymbol{m}})\rangle \in L^\infty(0,T; \mathcal{M}^+(\overline{\Omega})),$$

$$\mathfrak{R}_{cd} = \overline{1_{\varrho>0}\frac{\boldsymbol{m}\otimes\boldsymbol{m}}{\varrho} + p(\varrho)\mathbb{I}} - \left\langle \mathcal{V}; 1_{\widetilde{\varrho}>0}\frac{\widetilde{\boldsymbol{m}}\otimes\widetilde{\boldsymbol{m}}}{\widetilde{\varrho}} + p(\widetilde{\varrho})\mathbb{I}\right\rangle \in L^\infty(0,T; \mathcal{M}^+(\overline{\Omega}; R^{d\times d}_{\text{sym}})).$$

5.2 Dissipative Solutions to the Euler System

The measure-valued solutions introduced in the preceding section are probably the smallest class of objects that can accommodate limits of consistent approximations of the Euler system. There are, however, two conceptual difficulties related to representation of solutions via Young measures:

- The mapping
$$t \in [0, T] \mapsto \mathcal{V}_{t,x} \in L^\infty(\Omega; \mathcal{P}(R^{d+2}))$$
is *a priori* not continuous, not even in any weak sense. The continuous quantities are only its barycenter coordinates:
$$t \mapsto \left\langle \mathcal{V}_{t,x}; \widetilde{\varrho} \right\rangle \in C_{\rm weak}([0,T]; L^\gamma(\Omega)),$$
$$t \mapsto \left\langle \mathcal{V}_{t,x}; \widetilde{\boldsymbol{m}} \right\rangle \in C_{\rm weak}([0,T]; L^{\frac{2\gamma}{\gamma+1}}(\Omega; R^d)),$$
and, to certain extent,
$$t \mapsto \left\langle \mathcal{V}_{t,x}; \widetilde{S} \right\rangle \in BV_{\rm weak}([0,T]; L^\gamma(\Omega)).$$
- The specific shape (distribution) of the Young measure depends on the way how the consistent approximation is constructed.

The second problem is particularly relevant when reconstructing the approximate sequence from a discrete set of values produced by a numerical scheme. Consider a very simple example of a sequence $\{r_n\}_{n=1}^N$,

$$r_n = \begin{cases} 1 \text{ for odd } n \\ 0 \text{ for even } n. \end{cases},$$

and a family of discrete functions

$$g_N(x_n) = r_n, \ x_n = \frac{n}{N}, \ n = 0, 1 \dots, N.$$

Now, we extend g_n to a function defined in $[0, 1]$ in two different ways:

$$\widetilde{g}_N(x) = g_N(x_n) \text{ for } x \in [x_n, x_n + \frac{1}{N}],$$

and

$$\widehat{g}_N(x) \in C([0,1]), \ \widehat{g}_N(x_n) = g_N(x_n), \ g_N \text{ affine otherwise.}$$

It is easy to check that both $\{\widetilde{g}_N\}_{N=1}^\infty$ and $\{\widehat{g}_N\}_{N=1}^\infty$ generate a Young measure $\{\widetilde{\mathcal{V}}_x\}_{x\in(0,1)}$ and $\{\widehat{\mathcal{V}}_x\}_{x\in(0,1)}$, respectively, where

$$\widetilde{\mathcal{V}}_x = \frac{1}{2}\delta_0 + \frac{1}{2}\delta_1,\ \widehat{\mathcal{V}}_x = \mathcal{L}|_{[0,1]} - \text{the Lebesgue measure on } [0,1] \text{ for any } x \in (0,1).$$

Note, however, that the barycenter – $Y = \frac{1}{2}$ – is the same in both cases. Moreover, defining the oscillation defect as

$$|\widetilde{g}_N - \frac{1}{2}| = \frac{1}{2},\ |\widehat{g}_N - \frac{1}{2}| = \begin{cases} \frac{1}{2} \text{ for } x = x_n \\ 0 \text{ for } \frac{x_n + x_{n+1}}{2} \\ \text{affine continuous otherwise,} \end{cases}$$

we deduce

$$\left\langle \widetilde{\mathcal{V}}_x; \left| Y - \frac{1}{2} \right| \right\rangle = \frac{1}{2},\ \left\langle \widehat{\mathcal{V}}_x; \left| Y - \frac{1}{2} \right| \right\rangle = \frac{1}{4}.$$

This observation suggests that the only relevant piece of information to be retained for the limit problem is the value of some *observables* as the barycenter of the measure and possibly the value of the oscillation defect. This leads to the concept of *dissipative weak solution* introduced in this section. Very roughly indeed, a dissipative weak solution may be interpreted as a barycenter of a parametrized (Young) measure.

5.2.1 Dissipative Solutions to the Barotropic Euler System

We start with the barotropic Euler system, where the concept of dissipative weak solution can be introduced in an elegant way, eliminating completely the Young measure in the weak formulation of the problem. For the sake of simplicity, we suppose the pressure to be related to internal energy through the constitutive equation (5.78).

Definition 5.6 (DISSIPATIVE WEAK SOLUTION TO BAROTROPIC EULER SYSTEM)

Let $\Omega \subset R^d$, $d = 1, 2, 3$ be a bounded domain. Let the pressure $p = p(\varrho)$ be related to the internal energy e through the constitutive equation (5.78). We say that $[\varrho, \boldsymbol{m}]$ is a *dissipative weak (DW) solution* of the barotropic Euler system (5.71), with the impermeability boundary condition (5.72), and with the initial data $[\varrho_0, \boldsymbol{m}_0]$ if the following holds:

- (**weak continuity**)

$$\varrho \in C_{\text{weak}}([0,T]; L^{\gamma}(\Omega)),\ \boldsymbol{m} \in C_{\text{weak}}([0,T]; L^{\frac{2\gamma}{\gamma+1}}(\Omega; R^d)); \tag{5.84}$$

- (**energy inequality**) there is a defect measure $\mathfrak{E} \in L^{\infty}(0,T; \mathcal{M}^+(\overline{\Omega}))$ such that the energy inequality

$$\int_{\Omega} \left[\frac{1}{2} \frac{|\boldsymbol{m}|^2}{\varrho} + P(\varrho) \right] (\tau, \cdot) \, \mathrm{d}x + \int_{\overline{\Omega}} \mathrm{d}\mathfrak{E}(\tau) \leq \int_{\Omega} \left[\frac{1}{2} \frac{|\boldsymbol{m}_0|^2}{\varrho_0} + P(\varrho_0) \right] \, \mathrm{d}x \quad (5.85)$$

for a.a. $0 \leq \tau \leq T$;

- (**equation of continuity**) the integral identity

$$\left[\int_{\Omega} \varrho\varphi \, \mathrm{d}x \right]_{t=0}^{t=\tau} = \int_0^{\tau} \int_{\Omega} \left[\varrho\partial_t\varphi + \boldsymbol{m} \cdot \nabla_x\varphi \right] \, \mathrm{d}x \, \mathrm{d}t \quad (5.86)$$

for any $0 \leq \tau \leq T$, and any test function $\varphi \in W^{1,\infty}((0,T) \times \Omega)$;

- (**momentum equation**) the integral identity

$$\left[\int_{\Omega} \boldsymbol{m} \cdot \boldsymbol{\varphi} \, \mathrm{d}x \right]_{t=0}^{t=\tau}$$

$$= \int_0^{\tau} \int_{\Omega} \left[\boldsymbol{m} \cdot \partial_t\boldsymbol{\varphi} + 1_{\varrho>0} \frac{\boldsymbol{m} \otimes \boldsymbol{m}}{\varrho} : \nabla_x\boldsymbol{\varphi} + p(\varrho)\mathrm{div}_x\boldsymbol{\varphi} \right] \, \mathrm{d}x \, \mathrm{d}t \quad (5.87)$$

$$+ \int_0^{\tau} \int_{\overline{\Omega}} \nabla_x\boldsymbol{\varphi} : \mathrm{d}\mathfrak{R}(t) \, \mathrm{d}t$$

for any $0 \leq \tau \leq T$, any test function $\boldsymbol{\varphi} \in C^M([0,T] \times \overline{\Omega}; R^d)$, $\boldsymbol{\varphi} \cdot \boldsymbol{n}|_{\partial\Omega} = 0$, $M \geq 1$ with the Reynolds defect

$$\mathfrak{R} \in L^\infty(0,T; \mathcal{M}^+(\overline{\Omega}; R^{d\times d}_{\mathrm{sym}}));$$

- (**defect compatibility condition**)

$$\underline{d}\ \mathfrak{E} \leq \mathrm{tr}\,[\mathfrak{R}] \leq \overline{d}\ \mathfrak{E} \text{ for some constants } 0 \leq \underline{d} \leq \overline{d}. \quad (5.88)$$

Remark 5.16 It follows from the energy inequality (5.2) that

$$\varrho(t,x) \geq 0 \text{ for a.a. } (t,x) \in (0,T) \times \Omega, \text{ and } \boldsymbol{m}(\tau,x) = 0$$

for a.a. $x \in \{\varrho(\tau,x) = 0\}$, and any $0 \leq \tau \leq T$.

Remark 5.17 As the energy satisfied the *inequality* (5.85), we can always set

$$\mathfrak{E} = \frac{1}{\overline{d}}\mathrm{tr}[\mathfrak{R}]$$

eliminating entirely the energy defect $\mathfrak{E}$ from the weak formulation. Obviously, (5.88) remains valid. In addition, a short inspection of the convergence of consistent approximation reveal that the constant $\overline{d}$ can be expressed solely in terms of γ and the dimension d.

It turns out, at least for the barotropic Euler system, that a DW solution may be interpreted as a barycenter of a DMV solution and vice versa.

Theorem 5.4 (DW vs DMV)
(i) Let a parametrized measure $\{\mathcal{V}_{t,x}\}_{(t,x)\in(0,T)\times\Omega}$ be a DMV solution of the barotropic Euler system in the sense of Definition 5.5, with the associated concentration defect measures $\mathfrak{E}_{cd}$, $\mathfrak{R}_{cd}$.
Then

$$\varrho = \langle \mathcal{V}; \widetilde{\varrho} \rangle\ ,\ \boldsymbol{m} = \langle \mathcal{V}, \widetilde{\boldsymbol{m}} \rangle$$

is a DW solution in the sense of Definition 5.6, with the defect measures,

$$\mathfrak{E} = \mathfrak{E}_{cd} + \mathfrak{E}_{od},\ \mathfrak{E}_{od} \equiv \left\langle \mathcal{V}; \frac{1}{2}\frac{|\widetilde{\boldsymbol{m}}|^2}{\widetilde{\varrho}} - P(\widetilde{\varrho}) \right\rangle - \left(\frac{1}{2}\frac{|\boldsymbol{m}|^2}{\varrho} + P(\varrho) \right),$$

$$\mathfrak{R} = \mathfrak{R}_{cd} + \mathfrak{R}_{od},\ \mathfrak{R}_{od} = \left\langle \mathcal{V}; 1_{\widetilde{\varrho}>0}\frac{\widetilde{\boldsymbol{m}}\otimes\widetilde{\boldsymbol{m}}}{\widetilde{\varrho}} + p(\widetilde{\varrho})\mathbb{I} \right\rangle - \left(1_{\varrho>0}\frac{\boldsymbol{m}\otimes\boldsymbol{m}}{\varrho} + p(\varrho)\mathbb{I} \right).$$

(ii) Suppose that $[\varrho, \boldsymbol{m}]$ is a DW solution of the barotropic Euler system in the sense of Definition 5.6.
Then

$$\mathcal{V}_{t,x} = \delta_{\varrho(t,x),\boldsymbol{m}(t,x)} \in \mathcal{P}(R^{d+1})\ \textit{for a.a.}\ (t,x) \in (0,T)\times\Omega$$

is a DMV solution in the sense of Definition 5.5.

Proof The claim (ii) is obvious. To show (i), the only thing to observe is

$$\mathfrak{R}_{od} = \left\langle \mathcal{V}; 1_{\widetilde{\varrho}>0}\frac{\widetilde{\boldsymbol{m}}\otimes\widetilde{\boldsymbol{m}}}{\widetilde{\varrho}} + p(\widetilde{\varrho})\mathbb{I} \right\rangle - \left(1_{\varrho>0}\frac{\boldsymbol{m}\otimes\boldsymbol{m}}{\varrho} + p(\varrho)\mathbb{I} \right)$$
$$\in L^\infty(0,T; \mathcal{M}^+(\overline{\Omega}; R^{d\times d}_{\rm sym})),$$

meaning the defect $\mathfrak{R}_{od}$ is positively semidefinite matrix valued measure. However, this follows from Jensen's inequality, exactly as in (5.51). □

5.2.1.1 Conclusion

As we have seen in Theorem 5.3, a consistent approximation of the barotropic Euler system gives rise to a DMV solution that may be seen, in view of Theorem 5.4, as a DW solution of the same problem. This fact, however, *must not* be interpreted in the way that the associated Young measure is a Dirac mass! The corresponding oscillation defect is only "hidden" in the Reynolds defect $\mathfrak{R}$.

As observed in Remark 5.17, the energy defect $\mathfrak{E}$ can be replaced by $\mathrm{tr}[\mathfrak{R}]$ modulo a multiplicative constant. Thus the "turbulent" character of DW solutions is entirely captured by the *Reynolds defect*

$$\mathfrak{R} \in L^\infty(0,T; \mathcal{M}^+(\overline{\Omega}; R^{d\times d}_{\mathrm{sym}})).$$

Moreover, as shown in Theorem 5.4,

$$\begin{aligned}\mathfrak{R} = &\overline{\left[1_{\varrho>0}\frac{\boldsymbol{m}\otimes\boldsymbol{m}}{\varrho} + p(\varrho)\mathbb{I}\right]} - \left\langle \mathcal{V}; 1_{\widetilde{\varrho}>0}\frac{\widetilde{\boldsymbol{m}}\otimes\widetilde{\boldsymbol{m}}}{\widetilde{\varrho}} + p(\widetilde{\varrho})\mathbb{I}\right\rangle \\ &+ \left\langle \mathcal{V}; 1_{\widetilde{\varrho}>0}\frac{\widetilde{\boldsymbol{m}}\otimes\widetilde{\boldsymbol{m}}}{\widetilde{\varrho}} + p(\widetilde{\varrho})\mathbb{I}\right\rangle - \left(1_{\varrho>0}\frac{\boldsymbol{m}\otimes\boldsymbol{m}}{\varrho} + p(\varrho)\mathbb{I}\right),\end{aligned}$$

where

$$\left(1_{\varrho_n}\frac{\boldsymbol{m}_n\otimes\boldsymbol{m}_n}{\varrho_n} + p(\varrho_n)\mathbb{I}\right) \to \overline{\left[1_{\varrho>0}\frac{\boldsymbol{m}\otimes\boldsymbol{m}}{\varrho} + p(\varrho)\mathbb{I}\right]}$$
$$\text{weakly-(*) in } L^\infty(0,T; \mathcal{M}^+(\overline{\Omega}; R^{d\times d}_{\mathrm{sym}})),$$

and $\{\mathcal{V}_{t,x}\}_{(t,x)\in(0,T)\times\Omega}$ is the Young measure generated by the associated consistent approximation $\{\varrho_n, \boldsymbol{m}_n\}_{n=1}^\infty$.

5.2.2 *Dissipative Solutions to the Complete Euler System*

The concept of DW solution can be extended to the complete Euler system. Unfortunately, the resulting new formulation is less elegant as we have to retain the piece of information provided by the measure $\mathcal{V}$ because of the convective term in the entropy balance.

Definition 5.7 (Dissipative weak solution to complete Euler system)

Let $\Omega \subset R^d$, $d = 1, 2, 3$ be a bounded domain. We say that a trio $[\varrho, \boldsymbol{m}, S]$ is *dissipative weak (DW) solution* of the complete Euler system (5.16)–(5.18), with the boundary condition (5.19), and the initial condition $[\varrho_0, \boldsymbol{m}_0, S_0]$ if the following holds:

- (**weak continuity**)

$$
\begin{aligned}
&\varrho \in C_{\rm weak}([0,T]; L^\gamma(\Omega)),\\
&\boldsymbol{m} \in C_{\rm weak}([0,T]; L^{\frac{2\gamma}{\gamma+1}}(\Omega; R^d)),\\
&S \in L^\infty(0,T; L^\gamma(\Omega)) \cap BV_{\rm weak}([0,T]; L^\gamma(\Omega));
\end{aligned}
\tag{5.89}
$$

- (**energy inequality**) there exists a measure

$$\mathfrak{E} \in L^\infty(0,T; \mathcal{M}^+(\overline{\Omega}))$$

such that the inequality

$$\int_\Omega \left[\frac{1}{2}\frac{|\boldsymbol{m}|^2}{\varrho} + \varrho e(\varrho, S) \right](\tau,\cdot)\ {\rm d}x + \int_{\overline{\Omega}} {\rm d}\mathfrak{E}(\tau) \le \int_\Omega \left[\frac{1}{2}\frac{|\boldsymbol{m}_0|^2}{\varrho_0} + \varrho_0 e(\varrho_0, S_0) \right]\ {\rm d}x \tag{5.90}$$

holds for a.a. $0 \le \tau \le T$;
- (**equation of continuity**) the integral identity

$$\left[\int_\Omega \varrho\varphi\ {\rm d}x \right]_{t=0}^{t=\tau} = \int_0^\tau \int_\Omega \left[\varrho \partial_t \varphi + \boldsymbol{m} \cdot \nabla_x \varphi \right]\ {\rm d}x\,{\rm d}t \tag{5.91}$$

holds for any $0 \le \tau \le T$, any test function $\varphi \in W^{1,\infty}((0,T)\times\Omega)$;
- (**momentum equation**) the integral identity

$$
\begin{aligned}
&\left[\int_\Omega \boldsymbol{m}\cdot\boldsymbol{\varphi}\ {\rm d}x \right]_{t=0}^{t=\tau}\\
&= \int_0^\tau \int_\Omega \left[\boldsymbol{m}\cdot\partial_t\boldsymbol{\varphi} + 1_{\varrho>0}\frac{\boldsymbol{m}\otimes\boldsymbol{m}}{\varrho} : \nabla_x\boldsymbol{\varphi} + 1_{\varrho>0}p(\varrho,S){\rm div}_x\boldsymbol{\varphi} \right]\ {\rm d}x\,{\rm d}t\\
&+ \int_0^\tau \int_{\overline{\Omega}} \nabla_x\boldsymbol{\varphi} : {\rm d}\mathfrak{R}(t)\,{\rm d}t
\end{aligned}
\tag{5.92}
$$

holds for any $0 \le \tau \le T$, any test function $\boldsymbol{\varphi} \in C^M([0,T]\times\overline{\Omega}; R^d)$, $\boldsymbol{\varphi}\cdot\boldsymbol{n}|_{\partial\Omega} = 0$, $M \ge 1$ and a defect measure

$$\mathfrak{R} \in L^\infty(0,T; \mathcal{M}^+(\overline{\Omega}; R^{d\times d}_{\rm sym}));$$

- (**entropy balance**) the integral inequality

$$\left[\int_{\Omega} S\varphi \, dx\right]_{t=\tau_1-}^{t=\tau_2+} \geq \int_{\tau_1}^{\tau_2}\int_{\Omega} \left[S\partial_t\varphi + \left\langle \mathcal{V}; 1_{\widetilde{\varrho}>0}\left(\widetilde{S}\frac{\widetilde{\boldsymbol{m}}}{\widetilde{\varrho}}\right)\right\rangle \cdot \nabla_x\varphi\right] dx\, dt,$$
$$S(0-,\cdot) = S_0, \tag{5.93}$$

holds for any $0 \leq \tau_1 \leq \tau_2 < T$, any $\varphi \in W^{1,\infty}((0,T)\times\Omega)$, $\varphi \geq 0$, where

$$\{\mathcal{V}_{t,x}\}_{(t,x)\in(0,T)\times\Omega}$$

is a parametrized measure,

$$\mathcal{V} \in L^\infty((0,T)\times\Omega; \mathcal{P}(R^{d+2})),\ R^{d+2} = \left\{\widetilde{\varrho} \in R, \widetilde{\boldsymbol{m}} \in R^d, \widetilde{S} \in R\right\};$$

$$\langle\mathcal{V};\widetilde{\varrho}\rangle = \varrho,\ \langle\mathcal{V};\widetilde{\boldsymbol{m}}\rangle = \boldsymbol{m},\ \left\langle\mathcal{V};\widetilde{S}\right\rangle = S,$$
$$\mathcal{V}_{t,x}\left\{\widetilde{\varrho}\geq 0,\ \widetilde{S} \geq \underline{s}\widetilde{\varrho}\right\} = 1 \text{ for a.a. } (t,x) \in (0,T)\times\Omega; \tag{5.94}$$

- (**defect compatibility conditions**)

$$\underline{d}\ \mathfrak{E} \leq \mathrm{tr}\,[\mathfrak{R}] \leq \overline{d}\ \mathfrak{E} \text{ for some constants } 0 \leq \underline{d} \leq \overline{d}, \tag{5.95}$$

and

$$\mathfrak{E} \geq \left\langle \mathcal{V}; \frac{1}{2}\frac{|\widetilde{\boldsymbol{m}}|^2}{\widetilde{\varrho}} + \widetilde{\varrho}e(\widetilde{\varrho},\widetilde{S})\right\rangle - \left(\frac{1}{2}\frac{|\boldsymbol{m}|^2}{\varrho} + \varrho e(\varrho, S)\right) \tag{5.96}$$

Remark 5.18 As observed in Remark 5.17, we may set

$$\mathfrak{E} = 1(\overline{d})^{-1}\mathrm{tr}[\mathfrak{R}]$$

to eliminate $\mathfrak{E}$ in the definition.

Similarly to Theorem 5.4, we can show that the barycenter of any DMV solution to the Euler system represents a DW solution. The defect measures are exactly the same as in Theorem 5.4.

5.3 Measure-Valued Solutions to the Navier–Stokes System

The theory of measure-valued solutions for the Navier–Stokes system is technically more involved because of the gradient terms explicitly present in the field equations. At the level of a numerical scheme, they are replaced by discrete approximations that represent a kind of "independent" quantities to be handled in the asymptotic limit when the discretization step tends to zero. Another technical difficulty is that the momentum $\boldsymbol{m}$, which is a conservative variable, must be replaced by $\boldsymbol{m} = \varrho \boldsymbol{u}$, where $\boldsymbol{u}$ is the fluid velocity. Although this setting might seem quite natural, the hypothetical possibility of the vacuum state $\varrho = 0$ makes the analysis rather delicate.

We restrict ourselves to the space-periodic case $\Omega = \mathbb{T}^d$. The reason is rather simple. In the applications to the Navier–Stokes system, we shall assume that the latter admits a smooth solution. Of course, such a result relies essentially on smoothness of the boundary of the physical domain. In numerical experiments, however, the domain is usually a polyhedron with merely Lipschitz boundary.

5.3.1 *Consistent Approximation of the Navier–Stokes System*

Consider the Navier–Stokes system introduced in Chap. 3:

$$\partial_t \varrho + \mathrm{div}_x(\varrho \boldsymbol{u}) = 0, \tag{5.97}$$

$$\partial_t(\varrho \boldsymbol{u}) + \mathrm{div}_x(\varrho \boldsymbol{u} \otimes \boldsymbol{u}) + \nabla_x p(\varrho) = \mathrm{div}_x \mathbb{S}, \tag{5.98}$$

$$\mathbb{S} = \mu \left(\nabla_x \boldsymbol{u} + \nabla_x^T \boldsymbol{u} - \frac{2}{d} \mathrm{div}_x \boldsymbol{u} \mathbb{I} \right) + \lambda \mathrm{div}_x \boldsymbol{u} \mathbb{I}, \ \mu > 0, \ \lambda \geq 0. \tag{5.99}$$

As agreed, we consider the spatially periodic boundary conditions

$$\Omega = \mathbb{T}^d. \tag{5.100}$$

Formally, we may rewrite

$$\mathrm{div}_x \mathbb{S} = \mu \Delta_x \boldsymbol{u} + \nu \nabla_x \mathrm{div}_x \boldsymbol{u}, \ \nu = \mu \left(1 - \frac{2}{d} \right) + \lambda \geq 0 \text{ for } d = 2, 3. \tag{5.101}$$

Of course, the formula holds only if the viscosity coefficients are constant. As we shall see later, this splitting is convenient in numerical approximations as it separates $\nabla_x \boldsymbol{u}$ and $\mathrm{div}_x \boldsymbol{u}$.

Similarly to the Euler system discussed in the preceding part of this chapter, the DMV solutions to the Navier–Stokes system will be identified as limits of consistent approximations. There is a technical difficulty, however, due to the fact that the viscous stress usually acts on the gradient of the fluid velocity $\boldsymbol{u}$, while the conservative (weakly continuous) quantity is the momentum $\boldsymbol{m} = \varrho \boldsymbol{u}$. This gives rise to an ambiguous behavior of the velocity on the (hypothetical) vacuum zones $\{\varrho = 0\}$, where $\boldsymbol{m} = 0$, while, in general, $\boldsymbol{u} \neq 0$. Although the Navier–Stokes system was derived as a model of nondilute fluids, where vacuum is not expected, the available (known) regularity of the weak (or DMV) solutions is not strong enough to prevent the vacuum to appear. In contrast with the Euler system, we consider the density ϱ and the velocity $\boldsymbol{u}$ as the basic state variables.

Another problem is that numerical solutions are in general not differentiable, whereas the differential operators are replaced by suitable approximations. This requires further modification of the definition of consistent approximation at the level of the energy balance. The latter takes the form

$$\frac{\mathrm{d}}{\mathrm{d}t} \int_{\mathbb{T}^d} \left[\frac{1}{2} \varrho |\boldsymbol{u}|^2 + P(\varrho) \right] \mathrm{d}x + \int_{\mathbb{T}^d} \mathbb{S} : \nabla_x \boldsymbol{u} \, \mathrm{d}x = (\leq) 0,$$

where P is the pressure potential,

$$P'(\varrho)\varrho - P(\varrho) = p(\varrho).$$

Introducing the symmetric gradient

$$\mathbb{D}_x \boldsymbol{u} \equiv \frac{\nabla_x \boldsymbol{u} + \nabla_x^T \boldsymbol{u}}{2}, \quad \mathrm{tr}[\mathbb{D}_x \boldsymbol{u}] = \mathrm{div}_x \boldsymbol{u},$$

we can write

$$\mathbb{S} = 2\mu \left(\mathbb{D}_x \boldsymbol{u} - \frac{1}{d} \mathrm{tr}[\mathbb{D}_x \boldsymbol{u}] \mathbb{I} \right) + \lambda \mathrm{tr}[\mathbb{D}_x \boldsymbol{u}] \mathbb{I},$$

whereas the dissipation function reads

$$\mathbb{S} : \nabla_x \boldsymbol{u} = \mathbb{S} : \mathbb{D}_x \boldsymbol{u}.$$

Evoking the material of Sect. 1.3.2, we may write the constitutive law in an "implicit way"

$$\mathbb{S} : \mathbb{D}_x \boldsymbol{u} = F(\mathbb{D}_x \boldsymbol{u}) + F^*(\mathbb{S})$$

for a suitable convex potential F. Newton's rheological law (5.99) corresponds to the potential

$$F(\mathbb{D}) = \mu \left| \mathbb{D} - \frac{1}{d} \mathrm{tr}[\mathbb{D}] \mathbb{I} \right|^2 + \frac{\lambda}{2} |\mathrm{tr}[\mathbb{D}]|^2 \ \text{ for all } \mathbb{D} \in R^{d \times d}_{\mathrm{sym}}, \tag{5.102}$$

the conjugate function reads

$$F^*(\mathbb{S}) = \frac{1}{\mu} \left| \mathbb{S} - \frac{1}{d} \mathrm{tr}[\mathbb{S}] \mathbb{I} \right|^2 + \frac{1}{2\lambda} |\mathrm{tr}[\mathbb{S}]|^2 \tag{5.103}$$

if $\lambda > 0$,

$$F^*(\mathbb{S}) = \begin{cases} \frac{1}{\mu} |\mathbb{S}|^2 \text{ if } \mathrm{tr}[\mathbb{S}] = 0, \\ \\ \infty \text{ otherwise} \end{cases} \tag{5.104}$$

if $\lambda = 0$. Going back to formula (5.106) we may write

$$\int_{\mathbb{T}^d} \mu |\nabla_x \boldsymbol{u}|^2 + \nu |\mathrm{div}_x \boldsymbol{u}|^2 \, \mathrm{d}x = \int_{\mathbb{T}^d} \mathbb{S} : \nabla_x \boldsymbol{u} \, \mathrm{d}x = \int_{\mathbb{T}^d} F(\mathbb{D}_x \boldsymbol{u}) + F^*(\mathbb{S}) \, \mathrm{d}x. \tag{5.105}$$

Note that, in general,

$$\mu |\nabla_x \boldsymbol{u}|^2 + \nu |\mathrm{div}_x \boldsymbol{u}|^2 \neq F(\mathbb{D}_x \boldsymbol{u}) + F^*(\mathbb{S}),$$

however, the integrals over the physical domain of these quantities coincide as long as the spatially periodic functions are considered. It is easy to check that (5.105) holds whenever $\boldsymbol{u} \in W^{1,2}(\mathbb{T}^d; R^d)$.

We consider abstract approximate differential operators ∇_x^n, div_x^n. First, recall the Korn–Poincaré inequality that can be written in the context of spatially periodic functions as

$$\|\boldsymbol{u}\|^2_{L^2(\Omega)} \leq c(M, \Gamma) \int_{\mathbb{T}^d} \left[\varrho |\boldsymbol{u}|^2 + |\mathbb{D}_x \boldsymbol{u} - \frac{1}{d} \mathrm{div}_x \boldsymbol{u} \mathbb{I}|^2 \right] \mathrm{d}x,$$

$$\begin{aligned} \int_{\mathbb{T}^d} |\mathbb{D}_x \boldsymbol{u} - \frac{1}{d} \mathrm{div}_x \boldsymbol{u} \mathbb{I}|^2 \, \mathrm{d}x &= \int_{\mathbb{T}^d} \left[|\nabla_x \boldsymbol{u}|^2 - \frac{1}{d} |\mathrm{div}_x \boldsymbol{u}|^2 \right] \mathrm{d}x \\ &\leq \frac{d-1}{d} \int_{\mathbb{T}^d} |\nabla_x \boldsymbol{u}|^2 \, \mathrm{d}x, \end{aligned}$$

whenever

$$\varrho \geq 0, \ \int_{\mathbb{T}^d} \varrho \, \mathrm{d}x = M > 0, \ \int_{\mathbb{T}^d} \varrho^\gamma \, \mathrm{d}x < \Gamma.$$

This motivates the following definition.

Definition 5.8 (Compatibility of approximate differential operators)

Let

$$X_n \subset L^\infty(\mathbb{T}^d; R^d)$$

be a family of (finite-dimensional) spaces, along with a family of approximate differential operators

$$\nabla_x^n : X_n \to L^\infty(\mathbb{T}^d; R^{d \times d}), \ \mathrm{div}_x^n : X_n \to L^\infty(\mathbb{T}^d).$$

We say that ∇_x^n is *compatible* with ∇_x, and div_x^n *compatible* with div_x if the following is satisfied:

- (**Korn–Poincaré inequality**) The inequality

$$\|\boldsymbol{v}\|^2_{L^2(\Omega)} \leq c(M\,,\Gamma) \int_{\mathbb{T}^d} \left[\varrho |\boldsymbol{v}|^2 + \left|\nabla_x^n \boldsymbol{v}\right|^2 \right] \mathrm{d}x, \tag{5.106}$$

holds for any $\boldsymbol{v} \in X_n$, whenever

$$\varrho \geq 0, \int_{\mathbb{T}^d} \varrho \ \mathrm{d}x = M > 0, \int_{\mathbb{T}^d} \varrho^\gamma \ \mathrm{d}x < \Gamma, \ \gamma > 1,$$

with $c(M\,,\Gamma)$ independent of $n = 1, \dots$.

- (**Consistency (i)**) If

$$\begin{aligned} \boldsymbol{u}_n \in X_n, \ \boldsymbol{u}_n \to \boldsymbol{u} \text{ weakly in } L^2(\mathbb{T}^d; R^d), \\ \nabla_x^n[\boldsymbol{u}_n] \to \overline{\nabla_x \boldsymbol{u}} \text{ weakly in } L^2(\mathbb{T}^d; R^{d\times d}), \end{aligned}$$

then

$$\overline{\nabla_x \boldsymbol{u}} = \nabla_x \boldsymbol{u} \ \text{ in } \ \mathcal{D}'(\mathbb{T}^d; R^{d\times d}).$$

If

$$\begin{aligned} \boldsymbol{u}_n \in X_n, \ \boldsymbol{u}_n \to \boldsymbol{u} \ \text{ weakly in } \ L^2(\mathbb{T}^d; R^d), \\ \mathrm{div}_x^n[\boldsymbol{u}_n] \to \overline{\mathrm{div}_x \boldsymbol{u}} \ \text{ weakly in } \ L^2(\mathbb{T}^d), \end{aligned}$$

then

$$\overline{\mathrm{div}_x \boldsymbol{u}} = \mathrm{div}_x \boldsymbol{u} \text{ in } \mathcal{D}'(\mathbb{T}^d).$$

(**Consistency (ii)**) There exists a projection operator

$$\Pi_n : L^1(\mathbb{T}^d; R^d) \to X_n$$

such that

$$\begin{aligned} \Pi_n \boldsymbol{v} \to \boldsymbol{v} \text{ in } W^{1,\infty}((0,T) \times \mathbb{T}^d; R^d), \\ \nabla_x^n \left[\Pi_n \boldsymbol{v}\right] \to \nabla_x \boldsymbol{v} \text{ in } L^\infty(\mathbb{T}^d; R^{d\times d}) \text{ as } n \to \infty \text{ for any } \boldsymbol{v} \in C^1(\mathbb{T}^d; R^d), \\ \mathrm{div}_x^n \left[\Pi_n \boldsymbol{v}\right] \to \mathrm{div}_x \boldsymbol{v} \text{ in } L^\infty(\mathbb{T}^d) \text{ as } n \to \infty \text{ for any } \boldsymbol{v} \in C^1(\mathbb{T}^d; R^d). \end{aligned} \tag{5.107}$$

Now, we have everything at hand to introduce the concept of *consistent approximation* for the Navier–Stokes system.

Definition 5.9 (Consistent approximation of the Navier–Stokes system)

Let ∇_x^n be a family of linear operators compatible with ∇_x, and div_x^n compatible with div_x, defined on the associated sequence of function spaces $\{X_n\}_{n=1}^{\infty}$. A sequence $\{\varrho_n, \boldsymbol{u}_n\}_{n=1}^{\infty}$ is a *consistent approximation* of the Navier–Stokes system (5.97), (5.98), (5.101) in $(0,T) \times \mathbb{T}^d$, with the initial data $[\varrho_0, \boldsymbol{m}_0]$ if:

- **(energy inequality)** $\varrho_n \geq 0$, $\boldsymbol{u}_n \in X_n$, and the energy inequality

$$
\begin{aligned}
&\int_{\mathbb{T}^d} \left[\frac{1}{2} \varrho_n |\boldsymbol{u}_n|^2 + P(\varrho_n) \right] (\tau, \cdot) \, \mathrm{d}x \\
&+ \int_0^\tau \int_{\mathbb{T}^d} \left[\mu |\nabla_x^n \boldsymbol{u}_n|^2 + \nu |\mathrm{div}_x^n \boldsymbol{u}_n|^2 \, \mathrm{d}x \right] \mathrm{d}t \\
&\leq \int_{\mathbb{T}^d} \left[\frac{1}{2} \frac{|\boldsymbol{m}_{0,n}|^2}{\varrho_{0,n}} + P(\varrho_{0,n}) \right] \mathrm{d}x + e_n^1, \ P'(\varrho)\varrho - P(\varrho) = p(\varrho),
\end{aligned}
\tag{5.108}
$$

  holds for a.a. $0 \leq \tau \leq T$,

$$
\varrho_{0,n} \to \varrho_0 \text{ weakly in } L^1(\Omega), \ \boldsymbol{m}_{0,n} \to \boldsymbol{m}_0 \text{ weakly in } L^1(\Omega; R^d),
$$

  and

$$
\int_{\mathbb{T}^d} \left[\frac{1}{2} \frac{|\boldsymbol{m}_{0,n}|^2}{\varrho_{0,n}} + P(\varrho_{0,n}) \right] \mathrm{d}x \to \int_{\mathbb{T}^d} \left[\frac{1}{2} \frac{|\boldsymbol{m}_0|^2}{\varrho_0} + P(\varrho_0) \right] \mathrm{d}x < \infty,
$$

$$
e_n^1 \to 0
$$

  as $n \to \infty$;
- **(mass conservation)**

$$
\int_{\mathbb{T}^d} \varrho_n(\tau, \cdot) \, \mathrm{d}x = \int_{\mathbb{T}^d} \varrho_{0,n} \, \mathrm{d}x
\tag{5.109}
$$

  for a.a. $0 \leq \tau \leq T$;
- **(equation of continuity)** the integral identity

$$
\int_0^T \int_{\mathbb{T}^d} \left[\varrho_n \partial_t \varphi + \varrho_n \boldsymbol{u}_n \cdot \nabla_x \varphi \right] \mathrm{d}x \, \mathrm{d}t = - \int_{\mathbb{T}^d} \varrho_{0,n} \varphi(0, \cdot) \, \mathrm{d}x + e_n^2[\varphi]
\tag{5.110}
$$

holds for any $\varphi \in C_c^1([0,T) \times \mathbb{T}^d)$, where

$$e_n^2[\varphi] \to 0 \text{ as } n \to \infty \text{ for any } \varphi \in C_c^M([0,T) \times \mathbb{T}^d),$$

where $M \geq 1$ is a positive integer;

- (**momentum equation**) the integral identity

$$\begin{aligned} &\int_0^T \int_{\mathbb{T}^d} \left[\varrho_n \boldsymbol{u}_n \cdot \partial_t \boldsymbol{\varphi} + \varrho_n (\boldsymbol{u}_n \otimes \boldsymbol{u}_n) : \nabla_x \boldsymbol{\varphi} + p(\varrho_n) \mathrm{div}_x \boldsymbol{\varphi} \right] \mathrm{d}x \, \mathrm{d}t \\ &\quad = \int_0^T \int_{\mathbb{T}^d} \left(\mu \nabla_x^n \boldsymbol{u}_n : \nabla_x \boldsymbol{\varphi} + \nu \mathrm{div}_x^n \boldsymbol{u}_n \mathrm{div}_x \boldsymbol{\varphi} \right) \mathrm{d}x \, \mathrm{d}t \\ &\quad - \int_{\mathbb{T}^d} \boldsymbol{m}_{0,n} \cdot \boldsymbol{\varphi}(0, \cdot) \, \mathrm{d}x + e_n^3[\boldsymbol{\varphi}] \end{aligned} \tag{5.111}$$

holds for any $\boldsymbol{\varphi} \in C_c^1([0,T) \times \mathbb{T}^d; R^d) \cap L^2((0,T); W^{1,2}(\mathbb{T}^d; R^d))$, where

$$e_n^3[\boldsymbol{\varphi}] \to 0 \ \text{ as } \ n \to \infty \ \text{ for any } \ \boldsymbol{\varphi} \in C_c^M([0,T) \times \mathbb{T}^d; R^d), \ M \geq 1.$$

Remark 5.19 Note that (5.109) and (5.110) are compatible only if $e_n^2[\psi] = 0$ for any spatially homogeneous function $\psi = \psi(t)$.

5.3.2 Measure-Valued Solutions to the Navier–Stokes System

Similarly to the inviscid Euler models, the dissipative measure-valued (DMV) solutions of the Navier–Stokes system will be identified as asymptotic limit of consistent approximations. In order to derive the necessary uniform bounds, certain coercivity properties of the pressure are needed. For the sake of simplicity, we focus on the isentropic EOS:

$$p(\varrho) = a\varrho^\gamma, \ a > 0, \ \gamma > 1. \tag{5.112}$$

More general EOS can be considered retaining the basic asymptotic properties of (5.112):

$$p \in C(0,\infty) \cap C^2(0,\infty), \ p'(\varrho) > 0 \text{ for any } \varrho > 0, \ p'(\varrho) \approx \varrho^{\gamma-1} \text{ as } \varrho \to \infty. \tag{5.113}$$

Accordingly, the energy inequality (5.108) yield the uniform bound

$$\operatorname*{ess\,sup}_{t \in [0,T]} \|\varrho_n(t,\cdot)\|_{L^\gamma(\mathbb{T}^d)} \lesssim 1.$$

Similarly, by Hölder's inequality,

$$\|\varrho_n \boldsymbol{u}_n\|_{L^{\frac{2\gamma}{\gamma+1}}(\mathbb{T}^d;R^d)} \leq \left\|\sqrt{\varrho_n}\right\|_{L^{2\gamma}(\mathbb{T}^d)} \left\|\sqrt{\varrho_n}\boldsymbol{u}_n\right\|_{L^2(\mathbb{T}^d;R^d)}.$$

Consequently, in accordance with the approximate Korn–Poincaré inequality (5.106), we may suppose

$$\begin{aligned}
\varrho_n &\to \varrho \text{ weakly-(*) in } L^\infty(0,T;L^\gamma(\mathbb{T}^d)),\\
\boldsymbol{m}_n = (\varrho_n \boldsymbol{u}_n) &\to \boldsymbol{m} \text{ weakly-(*) in } L^\infty(0,T;L^{\frac{2\gamma}{\gamma+1}}(\mathbb{T}^d;R^d)),
\end{aligned}$$

and

$$\begin{aligned}
\boldsymbol{u}_n &\to \boldsymbol{u} \text{ weakly in } L^2((0,T)\times\mathbb{T}^d;R^d),\\
\nabla_x^n \boldsymbol{u}_n &\to \nabla_x \boldsymbol{u} \text{ weakly in } L^2((0,T)\times\mathbb{T}^d;R^d),\\
\mathrm{div}_x^n \boldsymbol{u}_n &\to \mathrm{div}_x \boldsymbol{u} \text{ weakly in } L^2((0,T)\times\mathbb{T}^d;R^d)).
\end{aligned}$$

passing to a suitable subsequence as the case may be. In particular,

$$\boldsymbol{u} \in L^2(0,T;W^{1,2}(\mathbb{T}^d;R^d)). \tag{5.114}$$

In addition, we may suppose that the sequence $\{\varrho_n, \boldsymbol{u}_n\}_{n=1}^\infty$ generates a Young measure

$$\{\mathcal{V}_{t,x}\}_{(t,x)\in(0,T)\times\mathbb{T}^d},\ \mathcal{V}_{t,x} \in \mathcal{P}\left\{[\widetilde{\varrho},\widetilde{\boldsymbol{u}}]\middle|\widetilde{\varrho}\in R, \widetilde{\boldsymbol{u}}\in R^d\right\},\ \mathrm{supp}[\mathcal{V}_{t,x}] \subset [0,\infty)\times R^d.$$

Note that $\mathcal{V}$ differs from its counterpart for the Euler system as it is generated by a different sequence, namely $[\varrho_n, \boldsymbol{u}_n]$ instead of $[\varrho_n, \boldsymbol{m}_n]$.

5.3.2.1 Equation of Continuity

Passing to the limit in (5.110) we deduce, by means of the same arguments as in Sect. 5.1.3,

$$t \mapsto \left\langle \mathcal{V}_{t,\cdot}; \widetilde{\varrho}\right\rangle \in C_{\rm weak}([0,T];L^\gamma(\mathbb{T}^d)), \tag{5.115}$$

and

$$\left[\int_{\mathbb{T}^d} \langle \mathcal{V};\widetilde{\varrho}\rangle \varphi \,\mathrm{d}x\right]_{t=0}^{t=\tau} = \int_0^\tau \int_{\mathbb{T}^d} \left[\langle \mathcal{V};\widetilde{\varrho}\rangle \partial_t \varphi + \langle \mathcal{V};\widetilde{\varrho}\widetilde{\boldsymbol{u}}\rangle \cdot \nabla_x \varphi\right] \mathrm{d}x\,\mathrm{d}t \tag{5.116}$$

for any $0 \leq \tau \leq T$. Using a simple density argument on $\mathbb{T}^d$, we can show that (5.116) holds for any test function $\varphi \in W^{1,\infty}((0,T)\times\mathbb{T}^d)$.

5.3.2.2 Momentum Equation

We start by the convective term,

$$\varrho_n \boldsymbol{u}_n \otimes \boldsymbol{u}_n \to \overline{\varrho \boldsymbol{u} \otimes \boldsymbol{u}} \text{ weakly-(*) in } L^\infty(0,T;\mathcal{M}(\mathbb{T}^d;R^{d\times d}_{\rm sym})).$$

Introducing the defect

$$\mathfrak{C}_{cd} = \overline{\varrho \boldsymbol{u} \otimes \boldsymbol{u}} - \langle \mathcal{V}; \widetilde{\varrho} \widetilde{\boldsymbol{u}} \otimes \widetilde{\boldsymbol{u}} \rangle$$

we apply Corollary 5.2 to

$$\overline{\varrho \boldsymbol{u} \otimes \boldsymbol{u}} : (\xi \otimes \xi),\ \xi \in R^d$$

to deduce

$$\mathfrak{C}_{cd} \in L^\infty(0,T;\mathcal{M}^+(\mathbb{T}^d;R^{d\times d}_{\rm sym})).$$

By the same token,

$$p(\varrho_n) \to \overline{p(\varrho)} \text{ weakly-(*) in } L^\infty(0,T;\mathcal{M}(\mathbb{T}^d)),$$

where

$$\mathfrak{P}_{cd} \equiv \overline{p(\varrho)} - \langle \mathcal{V}; p(\widetilde{\varrho}) \rangle \in L^\infty(0,T;\mathcal{M}^+(\mathbb{T}^d)).$$

Thus we may let $n \to \infty$ in the momentum equation (5.111) obtaining

$$t \mapsto \left\langle \mathcal{V}_{t,\cdot}; \widetilde{\varrho}\widetilde{\boldsymbol{u}} \right\rangle \in C_{\rm weak}([0,T];L^{\frac{2\gamma}{\gamma+1}}(\mathbb{T}^d;R^d)),\ \left\langle \mathcal{V}_{0,\cdot}; \widetilde{\varrho}\widetilde{\boldsymbol{u}} \right\rangle = \boldsymbol{m}_0, \tag{5.117}$$

and

$$\begin{aligned}
&\left[\int_{\mathbb{T}^d} \langle \mathcal{V}; \widetilde{\varrho}\widetilde{\boldsymbol{u}} \rangle \cdot \boldsymbol{\varphi} \ {\rm d}x \right]_{t=0}^{t=\tau} \\
&= \int_0^\tau \int_{\mathbb{T}^d} \Big[\langle \mathcal{V}; \widetilde{\varrho}\widetilde{\boldsymbol{u}} \rangle \cdot \partial_t \boldsymbol{\varphi} + \langle \mathcal{V}; \widetilde{\varrho}\widetilde{\boldsymbol{u}} \otimes \widetilde{\boldsymbol{u}} \rangle : \nabla_x \boldsymbol{\varphi} + \langle \mathcal{V}; p(\widetilde{\varrho}) \rangle {\rm div}_x \boldsymbol{\varphi} \Big] \ {\rm d}x \, {\rm d}t \\
&- \int_0^\tau \int_{\mathbb{T}^d} (\mu \nabla_x \boldsymbol{u} : \nabla_x \boldsymbol{\varphi} + \nu {\rm div}_x \boldsymbol{u} {\rm div}_x \boldsymbol{\varphi}) \ {\rm d}x \, {\rm d}t + \int_0^\tau \int_{\mathbb{T}^d} \nabla_x \boldsymbol{\varphi} : {\rm d}\mathfrak{R}(t) \, {\rm d}t
\end{aligned} \tag{5.118}$$

for any $0 \le \tau \le T$, and any test function $\boldsymbol{\varphi} \in C^1([0,T] \times \mathbb{T}^d; R^d)$. Here we have set

$$\mathfrak{R} = \mathfrak{C}_{cd} + \mathfrak{P}_{cd}\mathbb{I} \in L^\infty(0,T;\mathcal{M}^+(\mathbb{T}^d;R^{d\times d}_{\rm sym})).$$

5.3.2.3 Energy Inequality

Finally, we let $n \to \infty$ in the energy inequality (5.108). Using weak lower semicontinuity of convex functions, we obtain successively:

$$\varrho_n|\boldsymbol{u}_n|^2 + P(\varrho_n) \to \overline{\varrho|\boldsymbol{u}|^2 + P(\varrho)} \equiv \left\langle \mathcal{V}; \frac{1}{2}\widetilde{\varrho}|\widetilde{\boldsymbol{u}}|^2 + P(\widetilde{\varrho}) \right\rangle + \mathfrak{E}$$
$$\text{weakly-(*) in } L^\infty(0,T;\mathcal{M}(\mathbb{T}^d)),$$

$$\mu|\nabla_x^n \boldsymbol{u}_n|^2 + \lambda|\mathrm{div}_x \boldsymbol{u}_n|^2 \to \overline{\mu|\nabla_x \boldsymbol{u}|^2 + \lambda|\mathrm{div}_x \boldsymbol{u}|^2} \equiv \mu|\nabla_x \boldsymbol{u}|^2 + \lambda|\mathrm{div}_x \boldsymbol{u}|^2 + \mathfrak{D}$$
$$\text{weakly-(*) in } \mathcal{M}([0,T]\times\mathbb{T}^d),$$

and

$$\int_{\mathbb{T}^d} \left\langle \mathcal{V}; \frac{1}{2}\widetilde{\varrho}|\widetilde{\boldsymbol{u}}|^2 + P(\widetilde{\varrho}) \right\rangle (\tau,\cdot)\, \mathrm{d}x + \int_{\mathbb{T}^d} \mathrm{d}\mathfrak{E}(\tau)$$
$$+ \int_0^\tau \int_{\mathbb{T}^d} \left[\mu|\nabla_x \boldsymbol{u}|^2 + \nu|\mathrm{div}_x \boldsymbol{u}|^2 \right] \mathrm{d}x\,\mathrm{d}t + \int_0^\tau \int_{\mathbb{T}^d} \mathrm{d}\mathfrak{D} \tag{5.119}$$
$$\leq \int_{\mathbb{T}^d} \left[\frac{1}{2}\frac{|\boldsymbol{m}_0|^2}{\varrho_0} + P(\varrho_0) \right] \mathrm{d}x,$$

for a.a. $0 \leq \tau \leq T$, where the defect measures satisfy

$$\mathfrak{D} \in \mathcal{M}^+([0,T]\times\mathbb{T}^d),$$
$$\mathfrak{E} \in L^\infty(0,T;\mathcal{M}^+(\mathbb{T}^d)),\ \underline{d}\,\mathfrak{E} \leq \mathrm{tr}[\mathfrak{R}] \leq \overline{d}\,\mathfrak{E} \ \text{ for some constants } 0 \leq \underline{d} \leq \overline{d}. \tag{5.120}$$

Similarly to the Euler system, the measure $\mathfrak{E}$ can be interpreted as the energy defect, while $\mathfrak{D}$ represents the *dissipation defect* pertinent to models of viscous fluids.

Finally, we need certain relation between the limit Young measure and the Korn–Poincaré inequality. Consider a smooth vector field $\boldsymbol{U} \in C^1([0,T]\times\mathbb{T}^d;R^d)$. Using (5.106) and the hypothesis on compatibility of the differential operators ∇_x^n, div_x^n we deduce

$$
\begin{aligned}
&\int_0^\tau \int_{\mathbb{T}^d} |\boldsymbol{u}_n - \boldsymbol{U}|^2 \,\mathrm{d}x\,\mathrm{d}t \\
&\quad \lesssim \int_0^\tau \int_{\mathbb{T}^d} |\boldsymbol{u}_n - \Pi_n \boldsymbol{U}|^2 \,\mathrm{d}x\,\mathrm{d}t + \int_0^\tau \int_{\mathbb{T}^d} |\boldsymbol{U} - \Pi_n \boldsymbol{U}|^2 \,\mathrm{d}x\,\mathrm{d}t \\
&\quad \lesssim \int_0^\tau \int_{\mathbb{T}^d} \varrho_n |\boldsymbol{u}_n - \Pi_n \boldsymbol{U}|^2 \,\mathrm{d}x\,\mathrm{d}t \\
&\quad + \int_0^\tau \int_{\mathbb{T}^d} \left| \nabla_x^n \boldsymbol{u}_n - \nabla_x^n [\Pi_n \boldsymbol{U}] \right|^2 \,\mathrm{d}x\,\mathrm{d}t + \int_0^\tau \int_{\mathbb{T}^d} |\boldsymbol{U} - \Pi_n \boldsymbol{U}|^2 \,\mathrm{d}x\,\mathrm{d}t \\
&\quad \lesssim \int_0^\tau \int_{\mathbb{T}^d} \varrho_n |\boldsymbol{u}_n - \boldsymbol{U}|^2 \,\mathrm{d}x\,\mathrm{d}t + \int_0^\tau \int_{\mathbb{T}^d} \left| \nabla_x^n \boldsymbol{u}_n - \nabla_x \boldsymbol{U} \right|^2 \,\mathrm{d}x\,\mathrm{d}t \\
&\quad + \int_0^\tau \int_{\mathbb{T}^d} \left| \nabla_x \boldsymbol{U} - \nabla_x^n [\Pi_n \boldsymbol{U}] \right|^2 \,\mathrm{d}x\,\mathrm{d}t \\
&\quad + \int_0^\tau \int_{\mathbb{T}^d} |\boldsymbol{U} - \Pi_n \boldsymbol{U}|^2 \,\mathrm{d}x\,\mathrm{d}t + \int_0^\tau \int_{\mathbb{T}^d} \varrho_n |\boldsymbol{U} - \Pi_n \boldsymbol{U}|^2 \,\mathrm{d}x\,\mathrm{d}t
\end{aligned}
$$

It follows from the compatibility condition (5.107) that

$$
\begin{aligned}
&\int_0^\tau \int_{\mathbb{T}^d} \left| \nabla_x \boldsymbol{U} - \nabla_x^n [\Pi_n \boldsymbol{U}] \right|^2 \,\mathrm{d}x\,\mathrm{d}t \\
&+ \int_0^\tau \int_{\mathbb{T}^d} |\boldsymbol{U} - \Pi_n \boldsymbol{U}|^2 \,\mathrm{d}x\,\mathrm{d}t + \int_0^\tau \int_{\mathbb{T}^d} \varrho_n |\boldsymbol{U} - \Pi_n [\boldsymbol{U}]|^2 \,\mathrm{d}x\,\mathrm{d}t \to 0 \ \text{as} \ n \to \infty;
\end{aligned}
$$

whence

$$
\begin{aligned}
&\int_0^\tau \int_{\mathbb{T}^d} \left\langle \mathcal{V}; |\widetilde{\boldsymbol{u}} - \boldsymbol{U}|^2 \right\rangle \,\mathrm{d}x\,\mathrm{d}t \leq \liminf_{n\to\infty} \int_0^\tau \int_{\mathbb{T}^d} |\boldsymbol{u}_n - \boldsymbol{U}|^2 \,\mathrm{d}x\,\mathrm{d}t \\
&\quad \lesssim \int_0^\tau \int_{\mathbb{T}^d} \left\langle \mathcal{V}; \widetilde{\varrho} |\widetilde{\boldsymbol{u}} - \boldsymbol{U}|^2 \right\rangle \,\mathrm{d}x\,\mathrm{d}t \qquad (5.121) \\
&\quad + \int_0^\tau \int_{\mathbb{T}^d} |\nabla_x (\boldsymbol{u} - \boldsymbol{U})|^2 \,\mathrm{d}x\,\mathrm{d}t + \int_0^\tau \left(\int_{\mathbb{T}^d} \mathrm{d}\mathfrak{E}(t) \right) \mathrm{d}t + \int_0^\tau \int_{\mathbb{T}^d} \mathrm{d}\mathfrak{D}
\end{aligned}
$$

Relation (5.121) may be seen as a kind of compatibility condition relating the behavior of the Young measure to the differential of the limit velocity field $\boldsymbol{u}$.

Finally, going back to the original formula for the viscous stress

$$\mathbb{S} = 2\mu \left(\mathbb{D}_x \boldsymbol{u} - \frac{1}{d} \mathrm{tr}[\mathbb{D}_x \boldsymbol{u}] \mathbb{I} \right) + \lambda \mathrm{tr}[\mathbb{D}_x \boldsymbol{u}] \mathbb{I},$$

we may rewrite the corresponding integral in the momentum balance (5.118) as

$$\int_{\mathbb{T}^d} (\mu \nabla_x \boldsymbol{u} : \nabla_x \boldsymbol{\varphi} + \nu \mathrm{div}_x \boldsymbol{u} \mathrm{div}_x \boldsymbol{\varphi}) \, \mathrm{d}x = \int_{\mathbb{T}^d} \mathbb{S} : \nabla_x \boldsymbol{\varphi} \, \mathrm{d}x,$$

$\nu = \lambda - \frac{2}{d}\mu$, and the dissipation term in the energy inequality (5.119) as

$$\int_{\mathbb{T}^d} \mu |\nabla_x \boldsymbol{u}|^2 + \nu |\mathrm{div}_x \boldsymbol{u}|^2 \, \mathrm{d}x = \int_{\mathbb{T}^d} F(\mathbb{D}_x \boldsymbol{u}) + F^*(\mathbb{S}) \, \mathrm{d}x,$$

where the convex functions F and F^* have been introduced in (5.102) and (5.103), respectively.

Similarly to Sect. 5.1.5, we may define dissipative measure-valued solution for the Navier–Stokes system.

Definition 5.10 (**Dissipative measure-valued (DMV) solution to the Navier–Stokes system**)

Suppose that the pressure $p = p(\varrho)$ satisfies the isentropic EOS (5.112). A parametrized probability measure $\{\mathcal{V}_{t,x}\}_{(t,x) \in (0,T) \times \mathbb{T}^d}$,

$$\mathcal{V} \in L^\infty((0,T) \times \mathbb{T}^d; \mathcal{P}(R^{d+1})),\ R^{d+1} = \left\{ [\widetilde{\varrho}, \widetilde{\boldsymbol{u}}] \ \middle| \ \widetilde{\varrho} \in R,\ \widetilde{\boldsymbol{u}} \in R^d \right\},$$
$$\mathcal{V}_{t,x} \{\widetilde{\varrho} \geq 0\} = 1 \text{ for a.a. } (t,x),$$

is called *dissipative measure-valued (DMV) solution* of the Navier–Stokes system (5.97)–(5.100), with the initial conditions $[\varrho_0, \boldsymbol{m}_0]$ if the following holds:

- (**energy inequality**) the integral inequality

$$\begin{aligned} &\int_{\mathbb{T}^d} \left\langle \mathcal{V}_{\tau,x}; \frac{1}{2} \widetilde{\varrho} |\widetilde{\boldsymbol{u}}|^2 + P(\widetilde{\varrho}) \right\rangle \mathrm{d}x + \int_0^\tau \int_{\mathbb{T}^d} \left[F(\mathbb{D}_x \boldsymbol{u}) + F^*(\mathbb{S}) \right] \mathrm{d}x \, \mathrm{d}t \\ &+ \int_{\mathbb{T}^d} \mathrm{d}\mathfrak{E}(\tau) + \int_0^\tau \int_{\mathbb{T}^d} \mathrm{d}\mathfrak{D} \leq \int_{\mathbb{T}^d} \left[\frac{1}{2} \frac{|\boldsymbol{m}_0|^2}{\varrho_0} + P(\varrho_0) \right] \mathrm{d}x \end{aligned} \tag{5.122}$$

holds for a.a. $0 \le \tau \le T$, with the energy concentration defect

$$\mathfrak{E} \in L^\infty(0,T;\mathcal{M}^+(\mathbb{T}^d)),$$

and the dissipation defect

$$\mathfrak{D} \in \mathcal{M}^+([0,T] \times \mathbb{T}^d);$$

- (**equation of continuity**)

$$\langle \mathcal{V}; \widetilde{\varrho} \rangle \in C_{\rm weak}([0,T]; L^\gamma(\mathbb{T}^d)),\ \left\langle \mathcal{V}_{0,x}; \widetilde{\varrho} \right\rangle = \varrho_0(x) \ \text{ for a.a. } \ x \in \mathbb{T}^d,$$

and the integral identity

$$\left[\int_\Omega \langle \mathcal{V}; \widetilde{\varrho} \rangle \varphi \ {\rm d}x \right]_{t=0}^{t=\tau} = \int_0^\tau \int_{\mathbb{T}^d} \Big[\langle \mathcal{V}; \widetilde{\varrho} \rangle \partial_t \varphi + \langle \mathcal{V}; \widetilde{\varrho} \widetilde{\boldsymbol{u}} \rangle \cdot \nabla_x \varphi \Big] \ {\rm d}x \, {\rm d}t \tag{5.123}$$

for any $0 \le \tau \le T$, and any $\varphi \in W^{1,\infty}((0,T) \times \mathbb{T}^d)$;

- (**momentum equation**)

$$\langle \mathcal{V}; \widetilde{\varrho} \widetilde{\boldsymbol{u}} \rangle \in C_{\rm weak}([0,T]; L^{\frac{2\gamma}{\gamma+1}}(\mathbb{T}^d; R^d)),\ \left\langle \mathcal{V}_{0,x}; \widetilde{\varrho} \widetilde{\boldsymbol{u}} \right\rangle = \boldsymbol{m}_0(x) \text{ for a.a. } x \in \mathbb{T}^d,$$

and the integral identity

$$\begin{aligned}
&\left[\int_{\mathbb{T}^d} \langle \mathcal{V}; \widetilde{\varrho} \widetilde{\boldsymbol{u}} \rangle \cdot \boldsymbol{\varphi} \ {\rm d}x \right]_{t=0}^{t=\tau} \\
&= \int_0^\tau \int_{\mathbb{T}^d} \left[\langle \mathcal{V}; \widetilde{\varrho} \widetilde{\boldsymbol{u}} \rangle \cdot \partial_t \boldsymbol{\varphi} + \langle \mathcal{V}; \widetilde{\varrho} \widetilde{\boldsymbol{u}} \otimes \widetilde{\boldsymbol{u}} \rangle : \nabla_x \boldsymbol{\varphi} + \langle \mathcal{V}; p(\widetilde{\varrho}) \rangle {\rm div}_x \boldsymbol{\varphi} \right] \ {\rm d}x \, {\rm d}t \\
&- \int_0^\tau \int_{\mathbb{T}^d} \mathbb{S} : \nabla_x \boldsymbol{\varphi} \ {\rm d}x \, {\rm d}t + \int_0^\tau \int_{\mathbb{T}^d} \nabla_x \boldsymbol{\varphi} : {\rm d}\mathfrak{R}(t) \, {\rm d}t
\end{aligned} \tag{5.124}$$

holds for any $0 \le \tau \le T$, and any $\boldsymbol{\varphi} \in C^1([0,T] \times \mathbb{T}^d; R^d)$, with the Reynolds concentration defect

$$\mathfrak{R} \in L^\infty(0,T; \mathcal{M}^+(\mathbb{T}^d; R^{d \times d}_{\rm sym}))$$

satisfying

$$\underline{d} \ \mathfrak{E} \le {\rm tr}[\mathfrak{R}] \le \overline{d} \ \mathfrak{E} \text{ for some constants } 0 < \underline{d} \le \overline{d}; \tag{5.125}$$

- (**Korn–Poincaré inequality**)
the velocity field $\boldsymbol{u} = \langle \mathcal{V}; \widetilde{\boldsymbol{u}} \rangle \in L^2(0,T; W^{1,2}(\mathbb{T}^d; R^d))$ satisfies

$$
\begin{aligned}
&\int_0^\tau \int_{\mathbb{T}^d} \left\langle \mathcal{V}; |\widetilde{\boldsymbol{u}} - \boldsymbol{U}|^2 \right\rangle \, \mathrm{d}x \, \mathrm{d}t \\
&\quad \lesssim \int_0^\tau \int_{\mathbb{T}^d} \left\langle \mathcal{V}; \widetilde{\varrho} |\widetilde{\boldsymbol{u}} - \boldsymbol{U}|^2 \right\rangle \, \mathrm{d}x \, \mathrm{d}t \\
&\quad + \int_0^\tau \int_{\mathbb{T}^d} |\nabla_x (\boldsymbol{u} - \boldsymbol{U})|^2 \, \mathrm{d}x \, \mathrm{d}t + \int_0^\tau \left(\int_{\mathbb{T}^d} \mathrm{d}\mathfrak{E}(t) \right) \mathrm{d}t + \int_0^\tau \int_{\mathbb{T}^d} \mathrm{d}\mathfrak{D}
\end{aligned}
\tag{5.126}
$$

for a.a. $0 \leq \tau \leq T$, and any $\boldsymbol{U} \in C^1([0,T] \times \mathbb{T}^d; R^d)$.

Finally, we summarize the results obtained in this section.

Theorem 5.5 (Asymptotic limit of consistent approximation)
Let the pressure $p = p(\varrho)$ be given by (5.112). Let $\{\varrho_n, \boldsymbol{u}_n\}_{n=1}^\infty$ be a consistent approximation of the Navier–Stokes system (5.97), (5.98), (5.101) in the sense of Definition 5.9.

Then, for a suitable subsequence,

- *the sequence $\{\varrho_n, \boldsymbol{u}_n\}_{n=1}^\infty$ generates a Young measure*

$$
\{\mathcal{V}_{t,x}\}_{(t,x) \in (0,T) \times \mathbb{T}^d}, \ \mathcal{V} \in L^\infty_{\mathrm{weak}-(*)}((0,T) \times \mathbb{T}^d; \mathcal{P}(R^{d+1}));
$$

-

$$
\varrho_n \boldsymbol{u}_n \otimes \boldsymbol{u}_n + p(\varrho_n)\mathbb{I} \to \overline{\varrho \boldsymbol{u} \otimes \boldsymbol{u} + p(\varrho)\mathbb{I}} \ \textit{weakly-(*) in } L^\infty(0,T; \mathcal{M}(\mathbb{T}^d; R^{d \times d}));
$$

-

$$
\frac{1}{2}\varrho_n |\boldsymbol{u}_n|^2 + P(\varrho_n) \to \overline{\frac{1}{2}\varrho |\boldsymbol{u}|^2 + P(\varrho)} \ \textit{weakly-(*) in } L^\infty(0,T; \mathcal{M}(\mathbb{T}^d));
$$

-

$$
\mu |\nabla_x^n \boldsymbol{u}_n|^2 + \nu |\mathrm{div}_x^n \boldsymbol{u}_n|^2 \to \overline{\mu |\nabla_x \boldsymbol{u}|^2 + \nu |\mathrm{div}_x \boldsymbol{u}|^2} \ \textit{weakly-(*) in } \mathcal{M}([0,T] \times \mathbb{T}^d).
$$

The Young measure $\{\mathcal{V}_{t,x}\}_{(t,x) \in (0,T) \times \mathbb{T}^d}$ is a dissipative measure-valued (DMV) solution of the Navier–Stokes system in the sense of Definition 5.10, with the defect measures

$$
\mathfrak{E} = \overline{\frac{1}{2}\varrho |\boldsymbol{u}|^2 + P(\varrho)} - \left\langle \mathcal{V}; \frac{1}{2}\widetilde{\varrho} |\widetilde{\boldsymbol{u}}|^2 + P(\widetilde{\varrho}) \right\rangle \in L^\infty(0,T; \mathcal{M}^+(\mathbb{T}^d)),
$$

$$
\mathfrak{R} = \overline{\varrho \boldsymbol{u} \otimes \boldsymbol{u} + p(\varrho)\mathbb{I}} - \langle \mathcal{V}; \widetilde{\varrho} \widetilde{\boldsymbol{u}} \otimes \widetilde{\boldsymbol{u}} + p(\widetilde{\varrho})\mathbb{I} \rangle \in L^\infty(0,T; \mathcal{M}^+(\mathbb{T}^d; R^{d \times d}_{\mathrm{sym}})),
$$

$$
\mathfrak{D} = \overline{\mu |\nabla_x \boldsymbol{u}|^2 + \nu |\mathrm{div}_x \boldsymbol{u}|^2} - \left(\mu |\nabla_x \boldsymbol{u}|^2 + \nu |\mathrm{div}_x \boldsymbol{u}|^2 \right) \in \mathcal{M}^+([0,T] \times \mathbb{T}^d).
$$

5.3.3 *Dissipative Solutions to the Navier–Stokes System*

The concept of DMV solution for the Navier–Stokes system introduced in Definition 5.10 is rather awkward mixing up the nonconservative variable $\boldsymbol{u}$ with its differential. The situation simplifies a lot if additional compactness properties may be shown at the level of consistent approximation. Suppose that, in addition to the hypotheses listed in Definition 5.9, we are able to show that

$$\langle \mathcal{V}; \widetilde{\varrho \boldsymbol{u}} \rangle = \langle \mathcal{V}; \widetilde{\varrho} \rangle \, \langle \mathcal{V}; \widetilde{\boldsymbol{u}} \rangle \,, \tag{5.127}$$

or, in other words,

$$\begin{aligned} &\varrho_n \to \varrho \text{ weakly-(*) in } L^\infty(0,T; L^\gamma(\mathbb{T}^d)), \ \boldsymbol{u}_n \to \boldsymbol{u} \text{ weakly in } L^2((0,T)\times\mathbb{T}^d; R^d), \\ &\quad \text{and } \varrho_n \boldsymbol{u}_n \to \varrho \boldsymbol{u} \text{ weakly-(*) in } L^\infty(0,T; L^{\frac{2\gamma}{\gamma+1}}(\mathbb{T}^d; R^d)). \end{aligned} \tag{5.128}$$

Note that hypothesis (5.128) is quite realistic as boundedness of $\nabla_x^n[\boldsymbol{u}_n]$ may result in compactness of $\boldsymbol{u}_n$ in the space variable while ϱ_n is expected to be (weakly) compact in the time variable. The desired conclusion (5.128) may be then obtained applying the standard Aubin–Lions lemma.

Taking (5.128) for granted, we may simplify considerably the definition of the limit problem in the spirit of Sect. 5.2. Following the strategy of Sect. 5.2.1, we introduce *dissipative weak (DW) solution* of the Navier–Stokes system eliminating completely the Young measure in the definition.

Definition 5.11 (DISSIPATIVE WEAK (DW) SOLUTION TO THE NAVIER–STOKES SYSTEM)

Let the pressure $p = p(\varrho)$ be given by (5.112). We say that $[\varrho, \boldsymbol{u}]$ with the associated viscous stress $\mathbb{S}$ is *dissipative weak (DW) solution* to the Navier–Stokes system (5.97)–(5.100) with the initial data $[\varrho_0, \boldsymbol{m}_0]$ if the following holds:

- **(weak continuity)**

$$\begin{aligned} &\varrho \in C_{\rm weak}([0,T]; L^\gamma(\mathbb{T}^d)), \ \boldsymbol{u} \in L^2(0,T; W^{1,2}(\mathbb{T}^d; R^d)), \\ \boldsymbol{m} = (\varrho \boldsymbol{u}) &\in C_{\rm weak}([0,T]; L^{\frac{2\gamma}{\gamma+1}}(\mathbb{T}^d; R^d)), \\ &\mathbb{S} \in L^2((0,T)\times\mathbb{T}^d; R^{d\times d}_{\rm sym}); \end{aligned} \tag{5.129}$$

- (**energy inequality**) there is a defect measure $\mathfrak{E} \in L^\infty(0, T; \mathcal{M}^+(\mathbb{T}^d))$ such that the energy inequality

$$\int_{\mathbb{T}^d} \left[\frac{1}{2} \varrho |\boldsymbol{u}|^2 + P(\varrho) \right] (\tau, \cdot) \, \mathrm{d}x + \int_{\mathbb{T}^d} \mathrm{d}\mathfrak{E}(\tau) + \int_0^\tau \int_{\mathbb{T}^d} \left[F(\mathbb{D}_x \boldsymbol{u}) + F^*(\mathbb{S}) \right] \mathrm{d}x \, \mathrm{d}t \leq \int_{\mathbb{T}^d} \left[\frac{1}{2} \frac{|\boldsymbol{m}_0|^2}{\varrho_0} + P(\varrho_0) \right] \mathrm{d}x \tag{5.130}$$

for a.a. $0 \leq \tau \leq T$;
- (**equation of continuity**) the integral identity

$$\left[\int_{\mathbb{T}^d} \varrho \varphi \, \mathrm{d}x \right]_{t=0}^{t=\tau} = \int_0^\tau \int_{\mathbb{T}^d} \left[\varrho \partial_t \varphi + \varrho \boldsymbol{u} \cdot \nabla_x \varphi \right] \mathrm{d}x \, \mathrm{d}t \tag{5.131}$$

for any $0 \leq \tau \leq T$, and any test function $\varphi \in W^{1,\infty}((0, T) \times \mathbb{T}^d)$;
- (**momentum equation**) the integral identity

$$\left[\int_{\mathbb{T}^d} (\varrho \boldsymbol{u}) \cdot \boldsymbol{\varphi} \, \mathrm{d}x \right]_{t=0}^{t=\tau} = \int_0^\tau \int_{\mathbb{T}^d} \left[\varrho \boldsymbol{u} \cdot \partial_t \boldsymbol{\varphi} + \varrho \boldsymbol{u} \otimes \boldsymbol{u} : \nabla_x \boldsymbol{\varphi} + p(\varrho) \mathrm{div}_x \boldsymbol{\varphi} \right] \mathrm{d}x \, \mathrm{d}t - \int_0^\tau \int_{\mathbb{T}^d} \mathbb{S} : \nabla_x \boldsymbol{\varphi} \, \mathrm{d}x \, \mathrm{d}t + \int_0^\tau \int_{\mathbb{T}^d} \nabla_x \boldsymbol{\varphi} : \mathrm{d}\mathfrak{R}(t) \, \mathrm{d}t \tag{5.132}$$

for any $0 \leq \tau \leq T$, any test function $\boldsymbol{\varphi} \in C^1([0, T] \times \mathbb{T}^d; R^d)$, with the Reynolds defect

$$\mathfrak{R} \in L^\infty(0, T; \mathcal{M}^+(\mathbb{T}^d; R^{d \times d}_{\mathrm{sym}}));$$

- (**defect compatibility condition**)

$$\underline{d} \, \mathfrak{E} \leq \mathrm{tr} \, [\mathfrak{R}] \leq \overline{d} \, \mathfrak{E} \ \text{ for some constants } \ 0 \leq \underline{d} \leq \overline{d}. \tag{5.133}$$

Remark 5.20 In view (5.130), (5.133), we may eliminate $\mathfrak{E}$ setting

$$\mathfrak{E} = 1(\overline{d})^{-1} \mathrm{tr}[\mathfrak{R}].$$

Thus, similarly to the Euler system, the deviation of DW solutions from the standard weak (distributional) solution is encoded in a single quantity – the Reynolds defect $\mathfrak{R}$.

Finally, we may use the same arguments as for the Euler system in Sect. 5.2.1 to obtain the following result.

Theorem 5.6 (Asymptotic limit of consistent approximation)
Let the pressure $p = p(\varrho)$ be given by (5.112). *Let $\{\varrho_n, \boldsymbol{u}_n\}_{n=1}^{\infty}$ be a consistent approximation of the Navier–Stokes system* (5.97), (5.98), (5.101) *in the sense of Definition 5.9 generating a Young measure $\{\mathcal{V}_{t,x}\}_{(t,x)\in(0,T)\times\mathbb{T}^d}$ satisfying* (5.127).
Then, for a suitable subsequence,

$$\nabla_x^n \boldsymbol{u}_n \to \nabla_x \boldsymbol{u},\ \boldsymbol{u}_n \to \boldsymbol{u} \text{ weakly in } L^2((0,T)\times\mathbb{T}^d; R^d),$$

$$\varrho_n \boldsymbol{u}_n \otimes \boldsymbol{u}_n + p(\varrho_n)\mathbb{I} \to \overline{\varrho \boldsymbol{u}\otimes\boldsymbol{u} + p(\varrho)\mathbb{I}} \text{ weakly-(*) in } L^\infty(0,T;\mathcal{M}(\mathbb{T}^d;R^{d\times d})),$$

and

$$\frac{1}{2}\varrho_n|\boldsymbol{u}_n|^2 + P(\varrho_n) \to \overline{\frac{1}{2}\varrho|\boldsymbol{u}|^2 + P(\varrho)} \text{ weakly-(*) in } L^\infty(0,T;\mathcal{M}(\mathbb{T}^d)).$$

The quantities

$$\varrho = \langle \mathcal{V}; \widetilde{\varrho}\rangle,\ \boldsymbol{u} = \langle \mathcal{V}; \widetilde{\boldsymbol{u}}\rangle,$$

and the viscous stress

$$\mathbb{S} = \mu\left(\mathbb{D}_x\boldsymbol{u} - \frac{1}{d}\mathrm{div}_x\boldsymbol{u}\mathbb{I}\right) + \lambda\mathrm{div}_x\boldsymbol{u}\mathbb{I}$$

represent a dissipative weak (DW) solution of the Navier–Stokes system in the sense of Definition 5.11, with the defect measures

$$\mathfrak{E} = \overline{\frac{1}{2}\varrho|\boldsymbol{u}|^2 + P(\varrho)} - \left[\frac{1}{2}\varrho|\boldsymbol{u}|^2 + P(\varrho)\right] \in L^\infty(0,T;\mathcal{M}^+(\mathbb{T}^d)),$$

$$\mathfrak{R} = \overline{\varrho\boldsymbol{u}\otimes\boldsymbol{u} + p(\varrho)\mathbb{I}} - \left[\varrho\boldsymbol{u}\otimes\boldsymbol{u} + p(\varrho)\mathbb{I}\right] \in L^\infty(0,T;\mathcal{M}^+(\mathbb{T}^d;R^{d\times d}_{\mathrm{sym}})).$$

5.4 Compatibility

Compatibility of generalized solutions is the **property (C)** mentioned in the introduction to Chap. 5. A *compatible* generalized solution is a classical solution as long as it possess the necessary smoothness properties. Unlike the **weak-strong (WS) uniqueness** property asserting stability of strong solutions in the class of generalized (weak) solutions, compatibility is an intrinsic property of the generalized solution. In the framework of the mathematical theory, compatibility amounts to show that the defect measures used in the definition of generalized solutions vanish as long as the solution is smooth enough.

5.4.1 Compatibility for the Complete Euler System

Consider a dissipative weak (DW) solution $[\varrho, \boldsymbol{m}, S]$ of the complete Euler system in the sense of Definition 5.7. For the sake of simplicity, suppose that $M = 1$ in (5.92), meaning the function of class $C^1([0,T] \times \overline{\Omega}; R^d)$ can be used as test functions in the momentum balance (5.92). In addition, let $[\varrho, \boldsymbol{m}, S]$ belong to the regularity class

$$\varrho \in C^1([0,T] \times \overline{\Omega}), \ \inf_{(0,T)\times\Omega} \varrho > 0, \ \boldsymbol{u} \in C^1([0,T] \times \overline{\Omega}; R^d), \ S \in C^1([0,T] \times \overline{\Omega}).$$

Our goal is to show that $[\varrho, \boldsymbol{m}, S]$ is a classical solution of the complete Euler system.

First, we introduce the velocity field

$$\boldsymbol{u} \equiv \frac{\boldsymbol{m}}{\varrho} \in C^1([0,T] \times \overline{\Omega}; R^d).$$

Performing standard by parts integration in the equation of continuity (5.91) we get

$$\int_0^T \int_\Omega \left[\partial_t \varrho + \mathrm{div}_x(\varrho \boldsymbol{u}) \right] \varphi \, \mathrm{d}x \, \mathrm{d}t = 0$$

for any $\varphi \in C_c^\infty((0,T) \times \Omega)$. Consequently,

$$\partial_t \varrho + \mathrm{div}_x(\varrho \boldsymbol{u}) = 0 \text{ in } (0,T) \times \Omega. \tag{5.134}$$

Repeating the same process with a general $\varphi \in C_c^\infty((0,T) \times \overline{\Omega})$ and using (5.134), we get

$$\int_0^T \int_{\partial\Omega} \varrho \boldsymbol{u} \cdot \boldsymbol{n} \varphi \, \mathrm{d}S_x \, \mathrm{d}t = 0$$

yielding

$$\boldsymbol{u} \cdot \boldsymbol{n}|_{\partial\Omega} = 0. \tag{5.135}$$

Next, we use $\boldsymbol{u}$ as a test function in the momentum equation (5.92) obtaining

$$\left[\int_\Omega \varrho |\boldsymbol{u}|^2 \, \mathrm{d}x \right]_{t=0}^{t=\tau} = \int_0^\tau \int_\Omega \left[\varrho \boldsymbol{u} \cdot \partial_t \boldsymbol{u} + \varrho \boldsymbol{u} \otimes \boldsymbol{u} : \nabla_x \boldsymbol{u} + p(\varrho, S) \mathrm{div}_x \boldsymbol{u} \right] \mathrm{d}x \, \mathrm{d}t$$
$$+ \int_0^\tau \int_{\overline{\Omega}} \nabla_x \boldsymbol{u} : \mathrm{d}\mathfrak{R}(t) \, \mathrm{d}t.$$

With the help of (5.134), (5.135) we obtain

$$\left[\int_{\Omega} \frac{1}{2}\varrho|\boldsymbol{u}|^2 \,\mathrm{d}x\right]_{t=0}^{t=\tau} = \int_0^\tau \int_{\Omega} p(\varrho, S)\mathrm{div}_x\boldsymbol{u} \,\mathrm{d}x\,\mathrm{d}t + \int_0^\tau \int_{\overline{\Omega}} \nabla_x\boldsymbol{u} : \mathrm{d}\mathfrak{R}(t)\,\mathrm{d}t,$$

which, subtracted from the total energy balance (5.90), yields

$$\begin{aligned} &\int_{\Omega} \varrho e(\varrho, S)(\tau, \cdot) \,\mathrm{d}x + \int_{\overline{\Omega}} \mathrm{d}\mathfrak{E}(\tau) \\ &\leq \int_{\Omega} \varrho_0 e(\varrho_0, S_0) \,\mathrm{d}x - \int_0^\tau \int_{\Omega} p(\varrho, S)\mathrm{div}_x\boldsymbol{u} \,\mathrm{d}x\,\mathrm{d}t - \int_0^\tau \int_{\overline{\Omega}} \nabla_x\boldsymbol{u} : \mathrm{d}\mathfrak{R}(t)\,\mathrm{d}t. \end{aligned} \tag{5.136}$$

Finally, computing the temperature and the internal energy

$$s = \frac{S}{\varrho} = c_v \log(\vartheta) - \log(\varrho),$$

we use ϑ as test function in the entropy balance (5.93)

$$\begin{aligned} 0 &\geq \int_0^\tau \int_{\Omega} \left[-\partial_t S\vartheta - \mathrm{div}_x(\varrho s\boldsymbol{u})\vartheta + \left\langle \mathcal{V}; 1_{\widetilde{\varrho}>0}\left(\widetilde{S}\frac{\widetilde{\boldsymbol{m}}}{\widetilde{\varrho}}\right) - S\frac{\boldsymbol{m}}{\varrho}\right\rangle \cdot \nabla_x\vartheta\right] \mathrm{d}x\,\mathrm{d}t \\ &= \int_0^\tau \int_{\Omega} -\partial_t(\varrho e) + p\mathrm{div}_x\boldsymbol{u} + \left\langle \mathcal{V}; 1_{\widetilde{\varrho}>0}\left(\widetilde{S}\frac{\widetilde{\boldsymbol{m}}}{\widetilde{\varrho}}\right) - S\frac{\boldsymbol{m}}{\varrho}\right\rangle \cdot \nabla_x\vartheta \,\mathrm{d}x\,\mathrm{d}t. \end{aligned}$$

Adding the resulting expression to (5.136), we finally obtain

$$\int_{\overline{\Omega}} \mathrm{d}\mathfrak{E}(\tau) \leq \int_0^\tau \left|\int_{\Omega} \left\langle \mathcal{V}; 1_{\widetilde{\varrho}>0}\left(\widetilde{S}\frac{\widetilde{\boldsymbol{m}}}{\widetilde{\varrho}}\right) - S\frac{\boldsymbol{m}}{\varrho}\right\rangle \cdot \nabla_x\vartheta \,\mathrm{d}x\right| \mathrm{d}t - \int_0^\tau \int_{\overline{\Omega}} \nabla_x\boldsymbol{u} : \mathrm{d}\mathfrak{R}. \tag{5.137}$$

At this stage, we employ the defect compatibility conditions (5.94), (5.95), (5.96):

$$\begin{aligned} &\langle \mathcal{V}; \widetilde{\varrho}\rangle = \varrho,\ \langle \mathcal{V}; \widetilde{\boldsymbol{m}}\rangle = \boldsymbol{m},\ \left\langle \mathcal{V}; \widetilde{S}\right\rangle = S,\ \mathcal{V}_{t,x}\left\{\widetilde{\varrho} \geq 0,\ \widetilde{S} \geq \underline{s}\widetilde{\varrho}\right\} = 1 \\ &\quad \text{for a.a. } (t, x) \in (0, T) \times \Omega, \end{aligned} \tag{5.138}$$

$$\underline{d}\ \mathfrak{E} \leq \mathrm{tr}\,[\mathfrak{R}] \leq \overline{d}\ \mathfrak{E} \text{ for some constants } 0 < \underline{d} \leq \overline{d}, \tag{5.139}$$

and

$$\mathfrak{E} \geq \left\langle \mathcal{V}; \frac{1}{2} \frac{|\widetilde{\boldsymbol{m}}|^2}{\widetilde{\varrho}} + \widetilde{\varrho} e(\widetilde{\varrho}, \widetilde{S}) \right\rangle - \left(\frac{1}{2} \frac{|\boldsymbol{m}|^2}{\varrho} + \varrho e(\varrho, S) \right). \tag{5.140}$$

Our goal is to apply Gronwall's lemma to (5.137) to conclude that

$$\mathfrak{E} = \mathfrak{R} = 0, \ \mathcal{V}_{t,x} = \delta_{[\varrho(t,x), \boldsymbol{m}(t,x), S(t,x)]} \ (t, x) \in (0, T) \times \Omega. \tag{5.141}$$

To begin, we easily observe that the last integral in (5.137) is controlled by the left-hand side by virtue of (5.139).

Next, we decompose the domain of the measure $\mathcal{V}$ into two components:

$$\mathcal{R} = \left\{ [\widetilde{\varrho}, \widetilde{S}] \ \Big| \ \widetilde{S} \geq \underline{s} \widetilde{\varrho}, \ \frac{1}{2} \underline{\varrho} < \widetilde{\varrho} < 2 \overline{\varrho}, \ |S| < 2 \overline{S} \right\}, \ \mathcal{S} = \left\{ [\widetilde{\varrho}, \widetilde{S}] \ \Big| \ \widetilde{S} \geq \underline{s} \widetilde{\varrho} \right\} \setminus \mathcal{R},$$

where the constants $\underline{\varrho}, \overline{\varrho}, \overline{S}$ are chosen in such a way that

$$0 < \underline{\varrho} \leq \varrho(t, x) \leq \overline{\varrho}, \ |S(t, x)| \leq \overline{S} \text{ for all } (t, x) \in (0, T) \times \Omega.$$

As the energy

$$E(\widetilde{\varrho}, \widetilde{\boldsymbol{m}}, \widetilde{S}) = \frac{1}{2} \frac{|\widetilde{\boldsymbol{m}}|^2}{\widetilde{\varrho}} + \widetilde{\varrho} e(\widetilde{\varrho}, \widetilde{S})$$

is convex, we deduce from (5.138)–(5.140),

$$\mathfrak{E} \geq \left\langle \mathcal{V}; E(\widetilde{\varrho}, \widetilde{\boldsymbol{m}}, \widetilde{S}) - \partial_{\varrho, m, S} E(\widetilde{\varrho}, \widetilde{\boldsymbol{m}}, \widetilde{S}) \cdot [\widetilde{\varrho} - \varrho, \widetilde{\boldsymbol{m}} - \boldsymbol{m}, \widetilde{S} - S] - E(\varrho, \boldsymbol{m}, S) \right\rangle.$$

Accordingly, in view of the estimates derived in Lemma 5.2,

$$\left| \widetilde{S} \frac{\widetilde{\boldsymbol{m}}}{\widetilde{\varrho}} - S \frac{\boldsymbol{m}}{\varrho} \right| \lesssim E(\widetilde{\varrho}, \widetilde{\boldsymbol{m}}, \widetilde{S}) - \partial_{\varrho, m, S} E(\widetilde{\varrho}, \widetilde{\boldsymbol{m}}, \widetilde{S}) \cdot [\widetilde{\varrho} - \varrho, \widetilde{\boldsymbol{m}} - \boldsymbol{m}, \widetilde{S} - S] - E(\varrho, \boldsymbol{m}, S)$$

if $[\widetilde{\varrho}, \widetilde{\boldsymbol{m}}, \widetilde{S}] \in \mathcal{S}$. On the other hand, as $E = E(\widetilde{\varrho}, \widetilde{\boldsymbol{m}}, \widetilde{S})$ is strictly convex,

$$\begin{aligned} &\left\langle \mathcal{V}, 1_{[\widetilde{\varrho}, \widetilde{\boldsymbol{m}}, \widetilde{S}] \in \mathcal{R}} \left(\widetilde{S} \frac{\widetilde{\boldsymbol{m}}}{\widetilde{\varrho}} - S \frac{\boldsymbol{m}}{\varrho} \right) \right\rangle \\ &\lesssim \left\langle \mathcal{V}; E(\widetilde{\varrho}, \widetilde{\boldsymbol{m}}, \widetilde{S}) - \partial_{\varrho, m, S} E(\widetilde{\varrho}, \widetilde{\boldsymbol{m}}, \widetilde{S}) \cdot [\widetilde{\varrho} - \varrho, \widetilde{\boldsymbol{m}} - \boldsymbol{m}, \widetilde{S} - S] - E(\varrho, \boldsymbol{m}, S) \right\rangle. \end{aligned}$$

Thus applying Gronwall's lemma to (5.140) we obtain (5.141).

We have proved the following result.

Theorem 5.7 (Compatibility for complete Euler system)

Let $\Omega \subset R^d$ be a bounded Lipschitz domain. Let $[\varrho, \boldsymbol{m}, S]$ be a dissipative weak (DW) solution of the complete Euler system in the sense of Definition 5.7, with $M = 1$ in (5.92). In addition, let $[\varrho, \boldsymbol{m}, S]$ belong to the regularity class

$$\varrho \in C^1([0,T] \times \overline{\Omega}), \ \inf_{(0,T)\times\Omega} \varrho > 0, \ \boldsymbol{u} \in C^1([0,T] \times \overline{\Omega}; R^d), \ S \in C^1([0,T] \times \overline{\Omega}).$$

Then $[\varrho, \boldsymbol{m}, S]$ *is a classical solution of the complete Euler system. More specifically,*

$$\mathfrak{E} = \mathfrak{R} = 0, \ \textit{and} \ \mathcal{V}_{t,x} = \delta_{[\varrho(t,x),\boldsymbol{m}(t,x),S(t,x)]} \ \textit{for} \ (t,x) \in (0,T) \times \Omega.$$

5.4.2 Compatibility for the Navier–Stokes and the Barotropic Euler Systems

Following the ideas of the preceding section, we show the compatibility property for the dissipative weak (DW) solutions to the Navier–Stokes system specified in Definition 5.11.

Theorem 5.8 (Compatibility for the Navier–Stokes system)
Let $[\varrho, \boldsymbol{u}]$,

$$\varrho \in C^1([0,T] \times \mathbb{T}^d), \quad \inf_{(0,T)\times\mathbb{T}^d} \varrho > 0, \ \boldsymbol{u} \in C^1([0,T] \times \mathbb{T}^d; R^d)$$

be a dissipative weak solution of the Navier–Stokes system in the sense of Definition 5.11.

Then $[\varrho, \boldsymbol{u}]$ *is a classical solution, specifically,*

$$\mathfrak{E} = \mathfrak{R} = 0, \ \mathbb{S} \in \partial F(\mathbb{D}_x \boldsymbol{u}),$$

and the equations are satisfied in the classical sense.

Proof Following the arguments of the preceding section, we first show that

$$\partial_t \varrho + \mathrm{div}_x(\varrho \boldsymbol{u}) = 0 \text{ in } (0,T) \times \mathbb{T}^d.$$

Next, we use $\boldsymbol{u}$ as a test function in the momentum equation (5.132) obtaining

$$\begin{aligned} \left[\int_{\mathbb{T}^d} \varrho |\boldsymbol{u}|^2 \, \mathrm{d}x \right]_{t=0}^{t=\tau} &= \int_0^\tau \int_{\mathbb{T}^d} \left[\varrho \boldsymbol{u} \cdot \partial_t \boldsymbol{u} + \varrho \boldsymbol{u} \otimes \boldsymbol{u} : \nabla_x \boldsymbol{u} + p(\varrho) \mathrm{div}_x \boldsymbol{u} \right] \, \mathrm{d}x \, \mathrm{d}t \\ &\quad - \int_0^\tau \int_{\mathbb{T}^d} \mathbb{S} : \nabla_x \boldsymbol{u} \, \mathrm{d}x \, \mathrm{d}t + \int_0^\tau \int_{\mathbb{T}^d} \nabla_x \boldsymbol{u} : \mathrm{d}\mathfrak{R}(t) \, \mathrm{d}t. \end{aligned} \tag{5.142}$$

Relation (5.142) subtracted from the energy balance (5.130) gives rise to

$$\int_{\mathbb{T}^d} \mathrm{d}\mathfrak{E}(\tau) + \int_0^\tau \int_{\mathbb{T}^d} \Big[F(\mathbb{D}_x \boldsymbol{u}) + F^*(\mathbb{S}) - \mathbb{S} : \mathbb{D}_x \boldsymbol{u} \Big] \,\mathrm{d}x\,\mathrm{d}t \leq - \int_0^\tau \int_{\mathbb{T}^d} \nabla_x \boldsymbol{u} : \mathrm{d}\mathfrak{R}(t)\,\mathrm{d}t.$$

Thus the defect compatibility condition (5.133) combined with Gronwall's lemma yield first

$$\mathfrak{E} = \mathfrak{R} = 0,$$

and then, in accordance with Fenchel–Young inequality,

$$F(\mathbb{D}_x \boldsymbol{u}) + F^*(\mathbb{S}) = \mathbb{S} : \mathbb{D}_x \boldsymbol{u} \ \Rightarrow\ \mathbb{S} \in \partial F(\mathbb{D}_x \boldsymbol{u}).$$

□

Finally, we state the result for the barotropic Euler system, the proof of which is exactly the same as that of Theorem 5.8.

Theorem 5.9 (Compatibility for barotropic Euler system)
Let $[\varrho, \boldsymbol{m}]$,

$$\varrho \in C^1([0,T] \times \mathbb{T}^d), \quad \inf_{(0,T)\times\mathbb{T}^d} \varrho > 0, \ \boldsymbol{m} \in C^1([0,T] \times \mathbb{T}^d; R^d)$$

be a dissipative weak solution of the barotropic Euler system in the sense of Definition 5.6.

Then $[\varrho, \boldsymbol{m}]$ *is a classical solution, specifically,*

$$\mathfrak{E} = \mathfrak{R} = 0,$$

and the equations are satisfied in the classical sense.

5.5 Conclusion, Bibliographical Remarks

A general exposition of the theory of Young measures can be found in Ball [8] or in the monograph by Pedregal [178].

The measure-valued solutions as suitable objects to describe oscillations in systems of conservation laws were proposed in the pioneering work of DiPerna [65]. Later, DiPerna and Majda [66, 69, 70] developed a new approach based on measure-valued solutions to the Euler system describing the motion of incompressible fluids. Related results have been obtained by Greengard and Thomann [121].

The concept of measure-valued solutions has been further developed by Málek et al. [157] in the context of non-Newtonian fluids. The related mathematical theory of multipolar fluids was proposed by Nečas and Šilhavý [170], see also [168], [169], or [161]. Neustupa [171] developed the theory in the direction of the compressible

Euler and Navier–Stokes systems. The measure-valued solutions for the complete Euler system in the entropy formulation were obtained by Kröner and Zajaczkowski [144].

More recently, Székelyhidi and Wiedemann [188] have shown that *any* measure-valued solution to the incompressible Euler system can be approximated by a sequence of admissible weak solutions. A similar result, however, does not hold for the compressible barotropic Euler system, see [48].

Chapter 6
Weak-Strong Uniqueness Principle

This chapter is the heart of the theoretical part of this monograph. We discuss the *weak-strong uniqueness principle* for the class of generalized solutions introduced in the previous chapter. These results are absolutely indispensable for the numerical analysis in Part III. In particular, they can be interpreted in the spirit of the celebrated *Lax equivalence principle*

$$\text{stability} + \text{consistency} \Leftrightarrow \text{convergence}$$

extended to the nonlinear framework. Weak-strong uniqueness principle is a statement concerning stability of strong solutions in the class of generalized (weak) solutions: A weak solution coincides with the strong solution emanating from the same initial data as long as the latter exists. We examine the generalized solutions introduced in the previous chapter and establish the relevant results case by case. The crucial tool is the *relative energy inequality* introduced in the context of weak solutions in Chap. 4. The relative energy represents a kind of "metric" to evaluate the distance between a generalized solution and the smooth solution that is used as a "test function". More precisely, the relative energy is obtained via linearization of the energy functional around a given point in the phase space. As the energy is a convex function of the phase variables, the relative energy can be viewed as the so-called *Bregman distance*. The desired conclusion is obtained via a Gronwall type argument.

The relative energy is in fact a versatile tool, the applications of which go well beyond the weak-strong uniqueness principle. It can be used in problems involving stability of a particular smooth solution arising as singular limit, a long-time asymptotic regime, or a stationary state, among others. Another interesting application of the relative energy lies in the error analysis of numerical schemes as we will see in Part III.

E. Feireisl et al., *Numerical Analysis of Compressible Fluid Flows*, MS&A 20,
https://doi.org/10.1007/978-3-030-73788-7_6

6.1 Barotropic Euler System

We start with the simplest system of equations, for which the weak-strong uniqueness principle can be shown in an elegant way including at the same time the dissipative measure-valued (DMV) solutions introduced in Sect. 5.1.5 and the dissipative weak (DW) solutions from Sect. 5.2.1.

We start by recalling the classical formulation of the problem:

$$\begin{gathered}\partial_t \varrho + \mathrm{div}_x \boldsymbol{m} = 0 \text{ in } (0,T)\times\Omega,\ \varrho(0,\cdot) = \varrho_0,\\ \partial_t \boldsymbol{m} + \mathrm{div}_x \left(\frac{\boldsymbol{m}\otimes\boldsymbol{m}}{\varrho}\right) + \nabla_x p(\varrho) = 0 \text{ in } (0,T)\times\Omega,\ \boldsymbol{m}(0,\cdot) = \boldsymbol{m}_0,\\ \boldsymbol{m}\cdot\boldsymbol{n}|_{\partial\Omega} = 0.\end{gathered} \tag{6.1}$$

In Sect. 5.1.5, we have introduced the DMV solutions to problem (6.1) and in Sect. 5.2.1 the DW solutions. Moreover, we have shown that these two concepts are equivalent, see Theorem 5.4. Accordingly, we focus on the DW solutions, where the weak-strong uniqueness principle can be shown following the arguments of Sect. 4.1.3. We recall that the DW solutions solve the system of equations

$$\begin{gathered}\partial_t \varrho + \mathrm{div}_x \boldsymbol{m} = 0,\ \varrho(0,\cdot) = \varrho_0,\\ \partial_t \boldsymbol{m} + \mathrm{div}_x \left(\frac{\boldsymbol{m}\otimes\boldsymbol{m}}{\varrho}\right) + \nabla_x p(\varrho) = -\mathrm{div}_x \mathfrak{R},\ \boldsymbol{m}(0,\cdot) = \boldsymbol{m}_0\end{gathered} \tag{6.2}$$

in the sense of distributions, along with the total energy balance

$$\int_\Omega \left[\frac{1}{2}\frac{|\boldsymbol{m}|^2}{\varrho} + P(\varrho)\right](\tau,\cdot)\,\mathrm{d}x + \int_{\overline{\Omega}} \mathrm{d}\mathfrak{E}(\tau) \le \int_\Omega \left[\frac{1}{2}\frac{|\boldsymbol{m}_0|^2}{\varrho_0} + P(\varrho_0)\right]\,\mathrm{d}x. \tag{6.3}$$

The Reynolds defect $\mathfrak{R}$ and the energy defect $\mathfrak{E}$,

$$\mathfrak{R} \in L^\infty(0,T;\mathcal{M}^+(\overline{\Omega};R^{d\times d}_{\mathrm{sym}})),\ \mathfrak{E} \in L^\infty(0,T;\mathcal{M}^+(\overline{\Omega}))$$

satisfy the "compatibility condition"

$$\underline{d}\ \mathfrak{E} \le \mathrm{tr}[\mathfrak{R}] \le \overline{d}\ \mathfrak{E} \text{ for some constants } 0 < \underline{d} \le \overline{d}, \tag{6.4}$$

see Sect. 5.2.1 for details. As pointed out several times, the formulation can be simplified by setting

$$\mathfrak{E} = \frac{1}{\overline{d}}\mathrm{tr}[\mathfrak{R}].$$

6.1.1 Relative Energy Inequality

Given the apparent similarity of the system (6.2), (6.3) with the weak formulation introduced in Definition 4.2, the relative energy inequality for the barotropic Euler system can be derived via a straightforward modification of the arguments of Sect. 4.1.3. Indeed the energy balance (4.28) will contain and extra term due to the energy defect, while (4.29) is modified by the Reynolds defect. We should also keep in mind that the momentum equation in (6.2) contains a measure so that admissible test functions must be at least continuously differentiable in $\overline{\Omega}$.

Introducing the *relative energy*

$$E\left(\varrho, \boldsymbol{m} \,\Big|\, \widetilde{\varrho}, \widetilde{\boldsymbol{u}}\right) = \frac{1}{2}\left(\frac{|\boldsymbol{m}|^2}{\varrho} - 2\boldsymbol{m}\cdot\widetilde{\boldsymbol{u}} + \varrho|\widetilde{\boldsymbol{u}}|^2\right) + P(\varrho) - P'(\widetilde{\varrho})(\varrho - \widetilde{\varrho}) - P(\widetilde{\varrho})$$
$$= \frac{1}{2}\varrho\left|\frac{\boldsymbol{m}}{\varrho} - \widetilde{\boldsymbol{u}}\right|^2 + P(\varrho) - P'(\widetilde{\varrho})(\varrho - \widetilde{\varrho}) - P(\widetilde{\varrho})$$

we deduce the relative energy inequality in the form

$$\begin{aligned}
&\left[\int_\Omega E\left(\varrho, \boldsymbol{m} \,\Big|\, \widetilde{\varrho}, \widetilde{\boldsymbol{u}}\right)\,\mathrm{d}x\right]_{t=0}^{t=\tau} + \int_{\overline{\Omega}} \mathrm{d}\mathfrak{E}(\tau)\\
&\leq -\int_0^\tau\int_\Omega 1_{\varrho>0}\varrho\nabla_x\widetilde{\boldsymbol{u}}\cdot\left(\frac{\boldsymbol{m}}{\varrho} - \widetilde{\boldsymbol{u}}\right)\cdot\left(\frac{\boldsymbol{m}}{\varrho} - \widetilde{\boldsymbol{u}}\right)\,\mathrm{d}x\,\mathrm{d}t\\
&-\int_0^\tau\int_\Omega \Big[p(\varrho) - p'(\widetilde{\varrho})(\varrho - \widetilde{\varrho}) - p(\widetilde{\varrho})\Big]\mathrm{div}_x\widetilde{\boldsymbol{u}}\,\mathrm{d}x\,\mathrm{d}t\\
&+\int_0^\tau\int_\Omega \frac{1}{\widetilde{\varrho}}(\varrho\widetilde{\boldsymbol{u}} - \boldsymbol{m})\cdot\Big[\partial_t(\widetilde{\varrho}\widetilde{\boldsymbol{u}}) + \mathrm{div}_x\left(\widetilde{\varrho}\widetilde{\boldsymbol{u}}\otimes\widetilde{\boldsymbol{u}}\right) + \nabla_x p(\widetilde{\varrho})\Big]\,\mathrm{d}x\,\mathrm{d}t\\
&+\int_0^\tau\int_\Omega \left[\left(1 - \frac{\varrho}{\widetilde{\varrho}}\right)p'(\widetilde{\varrho}) + \frac{1}{\widetilde{\varrho}}\widetilde{\boldsymbol{u}}\cdot(\boldsymbol{m} - \varrho\widetilde{\boldsymbol{u}})\right]\Big[\partial_t\widetilde{\varrho} + \mathrm{div}_x(\widetilde{\varrho}\widetilde{\boldsymbol{u}})\Big]\,\mathrm{d}x\,\mathrm{d}t\\
&-\int_0^\tau\left(\int_{\overline{\Omega}} \nabla_x\widetilde{\boldsymbol{u}} : \mathrm{d}\mathfrak{R}(t)\right)\mathrm{d}t.
\end{aligned} \tag{6.5}$$

In view of (6.4), we may replace

$$\int_{\overline{\Omega}} \mathrm{d}\mathfrak{E}(\tau) \approx \frac{1}{d}\int_{\overline{\Omega}} \mathrm{d}\mathrm{tr}[\mathfrak{R}].$$

Inequality (6.5) holds for any DW solution $[\varrho, \boldsymbol{m}]$ specified in Definition 5.6 and any test functions $[\widetilde{\varrho}, \widetilde{\boldsymbol{u}}]$ in the class

$$\begin{gathered}\widetilde{\varrho} \in C^1([0,T] \times \overline{\Omega}),\ 0 < \underline{\varrho} \leq \widetilde{\varrho} \text{ in } [0,T] \times \overline{\Omega},\ \widetilde{\boldsymbol{u}} \in C^M([0,T] \times \overline{\Omega}; R^d),\\ M \geq 1,\ \widetilde{\boldsymbol{u}} \cdot \boldsymbol{n}|_{\partial\Omega} = 0,\end{gathered} \tag{6.6}$$

where M is the same as in the momentum balance (5.87).

6.1.2 *Weak-Strong Uniqueness for DW Solutions of Barotropic Euler System*

We have everything at hand to show the desired weak-strong uniqueness principle for DW solutions of the barotropic Euler system. Note that the proof is almost identical to that of Theorem 4.2.

Theorem 6.1 (Weak-strong uniqueness for DW solutions of barotropic Euler system)

Let $\Omega \subset R^d$, $d = 2, 3$, be a bounded domain of class C^N, $N = \min\{2, M\}$. Let the pressure $p = p(\varrho)$ satisfy

$$\begin{gathered}p \in C[0,\infty) \cap C^1(0,\infty),\ p(0) = 0,\ p'(\varrho) > 0 \textit{ for } \varrho > 0,\\ 0 < \liminf_{\varrho \to \infty} \frac{p'(\varrho)}{\varrho^{\gamma-1}} \leq \limsup_{\varrho \to \infty} \frac{p'(\varrho)}{\varrho^{\gamma-1}} < \infty\end{gathered} \tag{6.7}$$

for some $\gamma > 1$. Suppose that $[\varrho, \boldsymbol{m}]$ is a dissipative weak (DW) solution of the barotropic Euler system (6.1) in $(0,T) \times \Omega$ in the sense of Definition 5.6, with the initial data $[\varrho_0, \boldsymbol{m}_0]$.

Let $\widetilde{\varrho}$, $\widetilde{\boldsymbol{u}}$ be a strong solution of the same problem belonging to the class

$$\begin{gathered}\widetilde{\varrho} \in W^{1,\infty}((0,T) \times \Omega),\ \inf_{(t,x) \in (0,T) \times \Omega} \widetilde{\varrho}(t,x) > 0,\\ \widetilde{\boldsymbol{u}} \in W^{1,\infty}((0,T) \times \Omega; R^d),\ \widetilde{\boldsymbol{u}} \cdot \boldsymbol{n}|_{\partial\Omega} = 0,\end{gathered} \tag{6.8}$$

and such that

$$\widetilde{\varrho}(0,\cdot) = \varrho_0,\ \widetilde{\varrho}(0,\cdot)\widetilde{\boldsymbol{u}}(0,\cdot) = \boldsymbol{m}_0.$$

Then

$$\varrho = \widetilde{\varrho},\ \boldsymbol{m} = \widetilde{\varrho}\widetilde{\boldsymbol{u}} \textit{ in } (0,T) \times \Omega,$$

and

$$\mathfrak{E} = \mathfrak{R} = 0.$$

Proof The obvious idea is to use the strong solution as test functions in the relative energy inequality (6.5). Unfortunately, they do not meet the regularity requirement (6.6), thus some preliminary manipulation is needed. First, we realize that, by virtue of (6.4), (6.7),

$$\left| \int_0^\tau \int_\Omega 1_{\varrho>0} \varrho \nabla_x \widetilde{\boldsymbol{u}} \cdot \left(\frac{\boldsymbol{m}}{\varrho} - \widetilde{\boldsymbol{u}} \right) \cdot \left(\frac{\boldsymbol{m}}{\varrho} - \widetilde{\boldsymbol{u}} \right) \, \mathrm{d}x \, \mathrm{d}t \right|$$

$$\stackrel{<}{\sim} \|\nabla_x \widetilde{\boldsymbol{u}}\|_{L^\infty((0,T)\times\Omega; R^{d\times d})} \int_0^\tau \int_\Omega E\left(\varrho, \boldsymbol{m} \,\middle|\, \widetilde{\varrho}, \widetilde{\boldsymbol{u}}\right) \, \mathrm{d}x \, \mathrm{d}t,$$

$$\left| \int_0^\tau \int_\Omega \left[p(\varrho) - p'(\widetilde{\varrho})(\varrho - \widetilde{\varrho}) - p(\widetilde{\varrho}) \right] \mathrm{div}_x \widetilde{\boldsymbol{u}} \, \mathrm{d}x \, \mathrm{d}t \right|$$

$$\stackrel{<}{\sim} \|\nabla_x \widetilde{\boldsymbol{u}}\|_{L^\infty((0,T)\times\Omega; R^{d\times d})} \int_0^\tau \int_\Omega E\left(\varrho, \boldsymbol{m} \,\middle|\, \widetilde{\varrho}, \widetilde{\boldsymbol{u}}\right) \, \mathrm{d}x \, \mathrm{d}t,$$

and

$$\left| \int_0^\tau \left(\int_{\overline{\Omega}} \nabla_x \widetilde{\boldsymbol{u}} : \mathrm{d}\mathfrak{R}(t) \right) \mathrm{d}t \right| \stackrel{<}{\sim} \|\nabla_x \widetilde{\boldsymbol{u}}\|_{L^\infty((0,T)\times\Omega; R^{d\times d})} \int_0^\tau \int_{\overline{\Omega}} 1 \mathrm{d}\mathfrak{E}(t) \, \mathrm{d}t.$$

Accordingly, the relative energy inequality (6.5) reduces to

$$\begin{aligned}
&\left[\int_\Omega E\left(\varrho, \boldsymbol{m} \,\middle|\, \widetilde{\varrho}, \widetilde{\boldsymbol{u}}\right)(\tau, \cdot) \, \mathrm{d}x + \int_{\overline{\Omega}} \mathrm{d}\mathfrak{E}(\tau) \right] \stackrel{<}{\sim} \int_\Omega E\left(\varrho_0, \boldsymbol{m}_0 \,\middle|\, \widetilde{\varrho}(0,\cdot), \widetilde{\boldsymbol{u}}(0,\cdot)\right) \, \mathrm{d}x \\
&+ \int_0^\tau \int_\Omega \frac{1}{\widetilde{\varrho}} (\varrho \widetilde{\boldsymbol{u}} - \boldsymbol{m}) \cdot \left[\partial_t (\widetilde{\varrho} \widetilde{\boldsymbol{u}}) + \mathrm{div}_x (\widetilde{\varrho} \widetilde{\boldsymbol{u}} \otimes \widetilde{\boldsymbol{u}}) + \nabla_x p(\widetilde{\varrho}) \right] \mathrm{d}x \, \mathrm{d}t \\
&+ \int_0^\tau \int_\Omega \left[\left(1 - \frac{\varrho}{\widetilde{\varrho}} \right) p'(\widetilde{\varrho}) + \frac{1}{\widetilde{\varrho}} \widetilde{\boldsymbol{u}} \cdot (\boldsymbol{m} - \varrho \widetilde{\boldsymbol{u}}) \right] \left[\partial_t \widetilde{\varrho} + \mathrm{div}_x (\widetilde{\varrho} \widetilde{\boldsymbol{u}}) \right] \mathrm{d}x \, \mathrm{d}t \\
&+ \|\nabla_x \widetilde{\boldsymbol{u}}\|_{L^\infty((0,T)\times\Omega; R^{d\times d})} \int_0^\tau \left[\int_\Omega E\left(\varrho, \boldsymbol{m} \,\middle|\, \widetilde{\varrho}, \widetilde{\boldsymbol{u}}\right)(t, \cdot) \, \mathrm{d}x + \int_{\overline{\Omega}} \mathrm{d}\mathfrak{E}(t) \right] \mathrm{d}t
\end{aligned} \tag{6.9}$$

for the same class of test functions as in (6.6).

Now, we may approximate functions in the class (6.8) by those in (6.6) via Lemma 4.1. Consequently, we extend (6.9) to the class (6.8), specifically,

$$
\begin{aligned}
&\left[\int_\Omega E\left(\varrho, \boldsymbol{m} \,\middle|\, \widetilde{\varrho}, \widetilde{\boldsymbol{u}}\right)(\tau,\cdot)\,\mathrm{d}x + \int_{\overline{\Omega}} \mathrm{d}\mathfrak{E}(\tau)\right] \overset{<}{\sim} \int_\Omega E\left(\varrho_0, \boldsymbol{m}_0 \,\middle|\, \widetilde{\varrho}(0,\cdot), \widetilde{\boldsymbol{u}}(0,\cdot)\right)\,\mathrm{d}x \\
&+ \int_0^\tau \int_\Omega \frac{1}{\widetilde{\varrho}}(\varrho\widetilde{\boldsymbol{u}} - \boldsymbol{m}) \cdot \left[\partial_t(\widetilde{\varrho}\widetilde{\boldsymbol{u}}) + \mathrm{div}_x\,(\widetilde{\varrho}\widetilde{\boldsymbol{u}} \otimes \widetilde{\boldsymbol{u}}) + \nabla_x p(\widetilde{\varrho})\right]\,\mathrm{d}x\,\mathrm{d}t \\
&+ \int_0^\tau \int_\Omega \left[\left(1 - \frac{\varrho}{\widetilde{\varrho}}\right) p'(\widetilde{\varrho}) + \frac{1}{\widetilde{\varrho}}\widetilde{\boldsymbol{u}} \cdot (\boldsymbol{m} - \varrho\widetilde{\boldsymbol{u}})\right]\left[\partial_t\widetilde{\varrho} + \mathrm{div}_x(\widetilde{\varrho}\widetilde{\boldsymbol{u}})\right]\,\mathrm{d}x\,\mathrm{d}t \\
&+ c\left(\Omega, \|\nabla_x\widetilde{\boldsymbol{u}}\|_{L^\infty((0,T)\times\Omega;R^{d\times d})}\right)\int_0^\tau \left[\int_\Omega E\left(\varrho, \boldsymbol{m} \,\middle|\, \widetilde{\varrho}, \widetilde{\boldsymbol{u}}\right)(t,\cdot)\,\mathrm{d}x + \int_{\overline{\Omega}} \mathrm{d}\mathfrak{E}(t)\right]\mathrm{d}t
\end{aligned}
\tag{6.10}
$$

for any $[\widetilde{\varrho}, \widetilde{\boldsymbol{u}}]$ satisfying (6.8). In particular, we may plug the strong solution $[\widetilde{\varrho}, \widetilde{\boldsymbol{u}}]$ in (6.10) and apply Gronwall's lemma to obtain the desired conclusion

$$
\left[\int_\Omega E\left(\varrho, \boldsymbol{m} \,\middle|\, \widetilde{\varrho}, \widetilde{\boldsymbol{u}}\right)(\tau,\cdot)\,\mathrm{d}x + \int_{\overline{\Omega}} \mathrm{d}\mathfrak{E}(\tau)\right] \overset{<}{\sim} \int_\Omega E\left(\varrho_0, \boldsymbol{m}_0 \,\middle|\, \widetilde{\varrho}(0,\cdot), \widetilde{\boldsymbol{u}}(0,\cdot)\right)\,\mathrm{d}x (= 0).
\tag{6.11}
$$

□

Remark 6.1 (Stability)
Note that formula (6.11) provides also a piece information on *stability* of the strong solution with respect to the perturbation of the initial data. Similar idea will be used later in the numerical part to derive qualitative *error estimates*.

Some remarks are in order. Note that hypothesis (6.7) imposed on the pressure-density EOS is definitely less restrictive than in Definition 5.6 formulated for the *isentropic* Euler system, where

$$
p(\varrho) = a\varrho^\gamma.
$$

On the one hand, it can be shown that the conclusion of Theorem 6.1 remains valid under a more general restriction

$$
\begin{aligned}
&p \in C[0,\infty) \cap C^1(0,\infty),\ p(0) = 0,\ p'(\varrho) > 0 \text{ for } \varrho > 0, \\
&\limsup_{\varrho\to\infty} \frac{p(\varrho)}{P(\varrho)} < \infty,\ P'(\varrho)\varrho - P(\varrho) = p(\varrho).
\end{aligned}
\tag{6.12}
$$

On the other hand, however, the limit of a consistent approximation of the Euler system with p satisfying merely (6.12) *may not* be a DW solution as positivity of $\mathfrak{R}$ requires $p = p(\varrho)$ to be a convex function. From this point of view, the optimal pressure-density EOS should satisfy

$$\begin{aligned} &p \in C[0,\infty) \cap C^2(0,\infty),\ p(0)=0,\ p'(\varrho)>0 \text{ for } \varrho>0,\\ P'(\varrho)\varrho - P(\varrho) &= p(\varrho),\ P(0)=0, \\ &P - \underline{a}p,\ \overline{a}p - P \text{ convex for some } \underline{a}>0,\ \overline{a}>0. \end{aligned} \tag{6.13}$$

It is easy to check that (6.13) implies coercivity of the pressure potential, specifically,

$$P(\varrho) \geq a\varrho^{\gamma} \text{ whenever } \varrho \geq 1 \text{ for certain } a>0,\ \gamma = 1 + \frac{1}{\overline{a}} > 1. \tag{6.14}$$

Indeed

$$\overline{a}p''(\varrho) - P''(\varrho) = \overline{a}p''(\varrho) + \frac{p'(\varrho)}{\varrho} \geq 0;$$

whence

$$\left(\log(p'(\varrho))\right)' \geq \log'\left(\varrho^{\frac{1}{\overline{a}}}\right) \Rightarrow p'(\varrho) \geq p'(1)\varrho^{\frac{1}{\overline{a}}},\ \varrho \geq 1.$$

Note that (6.14), together with the energy inequality, guarantees the desired bounds

$$\varrho \in L^{\infty}(0,T;L^{\gamma}(\Omega)),\ \boldsymbol{m} \in L^{\infty}(0,T;L^{\frac{2\gamma}{\gamma+1}}(\Omega;R^d)).$$

In view of the equivalence of the concepts of dissipative weak and dissipative measure-valued solution stated in Theorem 5.4, we obtain the following corollary of Theorem 6.1.

Corollary 6.1 (Weak strong uniqueness for DMV solutions of barotropic Euler system)

Under the hypotheses of Theorem 6.1, suppose that a parametrized measure $\{\mathcal{V}_{t,x}\}_{(t,x)\in(0,T)\times\Omega}$ *is a DMV solution of the barotropic Euler system in the sense of Definition 5.5, with the initial data* $[\varrho_0, \boldsymbol{m}_0]$ *and the associated concentration defect measures* $\mathfrak{E}_{cd}$, $\mathfrak{R}_{cd}$. *Let* $\widetilde{\varrho}$, $\widetilde{\boldsymbol{u}}$ *be a strong solution of the same problem belonging to the class* (6.8) *and such that*

$$\widetilde{\varrho}(0,\cdot) = \varrho_0,\ \widetilde{\varrho}(0,\cdot)\widetilde{\boldsymbol{u}}(0,\cdot) = \boldsymbol{m}_0.$$

Then

$$\mathcal{V}_{t,x} = \delta_{\widetilde{\varrho}(t,x),\widetilde{\varrho}(t,x)\widetilde{\boldsymbol{u}}(t,x)} \text{ for a.a. } (t,x) \in (0,T)\times\Omega,$$

and

$$\mathfrak{E}_{cd} = \mathfrak{R}_{cd} = 0.$$

Remark 6.2 Both Theorem 6.1 and its Corollary 6.1 apply to the periodic boundary conditions $\Omega = \mathbb{T}^d$ with obvious modifications in the proof.

6.2 Complete the Euler System

Next, we consider the dissipative weak solutions to the complete Euler system introduced in Sect. 5.2.2. Recall that the classical formulation of the problem reads:

$$\begin{aligned}
\partial_t \varrho + \mathrm{div}_x \boldsymbol{m} &= 0 \text{ in } (0,T)\times\Omega,\ \varrho(0,\cdot)=\varrho_0,\\
\partial_t \boldsymbol{m} + \mathrm{div}_x\left(\frac{\boldsymbol{m}\otimes\boldsymbol{m}}{\varrho}\right) + \nabla_x p(\varrho,S) &= 0 \text{ in } (0,T)\times\Omega,\ \boldsymbol{m}(0,\cdot)=\boldsymbol{m}_0,\\
\partial_t\left[\frac{1}{2}\frac{|\boldsymbol{m}|^2}{\varrho} + \varrho e(\varrho,S)\right] + \mathrm{div}_x\left(\left[\frac{1}{2}\frac{|\boldsymbol{m}|^2}{\varrho} + \varrho e(\varrho,S) + p(\varrho,S)\right]\frac{\boldsymbol{m}}{\varrho}\right) &= 0 \text{ in } (0,T)\times\Omega,\\
S(0,\cdot) &= S_0,\\
\boldsymbol{m}\cdot\boldsymbol{n}|_{\partial\Omega} &= 0,
\end{aligned} \tag{6.15}$$

where $S=\varrho s$ is the total entropy, see Sect. 4.1.

The DW solutions satisfy the system of equations

$$\begin{aligned}
\partial_t \varrho + \mathrm{div}_x \boldsymbol{m} &= 0,\ \varrho(0,\cdot)=\varrho_0,\\
\partial_t \boldsymbol{m} + \mathrm{div}_x\left(1_{\varrho>0}\frac{\boldsymbol{m}\otimes\boldsymbol{m}}{\varrho}\right) + \nabla_x 1_{\varrho>0} p(\varrho,S) &= -\mathrm{div}_x \mathfrak{R},\ \boldsymbol{m}(0,\cdot)=\boldsymbol{m}_0,\\
\partial_t S + \mathrm{div}_x\left\langle \mathcal{V};\ 1_{\widehat{\varrho}>0}\widehat{S}\frac{\widehat{\boldsymbol{m}}}{\widehat{\varrho}}\right\rangle &\geq 0,\ S(0,\cdot)=S_0,\ \left\langle \mathcal{V}; [\widehat{\varrho},\widehat{\boldsymbol{m}},\widehat{S}]\right\rangle = [\varrho,\boldsymbol{m},S],\\
\mathcal{V}\left\{\widehat{\varrho}\geq 0,\ \widehat{S}\geq \underline{s}\widehat{\varrho}\right\} &= 1,
\end{aligned} \tag{6.16}$$

in the sense of distribution, together with the total energy balance

$$\int_\Omega \left[\frac{1}{2}\frac{|\boldsymbol{m}|^2}{\varrho} + \varrho e(\varrho,S)\right](\tau,\cdot)\,\mathrm{d}x + \int_{\overline{\Omega}} \mathrm{d}\mathfrak{E}(\tau) \leq \int_\Omega \left[\frac{1}{2}\frac{|\boldsymbol{m}_0|^2}{\varrho_0} + \varrho_0 e(\varrho_0,S_0)\right]\,\mathrm{d}x. \tag{6.17}$$

The energy defect $\mathfrak{E}\in L^\infty(0,T;\mathcal{M}^+(\overline{\Omega}))$ dominates both the Reynolds defect,

$$\underline{d}\mathfrak{E} \leq \mathrm{tr}[\mathfrak{R}] \leq \overline{d}\mathfrak{E},\ 0<\underline{d}\leq\overline{d}, \tag{6.18}$$

and the oscillation defect associated to the measure $\mathcal{V}$,

$$\mathfrak{E} \geq \left\langle \mathcal{V};\ \frac{1}{2}\frac{|\widehat{\boldsymbol{m}}|^2}{\widehat{\varrho}} + \widehat{\varrho}e(\widehat{\varrho},\widehat{S})\right\rangle - \left(\frac{1}{2}\frac{|\boldsymbol{m}|^2}{\varrho} + \varrho e(\varrho,S)\right), \tag{6.19}$$

cf. Definition 5.7.

Remark 6.3 Exceptionally, the dummy variables in the argument of the measure $\mathcal{V}$ are denoted $[\widehat{\varrho},\widehat{\boldsymbol{m}},\widehat{S}]$ instead of $[\widetilde{\varrho},\widetilde{\boldsymbol{m}},\widetilde{S}]$ as the latter symbol will be used for the test functions in the relative energy inequality.

6.2.1 Relative Energy Inequality

Similarly to the preceding section, we derive the relative energy inequality from its counterpart in Sect. 4.1.6. First we recall the relative energy in the conservative-entropy variables

$$
\begin{aligned}
& E\left(\varrho, \boldsymbol{m}, S \middle| \widetilde{\varrho}, \widetilde{\boldsymbol{u}}, \widetilde{S}\right) \\
& = \frac{1}{2}\varrho\left|\frac{\boldsymbol{m}}{\varrho} - \widetilde{\boldsymbol{u}}\right|^2 + \varrho e(\varrho, S) - \frac{\partial(\widetilde{\varrho} e(\widetilde{\varrho}, \widetilde{S}))}{\partial \varrho}(\varrho - \widetilde{\varrho}) - \frac{\partial(\widetilde{\varrho} e(\widetilde{\varrho}, \widetilde{S}))}{\partial S}(\widetilde{\varrho}, \widetilde{S})(S - \widetilde{S}) \\
& - \widetilde{\varrho} e(\widetilde{\varrho}, \widetilde{S}),
\end{aligned}
\tag{6.20}
$$

cf. (4.59).

The relative energy inequality stated in (4.66) requires only a minor modification to accommodate the defects. The final result reads

$$
\begin{aligned}
& \left[\int_{\Omega} E\left(\varrho, S, \boldsymbol{m} \middle| \widetilde{\varrho}, \widetilde{S}, \widetilde{\boldsymbol{u}}\right) \mathrm{d}x\right]_{t=0}^{t=\tau} + \int_{\overline{\Omega}} \mathrm{d}\mathfrak{E}(\tau) \\
& \leq - \int_0^\tau \int_{\Omega} 1_{\varrho > 0} \frac{(\varrho \widetilde{\boldsymbol{u}} - \boldsymbol{m}) \otimes (\varrho \widetilde{\boldsymbol{u}} - \boldsymbol{m})}{\varrho} : \nabla_x \widetilde{\boldsymbol{u}} \, \mathrm{d}x \, \mathrm{d}t \\
& - \int_0^\tau \int_{\Omega} \left[p(\varrho, S) - (\varrho - \widetilde{\varrho}) \frac{\partial p(\widetilde{\varrho}, \widetilde{S})}{\partial \varrho} - (S - \widetilde{S}) \frac{\partial p(\widetilde{\varrho}, \widetilde{S})}{\partial S} - p(\widetilde{\varrho}, \widetilde{S})\right] \mathrm{div}_x \widetilde{\boldsymbol{u}} \, \mathrm{d}x \, \mathrm{d}t \\
& + \int_0^\tau \int_{\Omega} (\varrho \widetilde{\boldsymbol{u}} - \boldsymbol{m}) \cdot \left[\partial_t \widetilde{\boldsymbol{u}} + \widetilde{\boldsymbol{u}} \cdot \nabla_x \widetilde{\boldsymbol{u}} + \frac{1}{\widetilde{\varrho}} \nabla_x \widetilde{p}\right] \mathrm{d}x \, \mathrm{d}t \\
& + \int_0^\tau \int_{\Omega} \left[(\widetilde{\varrho} - \varrho) \frac{1}{\widetilde{\varrho}} \frac{\partial p(\widetilde{\varrho}, \widetilde{S})}{\partial \varrho} \left(\partial_t \widetilde{\varrho} + \mathrm{div}_x (\widetilde{\varrho} \widetilde{\boldsymbol{u}})\right)\right] \mathrm{d}x \, \mathrm{d}t \\
& + \int_0^\tau \int_{\Omega} \left[(\widetilde{\varrho} - \varrho) \frac{1}{\widetilde{\varrho}} \frac{\partial p(\widetilde{\varrho}, \widetilde{S})}{\partial S} \left(\partial_t \widetilde{S} + \mathrm{div}_x (\widetilde{\boldsymbol{u}} \widetilde{S})\right)\right] \mathrm{d}x \, \mathrm{d}t \\
& + \int_0^\tau \int_{\Omega} \left[\left(\frac{\varrho}{\widetilde{\varrho}} \widetilde{S} - S\right) \left(\partial_t \widetilde{\vartheta} + \widetilde{\boldsymbol{u}} \cdot \nabla_x \widetilde{\vartheta} + \frac{\partial p(\widetilde{\varrho}, \widetilde{S})}{\partial S} \mathrm{div}_x \widetilde{\boldsymbol{u}}\right)\right] \mathrm{d}x \, \mathrm{d}t \\
& + \int_0^\tau \int_{\Omega} \left\langle \mathcal{V}_{t,x}; \left(\widehat{\varrho} \frac{\widetilde{S}}{\widetilde{\varrho}}(t,x) - \widehat{S}\right) \left(\frac{\widehat{\boldsymbol{m}}}{\widehat{\varrho}} - \widetilde{\boldsymbol{u}}(t,x)\right)\right\rangle \cdot \nabla_x \widetilde{\vartheta} \, \mathrm{d}x \, \mathrm{d}t \\
& - \int_0^\tau \left(\int_{\overline{\Omega}} \nabla_x \widetilde{\boldsymbol{u}} : \mathrm{d}\mathfrak{R}(t)\right) \mathrm{d}t
\end{aligned}
\tag{6.21}
$$

for any DW solution $[\varrho, \boldsymbol{m}, S]$ in the sense of Definition 5.7, and any trio of test functions $[\widetilde{\varrho}, \widetilde{\boldsymbol{m}}, \widetilde{S}]$ belonging to the class

$$\begin{aligned} &\widetilde{\varrho},\ \widetilde{S} \in W^{1,\infty}((0,T)\times\Omega),\ \widetilde{\varrho} \geq \underline{\varrho} > 0,\ \widetilde{\vartheta} \equiv \frac{1}{\gamma-1}\frac{\partial p(\widetilde{\varrho},\widetilde{S})}{\partial S} \geq \underline{\vartheta} > 0 \text{ in } (0,T)\times\Omega, \\ &\widetilde{\boldsymbol{u}} \in C^M([0,T]\times\overline{\Omega}; R^d),\ \widetilde{\boldsymbol{u}}\cdot\boldsymbol{n}|_{\partial\Omega} = 0,\ \ M \geq 1 \text{ is a positive integer.} \end{aligned} \tag{6.22}$$

6.2.2 Weak-Strong Uniqueness for DW Solutions of Complete the Euler System

We are ready to establish the weak-strong uniqueness principle for DW solutions of the complete Euler system following the line of arguments of the proof of Theorem 6.1.

Theorem 6.2 (Weak-strong uniqueness for DW solutions of complete Euler system)

Let $\Omega \subset R^d$, $d = 2, 3$, be a bounded domain of class C^N, $N = \min\{2, M\}$. Suppose that $[\varrho, \boldsymbol{m}, S]$ is a dissipative weak (DW) solution of the complete Euler system (6.15) in $(0,T)\times\Omega$ in the sense of Definition 5.7, with the initial data $[\varrho_0, \boldsymbol{m}_0, S_0]$.

Let a trio $[\widetilde{\varrho}, \widetilde{\boldsymbol{m}}, \widetilde{S}]$,

$$\text{with the velocity } \widetilde{\boldsymbol{u}} = \frac{\widetilde{\boldsymbol{m}}}{\widetilde{\varrho}},\ \text{ and the temperature } \widetilde{\vartheta} = \frac{1}{\gamma-1}\frac{1}{\widetilde{\varrho}}\frac{\partial p(\widetilde{\varrho},\widetilde{S})}{\partial S},$$

belonging to the class

$$\begin{aligned} &\widetilde{\varrho},\ \widetilde{S} \in W^{1,\infty}((0,T)\times\Omega),\ \widetilde{\varrho} \geq \underline{\varrho} > 0,\ \widetilde{\vartheta} \geq \underline{\vartheta} > 0 \text{ in } (0,T)\times\Omega, \\ &\widetilde{\boldsymbol{u}} \in W^{1,\infty}((0,T)\times\Omega; R^d),\ \widetilde{\boldsymbol{u}}\cdot\boldsymbol{n}|_{\partial\Omega} = 0. \end{aligned} \tag{6.23}$$

be a strong solution of the same problem,

$$\widetilde{\varrho}(0,\cdot) = \varrho_0,\ \widetilde{\varrho}(0,\cdot)\widetilde{\boldsymbol{u}}(0,\cdot) = \boldsymbol{m}_0,\ \widetilde{S}(0,\cdot) = S_0.$$

Then

$$\varrho = \widetilde{\varrho},\ \boldsymbol{m} = \widetilde{\varrho}\widetilde{\boldsymbol{u}},\ S = \widetilde{S} \text{ in } (0,T)\times\Omega,$$

and

$$\mathcal{V}_{t,x} = \delta_{\widetilde{\varrho}(t,x),\widetilde{\varrho}\widetilde{\boldsymbol{u}}(t,x),\widetilde{S}(t,x)} \text{ for a.a. } (t,x)\in(0,T)\times\Omega,\ \mathfrak{E} = \mathfrak{R} = 0.$$

Proof Exactly as in the proof of Theorem 6.1, we first enlarge the class of test functions (6.22) to accommodate the strong solution. Accordingly, we deduce

$$
\begin{aligned}
&\left[\int_{\Omega} E\left(\varrho, S, \boldsymbol{m}\Big|\widetilde{\varrho}, \widetilde{S}, \widetilde{\boldsymbol{u}}\right)(\tau, \cdot)\,\mathrm{d}x + \int_{\overline{\Omega}} \mathrm{d}\mathfrak{E}(\tau)\right] \\
&\leq \int_{\Omega} E\left(\varrho_0, S_0, \boldsymbol{m}_0\Big|\widetilde{\varrho}(0,\cdot), \widetilde{S}(0,\cdot), \widetilde{\boldsymbol{u}}(0,\cdot)\right)\,\mathrm{d}x \\
&+ \int_0^\tau \int_{\Omega} (\varrho\widetilde{\boldsymbol{u}} - \boldsymbol{m}) \cdot \left[\partial_t \widetilde{\boldsymbol{u}} + \widetilde{\boldsymbol{u}} \cdot \nabla_x \widetilde{\boldsymbol{u}} + \frac{1}{\widetilde{\varrho}} \nabla_x \widetilde{p}\right]\,\mathrm{d}x\,\mathrm{d}t \\
&+ \int_0^\tau \int_{\Omega} \left[(\widetilde{\varrho} - \varrho)\frac{1}{\widetilde{\varrho}}\frac{\partial p(\widetilde{\varrho}, \widetilde{S})}{\partial \varrho}\left(\partial_t \widetilde{\varrho} + \mathrm{div}_x(\widetilde{\varrho}\widetilde{\boldsymbol{u}})\right)\right]\,\mathrm{d}x\,\mathrm{d}t \\
&+ \int_0^\tau \int_{\Omega} \left[(\widetilde{\varrho} - \varrho)\frac{1}{\widetilde{\varrho}}\frac{\partial p(\widetilde{\varrho}, \widetilde{S})}{\partial S}\left(\partial_t \widetilde{S} + \mathrm{div}_x(\widetilde{\boldsymbol{u}}\widetilde{S})\right)\right]\,\mathrm{d}x\,\mathrm{d}t \\
&+ \int_0^\tau \int_{\Omega} \left[\left(\frac{\varrho}{\widetilde{\varrho}}\widetilde{S} - S\right)\left(\partial_t \widetilde{\vartheta} + \widetilde{\boldsymbol{u}} \cdot \nabla_x \widetilde{\vartheta} + \frac{\partial p(\widetilde{\varrho}, \widetilde{S})}{\partial S}\mathrm{div}_x \widetilde{\boldsymbol{u}}\right)\right]\,\mathrm{d}x\,\mathrm{d}t \\
&+ \int_0^\tau \int_{\Omega} \left\langle \mathcal{V}_{t,x}; \left(\widehat{\varrho}\frac{\widetilde{S}}{\widetilde{\varrho}}(t,x) - \widehat{S}\right)\left(\frac{\widehat{\boldsymbol{m}}}{\widehat{\varrho}} - \widetilde{\boldsymbol{u}}(t,x)\right)\right\rangle \cdot \nabla_x \widetilde{\vartheta}\,\mathrm{d}x\,\mathrm{d}t \\
&+ c\left(\Omega, \|\widetilde{\boldsymbol{u}}\|_{W^{1,\infty}((0,T)\times\Omega;R^d)}\right) \int_0^\tau \left(\int_{\Omega} E\left(\varrho, S, \boldsymbol{m}\Big|\widetilde{\varrho}, \widetilde{S}, \widetilde{\boldsymbol{u}}\right)(t,\cdot)\,\mathrm{d}x + \int_{\overline{\Omega}} \mathrm{d}\mathfrak{E}(t)\right)\mathrm{d}t
\end{aligned}
\tag{6.24}
$$

for any $[\widetilde{\varrho}, \widetilde{\boldsymbol{u}}, \widetilde{S}]$ as in (6.23). In particular, we may consider the strong solution $[\widetilde{\varrho}, \widetilde{\boldsymbol{u}}, \widetilde{S}]$ as a test function in (6.24) obtaining

$$
\begin{aligned}
&\left[\int_{\Omega} E\left(\varrho, S, \boldsymbol{m}\Big|\widetilde{\varrho}, \widetilde{S}, \widetilde{\boldsymbol{u}}\right)(\tau, \cdot)\,\mathrm{d}x + \int_{\overline{\Omega}} \mathrm{d}\mathfrak{E}(\tau)\right] \\
&\leq c\left(\Omega, \|\widetilde{\boldsymbol{u}}\|_{W^{1,\infty}((0,T)\times\Omega;R^d)}\right) \int_0^\tau \left(\int_{\Omega} E\left(\varrho, S, \boldsymbol{m}\Big|\widetilde{\varrho}, \widetilde{S}, \widetilde{\boldsymbol{u}}\right)(t,\cdot)\,\mathrm{d}x + \int_{\overline{\Omega}} \mathrm{d}\mathfrak{E}(t)\right)\mathrm{d}t \\
&+ \int_0^\tau \int_{\Omega} \left\langle \mathcal{V}_{t,x}; \left(\widehat{\varrho}\frac{\widetilde{S}}{\widetilde{\varrho}}(t,x) - \widehat{S}\right)\left(\frac{\widehat{\boldsymbol{m}}}{\widehat{\varrho}} - \widetilde{\boldsymbol{u}}(t,x)\right)\right\rangle \cdot \nabla_x \widetilde{\vartheta}\,\mathrm{d}x\,\mathrm{d}t.
\end{aligned}
\tag{6.25}
$$

By virtue of (6.19), we have

$$\int_{\Omega} \left\langle \mathcal{V}; E\left(\widehat{\varrho}, \widehat{S}, \widehat{\boldsymbol{m}} \Big| \widetilde{\varrho}, \widetilde{S}, \widetilde{\boldsymbol{u}}\right)\right\rangle (\tau, \cdot) \, \mathrm{d}x$$

$$\lesssim \left[\int_{\Omega} E\left(\varrho, S, \boldsymbol{m} \Big| \widetilde{\varrho}, \widetilde{S}, \widetilde{\boldsymbol{u}}\right)(\tau, \cdot) \, \mathrm{d}x + \int_{\overline{\Omega}} \mathrm{d}\mathfrak{E}(\tau) \right].$$

Consequently, we have only to show the algebraic inequality

$$\left| \left(\frac{\varrho}{\widetilde{\varrho}} \widetilde{S} - S \right) \left(\frac{\boldsymbol{m}}{\varrho} - \widetilde{\boldsymbol{u}} \right) \right| \lesssim E\left(\varrho, S, \boldsymbol{m} \Big| \widetilde{\varrho}, \widetilde{S}, \widetilde{\boldsymbol{u}}\right) \tag{6.26}$$

for any

$$\varrho > 0, \ S \geq \underline{s}\varrho, \tag{6.27}$$

cf. (5.94), and

$$0 < \underline{\varrho} \leq \widetilde{\varrho} \leq \overline{\varrho}, \ |\widetilde{S}| \leq \overline{S}, \ |\widetilde{\boldsymbol{u}}| \leq \overline{\boldsymbol{u}},$$

where the constant hidden in $\lesssim$ depends only on $\underline{\varrho}$, $\overline{\varrho}$, $\overline{S}$, and $\overline{\boldsymbol{u}}$.

To see (6.26), we decompose the set of $[\varrho, S]$ into two parts:

$$\mathcal{R} = \left\{ [\varrho, S] \ \Big| \ S \geq \underline{s}\varrho, \ \frac{1}{2}\underline{\varrho} < \varrho < 2\overline{\varrho}, \ |S| < 2\overline{S} \right\}, \ \mathcal{S} = \left\{ [\varrho, S] \ \Big| \ S \geq \underline{s}\varrho \right\} \setminus \mathcal{R},$$

cf. Sect. 5.4.1. One one hand, in view of strict convexity of the relative energy, we check easily that

$$E\left(\varrho, S, \boldsymbol{m} \Big| \widetilde{\varrho}, \widetilde{S}, \widetilde{\boldsymbol{u}}\right) \gtrsim \left| \frac{\boldsymbol{m}}{\varrho} - \widetilde{\boldsymbol{u}} \right|^2 + |\varrho - \widetilde{\varrho}|^2 + |S - \widetilde{S}|^2 \text{ whenever } [\varrho, S] \in \mathcal{R} \text{ for } [\varrho, S] \in \mathcal{S}.$$

Seeing that

$$\left| \left(\frac{\varrho}{\widetilde{\varrho}} \widetilde{S} - S \right) \left(\frac{\boldsymbol{m}}{\varrho} - \widetilde{\boldsymbol{u}} \right) \right| \lesssim |\varrho - \widetilde{\varrho}|^2 + |S - \widetilde{S}|^2 + \left| \frac{\boldsymbol{m}}{\varrho} - \widetilde{\boldsymbol{u}} \right|^2$$

we may infer that (6.26) holds whenever $[\varrho, S] \in \mathcal{R}$.

On the other hand, we can check by direct manipulation that

$$E\left(\varrho, S, \boldsymbol{m} \Big| \widetilde{\varrho}, \widetilde{S}, \widetilde{\boldsymbol{u}}\right) \gtrsim 1 + \frac{|\boldsymbol{m}|^2}{\varrho} + \varrho e(\varrho, S) \approx 1 + \frac{|\boldsymbol{m}|^2}{\varrho} + \varrho^{\gamma} \exp\left((\gamma - 1) \frac{S}{\varrho} \right), \tag{6.28}$$

while

$$\left| \left(\frac{\varrho}{\widetilde{\varrho}} \widetilde{S} - S \right) \left(\frac{\boldsymbol{m}}{\varrho} - \widetilde{\boldsymbol{u}} \right) \right| \lesssim \left| 1_{\varrho > 0} S \frac{|\boldsymbol{m}|}{\varrho} \right| + \varrho + |S| + |\boldsymbol{m}|$$

as soon as $[\varrho, S] \in \mathcal{S}$. Going back to the proof of Lemma 5.2 we can see that the term $\varrho + |S| + |\boldsymbol{m}|$ can be controlled by the right-hand side of (6.28) as required. Finally, we claim

$$\left| 1_{\varrho > 0} S \frac{|\boldsymbol{m}|}{\varrho} \right| \lesssim 1 + \frac{|\boldsymbol{m}|^2}{\varrho} + \varrho^\gamma \exp\left((\gamma - 1) \frac{S}{\varrho} \right).$$

Indeed we have

$$\left| 1_{\varrho > 0} S \frac{|\boldsymbol{m}|}{\varrho} \right| \lesssim \frac{|\boldsymbol{m}|^2}{\varrho} + \frac{S^2}{\varrho},$$

where

$$\frac{|S|^2}{\varrho} \leq -\underline{s} \varrho \text{ if } S \leq 0,$$

and

$$\left(\frac{|S|^2}{\varrho} \right)^\gamma \lesssim \varrho^\gamma \exp\left((\gamma - 1) \frac{S}{\varrho} \right) \text{ if } S > 0,$$

cf. (5.41). We have shown (6.26) so that the rest of the proof follows by direct application of Gronwall's lemma. □

At the level of DMV solutions introduced in Definition 5.3, Theorem 6.2 yields the following corollary.

Corollary 6.2 *Under the hypotheses of Theorem 6.2, let a parametrized measure*

$$\{\mathcal{V}_{t,x}\}_{(t,x) \in (0,T) \times \Omega}$$

be a dissipative measure-valued (DMV) solution of the complete Euler system (6.15) *in* $(0, T) \times \Omega$ *in the sense of Definition 5.3, with the initial data* $[\varrho_0, \boldsymbol{m}_0, S_0]$. *Let a trio* $[\widetilde{\varrho}, \widetilde{\boldsymbol{m}}, \widetilde{\boldsymbol{u}}]$, *with the temperature* $\widetilde{\vartheta}$,

$$\widetilde{\vartheta} = \frac{1}{\gamma - 1} \frac{1}{\widetilde{\varrho}} \frac{\partial p(\widetilde{\varrho}, \widetilde{S})}{\partial S},$$

belonging to the class (6.23) *be a strong solution of the same problem,*

$$\widetilde{\varrho}(0, \cdot) = \varrho_0, \ \widetilde{\varrho}(0, \cdot) \widetilde{\boldsymbol{u}}(0, \cdot) = \boldsymbol{m}_0, \ \widetilde{S}(0, \cdot) = S_0.$$

Then

$$\mathcal{V}_{t,x} = \delta_{\widetilde{\varrho}(t,x), \widetilde{\varrho}\widetilde{\boldsymbol{u}}(t,x), \widetilde{S}(t,x)} \text{ for a.a. } (t, x) \in (0, T) \times \Omega,$$

and

$$\mathfrak{E}_{cd} = \mathfrak{R}_{cd} = 0.$$

Remark 6.4 Both Theorem 6.2 and Corollary 6.2 remain valid for the periodic boundary conditions $\Omega = \mathbb{T}^d$.

6.3 Navier–Stokes System

We address the problem of weak-strong uniqueness for the Navier–Stokes system with the spatially periodic boundary conditions:

$$\begin{aligned} \partial_t \varrho + \mathrm{div}_x(\varrho \boldsymbol{u}) &= 0 \text{ in } (0,T) \times \mathbb{T}^d, \ \varrho(0,\cdot) = \varrho_0, \\ \partial_t(\varrho \boldsymbol{u}) + \mathrm{div}_x(\varrho \boldsymbol{u} \otimes \boldsymbol{u}) + \nabla_x p(\varrho) &= \mathrm{div}_x \mathbb{S} \text{ in } (0,T) \times \mathbb{T}^d, \ \varrho \boldsymbol{u}(0,\cdot) = \boldsymbol{m}_0, \\ \mathbb{S} = \mu \left(\nabla_x \boldsymbol{u} + \nabla_x^T \boldsymbol{u} - \frac{2}{d} \mathrm{div}_x \boldsymbol{u} \mathbb{I} \right) &+ \lambda \mathrm{div}_x \boldsymbol{u} \mathbb{I}, \ \mu > 0, \ \lambda \geq 0, \\ \Omega &= \mathbb{T}^d. \end{aligned} \tag{6.29}$$

As a matter of fact, we can accommodate more general viscous stress tensor $\mathbb{S}$ interrelated to $\mathbb{D}_x \boldsymbol{u}$ through the "implicit" rheological law,

$$\mathbb{S} : \mathbb{D}_x \boldsymbol{u} = F(\mathbb{D}_x \boldsymbol{u}) + F^*(\mathbb{S}),$$

as long as F enjoys certain coercivity, see Sect. 1.3.2.2. To simplify, however, we focus only on F with quadratic growth pertinent to the Navier–Stokes system.

Let us first consider the dissipative weak (DW) solutions introduced in Sect. 5.3.3, namely the system of field equations

$$\begin{aligned} \partial_t \varrho + \mathrm{div}_x(\varrho u) &= 0, \ \varrho(0,\cdot) = \varrho_0, \\ \partial_t(\varrho \boldsymbol{u}) + \mathrm{div}_x(\varrho \boldsymbol{u} \otimes \boldsymbol{u}) + \nabla_x p(\varrho) &= \mathrm{div}_x \mathbb{S} - \mathrm{div}_x \mathfrak{R}, \ \varrho \boldsymbol{u}(0,\cdot) = \boldsymbol{m}_0 \end{aligned} \tag{6.30}$$

satisfied in the sense of distributions and supplemented with the energy inequality

$$\begin{aligned} \int_{\mathbb{T}^d} \left[\frac{1}{2} \varrho |\boldsymbol{u}|^2 + P(\varrho) \right] (\tau,\cdot) \, \mathrm{d}x + \int_{\mathbb{T}^d} \mathrm{d}\mathfrak{E}(\tau) + \int_0^\tau \int_{\mathbb{T}^d} \left[F(\mathbb{D}_x \boldsymbol{u}) + F^*(\mathbb{S}) \right] \, \mathrm{d}x \, \mathrm{d}t \\ \leq \int_{\mathbb{T}^d} \left[\frac{1}{2} \frac{|\boldsymbol{m}_0|^2}{\varrho_0} + P(\varrho_0) \right] \, \mathrm{d}x. \end{aligned} \tag{6.31}$$

The defect measures

$$\mathfrak{E} \in L^\infty(0,T; \mathcal{M}^+(\mathbb{T}^d)), \ \mathfrak{R} \in L^\infty(0,T; \mathcal{M}^+(\mathbb{T}^d; R^{d \times d}_{\mathrm{sym}}))$$

satisfy the compatibility condition

$$\underline{d}\mathfrak{E} \leq \mathrm{tr}[\mathfrak{R}] \leq \overline{d}\mathfrak{E},\ 0 < \underline{d} \leq \overline{d}, \tag{6.32}$$

cf. Definition 5.11.

6.3.1 Relative Energy Inequality

The relative energy for the Navier–Stokes system is the same as for its inviscid Euler limit

$$E\left(\varrho, \boldsymbol{u} \,\middle|\, \widetilde{\varrho}, \widetilde{\boldsymbol{u}}\right) = \frac{1}{2}\varrho\left|\boldsymbol{u} - \widetilde{\boldsymbol{u}}\right|^2 + P(\varrho) - P'(\widetilde{\varrho})(\varrho - \widetilde{\varrho}) - P(\widetilde{\varrho}),$$

where we have replaced $\boldsymbol{m}$ by $\varrho\boldsymbol{u}$. The associated relative energy inequality may be recovered from (6.5) keeping in mind the extra contribution by the viscous stress. The resulting expression reads

$$\begin{aligned}
&\left[\int_{\mathbb{T}^d} E\left(\varrho, \boldsymbol{u} \,\middle|\, \widetilde{\varrho}, \widetilde{\boldsymbol{u}}\right)\,\mathrm{d}x\right]_{t=0}^{t=\tau} + \int_{\mathbb{T}^d} \mathrm{d}\mathfrak{E}(\tau) + \int_0^\tau \int_{\mathbb{T}^d} \left[F(\mathbb{D}_x \boldsymbol{u}) + F^*(\mathbb{S})\right] \mathrm{d}x\,\mathrm{d}t \\
&- \int_0^\tau \int_{\mathbb{T}^d} \mathbb{S} : \mathbb{D}_x \widetilde{\boldsymbol{u}}\,\mathrm{d}x\ \mathrm{d}t + \int_0^\tau \int_{\mathbb{T}^d} \widetilde{\mathbb{S}} : (\mathbb{D}_x \widetilde{\boldsymbol{u}} - \mathbb{D}_x \boldsymbol{u})\,\mathrm{d}x\,\mathrm{d}t \\
&\leq - \int_0^\tau \int_{\mathbb{T}^d} 1_{\varrho > 0} \varrho \nabla_x \widetilde{\boldsymbol{u}} \cdot (\boldsymbol{u} - \widetilde{\boldsymbol{u}}) \cdot (\boldsymbol{u} - \widetilde{\boldsymbol{u}})\ \mathrm{d}x\,\mathrm{d}t \\
&- \int_0^\tau \int_{\mathbb{T}^d} \left[p(\varrho) - p'(\widetilde{\varrho})(\varrho - \widetilde{\varrho}) - p(\widetilde{\varrho})\right] \mathrm{div}_x \widetilde{\boldsymbol{u}}\ \mathrm{d}x\,\mathrm{d}t \\
&+ \int_0^\tau \int_{\mathbb{T}^d} \frac{\varrho}{\widetilde{\varrho}} (\widetilde{\boldsymbol{u}} - \boldsymbol{u}) \cdot \left[\partial_t (\widetilde{\varrho}\widetilde{\boldsymbol{u}}) + \mathrm{div}_x\,(\widetilde{\varrho}\widetilde{\boldsymbol{u}} \otimes \widetilde{\boldsymbol{u}}) + \nabla_x p(\widetilde{\varrho}) - \mathrm{div}_x \widetilde{\mathbb{S}}\right] \mathrm{d}x\,\mathrm{d}t \\
&+ \int_0^\tau \int_{\mathbb{T}^d} \left[\left(1 - \frac{\varrho}{\widetilde{\varrho}}\right) p'(\widetilde{\varrho}) + \frac{\varrho}{\widetilde{\varrho}} \widetilde{\boldsymbol{u}} \cdot (\boldsymbol{u} - \widetilde{\boldsymbol{u}})\right] \left[\partial_t \widetilde{\varrho} + \mathrm{div}_x (\widetilde{\varrho}\widetilde{\boldsymbol{u}})\right] \mathrm{d}x\,\mathrm{d}t \\
&- \int_0^\tau \left(\int_{\mathbb{T}^d} \nabla_x \widetilde{\boldsymbol{u}} : \mathrm{d}\mathfrak{R}(t)\right) \mathrm{d}t + \int_0^\tau \int_{\mathbb{T}^d} (\varrho - \widetilde{\varrho})\,\frac{1}{\widetilde{\varrho}} \mathrm{div}_x \widetilde{\mathbb{S}} \cdot (\widetilde{\boldsymbol{u}} - \boldsymbol{u})\ \mathrm{d}x\,\mathrm{d}t
\end{aligned} \tag{6.33}$$

for any DW solution of the Navier–Stokes system and any trio of test functions $[\widetilde{\varrho}, \widetilde{\boldsymbol{u}}, \widetilde{\mathbb{S}}]$ belonging to the class

$$\begin{aligned} &\widetilde{\varrho} \in C^1([0,T] \times \mathbb{T}^d),\ 0 < \underline{\varrho} \leq \widetilde{\varrho} \text{ in } (0,T) \times \mathbb{T}^d, \\ &\widetilde{\boldsymbol{u}} \in C^1([0,T] \times \mathbb{T}^d; R^d),\ \widetilde{\mathbb{S}} \in L^2(0,T; W^{1,\infty}(\mathbb{T}^d; R^{d\times d}_{\rm sym})). \end{aligned} \tag{6.34}$$

Remark 6.5 Note that the integrals containing the tensor $\widetilde{\mathbb{S}}$ have been just added to the entropy inequality associated to the compressible Euler system. In particular, the regularity hypotheses imposed on $\widetilde{\mathbb{S}}$ may be relaxed given the available integrability of ϱ and $\boldsymbol{u}$.

6.3.2 Weak-Strong Uniqueness for DW Solutions of the Navier–Stokes System

Following the arguments of Sect. 6.1 we can establish the weak-strong uniqueness principle for the dissipative weak solutions to the Navier–Stokes system.

Theorem 6.3 (Weak-strong uniqueness for DW solutions of Navier–Stokes system)

Let the pressure $p = p(\varrho)$ satisfy

$$\begin{aligned} &p \in C^1[0,\infty) \cap C^2(0,\infty),\ p(0) = 0,\ p'(\varrho) > 0 \textit{ for } \varrho > 0, \\ &0 < \liminf_{\varrho\to\infty} \frac{p'(\varrho)}{\varrho^{\gamma-1}} \leq \limsup_{\varrho\to\infty} \frac{p'(\varrho)}{\varrho^{\gamma-1}} < \infty \end{aligned}$$

for some $\gamma > 1$. Let $[\varrho, \boldsymbol{u}]$, with the associated viscous stress $\mathbb{S}$, be a DW solution of the Navier–Stokes system (6.29) *with the initial data $[\varrho_0, \boldsymbol{m}_0 = \varrho\boldsymbol{u}(0,\cdot)]$. Suppose that $[\widetilde{\varrho}, \widetilde{\boldsymbol{u}}]$ is a strong solution of the same system, with*

$$\widetilde{\mathbb{S}} \in \partial F(\mathbb{D}_x \widetilde{\boldsymbol{u}}),\ \widetilde{\varrho}(0,\cdot) = \varrho_0,\ \widetilde{\varrho}\widetilde{\boldsymbol{u}}(0,\cdot) = \boldsymbol{m}_0,$$

belonging to the class (6.34).

Then

$$\varrho = \widetilde{\varrho},\ \boldsymbol{u} = \widetilde{\boldsymbol{u}},\ \mathbb{S} = \widetilde{\mathbb{S}} \textit{ in } (0,T) \times \mathbb{T}^d,$$

$$\mathfrak{E} = \mathfrak{R} = 0.$$

Proof Apparently, the only modification needed with respect to the proof of Theorem 6.1 consists in

- showing positivity of the integral

$$\int_0^\tau \int_{\mathbb{T}^d} \left[F(\mathbb{D}_x \boldsymbol{u}) + F^*(\mathbb{S}) - \mathbb{S} : \mathbb{D}_x \widetilde{\boldsymbol{u}} + \widetilde{\mathbb{S}} : (\mathbb{D}_x \widetilde{\boldsymbol{u}} - \mathbb{D}_x \boldsymbol{u}) \right] \mathrm{d}x\, \mathrm{d}t;$$

- controlling the integral

$$\int_0^\tau \int_{\mathbb{T}^d} (\varrho - \widetilde{\varrho}) \frac{1}{\widetilde{\varrho}} \mathrm{div}_x \widetilde{\mathbb{S}} \cdot (\widetilde{\boldsymbol{u}} - \boldsymbol{u})\, \mathrm{d}x\, \mathrm{d}t$$

on the right-hand side of the relative entropy inequality (6.33).

First observe that for $\widetilde{\mathbb{S}} \in \partial F(\mathbb{D}_x \widetilde{\boldsymbol{u}})$ we have

$$\begin{aligned} &F(\mathbb{D}_x \boldsymbol{u}) - \widetilde{\mathbb{S}} : (\mathbb{D}_x \boldsymbol{u} - \mathbb{D}_x \widetilde{\boldsymbol{u}}) - F(\mathbb{D}_x \widetilde{\boldsymbol{u}}) \\ &\approx \mu \left| \left(\mathbb{D}_x - \frac{1}{d} \mathrm{div}_x \mathbb{I} \right) (\boldsymbol{u} - \widetilde{\boldsymbol{u}}) \right|^2 + \lambda |\mathrm{div}_x (\boldsymbol{u} - \widetilde{\boldsymbol{u}})|^2, \end{aligned}$$

while, in view of the Fenchel–Young inequality,

$$F(\mathbb{D}_x \widetilde{\boldsymbol{u}}) + F^*(\mathbb{S}) - \mathbb{S} : \mathbb{D}_x \widetilde{\boldsymbol{u}} \geq 0.$$

Consequently, plugging the strong solution in the relative energy inequality (6.33) we obtain

$$\begin{aligned} &\left[\int_{\mathbb{T}^d} E\left(\varrho, \boldsymbol{u} \,\middle|\, \widetilde{\varrho}, \widetilde{\boldsymbol{u}} \right) (\tau, \cdot)\, \mathrm{d}x + \int_{\mathbb{T}^d} \mathrm{d}\mathfrak{E}(\tau) \right] \\ &+ \mu \int_0^\tau \int_{\mathbb{T}^d} \left| \left(\mathbb{D}_x - \frac{1}{d} \mathrm{div}_x \mathbb{I} \right) (\boldsymbol{u} - \widetilde{\boldsymbol{u}}) \right|^2 \mathrm{d}x\, \mathrm{d}t \\ &\lesssim \|\nabla_x \widetilde{\boldsymbol{u}}\|_{L^\infty((0,T)\times\mathbb{T}^d; R^{d\times d})} \int_0^\tau \int_{\mathbb{T}^d} \left[\mathcal{E}\left(\varrho, \boldsymbol{u} \,\middle|\, \widetilde{\varrho}, \widetilde{\boldsymbol{u}} \right) (t, \cdot) + \int_{\mathbb{T}^d} \mathrm{d}\mathfrak{E}(t) \right] \mathrm{d}x\, \mathrm{d}t \\ &+ \int_0^\tau \int_{\mathbb{T}^d} (\varrho - \widetilde{\varrho}) \frac{1}{\widetilde{\varrho}} \mathrm{div}_x \widetilde{\mathbb{S}} \cdot (\widetilde{\boldsymbol{u}} - \boldsymbol{u})\, \mathrm{d}x\, \mathrm{d}t \end{aligned} \tag{6.35}$$

To control the last integral in (6.35) we consider three sets:

$$\mathcal{R} = \left\{ 0 < \frac{1}{2}\underline{\varrho} < \varrho < 2\overline{\varrho} \right\}, \ \mathcal{S}^+ = \{ \varrho \geq 2\overline{\varrho} \}, \ \mathcal{S}^- = \left\{ 0 \leq \varrho \leq \frac{1}{2}\underline{\varrho} \right\}.$$

As P is strictly convex, we have

$$E\left(\varrho, \boldsymbol{u} \middle| \widetilde{\varrho}, \widetilde{\boldsymbol{u}}\right) \stackrel{>}{\sim} |\boldsymbol{u} - \widetilde{\boldsymbol{u}}|^2 + |\varrho - \widetilde{\varrho}|^2 \text{ whenever } \varrho \in \mathcal{R}.$$

Accordingly, we get

$$1_{\varrho \in \mathcal{R}} \left| (\varrho - \widetilde{\varrho}) \frac{1}{\widetilde{\varrho}} \mathrm{div}_x \widetilde{\mathbb{S}} \cdot (\widetilde{\boldsymbol{u}} - \boldsymbol{u}) \right| \stackrel{<}{\sim} 1_{\varrho \in \mathcal{R}} E\left(\varrho, \boldsymbol{u} \middle| \widetilde{\varrho}, \widetilde{\boldsymbol{u}}\right).$$

Next,

$$E\left(\varrho, \boldsymbol{u} \middle| \widetilde{\varrho}, \widetilde{\boldsymbol{u}}\right) \stackrel{>}{\sim} 1 + \varrho |\boldsymbol{u} - \widetilde{\boldsymbol{u}}|^2 + \varrho^\gamma \text{ whenever } \varrho \in \mathcal{S}^+ \cup \mathcal{S}^-.$$

Consequently,

$$1_{\varrho \in \mathcal{S}^+} \left| (\varrho - \widetilde{\varrho}) \frac{1}{\widetilde{\varrho}} \mathrm{div}_x \widetilde{\mathbb{S}} \cdot (\widetilde{\boldsymbol{u}} - \boldsymbol{u}) \right| \stackrel{<}{\sim} 1_{\varrho \in \mathcal{S}^+} \left(\varrho + \varrho |\widetilde{\boldsymbol{u}} - \boldsymbol{u}|^2 \right) \stackrel{<}{\sim} \mathcal{E}\left(\varrho, \boldsymbol{u} \middle| \widetilde{\varrho}, \widetilde{\boldsymbol{u}}\right)$$

for $\varrho \in \mathcal{S}^+$.

Finally,

$$1_{\varrho \in \mathcal{S}^-} \left| (\varrho - \widetilde{\varrho}) \frac{1}{\widetilde{\varrho}} \mathrm{div}_x \widetilde{\mathbb{S}} \cdot (\widetilde{\boldsymbol{u}} - \boldsymbol{u}) \right| \stackrel{<}{\sim} \delta |\boldsymbol{u} - \widetilde{\boldsymbol{u}}|^2 + 1_{\varrho \in \mathcal{S}^-} c(\delta) \text{ for any } \delta > 0.$$

Thus we may use the following variant of the Korn–Poincaré inequality

$$\int_{\mathbb{T}^d} |\boldsymbol{u} - \widetilde{\boldsymbol{u}}|^2 \, \mathrm{d}x \stackrel{<}{\sim} c(M, \Gamma) \left[\int_{\mathbb{T}^d} \varrho |\boldsymbol{u} - \widetilde{\boldsymbol{u}}|^2 \, \mathrm{d}x + \int_{\mathbb{T}^d} \left| \left(\mathbb{D}_x - \frac{1}{d} \mathrm{div}_x \mathbb{I} \right) (\boldsymbol{u} - \widetilde{\boldsymbol{u}}) \right|^2 \, \mathrm{d}x \right]$$
$$\text{whenever } 0 < M \leq \int_{\mathbb{T}^d} \varrho \, \mathrm{d}x, \ \|\varrho\|_{L^\gamma(\mathbb{T}^d)} \leq \Gamma \tag{6.36}$$

to conclude the proof. □

As we have observed, the proof does not depend essentially on the specific form of the viscous stress characterized by the potential F and the technique can be adapted to more general non-Newtonian constitutive relations for viscosity.

Theorem 6.3, combined with the local existence of strong solutions and the conditional regularity criterion of Sun, Wang, and Zhang (see Theorem 3.5), yields a remarkable corollary.

Corollary 6.3 (Conditional regularity of DW solutions of the Navier–Stokes system)
Let the pressure $p = p(\varrho)$ by given by the isentropic EOS,

$$p(\varrho) = a\varrho^{\gamma},\ a > 0,\ \gamma > 1.$$

Suppose the $[\varrho, \boldsymbol{u}]$ is a DW solution of the Navier–Stokes system in the sense of Definition 5.11 in $(0, T) \times \mathbb{T}^d$ with regular initial data

$$\varrho_0 \in W^{3,2}(\mathbb{T}^d),\ \boldsymbol{m}_0 = \varrho_0 \boldsymbol{u}_0,\ \boldsymbol{u}_0 \in W^{3,2}(\mathbb{T}^d, R^d),\ \inf_{x \in \mathbb{T}^d} \varrho_0(x) > 0.$$

(i) Suppose that the bulk viscosity coefficient $\lambda = 0$ and that $\varrho \in L^{\infty}((0, T) \times \mathbb{T}^d)$. Then $[\varrho, \boldsymbol{u}]$ is a strong solution and $\mathfrak{R} = \mathfrak{E} = 0$.

(ii) Suppose $\varrho \in L^{\infty}((0, T) \times \mathbb{T}^d)$, $\boldsymbol{u} \in L^{\infty}((0, T) \times \mathbb{T}^d; R^d)$. Then $[\varrho, \boldsymbol{u}]$ is a strong solution and $\mathfrak{R} = \mathfrak{E} = 0$.

Proof By virtue of the local existence results stated in Theorems 3.1, 3.4, there exists a time interval $[0, T_{\max})$ on which the problem admits a strong solution. Furthermore, it follows from the conditional regularity criterion of Sun, Wang, and Zhang (Theorem 3.5) that $T_{\max} = T$ in the case (i); while the same is true in the case (ii) in view of Remark 3.6. As pointed out in Remark 3.5, the solution remains in the same regularity class as the initial data, in particular, all relevant derivatives are continuous. □

6.3.3 Weak-Strong Uniqueness Principle for DMV Solutions to the Navier–Stokes System

The proof of weak-strong uniqueness for the dissipative measure-valued (DMV) solutions of the Navier–Stokes system requires only minor modification with respect to Theorem 6.3. In accordance with Definition 5.10, the relative energy reads

$$E\left(\varrho, \boldsymbol{u} \,\middle|\, \widetilde{\varrho}, \widetilde{\boldsymbol{u}}\right) = \left\langle \mathcal{V}; \frac{1}{2}\varrho\, |\boldsymbol{u} - \widetilde{\boldsymbol{u}}|^2 + P(\varrho) - P'(\widetilde{\varrho})(\varrho - \widetilde{\varrho}) - P(\widetilde{\varrho}) \right\rangle,$$

with the associated relative energy inequality

$$
\left[\int_{\mathbb{T}^d} E\left(\varrho, \boldsymbol{u} \middle| \widetilde{\varrho}, \widetilde{\boldsymbol{u}}\right) \, \mathrm{d}x \right]_{t=0}^{t=\tau} + \int_{\mathbb{T}^d} \mathrm{d}\mathfrak{E}(\tau) + \int_0^\tau \int_{\mathbb{T}^d} \mathfrak{D}
$$

$$
+ \int_0^\tau \int_{\mathbb{T}^d} \left[F(\mathbb{D}_x \boldsymbol{u}) + F^*(\mathbb{S}) \right] \mathrm{d}x \, \mathrm{d}t
$$

$$
- \int_0^\tau \int_{\mathbb{T}^d} \mathbb{S} : \mathbb{D}_x \widetilde{\boldsymbol{u}} \, \mathrm{d}x \ \mathrm{d}t + \int_0^\tau \int_{\mathbb{T}^d} \widetilde{\mathbb{S}} : (\mathbb{D}_x \widetilde{\boldsymbol{u}} - \mathbb{D}_x \boldsymbol{u}) \, \mathrm{d}x \, \mathrm{d}t
$$

$$
\leq - \int_0^\tau \int_{\mathbb{T}^d} \left\langle \mathcal{V}; 1_{\varrho > 0} \varrho \nabla_x \widetilde{\boldsymbol{u}} \cdot (\boldsymbol{u} - \widetilde{\boldsymbol{u}}) \cdot (\boldsymbol{u} - \widetilde{\boldsymbol{u}}) \right\rangle \, \mathrm{d}x \, \mathrm{d}t
$$

$$
- \int_0^\tau \int_{\mathbb{T}^d} \left\langle \mathcal{V}; p(\varrho) - p'(\widetilde{\varrho})(\varrho - \widetilde{\varrho}) - p(\widetilde{\varrho}) \right\rangle \mathrm{div}_x \widetilde{\boldsymbol{u}} \, \mathrm{d}x \, \mathrm{d}t
$$

$$
+ \int_0^\tau \int_{\mathbb{T}^d} \left\langle \mathcal{V}; \frac{\varrho}{\widetilde{\varrho}} (\widetilde{\boldsymbol{u}} - \boldsymbol{u}) \right\rangle \cdot \left[\partial_t (\widetilde{\varrho} \widetilde{\boldsymbol{u}}) + \mathrm{div}_x \, (\widetilde{\varrho} \widetilde{\boldsymbol{u}} \otimes \widetilde{\boldsymbol{u}}) + \nabla_x p(\widetilde{\varrho}) - \mathrm{div}_x \widetilde{\mathbb{S}} \right] \mathrm{d}x \, \mathrm{d}t
$$

$$
+ \int_0^\tau \int_{\mathbb{T}^d} \left\langle \mathcal{V}; \left(1 - \frac{\varrho}{\widetilde{\varrho}}\right) p'(\widetilde{\varrho}) + \frac{\varrho}{\widetilde{\varrho}} \widetilde{\boldsymbol{u}} \cdot (\boldsymbol{u} - \widetilde{\boldsymbol{u}}) \right\rangle \left[\partial_t \widetilde{\varrho} + \mathrm{div}_x (\widetilde{\varrho} \widetilde{\boldsymbol{u}}) \right] \mathrm{d}x \, \mathrm{d}t
$$

$$
- \int_0^\tau \left(\int_{\mathbb{T}^d} \nabla_x \widetilde{\boldsymbol{u}} : \mathrm{d}\mathfrak{R}(t) \right) \mathrm{d}t + \int_0^\tau \int_{\mathbb{T}^d} \left\langle \mathcal{V}; (\varrho - \widetilde{\varrho}) \frac{1}{\widetilde{\varrho}} \mathrm{div}_x \widetilde{\mathbb{S}} \cdot (\widetilde{\boldsymbol{u}} - \boldsymbol{u}) \right\rangle \, \mathrm{d}x \, \mathrm{d}t \tag{6.37}
$$

The proof of weak-strong uniqueness is now exactly the same as in Theorem 6.3, where the Korn–Poincaré inequality (6.36) is replaced by its Young measure counterpart (5.126). Let us state the result.

Theorem 6.4 (Weak-strong uniqueness for DMV solutions of the Navier–Stokes system)
Let the pressure $p = p(\varrho)$ *satisfy*

$$
p \in C[0, \infty) \cap C^1(0, \infty), \ p(0) = 0, \ p'(\varrho) > 0 \, \textit{for} \, \varrho > 0,
$$
$$
0 < \liminf_{\varrho \to \infty} \frac{p'(\varrho)}{\varrho^{\gamma - 1}} \leq \limsup_{\varrho \to \infty} \frac{p'(\varrho)}{\varrho^{\gamma - 1}} < \infty
$$

for some $\gamma > 1$*. Let a parametrized measure* $\{\mathcal{V}_{t,x}\}_{(t,x) \in (0,T) \times \mathbb{T}^d}$ *be a dissipative measure-valued solution of the Navier–Stokes system with the initial data* $[\varrho_0, \boldsymbol{m}_0 = \varrho \boldsymbol{u}(0, \cdot)]$ *introduced in Definition 5.10. Suppose that* $[\widetilde{\varrho}, \widetilde{\boldsymbol{u}}]$ *is a strong solution of the same system, with*

$$\widetilde{\mathbb{S}} \in \partial F(\mathbb{D}_x \boldsymbol{u}), \ \widetilde{\varrho}(0,\cdot) = \varrho_0, \ \widetilde{\varrho}\widetilde{\boldsymbol{u}}(0,\cdot) = \boldsymbol{m}_0,$$

belonging to the class (6.34).

Then

$$\mathcal{V}_{t,x} = \delta_{[\widetilde{\varrho}(t,x),\widetilde{\boldsymbol{u}}(t,x)]} \ \text{for a.a.} \ (t,x) \in (0,T) \times \mathbb{T}^d, \ \mathbb{S} = \widetilde{\mathbb{S}} \ \text{in} \ (0,T) \times \mathbb{T}^d,$$

$$\mathfrak{E} = \mathfrak{D} = \mathfrak{R} = 0.$$

Finally, we reformulate Corollary 6.3 in terms of the dissipative measure-valued solutions.

Corollary 6.4 (Conditional regularity of DMV solutions of the Navier–Stokes system)

Let the pressure $p = p(\varrho)$ *by given by the isentropic EOS,*

$$p(\varrho) = a\varrho^\gamma, \ a > 0, \ \gamma > 1.$$

Let a parametrized measure $\{\mathcal{V}_{t,x}\}_{(t,x)\in(0,T)\times\mathbb{T}^d}$ *be a DMV solution of the Navier–Stokes system in the sense of Definition 5.10 in* $(0,T) \times \mathbb{T}^d$ *with regular initial data*

$$\varrho_0 \in W^{3,2}(\mathbb{T}^d), \ \boldsymbol{m}_0 = \varrho_0 \boldsymbol{u}_0, \ \boldsymbol{u}_0 \in W^{3,2}(\mathbb{T}^d, R^d), \ \inf_{x\in\mathbb{T}^d} \varrho_0(x) > 0.$$

(i) Suppose that the bulk viscosity coefficient $\lambda = 0$ *and that*

$$\mathcal{V}_{t,x}\left\{0 \leq \widetilde{\varrho} \leq \overline{\varrho}\right\} = 1 \ \text{for a.a.} \ (t,x) \in (0,T) \times \mathbb{T}^d$$

and a certain constant $\overline{\varrho}$. *Then*

$$\mathcal{V} = \delta_{[\varrho(t,x),\boldsymbol{u}(t,x)]} \ \text{for a.a.} \ (t,x) \in (0,T) \times \mathbb{T}^d,$$

where $[\varrho, \boldsymbol{u}]$ *is a strong solution and* $\mathfrak{R} = \mathfrak{E} = \mathfrak{D} = 0$.

(ii) Suppose that

$$\mathcal{V}_{t,x}\left\{0 \leq \widetilde{\varrho} \leq \overline{\varrho}, \ |\boldsymbol{u}| \leq \overline{\boldsymbol{u}}\right\} = 1 \ \text{for a.a.} \ (t,x) \in (0,T) \times \mathbb{T}^d$$

and certain constants $\overline{\varrho}$, $\overline{\boldsymbol{u}}$.

Then

$$\mathcal{V} = \delta_{[\varrho(t,x),\boldsymbol{u}(t,x)]} \ \text{for a.a.} \ (t,x) \in (0,T) \times \mathbb{T}^d,$$

where $[\varrho, \boldsymbol{u}]$ *is a strong solution and* $\mathfrak{R} = \mathfrak{E} = \mathfrak{D} = 0$.

We point out that the presence of viscosity is crucial in this result. Similar conclusion is definitely false in the case of the Euler system.

6.4 Conclusion, Bibliographical Remarks

Weak strong uniqueness principle in the context of measure-valued solutions to the barotropic Euler system was shown by Gwiazda, Świerczewska–Gwiazda, and Wiedemann [126]. They used the generalized Young measures introduced by Alibert and Bouchitté [4]. The proof was simplified and extended to a larger class of measure-valued solutions for the Navier–Stokes system in [82]. A general result for systems of conservation laws without concentration effects was shown by Brenier et al. [26]. The complete Euler system including concentrations is treated in [40]. There is a nice survey of these results Wiedemann [199].

All results concerning weak-strong uniqueness for the Navier–Stokes system can be extended to more general "implicit" rheological relation

$$\mathbb{S} : \mathbb{D}_x \boldsymbol{u} = F(\mathbb{D}_x \boldsymbol{u}) + F^*(\mathbb{S})$$

under certain restrictions imposed on F. In view of the applications to the numerical schemes discussed in this monograph, we focused only on quadratic F corresponding to the standard Navier–Stokes system. The reader interested in general theory may consult [2].

Part III
Numerical Analysis

The ultimate and principal objective of this monograph is analysis of certain numerical schemes (methods) used in continuum fluid mechanics. Notably, we examine numerical approximations of the Euler and Navier–Stokes systems discussed in the previous chapters. There are three main issues to be discussed:

- **Stability**. The output of a numerical method is a finite set of numerical values that represents discrete approximation of the continuous (target) solution. These must be extended in a suitable way to the whole physical space $(0, T) \times \Omega$ to obtain a family of approximate solutions interpreted as numerical functions of the physical variables t and x. The approximate solutions are bounded, typically piecewise constant or piecewise polynomial and may enjoy certain properties inherited from the limit system. For instance, the approximate density and temperature may be *strictly* positive functions. A sequence of approximate solutions is *stable* if it admits bounds in certain function spaces that are *uniform* with respect to $n \to \infty$. The problem of stability is therefore intimately related to the *a priori* estimates available for the limit system. As we have seen in Part II, the (known) *a priori* bounds are based on the underlying laws of classical thermodynamics:

  (i) Conservation of mass yielding uniform integrability of the density.

  (ii) Conservation of energy, where the latter is a convex function of the density, momentum, and entropy.

  (iii) Entropy production yielding a lower bound on the total entropy.

  Besides, the problems with energy dissipation as the Navier–Stokes–Fourier system provide uniform bounds on the velocity and temperature gradient as a result of diffusive transport. A family of approximate solutions satisfying the relevant uniform bounds is termed *stable approximation.*

- **Consistency**. Stability itself is not sufficient for showing that a numerical method approaches the limit system. Consistent approximations discussed in detail in Chap. 5 solve the limit problem modulo an error that vanishes in the asymptotic limit. This is usually a sufficient piece of information for proving convergence to the continuous solution if the underlying system is linear (cf. the celebrated Lax equivalence theorem [149]). As we have seen in Chap. 5, consistent approximations converge to generalized (measure-valued or dissipative weak) solutions of nonlinear problems. In view of the weak–strong uniqueness principle derived for generalized solutions in Chap. 6, however, a generalized solution coincides with the (unique) strong solution of the limit problem as long as the latter exist. This property may be seen as an extension of Lax equivalence theorem to nonlinear problems. A family of approximate solutions consistent with the limit system of field equations is termed *consistent approximation.*

- **Convergence.**

The problem of *convergence* is fundamental in numerical analysis. There are two principal aspects discussed in chapters below:

(i) Identifying the limit of the approximate solutions and its relation to the exact solution of the target continuous problem.

(ii) The way how the limit is attained – an issue intimately related to *error estimates.*

Given the rather poor uniform bounds, the approximate solutions usually approach the limit only in the weak topology in the sense of Lebesgue spaces, meaning only integral means converge. The limit object is then characterized through the associated Young measure that captures statistical distribution of the oscillating sequence, cf. Chap. 5 and notably Sect. 5.1, where this issue is discussed for the Euler system. Even if we know that an approximate sequence $\{\mathbf{U}_n\}_{n=1}^{\infty}$ converges weakly to the exact solution $\mathbf{U}$, the output of a specific implementation of a numerical method would provide in this case only chaotic oscillations (wiggles) for which the desired limit object – barycenter of the associated Young measure – is hard to be identified.

Strong, meaning pointwise with respect to the physical variables (t, x), convergence is definitely more convenient as the limit object can be clearly identified at least at a sufficiently high resolution of the discretization parameters. Recall that pointwise convergence along with equi-integrability of a sequence $\{\mathbf{U}_n\}_{n=1}^{\infty}$ imply strong convergence in the integral norm $L^1((0, T) \times \Omega)$ but not vice versa, as the other implication requires to pass to a suitable subsequence as the case may be.

We propose a method how to convert weakly converging sequences to strongly converging by means of an averaging procedure mimicking the Strong Law of Large Numbers in probability. The resulting concept of convergence is commonly known as $\mathcal{K}$–convergence or convergence of Cesàro averages. Its functional analytic background goes back to the classical results of Banach and Saks; here, we use the refined more recent approach based on pointwise convergence due to Komlós. As we shall see, the method not only provides an efficient visualization of the asymptotic limits of numerical methods but also the associated Young measures that may be generated by them. In particular, we establish pointwise convergence in (t, x) of these quantities with respect to a suitable distance (Monge–Kantorowich or Wasserstein).

Chapter 7
Weak and Strong Convergence

We examine in detail the issue of weak and strong convergence of a sequence of approximate solutions resulting from a numerical scheme. Very roughly indeed, "weak" means the convergence in integral averages while "strong" refers to pointwise convergence with respect to the physical variables t and x. Here "pointwise" should be understood "pointwise a.a." with respect to the Lebesgue measure of the physical space $(0, T) \times \Omega$.

In particular, we consider sequences of approximate solutions to the compressible Euler system admitting uniform energy bounds (stable approximation) and/or satisfying the relevant field equations modulo an error vanishing in the asymptotic limit (consistent approximation). We discover a rather striking fact that either (i) the approximate sequence converges strongly in the energy norm, or (ii) the limit is not a weak solution of the associated Euler system. The latter alternative may be seen as a motivation for considering the more general classes of solutions introduced in Chap. 5.

7.1 Sharp Form of Jensen's Inequality

In Sect. 5.1 we have considered a general sequence $\{\boldsymbol{U}_n\}_{n=1}^{\infty}$ of measurable functions defined on a (bounded) set $Q \subset R^m$ and ranging in R^k, along with its superposition $E(\boldsymbol{U}_n)$ with a l.s.c. function

$$E : R^k \to [0, \infty].$$

If

$$\int_Q (|\boldsymbol{U}_n| + E(\boldsymbol{U}_n)) \, \mathrm{d}y \leq c$$

E. Feireisl et al., *Numerical Analysis of Compressible Fluid Flows*, MS&A 20,
https://doi.org/10.1007/978-3-030-73788-7_7

uniformly for $n \to \infty$, we may identify the limit

$$E(\boldsymbol{U}_n) \to \overline{E(\boldsymbol{U})} \text{ weakly-(*) in } \mathcal{M}^+(Q),$$

along with a Young measure $\mathcal{V}$ generated by $\{\boldsymbol{U}_n\}_{n=1}^{\infty}$, passing to a suitable subsequence as the case may be. Similarly to Sect. 5.1, we introduce the concentration defect

$$\overline{E(\boldsymbol{U})} - \left\langle \mathcal{V}; E(\widetilde{\boldsymbol{U}}) \right\rangle,$$

and the oscillation defect

$$\left\langle \mathcal{V}; E(\widetilde{\boldsymbol{U}}) \right\rangle - E\left(\left\langle \mathcal{V}; \widetilde{\boldsymbol{U}} \right\rangle \right).$$

We have also seen that the concentration defect is nonnegative and vanishes only if

$$\boldsymbol{U}_n \to \boldsymbol{U} \text{ weakly in } L^1(Q; R^k),$$

cf. Lemma 5.1.

If E is convex, then Jensen's inequality implies that also the oscillation defect is nonnegative. If, in addition, E is *strictly convex*, then, roughly speaking, the following implication holds:

$$\begin{aligned} &\text{vanishing oscillation defect} + E \text{ strictly convex} \\ &\quad \Rightarrow \mathcal{V}_y = \delta_{\boldsymbol{U}} \Rightarrow \boldsymbol{U}_n \to \boldsymbol{U} \text{ in measure (pointwise up to subsequence).} \end{aligned}$$

A rigorous statement reads as follows:

Lemma 7.1 (Sharp form of Jensen's inequality)

Suppose that $E : R^m \to [0, \infty]$ is a l.s.c. convex function satisfying:

- *E is strictly convex on its domain of positivity, meaning for any $y_1, y_2 \in R^m$ such that $0 < E(y_1) < \infty$, $E(y_2) < \infty$, $y_1 \neq y_2$, we have*

$$E\left(\frac{y_1 + y_2}{2}\right) < \frac{1}{2}E(y_1) + \frac{1}{2}E(y_2).$$

- *If $y \in \partial \mathrm{Dom}[E]$, then either $E(y) = \infty$ or $E(y) = 0$, in other words,*

$$E(y) = 0 \text{ whenever } y \in \mathrm{Dom}[E] \cap \partial \mathrm{Dom}[E]. \tag{7.1}$$

Let $\nu \in \mathcal{P}[R^m]$ be a (Borel) probability measure with finite first moment satisfying

$$E(\langle \nu; \widetilde{y} \rangle) = \langle \nu; E(\widetilde{y}) \rangle < \infty. \tag{7.2}$$

Then

(i) either

$$\nu = \delta_Y, \ Y = \langle \nu; \widetilde{y} \rangle \in \mathrm{Dom}[E], \ E(Y) > 0,$$

(ii) or

$$\mathrm{supp}[\nu] \subset \left\{ y \in R^m \;\middle|\; E(y) = 0 \right\}.$$

Proof First observe that, obviously, $\langle \nu; \widetilde{y} \rangle \in \mathrm{Dom}[E]$, and, by virtue of (7.2) and positivity of E,

$$\nu \left\{ R^m \setminus \mathrm{Dom}[E] \right\} = 0.$$

(i) Suppose first that $Y \equiv \langle \nu; \widetilde{y} \rangle \in \mathrm{int}[\mathrm{Dom}[E]]$, $E(Y) > 0$. Then there exists

$$\Lambda \in \partial E(Y)$$

such that

$$E(y) \geq E(Y) + \Lambda \cdot (y - Y) \text{ for any } y \in R^m.$$

As E is strictly convex in $\mathrm{Dom}[E] \cap \{E > 0\}$, however, we claim that the above inequality must be sharp:

$$E(y) - E(Y) - \Lambda \cdot (y - Y) > 0 \text{ for all } y \in R^d,\ y \neq Y.$$

Now it follows from (7.2) that

$$\left\langle \nu; E(\widetilde{y}) - E(Y) - \Lambda \cdot (\widetilde{y} - Y) \right\rangle = 0$$

which yields the desired conclusion (i).

(ii) Suppose that $Y = \langle \nu; \widetilde{y} \rangle \in \mathrm{Dom}[E] \cap \partial \mathrm{Dom}[E]$ or $E(Y) = 0$. In accordance with the hypothesis (7.1), we have in both cases

$$E(Y) = 0.$$

Consequently, we get from (7.2)

$$\langle \nu; E(\widetilde{y}) \rangle = 0$$

which implies that ν is supported by zero points of E as $E \geq 0$ which is alternative (ii). □

7.2 $\mathcal{K}$–Convergence

The basic idea behind the concept of $\mathcal{K}$*–convergence* is that averaging compactifies weakly converging sequences. A prominent example is the *Strong Law of Large Numbers* in probability,

$$\frac{1}{N}\sum_{n=1}^{N} f_n(y) \to \frac{1}{|Q|}\int_Q f_1(y)\, \mathrm{d}y \text{ as } N \to \infty \text{ for a.a. in } y \in Q, \tag{7.3}$$

where f_n are random variables (measurable functions) on a probability space $Q \subset R^m$ with the probability measure $\frac{1}{|Q|}$ dy that have the same distribution, meaning

$$|f_n^{-1}(B)| \text{ is independent of } n \text{ for any Borel set } B \subset R^1;$$

and are mutually independent, meaning

$$\left|\left\{y \,\middle|\, f_m(y) \in [a,b],\ f_n(y) \in [c,d]\right\}\right| = \left|\left\{y \,\middle|\, f_m(y) \in [a,b]\right\}\right| \left|\left\{y \,\middle|\, f_n(y) \in [c,d]\right\}\right|$$
$$\text{for any } m \neq n,\ a < b,\ c < d.$$

Komlós theorem discussed below represents a fundamental ingredient of our analysis of weakly convergent sequences and may be seen as a generalization of (7.3) to an arbitrary sequence of random variables.

7.2.1 Banach–Saks Theorem in Functional Analysis

The fundamental idea behind the Banach–Saks theorem and similar statements in functional analysis is that the weak and strong closure of a convex set coincide in an appropriate functional space. Let us recall the statement, see Banach and Saks [11], Kakutani [135].

Theorem 7.1 (Banach–Saks Theorem in L^p-spaces)
Let $Q \subset R^m$ be a bounded measurable set, $1 < p < \infty$. Let $\{U_n\}_{n=1}^{\infty}$ be a bounded sequence of functions in $L^p(Q)$,

$$\int_Q |U_n|^p \, \mathrm{d}y \lesssim 1 \textit{ uniformly for } n \to \infty.$$

Then there is an increasing sequence of indices $n_k \to \infty$ such that

$$U_{n_k} \to U \in L^p(Q) \textit{ weakly}, \quad \frac{1}{N}\sum_{k=1}^{N} U_{n_k} \to U \textit{ (strongly) in } L^p(Q).$$

In addition, Erdös and Magidor [75] showed that the subsequence in Theorem 7.1 can be chosen in such a way that any of its subsequences enjoys the same property:

Theorem 7.2 *Let $Q \subset R^m$ be a bounded measurable set, $1 < p < \infty$. Let $\{U_n\}_{n=1}^{\infty}$ be a bounded sequence of functions in $L^p(Q)$,*

$$\int_Q |U_n|^p \, dy \lesssim 1 \text{ uniformly for } n \to \infty.$$

Then there is an increasing sequence of indices $n_k \to \infty$ such that for each its subsequence $n_l \to \infty$ it holds

$$U_{n_l} \to U \in L^p(Q) \text{ weakly}, \; \frac{1}{N}\sum_{l=1}^{N} U_{n_l} \to U \text{ (strongly) in } L^p(Q).$$

As a matter of fact, there is a stronger version of Theorem 7.1 stating that

$$\left\| \sum_{k=1}^{N} (U_{n_k} - U) \right\|_{L^p(Q)} \lesssim N^{\max\left\{\frac{1}{p},\frac{1}{2}\right\}}, \tag{7.4}$$

see [10, Chap. 12, Theorem 2]. Extensions to weakly convergent sequences in L^1 are also available, see Szlenk [189].

The principal idea behind Banach–Saks theorem and related results is uniform convexity of the Banach norm in the spaces L^p. To simplify, consider the *Hilbert* space $L^2(0, 2\pi)$, together with its orthonormal basis $\left\{\frac{1}{\sqrt{2\pi}} \exp(inx)\right\}_{n=1}^{\infty}$. The sequence

$$U_n = \frac{1}{\sqrt{2\pi}} \exp(inx), \; n = 1, 2, \dots$$

is a standard example of a weak null sequence, meaning

$$U_n \to 0 \text{ weakly in } L^2(0, 2\pi) \text{ as } n \to \infty.$$

We have, because of orthonormality,

$$\|U_n\|_{L^2(0,2\pi)} = 1, \; \left\| \sum_{n=1}^{N} U_n \right\|_{L^2(0,2\pi)} = \left(\sum_{n=1}^{N} \|U_n\|^2_{L^2(0,2\pi)} \right)^{\frac{1}{2}} = N^{\frac{1}{2}}$$

in agreement with (7.4). A similar treatment may be applied to a general sequence

$$U_n \to U \text{ weakly in } L^2(Q).$$

First, we introduce

$$V_n = U_n - U \to 0 \text{ weakly in } L^2(Q), \; \|V_n\|_{L^2(Q)} = \overline{V}_n.$$

if the convergence is genuinely week (meaning not strong), there is a subsequence (not relabeled for simplicity) such that

$$\|V_n\| \to \overline{V} > 0 \text{ as } n \to \infty.$$

Normalizing $\{V_n\}$ we may suppose that $\|V_n\|_{L^2(Q)} = 1$ for all $n = 1, 2, \ldots$. Now, we fix V_1 and choose recursively V_{n_k} such that

$$\left| \int_Q V_{n_k} V_{n_j} \, \mathrm{d}y \right| < \chi(k), \; j = 1, \ldots, k-1,$$

where $\chi(k)$ is an arbitrary function such that $\chi(k) \to 0$ for $k \to \infty$. This is possible as V_n converges weakly to zero. Consequently,

$$\begin{aligned} S_N \equiv \left\| \sum_{k=1}^{N} V_{n_k} \right\|^2_{L^2(Q)} &= \left\| \sum_{k=1}^{N-1} V_{n_k} + V_{n_N} \right\|^2_{L^2(Q)} \\ &= \left\| \sum_{k=1}^{N-1} V_{n_k} \right\|^2_{L^2(Q)} + 1 + 2 \sum_{k=1}^{N-1} \int_Q V_{n_N} V_{n-k} \, \mathrm{d}y \le S_{N-1} + 1 + 2(N-1)\chi(N). \end{aligned}$$

We thus deduce recursively

$$S_N \le N + 2 \sum_{n=1}^{N} n \chi(n)$$

where $\chi(n) \to 0$ is arbitrary.

On the example of orthonormal basis $\{U_n\}_{n=1}^{\infty}$ we also see that

$$\left\| \sum_{k=1}^{N} U_{n_k} \right\|_{L^2(Q)} = N^{\frac{1}{2}}$$

independently of any chosen subsequence $n_k \to \infty$. Thus the convergence rate of the Cesàro averages

$$\left\| \frac{1}{N} \sum_{k=1}^{N} U_{n_k} \right\|_{L^2(Q)} = N^{-\frac{1}{2}}$$

is optimal and in fact independent of the choice of the subsequence. On the other hand, introducing the standard metric on the unit ball for the weak topology

$$d(U, V) = \sum_{n=1}^{\infty} \frac{1}{2^n} \left| \int_Q (U - V) U_n \, \mathrm{d}y \right| ,$$

we can see that the rate of weak convergence, measured in terms of d, can be made as fast as desired by choosing a suitable subsequence since

$$d(U_{n_k}, 0) = \frac{1}{2^{n_k}}.$$

The above observation indicate that the convergence rate of Cesàro averages of an oscillatory (weakly convergent) sequence $U_n \to 0$ depends on:

- the speed in which the oscillation defect $\|U_n\|_{L^2(Q)}$ converges to its (nonzero) asymptotic limit;
- the number N of terms in the Cesàro average.

Strangely enough, the frequency (rate of weak convergence in the metric d) may not have any influence on the convergence rate of Cesàro averages. Of course, one should keep in mind that all the above results require extracting a suitable but *a priori* unknown subsequence.

7.2.2 *Pointwise Convergence of Cesàro Averages – Komlós Theorem*

Komlós theorem is a variant of Banach–Saks theorem in the L^1-setting. It applies in the situation when the sequence is not equi-integrable and in general does not admit a weakly convergent subsequence. In particular, the version of Banach–Saks theorem of Szlenk [189] is not applicable.

Theorem 7.3 (Komlós Theorem)
Let $Q \subset R^m$ be a Borel set of finite Lebesgue measure. Suppose that

$$\int_Q |U_n| \, \mathrm{d}y \lesssim 1 \text{ for a sequence of measurable functions } \{U_n\}_{n=1}^{\infty}.$$

Then there exists a subsequence $n_k \to \infty$ such that for any subsequence $\{n_l\} \subset \{n_k\}$,

$$\frac{1}{N} \sum_{l=1}^{N} U_{n_l} \to U \in L^1(Q) \text{ a.a. in } Q \text{ as } N \to \infty.$$

Theorem 7.3 is due to Komlós [142]. There are several generalizations, notably by Berkes [19], who asserts validity of the above result not only for any subsequence $\{n_l\}_{l=1}^{\infty}$ but for arbitrary permutation of the sequence $\{n_k\}_{k=1}^{\infty}$.

7.2.3 Application to the Young Measures

The advantage of having pointwise convergence in Theorem 7.3 becomes imminent when the theory is extended to Young measures. Consider a sequence of measurable vector valued functions $\{\boldsymbol{U}_n\}_{n=1}^{\infty}$,

$$\boldsymbol{U}_n : Q \to R^m, \ Q \subset R^m \text{ a bounded domain,}$$

uniformly bounded in $L^1(Q; R^m)$,

$$\int_Q |\boldsymbol{U}_n| \, \mathrm{d}y \stackrel{<}{\sim} 1 \ \text{ for } \ n \to \infty.$$

In future applications to the Euler/Navier–Stokes system we consider

$$\boldsymbol{U}_n : (0, T) \times \Omega \subset R^{d+1} \mapsto [\varrho_n, \boldsymbol{m}_n, S_n] \in R^{d+2}.$$

As we have seen in Proposition 5.1, there is a subsequence (not relabeled) that generates a Young measure

$$\mathcal{V} : Q \to \mathcal{P}(R^k), \ \text{weakly-(*) measurable,}$$

meaning

$$y \in Q \mapsto \left\langle \mathcal{V}_y, g(\widetilde{\boldsymbol{U}}) \right\rangle \text{ is Lebesgue measurable for any } g \in BC(R^k).$$

By "generates" we mean that

$$g(\boldsymbol{U}_n) \to \left\langle \mathcal{V}; g(\widetilde{\boldsymbol{U}}) \right\rangle \ \text{ weakly-(*) in } \ L^{\infty}(Q) \ \text{ for any } \ g \in BC(R^k). \tag{7.5}$$

This can be equivalently stated in terms of the weak-(*) convergence in the space $L^{\infty}(Q; \mathcal{M}(R^k))$ if $\boldsymbol{U}_n$ is interpreted as the Dirac measure

$$\boldsymbol{U}_n(y) \approx \delta_{\boldsymbol{U}_n(y)} \ \text{ for a.a. } \ y \in Q.$$

Convergence claimed in (7.5) is very weak, meaning expressed only in terms of integral averages.

A better way to "visualize" the Young measure, inspired by Sects. 7.2.1, 7.2.2, is to consider the Cesàro averages

$$\frac{1}{N}\sum_{n=1}^{N}\delta_{U_n(y)} \in \mathcal{P}(R^k) \text{ for a.a. } y \in Q.$$

Such a process requires subtracting further subsequence(s), for which we are still keeping the original notation $\{\boldsymbol{U}_n\}_{n=1}^{\infty}$.

Similarly to the construction of the Young measure in Sect. 5.1, we consider a countable family of functions $\{g_j\}_{j=1}^{\infty}$ of compactly supported functions, dense in $C_c(R^k)$. In view of the Komlós theorem, see Theorem 7.3, applied to $\{g_j(\boldsymbol{U}_n)\}_{n=1}^{\infty}$, we may assume, having extracted several subsequences as the case may be,

$$\frac{1}{N}\sum_{n=1}^{N} g_j(\boldsymbol{U}_n(y)) \to \left\langle \mathcal{V}_y; g_j(\widetilde{\boldsymbol{U}})\right\rangle \text{ for a.a. } y \in Q,\ j = 1, 2, \dots \tag{7.6}$$

and

$$\frac{1}{N}\sum_{n=1}^{N} g_j(\boldsymbol{U}_n) \to \left\langle \mathcal{V}; g_j(\widetilde{\boldsymbol{U}})\right\rangle \text{ in } L^q(Q) \text{ for any } 1 \le q < \infty,\ j = 1, 2, \dots$$

Indeed we already know that $\{\boldsymbol{U}_n\}_{n=1}^{\infty}$ generates the Young measure $\mathcal{V}$; whence

$$g_j(\boldsymbol{U}_n) \to \left\langle \mathcal{V}; g_j(\widetilde{\boldsymbol{U}})\right\rangle \text{ weakly in } L^q(Q),\ 1 \le q < \infty,$$

which identifies the limit in (7.6).

The result can be extended to a larger class of sublinear nonlinearities if $\{\boldsymbol{U}_n\}_{n=1}^{\infty}$ is integrable.

Lemma 7.2 *Let $Q \subset R^m$ be a bounded domain,*

$$\boldsymbol{U}_n : Q \to R^k,\ \int_Q |\boldsymbol{U}_n| \,\mathrm{d}y \le c \text{ uniformly for } n \to \infty.$$

Then, extracting a suitable subsequence as the case may be, we have

$$\frac{1}{N}\sum_{n=1}^{N} g(\boldsymbol{U}_n)(y) \to \left\langle \mathcal{V}_y; g(\widetilde{\boldsymbol{U}})\right\rangle \text{ as } N \to \infty \text{ for a.a. } y \in Q$$

for any

$$g \in C(R^k),\ |g(\boldsymbol{U})| \lesssim 1 + |\boldsymbol{U}|.$$

In particular,

$$\frac{1}{N}\sum_{n=1}^{N}|\boldsymbol{U}_n(y)| \to \left\langle \mathcal{V}_y; |\widetilde{\boldsymbol{U}}|\right\rangle \text{ for a.a. } y \in Q. \tag{7.7}$$

Proof On one hand, by Biting Lemma, there is a subsequence (not relabeled) of measurable sets $M_k \subset Q$,

$$|M_k| \to 0 \text{ as } k \to \infty,$$

and a function $\overline{g(\boldsymbol{U})} \in L^1(Q)$ such that

$$g(\boldsymbol{U}_n) \to \overline{g(\boldsymbol{U})} \text{ weakly in } L^1(Q \setminus M_k) \text{ as } n \to \infty$$

for any $k = 1, 2, \dots$.

On the other hand, by the Komlós theorem, the Cesàro averages

$$\frac{1}{N}\sum_{n=1}^{N} g(\boldsymbol{U}_n) \text{ converge a.a. in } Q,$$

again for a suitable subsequence. Consequently,

$$\frac{1}{N}\sum_{n=1}^{N} g(\boldsymbol{U}_n)(y) \to \overline{g(\boldsymbol{U})}(y) \text{ for a.a. } y \in Q \setminus M_k$$

for any $k = 1, 2, \dots$. Thus we may infer that

$$\frac{1}{N}\sum_{n=1}^{N} g(\boldsymbol{U}_n)(y) \to \overline{g(\boldsymbol{U})}(y) \text{ for a.a. } y \in Q.$$

Finally, we observe that

$$\overline{g(\boldsymbol{U})}(y) = \lim_{m\to\infty} \overline{g_m(\boldsymbol{U})}(y) = \lim_{m\to\infty} \left\langle \mathcal{V}_y; b(\widetilde{\boldsymbol{U}})\right\rangle \text{ for a.a. } y \in Q$$

for a suitable approximating sequence $g_m \in BC(R^k)$, $g_m \to g$. We conclude

$$\overline{g(\boldsymbol{U})}(y) = \left\langle \mathcal{V}_y; g(\widetilde{\boldsymbol{U}})\right\rangle \text{ for a.a. } y \in Q.$$

□

Remark 7.1 ($\mathcal{K}$-limit vs. biting limit)

The quantity $\overline{g(\boldsymbol{U})}$ identified in the proof of Lemma 7.2 is called *biting limit* of the sequence $\{g(\boldsymbol{U}_n)\}_{n=1}^{\infty}$. We have shown that the biting limit coincides with the limit of Cesàro averages arising in the Komlós theorem. The relation between biting limits and the Young measures has been studied by Ball and Murat [9]. As somehow expected, the limit of Cesàro averages ignores concentrations in the generating sequence.

Resuming our discussion, we rewrite

$$\frac{1}{N}\sum_{n=1}^{N} g(\boldsymbol{U}_n(y)) = \frac{1}{N}\sum_{n=1}^{N}\left\langle \delta_{\boldsymbol{U}_n(y)}; g(\widetilde{\boldsymbol{U}})\right\rangle;$$

obtaining, by virtue of Theorem 4,

$$d_{W_1}\left[\frac{1}{N}\sum_{n=1}^{N}\delta_{\boldsymbol{U}_n(y)}; \mathcal{V}_y\right] \to 0 \text{ as } N \to \infty, \text{ for a.a. } y \in Q, \tag{7.8}$$

where d_{W_1} is the Wasserstein 1-distance. Moreover,

$$\int_Q \left| d_P\left[\frac{1}{N}\sum_{n=1}^{N}\delta_{\boldsymbol{U}_n(y)}; \mathcal{V}_y\right]\right|^q \mathrm{d}y \to 0 \text{ as } N \to \infty \text{ for any } 1 \le q < \infty, \tag{7.9}$$

where d_P is the Lévy–Prokhorov distance equivalent to $d_{w-(*)}$ distance. Note that $d_P(\nu, \mu) \le 1$ for any probability measures ν, μ.

If higher power integral bounds

$$\int_Q |\boldsymbol{U}_n|^r \,\mathrm{d}y \stackrel{<}{\sim} 1,\ r > 1$$

are available, then, in view of Lemma 7.2, (7.7) can be strengthened, for a suitable subsequence, to

$$\frac{1}{N}\sum_{n=1}^{N}|\boldsymbol{U}_n(y)|^r = \frac{1}{N}\sum_{n=1}^{N}\left\langle \delta_{\boldsymbol{U}_n(y)}; |\widetilde{\boldsymbol{U}}|^r\right\rangle \to \left\langle \mathcal{V}_y; |\widetilde{\boldsymbol{U}}|^r\right\rangle \text{ as } n \to \infty \text{ for a.a. } y \in Q, \tag{7.10}$$

and, in view of the Banach–Saks theorem, see Theorem 7.2,

$$\frac{1}{N}\sum_{n=1}^{N}|\boldsymbol{U}_n|^s = \frac{1}{N}\sum_{n=1}^{N}\left\langle \delta_{\boldsymbol{U}_n}; |\widetilde{\boldsymbol{U}}|^s\right\rangle \to \left\langle \mathcal{V}; |\widetilde{\boldsymbol{U}}|^s\right\rangle \text{ as } N \to \infty$$
$$\text{in } L^{\frac{r}{s}}(Q) \text{ for all } 1 \le s < r. \tag{7.11}$$

Thus we conclude

$$\int_Q \left| d_P \left[\frac{1}{N} \sum_{n=1}^{N} \delta_{U_n(y)}; \mathcal{V}_y \right] \right|^q \mathrm{d}y \to 0 \text{ as } N \to \infty \text{ for any } 1 \le q < \infty \tag{7.12}$$

and

$$\int_Q \left| d_{W_r} \left[\frac{1}{N} \sum_{n=1}^{N} \delta_{U_n(y)}; \mathcal{V}_y \right] \right|^s \mathrm{d}y \to 0 \text{ as } N \to \infty \text{ for any } 1 \le s < r, \tag{7.13}$$

where W_r denotes the Wasserstein r-distance.

The results obtained in this section are summarized in the following theorem.

Theorem 7.4 **($\mathcal{K}$–convergence of integrable sequences)**

Let $Q \subset R^m$ be a bounded domain. Let $\{\boldsymbol{U}_n\}_{n=1}^{\infty}$ be a sequence of measurable vector valued functions

$$\boldsymbol{U}_n : Q \to R^k, \int_{\Omega} |\boldsymbol{U}_n|^r \, \mathrm{d}x \lesssim 1 \textit{ for some } 1 \le r < \infty.$$

Then there is a subsequence $\{\boldsymbol{U}_{n_j}\}_{j=1}^{\infty}$ enjoying the following properties:

- *$\{\boldsymbol{U}_{n_j}\}_{j=1}^{\infty}$ generates a Young measure $\{\mathcal{V}_y\}_{y \in Q}$, specifically,*

$$\delta_{\boldsymbol{U}_{n_j}} \to \mathcal{V} \textit{ weakly-(*) in } L^\infty(Q; \mathcal{M}(R^k)) \textit{ as } j \to \infty.$$

- $$d_{W_1} \left[\frac{1}{N} \sum_{j=1}^{N} \delta_{\boldsymbol{U}_{n_j}(y)}; \mathcal{V}_y \right] \to 0 \textit{ as } N \to \infty \textit{ for a.a. } y \in Q.$$

- *If, in addition $r > 1$, then*

$$\int_Q \left| d_{W_r} \left[\frac{1}{N} \sum_{j=1}^{N} \delta_{\boldsymbol{U}_j(y)}; \mathcal{V}_y \right] \right|^s \mathrm{d}y \to 0 \textit{ as } N \to \infty \textit{ for any } 1 \le s < r.$$

7.3 Strong Convergence of Approximate Solutions to the Complete Euler System

We consider the complete Euler system introduced in Chap. 2 written in the conservative-entropy variables $[\varrho, \boldsymbol{m}, S]$

$$\partial_t \varrho + \mathrm{div}_x \boldsymbol{m} = 0,$$
$$\partial_t \boldsymbol{m} + \mathrm{div}_x \left(\frac{\boldsymbol{m} \otimes \boldsymbol{m}}{\varrho} \right) + \nabla_x p(\varrho, S) = 0,$$
$$\partial_t \left(\frac{1}{2} \frac{|\boldsymbol{m}|^2}{\varrho} + \varrho e(\varrho, S) \right) + \mathrm{div}_x \left[\left(\frac{1}{2} \frac{|\boldsymbol{m}|^2}{\varrho} + \varrho e(\varrho, S) + p(\varrho, S) \right) \frac{\boldsymbol{m}}{\varrho} \right] = 0, \tag{7.14}$$

supplemented with the (renormalized) entropy balance

$$\partial_t \left(\varrho Z \left(\frac{S}{\varrho} \right) \right) + \mathrm{div}_x \left(Z \left(\frac{S}{\varrho} \right) \boldsymbol{m} \right) \geq 0 \tag{7.15}$$

for any $Z \in C^1(R)$, $Z' \geq 0$. The boundary conditions are either periodic $\Omega = \mathbb{T}^d$ or impermeable, specifically $\Omega \subset R^d$ bounded and

$$\boldsymbol{m} \cdot \boldsymbol{n}|_{\partial \Omega} = 0.$$

In accordance with our choice of independent variables, we prescribe the initial conditions

$$\varrho(0, \cdot) = \varrho_0, \ \boldsymbol{m}(0, \cdot) = \boldsymbol{m}_0, \ S(0, \cdot) = S_0. \tag{7.16}$$

We consider a sequence of *approximate* solutions $\{\varrho_n, \boldsymbol{m}_n, S_n\}_{n=1}^{\infty}$, with the initial data $[\varrho_{0,n}, \boldsymbol{m}_{0,n}, S_{0,n}]$. Note that the notion of "initial data" is rather vague as we do not assume any continuity of the approximate sequence. Numerical solutions considered in this monograph are right continuous with respect to t so we may suppose

$$\varrho_n(0+, \cdot) = \varrho_{0,n}, \ \boldsymbol{m}_n(0+, \cdot) = \boldsymbol{m}_{0,n}, \ S(0+, \cdot) = S_{0,n}.$$

7.3.1 *Positivity of the Density and Total Mass Conservation*

Positivity (nonnegativity) of the approximate mass density ϱ_n is absolutely necessary not only because of its physical interpretation but also for all fields appearing in the equations to be well defined. Note that the equation of continuity written in terms of momentum itself is not sufficient to guarantee this property. From this point of view, it is more convenient to write the mass conservation in terms of the standard variables – the density and the velocity

$$\partial_t \varrho + \mathbf{u} \cdot \nabla_x \varrho = -\varrho \mathrm{div}_x \mathbf{u},$$

where, at least formally, the density remains positive as soon as it is so initially and $\mathrm{div}_x \mathbf{u}$ is controlled. If $\varrho \geq 0$ then the equation of continuity integrated over the spatial domain Ω gives rise to the total mass conservation

$$\int_\Omega \varrho(\tau, \cdot)\ \mathrm{d}x = \int_\Omega \varrho_0\ \mathrm{d}x \ \text{ for any } \ \tau \geq 0, \ \text{ in particular } \ \varrho \in L^\infty(0, T; L^1(\Omega)).$$

Formally, positivity can be enforced by setting the total energy

$$E = E(\varrho, \mathbf{m}, S) = \frac{1}{2} \frac{|\mathbf{m}|^2}{\varrho} + \varrho e(\varrho, S) = \infty \ \text{ whenever } \ \varrho < 0$$

and considering solutions with finite total energy. As for the approximate sequence, we shall therefore assume

$$\begin{aligned} &\varrho_n \geq 0 \ \text{ a.a. in } \ (0, T) \times \Omega, \\ &\sup_{\tau \in (0,T)} \|\varrho_n(\tau, \cdot)\|_{L^1(\Omega)} \lesssim 1 \text{ or, equivalently, } \sup_{\tau \in (0,T)} \int_\Omega \varrho_n(\tau, \cdot)\ \mathrm{d}x \leq M \\ &\text{uniformly for } n \to \infty. \end{aligned} \tag{7.17}$$

7.3.2 *Minimum Entropy Principle*

The minimum entropy principle for the Euler system states that

$$\min_{x \in \Omega} s(\tau, x) \geq \min_{x \in \Omega} s(0, x) = \min_{x \in \Omega} \frac{S_0}{\varrho_0}(x) \ \text{ for any } \ \tau \geq 0. \tag{7.18}$$

This property can be easily but still formally deduced from the entropy inequality (7.15) yielding for $Z(Y) = Y$ the transport equation (inequality)

$$\partial_t s + \mathbf{u} \cdot \nabla_x s \geq 0.$$

It is worth noting that we have divided by ϱ in this process so we have tacitly assumed that the density is strictly positive.

The minimum entropy principle can be rephrased as

$$S(t, x) \geq \underline{s} \varrho(t, x) \ \text{ for a.a. } \ (t, x), \ \text{ where } \ \underline{s} = \min_{x \in \Omega} s(0, x). \tag{7.19}$$

This condition has been introduced in Definition 5.2 of consistent approximation of the Euler system. Here, we start with a much weaker restriction imposed on the sequence of approximate solutions, namely

$$\int_{\Omega} S_n(\tau, x)\, \mathrm{d}x \geq \underline{S} \ \text{ for all } \ \tau \geq 0,\ n \to \infty. \tag{7.20}$$

In contrast with (7.19), verification of (7.20) requires only integration of (7.15) in the space variable. Moreover, unlike (7.19) the relation (7.20) holds for a large class of approximate solutions that comply with the Second law of thermodynamics – nonnegativity of the entropy production rate.

7.3.3 *Total Energy*

The total energy

$$E = e_{\text{kin}} + e_{\text{int}},\ e_{\text{kin}} \equiv \frac{1}{2}\frac{|\boldsymbol{m}|^2}{\varrho},\ e_{\text{int}} \equiv \frac{1}{\gamma - 1} p = c_v \varrho^{\gamma} \exp\left(\frac{S}{c_v \varrho}\right),\ c_v = \frac{1}{\gamma - 1}, \tag{7.21}$$

represents an absolutely crucial quantity in obtaining uniform bounds. We suppose, in accordance with the energy balance equation, that

$$\begin{aligned} \int_{\Omega} E(\varrho_n, \boldsymbol{m}_n, S_n)(\tau, \cdot)\, \mathrm{d}x &\equiv \int_{\Omega} \left(\frac{1}{2}\frac{|\boldsymbol{m}_n|^2}{\varrho_n} + \varrho_n e(\varrho_n, S_n)\right) \mathrm{d}x \\ &\leq \int_{\Omega} \left(\frac{1}{2}\frac{|\boldsymbol{m}_{0,n}|^2}{\varrho_{0,n}} + \varrho_{0,n} e(\varrho_{0,n}, S_{0,n})\right) \mathrm{d}x \stackrel{<}{\sim} 1 \text{ for } n \to \infty. \end{aligned} \tag{7.22}$$

Note that (7.21) includes the isentropic case $s = \overline{s}$ – a constant, for which

$$e_{\text{int}} = \frac{a}{\gamma - 1}\varrho^{\gamma},\ a = \exp\left(\frac{\overline{s}}{c_v}\right) > 0,\ p = a\varrho^{\gamma}.$$

The energy E as a function of the variables $\varrho, \boldsymbol{m}, S$ is correctly defined through (7.21) only out of the vacuum, meaning for $\varrho > 0$. As we have observed in Sect. 2.2.4, the total energy can be extended as a convex l.s.c. function to the whole phase space R^{d+2}

$$e_{\text{kin}}(\varrho, \boldsymbol{m}) = \begin{cases} 0 \ \text{if} \ \varrho = 0,\ \boldsymbol{m} = 0, \\ \infty \ \text{if} \ \varrho < 0 \ \text{or} \ \varrho = 0,\ \boldsymbol{m} \neq 0, \end{cases} \tag{7.23}$$

and

$$e_{\mathrm{int}}(\varrho, S) = \begin{cases} 0 \ \text{if} \ \varrho = 0, \ S \leq 0, \\ \infty \ \text{if} \ \varrho < 0 \ \text{or} \ \varrho = 0, \ S > 0. \end{cases} \tag{7.24}$$

Lemma 7.3 (Convexity of the total energy)
The function $E = E(\varrho, \boldsymbol{m}, S)$, defined for $\varrho > 0$ as

$$E(\varrho, \boldsymbol{m}, S) = e_{\mathrm{kin}}(\varrho, \boldsymbol{m}) + e_{\mathrm{int}}(\varrho, S) = \frac{1}{2}\frac{|\boldsymbol{m}|^2}{\varrho} + c_v \varrho^\gamma \exp\left(\frac{S}{c_v \varrho}\right),$$

and extended through (7.23), (7.24) *if* $\varrho \leq 0$, *is convex l.s.c. function of variables* $(\varrho, \boldsymbol{m}, S) \in R^{d+2}$ *ranging in* $[0, \infty]$. *Moreover, E is strictly convex at any point where* $\varrho > 0$, *in particular, it satisfies all hypotheses of Lemma 7.1.*

Next we claim that the uniform bounds (7.17), (7.20), and (7.22) imply bounds on the L^1 norm of all state variables $[\varrho_n, \boldsymbol{m}_n, S_n]$. To see this, we choose an arbitrary point $[\widetilde{\varrho}, 0, \widetilde{S}] \in R^{d+2}$, $\widetilde{\varrho} > 0$, and consider the function

$$\begin{aligned} 0 &\leq E(\varrho_n, \boldsymbol{m}_n, S_n) - \frac{\partial E(\widetilde{\varrho}, 0, \widetilde{S})}{\partial \varrho}(\varrho_n - \widetilde{\varrho}) - \frac{\partial E(\widetilde{\varrho}, 0, \widetilde{S})}{\partial \boldsymbol{m}} \cdot (\boldsymbol{m}_n - \widetilde{\boldsymbol{m}}) \\ &- \frac{\partial E(\widetilde{\varrho}, 0, \widetilde{S})}{\partial S}(S_n - \widetilde{S}) - E(\widetilde{\varrho}, 0, \widetilde{S}) \\ &= \frac{1}{2}\frac{|\boldsymbol{m}_n|^2}{\varrho_n} + \varrho_n e(\varrho_n, S_n) - \frac{\partial(\varrho e)(\widetilde{\varrho}, \widetilde{S})}{\partial \varrho}(\varrho - \widetilde{\varrho}) - \frac{\partial(\varrho e)(\widetilde{\varrho}, \widetilde{S})}{\partial S}(S - \widetilde{S}) \\ &- (\widetilde{\varrho} e(\widetilde{\varrho}, \widetilde{S})). \end{aligned}$$

Seeing that $\frac{\partial E}{\partial S} = \vartheta > 0$, we conclude

$$\begin{aligned} &\int_\Omega \Big[E(\varrho_n, \boldsymbol{m}_n, S_n) - \frac{\partial E(\widetilde{\varrho}, 0, \widetilde{S})}{\partial \varrho}(\varrho_n - \widetilde{\varrho}) - \frac{\partial E(\widetilde{\varrho}, 0, \widetilde{S})}{\partial \boldsymbol{m}} \cdot (\boldsymbol{m}_n - \widetilde{\boldsymbol{m}}) \\ &\quad - \frac{\partial E(\widetilde{\varrho}, 0, \widetilde{S})}{\partial S}(S_n - \widetilde{S}) - E(\widetilde{\varrho}, 0, \widetilde{S}) \Big] \mathrm{d}x \\ &\quad \leq c(\widetilde{\varrho}, \widetilde{S}) \left(1 + \int_\Omega E(\varrho_n, \boldsymbol{m}_n, S_n) \,\mathrm{d}x + \int_\Omega \varrho_n \,\mathrm{d}x - \int_\Omega S_n \,\mathrm{d}x \right) \\ &\quad \leq c(\widetilde{\varrho}, \widetilde{S}) \left(1 + \int_\Omega E(\varrho_{0,n}, \boldsymbol{m}_{0,n}, S_{0,n}) \,\mathrm{d}x + \int_\Omega \varrho_{0,n} \,\mathrm{d}x - \int_\Omega S_{0,n} \,\mathrm{d}x \right). \end{aligned}$$

As E is strictly convex at $[\widetilde{\varrho}, 0, \widetilde{S}]$, we have

$$
\begin{aligned}
E(\varrho_n, \boldsymbol{m}_n, S_n) &- \frac{\partial E(\widetilde{\varrho}, 0, \widetilde{S})}{\partial \varrho}(\varrho_n - \widetilde{\varrho}) \\
&- \frac{\partial E(\widetilde{\varrho}, 0, \widetilde{S})}{\partial \boldsymbol{m}} \cdot (\boldsymbol{m}_n - \widetilde{\boldsymbol{m}}) - \frac{\partial E(\widetilde{\varrho}, 0, \widetilde{S})}{\partial S}(S_n - \widetilde{S}) - \mathrm{e}(\widetilde{\varrho}, 0, \widetilde{S}) \\
&\gtrsim |\varrho_n - \widetilde{\varrho}| + |\boldsymbol{m}_n| + |S_n - \widetilde{S}|
\end{aligned}
$$

as soon as

$$
|\varrho_n - \widetilde{\varrho}| + |\boldsymbol{m}_n| + |S_n - \widetilde{S}| \geq 1.
$$

Definition 7.1 (STABLE APPROXIMATION TO COMPLETE EULER SYSTEM)
We say that a sequence $\{\varrho_n, \boldsymbol{m}_n, S_n\}_{n=1}^{\infty}$ is a *stable approximation* of the complete Euler system in $(0, T) \times \Omega$, with the initial data $[\varrho_0, \boldsymbol{m}_0, S_0]$, if

$$
\varrho_n \geq 0, \ \operatorname*{ess\ sup}_{\tau \in (0,T)} \int_{\Omega} \varrho_n(\tau, \cdot) \, dx \leq M,
$$

$$
\operatorname*{ess\ inf}_{\tau \in (0,T)} \int_{\Omega} S_n(\tau, \cdot) \, dx \geq \underline{S} \tag{7.25}
$$

uniformly for $n \to \infty$;

$$
\operatorname*{ess\ sup}_{\tau \in (0,T)} \int_{\Omega} E(\varrho_n, \boldsymbol{m}_n, S_n) \, dx \leq \int_{\Omega} E(\varrho_0, \boldsymbol{m}_0, S_0) \, dx + e_n \text{ for all } n = 1, 2, \ldots \tag{7.26}
$$

where $e_n \to 0$ as $n \to \infty$.

Summarizing the above discussion, we obtain the following result.

Proposition 7.1 (Energy bounds on stable approximation)
Let $\Omega \subset R^d$ be a bounded domain. Let a sequence $\{\varrho_n, \boldsymbol{m}_n, S_n\}_{n=1}^{\infty}$ be a stable approximation of the complete Euler system in the sense of Definition 7.1.
Then

$$
\operatorname*{ess\ sup}_{\tau \in (0,T)} \left[\|\varrho_n(\tau, \cdot)\|_{L^1(\Omega)} + \|\boldsymbol{m}_n(\tau, \cdot)\|_{L^1(\Omega;R^d)} + \|S_n(\tau, \cdot)\|_{L^1(\Omega)} \right] \lesssim 1
$$

uniformly for $n \to \infty$.

Remark 7.2 In contrast with Lemma 5.2, where uniform bounds on *consistent approximation* were established, Proposition 7.1 does not require validity of the minimum entropy principle (7.19).

We conclude that under very mild assumptions (7.17), (7.19), and (7.22), any stable approximation of the complete Euler system is bounded at least in the L^1-norm and as such, up to a suitable subsequence, generates a Young measure

$$\mathcal{V}_{t,x} \in L^\infty_{\rm weak}\left((0,T)\times\Omega; \mathcal{P}\left[R^{d+2} = \left\{\widetilde{\varrho}\in R,\ \widetilde{\boldsymbol{m}}\in R^d,\ \widetilde{S}\in R\right\}\right]\right).$$

Moreover, the measure $\mathcal{V}_{t,x}$ has finite moments for a.a. $(t,x)\in(0,T)\times\Omega$.

7.3.4 Strong Convergence of a Stable Approximation to a Weak Solution

The behavior of the total energy completely determines the convergence properties of the stable approximation. In accordance with (7.26), we may assume

$$E(\varrho_n, \boldsymbol{m}_n, S_n) \to \overline{E(\varrho,\boldsymbol{m},S)} \text{ weakly-(*) in } L^\infty(0,T;\mathcal{M}(\overline{\Omega}))$$

passing again to a suitable subsequence if necessary. The associated concentration defect is defined as

$$\mathfrak{E}_{cd} \equiv \overline{E(\varrho,\boldsymbol{m},S)} - \left\langle \mathcal{V}; E(\widetilde{\varrho},\widetilde{\boldsymbol{m}},\widetilde{S})\right\rangle,$$

while the oscillation defect reads

$$\mathfrak{E}_{od} \equiv \left\langle \mathcal{V}; E(\widetilde{\varrho},\widetilde{\boldsymbol{m}},\widetilde{S})\right\rangle - E(\varrho,\boldsymbol{m},S).$$

Here we have denoted

$$[\varrho(t,x), \boldsymbol{m}(t,x), S(t,x)] \equiv \left[\left\langle\mathcal{V}_{t,x};\widetilde{\varrho}\right\rangle, \left\langle\mathcal{V}_{t,x};\widetilde{\boldsymbol{m}}\right\rangle, \left\langle\mathcal{V}_{t,x};\widetilde{S}\right\rangle\right]$$

the barycenter of the Young measure generated by $\{\varrho_n, \boldsymbol{m}_n, S_n\}_{n=1}^\infty$. As pointed out in Remarks 5.4, 5.6, both concentration and oscillation defects are nonnegative, specifically,

$$\mathfrak{E}_{cd}\in L^\infty(0,T;\mathcal{M}^+(\overline{\Omega})),\ \mathfrak{E}_{od}\in L^1((0,T)\times\Omega),\ \mathfrak{E}_{od}\geq 0. \tag{7.27}$$

Our goal is to show a rather surprising result: A stable approximation converges strongly as soon as the barycenter $[\varrho,\boldsymbol{m},S]$ of the associated Young measure is a weak solution to the Euler system (7.14)–(7.16).

Theorem 7.5 (Asymptotic limit of stable approximation)
Let $\Omega\subset R^d$ be a bounded domain. Let $\{\varrho_n,\boldsymbol{m}_n,S_n\}_{n=1}^\infty$ be a stable approximation of the complete Euler system in the sense of Definition 7.1, with the initial data

$$\varrho_0 > 0,\ \boldsymbol{m}_0,\ S_0 \geq \varrho_0\underline{s},\ \text{where } \underline{s}\in R. \tag{7.28}$$

In addition, suppose that the sequence $\{\varrho_n,\boldsymbol{m}_n,S_n\}_{n=1}^\infty$ generates a Young measure $\{\mathcal{V}_{t,x}\}_{(t,x)\in(0,T)\times\Omega}$. Finally, let the barycenter

$$[\varrho(t,x), \boldsymbol{m}(t,x), S(t,x)] \equiv \left[\left\langle\mathcal{V}_{t,x};\widetilde{\varrho}\right\rangle, \left\langle\mathcal{V}_{t,x};\widetilde{\boldsymbol{m}}\right\rangle, \left\langle\mathcal{V}_{t,x};\widetilde{S}\right\rangle\right],\ (t,x)\in(0,T)\times\Omega,$$

be a weak solution of the Euler system (7.14)–(7.16) in the sense of Definition 2.4, satisfying the entropy inequality (2.70), and emanating form the initial data $[\varrho_0, \boldsymbol{m}_0, S_0]$.

Then

$$\mathcal{V}_{t,x} = \delta_{[\varrho(t,x), \boldsymbol{m}(t,x), S(t,x)]} \text{ for a.a. } (t,x) \in (0,T) \times \Omega,$$

and, up to a suitable subsequence,

$$\begin{aligned}\varrho_n &\to \varrho \text{ a.a. in } (0,T) \times \Omega,\\ \boldsymbol{m}_n &\to \boldsymbol{m} \text{ a.a. in } (0,T) \times \Omega,\\ S_n &\to S \text{ a.a. in } (0,T) \times \Omega.\end{aligned}$$

Proof It follows from the energy bound (7.26) that

$$E(\varrho_n, \boldsymbol{m}_n, S_n) \equiv \left[\frac{1}{2}\frac{|\boldsymbol{m}_n|^2}{\varrho_n} + \varrho_n e(\varrho_n, S_n)\right] \to \overline{E(\varrho, \boldsymbol{m}, S)}$$
$$\text{weakly-(*) in } L^\infty(0,T; \mathcal{M}^+(\overline{\Omega})),$$

where

$$\int_{\overline{\Omega}} \mathrm{d}\overline{E(\varrho, \boldsymbol{m}, S)}(\tau, \cdot) \leq \int_{\Omega} \left[\frac{1}{2}\frac{|\boldsymbol{m}_0|^2}{\varrho_0} + \varrho_0 e(\varrho_0, S_0)\right] \mathrm{d}x \text{ for a.a. } \tau \in (0,T). \quad (7.29)$$

On the other hand, it follows from the weak formulation of the energy equation (2.69) that

$$\int_{\Omega} \left[\frac{1}{2}\frac{|\boldsymbol{m}|^2}{\varrho} + \varrho e(\varrho, S)\right](\tau, \cdot)\, \mathrm{d}x = \int_{\Omega} \left[\frac{1}{2}\frac{|\boldsymbol{m}_0|^2}{\varrho_0} + \varrho_0 e(\varrho_0, S_0)\right] \mathrm{d}x$$

for a.a. $\tau \in (0,T)$. As

$$\left[\frac{1}{2}\frac{|\boldsymbol{m}|^2}{\varrho} + \varrho e(\varrho, S)\right] \leq \overline{E(\varrho, \boldsymbol{m}, S)},$$

we deduce from (7.29) that

$$\left[\frac{1}{2}\frac{|\boldsymbol{m}|^2}{\varrho} + \varrho e(\varrho, S)\right] = \overline{E(\varrho, \boldsymbol{m}, S)} \text{ a.a. in } (0,T) \times \Omega. \quad (7.30)$$

In particular,

$$\begin{aligned}\mathfrak{E}_{cd} &= \overline{E(\varrho, \boldsymbol{m}, S)} - \left\langle \mathcal{V}; E(\widetilde{\varrho}, \widetilde{\boldsymbol{m}}, \widetilde{S})\right\rangle = 0,\\ \mathfrak{E}_{od} &= \left\langle \mathcal{V}; E(\widetilde{\varrho}, \widetilde{\boldsymbol{m}}, \widetilde{S})\right\rangle - E(\varrho, \boldsymbol{m}, S) = 0.\end{aligned} \quad (7.31)$$

Now, we can use Lemma 7.1 to conclude that for a.a. (t, x) either

$$\mathcal{V}_{t,x} = \delta_{\varrho(t,x),\boldsymbol{m}(t,x),S(t,x)} \tag{7.32}$$

or

$$\text{supp}[\mathcal{V}_{t,x}] \subset \left\{ [\widetilde{\varrho}, \widetilde{\boldsymbol{m}}, \widetilde{S}] \;\middle|\; \widetilde{\varrho} = 0,\ \widetilde{\boldsymbol{m}} = 0,\ \widetilde{S} \leq 0 \right\}. \tag{7.33}$$

However, in accordance with hypothesis (7.28) and the minimum entropy principle established in Proposition 2.1, the total entropy S satisfies $S \geq \varrho \underline{s}$, in particular

$$\varrho(t, x) = 0 \text{ implies } S(t, x) = 0 \text{ for a.a. } (t, x).$$

Consequently, $\mathcal{V}$ reduces to a Dirac mass in the case (7.33) as well. In view of Proposition 7.1, this implies pointwise a.a. convergence of a suitable subsequence of $\{\varrho_n, \boldsymbol{m}_n, S_n\}_{n=1}^{\infty}$. □

The most striking aspect of the above result is that only weak convergence of the approximate sequence to a weak solution of the Euler system is required, without any consistency hypotheses imposed on the approximating sequence. The conclusion asserts that either (i) a sequence of approximate solutions converges pointwise or (ii) the limit is not a weak solution of the Euler system. The latter statement justifies the measure-valued solutions as a legitimate object to characterize the limit of numerical schemes. It should be pointed out, however, that the fact that an approximate sequence converges weakly does not exclude the possibility for the target problem to admit a weak solution.

7.3.5 Strong Convergence of Consistent Approximation

The pointwise convergence claimed in Theorem 7.5 can be considerably improved if the approximate sequence represents a consistent approximation of the Euler system in the sense of Definition 5.2. It turns out that the crucial property is the minimum entropy principle (5.29).

Theorem 7.6 (Asymptotic limit of stable approximation with minimum entropy principle)

Let $\{\varrho_n, \boldsymbol{m}_n, S_n\}_{n=1}^{\infty}$ be an approximate sequence satisfying all the hypotheses of Theorem 7.5. In addition, suppose that

$$S_n \geq \underline{s}\varrho_n \text{ for some constant } \underline{s} \in R \text{ uniformly for } n \to \infty.$$

Then

$$\varrho_n \to \varrho \text{ in } L^q(0, T; L^\gamma(\Omega)),$$

$$\boldsymbol{m}_n \to \boldsymbol{m} \text{ in } L^q(0, T; L^{\frac{2\gamma}{\gamma+1}}(\Omega; R^d)),$$

$$S_n \to S \text{ in } L^q(0, T; L^\gamma(\Omega))$$

for any finite $1 \le q < \infty$*, passing to suitable subsequences as the case may be.*

Proof As shown in Lemma 5.2, the approximate sequence admits the following uniform bounds:

$$\varrho_n(t,x) \ge 0, \text{ and } \varrho_n(t,x) = 0 \Rightarrow \boldsymbol{m}_n(t,x) = 0 \text{ and } S_n(t,x) = 0$$
$$\text{for a.a. } (t,x) \in (0,T) \times \Omega; \tag{7.34}$$

$$\operatorname*{ess\,sup}_{\tau \in (0,T)} \|\varrho_n(\tau,\cdot)\|_{L^\gamma(\Omega)} \lesssim 1; \tag{7.35}$$

$$\operatorname*{ess\,sup}_{\tau \in (0,T)} \|S_n(\tau,\cdot)\|_{L^\gamma(\Omega)} \lesssim 1; \tag{7.36}$$

$$\operatorname*{ess\,sup}_{\tau \in (0,T)} \|\boldsymbol{m}_n(\tau,\cdot)\|_{L^{\frac{2\gamma}{\gamma+1}}(\Omega;R^d)} \lesssim 1; \tag{7.37}$$

uniformly for $n \to \infty$. Accordingly, up to a subsequence, we may assume that

$$\begin{aligned} &\varrho_n \to \varrho \text{ as } n \to \infty \text{ weakly-(*) in } L^\infty(0,T;L^\gamma(\Omega)), \\ &\boldsymbol{m}_n \to \boldsymbol{m} \text{ as } n \to \infty \text{ weakly-(*) in } L^\infty(0,T;L^{\frac{2\gamma}{\gamma+1}}(\Omega;R^d)), \\ &S_n \to S \text{ as } n \to \infty \text{ weakly-(*) in } L^\infty(0,T;L^\gamma(\Omega)), \end{aligned} \tag{7.38}$$

where

$$\varrho \ge 0,\ S \ge \underline{s}\varrho \text{ a.a. in } (0,T) \times \Omega.$$

In addition, exactly as in the proof of Theorem 7.5, we have

$$E(\varrho_n, \boldsymbol{m}_n, S_n) \equiv \left[\frac{1}{2}\frac{|\boldsymbol{m}_n|^2}{\varrho_n} + \varrho_n e(\varrho_n, S_n)\right] \to \overline{E(\varrho, \boldsymbol{m}, S)}$$
$$\text{weakly-(*) in } L^\infty(0,T;\mathcal{M}^+(\overline{\Omega})),$$

where

$$\begin{aligned} \mathfrak{E}_{cd} &= \overline{E(\varrho, \boldsymbol{m}, S)} - \langle \mathcal{V}; E(\widetilde{\varrho}, \widetilde{\boldsymbol{m}}, \widetilde{S})\rangle = 0, \\ \mathfrak{E}_{od} &= \langle \mathcal{V}; E(\widetilde{\varrho}, \widetilde{\boldsymbol{m}}, \widetilde{S})\rangle - E(\varrho, \boldsymbol{m}, S) = 0. \end{aligned}$$

By virtue of Lemma 5.1 and the pointwise convergence established in Theorem 7.5, we may infer that

$$E(\varrho_n, \boldsymbol{m}_n, S_n) \equiv \left[\frac{1}{2}\frac{|\boldsymbol{m}_n|^2}{\varrho_n} + \varrho_n e(\varrho_n, S_n)\right] \to E(\varrho, \boldsymbol{m}, S) \text{ in } L^q(0,T;L^1(\Omega)), \tag{7.39}$$

for any $1 \leq q < \infty$. Consequently, keeping in mind the pointwise convergence already shown in Theorem 7.5, it is enough to observe that the total energy dominates the norms of $\{\varrho_n, \boldsymbol{m}_n, S_n\}_{n=1}^{\infty}$ in appropriate function spaces. Evoking the estimates obtained in the proof of Lemma 5.2, we get

$$
\begin{aligned}
|\varrho_n|^{\gamma} &\lesssim 1 + E(\varrho_n, \boldsymbol{m}_n, S_n),\\
|\boldsymbol{m}_n|^{\frac{2\gamma}{\gamma+1}} &\lesssim 1 + E(\varrho_n, \boldsymbol{m}_n, S_n),\\
|S_n|^{\gamma} &\lesssim 1 + E(\varrho_n, \boldsymbol{m}_n, S_n);
\end{aligned}
$$

whence the desired conclusion follows from (7.39) and the pointwise convergence established in Theorem 7.5. □

7.3.6 *Strong Convergence of Approximate Solutions to the Complete Euler System – Summary*

We summarize the results obtained in the previous two sections. We consider a general *stable approximation*, not necessarily consistent with the limit Euler system, that enjoys only the uniform mass and energy bounds accompanied, possibly, by the minimum entropy principle.

Theorem 7.7 (Asymptotic behavior of stable approximation to complete Euler system)

Let $\Omega \subset R^d$ be a bounded domain. Let $\{\varrho_n, \boldsymbol{m}_n, S_n\}_{n=1}^{\infty}$ be a stable approximation of the complete Euler system in the sense of Definition 7.1 with the initial data $[\varrho_0, \boldsymbol{m}_0, S_0]$,

$$\varrho_0,\ \boldsymbol{m}_0,\ \ S_0 \geq \varrho_0 \underline{s},\ \underline{s} \in R.$$

More specifically,

$$
\begin{aligned}
\varrho_n \geq 0,\ \operatorname*{ess\ sup}_{\tau \in (0,T)} \int_{\Omega} \varrho_n(\tau, \cdot)\ \mathrm{d}x &\leq M,\\
\operatorname*{ess\ inf}_{\tau \in (0,T)} \int_{\Omega} S_n(\tau, \cdot)\ \mathrm{d}x &\geq \underline{S}
\end{aligned}
$$

uniformly for $n \to \infty$;

$$\operatorname*{ess\ sup}_{\tau \in (0,T)} \int_{\Omega} E(\varrho_n, \boldsymbol{m}_n, S_n)\ \mathrm{d}x \leq \int_{\Omega} E(\varrho_0, \boldsymbol{m}_0, S_0)\ \mathrm{d}x + e_n \text{ for all } n = 1, 2, \ldots$$

where $e_n \to 0$ as $n \to \infty$.

Then

$$\operatorname*{ess\,sup}_{\tau\in(0,T)} \|\varrho_n(\tau,\cdot)\|_{L^1(\Omega)} \stackrel{<}{\sim} 1,$$
$$\operatorname*{ess\,sup}_{\tau\in(0,T)} \|\boldsymbol{m}_n(\tau,\cdot)\|_{L^1(\Omega)} \stackrel{<}{\sim} 1,$$
$$\operatorname*{ess\,sup}_{\tau\in(0,T)} \|S_n(\tau,\cdot)\|_{L^1(\Omega)} \stackrel{<}{\sim} 1, \tag{7.40}$$

and the following holds true for a suitable subsequence:

- *The sequence* $\{\varrho_n, \boldsymbol{m}_n, S_n\}_{n=1}^{\infty}$ *generates a Young measure* $\{\mathcal{V}_{t,x}\}_{(t,x)\in(0,T)\times\Omega}$, *with finite first moments*

$$\varrho = \langle \mathcal{V}; \widetilde{\varrho} \rangle, \ \boldsymbol{m} = \langle \mathcal{V}; \widetilde{\boldsymbol{m}} \rangle, \ S = \left\langle \mathcal{V}; \widetilde{S} \right\rangle;$$

-

$$\left(\frac{1}{2}\frac{|\boldsymbol{m}_n|^2}{\varrho_n} + \varrho_n e(\varrho_n, S_n) \right) \to \overline{\frac{1}{2}\frac{|\boldsymbol{m}|^2}{\varrho} + \varrho e(\varrho, S)} \ \textit{weakly-(*) in } L^\infty(0,T;\mathcal{M}(\overline{\Omega})), \tag{7.41}$$

and

$$\mathfrak{E}_{cd} = \overline{\frac{1}{2}\frac{|\boldsymbol{m}|^2}{\varrho} + \varrho e(\varrho, S)} - \left\langle \mathcal{V}; \frac{1}{2}\frac{|\widetilde{\boldsymbol{m}}|^2}{\widetilde{\varrho}} + \widetilde{\varrho} e(\widetilde{\varrho}, \widetilde{S}) \right\rangle \geq 0,$$
$$\mathfrak{E}_{od} = \left\langle \mathcal{V}; E(\widetilde{\varrho}, \widetilde{\boldsymbol{m}}, \widetilde{S}) \right\rangle - \frac{1}{2}\frac{|\boldsymbol{m}|^2}{\varrho} + \varrho e(\varrho, S) \geq 0. \tag{7.42}$$

- *If the limit* $[\varrho, \boldsymbol{m}, S]$ *is an admissible weak solution to the Euler system then*

$$\mathfrak{E}_{cd} = \mathfrak{E}_{od} = 0, \ \textit{and} \ \varrho(t,x) = 0 \Rightarrow S(t,x) = 0 \ \textit{for a.a.} \ (t,x),$$

and

$$\varrho_n \to \varrho \ \textit{a.a. in} \ (0,T)\times\Omega,$$
$$\boldsymbol{m}_n \to \boldsymbol{m} \ \textit{a.a. in} \ (0,T)\times\Omega,$$
$$S_n \to S \ \textit{a.a. in} \ (0,T)\times\Omega.$$

- *If the limit* $[\varrho, \boldsymbol{m}, S]$ *is an admissible weak solution to the Euler system, and, in addition,*

$$S_n \geq \underline{s}\varrho_n \ \textit{a.a. in} \ (0,T)\times\Omega \ \textit{for} \ n\to\infty,$$

then

$$\varrho_n \to \varrho \ \textit{in} \ L^q(0,T;L^\gamma(\Omega)),$$
$$\boldsymbol{m}_n \to \boldsymbol{m} \ \textit{in} \ L^q(0,T;L^{\frac{2\gamma}{\gamma+1}}(\Omega;R^d)),$$
$$S_n \to S \ \textit{in} \ L^q(0,T;L^\gamma(\Omega))$$

for any finite $1 \leq q < \infty$.

Remark 7.3 Theorem 7.7 remains valid for the periodic boundary conditions $\Omega = \mathbb{T}^d$.

Theorem 7.7 is a conditional result. A stable approximation converges strongly as soon as its limit is a weak solution of the Euler system. In view of the weak-strong uniqueness principle discussed in detail in Chap. 6, the convergence is strong on condition that (i) the sequence $\{\varrho_n, \boldsymbol{m}_n, S_n\}_{n=1}^{\infty}$ is a *consistent* approximation, and (ii) the limit system admits a smooth solution. We state the relevant results for the space periodic boundary conditions $\Omega = \mathbb{T}^d$. As observed in Chap. 6 certain regularity of the spatial domain is needed to guarantee the weak-strong uniqueness principle. As the Lipschitz domains pertinent to numerical experiments do not enjoy the necessary smoothness, it is more convenient to work on compact domains without boundary as $\mathbb{T}^d$.

Theorem 7.8 (Convergence of consistent approximation to strong solution) *Suppose that the thermodynamic functions p, e, and s satisfy the polytropic EOS (5.22). Let* $\{\varrho_n, \boldsymbol{m}_n, S_n\}_{n=1}^{\infty}$ *be a consistent approximation of the complete Euler system in* $(0, T) \times \mathbb{T}^d$*, with the initial data* $[\varrho_0, \boldsymbol{m}_0, S_0]$ *in the sense of Definition 5.2. Suppose that the limit Euler system admits a strong solution* $[\varrho, \boldsymbol{m}, S]$ *in the class*

$$\varrho,\ S \in W^{1,\infty}((0,T)\times\mathbb{T}^d),\ \boldsymbol{m} \in W^{1,\infty}((0,T)\times\mathbb{T}^d; R^d),\ 0 < \underline{\varrho} \leq \varrho \text{ in } [0,T)\times\mathbb{T}^d.$$

Then

$$\frac{1}{2}\frac{|\boldsymbol{m}_n|^2}{\varrho_n} + \varrho_n e(\varrho_n, S_n) \to \frac{1}{2}\frac{|\boldsymbol{m}|^2}{\varrho} + \varrho e(\varrho, S) \text{ in } L^q(0,T; L^1(\mathbb{T}^d)),$$

and

$$\begin{aligned} \varrho_n &\to \varrho \text{ in } L^q(0,T; L^\gamma(\Omega)), \\ \boldsymbol{m}_n &\to \boldsymbol{m} \text{ in } L^q(0,T; L^{\frac{2\gamma}{\gamma+1}}(\Omega; R^d)), \\ S_n &\to S \text{ in } L^q(0,T; L^\gamma(\Omega)) \end{aligned}$$

for any finite $1 \leq q < \infty$.

Proof It follows from Theorem 5.1 that the sequence $\{\varrho_n, \boldsymbol{m}_n, S_n\}_{n=1}^{\infty}$ generates (up to a subsequence) a Young measure $\{\mathcal{V}_{t,x}\}_{(t,x)\in(0,T)\times\mathbb{T}^d}$ that represents a DMV solution of the complete Euler system in the sense of Definition 5.3. As observed in Sect. 5.2.2, the barycenter

$$[\varrho, \boldsymbol{m}, S](t,x) = \left\langle \mathcal{V}_{t,x}; [\widetilde{\varrho}, \widetilde{\boldsymbol{m}}, \widetilde{S}] \right\rangle$$

is a DW solution of the same system specified in Definition 5.7. Moreover, the corresponding Reynolds defect $\mathfrak{R}$ satisfies

$$0 \leq \overline{\left[\frac{1}{2} \frac{|\boldsymbol{m}|^2}{\varrho} + \varrho e(\varrho, S) \right]} - \left(\frac{1}{2} \frac{|\boldsymbol{m}|^2}{\varrho} + \varrho e(\varrho, S) \right) \stackrel{<}{\sim} \mathrm{tr}[\mathfrak{R}],$$

where

$$\left(\frac{1}{2} \frac{|\boldsymbol{m}_n|^2}{\varrho_n} + \varrho_n e(\varrho_n, S_n) \right) \to \overline{\left[\frac{1}{2} \frac{|\boldsymbol{m}|^2}{\varrho} + \varrho e(\varrho, S) \right]} \text{ weakly-(*) in } L^\infty(0, T; \mathcal{M}(\mathbb{T}^d)).$$

Finally, applying the weak-strong uniqueness principle established in Theorem 6.2 (see also Remark 6.4), we conclude that

$$\mathcal{V}_{t,x} = \delta_{[\varrho(t,x), \boldsymbol{m}(t,x), S(t,x)]}, \ \mathfrak{R} = 0,$$

and the DW solution coincides with the strong solution. The desired convergence can be shown exactly as in Theorem 7.7. As the limit is unique, there is no need to consider subsequences. □

7.3.6.1 Conclusion

The major point of this section is that the approximate solutions to the complete Euler system converge pointwise a.a. if one of the following conditions is satisfied:

- the limit is a weak solution of the Euler system;
- the Euler system admits strong solution;
- the limit is continuously differentiable.

The third alternative follows from the compatibility principle discussed in Sect. 5.4.1.

7.4 Weak Convergence of Approximate Solutions to the Euler System

We use the abstract results obtained in Sect. 7.2.3 to describe the behavior of a weakly convergent sequences of approximate solutions to the complete Euler system. To begin, we consider stable approximations introduced in Definition 7.1. These are sequences of functions $\{\varrho_n, \boldsymbol{m}_n, S_n\}_{n=1}^\infty$ enjoying the following properties:

$$\varrho_n \geq 0, \ \int_\Omega \varrho_n \,\mathrm{d}x \stackrel{<}{\sim} 1; \tag{7.43}$$

$$-\int_{\Omega} S_n \, dx \lesssim 1; \tag{7.44}$$

$$\int_{\Omega} \left[\frac{1}{2} \frac{|\boldsymbol{m}_n|^2}{\varrho_n} + \varrho_n e(\varrho_n, S_n) \right] (\tau, \cdot) \, dx \leq E_{0,n} \to E_0 \text{ as } n \to \infty \tag{7.45}$$

for a.a. $0 \leq \tau \leq T$. These are only conditions reflecting the natural *a priori* bounds discussed in Sect. 7.3. We also consider approximate sequence that satisfy the minimum entropy principle

$$S_n \geq \underline{s} \varrho_n \text{ for some } \underline{s} \in R. \tag{7.46}$$

As we have seen in Theorem 7.5, the sequence of approximate solutions must converge pointwise (strongly) as long as its limit is a weak solution of the Euler system with the initial (total) energy E_0. In this section, we focus on the situation when the convergence is not strong, meaning the limit *is not* a weak solution to the complete Euler system but rather a dissipative measure-valued (DMV) solution in the sense of Definition 5.3. In particular, our goal will be to "compute" or "visualize" the associated Young measure via the method of averaging introduced in Sect. 7.2.3. We also focus on the barycenter of the Young measure that represents a DW solution of the Euler system. Recall that a DW solution satisfies the original system of equations modulo a Reynolds defect that is compensated by the corresponding defect in the total energy equation.

7.4.1 $\mathcal{K}$–Convergence

As we know from Theorem 7.5, even the rather poor a priori or stability estimates (7.43)–(7.45) give rise to uniform bounds on the integrals

$$\operatorname*{ess\,sup}_{\tau \in (0,T)} \int_{\Omega} \varrho_n(\tau, \cdot) \, dx \lesssim 1, \ \operatorname*{ess\,sup}_{\tau \in (0,T)} \int_{\Omega} |S_n|(\tau, \cdot) \, dx \lesssim 1,$$

$$\operatorname*{ess\,sup}_{\tau \in (0,T)} \int_{\Omega} |\boldsymbol{m}_n|(\tau, \cdot) \, dx \lesssim 1.$$

In particular, we may assume that $\{\varrho_n, \boldsymbol{m}_n, S_n\}_{n=1}^{\infty}$ generates a Young measure

$$\mathcal{V}_{(t,x)} \in \mathcal{P}(R^{d+2}), \ R^{d+2} = \left\{ [\widetilde{\varrho}, \widetilde{\boldsymbol{m}}, \widetilde{S}] \in R^{d+2} \right\}.$$

In accordance with (7.45) and weak lower semicontinuity of convex functions, we have

$$\operatorname{supp}[\mathcal{V}_{t,x}] \subset \{\widetilde{\varrho} \geq 0\}. \tag{7.47}$$

Moreover,

$$\mathcal{V}_{t,x}\left\{[\widetilde{\varrho},\widetilde{\boldsymbol{m}},\widetilde{S}]\,\middle|\,\widetilde{\varrho}=0,\ \widetilde{\boldsymbol{m}}\neq 0\right\}=0,$$
$$\mathcal{V}_{t,x}\left\{[\widetilde{\varrho},\widetilde{\boldsymbol{m}},\widetilde{S}]\,\middle|\,\widetilde{\varrho}=0,\ \widetilde{S}\geq 0\right\}=0,$$

meaning

$$\mathcal{V}_{t,x}\left\{\widetilde{\boldsymbol{m}}=0,\ \widetilde{S}\leq 0\,\middle|\,\widetilde{\varrho}=0\right\}=1. \tag{7.48}$$

Adapting the first part of Theorem 7.4 we obtain the following conclusion.

Proposition 7.2 (Weak convergence, part I)
Let $\{\varrho_n,\boldsymbol{m}_n,S_n\}_{n=1}^\infty$ be a stable approximation satisfying (7.43)–(7.45).
Then there exists a subsequence $\{\varrho_{n_k},\boldsymbol{m}_{n_k},S_{n_k}\}_{k=1}^\infty$ enjoying the following properties:

- **(Young measure)** *the sequence $\{\varrho_{n_k},\boldsymbol{m}_{n_k},S_{n_k}\}_{k=1}^\infty$ generates a Young measure $\{\mathcal{V}_{t,x}\}_{(t,x)\in(0,T)\times\Omega}$,*

$$\delta_{\varrho_{n,k},\boldsymbol{m}_{n_k},S_{n_k}}\to\mathcal{V}\ \text{as}\ k\to\infty\ \text{weakly-(*) in}\ L^\infty((0,T)\times\Omega;\mathcal{M}(R^{d+2}))$$

$$\mathrm{supp}[\mathcal{V}_{t,x}]\subset\{\widetilde{\varrho}\geq 0\},$$
$$\mathcal{V}_{t,x}\left\{[\widetilde{\varrho},\widetilde{\boldsymbol{m}},\widetilde{S}]\,\middle|\,\widetilde{\varrho}=0,\ \widetilde{\boldsymbol{m}}\neq 0\right\}=0,$$
$$\mathcal{V}_{t,x}\left\{[\widetilde{\varrho},\widetilde{\boldsymbol{m}},\widetilde{S}]\,\middle|\,\widetilde{\varrho}=0,\ \widetilde{S}\geq 0\right\}=0$$

for a.a. $(t,x)\in(0,T)\times\Omega$. The measure $\mathcal{V}_{t,x}$ possesses finite first moments,

$$\varrho(t,x)=\left\langle\mathcal{V}_{t,x};\widetilde{\varrho}\right\rangle,\ \boldsymbol{m}(t,x)=\left\langle\mathcal{V}_{t,x};\widetilde{\boldsymbol{m}}\right\rangle,\ S(t,x)=\left\langle\mathcal{V}_{t,x};\widetilde{S}\right\rangle\ \text{for a.a.}\ (t,x)\times\Omega.$$

- *(* **$\mathcal{K}$–convergence)**

$$\frac{1}{N}\sum_{k=1}^N\varrho_{n_k}\to\varrho\ \text{as}\ N\to\infty\ \text{a.a. in}\ (0,T)\times\Omega,$$
$$\frac{1}{N}\sum_{k=1}^N\boldsymbol{m}_{n_k}\to\boldsymbol{m}\ \text{as}\ N\to\infty\ \text{a.a. in}\ (0,T)\times\Omega,$$
$$\frac{1}{N}\sum_{k=1}^N S_{n_k}\to S\ \text{as}\ N\to\infty\ \text{a.a. in}\ (0,T)\times\Omega. \tag{7.49}$$

- **($\mathcal{K}$–convergence of Young measures)**

$$d_{W_1}\left[\frac{1}{N}\sum_{k=1}^N\delta_{[\varrho_{n_k}(t,x);\boldsymbol{m}_{n_k}(t,x),S_{n_k}(t,x)]};\mathcal{V}_{t,x}\right]\to 0\ \text{as}\ N\to\infty \tag{7.50}$$

a.a. in $(0, T) \times \Omega$.

Remark 7.4 ($\mathcal{K}$–convergence of energy)

In view of Lemma 7.2 and the convergences stated in (7.49), one is tempted to say

$$\frac{1}{N}\sum_{k=1}^{N}\left(\frac{1}{2}\frac{|\boldsymbol{m}_{n_k}|^2}{\varrho_{n_k}} + \varrho_{n_k} e(\varrho_{n_k}, S_{n_k})\right) \to \left\langle \mathcal{V}; \frac{1}{2}\frac{|\widetilde{\boldsymbol{m}}|^2}{\widetilde{\varrho}} + \widetilde{\varrho} e(\widetilde{\varrho}, \widetilde{S})\right\rangle \text{ as } N \to \infty \tag{7.51}$$

a.a. in $(0, T) \times \Omega$. This is, however, not true, in general, as the energy

$$[\widetilde{\varrho}, \widetilde{\boldsymbol{m}}, \widetilde{S}] \mapsto \frac{1}{2}\frac{|\widetilde{\boldsymbol{m}}|^2}{\widetilde{\varrho}} + \widetilde{\varrho} e(\widetilde{\varrho}, \widetilde{S}).$$

is not a *continuous* at $\widetilde{\varrho} = 0$ but only l.s.c. The correct conclusion is

$$\frac{1}{N}\sum_{k=1}^{N}\left(\frac{1}{2}\frac{|\boldsymbol{m}_{n_k}|^2}{\varrho_{n_k}} + \varrho_{n_k} e(\varrho_{n_k}, S_{n_k})\right) \to \overline{E} \text{ as } N \to \infty \tag{7.52}$$

a.a. in $(0, T) \times \Omega$, where

$$\overline{E} \geq \left\langle \mathcal{V}; \frac{1}{2}\frac{|\widetilde{\boldsymbol{m}}|^2}{\widetilde{\varrho}} + \widetilde{\varrho} e(\widetilde{\varrho}, \widetilde{S})\right\rangle.$$

In the analysis of certain schemes, we will *assume* that the sequence of approximate densities $\{\varrho_n\}_{n=1}^{\infty}$ is bounded below away from zero. If this is the case, the convergence stated in (7.51) holds true.

7.4.2 $\mathcal{K}$–Convergence with the Minimum Entropy Principle

If, in addition, the stable approximation satisfies the minimum entropy condition (7.46), the above result can be strengthened as follows.

Proposition 7.3 (Weak convergence, part II)

Under the hypotheses of Proposition 7.2, suppose that the approximate sequence $\{\varrho_n, \boldsymbol{m}_n, S_n\}_{n=1}^{\infty}$ *satisfies the minimum entropy principle*

$$S_n \geq \underline{s}\varrho_n \ \textit{for a constant}\ \underline{s} \in R.$$

Then, in addition to the conclusion of Proposition 7.3, we have:

- **(Support of Young measure)**

$$\operatorname{supp}[\mathcal{V}_{t,x}] \subset \left\{ \widetilde{\varrho} \geq 0,\ \widetilde{S} \geq \underline{s}\widetilde{\varrho} \right\}. \tag{7.53}$$

- **(Strong convergences of Cesàro averages)**

$$\frac{1}{N}\sum_{k=1}^{N} \varrho_{n_k} \to \varrho \ \ as \ \ N \to \infty \ in \ L^q(0,T; L^\gamma(\Omega)) \ for \ any \ 1 \leq q < \infty,$$

$$\frac{1}{N}\sum_{k=1}^{N} \boldsymbol{m}_{n_k} \to \boldsymbol{m} \ \ as \ \ N \to \infty \ in \ L^q(0,T; L^{\frac{2\gamma}{\gamma+1}}(\Omega; R^d)) \ \ for \ any \ \ 1 \leq q < \infty,$$

$$\frac{1}{N}\sum_{k=1}^{N} S_{n_k} \to S \ \ as \ \ N \to \infty \ in \ L^q(0,T; L^\gamma(\Omega)) \ for \ any \ 1 \leq q < \infty. \tag{7.54}$$

- **(L^q convergence to Young measure)**

$$d_{W_r}\left[\frac{1}{N}\sum_{k=1}^{N} \delta_{[\varrho_{n_k}; \boldsymbol{m}_{n_k}, S_{n_k}]}; \mathcal{V}\right] \to 0 \ as \ N \to \infty \ \ in \ L^s((0,T) \times \Omega) \tag{7.55}$$

for $r = \frac{2\gamma}{\gamma+1}$ *and any* $1 \leq s < r$.

Finally, keeping the notation of Proposition 7.3 we denote

$$\overline{\mathcal{V}}_N \equiv \frac{1}{N}\sum_{k=1}^{N} \delta_{[\varrho_{n_k}, \boldsymbol{m}_{n_k}, S_{n_k}]} \ \text{with the barycenter} \ [\overline{\mathcal{V}}_N] \equiv \frac{1}{N}\sum_{k=1}^{N} [\varrho_{n_k}, \boldsymbol{m}_{n,k}, S_{n_k}],$$

and

$$\mathcal{V} \ \text{with the barycenter} \ [\mathcal{V}] \equiv [\varrho, \boldsymbol{m}, S],$$

where all quantities are parametrized by (t, x). Next, we introduce the approximate *deviation*

$$\left\langle \overline{\mathcal{V}}_N; \left| [\widetilde{\varrho}, \widetilde{\boldsymbol{m}}, \widetilde{S}] - [\overline{\mathcal{V}}_N] \right| \right\rangle,$$

and the final deviation

$$\left\langle \mathcal{V}; \left| [\widetilde{\varrho}, \widetilde{\boldsymbol{m}}, \widetilde{S}] - [\mathcal{V}] \right| \right\rangle.$$

Combining (7.54), (7.55) with Theorem 4, we may infer that

$$\left\langle \overline{\mathcal{V}}_N; \left| [\widetilde{\varrho}, \widetilde{\boldsymbol{m}}, \widetilde{S}] - [\overline{\mathcal{V}}_N] \right| \right\rangle \to \left\langle \mathcal{V}; \left| [\widetilde{\varrho}, \widetilde{\boldsymbol{m}}, \widetilde{S}] - [\mathcal{V}] \right| \right\rangle \ \text{as}\ N \to \infty \ \text{in}\ L^1((0,T) \times \Omega). \tag{7.56}$$

7.4.2.1 Conclusion

For a general weakly convergent approximate sequence $\{\varrho_n, \boldsymbol{m}_n, S_n\}_{n=1}^{\infty}$, the Cesàro averages convergence strongly in the same topology as the strongly convergent consistent approximations discussed in Sect. 7.3. Such a process, however, requires extracting a subsequence as the case may be.

7.5 Convergence of Approximate Solutions to the Barotropic Euler System

The barotropic Euler system introduced in Sect. 2.3 reads

$$\begin{aligned} \partial_t \varrho + \mathrm{div}_x \boldsymbol{m} &= 0, \\ \partial_t \boldsymbol{m} + \mathrm{div}_x \left(\frac{\boldsymbol{m} \otimes \boldsymbol{m}}{\varrho} \right) + \nabla_x p(\varrho) &= 0, \end{aligned} \tag{7.57}$$

supplemented with the impermeability boundary condition

$$\boldsymbol{m} \cdot \boldsymbol{n}|_{\partial\Omega} = 0, \tag{7.58}$$

or, alternatively, the space periodic boundary conditions $\Omega = \mathbb{T}^d$. Accordingly, the system (7.57), (7.58) yields the total energy balance (inequality)

$$\int_{\Omega} \left[\frac{1}{2} \frac{|\boldsymbol{m}|^2}{\varrho} + P(\varrho) \right] (\tau, \cdot)\ \mathrm{d}x \leq \int_{\Omega} \left[\frac{1}{2} \frac{|\boldsymbol{m}_0|^2}{\varrho_0} + P(\varrho_0) \right]\ \mathrm{d}x,\ P'(\varrho)\varrho - P(\varrho) = p(\varrho). \tag{7.59}$$

We recall that strict inequality in (7.59) is pertinent to the weak solutions and may be seen as a "ghost" term obtained in the vanishing viscosity limit.

At least for the pressure in the iconic isentropic form $p(\varrho) = a\varrho^{\gamma}$, $a > 0$, $\varrho > 1$, the system (7.57) can be seen as a special case of the complete Euler system with constant entropy. Accordingly, the main convergence results stated in Theorem 7.7, Propositions 7.2, 7.3 transfer with only minor modifications. Unfortunately, the strong convergence to a weak solution stated for the complete system in Theorem 7.5 cannot be proven at the same level of generality as the weak solutions to the barotropic Euler system satisfy only the energy inequality (7.59) that may accommodate possible concentration/oscillation defect.

The approximate sequence $\{\varrho_n, \boldsymbol{m}_n\}_{n=1}^{\infty}$ is supposed to be measurable satisfying the energy bounds

$$\sup_{\tau \in (0,T)} \int_{\Omega} \left[\frac{1}{2} \frac{|\boldsymbol{m}_n|^2}{\varrho_n} + P(\varrho_n) \right] (\tau, \cdot)\ \mathrm{d}x \lesssim 1. \tag{7.60}$$

Recall that this already entails

$$\varrho_n(t,x) \geq 0,\ \varrho(t,x) = 0 \Rightarrow \boldsymbol{m}(t,x) = 0 \text{ for a.a. } (t,x) \in (0,T) \times \Omega,$$

and, in view of Hölder's inequality,

$$|\boldsymbol{m}|^{\frac{2\gamma}{\gamma+1}} + \varrho^\gamma \lesssim \varrho^\gamma + \frac{|\boldsymbol{m}|^2}{\varrho} \leq \frac{1}{2}\frac{|\boldsymbol{m}|^2}{\varrho} + P(\varrho), \tag{7.61}$$

the uniform bounds

$$\operatorname*{ess\ sup}_{\tau\in(0,T)} \|\boldsymbol{m}_n(\tau,\cdot)\|_{L^{\frac{2\gamma}{\gamma+1}}(\Omega;R^d)} \lesssim 1,\ \operatorname*{ess\ sup}_{\tau\in(0,T)} \|\varrho_n(\tau,\cdot)\|_{L^\gamma(\Omega)} \lesssim 1.$$

7.5.1 Weak Convergence of Approximate Solutions to the Barotropic Euler System

Rephrasing Propositions 7.2, 7.3 in terms of the barotropic Euler system we obtain the following results.

Theorem 7.9 (Weak convergence)
Let $\Omega \subset R^d$ be a bounded domain. Suppose the pressure $p = p(\varrho)$ satisfies the isentropic EOS,

$$p(\varrho) = a\varrho^\gamma,\ a > 0,\ \gamma > 1.$$

Let $\{\varrho_n, \boldsymbol{m}_n\}_{n=1}^\infty$ be an approximate sequence satisfying the uniform energy bound (7.60).

Then, extracting a suitable subsequence as the case may be, we have the following:

- **(Young measure, defect measures)** *The sequence generates a Young measure $\{\mathcal{V}_{t,x}\}_{(t,x)\in(0,T)\times\Omega}$. Moreover,*

$$\frac{1}{2}\frac{|\boldsymbol{m}_n|^2}{\varrho_n} + P(\varrho_n) \to \overline{\frac{1}{2}\frac{|\boldsymbol{m}|^2}{\varrho} + P(\varrho)} \text{ weakly-(*) in } L^\infty(0,T;\mathcal{M}(\overline{\Omega})), \tag{7.62}$$

where

$$\mathfrak{E}_{cd} \equiv \overline{\frac{1}{2}\frac{|\boldsymbol{m}|^2}{\varrho} + P(\varrho)} - \left\langle \mathcal{V}; \frac{1}{2}\frac{|\widetilde{\boldsymbol{m}}|^2}{\widetilde{\varrho}} + P(\widetilde{\varrho}) \right\rangle \in L^\infty(0,T;\mathcal{M}^+(\overline{\Omega})), \tag{7.63}$$

and

$$\mathfrak{E}_{od} \equiv \left\langle \mathcal{V}; \frac{1}{2}\frac{|\widetilde{\boldsymbol{m}}|^2}{\widetilde{\varrho}} + P(\widetilde{\varrho}) \right\rangle - \left[\frac{1}{2}\frac{|\boldsymbol{m}|^2}{\varrho} + P(\varrho)\right] \geq 0 \text{ a.a. in } (0,T)\times\Omega. \tag{7.64}$$

- **(Weak and $\mathcal{K}$–convergence)**

$$
\begin{aligned}
&\varrho_n \to \varrho \textit{ weakly-(*) as } n \to \infty \textit{ in } L^\infty(0,T;L^\gamma(\Omega)),\\
&\frac{1}{N}\sum_{n=1}^N \varrho_n \to \varrho \quad \textit{as } N \to \infty \textit{ in } L^q(0,T;L^\gamma(\Omega)),\ 1 \le q < \infty,\\
&\boldsymbol{m}_n \to \boldsymbol{m} \textit{ weakly-(*) as } n \to \infty \textit{ in } L^\infty(0,T;L^{\frac{2\gamma}{\gamma+1}}(\Omega;R^d)),\\
&\frac{1}{N}\sum_{n=1}^N \boldsymbol{m}_n \to \boldsymbol{m} \quad \textit{as } N \to \infty \textit{ in } L^q(0,T;L^{\frac{2\gamma}{\gamma+1}}(\Omega;R^d)),\ 1 \le q < \infty,\\
&\frac{1}{N}\sum_{n=1}^N \left(\frac{1}{2}\frac{|\boldsymbol{m}_n|^2}{\varrho_n} + P(\varrho_n)\right) \to \overline{E} \textit{ as } N \to \infty \textit{ a.a. in } (0,T)\times\Omega,
\end{aligned}
\tag{7.65}
$$

where

$$
\overline{E} \ge \left\langle \mathcal{V}; \frac{1}{2}\frac{|\widetilde{\boldsymbol{m}}|^2}{\widetilde{\varrho}} + P(\widetilde{\varrho})\right\rangle,
$$

and

$$
d_{W_r}\left[\frac{1}{N}\sum_{k=1}^N \delta_{[\varrho_{n_k}(t,x);\boldsymbol{m}_{n_k}(t,x)]};\mathcal{V}_{t,x}\right] \to 0 \textit{ as } N \to \infty \textit{ in } L^s((0,T)\times\Omega) \tag{7.66}
$$

for $r = \frac{2\gamma}{\gamma+1}$ *and any* $1 \le s < r$.

7.5.2 Strong Convergence of Approximate Solutions to the Barotropic Euler System

Our goal is to obtain an analogue of Theorem 7.5 in the context of the barotropic Euler equations. Specifically, the convergence of an approximate sequence is strong (a.a. pointwise) provided the limit object is a weak solution. The answer turns out to be different than for the complete Euler system, where this property is enforced by the fact that the limit system includes the energy conservation principle. We claim the following result.

Theorem 7.10 *Let* $\Omega \subset R^d$ *be a bounded domain. Suppose the pressure* $p = p(\varrho)$ *satisfies the isentropic EOS,*

$$
p(\varrho) = a\varrho^\gamma,\ a > 0,\ \gamma > 1.
$$

Let $\{\varrho_n, \boldsymbol{m}_n\}_{n=1}^\infty$ *be a consistent approximation of the barotropic Euler system in the sense of Definition 5.4. Suppose that the limit* $[\varrho, \boldsymbol{m}]$ *identified in Theorem 7.9 is a* *weak solution of the Euler system (7.57), (7.58). In addition, suppose there is an open neighborhood $\mathcal{U}$ of the boundary $\partial\Omega$ such that*

$$\int_0^T \int_{\mathcal{U}} \left[\frac{1}{2} \frac{|\boldsymbol{m}_n|^2}{\varrho_n} + P(\varrho_n) \right] \mathrm{d}x\,\mathrm{d}t \to \int_0^T \int_{\mathcal{U}} \left[\frac{1}{2} \frac{|\boldsymbol{m}|^2}{\varrho} + P(\varrho) \right] \mathrm{d}x\,\mathrm{d}t. \tag{7.67}$$

Then

$$\int_0^T \int_{\Omega} \left[\frac{1}{2} \frac{|\boldsymbol{m}_n|^2}{\varrho_n} + P(\varrho_n) \right] \mathrm{d}x\,\mathrm{d}t \to \int_0^T \int_{\Omega} \left[\frac{1}{2} \frac{|\boldsymbol{m}|^2}{\varrho} + P(\varrho) \right] \mathrm{d}x\,\mathrm{d}t, \tag{7.68}$$

in particular

$$\varrho_n \to \varrho \text{ in } L^q(0,T; L^\gamma(\Omega)),\ \boldsymbol{m}_n \to \boldsymbol{m} \text{ in } L^q(0,T; L^{\frac{2\gamma}{\gamma+1}}(\Omega; R^d)) \tag{7.69}$$

for any $1 \le q < \infty$.

Remark 7.5 In contrast with the complete Euler system considered in Theorem 7.5, the approximate sequence must be a consistent approximation, and the strong convergence of the total energy in a neighborhood of the boundary is required. On the other hand, strong convergence of the energy at the initial time is not necessary.

Proof It is enough to verify that

$$\mathfrak{E}_{cd} = \mathfrak{E}_{od} = 0. \tag{7.70}$$

Indeed it follows from Lemma 5.1, Proposition 5.3, and the estimate (7.61) that $\mathfrak{E}_{cd} = 0$ implies

$$\varrho_n^\gamma \to \varrho^\gamma \text{ weakly in } L^q(0,T; L^1(\Omega)),\ |\boldsymbol{m}_n|^{\frac{2\gamma}{\gamma+1}} \to |\boldsymbol{m}|^{\frac{2\gamma}{\gamma+1}} \tag{7.71}$$

weakly in $L^q(0,T; L^1(\Omega; R^d))$ for any $1 \le q < \infty$. Moreover, in view of Lemma 7.1, $\mathfrak{E}_{od} = 0$ yields

$$\mathcal{V}_{t,x} = \delta_{\varrho(t,x),\boldsymbol{m}(t,x)} \text{ for a.a. } (t,x) \in (0,T) \times \Omega. \tag{7.72}$$

Relations (7.71), (7.72) give rise to the desired conclusion.

To see (7.70), we evoke Theorems 5.3, 5.4 concerning the asymptotic limit of a consistent approximation. In particular, we recover the limit of the momentum equation

$$\left[\int_{\Omega} \boldsymbol{m} \cdot \boldsymbol{\varphi} \, \mathrm{d}x\right]_{t=0}^{t=\tau}$$
$$= \int_0^\tau \int_{\Omega} \left[\boldsymbol{m} \cdot \partial_t \boldsymbol{\varphi} + 1_{\varrho>0} \frac{\boldsymbol{m} \otimes \boldsymbol{m}}{\varrho} : \nabla_x \boldsymbol{\varphi} + p(\varrho) \mathrm{div}_x \boldsymbol{\varphi}\right] \mathrm{d}x \tag{7.73}$$
$$+ \int_0^\tau \int_{\overline{\Omega}} \nabla_x \boldsymbol{\varphi} : \mathrm{d}\mathfrak{R}(t) \, \mathrm{d}t$$

for any $0 \leq \tau \leq T$, and any test function $\boldsymbol{\varphi} \in C^1([0,T] \times \overline{\Omega}; R^d)$, $\boldsymbol{\varphi} \cdot \boldsymbol{n}|_{\partial\Omega} = 0$, with the Reynolds defect measure

$$\mathfrak{R} \in L^\infty(0,T; \mathcal{M}^+(\overline{\Omega}; R^{d\times d}_{\mathrm{sym}})),\ \underline{d}(\mathfrak{E}_{cd} + \mathfrak{E}_{od}) \leq \mathrm{tr}[\mathfrak{R}] \leq \overline{d}(\mathfrak{E}_{cd} + \mathfrak{E}_{od})$$

$\underline{d} = \min\{2, \gamma - 1\}$, $\overline{d} = \max\{2, \gamma - 1\}$. Consequently, (7.70) follows as soon as we can show $\mathfrak{R} = 0$.

It follows from hypothesis (7.67) that

$$\mathfrak{R}|_{(0,T)\times\mathcal{U}} = 0. \tag{7.74}$$

Moreover, as $[\varrho, \boldsymbol{m}]$ is a weak solution of the Euler system, we deduce from (7.73) that

$$\int_0^\tau \left(\int_{\overline{\Omega}} \nabla_x \boldsymbol{\varphi} : \mathrm{d}\mathfrak{R}\right) \mathrm{d}t = 0 \text{ for all } \boldsymbol{\varphi} \in C^1([0,T] \times \overline{\Omega}; R^d),\ \boldsymbol{\varphi} \cdot \boldsymbol{n}|_{\partial\Omega} = 0.$$

As $\mathfrak{R} \in L^\infty(0,T; \mathcal{M}^+(\overline{\Omega}; R^{d\times d}_{\mathrm{sym}}))$, it follows that

$$\int_{\overline{\Omega}} \nabla_x \boldsymbol{\varphi} : \mathrm{d}\mathfrak{R}(\tau) = 0 \text{ for any } \boldsymbol{\varphi} \in C^1(\overline{\Omega}; R^d),\ \boldsymbol{\varphi} \cdot \boldsymbol{n}|_{\partial\Omega} = 0 \text{ and a.a. } \tau \in (0,T). \tag{7.75}$$

Our ultimate goal is to show that (7.74), (7.75) imply $\mathfrak{R} = 0$ for any $\mathfrak{R} \in \mathcal{M}^+(R^{d\times d}_{\mathrm{sym}})$. To this end, we first observe that (7.75) remains valid for any $\boldsymbol{\varphi} \in C^1(\overline{\Omega}; R^d)$, meaning without the restriction $\boldsymbol{\varphi} \cdot \boldsymbol{n}|_{\partial\Omega} = 0$. To see this, write

$$\boldsymbol{\varphi} = \boldsymbol{\varphi}^1 + \boldsymbol{\varphi}^2,\ \boldsymbol{\varphi}^1 \in C^1_c(\Omega; R^d),\ \mathrm{supp}[\boldsymbol{\varphi}^2] \subset \mathcal{U}.$$

Using (7.74), (7.75) we obtain

$$\int_{\overline{\Omega}} \nabla_x \boldsymbol{\varphi} : \mathrm{d}\mathfrak{R} = \int_{\overline{\Omega}} \nabla_x \boldsymbol{\varphi}^1 : \mathrm{d}\mathfrak{R} + \int_{\mathcal{U}} \nabla_x \boldsymbol{\varphi}^2 : \mathrm{d}\mathfrak{R} = 0.$$

Consequently, we may infer that

$$\begin{aligned} &\int_{\overline{\Omega}} \nabla_x \boldsymbol{\varphi} : \mathrm{d}\mathfrak{R}(\tau) = 0 \ \text{ for any } \ \boldsymbol{\varphi} \in C^1(\overline{\Omega}; R^d), \\ &\mathfrak{R}(\tau) \in \mathcal{M}^+(\overline{\Omega}; R^{d\times d}_{\rm sym}) \ \text{ for a.a. } \ \tau \in (0, T). \end{aligned} \tag{7.76}$$

Finally, we choose

$$\boldsymbol{\varphi} = (\xi \otimes \xi) \cdot x, \ \xi \in R^d$$

obtaining

$$\int_{\overline{\Omega}} (\xi \otimes \xi) : \mathrm{d}\mathfrak{R}(\tau) = 0 \ \Rightarrow \ (\xi \otimes \xi) : \mathfrak{R}(\tau) = 0 \text{ for any } \xi \in R^d \ \Rightarrow \mathfrak{R}(\tau) = 0.$$

□

7.5.3 *Strong Convergence to Strong Solutions*

Assuming the limit Euler system admits a strong solution, we may show unconditional convergence of consistent approximation. The following result can be seen as an analogue of Theorem 7.8

Theorem 7.11 (Strong convergence to strong solution)
Let the pressure p be given,

$$p(\varrho) = a\varrho^\gamma, \ a > 0, \ \gamma > 1.$$

Let $\{\varrho_n, \boldsymbol{m}_n\}_{n=1}^\infty$ *be a consistent approximation of the barotropic Euler system in* $(0, T) \times \Omega$, *with the initial data* $[\varrho_0, \boldsymbol{m}_0]$ *in the sense of Definition 5.4. Suppose that the Euler system admits a strong solution* $[\varrho, \boldsymbol{m}]$ *in* $(0, T) \times \Omega$ *in the class*

$$\varrho \in W^{1,\infty}((0,T) \times \Omega), \ 0 < \underline{\varrho} \le \varrho \ in \ [0, T) \times \Omega, \ \boldsymbol{m} \in W^{1,\infty}((0,T) \times \Omega; R^d).$$

Then

$$\frac{1}{2}\frac{|\boldsymbol{m}_n|^2}{\varrho_n} + P(\varrho_n) \to \frac{1}{2}\frac{|\boldsymbol{m}|^2}{\varrho} + P(\varrho) \ in \ L^q(0, T; L^1(\Omega)),$$

and

$$\varrho_n \to \varrho \ in \ L^q(0,T;L^\gamma(\Omega)),\ \boldsymbol{m}_n \to \boldsymbol{m} \ in \ L^q(0,T;L^{\frac{2\gamma}{\gamma+1}}(\Omega;R^d))$$

for any $1 \leq q < \infty$.

Remark 7.6 The same result holds for the periodic boundary conditions $\Omega = \mathbb{T}^d$.

7.6 Convergence of Approximate Solutions to the Navier–Stokes System

We consider the Navier–Stokes system

$$\begin{aligned}
&\partial_t \varrho + \mathrm{div}_x(\varrho \boldsymbol{u}) = 0, \quad \varrho(0,\cdot) = \varrho_0, \\
&\partial_t(\varrho \boldsymbol{u}) + \mathrm{div}_x(\varrho \boldsymbol{u} \otimes \boldsymbol{u}) + \nabla_x p(\varrho) = \mathrm{div}_x \mathbb{S}, \quad \varrho \boldsymbol{u}(0,\cdot) = \boldsymbol{m}_0 \\
&\mathbb{S} = \mu \left(\nabla_x \boldsymbol{u} + \nabla_x^t \boldsymbol{u} - \frac{2}{d} \mathrm{div}_x \boldsymbol{u} \mathbb{I} \right) + \lambda \mathrm{div}_x \boldsymbol{u} \mathbb{I}.
\end{aligned} \tag{7.77}$$

For the sake of simplicity, we restrict ourselves to the periodic boundary conditions $\Omega = \mathbb{T}^d$.

Approximate solutions to the Navier–Stokes system usually enjoy slightly better regularity because of the presence of viscosity terms. Introducing $\boldsymbol{m} = \varrho \boldsymbol{u}$, the total energy is the same as for the Euler system,

$$E(\varrho, \boldsymbol{m}) = \frac{1}{2} \frac{|\boldsymbol{m}|^2}{\varrho} + P(\varrho),$$

while the energy inequality reads

$$\begin{aligned}
&\int_{\mathbb{T}^d} \left[\frac{1}{2} \frac{|\boldsymbol{m}|^2}{\varrho} + P(\varrho) \right] (\tau,\cdot) \,\mathrm{d}x + \int_0^\tau \int_{\mathbb{T}^d} \mathbb{S} : \nabla_x \boldsymbol{u} \,\mathrm{d}x \,\mathrm{d}t \\
&\leq \int_{\mathbb{T}^d} \left[\frac{1}{2} \frac{|\boldsymbol{m}_0|^2}{\varrho_0} + P(\varrho_0) \right] \mathrm{d}x.
\end{aligned}$$

Moreover, as we observed in Sect. 6.3, it is convenient to use the Fenchel–Young inequality and rewrite the energy inequality in the form

$$\begin{aligned}
\int_{\mathbb{T}^d} \left[\frac{1}{2} \frac{|\boldsymbol{m}|^2}{\varrho} + P(\varrho) \right] (\tau,\cdot) \,\mathrm{d}x + \int_0^\tau \int_{\mathbb{T}^d} \left[F(\mathbb{D}_x \boldsymbol{u}) + F^*(\mathbb{S}) \right] \mathrm{d}x \,\mathrm{d}t \\
\leq \int_{\mathbb{T}^d} \left[\frac{1}{2} \frac{|\boldsymbol{m}_0|^2}{\varrho_0} + P(\varrho_0) \right] \mathrm{d}x
\end{aligned}$$

for the convex potential F and its conjugate F^* specified in (5.102)–(5.104).

7.6.1 Momentum-Velocity Splitting

As the energy functional is the same as for the barotropic Euler system, Theorem 7.9 remains valid also for the Navier–Stokes system. We focus on the possibility of splitting the Young measure as in (5.127), meaning finding suitable conditions for the following relation to hold:

$$\begin{aligned}
&\varrho_n \to \varrho \text{ weakly-(*) in } L^\infty(0,T;L^\gamma(\mathbb{T}^d)),\\
&\boldsymbol{u}_n \to \boldsymbol{u} \text{ weakly in } L^2((0,T)\times\mathbb{T}^d;R^d),\\
&\varrho_n\boldsymbol{u}_n \to \boldsymbol{m} \text{ weakly-(*) in } L^\infty(0,T;L^{\frac{2\gamma}{\gamma+1}}(\mathbb{T}^d;R^d))\\
&\Rightarrow \boldsymbol{m} = \varrho\boldsymbol{u}.
\end{aligned} \tag{7.78}$$

The problem can be attacked by means of the following result.

Lemma 7.4 *Let $Q = (0,T)\times\Omega$, where $\Omega \subset R^d$ is a bounded domain. Suppose that*

$$r_n \to r \text{ weakly in } L^p(Q),\ v_n \to v \text{ weakly in } L^q(Q),\ p>1, q>1,$$

and

$$r_n v_n \to w \text{ weakly in } L^r(Q),\ r>1.$$

In addition, let

$$\begin{aligned}
&\partial_t r_n = \mathrm{div}_x \boldsymbol{g}_n + h_n \text{ in } \mathcal{D}'(Q),\ \|\boldsymbol{g}_n\|_{L^s(Q;R^d)} \stackrel{<}{\sim} 1,\ s>1,\\
&h_n \text{ precompact in } W^{-1,z},\ z>1,
\end{aligned}$$

and

$$\|\nabla_x v_n\|_{\mathcal{M}(Q;R^d)} \stackrel{<}{\sim} 1 \text{ uniformly for } n\to\infty.$$

Then

$$w = rv \text{ a.a. in } Q.$$

Proof First, we introduce a cut-off function

$$\begin{aligned}
&T_k(v) = kT\left(\frac{v}{k}\right),\ T \in C^\infty \cap BC(R),\\
&T(Z) = T(-Z),\ T(Z) = Z \text{ if } |Z| \le 1,\ 0 \le T'(Z) \le 1.
\end{aligned}$$

Next, write

$$v_n = T_k(v_n) + \Big(v_n - T_k(v)\Big),$$

and

$$r_n v_n = r_n T_k(v_n) + r_n\Big(v_n - T_k(v)\Big).$$

Passing to a subsequence (not relabeled) we may assume

$$T_k(v_n) \to \overline{T_k(v)} \text{ weakly-(*) in } L^\infty(Q),\ r_n T_k(v_n) \to w_k \text{ weakly in } L^r(Q) \text{ as } n \to \infty.$$

We claim that it is enough to show

$$w_k = r\overline{T_k(v)} \text{ a.a. in } Q \text{ for any } k \to \infty.$$

Indeed we have

$$\int_Q |v_n - T_k(v_n)| \,\mathrm{d}x\,\mathrm{d}t \le \int_{|v_n \ge k|} |v_n| \,\mathrm{d}x\,\mathrm{d}t$$
$$\le |\{v_n \ge k\}|^{\frac{1}{q'}} \|v_n\|_{L^q(Q)} \to 0 \text{ as } k \to \infty$$

uniformly for $n \to \infty$, and

$$\left\| v - \overline{T_k(v)} \right\|_{L^1(Q)} \le \liminf_{n\to\infty} \|v_n - T_k(v_n)\|_{L^1(Q)} \to 0 \text{ as } k \to \infty.$$

Note that this implies

$$w_k = r\overline{T_k(v)} \to rv \text{ in } L^1(Q)$$

by means of the Lebesgue dominance theorem, as

$$\|w_k\|_{L^r(Q)} \le \|w\|_{L^r(Q)} \text{ uniformly for } k \to \infty.$$

Similarly

$$\int_Q |r_n(v_n - T_k(v_n))| \ \mathrm{d}x\,\mathrm{d}t \le \int_{|v_n \ge k|} |r_n v_n| \,\mathrm{d}x\,\mathrm{d}t$$
$$\le |\{v_n \ge k\}|^{\frac{1}{r'}} \|r_n v_n\|_{L^r(Q)} \to 0 \text{ as } k \to \infty \text{ uniformly in } n.$$

Since

$$\|\nabla_x v_n\|_{\mathcal{M}(Q;R^d)} \stackrel{<}{\sim} 1 \Rightarrow \|\nabla_x T_k(v_n)\|_{\mathcal{M}(Q;R^d)} \stackrel{<}{\sim} 1 \text{ uniformly for } n \to \infty,$$

it is enough to show the conclusion under the assumption

$$v_n \to v \text{ weakly-(*) in } L^\infty(Q).$$

To this end, we apply Div–Curl lemma to the vector fields

$$\begin{aligned} &\boldsymbol{U}_n = [r_n, -\boldsymbol{g}_n] : Q \to R^{d+1}, \ \mathrm{div}_{t,x}\boldsymbol{U}_n = \partial_t r_n + \mathrm{div}_x \boldsymbol{g}_n = h_n, \\ &\boldsymbol{U}_n \to \boldsymbol{U} = [r, \boldsymbol{g}] \text{ weakly in } L^{\min\{s,p\}}(Q), \end{aligned}$$

and

$$\boldsymbol{V}_n = [v_n, 0] : Q \to R^{d+1}, \ \mathrm{curl}_{t,x}\boldsymbol{V}_n \approx \nabla_x v_n \text{ bounded in } \mathcal{M}(Q; R^{d\times d}).$$

Applying Div–Curl Lemma we obtain the desired conclusion. □

7.6.2 Strong Convergence to Strong Solutions

Similarly to the Euler system, a consistent approximation to the Navier–Stokes system converges strongly as soon as the limit problem admits a strong solution.

Theorem 7.12 (Convergence to strong solution)
Let $\{\varrho_n, \boldsymbol{u}_n\}_{n=1}^{\infty}$ be a consistent approximation of the Navier–Stokes system in $(0,T)\times\mathbb{T}^d$, with the initial data $[\varrho_0, \boldsymbol{m}_0 = \varrho_0\boldsymbol{u}_0]$ in the sense of Definition 5.9. Suppose that the Navier–Stokes system admits a strong solution $[\varrho, \boldsymbol{u}]$ in the class

$$\varrho \in C^1([0,T]\times\mathbb{T}^d; R^d), \ 0 < \underline{\varrho} \le \varrho, \ \boldsymbol{u} \in C^1([0,T]\times\mathbb{T}^d; R^d) \cap L^2(0,T; W^{2,\infty}(\mathbb{T}^d; R^d)).$$

Then

$$\varrho_n \to \varrho \text{ in } L^q(0,T; L^\gamma(\Omega)), \ \varrho_n\boldsymbol{u}_n \to \varrho\boldsymbol{u} \text{ in } L^q(0,T; L^{\frac{2\gamma}{\gamma+1}}(\Omega; R^d))$$

for any $1 \le q < \infty$, and

$$\boldsymbol{u}_n \to \boldsymbol{u} \text{ in } L^q((0,T)\times\mathbb{T}^d; R^d) \text{ for any } 1 \le q < 2 \text{ and weakly in } L^2((0,T)\times\mathbb{T}^d; R^d).$$

Proof By virtue of Theorem 5.5, the sequence $\{\varrho_n, \boldsymbol{u}_n\}_{n=1}^{\infty}$ generates a Young measure $\{\mathcal{V}_{t,x}\}_{(t,x)\in(0,T)\times\Omega}$ that represents a DMV solution of the Navier–Stokes system in the sense of Definition 5.10. More precisely,

$$\begin{aligned} \varrho_n &\to \varrho \text{ weakly-(*) in } L^\infty(0,T; L^\gamma(\mathbb{T}^d)), \\ \boldsymbol{m}_n = (\varrho_n\boldsymbol{u}_n) &\to \boldsymbol{m} \text{ weakly-(*) in } L^\infty(0,T; L^{\frac{2\gamma}{\gamma+1}}(\mathbb{T}^d; R^d)), \\ \boldsymbol{u}_n &\to \boldsymbol{u} \text{ weakly in } L^2((0,T)\times\mathbb{T}^d; R^d), \ \boldsymbol{u} \in L^2(0,T; W^{1,2}(\mathbb{T}^d; R^d)), \end{aligned}$$

$$[\varrho, \boldsymbol{u}] = \langle \mathcal{V}; [\widetilde{\varrho}, \widetilde{\boldsymbol{u}}] \rangle .$$

Moreover, the Reynolds defect is

$$\mathfrak{R} = \overline{\varrho \boldsymbol{u} \otimes \boldsymbol{u} + p(\varrho)\mathbb{I}} - \langle \mathcal{V}; \widetilde{\varrho}\widetilde{\boldsymbol{u}} \otimes \widetilde{\boldsymbol{u}} + p(\widetilde{\varrho})\mathbb{I} \rangle \in L^\infty(0, T; \mathcal{M}^+(\mathbb{T}^d; R^{d\times d}_{\rm sym})),$$

where

$$\varrho_n \boldsymbol{u}_n \otimes \boldsymbol{u}_n + p(\varrho_n)\mathbb{I} \to \overline{\varrho \boldsymbol{u} \otimes \boldsymbol{u} + p(\varrho)\mathbb{I}} \text{ weakly-(*) in } L^\infty(0, T; \mathcal{M}(\mathbb{T}^d; R^{d\times d}_{\rm sym})).$$

By virtue of the weak-strong uniqueness principle established in Theorem 6.4, we have

$$\mathfrak{R} = 0,$$

and

$$\mathcal{V}_{t,x} = \delta_{\varrho(t,x),\boldsymbol{u}(t,x)},$$

which yields the desired strong convergence. □

7.6.3 *Convergence of Bounded Consistent Approximations*

Probably the most convincing argument demonstrating the synergy effect between mathematical analysis and numerical experiments is the following result on convergence of bounded approximate solutions to the Navier–Stokes system. It turns out that uniformly bounded consistent approximation *always* converges strongly to a strong solution as soon as the initial data are sufficiently smooth.

Theorem 7.13 (Convergence of bounded approximate solutions)
Let $\{\varrho_n, \boldsymbol{u}_n\}_{n=1}^\infty$ *be a consistent approximation of the Navier–Stokes system in* $(0, T) \times \mathbb{T}^d$ *in the sense of Definition 5.9, with the initial data* $[\varrho_0, \boldsymbol{m}_0 = \varrho_0 \boldsymbol{u}_0]$,

$$\varrho_0 \in W^{3,2}(\mathbb{T}^d),\ \boldsymbol{m}_0 = \varrho_0 \boldsymbol{u}_0,\ \boldsymbol{u}_0 \in W^{3,2}(\mathbb{T}^d, R^d),\ \inf_{x\in\mathbb{T}^d} \varrho_0(x) > 0.$$

- *Suppose that the bulk viscosity* $\lambda = 0$ *and that*

$$0 \le \varrho_n \le \overline{\varrho} \text{ for all } n \to \infty.$$

Then

$$\varrho_n \to \varrho \text{ in } L^q((0, T) \times \mathbb{T}^d),\ \varrho_n \boldsymbol{u}_n \to \varrho \boldsymbol{u} \text{ in } L^q(0, T; L^{\frac{2\gamma}{\gamma+1}}(\mathbb{T}^d; R^d))$$

for any $1 \le q < \infty$, *and*

$$\boldsymbol{u}_n \to \boldsymbol{u} \text{ in } L^q((0, T) \times \mathbb{T}^d; R^d) \text{ for any } 1 \le q < 2$$
$$\text{and weakly in } L^2((0, T) \times \mathbb{T}^d; R^d),$$

where $[\varrho, \mathbf{u}]$ *is a strong solution of the Navier–Stokes system.*

- *Suppose that*

$$0 \leq \varrho_n \leq \overline{\varrho}, \quad |\mathbf{u}_n| \leq \overline{u} \text{ for all } n \to \infty.$$

Then

$$\varrho_n \to \varrho \text{ in } L^q((0,T) \times \mathbb{T}^d), \ \mathbf{u}_n \to \mathbf{u} \text{ in } L^q((0,T) \times \Omega; R^d))$$

for any $1 \leq q < \infty$*, where* $[\varrho, \mathbf{u}]$ *is a strong solution of the Navier–Stokes system.*

Proof The proof follows the same line of arguments as in Theorem 7.12. At the final step, use the conditional regularity for DMV solutions stated in Corollary 6.4. □

We point out that the existence of the strong solutions to the Navier–Stokes system is not *a priori* required in Theorem 7.13. A bounded consistent approximation converges always strongly and unconditionally to a smooth solution of the Navier–Stokes system! There is no counterpart of this result in the framework of the Euler (inviscid) fluid models.

Chapter 8
Numerical Methods

The principal objective of this monograph is to study *numerical methods* for models of compressible fluid flow. By a numerical method we mean a finite system of algebraic equations yielding approximate solutions to the target system of partial differential equations. The models are *evolutionary*—the state of the system depends on time.

We exclusively use the *implicit time discretization*. The resulting system has a structure closer to the target system so that rigorous convergence analysis may be performed. A nonlinear system that arise at each time step can be solved by a suitable iterative solver, such as the fixed point or Newton method. In some cases, we omit the process of time discretization and study semidiscrete schemes, where only spatially dependent operators are approximated. The resulting problem then represents a system of (nonlinear) ODEs to be attacked by purely analytical methods. From the analysis point of view, the semidiscrete schemes share the principal difficulties with the fully discrete implicit schemes avoiding the unnecessary technical difficulties connected with the time discretization.

The space discretization will be mostly performed via *finite volume approximation*. A finite volume (FV) method is very well suited for partial differential equations in the divergence form. The discretization is based on applying the Gauss–Green formula on each mesh element and applying a suitable approximation of the fluxes through the interfaces of the element. Consequently, the method is automatically in conservative form. Numerical solutions are piecewise discontinuous polynomials. This is to be compared with a finite element (FE) method, which is based on the variational formulation of the continuous problem and the Galerkin approximation. Numerical solutions are piecewise polynomial functions that satisfy some type of continuity along the element interfaces. This yields a higher connectivity in the resulting discrete system than by the finite volume methods.

Due to their flexibility and automatic conservativity property, the finite volume methods have been widely used as a basis of the CFD (Computational Fluid Dynamics) packages, in particular for the Euler and Navier–Stokes systems studied in this monograph. We focus on discretization via piecewise constant functions that yields

E. Feireisl et al., *Numerical Analysis of Compressible Fluid Flows*, MS&A 20,
https://doi.org/10.1007/978-3-030-73788-7_8

the first order approximation, but generalizations to higher order methods both in space and time are possible. We will also present examples where the basic finite volume strategy will be combined with the finite element approximation of diffusive parts of the Navier–Stokes system, see Chap. 13 or reformulated as the staggered finite difference method, see Chap. 14. Except Chap. 9, where, as an example, some standard finite volume methods for hyperbolic conservation laws will be analyzed for the barotropic Euler equation, we will apply a particular numerical flux function for the approximation of the convective terms.

The approximation process necessarily affects the boundary of the computational domain. While analytical statements are mostly stated for "sufficiently smooth" domains $\Omega \subset R^d$, the computational domain Ω_h is a polyhedral approximation of Ω. Accordingly, the computational domains are typically Lipschitz or, in the best case, convex approximation of the physical domain Ω. Such a low level of regularity does not guarantee the existence of regular (smooth) solutions of the target problem necessary to perform rigorous convergence analysis. Ignoring this problem is usually referred to as committing "variational crimes". The problem can be avoided or solved by considering the space periodic boundary conditions, meaning

$$\Omega = \mathbb{T}^d,$$

or approximating a smooth domain Ω by a family $\{\Omega_h\}_{h \searrow 0}$, respectively.

The finite volume methods yield an approximation of the *weak formulation* of the underlying system of PDEs in the corresponding discrete space. Clearly, the finite volume methods share many common properties with the PDEs. Still certain physically relevant principles may not be encoded directly in the approximation and must be either verified a posteriori or imposed as extra constraints. The prominent examples are:

- **Strict positivity of the mass density**. The mass density by its physical meaning is a nonnegative quantity. The models of viscous fluids as the Navier–Stokes system operating with velocity rather than momentum have been derived for *nondilute* fluids far from vacuum. Accordingly, the mass density should be strictly positive. To certain extent, the same stipulation may be applied to the Euler system as well. Note that we deal almost exclusively with the *weak solutions*, where the possibility of developing vacuum in finite time is an open problem (Navier–Stokes system) or not excluded (Euler system). Mainly because of this reason, we do not expect to keep the density bounded from below *uniformly* through the approximation process. Positivity of the density at a fixed level of approximation is, however, desirable for various physical quantities to be well defined and in many cases enhances stability of the method. Moreover, strong positivity of the mass density facilitates the passage from one type of state variables to another.
- **Strict positivity of the (absolute) temperature**. This is an indispensable property for several thermodynamic functions, notable the entropy to be well defined. Note that positivity of the absolute temperature is enforced by the Second law of thermodynamics. In the literature, positivity of the temperature is often replaced by

positivity of the pressure. Physically speaking, this is not the same but equivalence obviously holds for the EOS considered in this monograph. Note that the pressure vanishes on the vacuum together with the density while the (physical) meaning of the temperature in this area might be dubious.

- **Total energy balance**. The validity of the First law of thermodynamics, notably the control of the total energy is *absolutely necessary* for the analysis of all numerical methods in this monograph. Note that uniform boundedness of the energy at any level of numerical approximation is practically the only source of stability estimates indispensable for the method to converge at least in a weak sense. For barotropic systems the energy is identified with the "mathematical entropy" in the literature. In this context some authors call an energy stable scheme as entropy stable. This is the case of the barotropic Euler system studied in the forthcoming chapter.
- **Entropy balance and positivity of the entropy production rate**. The Second law of thermodynamics is encoded in the models involving thermal changes as the entropy balance. Although the latter may not be directly used in the *formulation* of a numerical method, its satisfaction is crucial for convergence. Indeed the convective terms in the energy equation are difficult to control by the available stability estimates and it is the entropy and the *total* energy balance that is used for showing convergence. At the level of the Euler system, this limit procedure gives rise to the dissipative solutions introduced and discussed in Sect. 5.2 of Chap. 5.

As we shall see below, incorporating the above properties in the numerical approximation is not only physically relevant but provides a valuable piece of information for the convergence analysis. In the context of the Euler system, the numerical methods satisfying the above properties are called *invariant domain preserving* methods.

8.1 Discrete Versus Continuous Time Representation

As agreed in Preliminary material, a numerical approximation of a field $\boldsymbol{U}$ will be denoted $\boldsymbol{U}_h^k$, where k refers to the time level and h to the spatial resolution. Specifically, given a time step $\Delta t > 0$ and the initial time t_0 (typically $t_0 = 0$), we denote

$$t_k = t_0 + k\Delta t,\ k = 1, 2, \dots .$$

Time derivative $\frac{\partial \boldsymbol{U}}{\partial t}$ is approximated by the backward Euler finite difference

$$\frac{\partial \boldsymbol{U}}{\partial t} \approx D_t \boldsymbol{U}_h^k \equiv \frac{\boldsymbol{U}_h^k - \boldsymbol{U}_h^{k-1}}{\Delta t};$$

while the associated evolutionary problem

$$\frac{\partial \boldsymbol{U}}{\partial t} = \boldsymbol{F}(\boldsymbol{U}, t)$$

is replaced by the *implicit time discretization*

$$\frac{\boldsymbol{U}_h^k - \boldsymbol{U}_h^{k-1}}{\Delta t} = \boldsymbol{F}_h(\boldsymbol{U}_h^k, t_k).$$

That being said, the numerical method is reduced to a (finite) system of nonlinear algebraic equations,

$$\boldsymbol{U}_h^k - \Delta t \boldsymbol{F}_h(\boldsymbol{U}_h^k, t_k) = \boldsymbol{U}_h^{k-1}, \ k = 1, 2, \ldots, N_T. \tag{8.1}$$

that can be solved recursively at any time level t_k, with $\boldsymbol{U}_h^0$ being a suitable discrete approximation of the initial data. Note that the specific form of the left hand side of equation (8.1) is suitable for a direct application of a fixed point theorem.

When using a *semidiscrete scheme* we consider a system of ordinary differential equations

$$\frac{\mathrm{d}}{\mathrm{d}t}\boldsymbol{U}_h(t) = \boldsymbol{F}_h(\boldsymbol{U}_h(t), t).$$

In practical applications a suitable ODE solver can be applied afterward.

To illustrate the proximity of semi and fully discrete schemes, we recall the definition of piecewise constant interpolation in time, cf. (47)

$$\boldsymbol{U}_h(t) = \sum_{k \geq 0} \boldsymbol{U}_h^k 1_{[t_k, t_{k+1})}(t), \ t \geq 0.$$

This can be seen as a function of time ranging in a suitable finite-dimensional space given by the type of space discretization. Let $\psi \in C_c^2[0, T)$, where, for the sake of simplicity, we suppose $T = t_{N_T}$ for some $N_T \geq 1$. Accordingly, following analogous arguments as in Lemma 8 we obtain

$$\begin{aligned}
\int_0^T \boldsymbol{U}_h \partial_t \psi \ \mathrm{d}t &= \sum_{k=0}^{N_T} \boldsymbol{U}_h^k \int_{t_k}^{t_{k+1}} \partial_t \psi \ \mathrm{d}t = \sum_{k=0}^{N_T} \boldsymbol{U}_h^k \Big(\psi(t_{k+1}) - \psi(t_k) \Big) \\
&= -\boldsymbol{U}_h^0 \psi(0) - \sum_{k=1}^{N_T} \psi(t_k) \Big(\boldsymbol{U}_h^k - \boldsymbol{U}_h^{k-1} \Big) \\
&= -\boldsymbol{U}_h^0 \psi(0) - \sum_{k=0}^{N_T} \int_{t_k}^{t_{k+1}} \psi(t_k) D_t \boldsymbol{U}_h \ \mathrm{d}t = -\boldsymbol{U}_h^0 \psi(0) \\
&- \int_0^T \psi D_t \boldsymbol{U}_h \ \mathrm{d}t + \sum_{k=1}^{N_T} \int_{t_k}^{t_{k+1}} \Big(\psi - \psi(t_k) \Big) D_t \boldsymbol{U}_h \ \mathrm{d}t.
\end{aligned} \tag{8.2}$$

As ψ is twice continuously differentiable, we may use the Taylor expansion writing the consistency error in the form

$$\sum_{k=1}^{N_T} \int_{t_k}^{t_{k+1}} \Big(\psi(t) - \psi(t_k)\Big) D_t \boldsymbol{U}_h \, \mathrm{d}t = e_1 + e_2,$$

where

$$e_1 \equiv \sum_{k=1}^{N_T} \int_{t_k}^{t_{k+1}} \Big(\psi(t) - \psi(t_k) - \partial_t \psi(t_k)(t - t_k)\Big) D_t \boldsymbol{U}_h \, \mathrm{d}t$$

$$e_2 \equiv \sum_{k=1}^{N_T} \int_{t_k}^{t_{k+1}} \partial_t \psi(t_k)(t - t_k) D_t \boldsymbol{U}_h \, \mathrm{d}t = \frac{\Delta t}{2} \sum_{k=1}^{N_T} \partial_t \psi(t_k) \Big(\boldsymbol{U}_h^k - \boldsymbol{U}_h^{k-1}\Big)$$

$$= -\frac{\Delta t}{2} \left(\boldsymbol{U}_h^0 \partial_t \psi(0) - \int_0^T \boldsymbol{U}_h \partial_t^2 \psi \, \mathrm{d}t \right).$$

Going back to (8.2) we may infer that

$$\int_0^T \boldsymbol{U}_h \partial_t \psi \, \mathrm{d}t = -\boldsymbol{U}_h^0 \psi(0) - \int_0^T \psi D_t \boldsymbol{U}_h \, \mathrm{d}t + e(\Delta t, T, \boldsymbol{U}_h, \psi), \tag{8.3}$$

with the consistency error

$$|e| \leq \Delta t \; c\left(\|\psi\|_{C^2[0,T]}\right) \left(|\boldsymbol{U}_h^0| + \int_0^T |\boldsymbol{U}_h| \, \mathrm{d}t \right). \tag{8.4}$$

Thus the distributional derivative $\frac{\mathrm{d}}{\mathrm{d}t}\boldsymbol{U}_h$ is well approximated by $D_t \boldsymbol{U}_h$ provided the right-hand side of (8.4) is controlled. For that we need very mild stability estimates, namely the uniform integrability of the approximate solutions $\boldsymbol{U}_h$ in $L^1(0, T)$. Note that such a bound is *necessary* to perform any kind of convergence analysis even in the framework of weak solutions.

In the light of the above arguments, the *analysis* of the semidiscrete and fully discrete methods concentrates on the spatial discretization and replacing D_t by $\frac{\mathrm{d}}{\mathrm{d}t}$ and vice versa does not create any extra difficulties.

8.2 Diffusive Upwind Numerical Flux

The Euler and Navier–Stokes systems considered in this monograph can be written in a general form

$$\partial_t \boldsymbol{U} + \mathrm{div}_x \mathbb{F}(\boldsymbol{U}) = \mathrm{div}_x \mathbb{T}(\boldsymbol{U}),$$

where $\mathbb{F}$ is the convective flux and $\mathbb{T}$ represents the diffusive stress.

Let us discuss shortly the discretization of the convective flux $\mathbb{F}(\boldsymbol{U})$ present in both the viscous and the inviscid model. Assume that the computational domain Ω is discretized by a regular mesh $\mathcal{T}_h$, and the approximate solutions $\boldsymbol{U}_h^k$ are piecewise constant (on any volume cell $K \subset \Omega$) at any time level k. Denoting $\boldsymbol{U}_K^k$ the value on the cell K, we may approximate

$$\partial_t \boldsymbol{U} + \mathrm{div}_x(\mathbb{F}(\boldsymbol{U})) \approx D_t \boldsymbol{U}_K^k + (\mathrm{div}_{\mathcal{D}} \mathbb{F}_h(\boldsymbol{U}_h^k))_K \ \text{ on } \ K \in \mathcal{T}_h.$$

In the finite volume methods the discrete divergence operator $\mathrm{div}_{\mathcal{D}}$, cf. (9.3), requires that the numerical flux $\mathbb{F}_h$ is evaluated on any face $\sigma \in \partial K$ using the values on the neighboring elements. A natural choice would be

$$\mathbb{F}_h(\boldsymbol{U}_h^k)\Big|_\sigma = \{\{\mathbb{F}(\boldsymbol{U}_h^k)\}\},$$

which is locally consistent but may create problems with stability estimates and consequently with the global consistency errors. As a result, a suitable stabilization term is needed. Accordingly, a general flux approximation

$$\boldsymbol{F}_\sigma(\boldsymbol{U}_h^k) = \{\{\mathbb{F}(\boldsymbol{U}_h^k)\}\}\,\boldsymbol{n} - \lambda\,[\![\boldsymbol{U}_h^k]\!], \ \text{ where } \mathbb{F}_h \boldsymbol{n} \equiv \boldsymbol{F}_\sigma \tag{8.5}$$

is commonly used, with a positive factor λ depending, in general, on $\boldsymbol{U}_h$. The term

$$\lambda\,[\![\boldsymbol{U}_h^k]\!]$$

represents *numerical diffusion*. Specific examples of numerical fluxes will be discussed in Chap. 9.

Formula (8.5) is simple, elegant, but not always convenient when the properties of transported quantities are analyzed. Indeed the crucial quantity for the transport is the fluid velocity $\boldsymbol{u}$. Although $\boldsymbol{u}$ is a natural state variable for the Navier–Stokes system, it is usually replaced by the (conservative) quantity – the momentum $\boldsymbol{m}$ – in the context of the Euler system. This leads to a rather awkward formulation of the equation of continuity and renders the problem of positivity of the density quite delicate. In Chap. 10, we propose a new discretization of the Euler system based on Brenner's idea discussed in Chap. 3, Sect. 3.2.2. The leading idea is to keep the velocity as a state variable at the discrete level. As a result, the convective term is discretized in the same way for both the Euler and Navier–Stokes system by means of the upwind discretization we shortly describe below.

First we realize that all convective terms take the form $\mathrm{div}_x(r\boldsymbol{u})$, where r stays for a transported quantity and $\boldsymbol{u}$ is the flow velocity. Integrating the convective term over an arbitrary element K and applying the Gauss–Green theorem we easily identify the leading term in the flux discretization $\int_\sigma r\boldsymbol{u}\cdot\boldsymbol{n}\,\mathrm{dS}_x$ for any $\sigma\in\mathcal{E}(K)$. A suitable stable approximation is obtained by following the flow along streamlines. This leads to the following upwind discretization of the convective term:

Given a velocity field $\boldsymbol{u}_h\in\boldsymbol{Q}_h$ or $\boldsymbol{V}_h$, the *upwind flux* on a face $\sigma\in\mathcal{E}$ is defined as

$$\begin{aligned} Up[r_h,\boldsymbol{u}_h] &= r_h^{\mathrm{in}}[\langle\boldsymbol{u}_h\rangle_\sigma\cdot\boldsymbol{n}]^+ + r_h^{\mathrm{out}}[\langle\boldsymbol{u}_h\rangle_\sigma\cdot\boldsymbol{n}]^- \\ &= \{\!\{r_h\}\!\}\ \langle\boldsymbol{u}_h\rangle_\sigma\cdot\boldsymbol{n} - \frac{1}{2}|\langle\boldsymbol{u}_h\rangle_\sigma\cdot\boldsymbol{n}|\,[\![r_h]\!], \end{aligned} \tag{8.6}$$

where we have denoted

$$[f]^{\pm} = \frac{f\pm|f|}{2}.$$

Recall that f^{in} and f^{out} are the inward and outward traces of a piecewise smooth function f on a face σ defined in (8). Introducing the upwind information of a function $r_h\in Q_h$ on a face $\sigma\in\mathcal{E}$ by

$$r_h^{\mathrm{up}} = \begin{cases} r_h^{\mathrm{in}} & \text{if } \langle\boldsymbol{u}_h\rangle_\sigma\cdot\boldsymbol{n}\geq 0, \\ r_h^{\mathrm{out}} & \text{if } \langle\boldsymbol{u}_h\rangle_\sigma\cdot\boldsymbol{n} < 0, \end{cases} \tag{8.7}$$

the upwind flux (8.6) can be rewritten equivalently as

$$Up[r_h,\boldsymbol{u}_h] = r_h^{\mathrm{up}}\ \langle\boldsymbol{u}_h\rangle_\sigma\cdot\boldsymbol{n}.$$

The resulting numerical flux may be augmented by additional numerical diffusion, which yields the following *diffusive upwind numerical flux function*

$$F_h^{\mathrm{up}}[r_h,\boldsymbol{u}_h] = Up[r_h,\boldsymbol{u}_h] - h^\varepsilon\,[\![r_h]\!], \qquad \varepsilon > -1. \tag{8.8}$$

Indeed the additional numerical diffusion term introduced above acts as an artificial diffusion of order $\mathcal{O}(h^{1+\varepsilon})$ since

$$\sum_{\sigma\in\mathcal{E}(K)}\frac{|\sigma|}{|K|}h^\varepsilon\,[\![r_h]\!] = h^{1+\varepsilon}(\Delta_h r_h)_K,$$

where Δ_h stays for the approximation of the Laplace operator with central differences. Compared with the general formula (8.5), the upwind flux retains the information concerning the fluid velocity. As we shall see below, this fact enables us to adopt the continuous methods for transport equations to obtain *unconditional* density positivity as well as the renormalized entropy equation yielding the minimum entropy principle at the discrete level.

Further, we define the upwind-type divergence operators

$$\mathrm{div}_h^{\mathrm{up}}(r_h, \boldsymbol{u}_h) = \sum_{K \in \mathcal{T}_h} 1_K \mathrm{div}_h^{\mathrm{up}}(r_h, \boldsymbol{u}_h)_K, \ \mathrm{div}_h^{\mathrm{up}}(r_h, \boldsymbol{u}_h)_K = \sum_{\sigma \in \mathcal{E}(K)} \frac{|\sigma|}{|K|} F_h^{\mathrm{up}}[r_h, \boldsymbol{u}_h]. \tag{8.9}$$

Note that the formulas make sense if the domain $\Omega = \mathbb{T}^d$ is periodic, or for the upwind flux (8.6) defined on a bounded domain with the impermeability condition

$$\langle \boldsymbol{u}_h \rangle_\sigma \cdot \boldsymbol{n} = 0 \text{ if } \sigma \subset \partial\Omega. \tag{8.10}$$

Moreover, we set

$$[\![r_h]\!]_\sigma = 0 \ \text{ if } \ \sigma \subset \partial\Omega, \tag{8.11}$$

which corresponds to the homogeneous Neumann boundary condition for r_h. The impermeability condition (8.10) together with the homogeneous Neumann boundary condition (8.11) are sometimes called in the literature *no-flux boundary conditions*.

Keeping the above convention in mind and realizing that $\boldsymbol{n}_{\sigma,K} = -\boldsymbol{n}_{\sigma,L}$ for all $\sigma = K|L \in \mathcal{E}$, we deduce the conservative property

$$\sum_{K \in \mathcal{T}_h} \sum_{\sigma \in \mathcal{E}(K)} F_h^{\mathrm{up}}[r_h, \boldsymbol{u}_h] = 0 \quad \text{and} \quad \int_\Omega \mathrm{div}_h^{\mathrm{up}}(r_h \boldsymbol{u}_h) \, \mathrm{d}x = 0 \tag{8.12}$$

for $r_h \in Q_h$ and $\boldsymbol{u}_h \in \boldsymbol{V}_{0,h}$ or $\boldsymbol{u}_h \in \boldsymbol{Q}_h \cup \boldsymbol{W}_h$.

The following technical result will be suitable in many future calculations.

Lemma 8.1 (Diffusive upwind numerical flux)

Let $r_h \in Q_h$, $\boldsymbol{u}_h \in \boldsymbol{Q}_h$ and $\phi \in C^1(\overline{\Omega})$. Let the periodic or the no-flux boundary conditions be imposed.

Then the following holds:

$$\begin{aligned} &\int_{\mathcal{E}_{int}} \left(Up[r_h, \boldsymbol{u}_h] \left[\!\!\left[\frac{|\boldsymbol{u}_h|^2}{2} \right]\!\!\right] - Up[r_h \boldsymbol{u}_h, \boldsymbol{u}_h] \cdot [\![\boldsymbol{u}_h]\!] \right) \mathrm{dS}_x \\ &= \frac{1}{2} \int_{\mathcal{E}_{int}} \varrho_h^{\mathrm{up}} \, |\langle \boldsymbol{u}_h \rangle_\sigma \cdot \boldsymbol{n}| \, [\![\boldsymbol{u}_h]\!]^2 \, \mathrm{dS}_x, \end{aligned} \tag{8.13a}$$

and

$$\begin{aligned} &\int_{\mathcal{E}_{int}} \left(F_h^{\mathrm{up}}[r_h, \boldsymbol{u}_h] \left[\!\!\left[\frac{|\boldsymbol{u}_h|^2}{2} \right]\!\!\right] - \boldsymbol{F}_h^{\mathrm{up}}[r_h \boldsymbol{u}_h, \boldsymbol{u}_h] \cdot [\![\boldsymbol{u}_h]\!] \right) \mathrm{dS}_x \\ &= \frac{1}{2} \int_{\mathcal{E}_{int}} \varrho_h^{\mathrm{up}} \, |\langle \boldsymbol{u}_h \rangle_\sigma \cdot \boldsymbol{n}| \, [\![\boldsymbol{u}_h]\!]^2 \, \mathrm{dS}_x + h^\varepsilon \int_{\mathcal{E}_{int}} \{\!\{ r_h \}\!\} \, |[\![\boldsymbol{u}_h]\!]|^2 \, \mathrm{dS}_x. \end{aligned} \tag{8.13b}$$

Moreover,

$$\int_{\Omega} r_h \boldsymbol{u}_h \cdot \nabla_x \phi \, \mathrm{d}x - \int_{\mathcal{E}_{int}} F_h^{\mathrm{up}}[r_h, \boldsymbol{u}_h] \, \llbracket \Pi_Q \phi \rrbracket \, \mathrm{dS}_x$$
$$= \int_{\mathcal{E}_{int}} \left(\frac{1}{2} | \langle \boldsymbol{u}_h \rangle_\sigma \cdot \boldsymbol{n}| + h^\varepsilon + \frac{1}{4} \llbracket \boldsymbol{u}_h \rrbracket \cdot \boldsymbol{n} \right) \llbracket r_h \rrbracket \, \llbracket \Pi_Q \phi \rrbracket \, \mathrm{dS}_x \tag{8.13c}$$
$$+ \int_{\Omega} r_h \boldsymbol{u}_h \cdot \left(\nabla_x \phi - \nabla_h (\Pi_Q \phi) \right) \, \mathrm{d}x.$$

Proof First, we recall the definition of the upwind flux (8.6) and notice $\langle \boldsymbol{u}_h \rangle_\sigma = \frac{\boldsymbol{u}_h^{\mathrm{in}} + \boldsymbol{u}_h^{\mathrm{out}}}{2}$ to derive (8.13b), i.e.

$$\int_{\mathcal{E}_{int}} Up[r_h, \boldsymbol{u}_h] \left\llbracket \frac{|\boldsymbol{u}_h|^2}{2} \right\rrbracket \mathrm{dS}_x - \int_{\mathcal{E}_{int}} Up[r_h \boldsymbol{u}_h, \boldsymbol{u}_h] \cdot \llbracket \boldsymbol{u}_h \rrbracket \, \mathrm{dS}_x$$
$$= \int_{\mathcal{E}_{int}} r_h^{\mathrm{up}} \, \langle \boldsymbol{u}_h \rangle_\sigma \cdot \boldsymbol{n} (\langle \boldsymbol{u}_h \rangle_\sigma - \boldsymbol{u}_h^{\mathrm{up}}) \cdot \llbracket \boldsymbol{u}_h \rrbracket \, \mathrm{dS}_x$$
$$= \int_{\mathcal{E}_{int}} r_h^{\mathrm{in}} [\langle \boldsymbol{u}_h \rangle_\sigma \cdot \boldsymbol{n}]^+ (\langle \boldsymbol{u}_h \rangle_\sigma - \boldsymbol{u}_h^{\mathrm{in}}) \cdot \llbracket \boldsymbol{u}_h \rrbracket \, \mathrm{dS}_x$$
$$+ \int_{\mathcal{E}_{int}} r_h^{\mathrm{out}} [\langle \boldsymbol{u}_h \rangle_\sigma \cdot \boldsymbol{n}]^- (\langle \boldsymbol{u}_h \rangle_\sigma - \boldsymbol{u}_h^{\mathrm{out}}) \cdot \llbracket \boldsymbol{u}_h \rrbracket \, \mathrm{dS}_x$$
$$= \frac{1}{2} \int_{\mathcal{E}_{int}} \left(r_h^{\mathrm{in}} [\langle \boldsymbol{u}_h \rangle_\sigma \cdot \boldsymbol{n}]^+ - r_h^{\mathrm{out}} [\langle \boldsymbol{u}_h \rangle_\sigma \cdot \boldsymbol{n}]^- \right) \llbracket \boldsymbol{u}_h \rrbracket^2 \, \mathrm{dS}_x$$
$$= \frac{1}{2} \int_{\mathcal{E}_{int}} r_h^{\mathrm{up}} \, |\langle \boldsymbol{u}_h \rangle_\sigma \cdot \boldsymbol{n}| \, \llbracket \boldsymbol{u}_h \rrbracket^2 \, \mathrm{dS}_x.$$

Next we recall the definition of the diffusive upwind flux (8.8) and equality (8.24a) to calculate

$$
\begin{aligned}
&\int_{\mathcal{E}_{int}} F_h^{\mathrm{up}}[r_h, \boldsymbol{u}_h] \left[\!\!\left[\frac{|\boldsymbol{u}_h|^2}{2} \right]\!\!\right] \mathrm{dS}_x - \int_{\mathcal{E}_{int}} F_h^{\mathrm{up}}[r_h \boldsymbol{u}_h, \boldsymbol{u}_h] \cdot [\![\boldsymbol{u}_h]\!] \, \mathrm{dS}_x \\
&= \int_{\mathcal{E}_{int}} \left(Up[r_h, \boldsymbol{u}_h] - h^{\varepsilon} [\![r_h]\!] \right) \left[\!\!\left[\frac{|\boldsymbol{u}_h|^2}{2} \right]\!\!\right] \mathrm{dS}_x \\
&- \int_{\mathcal{E}_{int}} \left(Up[r_h \boldsymbol{u}_h, \boldsymbol{u}_h] - h^{\varepsilon} [\![r_h \boldsymbol{u}_h]\!] \right) \cdot [\![\boldsymbol{u}_h]\!] \, \mathrm{dS}_x \\
&= \frac{1}{2} \int_{\mathcal{E}_{int}} r_h^{\mathrm{up}} \, |\langle \boldsymbol{u}_h \rangle_{\sigma} \cdot \boldsymbol{n}| \, [\![\boldsymbol{u}_h]\!]^2 \, \mathrm{dS}_x \\
&- h^{\varepsilon} \int_{\mathcal{E}_{int}} \left([\![r_h]\!] \left[\!\!\left[\frac{|\boldsymbol{u}_h|^2}{2} \right]\!\!\right] - [\![r_h \boldsymbol{u}_h]\!] \cdot [\![\boldsymbol{u}_h]\!] \right) \mathrm{dS}_x \\
&= \frac{1}{2} \int_{\mathcal{E}_{int}} r_h^{\mathrm{up}} \, |\langle \boldsymbol{u}_h \rangle_{\sigma} \cdot \boldsymbol{n}| \, [\![\boldsymbol{u}_h]\!]^2 \, \mathrm{dS}_x + h^{\varepsilon} \int_{\mathcal{E}_{int}} \{\!\{r_h\}\!\} \, [\![\boldsymbol{u}_h]\!]^2 \, \mathrm{dS}_x,
\end{aligned}
$$

which proves (8.13b). Further, we proceed to show (8.13c). Clearly,

$$
\begin{aligned}
\int_{\Omega} r_h \boldsymbol{u}_h \cdot \nabla_x \phi \, \mathrm{d}x &= \sum_{K \in \mathcal{T}_h} \int_K r_h \boldsymbol{u}_h \cdot \nabla_x \phi \, \mathrm{d}x \\
&= \sum_{K \in \mathcal{T}_h} \int_K r_h \boldsymbol{u}_h \cdot \left(\nabla_x \phi - \nabla_h \left(\Pi_Q \phi \right) \right) \mathrm{d}x + I_1,
\end{aligned}
$$

where the term I_1 can be reformulated as

$$
\begin{aligned}
I_1 &= \sum_{K \in \mathcal{T}_h} \int_K r_h \boldsymbol{u}_h \cdot \nabla_h \left(\Pi_Q \phi \right) \mathrm{d}x = \sum_{K \in \mathcal{T}_h} (r_h \boldsymbol{u}_h)_K \cdot \int_{\partial K} \boldsymbol{n} \, \{\!\{\Pi_Q \phi\}\!\} \, \mathrm{dS}_x \\
&= - \int_{\mathcal{E}_{int}} [\![r_h \boldsymbol{u}_h]\!] \cdot \boldsymbol{n} \, \{\!\{\Pi_Q \phi\}\!\} \, \mathrm{dS}_x = \int_{\mathcal{E}_{int}} \{\!\{r_h \boldsymbol{u}_h\}\!\} \cdot \boldsymbol{n} \, [\![\Pi_Q \phi]\!] \, \mathrm{dS}_x \\
&= \int_{\mathcal{E}_{int}} \left(\{\!\{r_h \boldsymbol{u}_h\}\!\} - \{\!\{r_h\}\!\} \, \{\!\{\boldsymbol{u}_h\}\!\} \right) \cdot \boldsymbol{n} \, [\![\Pi_Q \phi]\!] \, \mathrm{dS}_x \\
&+ \int_{\mathcal{E}_{int}} \{\!\{r_h\}\!\} \, \{\!\{\boldsymbol{u}_h\}\!\} \cdot \boldsymbol{n} \, [\![\Pi_Q \phi]\!] \, \mathrm{dS}_x \\
&= \int_{\mathcal{E}_{int}} \frac{1}{4} [\![r_h]\!] \, [\![\boldsymbol{u}_h]\!] \cdot \boldsymbol{n} \, [\![\Pi_Q \phi]\!] \, \mathrm{dS}_x + \int_{\mathcal{E}_{int}} F_h^{\mathrm{up}}[r_h, \boldsymbol{u}_h] \, [\![\Pi_Q \phi]\!] \, \mathrm{dS}_x \\
&+ \int_{\mathcal{E}_{int}} \left(\frac{1}{2} | \{\!\{\boldsymbol{u}_h\}\!\} \cdot \boldsymbol{n}| + h^{\varepsilon} \right) [\![r_h]\!] \, [\![\Pi_Q \phi]\!] \, \mathrm{dS}_x,
\end{aligned}
$$

thanks to (23), (39) and the definition of the numerical flux F_h^{up} in (8.8). Combining the above two steps concludes the proof.

Note that the upwind numerical flux (8.8) approximates the convective term in a scalar-like manner, meaning component by component, and depends only on the sign of the normal component of velocity. This is in contrast with some standard numerical flux functions used for systems of hyperbolic conservation laws, where the eigenvalues of the corresponding Jacobi matrices of the flux vector are considered, cf. (8.5) and Chap. 9 below.

8.3 Discrete Continuity Equation

The aim of this section is to demonstrate some fundamental properties of the discrete density obtained as a solution of a finite volume method based on the use of the numerical flux (8.8). In particular, we will show the conservation of mass and positivity of the density. To this end we start by a precise formulation of the numerical method for the continuity equation.

Let $\varrho_h^0 \in Q_h$ and $\boldsymbol{u}_h^k \in \boldsymbol{Q}_h$ or $\boldsymbol{V}_h$, $k = 1, 2, \ldots, N_T$, be given. In addition, if Ω is a domain with boundary, we suppose that the no-flux boundary conditions are applied, i.e. $\boldsymbol{u}_h^k \cdot \boldsymbol{n}|_{\partial\Omega} = 0$ and $[\![\varrho_h^k]\!]_\sigma = 0$, $\sigma \in \mathcal{E}_{ext}$. The numerical approximation of the density $\varrho_h^k \in Q_h$, $k = 1, 2, \ldots, N_T$, is computed by the following finite volume update

$$D_t \varrho_h^k\Big|_K + \sum_{\sigma \in \mathcal{E}(K) \cap \mathcal{E}_{int}} \frac{|\sigma|}{|K|} F_h^{\mathrm{up}}[\varrho_h^k, \boldsymbol{u}_h^k] = 0 \tag{8.14}$$

or equivalently,

$$\int_\Omega D_t \varrho_h^k \phi_h \,\mathrm{d}x - \int_{\mathcal{E}_{int}} F_h^{\mathrm{up}}[\varrho_h^k, \boldsymbol{u}_h^k] [\![\phi_h]\!] \,\mathrm{dS}_x = 0 \quad \text{for all } \phi_h \in Q_h. \tag{8.15}$$

8.3.1 Mass Conservation

Setting $\phi_h \equiv 1$ in (8.15), we obtain

$$\int_\Omega D_t \varrho_h^k \,\mathrm{d}x = 0$$

for all $k = 1, \dots, N_T$, which implies the total mass conservation

$$M_0 \equiv \int_\Omega \varrho_h^0 \, dx = \int_\Omega \varrho_h^k \, dx, \ k = 1, \dots, N_T. \tag{8.16}$$

8.3.2 Renormalized Continuity Equation and Positivity of Density

First, we derive a discrete version of the renormalized continuity equation introduced in Chap. 3, formula (3.34).

Lemma 8.2 (Renormalized continuity equation)
Let $\varrho_h^k \in Q_h$ be a solution to the discrete problem (8.14)*, $k = 1, \dots, N_T$, $\boldsymbol{u}_h^k \in \boldsymbol{Q}_h \cup \boldsymbol{V}_h$. We suppose that either the periodic or the no-flux boundary conditions hold.*

Then for any function $B \in C^1(R)$ and any $\phi_h \in Q_h$ it holds:

$$\begin{aligned}
&\int_\Omega D_t B(\varrho_h^k)\phi_h \, dx - \int_{\mathcal{E}_{int}} F_h^{\mathrm{up}}[B(\varrho_h^k), \boldsymbol{u}_h^k] \, [\![\phi_h]\!] \, dS_x \\
&+ \int_\Omega \phi_h \left(\varrho_h^k B'(\varrho_h^k) - B(\varrho_h^k)\right) \mathrm{div}_h \boldsymbol{u}_h^k \, dx \\
&= -\frac{1}{\Delta t} \int_\Omega \phi_h \left(B(\varrho_h^{k-1}) - B(\varrho_h^k) - B'(\varrho_h^k)(\varrho_h^{k-1} - \varrho_h^k)\right) \, dx, \\
&- \sum_{K \in \mathcal{T}_h} \int_K \phi_h \sum_{\sigma \in \mathcal{E}(K) \cap \mathcal{E}_{int}} \frac{|\sigma|}{|K|} \Big([\![B(\varrho_h^k)]\!] - B'(\varrho_h^k) [\![\varrho_h^k]\!] \Big)(h^\varepsilon - [\langle \boldsymbol{u}_h^k \rangle_\sigma \cdot \boldsymbol{n}]^-) \, dx.
\end{aligned} \tag{8.17}$$

Proof We use $\phi_h \approx B'(\varrho_h^k)\phi_h$ as a test function in the discrete continuity equation (8.15) obtaining

$$\int_\Omega D_t \varrho_h^k B'(\varrho_h^k)\phi_h \, dx + \int_\Omega \mathrm{div}_h^{\mathrm{up}} \left(\varrho_h^k \boldsymbol{u}_h^k\right) B'(\varrho_h^k)\phi_h \, dx = 0.$$

First, it is easy to check

$$\begin{aligned}
\int_\Omega D_t \varrho_h^k B'(\varrho_h^k)\phi_h \, dx &= \int_\Omega \frac{\varrho_h^k - \varrho_h^{k-1}}{\Delta t} B'(\varrho_h^k)\phi_h \, dx = \int_\Omega D_t B(\varrho_h^k)\phi_h \, dx \\
&+ \frac{1}{\Delta t} \int_\Omega \phi_h \left(B(\varrho_h^{k-1}) - B(\varrho_h^k) - B'(\varrho_h^k)(\varrho_h^{k-1} - \varrho_h^k)\right) \, dx.
\end{aligned}$$

Next, recalling the definition of the upwind flux (8.6) and denoting $u_\sigma = \left\langle \boldsymbol{u}_h^k \right\rangle_\sigma \cdot \boldsymbol{n}$, we may rewrite the convective term as

$$
\begin{aligned}
&\int_\Omega \operatorname{div}_h^{\mathrm{up}}(\varrho_h^k, \boldsymbol{u}_h^k) B'(\varrho_h^k)\phi_h \,\mathrm{d}x \\
&= \sum_{K\in\mathcal{T}_h} \int_K \phi_h B'(\varrho_h^k) \sum_{\sigma\in\mathcal{E}(K)\cap\mathcal{E}_{int}} \frac{|\sigma|}{|K|} \left(\varrho_h^{k,\mathrm{up}} u_\sigma - h^\varepsilon \left[\!\left[\varrho_h^k \right]\!\right] \right) \mathrm{d}x \\
&= \int_\Omega \phi_h \varrho_h^k B'(\varrho_h^k) \operatorname{div}_h \boldsymbol{u}_h^k \,\mathrm{d}x + \sum_{K\in\mathcal{T}_h} \int_K \phi_h B'(\varrho_h^k) \sum_{\sigma\in\mathcal{E}(K)} \frac{|\sigma|}{|K|} (\varrho_h^{k,\mathrm{up}} - \varrho_h^k) u_\sigma \mathrm{d}x \\
&\quad - h^\varepsilon \sum_{K\in\mathcal{T}_h} \int_K \phi_h B'(\varrho_h^k) \sum_{\sigma\in\mathcal{E}(K)\cap\mathcal{E}_{int}} \frac{|\sigma|}{|K|} \left[\!\left[\varrho_h^k \right]\!\right] \mathrm{d}x \\
&= \int_\Omega \phi_h \left(\varrho_h^k B'(\varrho_h^k) - B(\varrho_h^k) \right) \operatorname{div}_h \boldsymbol{u}_h^k \,\mathrm{d}x - \int_{\mathcal{E}_{int}} F_h^{\mathrm{up}}[B(\varrho_h^k), \boldsymbol{u}_h^k] \left[\!\left[\phi_h \right]\!\right] \mathrm{dS}_x \\
&\quad + \sum_{K\in\mathcal{T}_h} \int_K \phi_h B'(\varrho_h^k) \sum_{\sigma\in\mathcal{E}\cap\mathcal{E}_{int}} \frac{|\sigma|}{|K|} \left[\!\left[\varrho_h^k \right]\!\right] \left([u_\sigma]^- - h^\varepsilon \right) \mathrm{d}x + D,
\end{aligned}
$$

where

$$
D = \int_\Omega \phi_h B(\varrho_h^k) \operatorname{div}_h \boldsymbol{u}_h^k \,\mathrm{d}x + \int_{\mathcal{E}_{int}} F_h^{\mathrm{up}}[B(\varrho_h^k), \boldsymbol{u}_h^k] \left[\!\left[\phi_h \right]\!\right] \mathrm{dS}_x. \tag{8.18}
$$

Summing up the last two formulas we find that (8.17) holds as soon as (8.18) can be rewritten as

$$
D = \sum_{K\in\mathcal{T}_h} \int_K \phi_h \sum_{\sigma\in\mathcal{E}(K)\cap\mathcal{E}_{int}} \frac{|\sigma|}{|K|} \left[\!\left[B(\varrho_h^k) \right]\!\right] (h^\varepsilon - [\left\langle \boldsymbol{u}_h^k \right\rangle_\sigma \cdot \boldsymbol{n}]^-) \,\mathrm{d}x. \tag{8.19}
$$

For the first term on the right hand side of (8.18) we have

$$
\begin{aligned}
\int_\Omega \phi_h B(\varrho_h^k) \operatorname{div}_h \boldsymbol{u}_h^k \,\mathrm{d}x &= \sum_{K\in\mathcal{T}_h} \int_K \phi_h B(\varrho_h^k) \sum_{\sigma\in\mathcal{E}(K)} \frac{|\sigma|}{|K|} u_\sigma \,\mathrm{d}x \\
&= \sum_{K\in\mathcal{T}_h} \int_K \phi_h \sum_{\mathcal{E}(K)} \frac{|\sigma|}{|K|} \left(\{\!\{ B(\varrho_h^k) \}\!\} - \frac{1}{2} \left[\!\left[B(\varrho_h^k) \right]\!\right] \right) u_\sigma \,\mathrm{d}x.
\end{aligned} \tag{8.20}
$$

Further, concerning the second term on the right hand side of (8.18) we calculate

$$\int_{\mathcal{E}_{int}} F_h^{\rm up}[B(\varrho_h^k), \boldsymbol{u}_h^k] [\![\phi_h]\!] \ {\rm dS}_x$$

$$= - \sum_{K \in \mathcal{T}_h} \int_K \phi_h \sum_{\mathcal{E}(K) \cap \mathcal{E}_{int}} \frac{|\sigma|}{|K|} \Big(\{\!\{B(\varrho_h^k)\}\!\} \, u_\sigma - \frac{|u_\sigma|}{2} [\![B(\varrho_h^k)]\!] - h^\varepsilon [\![B(\varrho_h^k)]\!] \Big) \ {\rm d}x$$

$$= \sum_{K \in \mathcal{T}_h} \int_K \phi_h \sum_{\mathcal{E}(K) \cap \mathcal{E}_{int}} \frac{|\sigma|}{|K|} (h^\varepsilon - [u_\sigma]^-) [\![B(\varrho_h^k)]\!] \ {\rm d}x$$

$$- \sum_{K \in \mathcal{T}_h} \int_K \phi_h \sum_{\mathcal{E}(K) \cap \mathcal{E}_{int}} \frac{|\sigma|}{|K|} \Big(\{\!\{B(\varrho_h^k)\}\!\} - \frac{1}{2} [\![B(\varrho_h^k)]\!] \Big) u_\sigma \ {\rm d}x. \tag{8.21}$$

Finally, summing up (8.20) and (8.21) proves (8.19), which completes the proof of (8.17).

With the renormalized continuity equation at hand we are ready to show positivity of the discrete density.

Lemma 8.3 (Positivity of density)
Let $\varrho_h^0 > 0$. Then any solution of (8.14) *satisfies $\varrho_h^k > 0$ for all $k = 1, \ldots, N_T$.*

Proof We use mathematical induction and start with the induction hypothesis $\varrho_h^{k-1} > 0$. We use the renormalized continuity equation (8.17) for $\phi \equiv 1$ and B a continuously differentiable convex function obtaining

$$\int_\Omega D_t B(\varrho_h^k) \ {\rm d}x + \int_\Omega \Big(\varrho_h^k B'(\varrho_h^k) - B(\varrho_h^k) \Big) {\rm div}_h \boldsymbol{u}_h^k \ {\rm d}x \le 0. \tag{8.22}$$

Next, we aim to use $B(\varrho) = \max\{0, -\varrho\} \ge 0$. Strictly speaking, this is not a C^1 function, however, we may easily construct an approximate sequence

$$B_\delta : R \to [0, \infty), \ B_\delta \in C^1(R), \ B_\delta(\varrho) = 0 \text{ for } \varrho \ge -\delta, \ B_\delta'(0) = 0,$$
$$B_\delta(\varrho) \to \max\{0, -\varrho\} \text{ uniformly in } R, \ B_\delta'(\varrho) = -1 \text{ for all } \varrho < \delta.$$

Plugging B_δ in (8.17) and performing the limit $\delta \to 0$ we conclude

$$\int_\Omega B(\varrho_h^k) \ {\rm d}x \le 0.$$

As B is a nonnegative function we have $B(\varrho_h^k) \equiv 0$ holds for all $x \in \Omega$. Thus we have proved $\varrho_h^k \ge 0$.

Finally, we assume there exists a $K \in \mathcal{T}_h$ such that $\varrho_K^k = 0$. Then a straightforward calculation using the discrete equation for the density (8.14) yields

$$\frac{|K|}{\Delta t}(0 - \varrho_K^{k-1}) = - \int_K \mathrm{div}_h^{\mathrm{up}}(\varrho_h^k, \boldsymbol{u}_h^k)\, \mathrm{d}x \geq - \sum_{\sigma \in \mathcal{E}(K)} (\varrho_h^k)^{\mathrm{out}}[\langle \boldsymbol{u}_h^k \cdot \boldsymbol{n} \rangle_\sigma]^- \geq 0,$$

which implies $\varrho_K^{k-1} \leq 0$ in contrast with the assumption $\varrho_h^{k-1} > 0$. Thus necessarily $\varrho_h^k > 0$.

Note carefully that positivity of the density stated in Lemma 8.3 is *unconditional*; it holds independently of the length of the time step and the degree of spatial discretization. Standard numerical fluxes, such as those introduced in (8.5), in general do not enjoy this property unless certain restrictions are imposed on the length of the time step in the form of a CFL (Courant–Friedrichs–Lewy) condition. The crucial role in the diffusive upwind numerical flux is the presence of the flow velocity in the upwind discretization.

8.4 Stability and Consistency of Numerical Solutions

The notion of consistent approximation has been introduced and discussed in Sects. 7.3, 7.5 and 7.6. In this section, we revisit the topic in the context of numerical solutions introducing stable and consistent approximations to the Euler and Navier–Stokes systems.

8.4.1 *Stability*

Stability of a numerical method reflects uniform bounds imposed by natural constraints resulting from the laws of physics inherited by the system of field equations. Ideally, stability estimates coincide with the *a priori bounds* available for the continuous model. In addition, there is a common believe that the numerical viscosity omnipresent in numerical schemes may "pick up" the physically relevant solution even if the limit problem is ill-posed. This is particularly relevant for the Euler system, however, a rigorous proof at least in the multidimensional case is so far not available.

In the context of compressible fluid flow models, natural properties to be satisfied are the conservation of mass and positivity of the discrete density discussed in the preceding part. For models of viscous fluids, further important stability estimates result from the dissipation of the discrete energy. For the complete Euler system, the Second law of thermodynamics enforced through the discrete entropy inequality provides more bounds at the level of thermostatic variables. Its satisfaction entails other important properties as the minimum entropy principle. In what follows we will define more precisely the concept of a stable numerical method for the Euler and Navier–Stokes systems, respectively.

Consider the barotropic Euler system (7.57). Here, the conservative state variables are the mass density ϱ and the momentum $\boldsymbol{m}$. Alternatively, we may replace $\boldsymbol{m}$ by the velocity $\boldsymbol{u} = \frac{\boldsymbol{m}}{\varrho}$. Note, however, that strict positivity of ϱ is required to justify this choice.

Remark 8.1 Strictly speaking, numerical solutions $\{\varrho_h^k, \boldsymbol{m}_h^k\}_{h \searrow 0}$ resulting from the fully discrete scheme are defined on discrete time levels $t^k = t^0 + k\Delta t$, $k = 0, 1, \dots, N_T$. As introduced in Preliminary material, we work with the piecewise constant interpolation on the full time interval $t \in [0, T + \Delta t)$

$$\varrho_h(t, \cdot) = \sum_{k=0}^{N_T} \varrho_h^k 1_{[t_k, t_{k+1})}(t), \quad \boldsymbol{m}_h(t, \cdot) = \sum_{k=0}^{N_T} \boldsymbol{m}_h^k 1_{[t_k, t_{k+1})}(t).$$

Clearly, other suitable interpolations are possible as well. The solutions resulting from the semidiscrete (continuous time) schemes are defined naturally for any t. We tacitly assume this convention whenever omitting the upper time step index.

Definition 8.1 (STABLE NUMERICAL METHOD FOR THE BAROTROPIC EULER SYSTEM)

Let $\{\varrho_h, \boldsymbol{m}_h\}_{h \searrow 0}$ be a sequence obtained by a numerical method with the initial data $[\varrho_h^0, \boldsymbol{m}_h^0]$ and either periodic ($\Omega = \mathbb{T}^d$) or no-flux boundary conditions (8.10), (8.11) with $r_h = \varrho_h$. We say that a numerical method is *stable* for the barotropic Euler system (7.57), (7.58), if the resulting numerical solutions $(\varrho_h, \boldsymbol{m}_h)$ enjoy the following properties:

- positivity of density $\quad \varrho_h \geq 0$;
- conservation of total mass

$$\int_\Omega \varrho_h(t, \cdot) \,\mathrm{d}x = \int_\Omega \varrho_h^0 \,\mathrm{d}x, \quad t \in [0, T];$$

- momentum-density compatibility $\quad \varrho_h = 0 \Rightarrow \boldsymbol{m}_h = 0$;
- total energy inequality

$$\int_\Omega \left[\frac{1}{2}\frac{|\boldsymbol{m}_h|^2}{\varrho_h} + P(\varrho_h)\right](t, \cdot) \,\mathrm{d}x \leq \int_\Omega \left[\frac{1}{2}\frac{|\boldsymbol{m}_h^0|^2}{\varrho_h^0} + P(\varrho_h^0)\right] \mathrm{d}x \quad t \in [0, T],$$

where $P(\varrho) = \varrho \int_1^\varrho \frac{p(z)}{z^2} \mathrm{d}z$ is the pressure potential.

Remark 8.2 Recall that the kinetic energy $\frac{1}{2}\frac{|\boldsymbol{m}|^2}{\varrho}$ has been defined as a convex l.s.c. function for *any* $[\varrho, \boldsymbol{m}] \in R^{d+1}$ by (1.26). In particular, the stipulation that the kinetic energy is finite at each step of numerical approximation includes implicitly both nonnegativity of the density and the momentum-density compatibility property.

Next, we specify stability of a numerical method for the (complete) Euler system in terms of the conservative variables $[\varrho, \boldsymbol{m}, E]$. We also assume the existence of a physical entropy s, $\varrho s(\varrho, \boldsymbol{m}, E)$ being a concave function of the state variables. Recall that in the case of the standard Boyle–Mariotte EOS, the entropy takes the form

$$s = \frac{1}{\gamma - 1} \log \left(\frac{\gamma - 1}{\varrho^\gamma} \left(E - \frac{|\boldsymbol{m}|^2}{\varrho} \right) \right).$$

Definition 8.2 (STABLE NUMERICAL METHOD FOR THE EULER SYSTEM)
Let $\{\varrho_h, \boldsymbol{m}_h, E_h\}_{h \searrow 0}$ be a numerical solution of the Euler system (2.34)–(2.36), supplemented with either periodic ($\Omega = \mathbb{T}^d$) or the no-flux boundary conditions (8.10), (8.11) with $r_h = \varrho_h$, p_h and with the initial data $[\varrho_h^0, \boldsymbol{m}_h^0, E_h^0]$. Here p_h denotes the corresponding discrete pressure. We say that the numerical method is *stable* if the following properties hold:

- positivity of density $\quad \varrho_h \geq 0$;
- conservation of mass

$$\int_\Omega \varrho_h(t, \cdot) \, \mathrm{d}x = \int_\Omega \varrho_h^0 \, \mathrm{d}x, \quad t \in [0, T];$$

- conservation of total energy

$$\int_\Omega E_h(t, \cdot) \, \mathrm{d}x = \int_\Omega E_h^0 \, \mathrm{d}x, \quad t \in [0, T];$$

- renormalized entropy inequality

$$\int_\Omega \varrho_h \chi \Big(s(\varrho_h, \boldsymbol{m}_h, E_h) \Big)(t, \cdot) \, \mathrm{d}x \geq \int_\Omega \varrho_h^0 \chi \Big(s(\varrho_h^0, \boldsymbol{m}_h^0, E_h^0) \Big) \, \mathrm{d}x, \quad t \in [0, T], \tag{8.23}$$

  for any nondecreasing, concave, twice continuously differentiable function χ on R that is bounded from above;
- minimum entropy principle

$$s_h(t, \cdot) \geq \underline{s}, \quad t \in [0, T] \text{ and a.e. in } \Omega, \quad \underline{s} \in R.$$

Remark 8.3 As we have seen in Proposition 2.1, the minimum entropy principle follows from (8.23) as long as the initial entropy admits a uniform lower bound.

Remark 8.4 The minimum entropy principle represents a rather strong constraint that implies, at least in the case of Boyle–Mariotte EOS, other key stability estimates:

- momentum-density compatibility $\varrho = 0 \Rightarrow \boldsymbol{m} = 0$;
- strict positivity of the absolute temperature $\vartheta > 0$.

Finally, we consider the Navier–Stokes system (7.77). As viscosity plays a crucial role in this model, it is convenient to consider the mass density ϱ and the fluid velocity $\boldsymbol{u}$ as state variables.

Definition 8.3 (Stable numerical method for the Navier–Stokes system)
Let $\{\varrho_h, \boldsymbol{u}_h\}_{h\searrow 0}$ be an approximate solution resulting from a numerical method for the Navier–Stokes system (7.77), with the space periodic boundary condition ($\Omega = \mathbb{T}^d$) or the no-slip boundary conditions (1.54), and with the initial data $[\varrho_h^0, \boldsymbol{u}_h^0]$. We say that the numerical method is *stable* if the following holds:

- positivity of density $\varrho_h \geq 0$;
- conservation of mass

$$\int_{\Omega} \varrho_h(t, \cdot)\, \mathrm{d}x = \int_{\Omega} \varrho_h^0\, \mathrm{d}x, \quad t \in [0, T];$$

- total energy inequality

$$\int_{\Omega} \left[\frac{1}{2} \varrho_h |\boldsymbol{u}_h|^2 + P(\varrho_h) \right] (t, \cdot)\, \mathrm{d}x + \mu \, \|\nabla_h \boldsymbol{u}_h(t)\|^2_{L^2(\Omega)} + \nu \, \|\mathrm{div}_h \boldsymbol{u}_h(t)\|^2_{L^2(\Omega)}$$

$$\leq \int_{\Omega} \left[\frac{1}{2} \varrho_h^0 |\boldsymbol{u}_h^0|^2 + P(\varrho_h^0) \right] \mathrm{d}x, \quad t \in [0, T],$$

where $P(\varrho) = \varrho \int_1^{\varrho} \frac{p(z)}{z^2} \mathrm{d}z$ and $\nu = \frac{d-2}{d}\mu + \lambda$.

The symbols ∇_h and div_h are the discrete differential operators introduced in (15) for piecewise constant approximation of the velocity field. They may be replaced accordingly by their analogues in (15) if different velocity discretization is used.

8.4.2 Consistency

A numerical method is consistent if the numerical solution satisfies the continuous problem up to a consistency error that can be evaluated from the local truncation analysis. In what follows, we introduce consistent numerical methods for the Euler and Navier–Stokes systems. They should be compared to their continuous counterparts discussed in Chap. 5.

Definition 8.4 (CONSISTENT NUMERICAL METHOD FOR THE BAROTROPIC EULER SYSTEM)

Let $\{\varrho_h, \boldsymbol{m}_h\}_{h\searrow 0}$ be a sequence obtained by a numerical method for the barotropic Euler system with the no-flux boundary conditions (8.10), (8.11) with $r_h = \varrho_h$ or the periodic ($\Omega = \mathbb{T}^d$) boundary conditions, and with the initial data $[\varrho_h^0, \boldsymbol{m}_h^0]$. We say that a numerical method is *consistent* with the barotropic Euler system (5.71) if

$$-\int_\Omega \varrho_h^0 \varphi(0,\cdot)\,\mathrm{d}x = \int_0^T \int_\Omega \left[\varrho_h \partial_t \varphi + \boldsymbol{m}_h \cdot \nabla_x \varphi\right]\,\mathrm{d}x\,\mathrm{d}t + \int_0^T e_{1,h}(t,\varphi)\,\mathrm{d}t \tag{8.24a}$$

for any $\varphi \in C_c^2([0,T)\times\overline{\Omega})$;

$$\begin{aligned}
-\int_\Omega \boldsymbol{m}_h^0 \cdot \boldsymbol{\varphi}(0,\cdot)\,\mathrm{d}x = \\
&\int_0^T \int_\Omega \left[\boldsymbol{m}_h \cdot \partial_t \boldsymbol{\varphi} + \left(1_{\varrho_n>0} \frac{\boldsymbol{m}_h \otimes \boldsymbol{m}_h}{\varrho_h}\right) : \nabla_x \boldsymbol{\varphi} + p_h \mathrm{div}_x \boldsymbol{\varphi}\right]\,\mathrm{d}x\,\mathrm{d}t \\
&+ \int_0^T e_{2,h}(t,\boldsymbol{\varphi})\,\mathrm{d}t, \qquad p_h = p(\varrho_h)
\end{aligned} \tag{8.24b}$$

for any $\boldsymbol{\varphi} \in C_c^2([0,T)\times\overline{\Omega}; R^d)$, $\boldsymbol{\varphi}\cdot\boldsymbol{n}|_{\partial\Omega} = 0$;

$$\|e_{1,h}(\cdot,\varphi)\|_{L^1(0,T)} \lesssim h^\beta \|\varphi\|_{C^2},\ \|e_{2,h}(\cdot,\boldsymbol{\varphi})\|_{L^1(0,T)} \lesssim h^\beta \|\boldsymbol{\varphi}\|_{C^2} \text{ for some } \beta > 0.$$

Going back to Definition 5.4, Sect. 5.1.5, we easily check that the numerical solutions represent a consistent approximation in the sense of Definition 5.4 provided the method is both stable in the sense of Definition 8.1 and consistent in the sense of Definition 8.4. In particular, the abstract *convergence* results derived in Sect. 7.5 are applicable.

Next, we consider the (complete) Euler system written in the conservative variables $[\varrho, \boldsymbol{m}, E]$ in (2.34)–(2.36). As we have observed in Sects. 7.3, 7.4, the problems related to convergence of approximate solutions are easier to study in the framework of the conservative-entropy variables $[\varrho, \boldsymbol{m}, S]$. Note that both descriptions are equivalent at the discrete level as long as the mass density is strictly positive. Accordingly, we define consistency in terms of the entropy formulation (7.14), (7.15).

Definition 8.5 (CONSISTENT NUMERICAL METHOD FOR THE EULER SYSTEM)

Let $\{\varrho_h, \boldsymbol{m}_h, E_h\}_{h\searrow 0}$ be a sequence obtained by a numerical method for the Euler system (2.34)–(2.36), supplemented with either periodic ($\Omega = \mathbb{T}^d$) or the no-flux boundary conditions (8.10), (8.11) with $r_h = \varrho_h, p_h$, and with the initial data $[\varrho_h^0, \boldsymbol{m}_h^0, E_h^0]$. Here p_h and s_h denote the associated discrete pressure and entropy, respectively. We say that a numerical method is *consistent* with the Euler system (7.14), (7.15) if

$$-\int_{\Omega} \varrho_h^0 \varphi(0,\cdot)\,\mathrm{d}x = \int_0^T \int_{\Omega} \left[\varrho_h \partial_t \varphi + \boldsymbol{m}_h \cdot \nabla_x \varphi\right]\,\mathrm{d}x\,\mathrm{d}t + \int_0^T e_{1,h}(t,\varphi)\,\mathrm{d}t \tag{8.25}$$

for any $\varphi \in C_c^2([0,T) \times \overline{\Omega})$;

$$\begin{aligned}
-\int_{\Omega} \boldsymbol{m}_h^0 \cdot \boldsymbol{\varphi}(0,\cdot)\,\mathrm{d}x = \\
\int_0^T \int_{\Omega} \left[\boldsymbol{m}_h \cdot \partial_t \boldsymbol{\varphi} + \left(1_{\varrho_n > 0} \frac{\boldsymbol{m}_h \otimes \boldsymbol{m}_h}{\varrho_h}\right) : \nabla_x \boldsymbol{\varphi} + p_h \mathrm{div}_x \boldsymbol{\varphi}\right]\,\mathrm{d}x\,\mathrm{d}t \\
+ \int_0^T e_{2,h}(t,\boldsymbol{\varphi})\,\mathrm{d}t
\end{aligned} \tag{8.26}$$

for any $\boldsymbol{\varphi} \in C_c^2([0,T) \times \overline{\Omega}; R^d)$, $\boldsymbol{\varphi} \cdot \boldsymbol{n}|_{\partial\Omega} = 0$;

$$\begin{aligned}
\int_0^T \int_{\Omega} \left[\varrho_h \chi(s_h) \partial_t \varphi + (\chi(s_h) \boldsymbol{m}_n) \cdot \nabla_x \varphi\right]\,\mathrm{d}x\,\mathrm{d}t + \int_0^T e_{3,h}(t,\varphi)\,\mathrm{d}t \\
\leq -\int_{\Omega} \varrho_h^0 \chi(s_h^0) \varphi(0,\cdot)\,\mathrm{d}x
\end{aligned} \tag{8.27}$$

for any $\varphi \in C_c^2([0,T) \times \overline{\Omega})$, and χ a nondecreasing, concave, twice continuously differentiable function on R that is bounded from above.

The consistency errors satisfy

$$\|e_{j,h}(\cdot,\varphi)\|_{L^1(0,T)} \lesssim h^{\beta} \|\varphi\|_{C^2},\ j = 1,3, \quad \|e_{2,h}(\cdot,\boldsymbol{\varphi})\|_{L^1(0,T)} \lesssim h^{\beta} \|\boldsymbol{\varphi}\|_{C^2}, \quad \text{for some } \beta > 0.$$

One may add the conservation of the global energy

$$\int_{\Omega} E_h(t,\cdot)\,\mathrm{d}x = \int_{\Omega} E_h^0\,\mathrm{d}x, \quad t \in (0,T] \tag{8.28}$$

as another consistency requirement. Note, however, that this condition is already included in Definition 8.2 of stability of the Euler system.

It may seem that consistency for the (complete) Euler system requires actually much less than expected. In particular, the energy equation is replaced by the entropy inequality (8.27) and the total energy balance (8.28). As we shall see, convergence of a numerical method can be shown even under these "minimalist" conditions. The point is that suitable *a priori* bounds that would control the energy convection are simply not available given the present state-of-the-art of the mathematical theory.

Finally, we address consistency for the Navier–Stokes system.

Definition 8.6 (Consistent numerical method for the Navier–Stokes system)

Let $\{\varrho_h, \boldsymbol{u}_h\}_{h\searrow 0}$ be a sequence obtained by a numerical method for the Navier–Stokes system (7.77) with the initial data $[\varrho_h^0, \boldsymbol{u}_h^0]$, and the periodic boundary conditions ($\Omega = \mathbb{T}^d$) or the no-slip boundary conditions (1.41) and the Neumann boundary condition (8.11) for ϱ_h. We say that a numerical method is *consistent* with the Navier–Stokes system (7.77) if

$$
-\int_{\Omega} \varrho_h^0 \varphi(0,\cdot)\,\mathrm{d}x = \int_0^T \int_{\Omega} \left[\varrho_h \partial_t \varphi + \varrho_h \boldsymbol{u}_h \cdot \nabla_x \varphi\right]\,\mathrm{d}x\,\mathrm{d}t + \int_0^T e_{1,h}(t,\varphi)\,\mathrm{d}t, \tag{8.29a}
$$

for any $\varphi \in C_c^2([0,T)\times\overline{\Omega})$;

$$
\begin{aligned}
-\int_{\Omega} \varrho_h^0 \boldsymbol{u}_h^0 \cdot \boldsymbol{\varphi}(0,\cdot)\,\mathrm{d}x &= \int_0^T \int_{\Omega} \left[\varrho_h \boldsymbol{u}_h \cdot \partial_t \boldsymbol{\varphi} + (\varrho_h \boldsymbol{u}_h \otimes \boldsymbol{u}_h + p_h \mathbb{I}) : \nabla_x \boldsymbol{\varphi}\right]\,\mathrm{d}x\,\mathrm{d}t \\
&- \int_0^T \int_{\Omega} \left(\mu \nabla_h \boldsymbol{u}_h : \nabla_x \boldsymbol{\varphi} + \nu\, \mathrm{div}_h \boldsymbol{u}_h\, \mathrm{div}_x \boldsymbol{\varphi}\right)\,\mathrm{d}x\,\mathrm{d}t + \int_0^T e_{2,h}(t,\boldsymbol{\varphi})\,\mathrm{d}t,
\end{aligned} \tag{8.29b}
$$

where $\nu = \mu\frac{d-2}{d} + \lambda$ and $\boldsymbol{\varphi} \in C_c^2([0,T)\times\Omega; R^d)$;

$$
\|e_{1,h}(\cdot,\varphi)\|_{L^1(0,T)} \lesssim h^\beta \|\varphi\|_{C^2}, \quad \|e_{2,h}(\cdot,\boldsymbol{\varphi})\|_{L^1(0,T)} \lesssim h^\beta \|\boldsymbol{\varphi}\|_{C^2} \ \text{for some } \beta > 0.
$$

Note that the discrete operators div_h and ∇_h, introduced in Preliminary material, are compatible with div_x and ∇_x, cf. Definition 5.8.

As already pointed out in the context of the barotropic Euler system, a consistent approximation specified in Definition 5.9 includes also the energy inequality. Thus a numerical approximation is consistent in the sense of Definition 5.9, if the numerical method is stable in the sense of Definition 8.3 and consistent in the sense of Definition 8.6.

8.5 Convergence of Numerical Solutions

A common theme of this monograph is to establish convergence of certain numerical methods for compressible fluid flows in the spirit of the celebrated **Lax equivalence theorem**, see Lax and Richtmeyer, [149]:

Given a properly posed linear initial value problem and a linear finite difference approximation to it that satisfies the consistency condition, stability is a necessary and sufficient condition that a scheme be a convergent approximation of the initial value problem.

In the following chapters, we will present several numerical methods, based on the finite volume approximation and their combinations with the nonconforming finite elements or finite differences. In particular, the goal will be to show their stability and consistency in order to apply the results from Chap. 7 to obtain their convergence.

Mathematically speaking, showing convergence of a sequence of numerical approximations requires precompactness and passing to a *subsequence* as the case may be. This can be seen as a serious drawback from the practical view point as different subsequences may give rise to different solutions. This is indeed a serious issue in particular for the Euler system that is *known* to be ill posed in the class of admissible weak solutions. Indeed, convergence of the entire sequence can be rigorously justified only if the limit is unique, which translates to well posedness of the limit problem.

The method of averaging, incorporated in the theory via the concept of $\mathcal{K}$–convergence in Chap. 7, besides the original motivation to visualize the weak limits, may be seen as a tentative tool to eliminate possible fluctuations and "pick up" a unique limit. A rigorous proof of such a statement, however, is out of reach of the available analytical tools considered in this monograph.

8.6 Conclusion, Bibliographical Remarks

There is a large amount of literature on the finite element methods mostly devoted to the approximation of the elliptic or parabolic problems and on the finite volume methods that typically deals with the hyperbolic conservation laws. We assume that the reader is familiar with basic concepts of these methods. We refer to monographs Boffi et al. [23], Ciarlet [55, 56], Ern and Guermond [76] for the theory of finite element methods and to Godlewski, Raviart [119], Eymard, Gallouët, Herbin [78], Kröner [143], LeVeque [151], Toro [194] for the finite volume methods. A comprehensive overview of various numerical techniques suitable for the compressible and incompressible flow problems can be found in Feistauer et al. [99, 100] and John [132]. Among other methods available for numerical solution of partial differential equations, one should mention the so-called discontinuous Galerkin method. This numerical method can be viewed as an extension of the nonconforming finite element methods enriched by the concept of numerical fluxes borrowed from the finite volume methods, see, e.g., Di Pietro and Ern [63], Dolejší and Feistauer [71]. In fact, any finite volume method working with piecewise constant approximations can be seen as a low-order discontinuous Galerkin method.

A typical numerical flux used in the present monograph to approximate convective terms is the upwinding approximation (8.6). The basic idea is well-known and can be found both in the finite element as well as finite volume literature. The information that propagates along streamlines (or characteristics) is taken as a suitable candidate for a particular approximation. The numerical flux that we frequently use in what follows adds to the numerical diffusion due to the upwinding an additional numerical diffusion of order $h^{\varepsilon+1}$, $\varepsilon > -1$. The latter will allow us to obtain better stability estimates. We refer to Karper [140] and the monograph [88], where such a diffusive upwinding has been successfully applied to approximate weak solutions of the compressible Navier–Stokes equations. The interested reader may consult Karlsen, Karper [137–139] for the approximation of the Stokes problem and [87] for the convergence analysis to a weak solution of the Navier–Stokes–Fourier system. A recent result on the convergence analysis of the Navier–Stokes–Fourier system via the concept of DMV solution can be found in [95].

Chapter 9
Finite Volume Method for the Barotropic Euler System

The Euler equations belong to the class of hyperbolic conservation laws that can be written in a general form

$$
\begin{aligned}
\partial_t \boldsymbol{U} + \mathrm{div}_x \mathbb{F}(\boldsymbol{U}) &= 0, \qquad \text{in } \Omega \times (0, T) \\
\boldsymbol{U}(0, \cdot) &= \boldsymbol{U}_0, \qquad \text{in } \Omega.
\end{aligned}
\tag{9.1}
$$

Here $\boldsymbol{U}$, $\mathbb{F}(\boldsymbol{U})$ denote the vector of conservative variables and the flux matrix, respectively. The system (9.1) is usually accompanied with suitable boundary conditions, such as no-flux or periodic boundary conditions. System (9.1) is called *hyperbolic* if the Jacobian matrix $\frac{\mathrm{d}\mathbb{F}}{\mathrm{d}\boldsymbol{U}}$ is diagonalizable and has real eigenvalues. The methods of choice for hyperbolic conservation laws are the so-called *finite volume methods* that we now briefly describe. As discussed in the previous chapter there are different possibilities for time discretization. In this section we concentrate on the semidiscrete method, letting time continuous and applying piecewise constant approximation in space. We note by passing that in Chap. 12 we will introduce a new finite volume method for the barotropic Euler equations, where the implicit time discretization will be applied.

Let a computational domain Ω be discretized by a mesh $\mathcal{T}_h$ consisting of regular quadrilaterals ($d = 2$) or cuboids ($d = 3$), see Definition 3. The finite volume approximation of (9.1) is a piecewise constant vector valued function $\boldsymbol{U}_h, \boldsymbol{U}_h(t)_{|K} = \boldsymbol{U}_K(t)$, that satisfies

$$
\begin{aligned}
\frac{\mathrm{d}}{\mathrm{d}t}\boldsymbol{U}_K(t) + (\mathrm{div}_{\mathcal{D}} \mathbb{F}_h(t))_K &= 0, \quad t > 0, \quad K \in \mathcal{T}_h, \\
\boldsymbol{U}_K(0) &= (\Pi_Q(\boldsymbol{U}_0))_K, \quad K \in \mathcal{T}_h.
\end{aligned}
\tag{9.2}
$$

The numerical flux function $\mathbb{F}_h$ quantifies the flux across the interfaces $\sigma \in \mathcal{E}$. For $\sigma = K|L$ we set $\boldsymbol{F}_\sigma \equiv \mathbb{F}_h(\boldsymbol{U}_K, \boldsymbol{U}_L)\boldsymbol{n}_\sigma$. The discrete divergence operator $\mathrm{div}_{\mathcal{D}}$ represents the finite difference of the numerical flux on edges, i.e.

E. Feireisl et al., *Numerical Analysis of Compressible Fluid Flows*, MS&A 20,
https://doi.org/10.1007/978-3-030-73788-7_9

$$(\mathrm{div}_{\mathcal{D}}\mathbb{F}_h)_K = \sum_{\sigma\in\mathcal{E}(K)} \frac{|\sigma|}{|K|}\mathbb{F}_h \boldsymbol{n}_{\sigma K}. \tag{9.3}$$

In the literature, one can find a large variety of numerical flux functions. Most of them satisfy some suitable admissibility conditions. First, the numerical flux $\mathbb{F}_h$ should be *consistent* with the physical flux $\mathbb{F}$ in the sense that $\mathbb{F}_h(\boldsymbol{w},\boldsymbol{w}) = \mathbb{F}(\boldsymbol{w})$ for all $\boldsymbol{w} \in R^N$, N being the number of equations in the system (9.1). Further, the numerical flux is assumed to be *locally Lipschitz continuous*, i.e. for every compact set $D \subset R^N$ there exists $C > 0$ such that

$$\begin{aligned}&\|\boldsymbol{F}_\sigma - \mathbb{F}(\boldsymbol{U}_K)\boldsymbol{n}_\sigma\| \equiv \|\mathbb{F}_h(\boldsymbol{U}_K,\boldsymbol{U}_L)\boldsymbol{n}_\sigma - \mathbb{F}(\boldsymbol{U}_K)\boldsymbol{n}_\sigma\|\\ &\leq C\|\boldsymbol{U}_K - \boldsymbol{U}_L\|,\end{aligned}$$

whenever $\boldsymbol{U}_K, \boldsymbol{U}_L \in D$.

Certain numerical schemes are further required to preserve some important physical properties of the underlying continuous solutions, such as positivity preserving of some physical quantities. It turns out that for convergence analysis a discrete version of the entropy inequality plays a crucial role. Below, we will describe the concept of *entropy stable schemes* as introduced by E. Tadmor. The results are presented without proofs and the interested reader may consult Tadmor [190–192] for details.

Let $(\eta, \boldsymbol{g})$ be an *entropy-flux pair* associated with system (9.1), i.e., $(\eta, \boldsymbol{g}) : R^M \to R \times R^d$ such that $\eta = \eta(\boldsymbol{U})$ is a convex function of the state variables $\boldsymbol{U}$, and $\boldsymbol{g}$ is the associated entropy flux that satisfies for all $\boldsymbol{U} \in R^N$ the compatibility condition

$$\nabla_U g_s(\boldsymbol{U})^T = \nabla_U \eta(\boldsymbol{U})^T \nabla_U \boldsymbol{F}_s(\boldsymbol{U}), \quad s = 1, \ldots, d.$$

Accordingly, the entropy satisfies an additional conservation law

$$\partial_t \eta(\boldsymbol{U}) + \mathrm{div}_x \boldsymbol{g}(\boldsymbol{U}) = 0$$

that can be *derived* from the original system of equations (9.1) on condition that all quantities are smooth. Note carefully that in the context of the barotropic Euler system and/or the (complete) Euler system written down in the conservative-entropy variables $[\varrho, \boldsymbol{m}, S]$, the total energy is a mathematical entropy. If the complete Euler system is written in terms of the conservative variables $[\varrho, \boldsymbol{m}, E]$, then it is the total entropy $-S(\varrho, \boldsymbol{m}, E)$ that plays the role of mathematical entropy. As we have seen in Sect. 4.1.6, convexity of a (mathematical) entropy is closely related to the thermodynamic stability of the Euler system.

Finite volume method (9.2) is said to be *entropy stable* if it satisfies the discrete entropy inequality

$$\frac{\mathrm{d}}{\mathrm{d}t}\eta(\boldsymbol{U}_K(t)) + (\mathrm{div}_{\mathcal{D}}\boldsymbol{G}_h(t))_K \leq 0, \quad K \in \mathcal{T}_h,\ t > 0 \tag{9.4}$$

for a suitable numerical flux $\boldsymbol{G}_h$. If, in particular, equality holds in (9.4), we say the finite volume method (9.2) is *entropy conservative.* Similarly to $\mathbb{F}_h$, the numerical entropy flux $\boldsymbol{G}_h$ is a function of two neighboring values, i.e., $\boldsymbol{G}_\sigma \equiv \boldsymbol{G}_h(\boldsymbol{U}_K, \boldsymbol{U}_L)$ for $\sigma = K|L$. It is assumed to be *consistent* with the differential entropy flux $\boldsymbol{g}$, i.e., $\boldsymbol{G}_h(\boldsymbol{w}, \boldsymbol{w}) = \boldsymbol{g}(\boldsymbol{w})$ for all $\boldsymbol{w} \in R^N$. Introducing the entropy variables $\boldsymbol{V} = \nabla_U \eta(\boldsymbol{U})$ and the so-called potential function $\boldsymbol{\psi}(\boldsymbol{V})$, such that

$$\nabla_V \boldsymbol{\psi}(\boldsymbol{V}) = \mathbb{F}(\boldsymbol{U}(\boldsymbol{V}))$$

we can express the entropy flux $\boldsymbol{G}_h$ explicitly in terms of the numerical flux $\mathbb{F}_h$ as

$$\boldsymbol{G}_\sigma = \{\!\{\boldsymbol{V}_h\}\!\}_\sigma \boldsymbol{F}_\sigma - \{\!\{\boldsymbol{\psi}(\boldsymbol{V}_h)\}\!\}_\sigma . \tag{9.5}$$

Remark 9.1 (Entropy variables for the Euler system)
As it can be checked by rather lengthy but straightforward manipulation, the standard choice of entropy for the complete Euler system,

$$s = \log\left(\frac{p}{\varrho^\gamma}\right), \quad p = (\gamma - 1)\varrho e, \ e = c_v \vartheta,$$

yields the entropy variables in the form

$$\boldsymbol{V} = \frac{1}{e}\left[e(\gamma - s) - \frac{|\boldsymbol{u}|^2}{2}, \ \boldsymbol{u}, -1\right], \quad \boldsymbol{u} \equiv \frac{\boldsymbol{m}}{\varrho}.$$

In particular, the entropy variables $\boldsymbol{V}$ should not be confused with the conservative-entropy variables $[\varrho, \boldsymbol{m}, S]$ introduced in Sect. 2.2.4. The potential function for the Euler system reads $\psi(\boldsymbol{V}) = \boldsymbol{m}$.

Remark 9.2 (Entropy variables for the barotropic Euler system)
Direct calculation yields for the barotropic Euler system the entropy variables and the potential function

$$\boldsymbol{V} = \left[\frac{a\gamma}{\gamma - 1}\varrho^{\gamma-1} - \frac{|\boldsymbol{u}|^2}{2}, \boldsymbol{u}\right], \qquad \psi(\boldsymbol{V}) = a\gamma\varrho^{\gamma-1}\boldsymbol{m}.$$

9.1 Numerical Methods

In this chapter we will discuss some standard finite volume approximations of the Euler equations for barotropic fluids (2.34),(2.35), with the isentropic pressure-density EOS $p(\varrho) = a\varrho^\gamma$, $\gamma > 1$, cf. (2.71):

$$\partial_t \varrho + \operatorname{div}_x \boldsymbol{m} = 0, \tag{9.6}$$

$$\partial_t \boldsymbol{m} + \operatorname{div}_x \left(\frac{\boldsymbol{m} \otimes \boldsymbol{m}}{\varrho} \right) + \nabla_x p = 0. \tag{9.7}$$

The above system is written in the form (9.1), with the conservative variables $\boldsymbol{U} = (\varrho, \boldsymbol{m})$, and the flux $\mathbb{F}(\boldsymbol{U}) = (\boldsymbol{m}, \frac{\boldsymbol{m} \otimes \boldsymbol{m}}{\varrho} + p\mathbb{I})$. As pointed out above (cf. also (2.73)), the total energy E plays the role of mathematical entropy η for the barotropic Euler system

$$\eta \equiv E = \frac{1}{2} \frac{|\boldsymbol{m}|^2}{\varrho} + P(\varrho), \quad P(\varrho) = \frac{a}{\gamma - 1} \varrho^\gamma$$

with the associated flux $\boldsymbol{g} = \left(\frac{1}{2} \frac{|\boldsymbol{m}|^2}{\varrho} + P(\varrho) + p(\varrho) \right) \frac{\boldsymbol{m}}{\varrho}$.

In order to approximate (9.6), (9.7) we apply the *Lax–Friedrichs* or the *Rusanov* (local Lax–Friedrichs) numerical flux function

$$\boldsymbol{F}_\sigma \equiv \{\!\{ \mathbb{F}(\boldsymbol{U}_h) \}\!\} \cdot \boldsymbol{n} - \lambda_\sigma [\![\boldsymbol{U}_h]\!] . \tag{9.8}$$

In the case of the Rusanov flux we set $\lambda_\sigma \equiv \frac{1}{2} \max_{s=1,\ldots,d} (|\lambda^s(\boldsymbol{U}_K)|, |\lambda^s(\boldsymbol{U}_L)|)$, $\sigma = K|L$ and λ^s is the s-th eigenvalue of the corresponding Jacobian $\frac{\mathrm{d}\mathbb{F}}{\mathrm{d}\boldsymbol{U}}(\boldsymbol{U}_h)$. Setting

$$\lambda_\sigma \equiv \lambda \equiv \frac{1}{2} \max_{s=1,\ldots,d} \max_{K \in \mathcal{T}_h} |\lambda^s(\boldsymbol{U}_K)| \tag{9.9}$$

we obtain the Lax–Friedrichs finite volume scheme.

We consider either the periodic ($\Omega = \mathbb{T}^d$) or the impermeability boundary conditions

$$\{\!\{ \boldsymbol{m}_h \}\!\} \cdot \boldsymbol{n} = 0 \quad \text{for each } \sigma \in \mathcal{E}_{ext}. \tag{9.10}$$

In the latter case, we have to add the homogeneous Neumann boundary condition for ϱ_h, i.e.

$$[\![\varrho_h]\!] = 0 \quad \text{for each } \sigma \in \mathcal{E}_{ext} \tag{9.11}$$

yielding the no-flux boundary conditions to specify the numerical diffusion on the boundary. Plugging the above Rusanov flux function in the finite volume formulation (9.2) we obtain

$$\frac{\mathrm{d}}{\mathrm{d}t} \varrho_K(t) + \sum_{\sigma \in \mathcal{E}(K) \cap \mathcal{E}_{int}} \frac{|\sigma|}{|K|} \left(\{\!\{ \boldsymbol{m}_h(t) \}\!\} \boldsymbol{n} - \lambda_\sigma [\![\varrho_h(t)]\!] \right) = 0, \tag{9.12a}$$

$$\frac{\mathrm{d}}{\mathrm{d}t}\boldsymbol{m}_K(t) + \sum_{\sigma \in \mathcal{E}(K) \cap \mathcal{E}_{int}} \frac{|\sigma|}{|K|} \left(\left\{\!\!\left\{ \frac{\boldsymbol{m}_h(t) \otimes \boldsymbol{m}_h(t)}{\varrho_h(t)} + p_h(t) \right\}\!\!\right\} \boldsymbol{n} - \lambda_\sigma \left[\!\left[\boldsymbol{m}_h(t) \right]\!\right] \right) = 0,$$

$$t > 0, \ K \in \mathcal{T}_h. \tag{9.12b}$$

For the global numerical diffusion coefficient (9.9) gives

$$\frac{\mathrm{d}}{\mathrm{d}t}\varrho_K(t) + \sum_{\sigma \in \mathcal{E}(K) \cap \mathcal{E}_{int}} \frac{|\sigma|}{|K|} \left(\{\!\{ \boldsymbol{m}_h(t) \}\!\} \boldsymbol{n} - \lambda \left[\!\left[\varrho_h(t) \right]\!\right] \right) = 0, \tag{9.13a}$$

$$\frac{\mathrm{d}}{\mathrm{d}t}\boldsymbol{m}_K(t) + \sum_{\sigma \in \mathcal{E}(K) \cap \mathcal{E}_{int}} \frac{|\sigma|}{|K|} \left(\left\{\!\!\left\{ \frac{\boldsymbol{m}_h(t) \otimes \boldsymbol{m}_h(t)}{\varrho_h(t)} + p_h(t) \right\}\!\!\right\} \boldsymbol{n} - \lambda \left[\!\left[\boldsymbol{m}_h(t) \right]\!\right] \right) = 0,$$

$$t > 0, \ K \in \mathcal{T}_h. \tag{9.13b}$$

Recall that $p_h(t) = p(\varrho_h(t)) = a\varrho_h^\gamma(t)$, $\gamma > 1$, $a > 0$, cf. (2.71). The initial conditions for the schemes (9.12) and (9.13) are

$$(\varrho_K(0), \boldsymbol{m}_K(0)) = ((\Pi_Q \varrho_0)_K, (\Pi_Q \boldsymbol{m}_0)_K), \quad K \in \mathcal{T}_h. \tag{9.14}$$

In what follows we address the issues of stability, consistency, and, finally, convergence of the numerical schemes (9.12) and (9.13). As we shall see, the results will be *conditional*. Specifically, some additional hypotheses must be imposed on the family of approximate solutions that can not be derived from (9.12) and (9.13) in any direct manner. This is the price to pay for the relative simplicity of the standard schemes.

Clearly, positivity of the discrete density is necessary for the scheme to be properly defined. Starting from positive initial density $\varrho_h(0) > 0$, the semidiscrete scheme admits a unique solution defined on a maximal time interval $[0, T_{\max})$. Indeed, each of the finite volume methods (9.12) and (9.13) build an ODE system for the approximate solutions $(\varrho_h(t), \boldsymbol{m}_h(t))$. As the numerical flux functions are locally Lipschitz-continuous, the standard ODE theory implies that for a given initial state

$$\varrho_h(0) \in Q_h, \ \varrho_h(0) > 0, \ \boldsymbol{m}_h(0) \in \boldsymbol{Q}_h$$

each of the semidiscrete systems (9.12) and (9.13) admits a unique solution $(\varrho_h, \boldsymbol{m}_h)$ defined on a maximal time interval $[0, T_{\max})$, where

$$\varrho_h(t) > 0 \text{ for all } t \in [0, T_{\max}).$$

In general, $T_{\max}$ may even depend on the mesh parameter h and shrink to zero for $h \to 0$. In order to avoid this difficulty, suitable *a priori* bounds that would guarantee $\varrho_h(t)$ being bounded below away from zero must be established. At the level of limit system, such bounds may be derived from the existence of *invariant domains*—parts of the phase space preserved by the flow. For fully discrete methods based on the explicit time discretization some kind of the so-called CFL (Courant–Friedrichs–Lewy) stability condition are typically applied to preserve invariance at the discrete level. In general, this technique requires uniform bounds on numerical solutions that are not available. To eliminate this problem, we shall therefore impose positivity of ϱ_h as our principal working hypothesis:

$$\varrho_h(t) \geq \underline{\varrho} > 0 \text{ uniformly for } t \in [0, T], \ h \to 0 \tag{9.15}$$

for a positive constant $\underline{\varrho}$.

Remark 9.3 Strictly speaking, the method is well defined under a milder assumption

$$\varrho_h(t) \geq \underline{\varrho}_h > 0 \text{ uniformly for } t \in [0, T],$$

meaning the lower bound is allowed to vanish in the asymptotic limit $h \to 0$. This is, in turn, quite realistic if vacuum appears in the solution of the limit problem. As we shall see, condition (9.15) plays a crucial role not only for stability but also for consistency of the method.

9.2 Stability

We show the stability of the finite volume methods (9.12) and (9.13) by deriving a priori estimates and the discrete energy inequality (9.4).

9.2.1 Mass Conservation

First, we sum up the continuity equation (9.12a) (or (9.13a)) multiplied by $|K|$ for all $K \in \mathcal{E}$ and integrate in time to get

$$\int_\Omega \varrho_h(t) \, \mathrm{d}x = \int_\Omega \varrho_h(0) \, \mathrm{d}x.$$

The anticipated positivity of $\varrho_h(t)$ then implies $\varrho_h \in L^\infty(0, T; L^1(\Omega))$.

9.2.2 Energy Inequality

In what follows we will show that the finite volume methods (9.12) and (9.13) are *entropy stable*, cf. (9.4) bearing in mind that entropy $\approx$ energy in the present context. To this end let us multiply by $\boldsymbol{V}(\boldsymbol{U}) = \frac{\mathrm{d}E(U)}{\mathrm{d}U}$ the finite volume method (9.2). From the flux part we obtain

$$
\begin{aligned}
\boldsymbol{V}_K \mathrm{div}_{\mathcal{D}} \mathbb{F}_h &= \mathrm{div}_{\mathcal{D}}(\{\{\boldsymbol{V}_h\}\} \mathbb{F}_h) - \sum_{\sigma \in \mathcal{E}(K) \cap \mathcal{E}_{int}} \frac{1}{2d_\sigma} [\![\boldsymbol{V}_h]\!] \mathbb{F}_h \\
&= \mathrm{div}_{\mathcal{D}}(\{\{\boldsymbol{V}_h\}\} \mathbb{F}_h - \{\{\boldsymbol{\psi}_h\}\}) - \sum_{\sigma \in \mathcal{E}(K) \cap \mathcal{E}_{int}} \frac{1}{2d_\sigma} \left([\![\boldsymbol{V}_h]\!] \cdot \boldsymbol{F}_\sigma - [\![\boldsymbol{\psi}_h \cdot \boldsymbol{n}_\sigma]\!]\right).
\end{aligned}
$$

Hence, if we define the numerical entropy flux $\boldsymbol{G}_h \equiv \{\{\boldsymbol{V}_h\}\} \mathbb{F}_h - \{\{\boldsymbol{\psi}_h\}\}$, then (9.2) multiplied by $\boldsymbol{V}(\boldsymbol{U})_K$ gives

$$
\frac{\mathrm{d}}{\mathrm{d}t} E_K(t) + (\mathrm{div}_{\mathcal{D}} \boldsymbol{G}_h(t))_K = \sum_{\sigma \in \mathcal{E}(K) \cap \mathcal{E}_{int}} \frac{1}{2d_\sigma} r_\sigma, \quad K \in \mathcal{T}_h \tag{9.16}
$$

with the local energy residual $r_\sigma \equiv [\![\boldsymbol{V}_h]\!]_\sigma \cdot \boldsymbol{F}_\sigma - [\![\boldsymbol{\psi}_h \cdot \boldsymbol{n}]\!]_\sigma$. Direct calculation yields that the energy residual for the Rusanov and the Lax–Friedrichs numerical flux can be reformulated as

$$
r_\sigma = -\delta_\sigma \, [\![\boldsymbol{U}_h]\!]_\sigma \cdot [\![\boldsymbol{V}_h]\!]_\sigma .
$$

Here $\delta_\sigma \geq \lambda_\sigma = \frac{1}{2} \max(|\boldsymbol{u}_K| + c_K, |\boldsymbol{u}_L| + c_L)$ (Rusanov flux) or $\delta_\sigma \geq \lambda = \frac{1}{2}$ $\max(|\boldsymbol{u}_h| + c_h)$ (Lax–Friedrichs flux). Note that $\boldsymbol{u} \equiv \frac{\boldsymbol{m}}{\varrho}$ is the flow velocity, and the quantity $c_h \equiv \sqrt{\frac{\gamma p_h}{\varrho_h}} = \sqrt{a\gamma \varrho_h^{\gamma-1}}$ has the physical meaning of *sound speed* related directly to compressibility of the fluid.

Due to the mean value theorem we have for any interface σ

$$
[\![\boldsymbol{V}]\!] = \frac{\mathrm{d}\boldsymbol{V}(\boldsymbol{U}^*)}{\mathrm{d}\boldsymbol{U}} [\![\boldsymbol{U}]\!] = \frac{\mathrm{d}^2 E(\boldsymbol{U}^*)}{\mathrm{d}\boldsymbol{U}^2} [\![\boldsymbol{U}]\!],
$$

where $\boldsymbol{U}^* \in co\{\boldsymbol{U}_K, \boldsymbol{U}_L\}$ with $\sigma = K|L$.

Realizing that the energy $E(\boldsymbol{U})$ is a convex function of $\boldsymbol{U}$, see Remark 2.11 and also Lemma 9.1 below, we obtain $r_\sigma \leq 0$ and thus the desired energy inequality

$$
\frac{\mathrm{d}}{\mathrm{d}t} E_K(t) + (\mathrm{div}_{\mathcal{D}} \boldsymbol{G}_h(t))_K \leq 0, \quad K \in \mathcal{T}_h. \tag{9.17}
$$

Since the numerical energy flux given by (9.5) is conservative, which means $\sum_{K\in\mathcal{T}_h}(\mathrm{div}_{\mathcal{D}}\boldsymbol{G}_h(t))_K = 0$, the integral of (9.17) yields

$$\int_\Omega E_h(t)\,\mathrm{d}x \le \int_\Omega E_h(0)\,\mathrm{d}x.$$

Remark 9.4 We should not "forget" that the original form of the energy balance is (9.16). In particular, we also have

$$-\sum_{K\in\mathcal{T}_h}\frac{|K|}{2d_\sigma}\sum_{\sigma\in\mathcal{E}(K)\cap\mathcal{E}_{int}} r_\sigma \le \int_\Omega E_h(0)\,\mathrm{d}x$$

keeping in mind that $r_\sigma \le 0$. This is a valuable piece of information that will be exploited later to derive the so-called weak BV bounds.

Remark 9.5 As a matter of fact, the specific form of the energy (entropy) flux $\boldsymbol{G}_h$ will not be used in the analysis. The integrated version of (9.16) is sufficient.

Similarly to the above, the latter inequality gives rise to $E_h \in L^\infty(0,T;L^1(\Omega))$. Noting also (2.71) and (2.72), we obtain the following bounds of our finite volume solutions uniform with respect to the norms in the following spaces:

$$\begin{aligned}\varrho_h \in L^\infty(0,T;L^\gamma(\Omega)),\ \gamma>1, \qquad p_h \in L^\infty(0,T;L^1(\Omega)),\\ \sqrt{\varrho_h}\boldsymbol{u}_h \in L^\infty(0,T;L^2(\Omega)),\ \text{ and } \boldsymbol{m}_h = \varrho_h\boldsymbol{u}_h \in L^\infty(0,T;L^{\frac{2\gamma}{\gamma+1}}(\Omega)).\end{aligned} \tag{9.18}$$

9.2.3 *Weak BV Estimates*

Clearly, the uniform estimates obtained so far are quite limited in order to allow the limit passage $h \to 0$ in the nonlinear flux functions. On the other hand, as the Euler system is a model of inviscid fluid, there is no hope to obtain uniform estimates on the discrete gradients. But there is still one piece of information to be discussed mentioned in Remark 9.4. These are the so-called *weak BV estimates* arising from the discrete energy (entropy) inequality. Specifically, the precise form of the energy residual term r_σ allows us to control at least how fast the discrete gradients blow up. Unfortunately, the price to pay will be to augment the list of anticipated properties of the numerical solutions.

Lemma 9.1 (Weak BV estimates)

Let us assume that there exist two positive constants $\underline{\varrho}$, $\overline{\varrho}$ *such that*

$$0 < \underline{\varrho} \le \varrho_h(t) \le \overline{\varrho} \tag{9.19}$$

uniformly for $t \in [0,T]$, $h \to 0$.

Then the energy Hessian $\frac{d^2E(\boldsymbol{U})}{d\boldsymbol{U}^2}$ *is strictly positive definite and the following weak BV estimate*

$$\int_0^T \sum_{\sigma\in\mathcal{E}_{int}} \lambda_\sigma \left| [\![\boldsymbol{U}_h(t)]\!] \right| h^d \, \mathrm{d}t \lesssim h^{1/2} \quad as \quad h \to 0 \tag{9.20}$$

holds for the Rusanov scheme (9.2) *with* (9.8), *where*

$$\lambda_\sigma = \frac{1}{2} \max_{s=1,\ldots,d} (|\lambda^s(\boldsymbol{U}_K)|, |\lambda^s(\boldsymbol{U}_L)|), \; \sigma = K|L.$$

Proof For the barotropic Euler system the energy Hessian reads

$$\frac{\mathrm{d}^2E(\boldsymbol{U})}{\mathrm{d}\boldsymbol{U}^2} = \begin{pmatrix} a\gamma\varrho^{\gamma-2} + \frac{|\boldsymbol{m}|^2}{\varrho^3} & -\frac{\boldsymbol{m}^T}{\varrho^2} \\ -\frac{\boldsymbol{m}}{\varrho^2} & \frac{1}{\varrho}\mathbb{I} \end{pmatrix},$$

where $\mathbb{I}$ is a $d \times d$ unit matrix. The associated eigenvalues read

$$\lambda_{1,2} = \frac{1}{2\varrho^3}\left(a\gamma\varrho^{\gamma+1} + \varrho^2 + |\boldsymbol{m}|^2 \pm \sqrt{D}\right),$$

and $\lambda_{3,\ldots,d+1} = \frac{1}{\varrho}$ with $D = \left(a\gamma\varrho^{\gamma+1} + \varrho^2 + |\boldsymbol{m}|^2\right)^2 - 4a\gamma\varrho^{\gamma+3}$. Thus, the Hessian is uniformly strictly positive as soon as (9.19) holds. Consequently, there exists

$$\underline{E}(\underline{\varrho}, \overline{\varrho}) > 0 \quad \text{such that} \quad \frac{\mathrm{d}^2E(\boldsymbol{U})}{\mathrm{d}\boldsymbol{U}^2} \geq \underline{E}\mathbb{I}. \tag{9.21}$$

Next, as pointed out in Remark 9.4, the energy inequality (9.16), integrated over Ω and over the time interval $(0, T)$, yields

$$-\int_0^T \sum_{\sigma\in\mathcal{E}_{int}} \int_\sigma r_\sigma \, \mathrm{dS}_x \, \mathrm{d}t \lesssim \int_\Omega E_h(0) \, \mathrm{d}x. \tag{9.22}$$

Let us now reconsider the energy residual r_σ,

$$r_\sigma = -\delta_\sigma [\![\boldsymbol{U}_h]\!]_\sigma [\![\boldsymbol{V}_h]\!]_\sigma .$$

Positivity of the energy Hessian (9.21) and the mean value theorem imply

$$[\![\boldsymbol{U}]\!] = \boldsymbol{U}'(\boldsymbol{V}^*) [\![\boldsymbol{V}]\!] = \left(\frac{\mathrm{d}^2(E(\boldsymbol{U}^*))}{\mathrm{d}\boldsymbol{U}^2}\right)^{-1} [\![\boldsymbol{V}]\!]$$

and thus

$$\underline{E}\,[\![\boldsymbol{U}_h]\!] \leq [\![\boldsymbol{V}_h]\!].$$

Consequently, by virtue of (9.22) and the fact $|\sigma| \approx h^{d-1}$, we have

$$\int_0^T \sum_{\sigma\in\mathcal{E}_{int}} h^d \lambda_\sigma \, [\![\boldsymbol{U}_h]\!]^2 \, \mathrm{d}t \leq \frac{1}{\underline{E}} \int_0^T \sum_{\sigma\in\mathcal{E}_{int}} h^d \delta_\sigma \, [\![\boldsymbol{U}_h]\!] \, [\![\boldsymbol{V}_h]\!] \, \mathrm{d}t \stackrel{<}{\sim} h. \tag{9.23}$$

In order to control the weak BV estimates we apply first the Hölder inequality

$$\begin{aligned} &\int_0^T \sum_{\sigma\in\mathcal{E}_{int}} h^d \lambda_\sigma |\, [\![\boldsymbol{U}_h]\!] \,| \, \mathrm{d}t \\ &\leq \left(\int_0^T \sum_{\sigma\in\mathcal{E}_{int}} h^d \lambda_\sigma \, \mathrm{d}t \right)^{1/2} \left(\int_0^T \sum_{\sigma\in\mathcal{E}_{int}} h^d \lambda_\sigma |\, [\![\boldsymbol{U}_h]\!] \,|^2 \, \mathrm{d}t \right)^{1/2}. \end{aligned} \tag{9.24}$$

The second term on the right hand side is controlled by (9.23). To show boundedness of the first term we apply the discrete trace inequality that holds for arbitrary piecewise constant function $f_h \in Q_h$, cf. (28)

$$\| f_h \|_{L^p(\partial K)} \leq h^{-1/p} \| f_h \|_{L^p(K)}, \qquad 1 \leq p \leq \infty.$$

Thus, taking into account that $|K| \approx h^d$ yields

$$\begin{aligned} &\int_0^T \sum_{\sigma\in\mathcal{E}_{int}} h^d \lambda_\sigma \, \mathrm{d}t \leq h \int_0^T \sum_{K\in\mathcal{T}_h} \sum_{\sigma\in\mathcal{E}(K)\cap\mathcal{E}_{int}} \int_\sigma \lambda_\sigma \mathrm{d}S_x \, \mathrm{d}t \\ &\stackrel{<}{\sim} h \int_0^T \sum_{K\in\mathcal{T}_h} \frac{1}{h} \int_K |\lambda(\boldsymbol{U}_K)| \, \mathrm{d}x \, \mathrm{d}t \leq \text{const.}, \quad \lambda(\boldsymbol{U}_K) = |\boldsymbol{u}_K| + c_K. \end{aligned} \tag{9.25}$$

The last inequality follows from the assumption (9.15) and from a priori estimates (9.18). In conclusion, the weak BV estimate (9.20) holds for the finite volume scheme (9.12) provided there is no vacuum and the entropy Hessian is strictly positive definite.

Clearly, the above proof can be repeated step by step for the Lax–Friedrichs scheme up to inequality (9.25), except the last estimate. At this point, it is necessary to control the L^∞-norm of the approximate velocity, specifically the integral

$$\int_0^T \|\boldsymbol{u}_h\|_{L^\infty(\Omega; R^d)} \, \mathrm{d}t.$$

In view of the inverse L^p-estimates, cf. (26), we get

$$\int_0^T \|\boldsymbol{u}_h\|_{L^\infty(\Omega;R^d)}\,\mathrm{d}t \lesssim h^{-\frac{d}{2}} \int_0^T \|\boldsymbol{u}_h\|_{L^2(\Omega;R^d)}\,\mathrm{d}t \lesssim h^{-\frac{d}{2}}$$

uniformly for $h \to 0$. Going back to (9.24) we may infer that

$$\int_0^T \sum_{\sigma\in\mathcal{E}_{int}} \lambda_\sigma \left| [\![\boldsymbol{U}_h(t)]\!] \right| h^d\,\mathrm{d}t \lesssim h^{\frac{2-d}{4}} \quad \text{as} \quad h \to 0, \tag{9.26}$$

which yields an uniform bound for $d = 2$. In general, another extra hypothesis must be introduced to save the weak BV estimates

$$\int_0^T \lambda \sum_{\sigma\in\mathcal{E}_{int}} \left| [\![\boldsymbol{U}_h(t)]\!] \right| h^d\,\mathrm{d}t \to 0 \quad \text{as} \quad h \to 0 \tag{9.27}$$

for the Lax–Friedrichs scheme. To avoid technicalities, we simply assume that the *speed of propagation is controlled*, specifically there exists $\overline{\lambda}$

$$\lambda^s(\boldsymbol{U}_h(t)) \le \overline{\lambda} \text{ uniformly for } t \in [0, T] \text{ and } h \to 0. \tag{9.28}$$

In terms of the conservative variables $(\varrho, \boldsymbol{m})$ (9.28), together with (9.19), boil down to

$$0 < \underline{\varrho} \le \varrho_h(t) \le \overline{\varrho},\ |\boldsymbol{m}_h| \le \overline{m} \tag{9.29}$$

uniformly for $t \in [0, T]$ and $h \to 0$.

Remark 9.6 It might be of interest to compare the weak BV estimates (9.20) with the BV-norm of the piecewise constant function $\boldsymbol{U}_h$,

$$\|\boldsymbol{U}_h\|_{BV(\Omega)} = \|\boldsymbol{U}_h\|_{L^1(\Omega)} + h^{d-1} \sum_{\sigma\in\mathcal{E}_{int}} |[\![\boldsymbol{U}_h]\!]|$$

that, in view of (9.20), blows up as $h^{-\frac{1}{2}}$ for $h \to 0$.

Although the anticipated uniform bounds (9.19) and (9.29) may seem strong, they are still *weak stability estimates*. This means they guarantee compactness of the sequence of approximate solutions only in terms of integral averages. As we have observed in the first part of this monograph, this is a serious obstacle for convergence of the method in the context of nonlinear problems.

9.3 Consistency

Our aim is to show consistency of the finite volume methods (9.12), (9.13). To this end let us multiply the continuity equations (9.12a) or (9.13a) by $|K|(\Pi_Q\varphi(t))_K$, with $\varphi \in C_c^2([0,T) \times \overline{\Omega})$, and the momentum equations (9.12b) or (9.13b) with $\boldsymbol{\varphi} \in C_c^2([0,T) \times \overline{\Omega}; R^d)$. If the no-flux boundary condition (9.10) is imposed, the test function for the momentum equation should also satisfy

$$\boldsymbol{\varphi} \cdot \boldsymbol{n}|_{\partial\Omega} = 0. \tag{9.30}$$

This might be a bit ambiguous statement as the numerical domain in our case consists of a finite union of rectangles/cuboids. In particular, Ω need not be even Lipschitz if $d = 3$. Here and hereafter, we tacitly adopt the convention that $\boldsymbol{\varphi}$ satisfies (9.30) at any $x \in \partial\Omega$, where the outer normal vector $\boldsymbol{n}(x)$ exists. It is easy to check that some components of such a function must vanish on the certain edges (vertices) lying on $\partial\Omega$.

We sum the resulting equations over $K \in \mathcal{T}_h$ and integrate in time. A priori estimates (9.18) combined with the weak BV estimates (9.20), (9.27) allow us to show consistency.

Time derivative
Integration by parts with respect to time leads to

$$|K| \int_0^T \frac{\mathrm{d}}{\mathrm{d}t} \sum_{K\in\mathcal{T}_h} \varrho_K(t)(\Pi_Q\varphi(t))_K \,\mathrm{d}t = \int_0^T \frac{\mathrm{d}}{\mathrm{d}t} \int_\Omega \varrho_h(t)\varphi(t,x)\,\mathrm{d}x\,\mathrm{d}t$$
$$= \left[\int_\Omega \varrho_h(\tau)\varphi(\tau,\cdot)\,\mathrm{d}x \right]_{\tau=0}^{\tau=T} - \int_0^T \int_\Omega \varrho_h(t)\partial_t\varphi(t,x)\,\mathrm{d}x\,\mathrm{d}t$$

in the continuity equation, and, similarly, to

$$|K| \int_0^T \frac{\mathrm{d}}{\mathrm{d}t} \sum_{K\in\mathcal{T}_h} \boldsymbol{m}_K(t) \cdot (\Pi_Q\boldsymbol{\varphi}(t))_K \,\mathrm{d}t$$
$$= \left[\int_\Omega \boldsymbol{m}_h(\tau) \cdot \boldsymbol{\varphi}(\tau,x)\,\mathrm{d}x \right]_{\tau=0}^{\tau=T} - \int_0^T \int_{\Omega_h} \boldsymbol{m}_h(t) \cdot \partial_t\boldsymbol{\varphi}(t,x)\,\mathrm{d}x\,\mathrm{d}t$$

in the momentum equation.

Convective terms
To treat the convective terms in the continuity equations we use the discrete integration by parts (41) and the Taylor expansion to get

$$
\begin{aligned}
&|K| \int_0^T \sum_{K\in\mathcal{T}_h} \mathrm{div}_h \boldsymbol{m}_h(t) (\Pi_Q \varphi(t))_K \,\mathrm{d}t \\
&= -|K| \int_0^T \sum_{K\in\mathcal{T}_h} \sum_{s=1}^d m_K^s(t) \left(\int_K \frac{\varphi(t, x + h\mathbf{e}_s) - \varphi(t, x - h\mathbf{e}_s)}{2h} \,\mathrm{d}x \right) \mathrm{d}t \\
&= - \int_0^T \int_\Omega \boldsymbol{m}_h(t) \cdot \nabla_x \varphi(t, x) \,\mathrm{d}x \,\mathrm{d}t + r_1,
\end{aligned}
$$

where $\mathbf{e}_s$ is the unit basis vector in the s-th space direction, $s = 1, ..., d$, and the reminder term r_1 is estimated as follows

$$
r_1 \lesssim h \left\| \frac{\mathrm{d}^2 \varphi}{\mathrm{d}x^2}(\hat{x}) \right\|_{C(0,T)} \|\boldsymbol{m}_h\|_{L^\infty(L^1)}, \quad \text{where } \frac{\mathrm{d}^2\varphi}{\mathrm{d}x^2} \equiv \left(\frac{\partial^2 \varphi}{\partial x_i \partial x_j} \right)_{i,j=1}^d . \tag{9.31}
$$

Point $\hat{x}$ appears in the remainder of the Taylor expansion and lies either between the points $x + h\mathbf{e}_s$ and x or the points x and $x - h\mathbf{e}_s$.

We proceed analogously with the convective term in the momentum equations

$$
\begin{aligned}
&|K| \int_0^T \sum_{K\in\mathcal{T}_h} \mathrm{div}_h \left(\frac{\boldsymbol{m}_h(t) \otimes \boldsymbol{m}_h(t)}{\varrho_h(t)} + p_h(t)\mathbb{I} \right) (\Pi_Q \boldsymbol{\varphi}(t))_K \,\mathrm{d}t \\
&= -|K| \int_0^T \sum_{K\in\mathcal{T}_h} \sum_{s=1}^d \sum_{z=1}^d \left(\frac{m_h^s(t) m_h^z(t)}{\varrho_h(t)} + p_h(t) \right) \\
&\qquad \times \left(\int_K \frac{\varphi^z(t, x + h\mathbf{e}_s) - \varphi^z(t, x - h\mathbf{e}_s)}{2h} \,\mathrm{d}x \right) \mathrm{d}t \\
&= - \int_0^T \int_\Omega \left(\frac{\boldsymbol{m}_h(t) \otimes \boldsymbol{m}_h(t)}{\varrho_h(t)} + p_h(t)\mathbb{I} \right) \cdot \nabla_x \boldsymbol{\varphi}(t, x) \,\mathrm{d}x \,\mathrm{d}t + r_2,
\end{aligned}
$$

where term r_2 is bounded by

$$
r_2 \lesssim h \left\| \frac{\mathrm{d}^2 \varphi}{\mathrm{d}x^2}(\hat{x}) \right\|_{C(0,T)} \left\{ \|\sqrt{\varrho_h} \boldsymbol{u}_h\|_{L^\infty(L^2)} + \|p_h\|_{L^\infty(L^1)} \right\}.
$$

Note that we have used the fact that the normal trace of $\boldsymbol{\varphi}$ vanishes on the boundary of $\partial\Omega$.

Numerical diffusion

Now we estimate the numerical diffusion term of the Rusanov finite volume method (9.12), estimates for the Lax–Friedrichs method (9.13) can be obtained in an analogous way under the assumption (9.29) on the finite speed of propagation.

Considering the diffusion terms in (9.12) we obtain, cf. (19)

$$\int_0^T \sum_{K\in\mathcal{T}_h} \sum_{\sigma\in\mathcal{E}(K)\cap\mathcal{E}_{int}} \int_\sigma \lambda_\sigma [\![\boldsymbol{U}_h(t)]\!] \left(\Pi_Q \boldsymbol{\varphi}(t)\right)_K \,\mathrm{d}t. \tag{9.32}$$

The terms belonging to an arbitrary but fixed face $\sigma = K|L$ are

$$\frac{1}{d_\sigma}\int_0^T \left(\lambda_\sigma [\![\boldsymbol{U}_h(t)]\!] \int_K \boldsymbol{\varphi}(t)\,\mathrm{d}x - \lambda_\sigma [\![\boldsymbol{U}_h(t)]\!] \int_L \boldsymbol{\varphi}(t)\,\mathrm{d}x \right) \mathrm{d}t. \tag{9.33}$$

Let us now consider an arbitrary but fixed point $\tilde{x} \in \sigma$; without loss of generality let $\tilde{x} = (\tilde{x}_s, x')$, $x' \in R^{d-1}$, $s = 1, \ldots, d$. The Taylor expansion for $x = (x_s, x') \in K$ with respect to $(\tilde{x}_s, x')$ gives

$$\boldsymbol{\varphi}(x_s, x') = \boldsymbol{\varphi}(\tilde{x}_s, x') - \xi\,\partial_s \boldsymbol{\varphi}(\tilde{x}_s, x') + \mathcal{O}(h^2),$$

where $\xi \in (0, d_\sigma)$. Analogously, we have for $x = (\tilde{x}_s, x') \in L$

$$\boldsymbol{\varphi}(x_s, x') = \boldsymbol{\varphi}(\tilde{x}_s, x') + \xi\,\partial_s \boldsymbol{\varphi}(\tilde{x}_s, x') + \mathcal{O}(h^2).$$

Substituting the above Taylor expansions in (9.33) we directly see that the terms multiplied by $\boldsymbol{\varphi}(\tilde{x}_s, x')$ vanish. The resulting terms give

$$\begin{aligned}
&\Bigg| \int_0^T -\frac{1}{d_\sigma}\lambda_\sigma [\![\boldsymbol{U}_h]\!] \int_0^{d_\sigma}\int_\sigma \xi\,\partial_s\boldsymbol{\varphi}(\tilde{x}_s, x')\mathrm{d}S_{x'}\mathrm{d}\xi \\
&\qquad + \frac{1}{d_\sigma}\lambda_\sigma [\![\boldsymbol{U}_h]\!] \int_0^{d_\sigma}\int_\sigma -\xi\,\partial_s\boldsymbol{\varphi}(\tilde{x}_s, x')\mathrm{d}S_{x'}\mathrm{d}\xi \,\mathrm{d}t \Bigg| \\
&\le \frac{2}{d_\sigma}\int_0^T \left| \lambda_\sigma [\![\boldsymbol{U}_h]\!] \int_0^{d_\sigma}\int_\sigma \xi\,\partial_s\boldsymbol{\varphi}(\tilde{x}_s, x')\mathrm{d}S_{x'}\mathrm{d}\xi \right| \mathrm{d}t \\
&\lesssim h^d \int_0^T \lambda_\sigma \left| [\![\boldsymbol{U}_h]\!] \right| dt \, \|\boldsymbol{\varphi}\|_{C^1([0,T]\times\overline{\Omega})} \to 0 \qquad \text{for } h \to 0.
\end{aligned}$$

The last convergence follows from the weak BV property (9.20) and implies the consistency of the numerical diffusion term (9.32).

Let us summarize the obtained consistency result.

Lemma 9.2 (Consistency)
Let us assume that there exist two positive constants $\underline{\varrho}$, $\overline{\varrho}$ such that

$$0 < \underline{\varrho} \leq \varrho_h(t) \leq \overline{\varrho}$$

uniformly for $t \in [0, T]$, $h \to 0$.

Then the Rusanov finite volume method (9.12) is consistent with the barotropic Euler equations (9.6) *and* (9.7)*, specifically*

$$- \int_{\Omega} \varrho_h^0 \varphi(0, \cdot) \, \mathrm{d}x = \int_0^T \int_{\Omega} \left[\varrho_h \partial_t \varphi + \boldsymbol{m}_h \cdot \nabla_x \varphi \right] \, \mathrm{d}x \, \mathrm{d}t + \int_0^T e_{1,h}(t, \varphi) \, \mathrm{d}t, \quad (9.34a)$$

for any $\varphi \in C_c^2([0, T) \times \overline{\Omega})$;

$$- \int_{\Omega} \boldsymbol{m}_h^0 \boldsymbol{\varphi}(0, \cdot) \, \mathrm{d}x = \int_0^T \int_{\Omega} \left[\boldsymbol{m}_h \cdot \partial_t \boldsymbol{\varphi} + \frac{\boldsymbol{m}_h \otimes \boldsymbol{m}_h}{\varrho_h} : \nabla_x \boldsymbol{\varphi} + p_h \mathrm{div}_x \boldsymbol{\varphi} \right] \mathrm{d}x \, \mathrm{d}t + \int_0^T e_{2,h}(t, \boldsymbol{\varphi}) \, \mathrm{d}t \quad (9.34b)$$

for any $\boldsymbol{\varphi} \in C_c^2([0, T] \times \overline{\Omega})$, $\boldsymbol{\varphi} \cdot \boldsymbol{n}|_{\partial\Omega} = 0$;

$$\|e_{1,h}(\cdot, \varphi)\|_{L^1(0,T)} \lesssim h \|\varphi\|_{C^2}, \ \|e_{2,h}(\cdot, \boldsymbol{\varphi})\|_{L^1(0,T)} \lesssim h \|\boldsymbol{\varphi}\|_{C^2}.$$

If we assume that

$$0 < \underline{\varrho} \leq \varrho_h(t) \leq \overline{\varrho}, \ |\boldsymbol{m}_h(t)| \leq \overline{m}, \ t \in [0, T],$$

then the Lax–Friedrichs scheme (9.13) is consistent with the barotropic Euler equations and the above consistency formulation holds.

Remark 9.7 Note that the above formulation includes also the problem with space periodic boundary conditions as soon as $\mathbb{T}^d$ is interpreted as a d-dimensional manifold without boundary.

9.4 Convergence

In the previous two sections we have shown that the finite volume methods (9.12), (9.13) are stable and consistent. This implies that their numerical solutions $\{\varrho_h, \boldsymbol{m}_h\}_{h \searrow 0}$ yield consistent approximations of the barotropic Euler equations (9.6), (9.7), in the sense of Definition 5.4. We are ready to apply the theory of generalized solutions to study convergence.

To begin, we point out that our results so far have been conditioned by the hypotheses (9.19), (9.29), respectively. Accordingly, in the remaining part of this section, we suppose at least (9.19), specifically,

$$0 < \underline{\varrho} \leq \varrho_h(t) \leq \overline{\varrho} \ \text{for} \ t \in [0, T]$$

uniformly for $h \to 0$. Accordingly, there is a sequence $h_n \to 0$ such that

$$\varrho_n \equiv \varrho_{h_n} \to \varrho \ \text{weakly-(*) in} \ L^\infty((0,T) \times \Omega),$$
$$\boldsymbol{m}_n \equiv \boldsymbol{m}_{h_n} \to \boldsymbol{m} \ \text{weakly-(*) in} \ L^\infty(0,T; L^2(\Omega; R^d)).$$

In addition, the limit density satisfies

$$0 < \underline{\varrho} \leq \varrho \leq \overline{\varrho} \ \text{a.a. in} \ (0,T) \times \Omega.$$

Leaving apart the fundamental question in which sense the limit solves the Euler system, there are several issues to discuss even at this level:

- **Q1:** Does the limit depend on the choice of the sequence of numerical steps $h_n \to 0$?
- **Q2:** Is it possible to visualize the weak convergence and to identify the limit with some predictable error?

In the context of the barotropic Euler system, we can guarantee that the limit is unique, meaning independent of the choice of $h_n \to 0$ only if the limit system admits a unique solution. As the Euler system is in general ill posed in the class of (admissible) weak solutions, the only possibility is that the limit is a strong solution. As we have seen in Chap. 4, smooth initial data prescribed on smooth domain give rise to local-in-time strong solutions. In general, however, smooth solutions loose regularity in a finite time, see Chap. 2 for details. As soon as the limit problem is not uniquely solvable, we have to content ourselves with convergence *up to a subsequence*.

Weak convergence, or convergence of integral averages, provides very little piece of information concerning the limit. Even if we anticipate that concentrations are eliminated, which is actually the case under the hypothesis (9.29), oscillations or wiggles represent serious difficulties in identifying the weak limit. The concept of $\mathcal{K}$–convergence, discussed in Sect. 7.2, introduces statistical averaging in the limit process that may help to eliminate fluctuations and gives rise to *strong convergence*. Indeed, using Theorems 7.1, 7.3, we may infer that

$$\frac{1}{N}\sum_{n=1}^{N}\varrho_n \to \varrho \text{ (strongly) in } L^q((0,T)\times\Omega) \text{ as } N\to\infty, \text{ for any } 1\le q<\infty,$$
$$\frac{1}{N}\sum_{n=1}^{N}\boldsymbol{m}_n \to \boldsymbol{m} \text{ (strongly) in } L^2((0,T)\times\Omega;R^d)) \text{ as } N\to\infty$$

and

$$\frac{1}{N}\sum_{n=1}^{N}\varrho_n \to \varrho \text{ a.a. in } (0,T)\times\Omega \text{ as } N\to\infty,$$
$$\frac{1}{N}\sum_{n=1}^{N}\boldsymbol{m}_n \to \boldsymbol{m} \text{ a.a. in } (0,T)\times\Omega \text{ as } N\to\infty$$

passing again to a suitable subsequence as the case may be. Strong is definitely better than weak, however, the process is more costly in practical implementations as it requires the knowledge of all successive approximations up to order h_N, $h_N \to 0$. One may also worry that the method based on Cesàro averaging may actually slow down the real speed of convergence, if available. To illuminate this phenomenon, suppose that $U_n \to U$ in certain norm $\|\cdot\|$ with the rate

$$\|U_n - U\| \stackrel{<}{\sim} n^{-\beta},\ 0<\beta\le 1.$$

Then

$$\left\|\frac{1}{N}\sum_{n=1}^{N}U_n - U\right\| \le \frac{1}{N}\sum_{n=1}^{N}\|U_n - U\| \stackrel{<}{\sim} \frac{1}{N}\sum_{n=1}^{N}n^{-\beta} \approx \begin{cases}\frac{\log(N)}{N} & \text{if } \beta=1,\\ \frac{1}{1-\beta}N^{-\beta} & \text{if } 0<\beta<1.\end{cases}$$

In other words, if we assume that the convergence rate is $\mathcal{O}(h^\beta)$, i.e.

$$\|U_h - U\| \stackrel{<}{\sim} h^{\beta},\ 0<\beta\le 1,$$

then the convergence rate of the Cesàro averages for a subsequence $h_n \approx 1/n$ is

$$\left\|\frac{1}{N}\sum_{n=1}^{N}U_{h_n} - U\right\| \le \frac{1}{N}\sum_{n=1}^{N}\|U_{h_n} - U\| \stackrel{<}{\sim} \frac{1}{N}\sum_{n=1}^{N}h_n^{\beta} \approx \begin{cases}-\log(h_N)h_N & \text{if } \beta=1,\\ \frac{1}{1-\beta}h_N^{\beta} & \text{if } 0<\beta<1.\end{cases}$$

Thus the rate is of the same order if $\beta<1$ and slightly worse if $\beta=1$. Note that for piecewise constant approximations the value $\beta=1$ is optimal.

In addition, knowing the sequence of approximate solutions $\{\varrho_h, \boldsymbol{m}_h\}_{h\searrow 0}$ we may also consider a Young measure. Note that this step may require another subsequence even if the limits ϱ, $\boldsymbol{m}$ were already identified. More specifically, Theorem 5.3

implies that there is a subsequence of $\{\varrho_{h_n}, \boldsymbol{m}_{h_n}\}_{n=1}^{\infty}$ that generates a Young measure $\left\{\mathcal{V}_{t,x}\right\}_{(t,x)\in(0,T)\times\Omega}$,

$$\mathcal{V}_{t,x} \in \mathcal{P}(R^{d+1}), \ R^{d+1} = \left\{\widetilde{\varrho} \in R, \widetilde{\boldsymbol{m}} \in R^d\right\}$$

that is a DMV solution of the barotropic Euler equations in the sense of Definition 5.5. Moreover, in view of the available bounds, $\mathcal{V}_{t,x}$ possesses finite second moments, and

$$\mathcal{V}_{t,x}\left\{\underline{\varrho} \le \widetilde{\varrho} \le \overline{\varrho}, \ \widetilde{\boldsymbol{m}} \in R^d\right\} = 1$$

for a.a. (t, x). Finally, by virtue of Theorem 5.4, the weak limit $(\varrho, \boldsymbol{m})$ is the barycenter of $\mathcal{V}$,

$$\varrho = \langle \mathcal{V}; \widetilde{\varrho} \rangle , \ \boldsymbol{m} = \langle \mathcal{V}, \widetilde{\boldsymbol{m}} \rangle ,$$

and as such represents a DW solution in the sense of Definition 5.6.

9.4.1 Strong Convergence

We assume that the Euler system (9.6), (9.7), possibly with impermeability condition (9.10), admits a smooth solution $[\varrho, \boldsymbol{m}]$ in $[0, T]$. Our goal is to apply Theorem 7.11. Let us start with the periodic boundary conditions.

Theorem 9.1 (Strong convergence for Lax–Friedrichs/Rusanov scheme, periodic case)

Let $\{\varrho_h, \boldsymbol{m}_h\}_{h\searrow 0}$ be a family of approximate solutions obtained by the Rusanov finite volume method (9.12) or the Lax–Friedrichs finite volume method (9.13), with the periodic boundary conditions ($\Omega = \mathbb{T}^d$). In addition, suppose that $(\varrho_h, \boldsymbol{m}_h)$ satisfy

$$0 < \underline{\varrho} \le \varrho_h(t) \le \overline{\varrho} \ \textit{for} \ t \in [0, T]$$

for the Rusanov scheme, and

$$0 < \underline{\varrho} \le \varrho_h(t) \le \overline{\varrho}, \ |\boldsymbol{m}_h(t)| \le \overline{m}, \ t \in [0, T]$$

for the Lax–Friedrichs scheme.

Finally, suppose that the limit Euler system admits a strong solution $[\varrho, \boldsymbol{m}]$ in the class

$$\varrho \in W^{1,\infty}((0, T) \times \mathbb{T}^d), \ \inf_{(0,T)\times\mathbb{T}^d} \varrho > 0, \ \boldsymbol{m} \in W^{1,\infty}((0, T) \times \mathbb{T}^d; R^d).$$

Then

$$\varrho_h \to \varrho \text{ in } L^q((0,T) \times \mathbb{T}^d),\ \boldsymbol{m}_h \to \boldsymbol{m} \text{ in } L^q(0,T; L^2(\mathbb{T}^d; R^d))$$

as $h \to 0$ *for any* $1 \le q < \infty$.

To retain the same result for the problem endowed with the impermeability condition, we have to realize that the smooth solution $[\varrho, \boldsymbol{m}]$ must be eligible as a test function in the consistency formulation stated in Lemma 9.2, namely

$$\varrho \in C^2_c([0,T) \times \overline{\Omega}),\ \boldsymbol{m} \in C^2_c([0,T) \times \overline{\Omega}; R^d),\ \boldsymbol{m} \cdot \boldsymbol{n}|_{\partial\Omega} = 0. \tag{9.35}$$

At first glance, this may seem rather unrealistic, in view of the low regularity of the boundary of the numerical domain. However, smooth solutions of the barotropic Euler system admit the *finite speed of propagation*. Consider the Cauchy problem

$$\partial_t \varrho + \mathrm{div}_x \boldsymbol{m} = 0, \tag{9.36}$$

$$\partial_t \varrho + \mathrm{div}_x \left(\frac{\boldsymbol{m} \otimes \boldsymbol{m}}{\varrho} \right) + \nabla_x p(\varrho) = 0, \tag{9.37}$$

in $[0,T) \times R^d$, with the far-field conditions

$$\varrho \to \overline{\varrho} > 0,\ \boldsymbol{m} \to 0 \text{ as } |x| \to \infty,$$

and with the initial data confined to a bounded ball,

$$\varrho_0(x) = \overline{\varrho},\ \boldsymbol{m}_0(x) = 0 \text{ whenever } |x| > R. \tag{9.38}$$

Then any C^1 solution $[\varrho, \boldsymbol{m}]$ of the problem satisfies

$$\varrho(t,x) = \overline{\varrho},\ \boldsymbol{m}(t,x) = 0 \text{ whenever } |x| > R + t\sqrt{p'(\overline{\varrho})},$$

see e.g. Sideris [183, Lemma 3.2]. Clearly, the above solution is smooth if the initial data are smooth and it satisfies the impermeability condition on *any* sufficiently large domain $\Omega \subset R^d$ satisfying

$$\left\{ |x| \le R + T\sqrt{p'(\overline{\varrho})} \right\} \subset \Omega.$$

In view of this observation, let us reformulate Theorem 9.1 for the impermeability boundary conditions.

Theorem 9.2 (Strong convergence for Lax–Friedrich/Rusanov scheme, no-flux boundary)
Theorem 9.1 remains valid in the case of the no-flux boundary condition (9.10), (9.11) *provided:*

- *the numerical domain is bounded Lipschitz;*
- *the strong solution* $[\varrho, \boldsymbol{m}]$ *belongs to the class*

$$\varrho \in C_c^2([0,T) \times \overline{\Omega}),\ \boldsymbol{m} \in C_c^2([0,T) \times \overline{\Omega}; R^d),\ \boldsymbol{m} \cdot \boldsymbol{n}|_{\partial\Omega} = 0.$$

The weak-strong compatibility principle stated in Theorem 5.9 gives rise to another interesting situation when convergence is strong and unconditional. For the sake of simplicity, we consider the periodic boundary conditions.

Theorem 9.3 (Strong convergence for Lax–Friedrichs/Rusanov scheme, smooth limit)
Let $\{\varrho_h, \boldsymbol{m}_h\}_{h \searrow 0}$ *be a family of approximate solutions obtained by the Rusanov finite volume method (9.12) or the Lax–Friedrichs finite volume method (9.13), with the periodic boundary conditions* ($\Omega = \mathbb{T}^d$). *In addition, suppose that* $(\varrho_h, \boldsymbol{m}_h)$ *satisfy*

$$0 < \underline{\varrho} \leq \varrho_h(t) \leq \overline{\varrho} \ \text{for}\ t \in [0,T]$$

for the Rusanov scheme, and

$$0 < \underline{\varrho} \leq \varrho_h(t) \leq \overline{\varrho},\ |\boldsymbol{m}_h(t)| \leq \overline{m},\ t \in [0,T]$$

for the Lax–Friedrichs scheme.
Finally, suppose there is a sequence $h_n \to 0$ *such that*

$$\varrho_{h_n} \to \varrho \text{ weakly-(*) in } L^\infty((0,T) \times \mathbb{T}^d),$$
$$\boldsymbol{m}_{h_n} \to \boldsymbol{m} \text{ weakly-(*) in } L^\infty(0,T; L^2(\mathbb{T}^d; R^d)),$$

where

$$\varrho \in C^1([0,T] \times \mathbb{T}^d),\ \boldsymbol{m} \in C^1([0,T] \times \mathbb{T}^d; R^d).$$

Then $[\varrho, \boldsymbol{m}]$ *is a classical solution to the limit Euler system, and*

$$\varrho_h \to \varrho \ \text{in}\ L^q((0,T) \times \mathbb{T}^d),\ \boldsymbol{m}_h \to \boldsymbol{m} \ \text{in}\ L^q(0,T; L^2(\mathbb{T}^d; R^d))$$

as $h \to 0$ *for any* $1 \leq q < \infty$.

Note carefully the subtle difference between Theorems 9.1 and 9.3. The former postulates the *existence* of a smooth solution of the limit system; while the latter requires *smoothness* of the asymptotic limit. The convergence is unconditional, meaning for the full range $h \to 0$, in both cases.

The last situation when the strong convergence is guaranteed is a direct consequence of Theorem 7.10.

Theorem 9.4 (Strong convergence to weak solution)

Let $\{\varrho_h, \boldsymbol{m}_h\}_{h\searrow 0}$ *be a sequence of approximate solutions obtained by the Rusanov finite volume method (9.12) or the Lax–Friedrichs finite volume method (9.13), with the no-flux boundary condition* (9.10), (9.11) *on a bounded Lipschitz domain. In addition, suppose that* $(\varrho_h, \boldsymbol{m}_h)$ *satisfy*

$$0 < \underline{\varrho} \leq \varrho_h(t) \leq \overline{\varrho} \ \text{for} \ t \in [0, T]$$

for the Rusanov scheme, and

$$0 < \underline{\varrho} \leq \varrho_h(t) \leq \overline{\varrho}, \ |\boldsymbol{m}_h(t)| \leq \overline{m}, \ \ t \in [0, T]$$

for the Lax–Friedrichs scheme.

Let there exist a subsequence $h_n \to 0$*, such that*

$$\begin{aligned} \varrho_{h_n} &\to \varrho \ \text{weakly-(*) in} \ L^\infty((0, T) \times \Omega), \\ \boldsymbol{m}_{h_n} &\to \boldsymbol{m} \ \text{weakly-(*) in} \ L^\infty(0, T; L^2(\Omega; R^d)), \end{aligned}$$

where $[\varrho, \boldsymbol{m}]$ *is a weak solution to the limit Euler system in* $(0, T) \times \Omega$*. Finally, suppose that*

$$\varrho_{h_n} \to \varrho \ \text{in} \ L^1((0, T) \times \mathcal{U}), \ \boldsymbol{m}_{h_n} \to \boldsymbol{m} \ \text{in} \ L^2((0, T) \times \mathcal{U}; R^d), \tag{9.39}$$

where $\mathcal{U}$ *is an open neighborhood of the boundary* $\partial\Omega$.

Then

$$\varrho_{h_n} \to \varrho \ \text{in} \ L^q((0, T) \times \Omega), \ \boldsymbol{m}_{h_n} \to \boldsymbol{m} \ \text{in} \ L^q(0, T; L^2(\Omega; R^d))$$

for any $1 \leq q < \infty$.

Very roughly indeed, if the approximate sequence converges strongly in a neighborhood of the physical boundary to a weak solution, then the convergence must be strong in the full domain. Note that the scenario required in (9.39) may be quite realistic in view of the finite speed of propagation property for the Euler system discussed above. Indeed the weak solution may be constant in a neighborhood of the boundary and convergence may be strong in this neighborhood. Solutions with compact support are often chosen in the benchmark problems.

9.4.2 Weak Convergence

The weak convergence includes concentrations and oscillations in the approximate sequence. Here the concentrations have been (artificially) eliminated for the Lax–Friedrichs scheme by hypothesis (9.29) and largely eliminated by (9.19) for the Rusanov scheme. In addition, as the density is bounded below in both cases, the total energy

$$E = \frac{1}{2}\frac{|\boldsymbol{m}|^2}{\varrho} + \frac{a}{\gamma - 1}\varrho^\gamma$$

may be viewed as a continuous convex function of $[\varrho, \boldsymbol{m}]$.

In view of the convergence results stated in Theorems 9.1, 9.2, the convergence may be weak only if the limit problem does not admit a strong solution. Moreover, by virtue of Theorem 9.3, the weak limit $[\varrho, \boldsymbol{m}]$ is not continuously differentiable. Finally, extrapolating a bit the conclusion of Theorem 9.4, we may say that the limit is not a weak solution of the limit Euler system; whence a larger class of DW solutions must be used. Let us first summarize the general convergence result.

Theorem 9.5 (Weak convergence)

Let $\{\varrho_h, \boldsymbol{m}_h\}_{h\searrow 0}$ be a sequence of approximate solutions obtained by the Rusanov finite volume method (9.12) or the Lax–Friedrichs finite volume method (9.13) with the periodic boundary conditions ($\Omega = \mathbb{T}^d$) or with the no-flux boundary conditions (9.10), (9.11) on a bounded Lipschitz domain. In addition, suppose that $(\varrho_h, \boldsymbol{m}_h)$ satisfy

$$0 < \underline{\varrho} \le \varrho_h(t) \le \overline{\varrho},\ t \in [0, T]$$

for the Rusanov scheme, and

$$0 < \underline{\varrho} \le \varrho_h(t) \le \overline{\varrho},\ |\boldsymbol{m}_h(t)| \le \overline{m},\ t \in [0, T]$$

for the Lax–Friedrichs scheme.

Then there is a sequence $h_n \to 0$ such that

$$\begin{aligned} \varrho_{h_n} &\to \varrho \ \textit{weakly-(*) in } L^\infty((0,T)\times\Omega), \\ \boldsymbol{m}_{h_n} &\to \boldsymbol{m} \ \textit{weakly-(*) in } L^\infty(0,T; L^2(\Omega; R^d)), \end{aligned}$$

$$\frac{1}{N}\sum_{n=1}^{N} \varrho_{h_n} \to \varrho \ \textit{in } L^q((0,T)\times\Omega), \ \textit{and a.a. in } (0,T)\times\Omega \ \textit{as } N \to \infty,$$

$$\frac{1}{N}\sum_{n=1}^{N} \boldsymbol{m}_{h_n} \to \boldsymbol{m} \ \textit{in } L^2((0,T)\times\Omega; R^d)), \ \textit{and a.a. in } (0,T)\times\Omega \ \textit{as } N \to \infty,$$

for any $1 \le q < \infty$; and

$$\frac{\boldsymbol{m}_{h_n} \otimes \boldsymbol{m}_{h_n}}{\varrho_{h_n}} + a\varrho_{h_n}^{\gamma}\mathbb{I} \to \overline{\frac{\boldsymbol{m} \otimes \boldsymbol{m}}{\varrho} + a\varrho^{\gamma}\mathbb{I}} \ \text{weakly-(*) in } L^{\infty}(0,T; \mathcal{M}(\overline{\Omega}; R^{d\times d}_{\text{sym}})). \tag{9.40}$$

The limit $[\varrho, \boldsymbol{m}]$ *is a DW solution of the barotropic Euler system in the sense of Definition 5.6, with* $M = 1$ *in the case of periodic boundary conditions and* $M = 2$ *in the case of impermeability boundary conditions, and with the Reynolds defect*

$$\mathfrak{R} = \overline{\frac{\boldsymbol{m} \otimes \boldsymbol{m}}{\varrho} + a\varrho^{\gamma}\mathbb{I}} - \left(\frac{\boldsymbol{m} \otimes \boldsymbol{m}}{\varrho} + a\varrho^{\gamma}\mathbb{I}\right).$$

9.4.3 Weak Convergence – Young Measure

Finally, we reformulate the conclusion of the abstract result stated in Theorem 7.9 in the present context. As already pointed out, the family $\{\varrho_h, \boldsymbol{m}_h\}_{h\searrow 0}$ of approximate solutions generates a Young measure $\{\mathcal{V}_{t,x}\}_{(t,x)\in(0,T)\times\Omega}$. More specifically, there is a sequence $h_n \to 0$ such that

$$g(\varrho_{h_n}, \boldsymbol{m}_{h_n}) \to \langle \mathcal{V}; g(\widetilde{\varrho}, \widetilde{\boldsymbol{m}})\rangle \ \text{ weakly-(*) in } L^{\infty}((0,T)\times\Omega),$$
$$\frac{1}{N}\sum_{n=1}^{N} g(\varrho_{h_n}, \boldsymbol{m}_{h_n})(t,x) \to \left\langle \mathcal{V}_{t,x}; g(\widetilde{\varrho}, \widetilde{\boldsymbol{m}})\right\rangle \text{ as } N \to \infty \text{ for a.a. } (t,x) \in (0,T)\times\Omega \tag{9.41}$$

for any $g \in C_c(R^{d+1})$. Moreover, in view of the hypotheses (9.19), (9.29), validity (9.41) can be extended to any

$$g \in C(R^{d+1}),\ |g(\widetilde{\varrho}, \widetilde{\boldsymbol{m}})| \lesssim (1 + |\widetilde{\boldsymbol{m}}|^q),\ 1 \le q < 2 \ \text{ for all } \underline{\varrho} \le \widetilde{\varrho} \le \overline{\varrho}$$

in the case of Rusanov scheme, and to any

$$g \in C(R^{d+1})$$

in the case of the Lax–Friedrichs scheme. In particular, the Reynolds defect in Theorem 9.5 takes the form

$$\mathfrak{R} = \overline{\frac{\boldsymbol{m} \otimes \boldsymbol{m}}{\varrho}} + a\,\langle \mathcal{V}; \widetilde{\varrho}^{\gamma}\rangle\,\mathbb{I} - \left(\frac{\boldsymbol{m} \otimes \boldsymbol{m}}{\varrho} - a\varrho^{\gamma}\mathbb{I}\right)$$

for the Rusanov scheme, and

$$\mathfrak{R} = \left\langle \mathcal{V}; \frac{\widetilde{\boldsymbol{m}} \otimes \widetilde{\boldsymbol{m}}}{\widetilde{\varrho}} + a\widetilde{\varrho}^{\gamma}\mathbb{I}\right\rangle - \left(\frac{\boldsymbol{m} \otimes \boldsymbol{m}}{\varrho} - a\varrho^{\gamma}\mathbb{I}\right)$$

for the Lax–Friedrichs scheme. Recall that energy is continuous on the support of the Young measure $\mathcal{V}$ in both cases.

Summarizing the above observations, we may apply Theorem 7.9 (see also Theorem 6) to deduce the following conclusion.

Theorem 9.6 (Weak convergence – Young measure)

Under the hypotheses of Theorem 9.5, the subsequence $h_n \to 0$ can be chosen so that $\{\varrho_{h_n}, \boldsymbol{m}_{h_n}\}$ generates a Young measure $\mathcal{V}$. More specifically, there holds:

- **Weak convergence to Young measure**

$$\delta_{\varrho_{h_n}, \boldsymbol{m}_{h_n}} \to \mathcal{V} \text{ weakly-(*) in } L^\infty((0,T) \times \Omega; \mathcal{M}(R^{d+1}));$$

- **Strong convergence of Cesàro averages**

$$d_{W_s}\left[\frac{1}{N}\sum_{n=1}^{N} \delta_{\varrho_{h_n}(t,x), \boldsymbol{m}_{h_n}(t,x)}; \mathcal{V}_{t,x}\right] \to 0 \text{ as } N \to \infty$$
$$\text{in } L^s((0,T) \times \Omega) \text{ and for a.a. } (t,x) \in (0,T) \times \Omega,$$

where $1 \le s < 2$ for the Rusanov scheme, and $s \ge 1$ arbitrary finite for the Lax–Friedrichs scheme;

- **Strong convergence of deviations/variations**

$$\begin{aligned} &\left\langle \frac{1}{N}\sum_{n=1}^{N} \delta_{\varrho_{h_n}, \boldsymbol{m}_{h_n}}; \left| (\widetilde{\varrho}, \widetilde{\boldsymbol{m}}) - \left\langle \frac{1}{N}\sum_{n=1}^{N} \delta_{\varrho_{h_n}, \boldsymbol{m}_{h_n}}; (\widetilde{\varrho}, \widetilde{\boldsymbol{m}}) \right\rangle \right|^s \right\rangle \\ &\to \left\langle \mathcal{V}; \left| (\widetilde{\varrho}, \widetilde{\boldsymbol{m}}) - \langle \mathcal{V}; (\widetilde{\varrho}, \widetilde{\boldsymbol{m}}) \rangle \right|^s \right\rangle \text{ as } N \to \infty \text{ in } L^1((0,T) \times \Omega), \end{aligned} \tag{9.42}$$

where $1 \le s < 2$ for the Rusanov scheme, and $s \ge 1$ arbitrary finite for the Lax–Friedrichs scheme.

Note that $s = 1$ in (9.42) corresponds to convergence of deviations while $s = 2$ indicates convergence of variations.

9.5 Avoiding the Subsequence Principle – Statistical Convergence

The weak topology, discussed in detail in the preceding section, is not suitable for studying convergence properties of a numerical scheme. In order to capture the asymptotic behavior of an approximate sequence, we need to take into account the behavior of its last N terms, with a suitable large N. The resulting convergence results require substracting a subsequence that is not known *a priori*. In practice, we therefore assume that extracting of subsequence is not necessary and consider the whole sequence. This reflects the common belief, still not rigorously justified, that the scheme enjoys the same asymptotic properties in the vanishing discretization. We try to formulate this hypothesis in a rigorous way.

Suppose that for each $h \in (0, h_0)$, $h_0 < 1$, we have an approximate solution $\boldsymbol{U}_h : Q \subset R^L \to R^M$. In the context of the Euler system discussed in this chapter, $Q = (0, T) \times \Omega$ is the physical space, $\boldsymbol{U}_h = (\varrho_h, \boldsymbol{m}_h)$, and $L = M = d + 1$. In addition, suppose that $\boldsymbol{U}_h \in L^1(Q; R^M)$. Our aim is to introduce the concept of *statistical convergence* of the family $\{\boldsymbol{U}_h\}_{h \searrow 0}$ of approximate solutions. By this we mean, roughly speaking, that the statistical distribution of $\boldsymbol{U}_h$ in the phase space R^M should become stationary for $h \to 0$.

Definition 9.1 (STATIONARY FAMILY)

Let $\{\boldsymbol{U}^S_{h_n}\}_{n=1}^\infty$ be a family of functions, $\boldsymbol{U}_{h_n} \in L^1(Q; R^M)$ for any $n = 1, 2, \dots$ We say that $\{\boldsymbol{U}^S_{h_n}\}_{n=1}^\infty$ is *stationary* if

$$\int_Q F\left(\boldsymbol{U}^S_{h_{k_1}}, \boldsymbol{U}^S_{h_{k_2}} \dots, \boldsymbol{U}^S_{h_{k_j}}\right) \,\mathrm{d}y = \int_Q F\left(\boldsymbol{U}^S_{h_{k_1+n}}, \dots, \boldsymbol{U}^S_{h_{k_j+n}}\right) \,\mathrm{d}y,$$

for all integers $k_1, \dots, k_j$, $j \geq 0$, all $n > 0$, and any $F \in C_c(R^{Nj})$.

The definition of stationary family is reminiscent of that of the stochastic process ranging in R^M, with the probability space Q endowed with the (normalized) Lebesgue measure. Of course, we do not expect a family of numerical solutions to be stationary. However, we desire that any sequence of approximate solutions $\{\boldsymbol{U}_{h_n}\}_{n=1}^\infty$ behaves in the regime of small numerical steps as a stationary family. More precisely, if $j > 0$ is given and $B_1, \dots, B_j$ are Borel sets in R^M, then the measure of the sets

$$\left\{(t, x) \in Q \,\middle|\, \boldsymbol{U}_{h_{k_1}} \in B_1, \ \dots, \ \boldsymbol{U}_{h_{k_j}} \in B_j\right\}$$

$$\left\{(t, x) \in Q \,\middle|\, \boldsymbol{U}_{h_{k_1+n}} \in B_1, \ \dots, \ \boldsymbol{U}_{h_{k_j+n}} \in B_j\right\}, \ n > 0,$$

is asymptotically the same when $\min_j \{k_j\} \to \infty$. A suitable concept is the (S)–convergence introduced below.

Definition 9.2 ((S)–CONVERGENCE)

Let $\{\boldsymbol{U}_{h_n}\}_{n=1}^\infty$ be a sequence of approximate solutions. We say that $\{\boldsymbol{U}_{h_n}\}_{n=1}^\infty$ is *statistically (S)–convergent* if for any $b \in C_c(R^M)$ the following hold:

- **Correlation limit**

$$\lim_{n \to \infty} \int_Q b(\boldsymbol{U}_{h_n}) b(\boldsymbol{U}_{h_m}) \,\mathrm{d}y \text{ exists for any fixed } m; \tag{9.43}$$

- **Disintegration for correlations**

$$\lim_{N\to\infty} \frac{1}{N^2} \sum_{n=1,m=1}^{N} \int_Q b(\boldsymbol{U}_{h_n}) b(\boldsymbol{U}_{h_m})\,\mathrm{d}y$$
$$= \lim_{M\to\infty} \frac{1}{M} \sum_{m=1}^{M} \left(\lim_{N\to\infty} \frac{1}{N} \sum_{n=1}^{N} \int_Q b(\boldsymbol{U}_{h_n}) b(\boldsymbol{U}_{h_m})\,\mathrm{d}y \right). \tag{9.44}$$

Both (9.43) and (9.44) concern the asymptotic behavior of correlations of parts of the "tail" of the approximate sequence. The first crucial observation is that (9.43) yields weak-(*) convergence of the sequence $\{b(\boldsymbol{U}_{h_n})\}_{n=1}^{\infty}$.

Lemma 9.3 *Let $\{\boldsymbol{U}_{h_n}\}_{n=1}^{\infty}$ be a sequence of approximate solutions and $b \in C_c(R^M)$. The following statements are equivalent:*

- $\{b(\boldsymbol{U}_{h_n})\}_{n=1}^{\infty}$ *satisfies* (9.43);
- $$b(\boldsymbol{U}_{h_n}) \to \overline{b(\boldsymbol{U})} \text{ weakly-(*) in } L^\infty(Q) \text{ as } n \to \infty; \tag{9.45}$$
- *the limit*
$$\lim_{N\to\infty} \frac{1}{N} \sum_{k=1}^{N} \int_Q b(\boldsymbol{U}_{h_{n_k}}) b(\boldsymbol{U}_{h_m})\,\mathrm{d}y \tag{9.46}$$

exists for any m and any subsequence $\{\boldsymbol{U}_{h_{n_k}}\}_{k=1}^{\infty}$.

Proof Obviously, the weak-(*) convergence in (9.45) yields both (9.43) and (9.46).

Next, it is easy to check that if the limit in (9.46) exists for any subsequence, then (9.43) holds. Indeed, should there be two subsequences such that

$$\lim_{k\to\infty} \int_Q b(\boldsymbol{U}_{h_{n_k}}) b(\boldsymbol{U}_{h_m})\mathrm{d}y \neq \lim_{l\to\infty} \int_Q b(\boldsymbol{U}_{h_{n_l}}) b(\boldsymbol{U}_{h_m})\mathrm{d}y,$$

one could construct a new subsequence for which the limit (9.46) would not exist. This follows from the fact that the limit in (9.46) is the same if we change a finite number of terms in $\{\boldsymbol{U}_{h_{n_k}}\}_{k=1}^{\infty}$.

Consequently, it remains to show that (9.43) implies the weak-(*) convergence in (9.45). To this end, consider

$$W = \mathrm{closure}_{L^2(Q)} \left[\mathrm{span} \left\{ b(\boldsymbol{U}_{h_m}) \;\middle|\; m = 1, 2, \dots \right\} \right].$$

W is a closed subspace of the Hilbert space $L^2(Q)$ that can be written as

$$L^2(Q) = W \oplus W^{\perp}.$$

In view of (9.43), we get

$$\lim_{n\to\infty} \int_Q b(\boldsymbol{U}_{h_n}) w \,\mathrm{d}y \ \text{ for any } \ w \in W,$$

while

$$\int_Q b(\boldsymbol{U}_{h_n}) v \,\mathrm{d}y = 0 \ \text{ for any } \ n \ \text{ as soon as } \ v \in W^{\perp}.$$

Consequently

$$b(\boldsymbol{U}_{h_n}) \to \overline{b(\boldsymbol{U})} \ \text{ weakly in } \ L^2(Q),$$

which yields (9.45).

Obviously, any convergent family satisfying

$$\boldsymbol{U}_{h_n} \to \boldsymbol{U} \text{ in } L^1(Q; R^M) \text{ as } \ h_n \to 0$$

is (S)–convergent. In particular, the approximate families obtained in Theorems 9.1, 9.2, 9.3, and 9.4 are strongly convergent; whence (S)–convergent. Note that strongly convergent approximate solutions obviously satisfy

$$\int_Q b(\boldsymbol{U}_{h_n}) b(\boldsymbol{U}_{h_m}) \,\mathrm{d}y \to \int_Q b(\boldsymbol{U}) b(\boldsymbol{U}) \,\mathrm{d}y, \ h_n \to 0, h_m \to 0,$$

where $\boldsymbol{U}$ is the (strong) limit.

On the other hand, the (S)–convergent approximate families are well behaved in the statistical sense, in particular, they generate a single Young measure and their Cesàro means converge strongly as following result shows.

Theorem 9.7 (Unconditional weak convergence)
Suppose that $\{\boldsymbol{U}_{h_n}\}_{n=1}^{\infty}$ is an (S)–convergent approximate family,

$$\int_Q |\boldsymbol{U}_{h_n}| \,\mathrm{d}y \overset{<}{\sim} 1.$$

Then for any $b \in C_c(R^M)$ there is a unique $\overline{b(\boldsymbol{U})}$ such that

$$\frac{1}{N} \sum_{n=1}^{N} b(\boldsymbol{U}_{h_n}) \to \overline{b(\boldsymbol{U})} \ \text{ in } \ L^1(Q) \ \text{ as } \ N \to \infty. \tag{9.47}$$

If, in addition,

$$\|\boldsymbol{U}_h\|_{L^q(Q)} \overset{<}{\sim} 1 \ \text{ for some } \ q > 1,$$

then

$$\frac{1}{N}\sum_{n=1}^{N} \boldsymbol{U}_{h_n} \to \boldsymbol{U} \ \ in \ \ L^1(Q; R^M) \ \ as \ \ N \to \infty. \tag{9.48}$$

Proof As we know from Lemma 9.3,

$$b(\boldsymbol{U}_{h_n}) \to \overline{b(\boldsymbol{U})} \text{ weakly-(*) in } L^\infty(Q) \text{ as } n \to \infty;$$

in particular,

$$\frac{1}{N}\sum_{n=1}^{N} b(\boldsymbol{U}_{h_n}) \to \overline{b(\boldsymbol{U})} \text{ weakly-(*) in } L^\infty(Q) \text{ as } N \to \infty.$$

Next, it follows from (9.46) that

$$\begin{aligned}\left\| \tfrac{1}{N}\textstyle\sum_{n=1}^{N} b(\boldsymbol{U}_{h_n})\right\|^2_{L^2(Q)} &= \tfrac{1}{N^2}\textstyle\sum_{n,m=1}^{N}\int_Q b(\boldsymbol{U}_{h_n})b(\boldsymbol{U}_{h_m})\,\mathrm{d}y \\ &= \lim_{M\to\infty}\tfrac{1}{M}\textstyle\sum_{m=1}^{M}\int_Q b(\boldsymbol{U}_{h_m})\overline{b(\boldsymbol{U})}\,\mathrm{d}y = \left\|\overline{b(\boldsymbol{U})}\right\|^2_{L^2(Q)},\end{aligned} \tag{9.49}$$

which yields the strong convergence claimed in (9.47).

Finally, relation (9.48) follows under the hypothesis of higher integrability of $\boldsymbol{U}_h$.

Theorem 9.7 shows that any (S)–convergent approximate sequence generates a single Young measure that can be strongly approximated via Cesàro averages, in particular there is no need to pick up a subsequence in Theorem 9.6. Whether or not a given numerical scheme is (S)–convergent remains an interesting open question. In practical implementations, however, the (S)–convergence is tacitly assumed as the experimental convergence analysis is performed directly on a sequence (not a subsequence) of approximate solutions.

9.6 Conclusion, Bibliographical Remarks

The aim of this chapter was to present convergence analysis of some standard finite volume methods for multidimensional Euler equations describing motion of barotropic fluids, i.e. $p = p(\varrho)$. In particular, we have analyzed the Rusanov and Lax–Friedrichs finite volume methods, see, e.g., Feistauer, Felcman, Straškraba [100], Kröner [143], Toro [194]. Both of them belong to the class of so-called invariant domain preserving methods that have been introduced in Guermond and Popov [123, 124]. It means that the finite volume methods preserve some important (physical) properties of the underlying continuous problem, such as the positivity of density and pressure as well as the entropy/energy inequality. Indeed, a crucial property in order to prove convergence of a numerical scheme is the entropy stability, that means that the discrete entropy inequality holds on each mesh cell. The concept of entropy stable/entropy conservative methods has been introduced by Tadmor [190], see also [102, 105, 191, 192]. We should also mention the results of Bouchut et al. [20, 21, 24], where the kinetic flux-splitting method has been used. Relying on the fully discrete entropy inequality and applying the method of DiPerna [64] and Tartar's results on compensated compactness the authors proved strong convergence of fully discrete kinetic flux-splitting scheme to the bounded weak entropy solution of barotropic Euler equations provided numerical solutions satisfy L^∞-bounds and the vacuum does not appear.

In [134] Jovanović and Rohde assumed the existence of a classical solution to the Cauchy problem of a general multidimensional hyperbolic conservation law. Applying the stability result for classical solutions in the class of entropy solutions due to Dafermos [59] and DiPerna's method [64, 67], they derived error estimates for the explicit finite volume schemes satisfying the discrete entropy inequality and thus proved that the numerical solutions converge strongly to the exact classical solution.

In view of the results on nonuniqueness of weak entropy solutions Fjordholm, Mishra and Tadmor revisited the question of convergence and proved that the semidiscrete entropy stable finite volume schemes converge to the measure-valued solutions provided numerical solutions satisfy L^∞-bounds, coefficients of numerical viscosity are uniformly bounded from below by a positive constant and the entropy Hessian is strictly positive definite, see [101, 106]. In this context a further generalization, the so-called statistical solutions have been introduced by Fjordholm, Mishra et al. for general hyperbolic conservation laws, [103, 104]. Analogously to the DMV or DW solutions the statistical solutions are probabilistic-type solutions. In fact, they are time-parametrized probability measures satisfying an infinite set of partial differential equations consistent with the underlying hyperbolic conservation laws. Thus, they are the measure-valued solutions augmented by information on multipoint spatial correlations. In order to obtain the strong convergence of the entropy stable finite volume solutions to a statistical solution one needs to assume that a special condition on an approximate scaling of structure factors holds. The latter is related to the Kolmogorov compactness criterium.

In this chapter we have showed that under the assumption on boundedness of the density (and the velocity for the Lax–Friedrichs method) the finite volume solutions generated by the Rusanov and the Lax–Friedrichs methods converge in the following way

- weak convergence to a DW solution
- $\mathcal{K}$–convergence to a DW solution, i.e. strong convergence of Cesàro averages of a suitable subsequence
- if a limit identified above is a weak solution and condition (9.39) on the behavior around boundary holds, then the convergence is strong
- strong convergence to the strong solution, provided the strong solution exists.

As we have already observed the concept of $\mathcal{K}$–convergence based on the averaging over different meshes naturally inherits compactness. Consequently, we have obtained the strong solution to a generalized DW solution without any additional assumptions. Note that this convergence result holds in general only for a subsequence. However, if the sequence of approximate solutions is (S)–convergent then the strong convergence of the Cesàro averages holds for the whole sequence. The aim of this chapter was to demonstrate the application of theory developed in the previous parts of this book, in particular Chaps. 5–7 to some standard first order finite volume methods. For this purpose we have chosen the Lax–Friedrichs finite volume method and its local version, the Rusanov method. Generalization to other well-known first order finite volume methods as well as their higher order versions is possible and it is left to an interested reader.

Chapter 10
Finite Volume Method for the Complete Euler System

Having analyzed some standard finite volume methods for the barotropic Euler equations we continue with the convergence analysis of the complete Euler system of equations of gas dynamics (2.1)–(2.3), see also (2.34)–(2.36). Following the general approach described in Chap. 8, we introduce a new finite volume method based on the Brenner model discussed in Sect. 3.2.2. Specifically, H. Brenner proposed the following alternative to the complete Navier–Stokes–Fourier system as a model of viscous and heat conductive fluids:

$$\begin{aligned}
&\partial_t \varrho + \mathrm{div}_x(\varrho \boldsymbol{u}_m) = 0,\\
&\partial_t(\varrho \boldsymbol{u}) + \mathrm{div}_x\left(\varrho \boldsymbol{u} \otimes \boldsymbol{u}_m\right) + \nabla_x p(\varrho) = \mathrm{div}_x \mathbb{S}(\nabla_x \boldsymbol{u}),\\
&\partial_t E + \mathrm{div}_x(E \boldsymbol{u}_m) + \mathrm{div}_x(p\boldsymbol{u}) + \mathrm{div}_x \boldsymbol{q} = \mathrm{div}_x(\mathbb{S}(\nabla_x \boldsymbol{u}) \cdot \boldsymbol{u})
\end{aligned} \tag{10.1}$$

with the Fourier heat flux $\boldsymbol{q} = -\kappa \nabla_x \vartheta$ and Newton's rheological relation

$$\mathbb{S}(\nabla_x \boldsymbol{u}) = \mu \left(\nabla_x \boldsymbol{u} + \nabla_x^T \boldsymbol{u} - \frac{2}{d} \mathrm{div}_x \boldsymbol{u} \mathbb{I} \right) + \lambda \mathrm{div}_x \boldsymbol{u} \mathbb{I}. \tag{10.2}$$

The model contains two velocities, $\boldsymbol{u}$ and $\boldsymbol{u}_m$ interrelated, as Brenner suggested, through the following phenomenological relation:

$$\boldsymbol{u} - \boldsymbol{u}_m = K \nabla_x \log(\varrho). \tag{10.3}$$

Although the model suffers numerous physical deficiencies and as such has been thoroughly criticized, it exhibits striking similarity with certain *numerical* approximations. Indeed setting $\kappa = c_v \varrho K = c_v h \varrho \lambda,\ \lambda \geq 0,\ h > 0$ and

$$\mathbb{S}(\nabla_x \boldsymbol{u}) = h\lambda\varrho \nabla_x \boldsymbol{u} + h^\alpha \nabla_x \boldsymbol{u},$$

E. Feireisl et al., *Numerical Analysis of Compressible Fluid Flows*, MS&A 20,
https://doi.org/10.1007/978-3-030-73788-7_10

the system (10.1) rewrites as

$$\partial_t \varrho + \mathrm{div}_x(\boldsymbol{m}) = h\mathrm{div}_x(\lambda \nabla_x \varrho), \tag{10.4}$$

$$\partial_t \boldsymbol{m} + \mathrm{div}_x(\boldsymbol{m} \otimes \boldsymbol{u}) + \nabla_x p = h\mathrm{div}_x\,(\lambda \nabla_x \boldsymbol{m}) + h^\alpha \Delta_x \boldsymbol{u}, \tag{10.5}$$

$$\partial_t E + \mathrm{div}_x((E+p)\boldsymbol{u}) = h\mathrm{div}_x(\lambda \nabla_x E) + h^\alpha \mathrm{div}_x(\nabla_x \boldsymbol{u} \cdot \boldsymbol{u}). \tag{10.6}$$

One can directly observe that without the h^α-dependent terms, the system (10.4)–(10.6) is strongly reminiscent of the Rusanov or the Lax–Friedrichs numerical scheme where $h\mathrm{div}_x(\lambda\nabla_x \boldsymbol{U})$, $\boldsymbol{U} = (\varrho, \boldsymbol{m}, E)$, plays the role of the numerical diffusion, cf. (9.12), (9.13).
We set $\lambda \equiv \lambda_h = \{\{\boldsymbol{u}_h\}\} \cdot \boldsymbol{n} + h^\varepsilon$, $\varepsilon > -1$, where $\boldsymbol{u}_h \in \boldsymbol{Q}_h$ is a numerical approximation of the velocity $\boldsymbol{u}$. Approximating the divergence operator with the central differences leads to the following (vector valued) numerical flux function, cf. (8.8)

$$\boldsymbol{F}_h^{\mathrm{up}} = Up[\boldsymbol{U}_h, \boldsymbol{u}_h] - h^\varepsilon \left[\!\left[\boldsymbol{U}_h \right]\!\right], \tag{10.7}$$

where $\boldsymbol{U}_h = (\varrho_h, \boldsymbol{m}_h, E_h)$ is a piecewise constant approximation of $\boldsymbol{U}$ and $\varepsilon > -1$. Further,

$$\begin{aligned} Up[\boldsymbol{U}_h, \boldsymbol{u}_h] &= \{\{\boldsymbol{U}_h\}\}\ \{\{\boldsymbol{u}_h\}\} \cdot \boldsymbol{n} - \frac{1}{2} | \,\{\{\boldsymbol{u}_h\}\} \cdot \boldsymbol{n}|\, [\![\boldsymbol{U}_h]\!] \\ &= \boldsymbol{U}_h^{\mathrm{in}}[\{\{\boldsymbol{u}_h\}\} \cdot \boldsymbol{n}]^+ + \boldsymbol{U}_h^{\mathrm{out}}[\{\{\boldsymbol{u}_h\}\} \cdot \boldsymbol{n}]^-, \end{aligned} \tag{10.8}$$

see also (8.6) with $\langle \boldsymbol{u}_h \rangle_\sigma \equiv \{\{\boldsymbol{u}_h\}\}$ for $\boldsymbol{u}_h \in \boldsymbol{Q}_h$. The vector valued flux function can be written as $\boldsymbol{F}_h^{\mathrm{up}} = (F_{h,\varrho}^{\mathrm{up}}, \boldsymbol{F}_{h,\boldsymbol{m}}^{\mathrm{up}}, F_{h,E}^{\mathrm{up}})$ with the corresponding components for the density, momentum and energy equations.

10.1 Numerical Method

Let $\Omega \subset R^d, d = 2, 3$ be a bounded domain. We are now ready to formulate a semidiscrete finite volume method based on the numerical flux function (10.7) to approximate the complete Euler system

$$\begin{aligned} \partial_t \varrho + \mathrm{div}_x \boldsymbol{m} &= 0, \\ \partial_t \boldsymbol{m} + \mathrm{div}_x \left(\frac{\boldsymbol{m} \otimes \boldsymbol{m}}{\varrho} \right) + \nabla_x p &= 0, \\ \partial_t E + \mathrm{div}_x \left((E+p) \frac{\boldsymbol{m}}{\varrho} \right) &= 0, \\ \boldsymbol{m} \cdot \boldsymbol{n}|_{\partial\Omega} &= 0. \end{aligned} \tag{10.9}$$

We consider an unstructured mesh in the sense of Definition 1. Given the initial values $(\varrho_{0,h}, \boldsymbol{m}_{0,h}, E_{0,h}) \in Q_h \times \boldsymbol{Q}_h \times Q_h$, we seek a piecewise constant approximation $(\varrho_h(t), \boldsymbol{m}_h(t), E_h(t)) \in Q_h \times \boldsymbol{Q}_h \times Q_h$ that solves at any time $t \in (0, T)$ the following system of ordinary differential equations

$$
\begin{aligned}
&\frac{\mathrm{d}}{\mathrm{d}t}\varrho_K + \sum_{\sigma\in\mathcal{E}(K)\cap\mathcal{E}_{int}} \frac{|\sigma|}{|K|} F_{h,\varrho}^{\mathrm{up}} = 0,\\
&\frac{\mathrm{d}}{\mathrm{d}t}\boldsymbol{m}_K + \sum_{\sigma\in\mathcal{E}(K)\cap\mathcal{E}_{int}} \frac{|\sigma|}{|K|} \left(\boldsymbol{F}_{h,\boldsymbol{m}}^{\mathrm{up}} + \{\!\{p_h\}\!\}\boldsymbol{n}\right) = h^{\alpha-1} \sum_{\sigma\in\mathcal{E}(K)\cap\mathcal{E}_{int}} \frac{|\sigma|}{|K|} [\![\boldsymbol{u}_h]\!],\\
&\frac{\mathrm{d}}{\mathrm{d}t}E_K + \sum_{\sigma\in\mathcal{E}(K)\cap\mathcal{E}_{int}} \frac{|\sigma|}{|K|} \left(F_{h,E}^{\mathrm{up}} + \left(\{\!\{p_h\}\!\}\boldsymbol{u}_h + p_h\{\!\{\boldsymbol{u}_h\}\!\}\right)\cdot\boldsymbol{n}\right)\\
&\quad = \frac{h^{\alpha-1}}{2} \sum_{\sigma\in\mathcal{E}(K)\cap\mathcal{E}_{int}} \frac{|\sigma|}{|K|} \left[\!\left[|\boldsymbol{u}_h|^2\right]\!\right],
\end{aligned}
\tag{10.10}
$$

for any $K \in \mathcal{T}_h$. Analogously as in Chap. 9

$$
p_h = (\gamma - 1)\left(E_h - \frac{1}{2}\frac{|\boldsymbol{m}_h|^2}{\varrho_h}\right), \quad \vartheta_h = \frac{p_h}{\varrho_h}, \quad e_h = c_v\vartheta_h, \quad s_h = \log\left(\frac{\vartheta_h^{c_v}}{\varrho_h}\right)
$$

are the approximations of the pressure p, the temperature ϑ, the internal energy e and the entropy s, respectively. We accompany (10.10) with the no-flux boundary conditions that can be written as

$$
\{\!\{\boldsymbol{u}_h\}\!\}\cdot\boldsymbol{n} = 0, \ \left[\!\left[\varrho_h\right]\!\right] = 0 = \left[\!\left[p_h\right]\!\right] \quad \text{for any } \sigma \in \mathcal{E}_{ext}.
$$

Note that the boundary conditions for the density and the pressure are enforced by the artificial viscosity contained in the approximation. In the target Euler system only the impermeability condition $\{\!\{\boldsymbol{u}\}\!\}\cdot\boldsymbol{n} = 0$ is required.

In what follows we will refer to the scheme (10.10) as *viscous finite volume (VFV) method*. Indeed the h^α-terms in the momentum and energy equations can be identified with a vanishing viscosity regularization. Formulation (10.10) can be equivalently rewritten in the integral form

$$
\int_\Omega \frac{\mathrm{d}}{\mathrm{d}t}\varrho_h\phi_h \,\mathrm{d}x - \sum_{\sigma\in\mathcal{E}_{int}} \int_\sigma F_{h,\varrho}^{\mathrm{up}}[\varrho_h, \boldsymbol{u}_h]\,[\![\phi_h]\!]\,\mathrm{dS}_x = 0, \quad \text{for all } \phi_h \in Q_h, \tag{10.11}
$$

$$
\begin{aligned}
&\int_\Omega \frac{\mathrm{d}}{\mathrm{d}t}\boldsymbol{m}_h\cdot\boldsymbol{\phi}_h \,\mathrm{d}x - \sum_{\sigma\in\mathcal{E}_{int}} \int_\sigma \left(\boldsymbol{F}_{h,\boldsymbol{m}}^{\mathrm{up}}[\boldsymbol{m}_h, \boldsymbol{u}_h] - \{\!\{p_h\}\!\}\boldsymbol{n}\right)\cdot\left[\!\left[\boldsymbol{\phi}_h\right]\!\right]\mathrm{dS}_x\\
&\quad = -h^{\alpha-1}\sum_{\sigma\in\mathcal{E}_{int}} \int_\sigma [\![\boldsymbol{u}_h]\!]\cdot\left[\!\left[\boldsymbol{\phi}_h\right]\!\right]\mathrm{dS}_x,
\end{aligned}
\tag{10.12}
$$

$$
\text{for all } \boldsymbol{\phi}_h \in \boldsymbol{Q}_h, \left\{\!\!\left\{\boldsymbol{\phi}_h\right\}\!\!\right\} \cdot \boldsymbol{n} = 0 \text{ for any } \sigma \in \mathcal{E}_{ext},
$$

$$
\begin{aligned}
\int_{\Omega} \frac{\mathrm{d}}{\mathrm{d}t} E_h \phi_h \,\mathrm{d}x &- \sum_{\sigma \in \mathcal{E}_{int}} \int_{\sigma} F^{\mathrm{up}}_{h,E}[E_h, \boldsymbol{u}_h] \, [\![\phi_h]\!] \,\mathrm{dS}_x \\
&- \sum_{\sigma \in \mathcal{E}_{int}} \int_{\sigma} \Big(\{\!\{p_h\}\!\} \, [\![\phi_h \boldsymbol{u}_h]\!] - \{\!\{p_h \phi_h\}\!\} \, [\![\boldsymbol{u}_h]\!] \Big) \cdot \boldsymbol{n} \,\mathrm{dS}_x \\
&= -h^{\alpha-1} \sum_{\sigma \in \mathcal{E}_{int}} \int_{\sigma} [\![\boldsymbol{u}_h]\!] \cdot \{\!\{\boldsymbol{u}_h\}\!\} \, [\![\phi_h]\!] \,\mathrm{dS}_x, \quad \text{for all } \phi_h \in Q_h.
\end{aligned}
\tag{10.13}
$$

We point out that the pressure term in the energy equation can be rewritten as

$$
\begin{aligned}
&\sum_{\sigma \in \mathcal{E}_{int}} \int_{\sigma} \Big(\{\!\{p_h\}\!\} \, [\![\phi_h \boldsymbol{u}_h]\!] - \{\!\{p_h \phi_h\}\!\} \, [\![\boldsymbol{u}_h]\!] \Big) \cdot \boldsymbol{n} \,\mathrm{dS}_x \\
&= \sum_{\sigma \in \mathcal{E}_{int}} \int_{\sigma} \{\!\{p_h\}\!\} \, \{\!\{\boldsymbol{u}_h\}\!\} \cdot \boldsymbol{n} \, [\![\phi_h]\!] \,\mathrm{dS}_x - \frac{1}{4} \sum_{\sigma \in \mathcal{E}_{int}} \int_{\sigma} [\![p_h]\!] \, [\![\boldsymbol{u}_h]\!] \cdot \boldsymbol{n} \, [\![\phi_h]\!] \,\mathrm{dS}_x.
\end{aligned}
\tag{10.14}
$$

The numerical approximations $(\varrho_h, \boldsymbol{m}_h, E_h)$ provided by a semidiscrete scheme are continuous functions of time. Thus, the VFV method (10.11)–(10.13) may be interpreted as a finite system of ODEs. Since the flux terms are locally Lipschitz-continuous it follows from the standard ODE theory that for a given initial state

$$
\begin{aligned}
&\varrho_h(0) = \varrho_{0,h} \in Q_h, \ \varrho_{0,h} > 0, \ \boldsymbol{m}(0) = \boldsymbol{m}_{0,h} \in \boldsymbol{Q}_h, \ E_h(0) = E_{0,h} \in Q_h, \\
&E_{0,h} - \frac{1}{2} \frac{|\boldsymbol{m}_{0,h}|^2}{\varrho_{0,h}} > 0,
\end{aligned}
$$

the semidiscrete system (10.11)–(10.13) admits a unique solution $(\varrho_h, \boldsymbol{m}_h, E_h)$ defined on a maximal time interval $[0, T_{\max})$, where

$$
\varrho_h(t) > 0, \ p_h(t) = (\gamma - 1) \left(E_h(t) - \frac{1}{2} \frac{|\boldsymbol{m}_h(t)|^2}{\varrho_h(t)} \right) > 0 \text{ for all } t \in [0, T_{\max}).
\tag{10.15}
$$

In particular, the absolute temperature ϑ_h can be defined,

$$
\vartheta_h(t) = \frac{p_h(t)}{\varrho_h(t)} = \frac{\gamma - 1}{\varrho_h(t)} \left(E_h(t) - \frac{1}{2} \frac{|\boldsymbol{m}_h(t)|^2}{\varrho_h(t)} \right).
\tag{10.16}
$$

We will show in Sect. 10.2 that the system (10.11)–(10.13) admits sufficiently strong *a priori* bounds that will guarantee **(i)** $T_{\max} = \infty$, **(ii)** validity of (10.15) for any $t \geq 0$.

In what follows we proceed with a discussion on invariant domain preserving properties and stability of the VFV method (10.11)–(10.13). In particular, we show

- positivity of the discrete density, pressure and temperature
- entropy stability
- minimum entropy principle
- weak BV estimates.

10.2 Stability

We start by deriving some standard a priori bounds that directly follow from the finite volume formulation (10.11)–(10.13). First, taking $\phi_h \equiv 1$ in the equation of continuity (10.11) yields the total mass conservation

$$\int_{\Omega} \varrho_h(t,\cdot)\, \mathrm{d}x = \int_{\Omega} \varrho_{0,h}\, \mathrm{d}x = M_0 > 0,\ t \geq 0. \tag{10.17}$$

A similar argument applied to the total energy balance yields

$$\int_{\Omega} E_h(t,\cdot)\, \mathrm{d}x = \int_{\Omega} E_{0,h}\, \mathrm{d}x = E_0 > 0,\ t \geq 0. \tag{10.18}$$

Further important source of a priori estimates is the discrete entropy inequality that we will derive in what follows.

10.2.1 Entropy Stability

We start by introducing some notation:

$$r_h^{\mathrm{up}} = \begin{cases} r_h^{\mathrm{in}} \text{ if } \{\!\{\boldsymbol{u}_h\}\!\} \cdot \boldsymbol{n} \geq 0 \\ r_h^{\mathrm{out}} \text{ if } \{\!\{\boldsymbol{u}_h\}\!\} \cdot \boldsymbol{n} < 0, \end{cases} \qquad r_h^{\mathrm{down}} = \begin{cases} r_h^{\mathrm{out}} \text{ if } \{\!\{\boldsymbol{u}_h\}\!\} \cdot \boldsymbol{n} \geq 0 \\ r_h^{\mathrm{in}} \text{ if } \{\!\{\boldsymbol{u}_h\}\!\} \cdot \boldsymbol{n} < 0, \end{cases} \tag{10.19}$$

and

$$\widetilde{[\![r_h]\!]} = r_h^{\mathrm{up}} - r_h^{\mathrm{down}} = -\, [\![r_h]\!]\ \mathrm{sgn}(\{\!\{\boldsymbol{u}_h\}\!\} \cdot \boldsymbol{n}), \tag{10.20}$$

for any $r_h \in Q_h$, $\boldsymbol{u}_h \in \boldsymbol{Q}_h$.

Next, we derive renormalized versions of the discrete continuity equation and the transport equation.

Lemma 10.1 *Let $b \in C^2(R)$, $\chi \in C^2(R)$, and let $(\varrho_h, \boldsymbol{u}_h)$ solve the VFV method* (10.11)–(10.13).

Then there hold:

- **Renormalized discrete continuity equation**

$$\begin{aligned}
&\int_{\Omega} \frac{d}{dt} b(\varrho_h)\phi_h \,\mathrm{dx} - \sum_{\sigma\in\mathcal{E}_{int}} \int_{\sigma} Up[b(\varrho_h), \boldsymbol{u}_h] \,[\![\phi_h]\!] \,\mathrm{dS}_x \\
&+ \sum_{\sigma\in\mathcal{E}_{int}} \int_{\sigma} \{\!\{\boldsymbol{u}_h\}\!\} \cdot \boldsymbol{n} \left[\!\!\left[\Big(b(\varrho_h) - b'(\varrho_h)\varrho_h\Big)\phi_h\right]\!\!\right] \mathrm{dS}_x \\
&= - \sum_{\sigma\in\mathcal{E}_{int}} h^{\varepsilon} \int_{\sigma} [\![\varrho_h]\!] \,[\![b'(\varrho_h)\phi_h]\!] \,\mathrm{dS}_x \\
&- \sum_{\sigma\in\mathcal{E}_{int}} \int_{\sigma} \phi_h^{\mathrm{down}} \Big(\widetilde{[\![b(\varrho_h)]\!]} - b'(\varrho_h^{\mathrm{down}})\widetilde{[\![\varrho_h]\!]}\Big) |\, \{\!\{\boldsymbol{u}_h\}\!\} \cdot \boldsymbol{n}| \mathrm{dS}_x,
\end{aligned} \tag{10.21}$$

for any $\phi_h \in Q_h$.

- **Renormalized discrete transport equation**

$$\begin{aligned}
&\int_{\Omega} \frac{d}{dt}(\varrho_h g_h)\chi'(g_h)\phi_h \,\mathrm{dx} - \sum_{\sigma\in\mathcal{E}_{int}} \int_{\sigma} Up[\varrho_h g_h, \boldsymbol{u}_h] \left[\!\left[\chi'(g_h)\phi_h\right]\!\right] \mathrm{dS}_x \\
&= \int_{\Omega} \frac{d}{dt}(\varrho_h \chi(g_h))\phi_h \,\mathrm{dx} - \sum_{\sigma\in\mathcal{E}_{int}} \int_{\sigma} Up[\varrho_h \chi(g_h), \boldsymbol{u}_h] \,[\![\phi_h]\!] \,\mathrm{dS}_x \\
&+ h^{\varepsilon} \sum_{\sigma\in\mathcal{E}_{int}} \int_{\sigma} [\![\varrho_h]\!] \,\left[\!\left[\big(\chi(g_h) - \chi'(g_h)g_h\big)\, \phi_h\right]\!\right] \mathrm{dS}_x \\
&+ \sum_{\sigma\in\mathcal{E}_{int}} \int_{\sigma} \phi_h^{\mathrm{down}} \varrho_h^{\mathrm{up}} \left(\widetilde{[\![\chi(g_h)]\!]} - \chi'(g_h^{\mathrm{down}})\widetilde{[\![g_h]\!]}\right) |\, \{\!\{\boldsymbol{u}_h\}\!\} \cdot \boldsymbol{n}| \mathrm{dS}_x
\end{aligned} \tag{10.22}$$

for any $g_h \in Q_h$, $\phi_h \in Q_h$.

Proof The discrete renormalized continuity equation follows from Lemma 8.2, with an obvious modification to accommodate the continuous time derivative. In order to derive (10.22), we consider first the upwind term. Direct calculations yield

$$\begin{aligned}
&- \sum_{\sigma\in\mathcal{E}_{int}} \int_{\sigma} Up[\varrho_h g_h, \boldsymbol{u}_h] \left[\!\left[\chi'(g_h)\phi_h\right]\!\right] \mathrm{dS}_x \\
&= - \sum_{\sigma\in\mathcal{E}_{int}} \int_{\sigma} Up[\varrho_h \chi(g_h), \boldsymbol{u}_h] \,[\![\phi_h]\!] - Up[\varrho_h, \boldsymbol{u}_h] \,[\![\chi(g_h)\phi_h]\!] \,\mathrm{dS}_x \\
&- \sum_{\sigma\in\mathcal{E}_{int}} \int_{\sigma} Up[\varrho_h, \boldsymbol{u}_h] \left[\!\left[\chi'(g_h)g_h\phi_h\right]\!\right] \mathrm{dS}_x \\
&+ \sum_{\sigma\in\mathcal{E}_{int}} \int_{\sigma} \phi_h^{\mathrm{down}} \varrho_h^{\mathrm{up}} \left(\widetilde{[\![\chi(g_h)]\!]} - \chi'(g_h^{\mathrm{down}})\widetilde{[\![g_h]\!]}\right) |\, \{\!\{\boldsymbol{u}_h\}\!\} \cdot \boldsymbol{n}| \mathrm{dS}_x.
\end{aligned}$$

Combining the above with the continuity equation (10.11) leads to

$$\begin{aligned}
&\int_{\Omega} \frac{\mathrm{d}}{\mathrm{d}t}(\varrho_h g_h)\chi'(g_h)\phi_h \,\mathrm{d}x - \sum_{\sigma\in\mathcal{E}_{int}} \int_{\sigma} Up[\varrho_h g_h, \boldsymbol{u}_h] \left[\!\left[\chi'(g_h)\phi_h\right]\!\right] \mathrm{dS}_x \\
&= \int_{\Omega} \varrho_h \frac{\mathrm{d}}{\mathrm{d}t}\chi(g_h)\phi_h \,\mathrm{d}x - \sum_{\sigma\in\mathcal{E}_{int}} \int_{\sigma} Up[\varrho_h \chi(g_h), \boldsymbol{u}_h] \left[\!\left[\phi_h\right]\!\right] \mathrm{dS}_x \\
&+ \sum_{\sigma\in\mathcal{E}_{int}} \int_{\sigma} Up[\varrho_h, \boldsymbol{u}_h] \left[\!\left[\chi(g_h)\phi_h\right]\!\right] \mathrm{dS}_x - h^{\varepsilon} \sum_{\sigma\in\mathcal{E}_{int}} \int_{\sigma} \left[\!\left[\varrho_h\right]\!\right] \left[\!\left[\chi'(g_h) g_h \phi_h\right]\!\right] \mathrm{dS}_x \\
&+ \sum_{\sigma\in\mathcal{E}_{int}} \int_{\sigma} \phi_h^{\rm down} \varrho_h^{\rm up} \left(\widetilde{\left[\!\left[\chi(g_h)\right]\!\right] - \chi'(g_h^{\rm down})\left[\!\left[g_h\right]\!\right]} \right) | \{\!\{\boldsymbol{u}_h\}\!\} \cdot \boldsymbol{n} | \mathrm{dS}_x .
\end{aligned}$$

Finally, we consider $\chi(g_h)\phi_h$ as a test function in the discrete continuity equation (10.11) and deduce the desired result.

We proceed to prove that the VFV method satisfies the discrete entropy equation as well as its renormalized version.

Theorem 10.1 (Entropy stability)
The VFV method (10.11)–(10.13) *is entropy stable, i.e. the discrete entropy equation and the renormalized discrete entropy inequality hold:*

- **Discrete entropy equation**

$$\begin{aligned}
&\frac{d}{dt} \int_{\Omega} \varrho_h s_h \phi_h \,\mathrm{d}x - \sum_{\sigma\in\mathcal{E}_{int}} \int_{\sigma} Up[\varrho_h s_h, \boldsymbol{u}_h] \left[\!\left[\phi_h\right]\!\right] \mathrm{dS}_x \\
&= h^{\alpha-1} \sum_{\sigma\in\mathcal{E}_{int}} \int_{\sigma} \left[\!\left[\boldsymbol{u}_h\right]\!\right]^2 \left\{\!\!\left\{ \frac{\phi_h}{\vartheta_h} \right\}\!\!\right\} \mathrm{dS}_x \\
&+ \frac{1}{2} \sum_{\sigma\in\mathcal{E}_{int}} \int_{\sigma} \left(\frac{\phi_h}{\vartheta_h} \right)^{\rm down} \varrho_h^{\rm up} | \{\!\{\boldsymbol{u}_h\}\!\} \cdot \boldsymbol{n} | \left[\!\left[\boldsymbol{u}_h\right]\!\right]^2 \mathrm{dS}_x \\
&+ \sum_{\sigma\in\mathcal{E}_{int}} \int_{\sigma} h^{\varepsilon} \left(\{\!\{\varrho_h\}\!\} \left\{\!\!\left\{ \frac{\phi_h}{\vartheta_h} \right\}\!\!\right\} - \frac{1}{4} \left[\!\left[\varrho_h\right]\!\right] \left[\!\!\left[\frac{\phi_h}{\vartheta_h} \right]\!\!\right] \right) \left[\!\left[\boldsymbol{u}_h\right]\!\right]^2 \mathrm{dS}_x \\
&+ \sum_{\sigma\in\mathcal{E}_{int}} \int_{\sigma} \phi_h^{\rm down} \left(\widetilde{\left[\!\left[b(\varrho_h)\right]\!\right] - b'(\varrho_h^{\rm down})\left[\!\left[\varrho_h\right]\!\right]} \right) | \{\!\{\boldsymbol{u}_h\}\!\} \cdot \boldsymbol{n} | \mathrm{dS}_x \\
&- c_v \sum_{\sigma\in\mathcal{E}_{int}} \int_{\sigma} \phi_h^{\rm down} \varrho_h^{\rm up} \left(\widetilde{\left[\!\left[\log(\vartheta_h)\right]\!\right] - \frac{1}{\vartheta_h^{\rm down}}\left[\!\left[\vartheta_h\right]\!\right]} \right) | \{\!\{\boldsymbol{u}_h\}\!\} \cdot \boldsymbol{n} | \mathrm{dS}_x
\end{aligned}$$

$$
\begin{aligned}
&- c_v \sum_{\sigma\in\mathcal{E}_{int}} \int_\sigma h^\varepsilon \left[\!\left[\varrho_h \vartheta_h\right]\!\right] \left[\!\left[\frac{\phi_h}{\vartheta_h}\right]\!\right] \mathrm{dS}_x \\
&- c_v \sum_{\sigma\in\mathcal{E}_{int}} \int_\sigma h^\varepsilon \left[\!\left[\varrho_h\right]\!\right] \left[\!\left[(\log(\vartheta_h) - 1)\,\phi_h\right]\!\right] \mathrm{dS}_x \\
&+ \sum_{\sigma\in\mathcal{E}_{int}} \int_\sigma h^\varepsilon \left[\!\left[\varrho_h\right]\!\right] \left[\!\left[b'(\varrho_h)\phi_h\right]\!\right] \mathrm{dS}_x, \ \textit{where}\ \ b(\varrho) = \varrho\log(\varrho)
\end{aligned}
\tag{10.23}
$$

holds for any $\phi_h \in Q_h$.

- **Renormalized discrete entropy inequality** *Let* χ *be a nondecreasing, concave, twice continuously differentiable function on R that is bounded from above. Then we have for* $\phi_h \in Q_h$, $\phi_h \geq 0$

$$
\begin{aligned}
&\frac{d}{dt}\int_\Omega \varrho_h \chi(s_h)\phi_h \,\mathrm{d}x - \sum_{c}\int_\sigma \Big(Up[\varrho_h\chi(s_h), \boldsymbol{u}_h] \\
&\qquad + h^\varepsilon \left(\left\{\!\left\{\nabla_\varrho(\varrho_h\chi(s_h))\right\}\!\right\} \left[\!\left[\varrho_h\right]\!\right] + \left\{\!\left\{\nabla_p(\varrho_h\chi(s_h))\right\}\!\right\} \left[\!\left[p_h\right]\!\right] \right) \Big) \left[\!\left[\phi_h\right]\!\right] \mathrm{dS}_x \\
&\geq \sum_{\sigma\in\mathcal{E}_{int}} \int_\sigma h^{\alpha-1} \left[\!\left[\boldsymbol{u}_h\right]\!\right]^2 \left\{\!\left\{\frac{\chi'(s_h)\phi_h}{\vartheta_h}\right\}\!\right\} \mathrm{dS}_x \\
&+ \sum_{\sigma\in\mathcal{E}_{int}} \int_\sigma h^\varepsilon \left(\left\{\!\left\{\varrho_h\right\}\!\right\} \left\{\!\left\{\frac{\chi'(s_h)\phi_h}{\vartheta_h}\right\}\!\right\} - \frac{1}{4}\left[\!\left[\varrho_h\right]\!\right] \left[\!\left[\frac{\chi'(s_h)\phi_h}{\vartheta_h}\right]\!\right] \right) \left[\!\left[\boldsymbol{u}_h\right]\!\right]^2 \mathrm{dS}_x \\
&+ \frac{1}{2}\sum_{\sigma\in\mathcal{E}_{int}} \int_\sigma \left(\frac{\chi'(s_h)\phi_h}{\vartheta_h}\right)^{\mathrm{down}} \varrho_h^{\mathrm{up}} \left| \left\{\!\left\{\boldsymbol{u}_h\right\}\!\right\} \cdot \boldsymbol{n}\right| \left[\!\left[\boldsymbol{u}_h\right]\!\right]^2 \mathrm{dS}_x \\
&+ \sum_{\sigma\in\mathcal{E}_{int}} \int_\sigma (\chi'(s_h)\phi_h)^{\mathrm{down}} \left(\widetilde{\left[\!\left[b(\varrho_h)\right]\!\right]} - b'(\varrho_h^{\mathrm{down}})\widetilde{\left[\!\left[\varrho_h\right]\!\right]} \right) \left| \left\{\!\left\{\boldsymbol{u}_h\right\}\!\right\} \cdot \boldsymbol{n}\right| \mathrm{dS}_x \\
&- c_v \sum_{\sigma\in\mathcal{E}_{int}} \int_\sigma (\chi'(s_h)\phi_h)^{\mathrm{down}} \varrho_h^{\mathrm{up}} \left(\widetilde{\left[\!\left[\log(\vartheta_h)\right]\!\right]} - \frac{1}{\vartheta_h^{\mathrm{down}}}\widetilde{\left[\!\left[\vartheta_h\right]\!\right]} \right) \left| \left\{\!\left\{\boldsymbol{u}_h\right\}\!\right\} \cdot \boldsymbol{n}\right| \mathrm{dS}_x \\
&- \sum_{\sigma\in\mathcal{E}_{int}} \int_\sigma \phi_h^{\mathrm{down}} \varrho_h^{\mathrm{up}} \left(\widetilde{\left[\!\left[\chi(s_h)\right]\!\right]} - \chi'(s_h^{\mathrm{down}})\widetilde{\left[\!\left[s_h\right]\!\right]} \right) \left| \left\{\!\left\{\boldsymbol{u}_h\right\}\!\right\} \cdot \boldsymbol{n}\right| \mathrm{dS}_x,
\end{aligned}
\tag{10.24}
$$

where $b(\varrho) = \varrho\log(\varrho)$.

Proof In order to obtain the discrete entropy equation, we first derive a discrete kinetic energy equation which will be subtracted from the discrete energy equation (10.13) to obtain a discrete internal energy equation.

To this end we first take a test function $\boldsymbol{\phi}_h = \boldsymbol{u}_h \phi_h$ in (10.12):

$$\frac{\mathrm{d}}{\mathrm{d}t} \int_\Omega \boldsymbol{m}_h \cdot \boldsymbol{u}_h \phi_h \,\mathrm{d}x - \sum_{\sigma \in \mathcal{E}_{int}} \int_\sigma \boldsymbol{F}_h^{\mathrm{up}}[\boldsymbol{m}_h, \boldsymbol{u}_h] \cdot [\![\boldsymbol{u}_h \phi_h]\!] \,\mathrm{dS}_x$$

$$- \sum_{\sigma \in \mathcal{E}_{int}} \int_\sigma \{\!\{p_h\}\!\} \boldsymbol{n} \cdot [\![\boldsymbol{u}_h \phi_h]\!] \,\mathrm{dS}_x = -h^{\alpha-1} \sum_{\sigma \in \mathcal{E}_{int}} \int_\sigma [\![\boldsymbol{u}_h]\!] \cdot [\![\boldsymbol{u}_h \phi_h]\!] \,\mathrm{dS}_x.$$

Next, we use relation (10.22) for each component of $\boldsymbol{u}_h$, i.e. for $g_h = u_{h,k}$, $k = 1, \ldots, d$ and $\chi(u_{h,k}) = \frac{1}{2} u_{h,k}^2$. Summing the resulting equations for $k = 1, \ldots, d$ yields

$$\begin{aligned}
&\frac{\mathrm{d}}{\mathrm{d}t} \int_\Omega \boldsymbol{m}_h \cdot \boldsymbol{u}_h \phi_h \,\mathrm{d}x - \sum_{\sigma \in \mathcal{E}_{int}} \int_\sigma Up[\boldsymbol{m}_h, \boldsymbol{u}_h] \cdot [\![\boldsymbol{u}_h \phi_h]\!] \,\mathrm{dS}_x \\
&= \frac{\mathrm{d}}{\mathrm{d}t} \int_\Omega \varrho_h \boldsymbol{u}_h \cdot \boldsymbol{u}_h \phi_h \,\mathrm{d}x - \sum_{\sigma \in \mathcal{E}_{int}} \int_\sigma Up[\varrho_h \boldsymbol{u}_h, \boldsymbol{u}_h] \cdot [\![\boldsymbol{u}_h \phi_h]\!] \,\mathrm{dS}_x \\
&= \frac{\mathrm{d}}{\mathrm{d}t} \int_\Omega \frac{1}{2} \varrho_h |\boldsymbol{u}_h|^2 \phi_h \,\mathrm{d}x - \sum_{\sigma \in \mathcal{E}_{int}} \int_\sigma Up\left[\frac{1}{2} \varrho_h |\boldsymbol{u}_h|^2, \boldsymbol{u}_h\right] [\![\phi_h]\!] \,\mathrm{dS}_x \\
&- \sum_{\sigma \in \mathcal{E}_{int}} \int_\sigma h^\varepsilon \, [\![\varrho_h]\!] \left[\!\!\left[\frac{1}{2} |\boldsymbol{u}_h|^2 \phi_h\right]\!\!\right] + \frac{1}{2} \phi_h^{\mathrm{down}} \varrho_h^{\mathrm{up}} | \{\!\{\boldsymbol{u}_h\}\!\} \cdot \boldsymbol{n} | \, [\![\boldsymbol{u}_h]\!]^2 \,\mathrm{dS}_x.
\end{aligned}$$

Consequently, summing up the previous two relations we derive the *discrete kinetic energy equation*

$$\begin{aligned}
&\frac{\mathrm{d}}{\mathrm{d}t} \int_\Omega \frac{1}{2} \varrho_h |\boldsymbol{u}_h|^2 \phi_h \,\mathrm{d}x - \sum_{\sigma \in \mathcal{E}_{int}} \int_\sigma Up\left[\frac{1}{2} \varrho_h |\boldsymbol{u}_h|^2, \boldsymbol{u}_h\right] [\![\phi_h]\!] \,\mathrm{dS}_x \\
&= -h^{\alpha-1} \sum_{\sigma \in \mathcal{E}_{int}} \int_\sigma [\![\boldsymbol{u}_h]\!] \cdot [\![\boldsymbol{u}_h \phi_h]\!] \,\mathrm{dS}_x + \sum_{\sigma \in \mathcal{E}_{int}} \int_\sigma \{\!\{p_h\}\!\} \boldsymbol{n} \cdot [\![\boldsymbol{u}_h \phi_h]\!] \,\mathrm{dS}_x \\
&+ h^\varepsilon \sum_{\sigma \in \mathcal{E}_{int}} \int_\sigma - [\![\boldsymbol{m}_h]\!] \, [\![\boldsymbol{u}_h \phi_h]\!] + [\![\varrho_h]\!] \left[\!\!\left[\frac{1}{2} |\boldsymbol{u}_h|^2 \phi_h\right]\!\!\right] \mathrm{dS}_x \\
&- \frac{1}{2} \sum_{\sigma \in \mathcal{E}_{int}} \int_\sigma \phi_h^{\mathrm{down}} \varrho_h^{\mathrm{up}} | \{\!\{\boldsymbol{u}_h\}\!\} \cdot \boldsymbol{n} | \, [\![\boldsymbol{u}_h]\!]^2 \,\mathrm{dS}_x.
\end{aligned} \tag{10.25}$$

Now, let us subtract (10.25) from the total energy balance (10.13)

$$
\begin{aligned}
&\frac{\mathrm{d}}{\mathrm{d}t}\int_{\Omega} \varrho_h e_h \phi_h \,\mathrm{d}x - \sum_{\sigma\in\mathcal{E}_{int}}\int_{\sigma}\Big(Up[\varrho_h e_h, \boldsymbol{u}_h] - h^{\varepsilon}\left[\!\left[\varrho_h e_h\right]\!\right]\Big)[\![\phi_h]\!]\,\mathrm{dS}_x \\
&= h^{\alpha-1}\sum_{\sigma\in\mathcal{E}_{int}}\int_{\sigma} [\![\boldsymbol{u}_h]\!]^2 \{\!\{\phi_h\}\!\}\,\mathrm{dS}_x + \frac{1}{2}\sum_{\sigma\in\mathcal{E}_{int}}\int_{\sigma}\phi_h^{\mathrm{down}}\varrho_h^{\mathrm{up}}|\{\!\{\boldsymbol{u}_h\}\!\}\cdot\boldsymbol{n}|\,[\![\boldsymbol{u}_h]\!]^2\,\mathrm{dS}_x \\
&- \sum_{\sigma\in\mathcal{E}_{int}}\int_{\sigma}\{\!\{p_h\phi_h\}\!\}[\![\boldsymbol{u}_h]\!]\cdot\boldsymbol{n}\mathrm{dS}_x + \sum_{\sigma\in\mathcal{E}_{int}}\int_{\sigma} h^{\varepsilon}\left[\!\left[\varrho_h\boldsymbol{u}_h\right]\!\right][\![\boldsymbol{u}_h\phi_h]\!]\,\mathrm{dS}_x \\
&- \sum_{\sigma\in\mathcal{E}_{int}}\int_{\sigma} h^{\varepsilon}\left[\!\left[\varrho_h\right]\!\right]\left[\!\left[\frac{1}{2}|\boldsymbol{u}_h|^2\phi_h\right]\!\right]\mathrm{dS}_x - \sum_{\sigma\in\mathcal{E}_{int}}\int_{\sigma} h^{\varepsilon}\left[\!\left[\frac{1}{2}\varrho_h|\boldsymbol{u}_h|^2\right]\!\right][\![\phi_h]\!]\,\mathrm{dS}_x.
\end{aligned}
$$

Applying the product rule we obtain

$$
\begin{aligned}
&\left[\!\left[\varrho_h\boldsymbol{u}_h\right]\!\right][\![\boldsymbol{u}_h\phi_h]\!] - \frac{1}{2}\left[\!\left[\varrho_h\right]\!\right]\left[\!\left[|\boldsymbol{u}_h|^2\phi_h\right]\!\right] - \frac{1}{2}\left[\!\left[\varrho_h|\boldsymbol{u}_h|^2\right]\!\right][\![\phi_h]\!] \\
&= \{\!\{\varrho_h\}\!\}[\![\boldsymbol{u}_h]\!]\cdot[\![\boldsymbol{u}_h]\!]\{\!\{\phi_h\}\!\} + \{\!\{\varrho_h\}\!\}\ \{\!\{\boldsymbol{u}_h\}\!\}\cdot[\![\boldsymbol{u}_h]\!][\![\phi_h]\!] \\
&+ \frac{1}{2}\left[\!\left[\varrho_h\right]\!\right]\{\!\{\boldsymbol{u}_h\}\!\}\cdot[\![\boldsymbol{u}_h\phi_h]\!] - \frac{1}{2}\left[\!\left[\varrho_h\right]\!\right][\![\boldsymbol{u}_h]\!]\cdot\{\!\{\boldsymbol{u}_h\phi_h\}\!\} - \frac{1}{2}\left[\!\left[\varrho_h|\boldsymbol{u}_h|^2\right]\!\right][\![\phi_h]\!] \\
&= \{\!\{\varrho_h\}\!\}[\![\boldsymbol{u}_h]\!]\cdot[\![\boldsymbol{u}_h]\!]\{\!\{\phi_h\}\!\} + \{\!\{\varrho_h\}\!\}\ \{\!\{\boldsymbol{u}_h\}\!\}\cdot[\![\boldsymbol{u}_h]\!][\![\phi_h]\!] \\
&+ \frac{1}{2}\left[\!\left[\varrho_h\right]\!\right]|\{\!\{\boldsymbol{u}_h\}\!\}|^2[\![\phi_h]\!] - \frac{1}{2}\left[\!\left[\varrho_h|\boldsymbol{u}_h|^2\right]\!\right][\![\phi_h]\!] - \frac{1}{8}\left[\!\left[\varrho_h\right]\!\right][\![\boldsymbol{u}_h]\!]^2[\![\phi_h]\!] \\
&= \{\!\{\varrho_h\}\!\}[\![\boldsymbol{u}_h]\!]\cdot[\![\boldsymbol{u}_h]\!]\{\!\{\phi_h\}\!\} + \{\!\{\varrho_h\}\!\}\ \{\!\{\boldsymbol{u}_h\}\!\}\cdot[\![\boldsymbol{u}_h]\!][\![\phi_h]\!] - \frac{1}{2}\{\!\{\varrho_h\}\!\}[\![\boldsymbol{u}_h\cdot\boldsymbol{u}_h]\!][\![\phi_h]\!] \\
&- \frac{1}{4}\left[\!\left[\varrho_h\right]\!\right][\![u_h]\!]^2[\![\phi_h]\!] = \{\!\{\varrho_h\}\!\}[\![\boldsymbol{u}_h]\!]^2\{\!\{\phi_h\}\!\} - \frac{1}{4}\left[\!\left[\varrho_h\right]\!\right][\![\boldsymbol{u}_h]\!]^2[\![\phi_h]\!].
\end{aligned}
$$

Consequently, we derive the *discrete internal energy equation* for $e_h = c_v\vartheta_h$

$$
\begin{aligned}
&\frac{\mathrm{d}}{\mathrm{d}t}\int_{\Omega} \varrho_h e_h \phi_h \,\mathrm{d}x - \sum_{\sigma\in\mathcal{E}_{int}}\int_{\sigma}\Big(Up[\varrho_h e_h, \boldsymbol{u}_h] - h^{\varepsilon}\left[\!\left[\varrho_h e_h\right]\!\right]\Big)[\![\phi_h]\!]\,\mathrm{dS}_x \\
&= h^{\alpha-1}\sum_{\sigma\in\mathcal{E}_{int}}\int_{\sigma} [\![\boldsymbol{u}_h]\!]^2 \{\!\{\phi_h\}\!\}\,\mathrm{dS}_x + \frac{1}{2}\sum_{\sigma\in\mathcal{E}_{int}}\int_{\sigma}\phi_h^{\mathrm{down}}\varrho_h^{\mathrm{up}}|\{\!\{\boldsymbol{u}_h\}\!\}\cdot\boldsymbol{n}|\,[\![\boldsymbol{u}_h]\!]^2\,\mathrm{dS}_x \\
&+ \sum_{\sigma\in\mathcal{E}_{int}}\int_{\sigma} h^{\varepsilon}\left(\{\!\{\varrho_h\}\!\}\{\!\{\phi_h\}\!\} - \frac{1}{4}\left[\!\left[\varrho_h\right]\!\right][\![\phi_h]\!]\right)[\![\boldsymbol{u}_h]\!]^2\,\mathrm{dS}_x \\
&- \sum_{\sigma\in\mathcal{E}_{int}}\int_{\sigma}\{\!\{p_h\phi_h\}\!\}[\![\boldsymbol{u}_h]\!]\cdot\boldsymbol{n}\mathrm{dS}_x.
\end{aligned}
\tag{10.26}
$$

$$- c_v \sum_{\sigma\in\mathcal{E}_{int}} \int_\sigma \phi_h^{\rm down} \varrho_h^{\rm up} \left(\widetilde{[\![\log(\vartheta_h)]\!]} - \frac{1}{\vartheta_h^{\rm down}} \widetilde{[\![\vartheta_h]\!]} \right) |\, \{\!\{\boldsymbol{u}_h\}\!\} \cdot \boldsymbol{n}| {\rm dS}_x \tag{10.30}$$

$$- c_v \sum_{\sigma\in\mathcal{E}_{int}} h^\varepsilon \int_\sigma \left([\![\varrho_h\vartheta_h]\!] \left[\!\!\left[\frac{\phi_h}{\vartheta_h}\right]\!\!\right] + [\![\varrho_h]\!]\, [\![(\log(\vartheta_h) - 1)\,\phi_h]\!] \right) {\rm dS}_x$$

$$+ \sum_{\sigma\in\mathcal{E}_{int}} \int_\sigma h^\varepsilon\, [\![\varrho_h]\!]\, [\![b'(\varrho_h)\phi_h]\!]\, {\rm dS}_x, \ \text{ where } b(\varrho) = \varrho\log(\varrho).$$

In order to control the sign of the last two integrals in (10.30) we proceed by the derivation of the renormalized discrete entropy inequality. Applying formula (10.22) in (10.30) and taking $b(\varrho) = \varrho\log(\varrho)$ we get

$$\frac{\rm d}{{\rm d}t} \int_\Omega \varrho_h \chi(s_h)\phi_h \,{\rm d}x - \sum_{\sigma\in\mathcal{E}_{int}} \int_\sigma Up[\varrho_h\chi(s_h), \boldsymbol{u}_h]\, [\![\phi_h]\!]\, {\rm dS}_x$$

$$= \sum_{\sigma\in\mathcal{E}_{int}} \int_\sigma h^{\alpha-1} [\![\boldsymbol{u}_h]\!]^2 \left\{\!\!\left\{ \frac{\chi'(s_h)\phi_h}{\vartheta_h} \right\}\!\!\right\} {\rm dS}_x$$

$$+ \sum_{\sigma\in\mathcal{E}_{int}} \int_\sigma h^\varepsilon \left(\{\!\{\varrho_h\}\!\} \left\{\!\!\left\{ \frac{\chi'(s_h)\phi_h}{\vartheta_h} \right\}\!\!\right\} - \frac14 [\![\varrho_h]\!] \left[\!\!\left[\frac{\chi'(s_h)\phi_h}{\vartheta_h} \right]\!\!\right] \right) [\![\boldsymbol{u}_h]\!]^2 {\rm dS}_x$$

$$+ \frac12 \sum_{\sigma\in\mathcal{E}_{int}} \int_\sigma \left(\frac{\chi'(s_h)\phi_h}{\vartheta_h} \right)^{\rm down} \varrho_h^{\rm up} |\, \{\!\{\boldsymbol{u}_h\}\!\} \cdot \boldsymbol{n}|\, [\![\boldsymbol{u}_h]\!]^2 {\rm dS}_x$$

$$\sum \int_\sigma (\chi'(s_h)\phi_h)^{\rm down} \left(\widetilde{[\![b(\varrho_h)]\!]} - b'(\varrho_h^{\rm down}) \widetilde{[\![\varrho_h]\!]} \right) |\, \{\!\{\boldsymbol{u}_h\}\!\} \cdot \boldsymbol{n}| {\rm dS}_x$$

$$(\chi'(s_h)\phi_h)^{\rm down} \varrho_h^{\rm up} \left(\widetilde{[\![\log(\vartheta_h)]\!]} - \frac{1}{\vartheta_h^{\rm down}} \widetilde{[\![\vartheta_h]\!]} \right) |\, \{\!\{\boldsymbol{u}_h\}\!\} \cdot$$

$$\widetilde{[\![\chi(s_h)]\!]} - \chi'(s_h^{\rm down}) \widetilde{[\![s_h]\!]} \big) |\, \{\!\{\boldsymbol{u}_h\}\!\} \cdot \boldsymbol{n}| {\rm dS}_x$$

$$\frac{\chi'(s_h)\phi_h}{\vartheta_h} \Big]\!\!\Big] + [\![\varrho_h]\!]\, [\![(\log(\vartheta$$

$$]\!] - [\![\varrho_h]\!]\, [\![$$

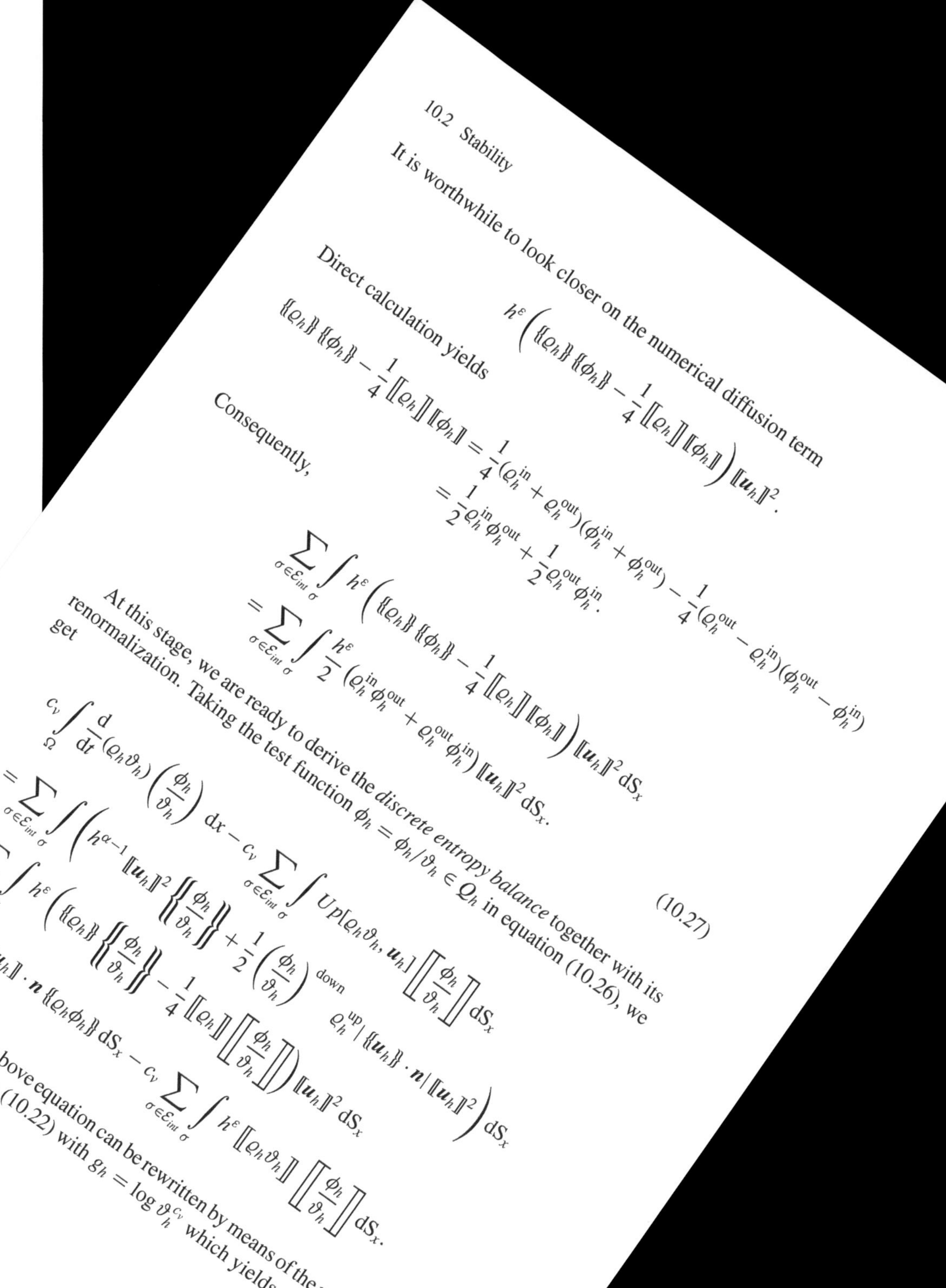

10.2 Stability 317

It is worthwhile to look closer on the numerical diffusion term

$$h^\varepsilon \left(\{\!\{\varrho_h\}\!\}\{\!\{\phi_h\}\!\} - \frac14 [\![\varrho_h]\!][\![\phi_h]\!] \right) [\![\boldsymbol{u}_h]\!]^2.$$

Direct calculation yields

$$\{\!\{\varrho_h\}\!\}\{\!\{\phi_h\}\!\} - \frac14 [\![\varrho_h]\!][\![\phi_h]\!] = \frac14 (\varrho_h^{\rm in} + \varrho_h^{\rm out})(\phi_h^{\rm in} + \phi_h^{\rm out}) - \frac14 (\varrho_h^{\rm out} - \varrho_h^{\rm in})(\phi_h^{\rm out} - \phi_h^{\rm in})$$
$$= \frac12 \varrho_h^{\rm in}\phi_h^{\rm out} + \frac12 \varrho_h^{\rm out}\phi_h^{\rm in}.$$

Consequently,

$$\sum_{\sigma\in\mathcal{E}_{int}} \int_\sigma h^\varepsilon \left(\{\!\{\varrho_h\}\!\}\{\!\{\phi_h\}\!\} - \frac14 [\![\varrho_h]\!][\![\phi_h]\!] \right) [\![\boldsymbol{u}_h]\!]^2 {\rm dS}_x$$
$$= \sum_{\sigma\in\mathcal{E}_{int}} \int_\sigma \frac{h^\varepsilon}{2} (\varrho_h^{\rm in}\phi_h^{\rm out} + \varrho_h^{\rm out}\phi_h^{\rm in}) [\![\boldsymbol{u}_h]\!]^2 {\rm dS}_x. \tag{10.27}$$

At this stage, we are ready to derive the *discrete entropy balance* together with its renormalization. Taking the test function $\phi_h = \phi_h/\vartheta_h \in Q_h$ in equation (10.26), we get

$$c_v \int_\Omega \frac{\rm d}{{\rm d}t}(\varrho_h\vartheta_h) \left(\frac{\phi_h}{\vartheta_h}\right) {\rm d}x - c_v \sum_{\sigma\in\mathcal{E}_{int}} \int_\sigma Up[\varrho_h\vartheta_h, \boldsymbol{u}_h] \left[\!\!\left[\frac{\phi_h}{\vartheta_h}\right]\!\!\right] {\rm dS}_x$$
$$= \sum_{\sigma\in\mathcal{E}_{int}} \int_\sigma \left(h^{\alpha-1} [\![\boldsymbol{u}_h]\!]^2 \left\{\!\!\left\{\frac{\phi_h}{\vartheta_h}\right\}\!\!\right\} + \frac12 \left(\frac{\phi_h}{\vartheta_h}\right)^{\rm down} \varrho_h^{\rm up} |\, \{\!\{\boldsymbol{u}_h\}\!\} \cdot \boldsymbol{n}|\, [\![\boldsymbol{u}_h]\!]^2 \right) {\rm dS}_x$$
$$+ \sum_{\sigma\in\mathcal{E}_{int}} \int_\sigma h^\varepsilon \left(\{\!\{\varrho_h\}\!\} \left\{\!\!\left\{\frac{\phi_h}{\vartheta_h}\right\}\!\!\right\} - \frac14 [\![\varrho_h]\!] \left[\!\!\left[\frac{\phi_h}{\vartheta_h}\right]\!\!\right] \right) [\![\boldsymbol{u}_h]\!]^2 {\rm dS}_x$$
$$- \sum_{\sigma\in\mathcal{E}_{int}} \int_\sigma [\![\boldsymbol{u}_h]\!] \cdot \boldsymbol{n}\, \{\!\{\varrho_h\phi_h\}\!\} {\rm dS}_x - c_v \sum_{\sigma\in\mathcal{E}_{int}} \int_\sigma h^\varepsilon [\![\varrho_h\vartheta_h]\!] \left[\!\!\left[\frac{\phi_h}{\vartheta_h}\right]\!\!\right] {\rm dS}_x.$$

The left hand side of the above equation can be rewritten by means of the renormalized discrete transport equation (10.22) with $g_h = \log\vartheta_h^{c_v}$ which yields

Next, we compute

$$
\begin{aligned}
&- c_v \sum_{\sigma \in \mathcal{E}_{int}} h^\varepsilon \int_\sigma \left([\![\varrho_h \vartheta_h]\!] \left[\!\!\left[\frac{\chi'(s_h)\phi_h}{\vartheta_h} \right]\!\!\right] + [\![\varrho_h]\!] \, [\![(\log(\vartheta_h) - 1)\, \chi'(s_h)\phi_h]\!] \right) \mathrm{dS}_x \\
&+ \sum_{\sigma \in \mathcal{E}_{int}} h^\varepsilon \int_\sigma \left([\![\varrho_h]\!] \, [\![b'(\varrho_h)\chi'(s_h)\phi_h]\!] - [\![\varrho_h]\!] \, [\![(\chi(s_h) - \chi'(s_h)s_h)\, \phi_h]\!] \right) \mathrm{dS}_x \\
&= -c_v \sum_{\sigma \in \mathcal{E}_{int}} h^\varepsilon \int_\sigma \left([\![\varrho_h \vartheta_h]\!] \left[\!\!\left[\frac{\chi'(s_h)\phi_h}{\vartheta_h} \right]\!\!\right] + [\![\varrho_h]\!] \, [\![\log(\vartheta_h)\chi'(s_h)\phi_h]\!] \right) \mathrm{dS}_x \\
&+ \sum_{\sigma \in \mathcal{E}_{int}} h^\varepsilon \int_\sigma \left([\![\varrho_h]\!] \, [\![\log(\varrho_h)\chi'(s_h)\phi_h]\!] - [\![\varrho_h]\!] \, [\![(\chi(s_h) - \chi'(s_h)s_h)\, \phi_h]\!] \right) \mathrm{dS}_x \\
&+ (c_v + 1) \sum_{\sigma \in \mathcal{E}_{int}} \int_\sigma h^\varepsilon \, [\![\varrho_h]\!] \, [\![\chi'(s_h)\phi_h]\!] \, \mathrm{dS}_x \\
&= -c_v \sum_{\sigma \in \mathcal{E}_{int}} \int_\sigma h^\varepsilon \, [\![\varrho_h \vartheta_h]\!] \left[\!\!\left[\varrho_h \frac{\chi'(s_h)\phi_h}{\varrho_h \vartheta_h} \right]\!\!\right] \mathrm{dS}_x \\
&- \sum_{\sigma \in \mathcal{E}_{int}} \int_\sigma h^\varepsilon \, [\![\varrho_h]\!] \left[\!\!\left[\Big(\chi(s_h) - (c_v + 1)\chi'(s_h) \Big) \phi_h \right]\!\!\right] \mathrm{dS}_x \\
&= -c_v \sum_{\sigma \in \mathcal{E}_{int}} \int_\sigma h^\varepsilon \, [\![p_h]\!] \left[\!\!\left[\varrho_h \frac{\chi'(s_h)\phi_h}{p_h} \right]\!\!\right] \mathrm{dS}_x \\
&- \sum_{\sigma \in \mathcal{E}_{int}} h^\varepsilon \int_\sigma [\![\varrho_h]\!] \left[\!\!\left[\Big(\chi(s_h) - (c_v + 1)\chi'(s_h) \Big) \phi_h \right]\!\!\right] \mathrm{dS}_x \\
&= - \sum_{\sigma \in \mathcal{E}_{int}} h^\varepsilon \int_\sigma \left([\![\phi_h \nabla_\varrho (\varrho_h \chi(s_h))]\!] \, [\![\varrho_h]\!] + [\![\phi_h \nabla_p (\varrho_h \chi(s_h))]\!] \, [\![p_h]\!] \right) \mathrm{dS}_x .
\end{aligned}
$$

Finally, we obtain the general entropy balance

$$
\begin{aligned}
&\frac{\mathrm{d}}{\mathrm{d}t}\int_{\Omega} \varrho_h \chi(s_h)\phi_h \,\mathrm{d}x - \sum_{\sigma\in\mathcal{E}_{int}}\int_{\sigma} Up[\varrho_h\chi(s_h), \boldsymbol{u}_h]\,[\![\phi_h]\!]\,\mathrm{dS}_x \\
&= \sum_{\sigma\in\mathcal{E}_{int}}\int_{\sigma} h^{\alpha-1}\,[\![\boldsymbol{u}_h]\!]^2 \left\{\!\!\left\{\frac{\chi'(s_h)\phi_h}{\vartheta_h}\right\}\!\!\right\}\mathrm{dS}_x \\
&+ \sum_{\sigma\in\mathcal{E}_{int}}\int_{\sigma} h^{\varepsilon}\left(\{\!\{\varrho_h\}\!\}\left\{\!\!\left\{\frac{\chi'(s_h)\phi_h}{\vartheta_h}\right\}\!\!\right\} - \frac{1}{4}[\![\varrho_h]\!]\left[\!\!\left[\frac{\chi'(s_h)\phi_h}{\vartheta_h}\right]\!\!\right]\right)[\![\boldsymbol{u}_h]\!]^2\,\mathrm{dS}_x \\
&+ \frac{1}{2}\sum_{\sigma\in\mathcal{E}_{int}}\int_{\sigma}\left(\frac{\chi'(s_h)\phi_h}{\vartheta_h}\right)^{\rm down}\varrho_h^{\rm up}\,|\,\{\!\{\boldsymbol{u}_h\}\!\}\cdot\boldsymbol{n}|\,[\![\boldsymbol{u}_h]\!]^2\,\mathrm{dS}_x \\
&+ \sum_{\sigma\in\mathcal{E}_{int}}\int_{\sigma}\left(\chi'(s_h)\phi_h\right)^{\rm down}\left(\widetilde{[\![b(\varrho_h)]\!]} - b'(\varrho_h^{\rm down})\widetilde{[\![\varrho_h]\!]}\right)|\,\{\!\{\boldsymbol{u}_h\}\!\}\cdot\boldsymbol{n}|\mathrm{dS}_x \\
&- c_v\sum_{\sigma\in\mathcal{E}_{int}}\int_{\sigma}\left(\chi'(s_h)\phi_h\right)^{\rm down}\varrho_h^{\rm up}\left(\widetilde{[\![\log(\vartheta_h)]\!]} - \frac{1}{\vartheta_h^{\rm down}}\widetilde{[\![\vartheta_h]\!]}\right)|\,\{\!\{\boldsymbol{u}_h\}\!\}\cdot\boldsymbol{n}|\mathrm{dS}_x \\
&- \sum_{\sigma\in\mathcal{E}_{int}}\int_{\sigma}\phi_h^{\rm down}\varrho_h^{\rm up}\left(\widetilde{[\![\chi(s_h)]\!]} - \chi'(s_h^{\rm down})\widetilde{[\![s_h]\!]}\right)|\,\{\!\{\boldsymbol{u}_h\}\!\}\cdot\boldsymbol{n}|\mathrm{dS}_x \\
&- \sum_{\sigma\in\mathcal{E}_{int}} h^{\varepsilon}\int_{\sigma}\left([\![\phi_h\nabla_{\varrho}(\varrho_h\chi(s_h))]\!]\,[\![\varrho_h]\!] + [\![\phi_h\nabla_{p}(\varrho_h\chi(s_h))]\!]\,[\![p_h]\!]\right)\mathrm{dS}_x,
\end{aligned}
\tag{10.32}
$$

where $b(\varrho) = \varrho\log(\varrho)$. The last integral can be rewritten using the product rule as

$$
\begin{aligned}
&- \sum_{\sigma\in\mathcal{E}_{int}}\int_{\sigma} h^{\varepsilon}\,[\![\phi_h]\!]\left(\{\!\{\nabla_{\varrho}(\varrho_h\chi(s_h))\}\!\}\,[\![\varrho_h]\!] + \{\!\{\nabla_{p}(\varrho_h\chi(s_h))\}\!\}\,[\![p_h]\!]\right)\mathrm{dS}_x \\
&- \sum_{\sigma\in\mathcal{E}_{int}}\int_{\sigma} h^{\varepsilon}\,\{\!\{\phi_h\}\!\}\left([\![\nabla_{\varrho}(\varrho_h\chi(s_h))]\!]\,[\![\varrho_h]\!] + [\![\nabla_{p}(\varrho_h\chi(s_h))]\!]\,[\![p_h]\!]\right)\mathrm{dS}_x.
\end{aligned}
\tag{10.33}
$$

The first sum in (10.33), together with the upwind term in (10.32), represents the numerical entropy flux $\boldsymbol{G}_h$

$$
\begin{aligned}
\sum_{\sigma\in\mathcal{E}_{int}}\int_{\sigma}\boldsymbol{G}_h\cdot\boldsymbol{n}\,[\![\phi_h]\!]\,\mathrm{dS}_x = \sum_{\sigma\in\mathcal{E}_{int}}\int_{\sigma}\Big(&Up[\varrho_h\chi(s_h), \boldsymbol{u}_h] \\
&- h^{\varepsilon}\left(\{\!\{\nabla_{\varrho}(\varrho_h\chi(s_h))\}\!\}\,[\![\varrho_h]\!] + \{\!\{\nabla_{p}(\varrho_h\chi(s_h))\}\!\}\,[\![p_h]\!]\right)\Big)[\![\phi_h]\!]\,\mathrm{dS}_x,
\end{aligned}
$$

The rest in (10.32) and (10.33) gives the numerical entropy production $r_h \equiv \sum_{\sigma\in\mathcal{E}_{int}} r_\sigma$, cf. (9.22). Indeed, for $\varrho_h > 0$ and $\vartheta_h > 0$ we have $r_h \geq 0$.

To see this property, let us firstly realize that due to (10.27) the numerical diffusion term can be rewritten in the following way

$$
\sum_{\sigma\in\mathcal{E}_{int}} \int_\sigma h^\varepsilon \left(\{\!\{\varrho_h\}\!\} \left\{\!\!\left\{ \frac{\chi'(s_h)\phi_h}{\vartheta_h} \right\}\!\!\right\} - \frac{1}{4} [\![\varrho_h]\!] \left[\!\!\left[\frac{\chi'(s_h)\phi_h}{\vartheta_h} \right]\!\!\right] \right) [\![\boldsymbol{u}_h]\!]^2 \, \mathrm{dS}_x
$$

$$
= \sum_{\sigma\in\mathcal{E}_{int}} \int_\sigma \frac{h^\varepsilon}{2} \left(\varrho_h^{\rm in} \frac{\chi'(s_h^{\rm out})\phi_h^{\rm out}}{\vartheta_h^{\rm out}} + \varrho_h^{\rm out} \frac{\chi'(s_h^{\rm in})\phi_h^{\rm in}}{\vartheta_h^{\rm in}} \right) [\![\boldsymbol{u}_h]\!]^2 \, \mathrm{dS}_x \geq 0.
$$

Further, recall that the total entropy

$$
(\varrho, p) \mapsto -\varrho\chi(s(\varrho,p)) = -\varrho\chi\left(\log\left(\frac{\vartheta^{c_v}}{\varrho}\right)\right) = -\varrho\chi\left(\frac{1}{\gamma-1}\log\left(\frac{p}{\varrho^\gamma}\right)\right)
$$

is a convex function of the variables ϱ and p. Indeed, direct calculations show that the function $\mathcal{S}(Z) = \chi(\log(Z))$ satisfies the conditions (2.49), i.e.

$$
\mathcal{S}'(Z) > 0, \ (\gamma-1)\mathcal{S}'(Z) + \gamma\mathcal{S}''(Z)Z < 0 \quad \text{for all } Z > 0.
$$

Now, it follows from Sect. 2.2.3 that $-\varrho\chi\left(\frac{1}{\gamma-1}\log\left(\frac{p}{\varrho^\gamma}\right)\right)$ is a convex function of ϱ, p. In particular, $-\nabla_{\varrho,p}(\varrho\chi(s(\varrho_h,p_h)))$ is monotone and the term in the second line of (10.33) is nonnegative. This concludes the proof.

We will discuss the positivity of discrete density and temperature in what follows.

10.2.2 Positivity of the Discrete Density

Our aim is to show that the discrete density obtained by the VFV method (10.11)–(10.13) remains strictly positive on $[0,T]$ provided $\varrho_0 > 0$. To this end, we prove the following discrete version of the comparison theorem.

Lemma 10.2 (Positivity of the discrete density)
Let $(\varrho_h, \boldsymbol{u}_h)$ be the discrete density and velocity obtained by the VFV method, specifically,

$$
\frac{\mathrm{d}}{\mathrm{d}t}\int_\Omega \varrho_h\phi_h \, \mathrm{d}x - \sum_{\sigma\in\mathcal{E}_{int}} \int_\sigma F_h^{\rm up}[\varrho_h, \boldsymbol{u}_h] [\![\phi_h]\!] \, \mathrm{dS}_x = 0 \ \textit{for any } \phi_h \in Q_h.
$$

Let $\underline{\varrho}$ be a subsolution of the same equation, meaning,

$$
\frac{\mathrm{d}}{\mathrm{d}t}\int_\Omega \underline{\varrho}\phi_h \, \mathrm{d}x - \sum_{\sigma\in\mathcal{E}_{int}} \int_\sigma F_h^{\rm up}[\underline{\varrho}, \boldsymbol{u}_h] [\![\phi_h]\!] \, \mathrm{dS}_x \leq 0 \, \textit{for any } \phi_h \in Q_h, \ \phi_h \geq 0.
\tag{10.34}
$$

In addition, suppose that

$$
\varrho_h(0) \geq \underline{\varrho}(0).
$$

Then

$$\varrho_h(t) \geq \underline{\varrho}(t) \text{ for all } 0 \leq t \leq T.$$

Proof It is easy to check that the difference $r = \varrho_h - \underline{\varrho}$ is a supersolution, meaning

$$\frac{\mathrm{d}}{\mathrm{d}t} \int_\Omega r\phi_h \,\mathrm{d}x - \sum_{\sigma \in \mathcal{E}_{int}} \int_\sigma F_h^{\rm up}[r, \boldsymbol{u}_h] [\![\phi_h]\!] \,\mathrm{dS}_x \geq 0 \text{ for any } \phi_h \in Q_h, \ \phi_h \geq 0.$$

Let $b : R \to R$ be a convex function such that $b'(r) \leq 0$ for a.a. $r \in R$. Similarly to (10.21), we obtain the integrated renormalized inequality

$$\int_\Omega \frac{\mathrm{d}}{\mathrm{d}t} b(r) \,\mathrm{d}x + \sum_{\sigma \in \mathcal{E}_{int}} \int_\sigma \{\!\{\boldsymbol{u}_h\}\!\} \cdot \boldsymbol{n} \left[\!\!\left[\left(b(r) - b'(r) r \right) \right]\!\!\right] \mathrm{dS}_x \leq 0.$$

Taking $b(r) = -r^- = -\min\{r, 0\}$ gives rise to

$$\int_\Omega \frac{\mathrm{d}}{\mathrm{d}t} (\varrho_h - \underline{\varrho})^- \,\mathrm{d}x \geq 0$$

yielding the desired conclusion

$$r = \varrho_h - \underline{\varrho} \geq 0 \text{ for any } t \in [0, T].$$

Now we choose $\underline{\varrho}_0 > 0$, such that

$$\varrho_h(0, \cdot) \geq \underline{\varrho}_0,$$

and consider

$$\underline{\varrho}(t) = \exp(-L(t)) \underline{\varrho}_0, \ t \geq 0, \ L(0) = 0.$$

Seeing that

$$\left[\!\!\left[\underline{\varrho}(t) \right]\!\!\right] = 0, \ \left\{\!\!\left\{ \underline{\varrho}(t) \right\}\!\!\right\} = \underline{\varrho}(t),$$

we easily deduce that

$$\begin{aligned}
&\frac{\mathrm{d}}{\mathrm{d}t} \int_\Omega \underline{\varrho}\phi_h \,\mathrm{d}x - \sum_{\sigma \in \mathcal{E}_{int}} \int_\sigma F_h^{\rm up}[\underline{\varrho}, \boldsymbol{u}_h] [\![\phi_h]\!] \,\mathrm{dS}_x \\
&= -L'(t) \int_\Omega \exp(-L(t)) \underline{\varrho}_0 \phi_h \,\mathrm{d}x + \sum_{\sigma \in \mathcal{E}_{int}} \int_\sigma \exp(-L(t)) \underline{\varrho}_0 \{\!\{\boldsymbol{u}_h\}\!\} \cdot \boldsymbol{n} [\![\phi_h]\!] \,\mathrm{dS}_x
\end{aligned}$$

$$\leq -L'(t) \int_{\Omega} \exp(-L(t))\underline{\varrho}_0 \phi_h \, \mathrm{d}x + Z(t,h) \int_{\Omega} \exp(-L(t))\underline{\varrho}_0 \phi_h \, \mathrm{d}x. \tag{10.35}$$

Using the discrete kinetic energy equation (10.25) we may derive that

$$\int_0^T \left(\sup_{\sigma \in \mathcal{E}_{int}} [\![\boldsymbol{u}_h]\!]^2 \right) \mathrm{d}t \lesssim \omega(h),$$

where $\omega(h)$ denotes a generic function that may blow up in the asymptotic regime $h \to 0$. Indeed, let us consider (10.25) with a test function $\phi_h = 1$. Seeing that

$$- \sum_{\sigma \in \mathcal{E}_{int}} \int_{\sigma} h^{\varepsilon} [\![\boldsymbol{m}_h]\!] [\![\boldsymbol{u}_h]\!] \, \mathrm{dS}_x + \sum_{\sigma \in \mathcal{E}_{int}} \int_{\sigma} h^{\varepsilon} [\![\varrho_h]\!] \left[\!\!\left[\frac{1}{2} |\boldsymbol{u}_h|^2 \right]\!\!\right] \mathrm{dS}_x =$$
$$- \sum_{\sigma \in \mathcal{E}_{int}} \int_{\sigma} h^{\varepsilon} \{\!\{ \varrho_h \}\!\} [\![\boldsymbol{u}_h]\!]^2 \, \mathrm{dS}_x,$$

we may integrate (10.25) in time and use the energy bound (10.18) to deduce

$$h^{\alpha-1} \int_0^T \sum_{\sigma \in \mathcal{E}_{int}} \int_{\sigma} [\![\boldsymbol{u}_h]\!]^2 \, \mathrm{dS}_x \, \mathrm{d}t + h^{\varepsilon} \int_0^T \sum_{\sigma \in \mathcal{E}_{int}} \int_{\sigma} \{\!\{ \varrho_h \}\!\} [\![\boldsymbol{u}_h]\!]^2 \, \mathrm{dS}_x \, \mathrm{d}t$$
$$+ \frac{1}{2} \int_0^T \sum_{\sigma \in \mathcal{E}_{int}} \int_{\sigma} \varrho_h^{\mathrm{up}} | \{\!\{ \boldsymbol{u}_h \}\!\} \cdot \boldsymbol{n} | [\![\boldsymbol{u}_h]\!]^2 \, \mathrm{dS}_x \, \mathrm{d}t$$
$$\lesssim 1 + \sum_{\sigma \in \mathcal{E}_{int}} \int_0^T \int_{\sigma} \{\!\{ p_h \}\!\} \boldsymbol{n} \cdot [\![\boldsymbol{u}_h]\!] \, \mathrm{dS}_x \, \mathrm{d}t.$$

Finally, we again use (10.18) combined with the inverse L^p–estimates (26) and Hölder's inequality to conclude

$$\int_0^T \sum_{\sigma \in \mathcal{E}_{int}} \int_{\sigma} [\![\boldsymbol{u}_h]\!]^2 \, \mathrm{dS}_x \, \mathrm{d}t \lesssim \omega(h). \tag{10.36}$$

In particular, relation (10.36) implies

$$\int_0^T \left(\sup_{\sigma \in \mathcal{E}_{int}} [\![\boldsymbol{u}_h]\!]^2 \right) \mathrm{d}t \lesssim \omega(h), \tag{10.37}$$

with another $\omega(h)$ generally different from its counterpart in (10.36).

Going back to (10.35) the latter implies

$$Z(t,h) \geq 0,\ Z(\cdot,h) \in L^1(0,T).$$

Taking $L(t) = \int_0^t Z(h,s)\ \mathrm{d}s$, we conclude that $\underline{\varrho}(t) = \underline{\varrho}(h,t)$ is a subsolution and strict positivity of ϱ_h follows from Lemma 10.2:

$$\varrho_h(t) \geq \underline{\varrho}(h,t) > 0 \text{ for all } t \in [0,T]. \tag{10.38}$$

Clearly, the estimate (10.38) is not uniform, neither with respect to T nor for $h \to 0$. But for each fixed mesh step h we have a strictly positive discrete density ϱ_h on any given time interval $[0,T]$.

10.2.3 *Minimum Entropy Principle and Positivity of the Discrete Pressure and Temperature*

In what follows we want to derive the minimum entropy principle from the renormalized discrete entropy inequality (10.22), where we consider

$$\phi_h = 1,\ \chi(s_h) = (s_h - \underline{s})^-,\ -\infty < \underline{s} < \min s_h(0).$$

Realizing that

$$\varrho \mapsto \varrho \log(\varrho) \text{ is convex, } \vartheta \mapsto \log(\vartheta) \text{ is concave, } s \mapsto \chi(s) \text{ is concave,}$$
$$\text{and } (\varrho, p) \mapsto \varrho\chi(s(\varrho, p)) \text{ is concave,}$$

the entropy production term r on the right hand side of (10.22) is nonnegative, and we may infer that

$$\int_\Omega \varrho_h(t)(s_h(t) - \underline{s})^-\ \mathrm{d}x \geq 0 \ \text{ for any } \ t \geq 0.$$

Consequently, we have obtained *the minimum entropy principle*

$$s_h(t) \geq \underline{s} \ \text{ for all } \ t \geq 0. \tag{10.39}$$

We proceed by considering the entropy as a function of ϱ and p,

$$s = \frac{1}{\gamma - 1}\log\left(\frac{p}{\varrho^\gamma}\right).$$

Minimum entropy principle (10.39) immediately implies

$$0 < \exp\{(\gamma - 1)\underline{s}\} \le \frac{p_h(t)}{\varrho_h^\gamma(t)} \text{ for all } t \ge 0. \tag{10.40}$$

In particular, the pressure is positive and (10.16), meaning the relation

$$\vartheta_h(t) = \frac{p_h(t)}{\varrho_h(t)}$$

is valid for any $t \ge 0$. Evoking the energy bound (10.18) we get

$$\frac{1}{2}\int_\Omega \frac{|\boldsymbol{m}_h(t)|^2}{\varrho_h(t)} \, \mathrm{d}x + c_v \int_\Omega \varrho_h(t)\vartheta_h(t) \, \mathrm{d}x \le E_0 \text{ for all } t \ge 0. \tag{10.41}$$

Thus going back to (10.40) we obtain

$$\int_\Omega \varrho_h^\gamma(t) \, \mathrm{d}x \lesssim \int_\Omega p_h(t) \, \mathrm{d}x \lesssim E_0 \text{ for all } t \ge 0. \tag{10.42}$$

The following lemma summarizes the properties of discrete solution $(\varrho_h, \boldsymbol{m}_h, E_h)$ obtained by the VFV method (10.11)–(10.13).

Lemma 10.3
Suppose that the initial data $(\varrho_{0,h}, \boldsymbol{m}_{0,h}, E_{0,h})$ satisfy

$$\varrho_{0,h} \ge \underline{\varrho} > 0, \ E_{0,h} - \frac{1}{2}\frac{|\boldsymbol{m}_{0,h}|^2}{\varrho_{0,h}} > 0. \tag{10.43}$$

Then the semidiscrete approximate system (10.11)–(10.13) admits a unique global-in-time solution $(\varrho_h, \boldsymbol{m}_h, E_h)$ such that

$$\varrho_h(t) > 0, \ E_h(t) - \frac{1}{2}\frac{|\boldsymbol{m}_h(t)|^2}{\varrho_h(t)} > 0 \textit{ for any } t \ge 0$$

and the following a priori bounds hold uniformly for $h \to 0$:

$$\|\varrho_h\|_{L^\infty(0,T;L^\gamma(\Omega))} \lesssim 1, \ \ \|\boldsymbol{m}_h\|_{L^\infty(0,T;L^{2\gamma/(\gamma+1)}(\Omega))} \lesssim 1, \ \ \|E_h\|_{L^\infty(0,T;L^1(\Omega))} \lesssim 1. \tag{10.44}$$

Moreover, the renormalized entropy balance (10.24) holds.

10.2.4 Weak BV Estimates

We conclude the discussion on the stability properties of the VFV method by deriving some suitable weak BV estimates. To this end we examine more closely the dissipation mechanism hidden in the entropy production term r_h, cf. (10.22).

Theorem 10.2 (Weak BV estimates)
Let $(\varrho_h, \boldsymbol{m}_h, E_h)$ be the family of approximate solutions generated by the VFV method. Let the hypothesis (10.43) *on the initial data hold.*
Then we have

$$\int_\Omega \varrho_h s_h(t)\,\mathrm{d}x \lesssim 1 + \int_\Omega E_h(t)\,\mathrm{d}x \le 1 + E_0, \tag{10.45}$$

and the following weak BV estimates

$$\begin{aligned}
&\int_0^T \sum_{\sigma\in\mathcal{E}_{int}} \int_\sigma h^{\alpha-1} [\![\boldsymbol{u}_h]\!]^2 \left\{\!\!\left\{\frac{1}{\vartheta_h}\right\}\!\!\right\} \mathrm{dS}_x\,\mathrm{d}t \\
&+ \int_0^T \sum_{\sigma\in\mathcal{E}_{int}} \int_\sigma \frac{h^\varepsilon}{2}\left(\frac{\varrho_h{}^{\mathrm{in}}}{\vartheta_h^{\mathrm{out}}} + \frac{\varrho_h{}^{\mathrm{out}}}{\vartheta_h^{\mathrm{in}}}\right) [\![\boldsymbol{u}_h]\!]^2\,\mathrm{dS}_x\,\mathrm{d}t \\
&+ \frac{1}{2}\int_0^T \sum_{\sigma\in\mathcal{E}_{int}} \int_\sigma \left(\frac{1}{\vartheta_h}\right)^{\mathrm{down}} \varrho_h^{\mathrm{up}} |\, \{\!\{\boldsymbol{u}_h\}\!\}\cdot\boldsymbol{n}|\, [\![\boldsymbol{u}_h]\!]^2\,\mathrm{dS}_x\,\mathrm{d}t \\
&+ \int_0^T \sum_{\sigma\in\mathcal{E}_{int}} \int_\sigma \left(\widetilde{[\![b(\varrho_h)]\!]} - b'(\varrho_h^{\mathrm{down}})\widetilde{[\![\varrho_h]\!]}\right) |\, \{\!\{\boldsymbol{u}_h\}\!\}\cdot\boldsymbol{n}|\mathrm{dS}_x\,\mathrm{d}t \\
&- c_v \int_0^T \sum_{\sigma\in\mathcal{E}_{int}} \int_\sigma \varrho_h^{\mathrm{up}} \left(\widetilde{[\![\log(\vartheta_h)]\!]} - \frac{1}{\vartheta_h^{\mathrm{down}}}\widetilde{[\![\vartheta_h]\!]}\right) |\, \{\!\{\boldsymbol{u}_h\}\!\}\cdot\boldsymbol{n}|\mathrm{dS}_x\,\mathrm{d}t \\
&- c_v \int_0^T \sum_{\sigma\in\mathcal{E}_{int}} \int_\sigma h^\varepsilon \min\{\varrho_h^{\mathrm{in}}, \varrho_h^{\mathrm{out}}\}\, [\![\vartheta_h]\!] \left[\!\!\left[\frac{1}{\vartheta_h}\right]\!\!\right] \mathrm{dS}_x\,\mathrm{d}t \\
&+ \int_0^T \sum_{\sigma\in\mathcal{E}_{int}} \int_\sigma h^\varepsilon\, [\![\varrho_h]\!]\, [\![\log(\varrho_h)]\!]\,\mathrm{dS}_x\,\mathrm{d}t \lesssim (1+E_0).
\end{aligned} \tag{10.46}$$

Proof In view of the minimum entropy principle established in (10.39), it is enough to observe that

$$\varrho_h \log\left(\frac{\vartheta_h^{c_v}}{\varrho_h}\right) \stackrel{<}{\sim} 1 + \varrho_h\vartheta_h \text{ provided } 0 < \varrho_h \stackrel{<}{\sim} \vartheta_h^{c_v}.$$

Seeing that $\varrho_h \log(\varrho_h)$ is controlled by (10.42) we restrict ourselves to $\varrho_h \log(\vartheta_h^{c_v})$. Here,

$$\varrho_h \log(\vartheta_h^{c_v}) \stackrel{<}{\sim} \varrho_h\vartheta_h \stackrel{<}{\sim} E_0 \text{ if } \vartheta_h \geq 1,$$

while

$$|\varrho_h \log(\vartheta_h^{c_v})| \leq \vartheta_h^{c_v} |\log(\vartheta_h^{c_v})| \stackrel{<}{\sim} 1 \text{ for } \vartheta_h \leq 1.$$

Thus we have shown (10.45).

Now, we take $\phi_h = 1$, $\chi_\varepsilon(s) = \min\{s, \frac{1}{\varepsilon}\}$ in the renormalized entropy balance (10.32). Letting $\varepsilon \to 0$ and recalling (10.27) we obtain the uniform estimate:

$$\begin{aligned}
&\int_0^T \sum_{\sigma\in\mathcal{E}_{int}} \int_\sigma h^{\alpha-1} [\![\boldsymbol{u}_h]\!]^2 \left\{\!\!\left\{\frac{1}{\vartheta_h}\right\}\!\!\right\} \mathrm{dS}_x \,\mathrm{d}t \\
&+ \int_0^T \sum_{\sigma\in\mathcal{E}_{int}} \int_\sigma \frac{h^\varepsilon}{2} \left(\frac{\varrho_h{}^{\mathrm{in}}}{\vartheta_h^{\mathrm{out}}} + \frac{\varrho_h{}^{\mathrm{out}}}{\vartheta_h^{\mathrm{in}}}\right) [\![\boldsymbol{u}_h]\!]^2 \,\mathrm{dS}_x \,\mathrm{d}t \\
&+ \frac{1}{2}\int_0^T \sum_{\sigma\in\mathcal{E}_{int}} \int_\sigma \left(\frac{1}{\vartheta_h}\right)^{\mathrm{down}} \varrho_h^{\mathrm{up}} |\{\!\{\boldsymbol{u}_h\}\!\} \cdot \boldsymbol{n}| \, [\![\boldsymbol{u}_h]\!]^2 \,\mathrm{dS}_x \,\mathrm{d}t \\
&+ \int_0^T \sum_{\sigma\in\mathcal{E}_{int}} \int_\sigma \left(\widetilde{[\![b(\varrho)]\!]} - b'(\varrho_h^{\mathrm{down}})\widetilde{[\![\varrho_h]\!]}\right) |\{\!\{\boldsymbol{u}_h\}\!\} \cdot \boldsymbol{n}| \mathrm{dS}_x \,\mathrm{d}t \\
&- c_v \int_0^T \sum_{\sigma\in\mathcal{E}_{int}} \int_\sigma \varrho_h^{\mathrm{down}} \left(\widetilde{[\![\log(\vartheta_h)]\!]} - \frac{1}{\vartheta_h^{\mathrm{down}}}\widetilde{[\![\vartheta_h]\!]}\right) |\{\!\{\boldsymbol{u}_h\}\!\} \cdot \boldsymbol{n}| \mathrm{dS}_x \,\mathrm{d}t \\
&- \int_0^T \sum_{\sigma\in\mathcal{E}_{int}} h^\varepsilon \int_\sigma \left([\![\nabla_\varrho(\varrho_h s_h)]\!] \; [\![\varrho_h]\!] + [\![\nabla_p(\varrho_h s_h)]\!] \; [\![p_h]\!]\right) \mathrm{dS}_x \,\mathrm{d}t \stackrel{<}{\sim} (1 + E_0).
\end{aligned} \tag{10.47}$$

Direct manipulation of the last integral in (10.47) yields

$$
\begin{aligned}
&-\int_0^T \sum_{\sigma\in\mathcal{E}_{int}} h^{\varepsilon} \int_\sigma [\![\nabla_\varrho(\varrho_h s_h)]\!]\, [\![\varrho_h]\!] + [\![\nabla_p(\varrho_h s_h)]\!]\, [\![p_h]\!]\, \mathrm{dS}_x\, \mathrm{d}t \\
&= -c_v \int_0^T \sum_{\sigma\in\mathcal{E}_{int}} \int_\sigma h^{\varepsilon} \{\!\{\varrho_h\}\!\} [\![\vartheta_h]\!] \left[\!\!\left[\frac{1}{\vartheta_h}\right]\!\!\right] \mathrm{dS}_x\, \mathrm{d}t \\
&+ \int_0^T \sum_{\sigma\in\mathcal{E}_{int}} \int_\sigma h^{\varepsilon} [\![\varrho_h]\!]\, [\![\log(\varrho_h)]\!]\, \mathrm{dS}_x\, \mathrm{d}t \\
&- c_v \int_0^T \sum_{\sigma\in\mathcal{E}_{int}} \int_\sigma h^{\varepsilon} \left([\![\log(\vartheta_h)]\!] + \{\!\{\vartheta_h\}\!\} \left[\!\!\left[\frac{1}{\vartheta_h}\right]\!\!\right] \right) [\![\varrho_h]\!]\, \mathrm{dS}_x\, \mathrm{d}t.
\end{aligned}
$$

Next, we show that

$$
-[\![\varrho_h]\!] \left([\![\log(\vartheta_h)]\!] + \{\!\{\vartheta_h\}\!\} \left[\!\!\left[\frac{1}{\vartheta_h}\right]\!\!\right] \right) \le -\frac{1}{2} |\, [\![\varrho_h]\!]\, |\, [\![\vartheta_h]\!] \left[\!\!\left[\frac{1}{\vartheta_h}\right]\!\!\right]. \tag{10.48}
$$

First realize that both expressions in the above inequality are invariant with respect to the change "in" and "out". Moreover, the right-hand side is invariant with respect to the same operation in ϱ_h and ϑ_h separately. Thus, it is enough to show (10.48) assuming $\varrho_h^{\rm in} \ge \varrho_h^{\rm out}$. In other words, $-[\![\varrho_h]\!] = |\, [\![\varrho_h]\!]\, | \ge 0$. Consequently, the proof of (10.48) reduces to the inequality

$$
[\![\log(\vartheta_h)]\!] + \{\!\{\vartheta_h\}\!\} \left[\!\!\left[\frac{1}{\vartheta_h}\right]\!\!\right] \le -\frac{1}{2} [\![\vartheta_h]\!] \left[\!\!\left[\frac{1}{\vartheta_h}\right]\!\!\right].
$$

Denoting $Z = \frac{\vartheta_h^{\rm out}}{\vartheta_h^{\rm in}}$, we have to show

$$
\log(Z) - \frac{1}{2}\left(Z - \frac{1}{Z}\right) \le \frac{1}{2}\left(Z + \frac{1}{Z}\right) - 1
$$

or

$$
\log(Z) \le Z - 1.
$$

The latter is clear since log is a concave function. Relations (10.48) and (10.47) yield (10.46) and conclude the proof.

10.3 Consistency

The next step towards the convergence is to prove that the VFV scheme (10.11)–(10.13) is consistent with the Euler system (7.14)–(7.15) in the sense of Definition 8.5. Careful analysis of consistency error below shows that this can be done under additional hypotheses on the boundedness of discrete density from below and discrete temperature from above uniformly for $h \to 0$.

Theorem 10.3 (Consistency of VFV scheme)
Suppose that the initial data $(\varrho_{0,h}, \boldsymbol{m}_{0,h}, E_{0,h})$ *satisfy*

$$\varrho_{0,h} \geq \underline{\varrho} > 0,\ E_{0,h} - \frac{1}{2}\frac{|\boldsymbol{m}_{0,h}|^2}{\varrho_{0,h}} > 0.$$

Let $(\varrho_h, \boldsymbol{m}_h, E_h)$ *be the unique solution of the VFV scheme* (10.11)–(10.13) *on the time interval* $[0, T]$.
Then

$$\left[\int_\Omega \varrho_h \varphi \, \mathrm{d}x\right]_{t=0}^{t=\tau} = \int_0^\tau \int_\Omega \left[\varrho_h \partial_t \varphi + \boldsymbol{m}_h \cdot \nabla_x \varphi\right] \, \mathrm{d}x \, \mathrm{d}t + \int_0^\tau e_{1,h}(t, \varphi) \, \mathrm{d}t \quad (10.49)$$

for any $\varphi \in C^2([0, T] \times \overline{\Omega})$;

$$\left[\int_\Omega \boldsymbol{m}_h \cdot \boldsymbol{\varphi} \, \mathrm{d}x\right]_{t=0}^{t=\tau} = \int_0^\tau \int_\Omega \left[\boldsymbol{m}_h \cdot \partial_t \boldsymbol{\varphi} + \frac{\boldsymbol{m}_h \otimes \boldsymbol{m}_h}{\varrho_h} : \nabla_x \boldsymbol{\varphi} + p_h \mathrm{div}_x \boldsymbol{\varphi}\right] \mathrm{d}x \, \mathrm{d}t + \int_0^\tau e_{2,h}(t, \boldsymbol{\varphi}) \, \mathrm{d}t$$

for any $\boldsymbol{\varphi} \in C^2([0, T] \times \overline{\Omega}; R^d)$, $\boldsymbol{\varphi} \cdot \boldsymbol{n}|_{\partial\Omega} = 0$;

$$\int_\Omega E_h(t) \, \mathrm{d}x = \int_\Omega E_{0,h} \, \mathrm{d}x; \quad (10.50)$$

$$\left[\int_\Omega \varrho_h \chi(s_h) \varphi \, \mathrm{d}x\right]_{t=0}^{t=\tau} \geq \int_0^\tau \int_\Omega \left[\varrho_h \chi(s_h) \partial_t \varphi + \chi(s_h) \boldsymbol{m}_h \cdot \nabla_x \varphi\right] \, \mathrm{d}x \, \mathrm{d}t + \int_0^\tau e_{3,h}(t, \varphi) \, \mathrm{d}t \quad (10.51)$$

for any $\varphi \in C^2([0, T] \times \overline{\Omega})$, $\varphi \geq 0$, *and any* χ,

$$\chi : R \to R \text{ a nondecreasing concave function},\ \chi(s) \leq \overline{\chi} \text{ for all } s \in R.$$

If, in addition,

$$0 < \alpha < \frac{4}{3}, \quad (10.52)$$

and

$$0 < \underline{\varrho} \le \varrho_h(t),\ \vartheta_h(t) \le \overline{\vartheta} \ \textit{for all} \ t \in [0, T] \ \textit{uniformly for} \ h \to 0, \tag{10.53}$$

then

$$\|e_{j,h}(\cdot, \varphi)\|_{L^1(0,T)} \lesssim h^{\beta} \|\varphi\|_{C^2},\ j = 1, 3, \quad \|e_{2,h}(\cdot, \boldsymbol{\varphi})\|_{L^1(0,T)} \lesssim h^{\beta} \|\boldsymbol{\varphi}\|_{C^2}, \quad \textit{for some} \ \beta > 0.$$

Proof First of all we note that in view of the minimum entropy principle (10.39) the hypothesis (10.53) yields the uniform boundedness of the discrete density and temperature as $h \to 0$. Indeed, under the assumption

$$0 < \vartheta_h(t) \le \overline{\vartheta} \text{ for all } t \in [0, T] \text{ uniformly for } h \to 0, \tag{10.54}$$

the minimum entropy principle (10.39) yields a similar bound on the discrete density,

$$0 < \varrho_h(t) \le \overline{\varrho} \text{ for all } t \in [0, T] \text{ uniformly for } h \to 0. \tag{10.55}$$

On the other hand, in view of (10.39) assuming

$$0 < \underline{\varrho} \le \varrho_h(t) \text{ for all } t \in [0, T] \text{ uniformly for } h \to 0 \tag{10.56}$$

implies a similar lower bound on the discrete temperature,

$$0 < \underline{\vartheta} \le \vartheta_h(t) \text{ for all } t \in [0, T] \text{ uniformly for } h \to 0. \tag{10.57}$$

Under these circumstances, we easily deduce from the total energy inequality (10.41) and weak BV estimates (10.46) the following bounds:

$$\sup_{t \in [0,T]} \|\boldsymbol{u}_h(t)\|_{L^2(\Omega)} \lesssim 1, \tag{10.58}$$

$$\int_0^T \sum_{\sigma \in \mathcal{E}_{int}} \int_\sigma |\{\!\{\boldsymbol{u}_h\}\!\} \cdot \boldsymbol{n}|\, [\![\boldsymbol{u}_h]\!]^2 \, \mathrm{dS}_x \, \mathrm{d}t \lesssim 1, \tag{10.59}$$

$$\int_0^T \sum_{\sigma \in \mathcal{E}_{int}} \int_\sigma |\{\!\{\boldsymbol{u}_h\}\!\} \cdot \boldsymbol{n}|\, [\![\varrho_h]\!]^2 \, \mathrm{dS}_x \, \mathrm{d}t \lesssim 1, \tag{10.60}$$

$$\int_0^T \sum_{\sigma \in \mathcal{E}_{int}} \int_\sigma |\{\!\{\boldsymbol{u}_h\}\!\} \cdot \boldsymbol{n}|\, [\![\vartheta_h]\!]^2 \, \mathrm{dS}_x \, \mathrm{d}t \lesssim 1, \tag{10.61}$$

$$\int_0^T \sum_{\sigma \in \mathcal{E}_{int}} \int_\sigma h^\varepsilon \llbracket \boldsymbol{u}_h \rrbracket^2 \, \mathrm{dS}_x \, \mathrm{d}t \stackrel{<}{\sim} 1. \tag{10.62}$$

In particular, we obtain the estimates

$$h^{\alpha-1} \int_0^T \sum_{\sigma \in \mathcal{E}_{int}} \int_\sigma \llbracket \boldsymbol{u}_h \rrbracket^2 \, \mathrm{dS}_x \, \mathrm{d}t \stackrel{<}{\sim} 1, \tag{10.63}$$

$$\int_0^T \sum_{\sigma \in \mathcal{E}_{int}} \int_\sigma \lambda_h \llbracket \varrho_h \rrbracket^2 \, \mathrm{dS}_x \, \mathrm{d}t \stackrel{<}{\sim} 1, \tag{10.64}$$

$$\int_0^T \sum_{\sigma \in \mathcal{E}_{int}} \int_\sigma \lambda_h \llbracket \vartheta_h \rrbracket^2 \, \mathrm{dS}_x \, \mathrm{d}t \stackrel{<}{\sim} 1, \tag{10.65}$$

where $\lambda_h \approx |\{\!\{\boldsymbol{u}_h\}\!\} \cdot \boldsymbol{n}| + h^\varepsilon$. In view of our hypothesis (10.53), the product rule yields

$$\llbracket p_h \rrbracket \approx \llbracket \varrho_h \rrbracket + \llbracket \vartheta_h \rrbracket, \tag{10.66}$$

and the estimates (10.64), (10.65) imply

$$\int_0^T \sum_{\sigma \in \mathcal{E}_{int}} \int_\sigma h^\varepsilon \llbracket \varrho_h \rrbracket^2 \, \mathrm{dS}_x \stackrel{<}{\sim} 1, \tag{10.67}$$

$$\int_0^T \sum_{\sigma \in \mathcal{E}_{int}} \int_\sigma h^\varepsilon \llbracket \vartheta_h \rrbracket^2 \, \mathrm{dS}_x \stackrel{<}{\sim} 1. \tag{10.68}$$

Moreover, by virtue of the minimum entropy principle (10.39), the entropy s_h is bounded below uniformly for $h \to 0$. As the cut-off function χ is supposed to be bounded from above, we may assume

$$|\chi(s_h)| \stackrel{<}{\sim} 1 \text{ for } h \to 0. \tag{10.69}$$

Having derived the necessary bounds to control the consistency error we proceed step by step with the estimates.

Let $\varphi \in C^2([0,T] \times \overline{\Omega})$ and $\boldsymbol{\varphi} \in C^2([0,T] \times \overline{\Omega}; R^d)$, $\boldsymbol{\varphi} \cdot \boldsymbol{n}|_{\partial\Omega} = 0$. We take $\phi_h = \Pi_Q \phi$ and $\boldsymbol{\varphi}_h = \Pi_Q \boldsymbol{\varphi}$ in the continuity equation (10.11) and the momentum equation (10.12), respectively. Moreover, we consider $\varphi_h = \Pi_Q \varphi$ in the renormalized discrete entropy inequality (10.24) for a function $\varphi \in C^2([0,T] \times \overline{\Omega})$ such that $\varphi \geq 0$. In what follows we show the consistency of the VFV scheme in three steps.

(1) Convective terms:
To show the consistency of the numerical fluxes in the continuity and momentum equations, as well as the numerical entropy flux we recall the identity (8.13c) from Lemma 8.1,

$$\int_{\Omega} r_h \boldsymbol{u}_h \cdot \nabla_x \varphi \, \mathrm{d}x - \int_{\mathcal{E}_{int}} F_h^{\mathrm{up}}[r_h, \boldsymbol{u}_h] \left[\!\left[\Pi_Q \varphi \right]\!\right] \mathrm{dS}_x = \sum_{j=1}^{4} E_j(r_h),$$

where

$$E_1(r_h) = \frac{1}{2} \int_{\mathcal{E}_{int}} |\{\!\{\boldsymbol{u}_h\}\!\} \cdot \boldsymbol{n}| \, [\![r_h]\!] \left[\!\left[\Pi_Q \varphi \right]\!\right] \mathrm{dS}_x,$$

$$E_2(r_h) = \frac{1}{4} \int_{\mathcal{E}_{int}} [\![\boldsymbol{u}_h]\!] \cdot \boldsymbol{n} \, [\![r_h]\!] \left[\!\left[\Pi_Q \varphi \right]\!\right] \mathrm{dS}_x,$$

$$E_3(r_h) = \int_{\Omega} r_h \boldsymbol{u}_h \cdot \Big(\nabla_x \varphi - \nabla_h (\Pi_Q \varphi) \Big) \, \mathrm{d}x,$$

$$E_4(r_h) = \int_{\mathcal{E}_{int}} h^{\varepsilon} \, [\![r_h]\!] \left[\!\left[\Pi_Q \varphi \right]\!\right] \mathrm{dS}_x$$

are the errors to be controlled. Recall $\int_{\mathcal{E}_{int}} \mathrm{dS}_x \equiv \sum_{\sigma \in \mathcal{E}_{int}} \int_{\sigma} \mathrm{dS}_x$. The first three of them correspond to the upwind part of the numerical flux and hence need to be controlled for the consistency of the numerical entropy flux as well, specifically for r_h either equal to ϱ_h, $\varrho_h u_{j,h}$, $j = 1, \ldots, d$, or $\varrho_h \chi(s_h)$. The last term only appears in the numerical flux of the continuity and momentum equations, and thus needs to be handled for $r_h = \varrho_h$ and $r_h = \varrho_h u_{j,h}$, $j = 1, \ldots, d$. The h^{ε}-term in the numerical entropy flux reads

$$E_5 = h^{\varepsilon} \int_{\mathcal{E}_{int}} \left(\left\{\!\!\left\{ \nabla_{\varrho} (\varrho_h \chi(s_h)) \right\}\!\!\right\} \left[\!\left[\varrho_h \right]\!\right] + \left\{\!\!\left\{ \nabla_p (\varrho_h \chi(s_h)) \right\}\!\!\right\} \left[\!\left[p_h \right]\!\right] \right) \left[\!\left[\Pi_Q \varphi \right]\!\right] \mathrm{dS}_x.$$

In what follows we show E_j, $j = 1, \ldots, 5$ vanish as $h \to 0$.

- **Term** E_1: In view of (19) it is enough to show that

$$E_{1,h}(r_h) = h \int_{\mathcal{E}_{int}} |\{\!\{\boldsymbol{u}_h\}\!\} \cdot \boldsymbol{n}| \, |[\![r_h]\!]| \, \mathrm{dS}_x \to 0, \tag{10.70}$$

  as $h \to 0$ for any fixed $\varphi \in C^2(\overline{\Omega})$.

Let r_h be either equal to ϱ_h or $\varrho_h \chi(s_h)$. The error can be handled as

$$h \int_{\mathcal{E}_{int}} | \{\!\{ \boldsymbol{u}_h \}\!\} \cdot \boldsymbol{n}| \ |[\![r_h]\!]| \ \mathrm{dS}_x \stackrel{<}{\sim} h \int_{\mathcal{E}_{int}} | \{\!\{ \boldsymbol{u}_h \}\!\} \cdot \boldsymbol{n}| \left(| [\![\varrho_h]\!] | + | [\![\vartheta_h]\!] | \right) \mathrm{dS}_x$$

$$\stackrel{<}{\sim} h \left(\int_{\mathcal{E}_{int}} | \{\!\{ \boldsymbol{u}_h \}\!\} | \, \mathrm{dS}_x \right)^{1/2} \left(\int_{\mathcal{E}_{int}} | \{\!\{ \boldsymbol{u}_h \}\!\} \cdot \boldsymbol{n}| \left([\![\varrho_h]\!]^2 + [\![\vartheta_h]\!]^2 \right) \mathrm{dS}_x \right)^{1/2}$$

$$\stackrel{<}{\sim} \sqrt{h} \| \boldsymbol{u}_h \|_{L^1(\Omega)}^{1/2} F_h^1 \stackrel{<}{\sim} \sqrt{h} F_h^1, \ \| F_h^1 \|_{L^2(0,T)} \stackrel{<}{\sim} 1.$$

We are left with $E_{1,h}(\varrho_h u_{j,h})$, specifically,

$$h \int_{\mathcal{E}_{int}} | \{\!\{ \boldsymbol{u}_h \}\!\} \cdot \boldsymbol{n}| \left| [\![\varrho_h u_{j,h}]\!] \right| \, \mathrm{dS}_x$$

$$\stackrel{<}{\sim} h \int_{\mathcal{E}_{int}} | \{\!\{ \boldsymbol{u}_h \}\!\} \cdot \boldsymbol{n}| | [\![\boldsymbol{u}_h]\!] | \, \mathrm{dS}_x + h \int_{\mathcal{E}_{int}} | \{\!\{ \boldsymbol{u}_h \}\!\} \cdot \boldsymbol{n}| | \{\!\{ \boldsymbol{u}_h \}\!\} | | [\![\varrho_h]\!] | \, \mathrm{dS}_x.$$

The first integral can be estimated as

$$\begin{aligned} h \int_{\mathcal{E}_{int}} | \{\!\{ \boldsymbol{u}_h \}\!\} \cdot \boldsymbol{n}| | [\![\boldsymbol{u}_h]\!] | \, \mathrm{dS}_x &\stackrel{<}{\sim} h \left(\int_{\mathcal{E}_{int}} [\![\boldsymbol{u}_h]\!]^2 \, \mathrm{dS}_x \right)^{1/2} \left(\int_{\mathcal{E}_{int}} | \{\!\{ \boldsymbol{u}_h \}\!\} |^2 \, \mathrm{dS}_x \right)^{1/2} \\ &\stackrel{<}{\sim} h^{\frac{3}{2} - \frac{\alpha}{2}} F_h^2 h^{-\frac{1}{2}} \| \boldsymbol{u}_h \|_{L^2(\Omega)} \leq h^{1 - \frac{\alpha}{2}} F_h^2, \ \| F_h^2 \|_{L^2(0,T)} \stackrel{<}{\sim} 1, \end{aligned} \tag{10.71}$$

where we have used the trace inequality, (10.58) and (10.63). Next, by Hölder's inequality, the trace inequality, (10.60), we get

$$h \int_{\mathcal{E}_{int}} | \{\!\{ \boldsymbol{u}_h \}\!\} \cdot \boldsymbol{n}| | \{\!\{ \boldsymbol{u}_h \}\!\} | | [\![\varrho_h]\!] | \, \mathrm{dS}_x$$

$$\stackrel{<}{\sim} h \left(\int_{\mathcal{E}_{int}} | \{\!\{ \boldsymbol{u}_h \}\!\} |^3 \, \mathrm{dS}_x \right)^{1/2} \left(\int_{\mathcal{E}_{int}} | \{\!\{ \boldsymbol{u}_h \}\!\} \cdot \boldsymbol{n}| \, [\![\varrho_h]\!]^2 \, \mathrm{dS}_x \right)^{1/2}$$

$$\stackrel{<}{\sim} \sqrt{h} \| \boldsymbol{u}_h \|_{L^3(\Omega)}^{3/2} F_h^3, \ \| F_h^3 \|_{L^2(0,T)} \stackrel{<}{\sim} 1.$$

Now, in view of the interpolation inequality

$$\| \boldsymbol{u}_h \|_{L^3(\Omega)} \stackrel{<}{\sim} \| \boldsymbol{u}_h \|_{L^2(\Omega)}^{1/2} \| \boldsymbol{u}_h \|_{L^6(\Omega)}^{1/2},$$

combined with (10.58), we obtain

$$h\int_{\mathcal{E}_{int}} |\{\!\{\boldsymbol{u}_h\}\!\}\cdot\boldsymbol{n}|\,|\{\!\{\boldsymbol{u}_h\}\!\}|\,|[\![\varrho_h]\!]|\,\mathrm{dS}_x \lesssim \sqrt{h}\|\boldsymbol{u}_h\|_{L^6(\Omega)}^{3/4} F_h^3.$$

Finally, we apply the discrete Sobolev embedding (10.34) and (10.63) to conclude

$$\begin{aligned} h\int_{\mathcal{E}_{int}} |\{\!\{\boldsymbol{u}_h\}\!\}\cdot\boldsymbol{n}|\,|\{\!\{\boldsymbol{u}_h\}\!\}|\,|[\![\varrho_h]\!]|\,\mathrm{dS}_x &\lesssim \sqrt{h}F_h^3\left(1+\left(\int_{\mathcal{E}_{int}}\frac{[\![\boldsymbol{u}_h]\!]^2}{h}\,\mathrm{dS}_x\right)^{1/2}\right)^{3/4} \\ &\lesssim h^{\frac{4-3\alpha}{8}} F_h^3 F_h^4,\ \|F_h^4\|_{L^{8/3}(0,T)} \lesssim 1. \end{aligned}$$

Note that $(4-3\alpha)/8 > 0$ due to (10.52).

- **Term** E_2: Using (19) we only need to show that

$$E_{2,h}(r_h) = h\int_{\mathcal{E}_{int}} |[\![\boldsymbol{u}_h]\!]\cdot\boldsymbol{n}|\,|[\![r_h]\!]|\,\mathrm{dS}_x \to 0 \tag{10.72}$$

as $h\to 0$ for any fixed $\varphi\in C^2(\overline{\Omega})$.
With (10.55), (10.56), and (10.69) at hand, the convergence of these errors for r_h either equal to ϱ_h or $\varrho_h\chi(s_h)$ reduces to showing

$$h\int_{\mathcal{E}_{int}} |[\![\boldsymbol{u}_h]\!]|\,\mathrm{dS}_x \to 0.$$

To see this, we use Hölder's inequality,

$$\begin{aligned} h\int_{\mathcal{E}_{int}} |[\![\boldsymbol{u}_h]\!]|\,\mathrm{dS}_x &\le h\left(\int_{\mathcal{E}_{int}}[\![\boldsymbol{u}_h]\!]^2\,\mathrm{dS}_x\right)^{1/2}\left(\int_{\mathcal{E}_{int}}1\,\mathrm{dS}_x\right)^{1/2} \\ &\lesssim \sqrt{h}\left(\int_{\mathcal{E}_{int}}[\![\boldsymbol{u}_h]\!]^2\,\mathrm{dS}_x\right)^{1/2} \lesssim h^{1-\frac{\alpha}{2}}F_h^2,\ \|F_h^2\|_{L^2(0,T)}\lesssim 1, \end{aligned}$$

where the last inequality follows from (10.63).

Analogously as above, we need to show $E_{2,h}(\varrho_h u_{j,h}) \to 0$ as $h \to 0$. It rewrites as

$$
\begin{aligned}
& h \int_{\mathcal{E}_{int}} |[\![\varrho_h u_{j,h}]\!]| \, |[\![\boldsymbol{u}_h]\!] \cdot \boldsymbol{n}| \, \mathrm{dS}_x \\
& \lesssim h \int_{\mathcal{E}_{int}} [\![\boldsymbol{u}_h]\!]^2 \, \mathrm{dS}_x + h \int_{\mathcal{E}_{int}} | \{\!\{\boldsymbol{u}_h\}\!\} | \, | [\![\boldsymbol{u}_h]\!] | \, \mathrm{dS}_x \\
& \lesssim h^{2-\alpha} F_h^5 + h \left(\int_{\mathcal{E}_{int}} [\![\boldsymbol{u}_h]\!]^2 \, \mathrm{dS}_x \right)^{1/2} \left(\int_{\mathcal{E}_{int}} | \{\!\{\boldsymbol{u}_h\}\!\} |^2 \, \mathrm{dS}_x \right)^{1/2} \\
& \lesssim h^{2-\alpha} F_h^5 + h^{\frac{3}{2}-\frac{\alpha}{2}} F_h^2 h^{-\frac{1}{2}} \|\boldsymbol{u}_h\|_{L^2(\Omega)} \lesssim h^{2-\alpha} F_h^5 + h^{1-\frac{\alpha}{2}} F_h^2, \ \|F_h^5\|_{L^1(0,T)} \lesssim 1.
\end{aligned}
$$

- **Term** E_3: For this integral with r_h either equal to ϱ_h, $\varrho_h \boldsymbol{u}_h$ or $\varrho_h \chi(s_h)$ we directly get

$$
\int_{\Omega} r_h \boldsymbol{u}_h \cdot \Big(\nabla_x \varphi - \nabla_h \big(\Pi_Q \varphi \big) \Big) \, \mathrm{d}x \lesssim h \, \|r_h\|_{L^2} \, \|\boldsymbol{u}_h\|_{L^2} \lesssim h
$$

due to (10.20), (10.55), (10.56), (10.58), and (10.69).

- **Term** E_4: Firstly, let $r_h = \varrho_h$. Hölder's inequality with (10.19) and (10.67) directly yield

$$
\begin{aligned}
& \int_{\mathcal{E}_{int}} h^{\varepsilon} \, [\![\varrho_h]\!] \, [\![\Pi_Q \varphi]\!] \, \mathrm{dS}_x \lesssim \left(\int_{\mathcal{E}_{int}} h^{\varepsilon} \, [\![\varrho_h]\!]^2 \, \mathrm{dS}_x \right)^{1/2} \left(\int_{\mathcal{E}_{int}} h^{2+\varepsilon} \, \mathrm{dS}_x \right)^{1/2} \\
& \lesssim h^{\frac{1+\varepsilon}{2}} F_h^6, \ \|F_h^6\|_{L^2(0,T)} \lesssim 1.
\end{aligned}
$$

Secondly, for $r_h = \varrho_h \boldsymbol{u}_h$, analogously as above, using the product rule, the trace inequality, and bounds (10.55), (10.62), (10.63), (10.67) we get

$$
\begin{aligned}
& \int_{\mathcal{E}_{int}} h^{\varepsilon} \, [\![\varrho_h \boldsymbol{u}_h]\!] \cdot [\![\Pi_Q \boldsymbol{\varphi}_h]\!] \, \mathrm{dS}_x \\
& \lesssim h \left(\int_{\mathcal{E}_{int}} h^{\varepsilon} \, [\![\varrho_h]\!]^2 \, \mathrm{dS}_x \right)^{1/2} \left(\int_{\mathcal{E}_{int}} h^{\varepsilon} | \{\!\{\boldsymbol{u}_h\}\!\} |^2 \, \mathrm{dS}_x \right)^{1/2} \\
& + h^{\frac{1+\varepsilon}{2}} \left(\int_{\mathcal{E}_{int}} h^{\varepsilon} \, [\![\boldsymbol{u}_h]\!]^2 \, \mathrm{dS}_x \right)^{1/2} \\
& \lesssim h^{\frac{1+\varepsilon}{2}} \|\boldsymbol{u}_h\|_{L^2(\Omega)} F_h^6 + h^{\frac{1+\varepsilon}{2}} F_h^7 \lesssim h^{\frac{1+\varepsilon}{2}} F_h^6 + h^{\frac{1+\varepsilon}{2}} F_h^7, \ \|F_h^7\|_{L^2(0,T)} \lesssim 1.
\end{aligned}
$$

Both exponents of h are positive as $\varepsilon > -1$.

- **Term** E_5: Finally, (10.66), (10.67) and (10.68) imply

$$\int_{\mathcal{E}_{int}} h^{\varepsilon}\Big(\big\{\!\!\big\{ \nabla_{\varrho}(\varrho_h\chi(s_h)) \big\}\!\!\big\} \big[\!\big[\varrho_h\big]\!\big] + \big\{\!\!\big\{ \nabla_{p}(\varrho_h\chi(s_h)) \big\}\!\!\big\} \big[\!\big[p_h\big]\!\big] \Big) \big[\!\big[\Pi_Q\varphi\big]\!\big] \, \mathrm{dS}_x$$

$$\lesssim \sqrt{h} \left(\int_{\mathcal{E}_{int}} h^{\varepsilon} \big[\!\big[\varrho_h\big]\!\big]^2 \, \mathrm{dS}_x \right)^{1/2} + \sqrt{h} \left(\int_{\mathcal{E}_{int}} h^{\varepsilon} \big[\!\big[\vartheta_h\big]\!\big]^2 \, \mathrm{dS}_x \right)^{1/2}$$

$$\lesssim \sqrt{h}(F_h^6 + F_h^8), \ \|F_h^8\|_{L^2(0,T)} \lesssim 1.$$

(2) Artificial viscosity term:
In accordance with (10.19), for the artificial viscosity term in the momentum equation (10.12), we get

$$h^{\alpha-1} \int_{\mathcal{E}_{int}} \big[\!\big[\boldsymbol{u}_h\big]\!\big] \cdot \big[\!\big[\Pi_Q\boldsymbol{\varphi}\big]\!\big] \, \mathrm{dS}_x \lesssim h^{\alpha-1} \left(\int_{\mathcal{E}_{int}} \big[\!\big[\boldsymbol{u}_h\big]\!\big]^2 \, \mathrm{dS}_x \right)^{1/2} \left(\int_{\mathcal{E}_{int}} h^2 \, \mathrm{dS}_x \right)^{1/2}$$

$$\lesssim h^{\alpha-\frac{1}{2}} \left(\int_{\mathcal{E}_{int}} \big[\!\big[\boldsymbol{u}_h\big]\!\big]^2 \, \mathrm{dS}_x \right)^{1/2} \lesssim h^{\frac{\alpha}{2}} F_h^2.$$

(3) Pressure term:
In order to control the last remaining term in the momentum equation (10.12) we recall the definition of the discrete divergence operator (15) and compute

$$\int_{\mathcal{E}_{int}} \{\!\!\{p_h\}\!\!\} \boldsymbol{n} \cdot \big[\!\big[\Pi_Q\boldsymbol{\varphi}\big]\!\big] \, \mathrm{dS}_x = \int_{\Omega} p_h \mathrm{div}_h(\Pi_Q\boldsymbol{\varphi}) \, \mathrm{d}x$$

$$= \int_{\Omega} p_h\big(\mathrm{div}_h(\Pi_Q\boldsymbol{\varphi}) - \mathrm{div}_x\boldsymbol{\varphi}\big) \, \mathrm{d}x + \int_{\Omega} p_h \mathrm{div}_x\boldsymbol{\varphi} \, \mathrm{d}x.$$

In view of the hypotheses on uniform boundedness of the approximate density and temperature the corresponding consistency error can be easily estimated using (20),

$$\int_{\Omega} p_h\big(\mathrm{div}_h(\Pi_Q\boldsymbol{\varphi}) - \mathrm{div}_x\boldsymbol{\varphi}\big) \, \mathrm{d}x \lesssim h \, \|p_h\|_{L^1} \, \|\boldsymbol{\varphi}\|_{C^2},$$

and tends to 0 for $h \to 0$ as $p_h \in L^{\infty}(0, T; L^1(\Omega))$.

Summing up the latter results we get the consistency formulation of the VFV scheme (10.11)–(10.13).

10.4 Convergence

Based on the results of the previous two sections and applying the convergence results for consistent approximations from Sects. 7.3, 7.4 we are now ready to prove the convergence of the VFV method. Although the approximate problem was stated and solved in terms of the field equations for the *conservative variables* $(\varrho_h, \boldsymbol{m}_h, E_h)$, the convergence will be studied in terms of the conservative-entropy variables $(\varrho_h, \boldsymbol{m}_h, S_h)$. The reason is that the uniform bounds obtained in the previous part are not strong enough to control integrability of the convective terms in the energy equation (10.13) that must be replaced by the entropy balance.

As the approximate density and temperature are strictly positive, there is a bijective mapping between the conservative variables $(\varrho_h, \boldsymbol{m}_h, E_h)$ computed by the VFV method (10.11)–(10.13) and the conservative-entropy variables $(\varrho_h, \boldsymbol{m}_h, S_h)$, where $E_h = \dfrac{|\boldsymbol{m}_h|^2}{2\varrho_h} + e(\varrho_h, S_h)$, the internal energy $e(\varrho, S)$ is given by (5.25), and equivalently $S_h = \dfrac{1}{\gamma - 1}\varrho_h \log\left(\dfrac{\gamma-1}{\varrho_h^\gamma}\left(E_h - \dfrac{|\boldsymbol{m}_h|^2}{\varrho_h}\right)\right)$, cf. Sections 4.1, 5.1.2. Thus, for the purpose of convergence analysis we may consider a sequence of numerical solutions in the conservative-entropy variables $\{\varrho_h, \boldsymbol{m}_h, S_h\}_{h\searrow 0}$ which yields a consistent approximation of the Euler system in the sense of Definition 5.2. Consequently, there exists a subsequence $\{\varrho_{h_n}, \boldsymbol{m}_{h_n}, S_{h_n}\}_{h_n\searrow 0}$ that generates a Young measure $\left\{\mathcal{V}_{t,x}\right\}_{(t,x)\in(0,T)\times\Omega}$,

$$
\begin{aligned}
&\mathcal{V}_{t,x} \in \mathcal{P}(R^{d+2}),\ R^{d+2} = \left\{\widetilde{\varrho} \in R, \widetilde{\boldsymbol{m}} \in R^d,\ \widetilde{S} \in R\right\} \\
&\mathcal{V}_{t,x}\left\{\widetilde{\varrho} > 0,\ \widetilde{S} \geq \underline{s}\widetilde{\varrho}\right\} = 1 \text{ for a.a. } (t,x) \in (0,T) \times \Omega,
\end{aligned}
$$

which is a DMV solution in the sense of Definition 5.3. Moreover, in accordance with the extra hypotheses imposed on the sequence of numerical solutions, we obtain:

$$
\mathcal{V}_{t,x}\left\{0 < \underline{\varrho} \leq \widetilde{\varrho} \leq \overline{\varrho},\ \underline{s}\widetilde{\varrho} \leq \widetilde{S} \leq \widetilde{\varrho}\,\overline{s}\right\} = 1 \text{ for a.a. } (t,x) \in (0,T) \times \Omega.
$$

The first two coordinates of the barycenter of the Young measure

$$
\langle \mathcal{V}; \widetilde{\varrho}\rangle = \varrho,\ \langle \mathcal{V}; \widetilde{\boldsymbol{m}}\rangle = \boldsymbol{m} \tag{10.73}
$$

are weakly continuous functions, specifically,

$$
\varrho \in C_{\text{weak}}([0,T]; L^\gamma(\Omega)),\quad \boldsymbol{m} \in C_{\text{weak}}([0,T]; L^{\frac{2\gamma}{\gamma+1}}(\Omega; R^d))
$$

while $S = \left\langle \mathcal{V}; \widetilde{S}\right\rangle$ satisfies

$$
S \in L^\infty(0,T; L^\gamma(\Omega)) \cap BV_{\text{weak}}([0,T]; L^\gamma(\Omega)).
$$

More importantly, $[\varrho, \boldsymbol{m}, S]$ is a DW solution in the sense of Definition 5.7. Applying Theorem 5.1 we have the following convergence result for the VFV method.

Theorem 10.4 (Weak convergence of the VFV method)
Let $\{\varrho_h, \boldsymbol{m}_h, S_h\}_{h\searrow 0}$ be the family of numerical solutions generated by the VFV method (10.11)–(10.13) *with* $0 < \alpha < 4/3$ *and* $\varepsilon > -1$. *In addition, assume that the numerical solutions remain in the gas nondegenerate region, i.e.* $0 < \underline{\varrho} \leq \varrho_h$ *and* $\vartheta_h \leq \overline{\vartheta}$, *uniformly for* $h \to 0$.
Then there exists a subsequence $\{\varrho_{h_n}, \boldsymbol{m}_{h_n}, S_{h_n}\}_{h_n\searrow 0}$ *such that:*

- ***weak convergence and defect measures***

$$\begin{aligned} \varrho_{h_n} &\to \varrho \ \text{ weakly-}(*) \text{ in } L^\infty((0,T)\times\Omega), \\ S_{h_n} &\to S \ \text{ weakly-}(*) \text{ in } L^\infty((0,T)\times\Omega), \\ \boldsymbol{m}_{h_n} &\to \boldsymbol{m} \ \text{ weakly-}(*) \text{ in } L^\infty(0,T;L^2(\Omega;R^d)) \end{aligned}$$

as $h_n \to 0$, *where* $[\varrho, \boldsymbol{m}, S]$ *is a DW solution of the Euler system (7.14)–(7.16). In addition,*

$$\varrho \geq \underline{\varrho} > 0, \quad S \geq \underline{\varrho s} \ \text{ a.a. in } (0,T)\times\Omega.$$

Moreover,

$$E(\varrho_{h_n}, \boldsymbol{m}_{h_n}, S_{h_n}) = \left[\frac{1}{2}\frac{|\boldsymbol{m}_{h_n}|^2}{\varrho_{h_n}} + \varrho_{h_n} e(\varrho_{h_n}, \boldsymbol{m}_{h_n})\right] \to \overline{E(\varrho, \boldsymbol{m}, S)}$$

weakly-(*) in $L^\infty(0,T;\mathcal{M}(\overline{\Omega}))$ *and the energy defect measure* $\mathfrak{E}$ *is a sum of the energy concentration defect* $\mathfrak{E}_{cd}$ *and the energy oscillation defect* $\mathfrak{E}_{od}$, $\mathfrak{E} = \mathfrak{E}_{cd} + \mathfrak{E}_{od}$

$$\begin{aligned} \mathfrak{E}_{cd} &\equiv \overline{E(\varrho, \boldsymbol{m}, S)} - \left\langle \mathcal{V}; E(\widetilde{\varrho}, \widetilde{\boldsymbol{m}}, \widetilde{S})\right\rangle \in L^\infty(0,T;\mathcal{M}^+(\overline{\Omega})), \\ \mathfrak{E}_{od} &\equiv \left\langle \mathcal{V}; E(\widetilde{\varrho}, \widetilde{\boldsymbol{m}}, \widetilde{S})\right\rangle - E(\varrho, \boldsymbol{m}, S) \geq 0. \end{aligned}$$

Furthermore, we have for the momentum flux

$$\left(\frac{\boldsymbol{m}_{h_n} \otimes \boldsymbol{m}_{h_n}}{\varrho_{h_n}} + p(\varrho_{h_n}, S_{h_n})\mathbb{I}\right) \to \overline{\left(\frac{\boldsymbol{m}\otimes\boldsymbol{m}}{\varrho} + p(\varrho, S)\mathbb{I}\right)}$$

weakly-(*) in $L^\infty(0,T;\mathcal{M}(\overline{\Omega};R^{d\times d}_{\text{sym}}))$ *and the Reynolds defect* $\mathfrak{R}$ *is a sum of the concentration defect* $\mathfrak{R}_{cd}$ *and the oscillation defect* $\mathfrak{R}_{od}$, $\mathfrak{R} = \mathfrak{R}_{cd} + \mathfrak{R}_{od}$

$$\begin{aligned} \mathfrak{R}_{cd} &\equiv \overline{\frac{\boldsymbol{m}\otimes\boldsymbol{m}}{\varrho}} - \left\langle \mathcal{V}; \frac{\widetilde{\boldsymbol{m}}\otimes\widetilde{\boldsymbol{m}}}{\widetilde{\varrho}}\right\rangle \in L^\infty(0,T;\mathcal{M}^+(\overline{\Omega};R^{d\times d}_{\text{sym}})), \\ \mathfrak{R}_{od} &\equiv \left\langle \mathcal{V}; \left(\frac{\widetilde{\boldsymbol{m}}\otimes\widetilde{\boldsymbol{m}}}{\widetilde{\varrho}} + p(\widetilde{\varrho}, \widetilde{S})\mathbb{I}\right)\right\rangle - \left(\frac{\boldsymbol{m}\otimes\boldsymbol{m}}{\varrho} + p(\varrho, S)\mathbb{I}\right). \end{aligned}$$

Specifically, the Reynolds defect is controlled by the energy defect in the following way

$$\underline{d}\ \mathfrak{E} \le \mathrm{tr}[\mathfrak{R}] \le \overline{d}\ \mathfrak{E} \quad \textit{for certain constants } 0 < \underline{d} \le \overline{d}. \tag{10.74}$$

Proof follows from Theorem 5.1, the detailed discussion of the limiting process is presented in Sect. 5.1.3. As pointed in Remarks 5.4 and 5.6 both the concentration and oscillation energy defects are nonnegative. The control of the Reynolds defect by the energy defect (10.74) follows from (5.53). Note that there is no defect measure in the entropy inequality, since the limiting process in the entropy convective term yields

$$S_{h_n}\frac{\boldsymbol{m}_{h_n}}{\varrho_{h_n}} \to \overline{S\frac{\boldsymbol{m}}{\varrho}} = \left\langle \mathcal{V}; \widetilde{S}\frac{\widetilde{\boldsymbol{m}}}{\widetilde{\varrho}} \right\rangle \text{ weakly-}(*) \text{ in } L^\infty(0,T;L^2(\Omega;R^d)),$$

see Sect. 5.1.3.4.

Clearly, the weak convergence of the VFV method is not a very suitable indication in practical numerical simulations. It is *the strong convergence* that is more desirable. As elaborated in the analytical part, it is therefore more convenient to pass to the corresponding DMV solution that can be visualized by computing its Cesàro means. A suitable tool to achieve this goal is the $\mathcal{K}$–convergence that provides strong convergence of the Cesàro averages to a DW solution (the barycenter of the DMV solution) as well as strong convergence of the approximate deviation of the associated Young measure.

Theorem 10.5 **($\mathcal{K}$–convergence of the VFV method)**
Under the hypothesis of Theorem 10.4 there exists a subsequence $\{\varrho_{h_n}, \boldsymbol{m}_{h_n}, S_{h_n}\}_{h_n \searrow 0}$ *such that*

- ***strong convergences of Cesàro averages***

$$\frac{1}{N}\sum_{n=1}^{N}\varrho_{h_n} \to \varrho \textit{ as } N \to \infty \textit{ in } L^q((0,T)\times\Omega), \quad \textit{for any } 1 \le q < \infty,$$

$$\frac{1}{N}\sum_{n=1}^{N}\boldsymbol{m}_{h_n} \to \boldsymbol{m} \textit{ as } N \to \infty \textit{ in } L^q(0,T;L^2(\Omega;R^d)) \textit{ for any } 1 \le q < \infty,$$

$$\frac{1}{N}\sum_{n=1}^{N}S_{h_n} \to S \textit{ as } N \to \infty \textit{ in } L^q((0,T)\times\Omega) \textit{ for any } 1 \le q < \infty.$$

- ***L^q convergencetoYoungmeasure***

$$d_{W_s}\left[\frac{1}{N}\sum_{n=1}^{N}\delta_{[\varrho_{h_n},\boldsymbol{m}_{h_n},S_{h_n}]};\mathcal{V}\right] \to 0 \textit{ as } N \to \infty \textit{ in } L^q((0,T)\times\Omega)$$

for any $1 \le q < s \le 2$.

- L^1 **convergenceofthedeviation**

$$\frac{1}{N}\sum_{n=1}^{N}\left\langle[\varrho_{h_n},\boldsymbol{m}_{h_n},S_{h_n}]-\frac{1}{N}\sum_{k=1}^{N}[\varrho_{h_n},\boldsymbol{m}_{h_n},S_{h_n}]\right\rangle\equiv$$
$$\left\langle\frac{1}{N}\sum_{n=1}^{N}\delta_{[\varrho_{h_n},\boldsymbol{m}_{h_n},S_{h_n}]};\left|[\widetilde{\varrho},\widetilde{\boldsymbol{m}},\widetilde{S}]-\left\langle\frac{1}{N}\sum_{n=1}^{N}\delta_{[\varrho_{h_n},\boldsymbol{m}_{h_n},S_{h_n}]};[\widetilde{\varrho},\widetilde{\boldsymbol{m}},\widetilde{S}]\right\rangle\right|\right\rangle\to$$
$$\left\langle\mathcal{V};\left|[\widetilde{\varrho},\widetilde{\boldsymbol{m}},\widetilde{S}]-[\varrho,\boldsymbol{m},S]\right|\right\rangle\quad \text{as } N\to\infty \text{ in } L^1((0,T)\times\Omega).$$

Proof follows from (7.54), (7.55) and Theorem 10.6.

Finally, we are ready to show that a sequence of finite volume solutions itself converges strongly, not only its Cesàro averages. Indeed, applying Theorems 7.5, 7.6 and 7.8 we obtain that the finite volume solutions converge strongly. Consequently, the limit is a weak (or strong) solution of the Euler system (5.16)–(5.19) with the entropy inequality (5.21) (or the entropy equality (5.20)).

Theorem 10.6 (Strong convergence of the VFV method) *Let* $\{\varrho_h,\boldsymbol{m}_h,S_h\}_{h\searrow 0}$ *be finite volume solutions generated by the VFV method* (10.11)–(10.13) *with* $0<\alpha<4/3$, $\varepsilon>-1$ *and with the initial data* $\varrho_{0,h}=\Pi_Q\varrho_0$, $\boldsymbol{m}_{0,h}=\Pi_Q\boldsymbol{m}_0$, $S_{0,h}=\Pi_Q S_0$, $\varrho_0\geq\underline{\varrho}>0$, $S_0\geq\underline{\varrho s}$. *Assume that the numerical solutions remain in a gas nondegenerate region, i.e.* $0<\underline{\varrho}\leq\varrho_h$ *and* $\vartheta_h\leq\overline{\vartheta}$.

Let

$$(\varrho_{h_n},\boldsymbol{m}_{h_n},S_{h_n})\to[\varrho,\boldsymbol{m},S]\ \text{as}\ n\to\infty$$

in the sense specified in Theorem 10.4.

Then the following holds:

- ***weak solution***
  If $[\varrho,\boldsymbol{m},S]$ *is a weak solution of the Euler system (7.14)–(7.16) in the sense of Definition 2.4, satisfying the entropy inequality (2.70), and emanating from the initial data* $[\varrho_0,\boldsymbol{m}_0,S_0]$, *then*

$$\mathcal{V}_{t,x}=\delta_{[\varrho(t,x),\boldsymbol{m}(t,x),S(t,x)]}\ \text{for a.a.}\ (t,x)\in(0,T)\times\Omega,$$

and

$$\begin{aligned}\varrho_{h_n}&\to\varrho\ \text{in}\ L^q((0,T)\times\Omega),\\ \boldsymbol{m}_{h_n}&\to\boldsymbol{m}\ \text{in}\ L^q(0,T;L^2(\Omega;R^d)),\\ S_{h_n}&\to S\ \text{in}\ L^q((0,T)\times\Omega),\\ \frac{1}{2}\frac{|\boldsymbol{m}_{h_n}|^2}{\varrho_{h_n}}+\varrho_{h_n}e(\varrho_{h_n},S_{h_n})&\to\frac{1}{2}\frac{|\boldsymbol{m}|^2}{\varrho}+\varrho e(\varrho,S)\ \text{in}\ L^q(0,T;L^1(\Omega))\end{aligned}$$

for any $1\leq q<\infty$.

- ***strong solution***
  Suppose that the Euler system admits a strong solution $[\varrho, \boldsymbol{m}, S]$ *in the class*

$$\varrho,\ S \in W^{1,\infty}((0,T)\times\Omega),\ \boldsymbol{m} \in W^{1,\infty}((0,T)\times\Omega; R^d),\ \varrho \geq \underline{\varrho} > 0\ in\ [0,T)\times\Omega$$

emanating from the initial data $[\varrho_0, \boldsymbol{m}_0, S_0]$. *Then for any* $1 \leq q < \infty$ *and* $h \to 0$

$$\frac{1}{2}\frac{|\boldsymbol{m}_h|^2}{\varrho_h} + \varrho_h e(\varrho_h, S_h) \to \frac{1}{2}\frac{|\boldsymbol{m}|^2}{\varrho} + \varrho e(\varrho, S)\ \ in\ \ L^q(0,T; L^1(\Omega))$$

and

$$\begin{aligned} \varrho_h &\to \varrho\ in\ \ L^q((0,T)\times\Omega), \\ \boldsymbol{m}_h &\to \boldsymbol{m}\ in\ \ L^q(0,T; L^2(\Omega; R^d)), \\ S_h &\to S\ in\ \ L^q((0,T)\times\Omega). \end{aligned}$$

- ***classical solution***
  Let $\Omega \subset R^d$ *be a bounded Lipschitz domain and* $[\varrho, \boldsymbol{m}, S]$ *such that*

$$\varrho \in C^1([0,T], \times\overline{\Omega}),\ \varrho \geq \underline{\varrho} > 0,\ \boldsymbol{m} \in C^1([0,T]\times\overline{\Omega}; R^d),\ \ S \in C^1([0,T]\times\overline{\Omega}).$$

Then $[\varrho, \boldsymbol{m}, S]$ *is a classical solution to the Euler system and*

$$\begin{aligned} \varrho_h &\to \varrho\ in\ \ L^q((0,T)\times\Omega), \\ \boldsymbol{m}_h &\to \boldsymbol{m}\ in\ \ L^q(0,T; L^2(\Omega; R^d)), \\ S_h &\to S\ in\ \ L^q((0,T)\times\Omega) \end{aligned}$$

as $h \to 0$ *for any* $q \geq 1$.

Proof of the strong convergence to a weak solution follows from Theorem 7.6, see also the proof of Theorem 7.5. The crucial point is to realize that the energy defect measures $\mathfrak{E}_{cd}$ and $\mathfrak{E}_{od}$ vanish and the strong convergence of $E(\varrho_{h_n}, \boldsymbol{m}_{h_n}, S_{h_n}) \to E(\varrho, \boldsymbol{m}, S)$ in $L^q(0,T; L^1(\Omega))$ follows. Due to the sharp form of the Jensen inequality, cf. Lemma 7.1 we conclude that

$$\mathcal{V}_{t,x} = \delta_{[\varrho(t,x),\boldsymbol{m}(t,x),S(t,x)]}\ \text{ for a.a. } (t,x) \in (0,T)\times\Omega,$$

and in view of Proposition 7.1 and a priori bounds (10.44) we obtain the strong convergence of $\{\varrho_{h_n}, \boldsymbol{m}_{h_n}, S_{h_n}\}_{h\searrow 0}$ to a weak solution $[\varrho, \boldsymbol{m}, S]$.

The strong convergence to the strong solution follows from Theorem 7.8. Indeed, if the strong solution to the Euler system exists, we apply the weak-strong uniqueness principle established in Theorem 6.2 (see also Remark 6.4). Consequently, we conclude that

$$\mathcal{V}_{t,x} = \delta_{[\varrho(t,x),\boldsymbol{m}(t,x),S(t,x)]}, \ \mathfrak{R} = 0,$$

and a DW solution $[\varrho, \boldsymbol{m}, S]$ is the strong solution. As the limit is unique, the whole sequence of finite volume solutions $\{\varrho_h, \boldsymbol{m}_h, S_h\}_{h\searrow 0}$ converges strongly to the strong solution. We note that the results of Theorem 7.8 hold not only for the periodic boundary conditions but also for the impermeability boundary conditions $\boldsymbol{m} \cdot \boldsymbol{n} = 0$, as long as the strong solution exists.

The last statement follows from the weak-strong uniqueness principle between a DW solution and the classical solution, cf. Theorem 6.2.

Remark 10.1 It is crucial to notice the substantial difference between the "weak" and "strong" part of the conclusion of the above theorem. In the "weak" statement, we suppose that the approximate solutions, expressed in terms of the conservative-entropy variables, *converge* (weakly) to a weak solution of the limit problem. The statement need not hold if we instead suppose the limit problem *admits a weak solution*. In the "strong" statement (the second statement), we suppose that the limit problem *admits a (unique) strong solution* and deduce strong convergence to it.

There is also a delicate difference between the convergence result in the second and third statements. While in the convergence result to the strong solution its existence is postulated, in the convergence result to the classical solution we only require smoothness of the limit $[\varrho, \boldsymbol{m}, S]$. In both cases the convergence holds for the whole sequence $\{\varrho_h, \boldsymbol{m}_h, S_h\}_{h\searrow 0}$ which is advantageous to result on the $\mathcal{K}$–convergence presented in Theorem (10.5).

10.5 Numerical Experiments

To illustrate the concept of $\mathcal{K}$–convergence we consider the Kelvin–Helmholtz problem with shear flow over two interfaces. The basic mechanism of the *Kelvin–Helmholtz problem* lies in the existence of a uniform velocity shear and fluids of different densities superposed one over the other. Consequently, the instability develops that is characterized as unstable small scale motions occurring vertically and laterally. It is a well-known fact that numerical methods for inviscid fluids do not converge for this test problem as we will also see below. Our aim is to demonstrate the role of $\mathcal{K}$–convergence and to show experimentally that strong convergence of the coarse-grained quantities, such as the mean and the first deviation.

Clearly, the VFV method works with the conservative variables $(\varrho_h, \boldsymbol{m}_h, E_h)$. Under the hypothesis of strict positivity of the discrete density there is a bijective mapping between the conservative variables $(\varrho_h, \boldsymbol{m}_h, E_h)$ and the conservative-entropy variables $(\varrho_h, \boldsymbol{m}_h, S_h)$ in which we have proved the convergence theoretically. In numerical simulations we will investigate convergence of the conservative-entropy variables $(\varrho_h, \boldsymbol{m}_h, S_h)$ as well as of the energy E_h. Although our theoretical results of the $\mathcal{K}$–convergence to a DW solution presented in Theorem 10.5 do not yield the strong convergence for the energy, we will observe in numerical simulations below that the Cesàro averages of the energy converge strongly. Note however that the limit of the Cesàro averages of the energy is in general neither equal to $E(\varrho, \boldsymbol{m}, S)$ being the function of the limiting DW solution $[\varrho, \boldsymbol{m}, S]$ nor to $\langle \mathcal{V}; E(\tilde{\varrho}, \tilde{\boldsymbol{m}}, \tilde{S}) \rangle$.

Let us first introduce the following abbreviations for averages

$$\widetilde{U}_{h_n} = \frac{1}{n}\sum_{j=1}^{n} U_{h_j}, \quad U^{\dagger}_{h_n} = \frac{1}{n}\sum_{j=1}^{n} \left| U_{h_j} - \widetilde{U}_{h_n} \right|, \quad \overline{\mathcal{V}}^{n}_{t,x} \equiv \frac{1}{n}\sum_{j=1}^{n} \delta_{U_{h_j}(t,x)},$$

where $U \in \{\varrho, \boldsymbol{m}, S, E\}$. We denote by E_1, E_2, E_3, and E_4 the L^1-error of the difference between the numerical solution and the reference solution computed on the finest grid, the L^1-error of the Cesàro averages, the L^1-error of the deviation, and the L^1-error of the Wasserstein distance for the Cesàro average of the Dirac measures concentrated on numerical solutions, respectively. More precisely, we have

$$\begin{aligned} E_1 &= \left\| U_{h_n} - U_{h_N} \right\|, \; E_2 = \left\| \widetilde{U}_{h_n} - \widetilde{U}_{h_N} \right\|, \\ E_3 &= \left\| U^{\dagger}_{h_n} - U^{\dagger}_{h_N} \right\|, \; E_4 = \left\| W_1(\overline{\mathcal{V}}^{n}_{t,x}, \overline{\mathcal{V}}^{N}_{t,x}) \right\|, \end{aligned} \tag{10.75}$$

where $h_n = \frac{1}{n}$, n is the number of rectangular mesh cells in each direction, $N = 2048$. Thus, the reference solution $U_{\text{ref}} \equiv U_{h_N}$. In all of the following tests we set $\Omega = [0, 1]^2$ and apply periodic boundary conditions. We consider the following initial data for the Kelvin–Helmholtz problem

$$(\varrho, u_1, u_2, p)(x) = \begin{cases} (2, -0.5, 0, 2.5), & \text{if } I_1 < x_2 < I_2 \\ (1, 0.5, 0, 2.5), & \text{otherwise.} \end{cases}$$

Here the interface profiles $I_j = I_j(x) \equiv J_j + \epsilon Y_j(x)$, $j = 1, 2$, are chosen to be small perturbations around the lower $x_2 = J_1 = 0.25$ and the upper $x_2 = J_2 = 0.75$ interface, respectively. Further,

$$Y_j(x) = \sum_{k=1}^{m} a_j^k \cos(b_j^k + 2k\pi x_1), \quad j = 1, 2,$$

where $a_j^k \in [0, 1]$ and $b_j^k, j = 1, 2, k = 1, \ldots, m$ are fixed numbers. The coefficients a_j^k have been normalized such that $\sum_{k=1}^{m} a_j^k = 1$ to guarantee that $|I_j(x) - J_j| \le \epsilon$ for $j = 1, 2$. We have set $m = 10$ and $\epsilon = 0.01$.

In what follows we present the numerical simulations obtained by the VFV method with $\alpha = 1.8$, $\varepsilon = 0.8$, upwind finite volume method, i.e. the VFV method without the artificial diffusion terms of the order h^{ε} and h^{α}, and the generalized Riemann problem (GRP) finite volume method. The latter is chosen as an example of higher order finite volume method. The GRP method is a second order finite volume method based on the use of a solution to a generalized Riemann problem with piecewise linear data. In the VFV method and the upwind finite volume method we use the

Table 10.1 Convergence study for the Kelvin–Helmholtz problem: E_1, E_2, E_3, and E_4 errors for density (left to right)

h_n	E_1		E_2		E_3		E_4	
	error	EOC	error	EOC	error	EOC	error	EOC
(a) VFV scheme								
1/64	3.09e−01	–	1.81e−01	–	2.00e−01	–	2.45e−01	–
1/128	3.13e−01	−0.02	1.26e−01	0.52	1.06e−01	0.92	1.64e−01	0.58
1/256	3.27e−01	−0.06	9.83e−02	0.36	7.44e−02	0.51	1.20e−01	0.45
1/512	3.54e−01	−0.12	6.94e−02	0.50	4.99e−02	0.58	8.62e−02	0.48
1/1024	3.58e−01	−0.01	4.59e−02	0.59	2.97e−02	0.75	5.54e−02	0.64
(b) upwind FV scheme								
1/64	3.24e−01	–	2.41e−01	–	2.40e−01	–	2.97e−01	–
1/128	3.41e−01	−0.08	1.59e−01	0.61	1.21e−01	0.99	1.95e−01	0.61
1/256	3.64e−01	−0.09	1.12e−01	0.50	7.96e−02	0.60	1.34e−01	0.54
1/512	3.90e−01	−0.10	7.36e−02	0.60	5.23e−02	0.61	9.37e−02	0.52
1/1024	3.87e−01	0.01	4.75e−02	0.63	3.00e−02	0.80	6.02e−02	0.64
(c) GRP scheme								
1/64	2.03e−01	–	1.28e−01	–	1.06e−01	–	1.44e−01	–
1/128	1.49e−01	0.45	6.95e−02	0.88	5.21e−02	1.03	9.98e−02	0.53
1/256	1.40e−01	0.08	4.45e−02	0.64	3.41e−02	0.61	6.52e−02	0.61
1/512	1.86e−01	−0.41	3.81e−02	0.22	2.48e−02	0.46	5.65e−02	0.21
1/1024	1.31e−01	0.50	2.33e−02	0.71	1.60e−02	0.63	2.18e−02	1.37

forward Euler method to solve the resulting ODE system which yields the first order explicit approximation in time. On the other hand the GRP method profits from the construction of an approximate Riemann solution at time $t_k + \Delta t/2$ and applies the second order Runge–Kutta method.

At the final time $T = 2$ we can observe small-scales vortex sheets that already been formed at the interfaces. In Table 10.1 we show the results of the convergence study for the errors $E_1, \dots, E_4$ in density. The experimental convergence order (EOC) is determined by

$$\mathrm{EOC} = \log_2(\mathrm{err_n}/\mathrm{err_{2n}}), \tag{10.76}$$

where $\mathrm{err_n}$ and $\mathrm{err_{2n}}$ stand for the corresponding errors $E_1, \dots, E_4$ evaluated at the grids with h_n and h_{2n} mesh steps, respectively. We can clearly recognize that none of the methods converges in the classical sense, i.e. single numerical solutions do not converge, see the first column. This behavior is also demonstrated in the Figs. 10.1, 10.2.

The second, third and fourth columns in Table 10.1 show the convergence results for the Cesàro averages of numerical solutions and their first deviation, as well as $\mathcal{K}$–convergence of the Wasserstein distance of the corresponding Dirac measures. We should point out that the convergence is strong in the L^1-norm as proved above, cf. Theorem 10.5. The graphs in Fig. 10.1 show the results of the convergence study for all variables $\varrho, \boldsymbol{m}, S, E$. As expected, these variables behave in a similar manner.

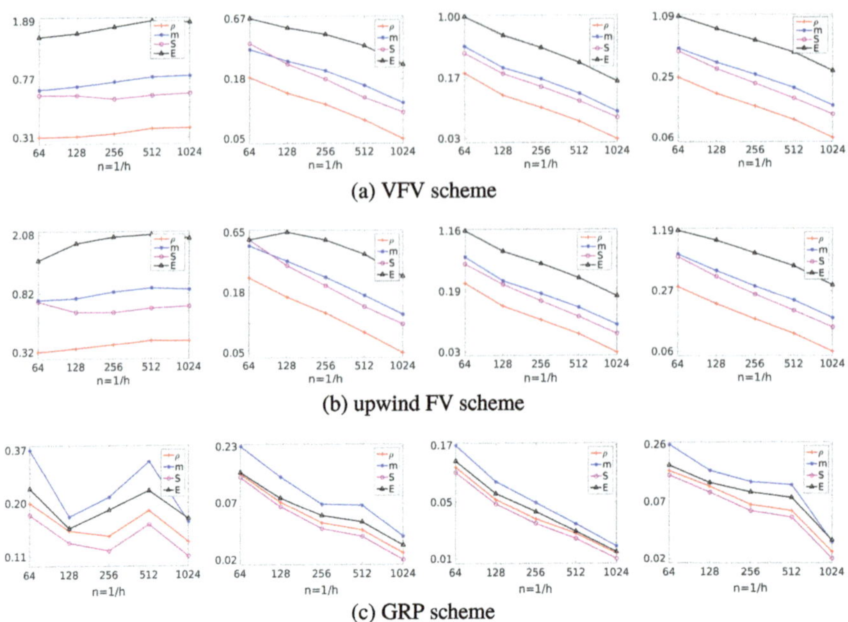

Fig. 10.1 Convergence study for the Kelvin–Helmholtz problem: E_1, E_2, E_3, and E_4 errors (left to right)

Approximate solutions for the density computed by the second order GRP scheme are presented in Fig. 10.2, which also clearly indicates that by refining the mesh we recover finer and finer vortex structures and the method does not converge in the classical sense. However, as already pointed out we have $\mathcal{K}$–convergence of the numerical solution and its first deviation, see Figs. 10.3, 10.4. The results obtained by the VFV scheme are similar, but more diffusive due to the first order accuracy.

10.6 Conclusion, Bibliographical Remarks

The aim of this chapter was to present the convergence of a new finite volume method, the so-called VFV method, that is based on the two-velocity model proposed by Brenner, see [27–29]. Brenner's approach to model dynamics of viscous and heat conducting fluids is based on two velocity fields distinguishing the bulk mass transport from the purely microscopic motion. On the one hand side this model has been subjected in literature to thorough criticism due to the incompatibility with certain physical principles. Nevertheless, on the other hand side some computational simulations have been performed by Greenschields and Reese [122], Bardow and Öttinger [12], Guo and Xu [125], who showed suitability of the model in specific situations. More recently, Guermond and Popov [124] rediscovered the Brenner model

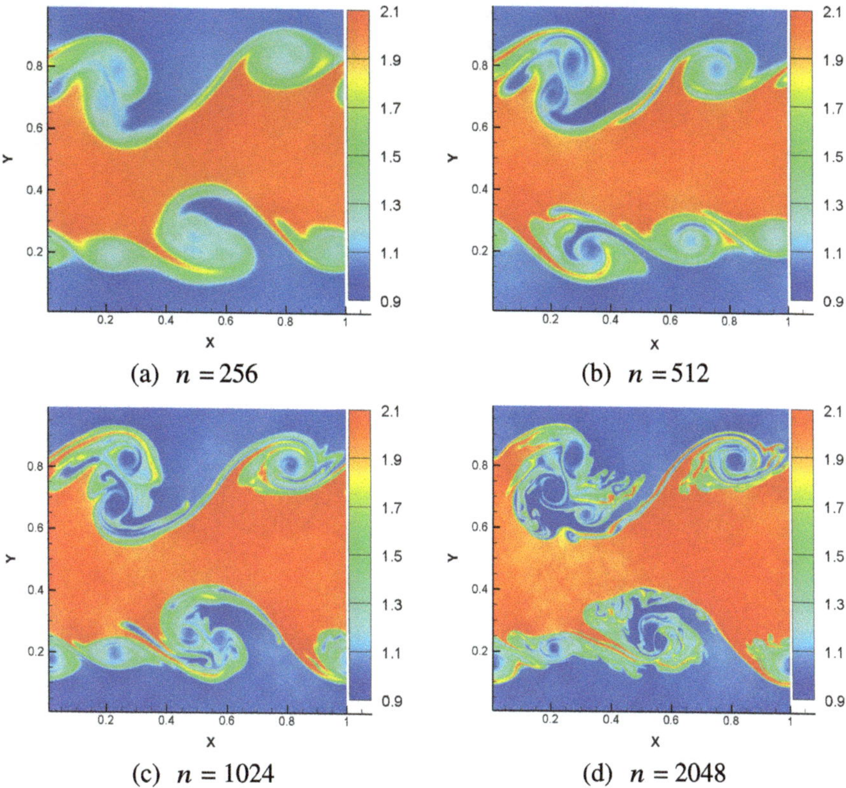

Fig. 10.2 Density computed by the GRP scheme at $T = 2$ for the Kelvin–Helmholtz problem on a mesh with $n \times n$ cells

pointing out its striking similarity with certain numerical methods based on the finite volume approximation of the inviscid fluids. In particular, unlike the conventional and well accepted Navier–Stokes–Fourier system, the Brenner model reflects the basic physical properties of the complete Euler system in the asymptotic limit of vanishing viscosity coefficients. Thus, the Brenner model can be seen as a physically admissible viscous regularization of the Euler system.

Consequently, the VFV method (10.11)–(10.13), that is based on the Brenner model, enjoys some crucial properties of the Euler system, such as the positivity of the approximate density and internal energy, the minimal entropy principle and the Second law of thermodynamics. Thus, the VFV method that was proposed in [93] belongs to the class of invariant domain preserving methods.

The analysis presented here has been done for the first order semidiscrete finite volume method. The numerical solutions are piecewise constant function in space and continuous in time. A generalization to a higher order method, where a suitable piecewise bilinear reconstruction is applied, is in principle possible and it is

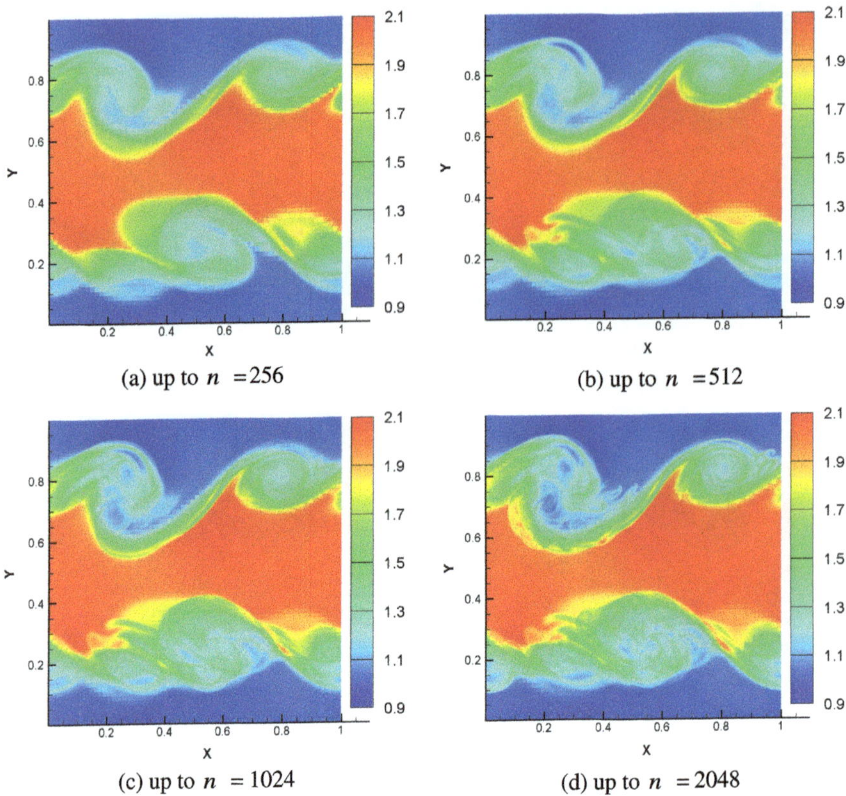

Fig. 10.3 Cesàro averages of the density computed by the GRP method on meshes with $n \times n$ cells, $n = 64, 128, \ldots, N$, for the Kelvin–Helmholtz problem

an interesting question for future study. In practical simulations the ODE system (10.11)–10.13) can be approximated by any suitable ODE solver. Due to the nonlinearity of the ODE system one typically opts for an explicit ODE solver, such as the explicit Euler or SSP (strong stability preserving) higher order time discretizations [120].

In Sect. 10.5 we have presented simulations obtained not only by the first order VFV method but also by the second order GRP finite volume method. The generalized Riemann problem finite volume method is one of successful standard numerical methods to simulate the Euler equations. It was developed as an analytical second order accurate extension of the classical Godunov finite volume method Ben–Artzi, Falcovitz [15, 16]. A direct Eulerian GRP scheme was presented in Ben–Artzi, Li [17], Li, Sun [153] by employing the regularity property of the Riemann invariants. Theoretically, a close coupling between the spatial and temporal evolution is recovered through the analysis of detailed wave interactions in the GRP scheme. Numerical results presented in Sect. 10.5 for the GRP finite volume method confirm

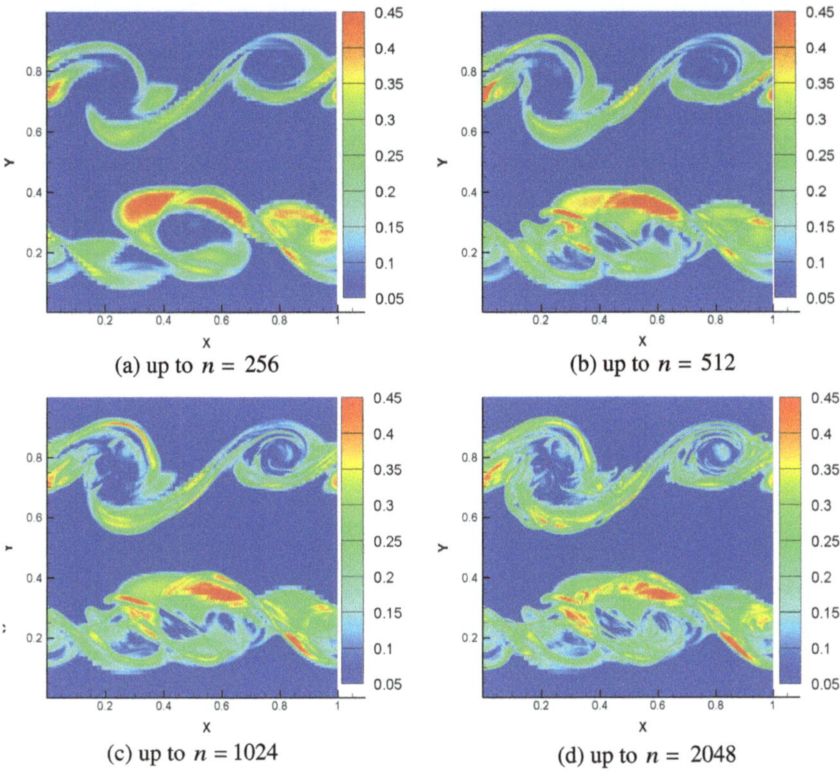

(a) up to $n = 256$

(b) up to $n = 512$

(c) up to $n = 1024$

(d) up to $n = 2048$

Fig. 10.4 Deviation of the density computed by the GRP method on meshes with $n \times n$ cells, $n = 64, 128, \ldots, N$, for the Kelvin–Helmholtz problem

that analogous convergence results for the Cesàro averages, deviation as well as for the Cesàro average of the corresponding Young measures hold. Consequently, we believe that the convergence can be proved rigorously for the GRP method or other standard finite volume methods applying the same tools as presented in Chaps. 9, 10.

As already pointed out before the results presented here can be seen as a nonlinear generalization of the celebrated Lax equivalence principle. Indeed, showing that the VFV method is stable and consistent already implies its convergence. In summary, we have shown the following convergence results under the hypothesis that the numerical solutions remain in a gas nondegenerate region, i.e. $0 < \underline{\varrho} \leq \varrho_h$ and $\vartheta_h \leq \overline{\vartheta}$:

- the VFV method converges weakly to a DW solution of the Euler system
- the Cesàro averages of the VFV solutions converge strongly to a DW solution of the Euler system
- if the above DW solution is a weak solution of the Euler system, then the VFV method converges strongly to this weak solution

- if the above DW solution belongs to the class of $C^1([0, T] \times \overline{\Omega})$ functions, then it is a classical solution and the convergence of the VFV method is strong
- the VFV method converges strongly to the strong solution of the Euler system on the lifespan of the latter.

Whereas the first three convergence results hold (up to a suitable subsequence), the convergence to the strong or classical solution holds for the whole finite volume sequence generated by the VFV method as the mesh step $h \to 0$.

Chapter 11
Finite Volume Method for the Navier–Stokes System

Having studied finite volume approximation of the inviscid compressible fluids in preceding chapters we now proceed with a finite volume approximation of the Navier–Stokes system, cf. (3.14)–(3.16):

$$
\begin{aligned}
&\partial_t \varrho + \mathrm{div}_x(\varrho \boldsymbol{u}) = 0 \\
&\partial_t(\varrho \boldsymbol{u}) + \mathrm{div}_x(\varrho \boldsymbol{u} \otimes \boldsymbol{u}) + \nabla_x p(\varrho) = \mathrm{div}_x \mathbb{S}(\nabla_x \boldsymbol{u})
\end{aligned}
\tag{11.1}
$$

where $\mathbb{S}$ is the viscous stress tensor

$$
\mathbb{S}(\nabla_x \boldsymbol{u}) = \mu \left(\nabla_x \boldsymbol{u} + \nabla_x^T \boldsymbol{u} - \frac{2}{d} \mathrm{div}_x \boldsymbol{u} \mathbb{I} \right) + \lambda \mathrm{div}_x \boldsymbol{u} \mathbb{I},\ \mu > 0,\ \lambda \geq -\frac{d-2}{d}\mu,
$$

and p is the pressure given by the isentropic EOS

$$
p = a\varrho^{\gamma},\ a > 0,\ \gamma > 1, \tag{11.2}
$$

cf. (2.71). The system is supplemented with the periodic boundary conditions ($\Omega = \mathbb{T}^d$) and the initial data

$$
\varrho(0) = \varrho_0 \geq \underline{\varrho} > 0, \quad \varrho_0 \in L^{\infty}(\mathbb{T}^d), \quad \boldsymbol{u}(0) = \boldsymbol{u}_0 \in L^{\infty}(\mathbb{T}^d; R^d). \tag{11.3}
$$

The case of Dirichlet boundary conditions is discussed in Chap. 13, where a mixed finite volume – finite element method is studied.

The aim of this chapter is to show that the numerical solution is a consistent approximation of the Navier–Stokes system in the sense of Definition 5.9. Then, applying the theoretical results stated in Theorems 5.5, 7.12, and 7.13, we obtain convergence of the scheme. More precisely, we need the numerical solution to be stable in the sense of Definition 8.3 and consistent in the sense of Definition 8.6. Our scheme is solvable at any step of discretization and preserves the total mass as

E. Feireisl et al., *Numerical Analysis of Compressible Fluid Flows*, MS&A 20,
https://doi.org/10.1007/978-3-030-73788-7_11

well as positivity of the discrete densities. This is due to the presence of viscosity in the approximated Navier–Stokes system. In particular, we may use the velocity $\boldsymbol{u}$ rather than the momentum $\boldsymbol{m}$ as the main phase variable. Accordingly, the equation of continuity can be exploited directly to avoid the arguments based on invariant domains necessary in many standard schemes for the Euler system.

11.1 Numerical Method

We consider a fully discrete FV method approximating the Navier–Stokes system (11.1) on a structured mesh in the sense of Definition 3, cf. also the notation introduced in Preliminary material and Chap. 8. Clearly, the grid satisfies the mesh regularity assumption (33). The computational domain is the flat torus $\Omega = \mathbb{T}^d$ consistent with the periodic boundary conditions. Accordingly, we have $\mathcal{E}_{ext} = \emptyset$ and $\mathcal{E} = \mathcal{E}_{int}$. We use the upwinding flux functions introduced in (8.6) to approximate the convective terms.

Definition 11.1 **(FV METHOD FOR THE NAVIER–STOKES SYSTEM)**
Given the initial data $(\varrho_h^0, \boldsymbol{u}_h^0) = (\Pi_Q \varrho_0, \Pi_Q \boldsymbol{u}_0) \in Q_h \times \boldsymbol{Q}_h$, the FV approximation $\{\varrho_h^k, \boldsymbol{u}_h^k\}_{k=1}^{N_T} \in Q_h \times \boldsymbol{Q}_h$ is a solution of the following system of algebraic equations:

$$
\begin{aligned}
D_t \varrho_h^k + \mathrm{div}_h^{\mathrm{up}}(\varrho_h^k, \boldsymbol{u}_h^k) &= 0, \\
D_t \boldsymbol{m}_h^k + \mathrm{div}_h^{\mathrm{up}}(\boldsymbol{m}_h^k, \boldsymbol{u}_h^k) + \nabla_h p_h^k &= \mu \Delta_h \boldsymbol{u}_h^k + \nu \nabla_h \mathrm{div}_h \boldsymbol{u}_h^k.
\end{aligned}
\tag{11.4}
$$

where $\nu = \frac{d-2}{d}\mu + \lambda$, $\boldsymbol{m}_h^k = \varrho_h^k \boldsymbol{u}_h^k$, $\boldsymbol{Q}_h = (Q_h)^d$ and $p_h = p(\varrho_h)$. Here the discrete operators ∇_h, div_h, Δ_h are given in (11.15), and $\mathrm{div}_h^{\mathrm{up}}$ is given in (8.9).

We may interpret the above FV method elementwisely for all $K \in \mathcal{T}_h$ as

$$
D_t \varrho_K^k + \sum_{\sigma \in \mathcal{E}(K)} \frac{|\sigma|}{|K|} F_h^{\mathrm{up}}[\varrho_h^k, \boldsymbol{u}_h^k] = 0,
$$

$$
D_t(\varrho_h^k \boldsymbol{u}_h^k)_K + \sum_{\sigma \in \mathcal{E}(K)} \frac{|\sigma|}{|K|} \left(\boldsymbol{F}_h^{\mathrm{up}}[\varrho_h^k \boldsymbol{u}_h^k, \boldsymbol{u}_h^k] + \{\{p_h^k - \nu \mathrm{div}_h \boldsymbol{u}_h^k\}\} \boldsymbol{n} - \mu \frac{[[\boldsymbol{u}_h^k]]}{d_\sigma} \right) = 0,
$$

where $F_h^{\mathrm{up}}[r, \boldsymbol{u}]$ is the diffusive upwind flux given in (8.8), $\{\{\cdot\}\}$ and $[[\cdot]]$ are the average and jump operators defined in (9), respectively.

For the purpose of numerical analysis it is convenient to rewrite the method in a weak form.

Lemma 11.1 **(FV method for the Navier–Stokes system: weak formulation)**
Let $(\varrho_h^k, \boldsymbol{u}_h^k) \in Q_h \times \boldsymbol{Q}_h$ *be a finite volume solution of* (11.4).
Then $(\varrho_h^k, \boldsymbol{u}_h^k)$ *fulfills the following weak formulation*

$$\int_{\mathbb{T}^d} D_t \varrho_h^k \phi_h \,\mathrm{d}x - \int_{\mathcal{E}} F_h^{\mathrm{up}}[\varrho_h^k, \boldsymbol{u}_h^k] [\![\phi_h]\!] \,\mathrm{dS}_x = 0 \quad \textit{for all } \phi_h \in Q_h, \tag{11.5a}$$

$$\int_{\mathbb{T}^d} D_t(\varrho_h^k \boldsymbol{u}_h^k) \cdot \boldsymbol{\phi}_h \,\mathrm{d}x - \int_{\mathcal{E}} \boldsymbol{F}_h^{\mathrm{up}}[\varrho_h^k \boldsymbol{u}_h^k, \boldsymbol{u}_h^k] \cdot [\![\boldsymbol{\phi}_h]\!] \,\mathrm{dS}_x - \int_{\mathbb{T}^d} p_h^k \mathrm{div}_h \boldsymbol{\phi}_h \,\mathrm{d}x$$

$$= -\mu \int_{\mathbb{T}^d} \nabla_{\mathcal{D}} \boldsymbol{u}_h^k : \nabla_{\mathcal{D}} \boldsymbol{\phi}_h \,\mathrm{d}x - \nu \int_{\mathbb{T}^d} \mathrm{div}_h \boldsymbol{u}_h^k \,\mathrm{div}_h \boldsymbol{\phi}_h \,\mathrm{d}x \quad \textit{for all } \boldsymbol{\phi}_h \in \boldsymbol{Q}_h. \tag{11.5b}$$

Note that the discrete gradient $\nabla_{\mathcal{D}}$ is different from ∇_h. The former is based on the dual element while the latter is based on the primary element, see in (11.16) and (11.15) for their definitions.

11.1.1 Fundamental Properties

In this subsection we show some important properties of the FV method (11.5): the conservation of total mass, positivity of the density, and the internal energy balance. Finally, we establish existence of numerical solutions. We recall the notation

$$c \in \mathrm{co}\{a, b\} \text{ if } \min\{a, b\} \le c \le \max\{a, b\}. \tag{11.6}$$

Lemma 11.2 (Positivity of density, mass conservation, and internal energy balance)

Let $(\varrho_h, \boldsymbol{u}_h)$ be a solution to the FV method (11.5) with the initial density $\varrho_0 > 0$. Then it enjoys the following properties:

1. **Discrete conservation of mass.**

$$\int_{\mathbb{T}^d} \varrho_h(t) \,\mathrm{d}x = \int_{\mathbb{T}^d} \varrho_0 \,\mathrm{d}x \equiv M_0, \ t \in (0, T). \tag{11.7}$$

2. **Positivity of discrete density.** $\varrho_h(t) > 0$ *for all* $t \in (0, T)$.
3. **Internal energy balance (integrated renormalization).** *For the pressure potential* $P(\varrho) = \frac{a\varrho^\gamma}{\gamma - 1}$ *we have*

$$\begin{aligned} &\int_{\mathbb{T}^d} D_t P(\varrho_h^k) \,\mathrm{d}x + \int_{\mathbb{T}^d} p(\varrho_h^k) \mathrm{div}_h \boldsymbol{u}_h^k \,\mathrm{d}x = -\frac{\Delta t}{2} \int_{\mathbb{T}^d} P''(\xi) |D_t \varrho_h^k|^2 \,\mathrm{d}x - \\ &- \int_{\mathcal{E}} P''(\varrho_{h,\dagger}^k) [\![\varrho_h^k]\!]^2 \left(h^\varepsilon + \frac{1}{2} | \langle \boldsymbol{u}_h^k \rangle_\sigma \cdot \boldsymbol{n} | \right) \mathrm{dS}_x, \end{aligned} \tag{11.8}$$

for some $\varrho_{h,\dagger}^k \in \mathrm{co}\{(\varrho_h^k)^{\mathrm{in}}, (\varrho_h^k)^{\mathrm{out}}\}$ *for any* $\sigma \in \mathcal{E}$ *and* $\xi \in \mathrm{co}\{\varrho_h^{k-1}, \varrho_h^k\}$.

Proof We refer to equation (8.16) and Lemma 8.3, respectively, for the proof of mass conservation and positivity of density. Moreover, the internal energy balance is a consequence of the renormalized continuity equation stated in Lemma 8.2, which can be derived by setting $\phi_h = 1$ and $B = P(\varrho_h)$ in Lemma 8.2.

Remark 11.1 Note that $\langle f_h \rangle_\sigma \equiv \{\!\{ f_h \}\!\}_\sigma$ for $f_h \in Q_h$. To keep consistent with the notations of other chapters, we have decided to use $\{\!\{ \varrho_h \}\!\}_\sigma$ and $\langle \boldsymbol{u}_h \rangle_\sigma$ respectively for the mean value of density and velocity on a face σ.

The FV method (11.5) is a system of nonlinear algebraic equations. Next, we show that it admits at least one solution.

Lemma 11.3 (Existence of a numerical solution)
Suppose that $\varrho_0 > 0$. Then for any $k = 1, \dots, N_T$, there exists a solution $(\varrho_h^k, \boldsymbol{u}_h^k) \in Q_h \times \boldsymbol{Q}_h$ to the discrete problem (11.5). Moreover, $\varrho_h^k > 0$.

Proof The proof is done via the topological degree theory, specifically Theorem 15. First, let us denote

$$V = \left\{ (\varrho_h^k, \boldsymbol{u}_h^k) \in Q_h \times \boldsymbol{Q}_h,\ \varrho_h^k > 0 \right\},$$
$$W = \left\{ (\varrho_h^k, \boldsymbol{u}_h^k) \in Q_h \times \boldsymbol{Q}_h,\ \epsilon < \varrho_h^k < C_1,\ \left\| \boldsymbol{u}_h^k \right\| \le C_2 \right\},$$

where $\left\| \boldsymbol{u}_h^k \right\| \equiv \left\| \nabla_{\mathcal{D}} \boldsymbol{u}_h^k \right\|_{L^2(\mathbb{T}^d)} + \left\| \boldsymbol{u}_h^k \right\|_{L^2(\mathbb{T}^d)}$. Note that by writing $\varrho_h^k > c$ we mean $\varrho_K^k > c$ for all $K \in \mathcal{T}_h$.

It is obvious that the dimension of the spaces Q_h and $\boldsymbol{Q}_h$ is finite. Indeed the space Q_h can be identified as $Q_h \subset R^M$, where M is the total number of elements of $\mathcal{T}_h$. Analogously, $\boldsymbol{Q}_h \subset R^N$, where $N = d\,M$.

Second, we define the following mapping

$$F : V \times [0,1] \to Q_h \times \boldsymbol{Q}_h, \quad (\varrho_h^k, \boldsymbol{u}_h^k, \zeta) \longmapsto (\varrho^\star, \boldsymbol{u}^\star) = F(\varrho_h^k, \boldsymbol{u}_h^k, \zeta),$$

where $\zeta \in [0,1]$ and $(\varrho^\star, \boldsymbol{u}^\star) \in Q_h \times \boldsymbol{Q}_h$ satisfy

$$\int_{\mathbb{T}^d} \varrho^\star \phi_h \,\mathrm{d}x = \int_{\mathbb{T}^d} \frac{\varrho_h^k - \varrho_h^{k-1}}{\Delta t} \phi_h \,\mathrm{d}x + \zeta \int_{\mathbb{T}^d} \mathrm{div}_h^{\mathrm{up}}(\varrho_h^k, \boldsymbol{u}_h^k) \phi_h \,\mathrm{d}x, \tag{11.9}$$

$$\int_{\mathbb{T}^d} \boldsymbol{u}^\star \cdot \boldsymbol{\phi}_h \,\mathrm{d}x = \int_{\mathbb{T}^d} \frac{\varrho_h^k \boldsymbol{u}_h^k - \varrho_h^{k-1} \boldsymbol{u}_h^{k-1}}{\Delta t} \cdot \boldsymbol{\phi}_h \,\mathrm{d}x + \mu \int_{\mathbb{T}^d} \nabla_{\mathcal{D}} \boldsymbol{u}_h^k : \nabla_{\mathcal{D}} \boldsymbol{\phi}_h \,\mathrm{d}x$$
$$+ \zeta \int_{\mathbb{T}^d} \mathrm{div}_h^{\mathrm{up}}(\varrho_h^k \boldsymbol{u}_h^k, \boldsymbol{u}_h^k) \cdot \boldsymbol{\phi}_h \,\mathrm{d}x + \zeta \int_{\mathbb{T}^d} \left(\mathrm{div}_h \boldsymbol{u}_h^k - p(\varrho_h^k) \right) \mathrm{div}_h \boldsymbol{\phi}_h \,\mathrm{d}x, \tag{11.10}$$

for any $\phi_h \in Q_h$ and $\boldsymbol{\phi} = (\phi_{1,h}, \dots, \phi_{d,h}) \in \boldsymbol{Q}_h$.

Obviously F is well defined and continuous since the values of $\varrho^\star$ and $\boldsymbol{u}^\star$ can be determined by setting $\phi_h = 1_K$ in (11.9), and $\phi_{i,h} = 1_K, \phi_{j,h} = 0$ for $j \neq i,\ i,j \in (1,\ldots,d)$ in (11.10).

Next, we aim to show that the first hypothesis of Theorem 15 holds. To this end, we suppose $(\varrho_h^k, \boldsymbol{u}_h^k) \in Q_h \times \boldsymbol{Q}_h$ is a solution to $F(\varrho_h^k, \boldsymbol{u}_h^k, \zeta) = (0,0)$ for any $\zeta \in [0,1]$. Then the system (11.9)–(11.10) becomes

$$\int_{\mathbb{T}^d} \frac{\varrho_h^k - \varrho_h^{k-1}}{\Delta t}\phi_h \,\mathrm{d}x + \zeta \int_{\mathbb{T}^d} \mathrm{div}_h^{\mathrm{up}}(\varrho_h^k, \boldsymbol{u}_h^k)\phi_h \,\mathrm{d}x = 0 \tag{11.11a}$$

$$\int_{\mathbb{T}^d} \frac{\varrho_h^k\boldsymbol{u}_h^k - \varrho_h^{k-1}\boldsymbol{u}_h^{k-1}}{\Delta t}\cdot\boldsymbol{\phi}_h \,\mathrm{d}x + \mu\int_{\mathbb{T}^d} \nabla_{\mathcal{D}}\boldsymbol{u}_h^k : \nabla_{\mathcal{D}}\boldsymbol{\phi}_h \,\mathrm{d}x$$
$$+\zeta\int_{\mathbb{T}^d} \mathrm{div}_h^{\mathrm{up}}(\varrho_h^k\boldsymbol{u}_h^k, \boldsymbol{u}_h^k)\cdot\boldsymbol{\phi}_h \,\mathrm{d}x + \zeta\int_{\mathbb{T}^d}\left(\mathrm{div}_h\boldsymbol{u}_h^k - p(\varrho_h^k)\right)\mathrm{div}_h\boldsymbol{\phi}_h \,\mathrm{d}x = 0. \tag{11.11b}$$

Taking $\phi_h = 1$ as a test function in (11.11a) and using the conservativity of the numerical flux (8.12) we obtain

$$\|\varrho_h^k\|_{L^1(\mathbb{T}^d)} = \int_{\mathbb{T}^d}\varrho_h^k \,\mathrm{d}x = \int_{\mathbb{T}^d}\varrho_h^{k-1}\,\mathrm{d}x \equiv M_0 > 0. \tag{11.12}$$

Further, setting $\phi_h = P'(\varrho_h^k) - \frac{|\boldsymbol{u}_h^k|^2}{2}$ and $\boldsymbol{\phi}_h = \boldsymbol{u}_h^k$ as the test functions in (11.11a) and (11.11b), respectively, it follows from the standard proof of energy estimates, cf. also Theorem 11.1 below that the velocity gradient is bounded. Moreover, applying the Sobolev–Poincaré inequality, we have the following bound

$$\|\boldsymbol{u}_h^k\| \equiv \|\nabla_{\mathcal{D}}\boldsymbol{u}_h^k\|_{L^2(\mathbb{T}^d)} + \|\boldsymbol{u}_h^k\|_{L^2(\mathbb{T}^d)} \leq C_2, \tag{11.13}$$

where C_2 depends on the data of the problem.

Further, let $K \in \mathcal{T}_h$ be such that $\varrho_K^k = \min_{L\in\mathcal{T}_h}\varrho_h^k$. Then $[\![\varrho_h^k]\!]_{\sigma\in\mathcal{E}(K)} \geq 0$. Now setting $\phi_h = 1_K$ leads to

$$\begin{aligned}
&\frac{|K|}{\Delta t\zeta}(\varrho_K^k - \varrho_K^{k-1}) = -\int_K \mathrm{div}_h^{\mathrm{up}}(\varrho_h^k, \boldsymbol{u}_h^k)\\
&= -\sum_{\sigma\in\mathcal{E}(K)}|\sigma|\,\varrho_h^{k,\mathrm{up}}\left\langle\boldsymbol{u}_h^k\right\rangle_\sigma\cdot\boldsymbol{n} + \sum_{\sigma\in\mathcal{E}(K)}|\sigma|\,h^\varepsilon\,[\![\varrho_h^k]\!]\\
&\geq -\sum_{\sigma\in\mathcal{E}(K)}|\sigma|\,\varrho_K^k\left\langle\boldsymbol{u}_h^k\right\rangle_\sigma\cdot\boldsymbol{n} + \sum_{\sigma\in\mathcal{E}(K)}|\sigma|\,(\varrho_K^k - \varrho_h^{k,\mathrm{up}})\left\langle\boldsymbol{u}_h^k\right\rangle_\sigma\cdot\boldsymbol{n}\\
&= -|K|\,\varrho_K^k(\mathrm{div}_h\boldsymbol{u}_h^k)_K - \sum_{\sigma\in\mathcal{E}(K)}|\sigma|\,[\![\varrho_h^k]\!]\,[\left\langle\boldsymbol{u}_h^k\right\rangle_\sigma\cdot\boldsymbol{n}]^- \geq -|K|\,\varrho_K^k(\mathrm{div}_h\boldsymbol{u}_h^k)_K\\
&\geq -|K|\,\varrho_K^k\left|\mathrm{div}_h\boldsymbol{u}_h^k\right|_K.
\end{aligned}$$

Thus $\varrho_h^k \geq \varrho_K^k \geq \frac{\varrho_K^{k-1}}{1+\Delta t\zeta|(\mathrm{div}_h \boldsymbol{u}_h^k)_K|} > 0$. Consequently, by virtue of (11.13) $\varrho_h^k > \epsilon$, where ϵ depends only on the data of the problem. Further, we get from (11.12) that $\varrho_h^k \leq \frac{M_0}{\min_{K\in\mathcal{T}_h}|K|}$, which indicates the existence of $C_1 > 0$ such that $\varrho_h^k < C_1$. Therefore, the first hypothesis of Theorem 15 is satisfied.

Next, we proceed to show that the second hypothesis of Theorem 15 is satisfied. Let $\zeta = 0$ then the system $F(\varrho_h^k, \boldsymbol{u}_h^k, 0) = 0$ reads

$$\varrho_h^k = \varrho_h^{k-1} \tag{11.14a}$$

$$\int_{\mathbb{T}^d} \frac{\varrho_h^k \boldsymbol{u}_h^k - \varrho_h^{k-1}\boldsymbol{u}_h^{k-1}}{\Delta t} \cdot \boldsymbol{\phi}_h \,\mathrm{d}x + \mu \int_{\mathbb{T}^d} \nabla_{\mathcal{D}} \boldsymbol{u}_h^k : \nabla_{\mathcal{D}} \boldsymbol{\phi}_h \,\mathrm{d}x = 0. \tag{11.14b}$$

From (11.14a) it is obvious $\varrho_h^k = \varrho_h^{k-1} > 0$. Substituting (11.14a) into (11.14b) we arrive at a linear system on $\boldsymbol{u}_h^k$ with a symmetric positive definite matrix. Thus (11.14b) admits a unique solution. Consequently, the second hypothesis of Theorem 15 is satisfied.

We have shown that both hypotheses of Theorem 15 hold. Applying Theorem 15 finishes the proof.

11.2 Stability

The scheme (11.5) is stable in the sense of Definition 8.3. Indeed, we have the following energy estimates.

Theorem 11.1 (Energy stability of the FV method)

Let the pressure p satisfy the equation of state (11.2) and let $(\varrho_h, \boldsymbol{u}_h) = \{\varrho_h^k, \boldsymbol{u}_h^k\}_{k=1}^{N_T}$ be a family of numerical solutions obtained by the FV method (11.5).

Then there exist $\xi \in \mathrm{co}\{\varrho_h^{k-1}, \varrho_h^k\}$ and for any $\sigma \in \mathcal{E}_{int}$ $\varrho_{h,\dagger}^k \in \mathrm{co}\{(\varrho_h^k)^{\mathrm{in}}, (\varrho_h^k)^{\mathrm{out}}\}$ such that

$$\begin{aligned}
&D_t \int_{\mathbb{T}^d} \left(\frac{1}{2}\varrho_h^k |\boldsymbol{u}_h^k|^2 + P(\varrho_h^k)\right) \mathrm{d}x + \mu \left\|\nabla_{\mathcal{D}} \boldsymbol{u}_h^k\right\|_{L^2}^2 + \nu \left\|\mathrm{div}_h \boldsymbol{u}_h^k\right\|_{L^2}^2 \\
&= -h^{\varepsilon} \int_{\mathcal{E}} \{\{\varrho_h^k\}\} \left|[[\boldsymbol{u}_h^k]]\right|^2 \mathrm{dS}_x - \frac{\Delta t}{2} \int_{\mathbb{T}^d} P''(\xi)|D_t \varrho_h^k|^2 \,\mathrm{d}x \\
&- \int_{\mathcal{E}} P''(\varrho_{h,\dagger}^k) \left[[\varrho_h^k]\right]^2 \left(h^{\varepsilon} + \frac{1}{2}\left|\langle \boldsymbol{u}_h^k\rangle_\sigma \cdot \boldsymbol{n}\right|\right) \mathrm{dS}_x \\
&- \frac{\Delta t}{2} \int_{\mathbb{T}^d} \varrho_h^{k-1} |D_t \boldsymbol{u}_h^k|^2 \,\mathrm{d}x - \frac{1}{2}\int_{\mathcal{E}} (\varrho_h^k)^{\mathrm{up}} \left|\langle \boldsymbol{u}_h^k\rangle_\sigma \cdot \boldsymbol{n}\right| \left|[[\boldsymbol{u}_h^k]]\right|^2 \mathrm{dS}_x.
\end{aligned} \tag{11.15}$$

Proof First, taking $\phi_h = -\frac{|\boldsymbol{u}_h^k|^2}{2}$ in (11.5a) we get

$$-\int_{\mathbb{T}^d} D_t \varrho_h^k \frac{|\boldsymbol{u}_h^k|^2}{2} \, \mathrm{d}x + \int_{\mathcal{E}} F_h^{\mathrm{up}}[\varrho_h^k, \boldsymbol{u}_h^k] \left[\!\!\left[\frac{|\boldsymbol{u}_h^k|^2}{2} \right]\!\!\right] \mathrm{dS}_x = 0$$

Next, by taking $\boldsymbol{\phi}_h = \boldsymbol{u}_h^k$ in (11.5b) we derive

$$\int_{\mathbb{T}^d} D_t(\varrho_h^k \boldsymbol{u}_h^k) \cdot \boldsymbol{u}_h^k \, \mathrm{d}x - \int_{\mathcal{E}} F_h^{\mathrm{up}}[\varrho_h^k \boldsymbol{u}_h^k, \boldsymbol{u}_h^k] \cdot [\![\boldsymbol{u}_h^k]\!] \, \mathrm{dS}_x - \int_{\mathbb{T}^d} p(\varrho_h^k) \mathrm{div}_h \boldsymbol{u}_h^k \, \mathrm{d}x$$
$$= -\mu \left\| \nabla_{\mathcal{D}} \boldsymbol{u}_h^k \right\|_{L^2}^2 - \nu \left\| \mathrm{div}_h \boldsymbol{u}_h^k \right\|_{L^2}^2 .$$

Further, summing up the previous two equalities and recalling (8.13b) we infer that

$$\begin{aligned} D_t \int_{\mathbb{T}^d} \frac{1}{2} \varrho_h |\boldsymbol{u}_h^k|^2 \, \mathrm{d}x + \mu \left\| \nabla_{\mathcal{D}} \boldsymbol{u}_h^k \right\|_{L^2}^2 + \nu \left\| \mathrm{div}_h \boldsymbol{u}_h^k \right\|_{L^2}^2 \\ = -\frac{\Delta t}{2} \int_{\mathbb{T}^d} \varrho_h^{k-1} |D_t \boldsymbol{u}_h^k|^2 \, \mathrm{d}x - \int_{\mathbb{T}^d} p_h^k \mathrm{div}_h \boldsymbol{u}_h^k \, \mathrm{d}x - h^{\varepsilon} \int_{\mathcal{E}} \{\!\{\varrho_h^k\}\!\} \left| [\![\boldsymbol{u}_h^k]\!] \right|^2 \mathrm{dS}_x \\ - \frac{1}{2} \int_{\mathcal{E}} (\varrho_h^k)^{\mathrm{up}} \left| \langle \boldsymbol{u}_h^k \rangle_{\sigma} \cdot \boldsymbol{n} \right| \left| [\![\boldsymbol{u}_h^k]\!] \right|^2 \mathrm{dS}_x. \end{aligned} \tag{11.16}$$

Finally, combining (11.16) with (11.8) we get

$$\begin{aligned} D_t \int_{\mathbb{T}^d} \left(\frac{1}{2} \varrho_h^k |\boldsymbol{u}_h^k|^2 + P(\varrho_h^k) \right) \, \mathrm{d}x + \mu \left\| \nabla_{\mathcal{D}} \boldsymbol{u}_h^k \right\|_{L^2}^2 + \nu \left\| \mathrm{div}_h \boldsymbol{u}_h^k \right\|_{L^2}^2 \\ = -\frac{\Delta t}{2} \int_{\mathbb{T}^d} \varrho_h^{k-1} |D_t \boldsymbol{u}_h^k|^2 \, \mathrm{d}x - \frac{\Delta t}{2} \int_{\mathbb{T}^d} P''(\xi) |D_t \varrho_h^k|^2 \, \mathrm{d}x \\ - \int_{\mathcal{E}} P''(\varrho_{h,\dagger}^k) \, [\![\varrho_h^k]\!]^2 \left(h^{\varepsilon} + \frac{1}{2} \left| \langle \boldsymbol{u}_h^k \rangle_{\sigma} \cdot \boldsymbol{n} \right| \right) \mathrm{dS}_x \\ - h^{\varepsilon} \int_{\mathcal{E}} \{\!\{\varrho_h^k\}\!\} \left| [\![\boldsymbol{u}_h^k]\!] \right|^2 \mathrm{dS}_x - \frac{1}{2} \int_{\mathcal{E}} (\varrho_h^k)^{\mathrm{up}} \left| \langle \boldsymbol{u}_h^k \rangle_{\sigma} \cdot \boldsymbol{n} \right| \left| [\![\boldsymbol{u}_h^k]\!] \right|^2 \mathrm{dS}_x, \end{aligned}$$

which completes the proof.

11.2.1 Uniform Bounds

In this subsection we derive the uniform bounds that are necessary to show the consistency of the FV method (11.5).

First, as a consequence of Theorem 11.1 and the Sobolev–Poincaré inequality (37), we have the following corollary, that gives *a priori* estimates for numerical solutions of the FV method (11.5).

Corollary 11.1 (Uniform bounds)

Let the pressure p satisfy the equation of state (11.2) *and* $(\varrho_h, \boldsymbol{u}_h)$ *be a solution of the FV method (11.5) in the sense of (47). Then the following estimates hold*

$$\left\|\varrho_h \left|\boldsymbol{u}_h\right|^2\right\|_{L^\infty L^1} \lesssim 1, \quad \|\varrho_h\|_{L^\infty L^\gamma} \lesssim 1, \quad \|\varrho_h \boldsymbol{u}_h\|_{L^\infty L^{\frac{2\gamma}{\gamma+1}}} \lesssim 1, \tag{11.17a}$$

$$\|\nabla_{\mathcal{D}} \boldsymbol{u}_h\|_{L^2 L^2} \lesssim 1, \quad \|\mathrm{div}_h \boldsymbol{u}_h\|_{L^2 L^2} \lesssim 1, \quad \|\boldsymbol{u}_h\|_{L^2 L^6} \lesssim 1, \tag{11.17b}$$

$$h^\varepsilon \int_0^T \int_{\mathcal{E}} \{\!\{\varrho_h\}\!\} \left|[\![\boldsymbol{u}_h]\!]\right|^2 \mathrm{dS}_x \, \mathrm{d}t \lesssim 1, \tag{11.17c}$$

$$\int_0^T \int_{\mathcal{E}} P''(\varrho_{h,\dagger}) \left[\!\left[\varrho_h\right]\!\right]^2 (h^\varepsilon + |\{\!\{\boldsymbol{u}_h\}\!\} \cdot \boldsymbol{n}|) \mathrm{dS}_x \, \mathrm{d}t \lesssim 1, \tag{11.17d}$$

where $\varrho_{h,\dagger} \in \mathrm{co}\{\varrho_h^{\mathrm{in}}, \varrho_h^{\mathrm{out}}\}$ *for any* $\sigma \in \mathcal{E}_{int}$.

Remark 11.2 Obviously, the estimate $\|\boldsymbol{u}_h\|_{L^2 L^6} \lesssim 1$ in (11.17b) follows from the Sobolev embeddings for $d = 3$. If $d = 2$ we have instead a better estimate $\|\boldsymbol{u}_h\|_{L^2 L^p} \lesssim 1$ for any $1 \leq p < \infty$.

We proceed with deriving negative estimates that are less obvious but follow also from the energy estimates (11.15).

Lemma 11.4 (Negative estimates of density and momentum)

Let the pressure p satisfy (11.2), *and let* $(\varrho_h, \boldsymbol{u}_h)$ *be a solution to the FV method (11.5) in the sense of (47) with* $h \in (0, 1)$. *Then we have the following negative estimates*

$$\|\varrho_h\|_{L^2 L^2} \lesssim h^{\beta_D}, \quad \beta_D = \begin{cases} \max\left\{-\frac{3\varepsilon+3+d}{6\gamma}, \frac{\gamma-2}{2\gamma}d\right\}, & \text{if } \gamma \in (1,2), \\ 0, & \text{if } \gamma \geq 2, \end{cases} \tag{11.18a}$$

$$\|\varrho_h\|_{L^2 L^{6/5}} \lesssim h^{\beta_R}, \quad \beta_R = \begin{cases} \max\left\{-\frac{3\varepsilon+3+d}{6\gamma}, \frac{5\gamma-6}{6\gamma}d\right\}, & \text{if } \gamma \in (1, \frac{6}{5}), \\ 0, & \text{if } \gamma \geq \frac{6}{5}, \end{cases} \tag{11.18b}$$

$$\|\varrho_h \boldsymbol{u}_h\|_{L^2L^2} \lesssim h^{\beta_M}, \quad \beta_M = \begin{cases} -\frac{3\varepsilon+3+d}{6\gamma}, & \text{if } \gamma \in (1,2), \\ \frac{\gamma-3}{3\gamma} d, & \text{if } \gamma \in [2,3), \\ 0, & \text{if } \gamma \geq 3 \text{ for } d=3, \\ 0, & \text{if } \gamma > 2 \text{ for } d=2. \end{cases} \tag{11.18c}$$

Proof We start with the proof of (11.18a). Firstly, for $\gamma \geq 2$ it is clear that

$$\|\varrho_h\|_{L^2L^2} \lesssim \|\varrho_h\|_{L^\infty L^\gamma} \lesssim 1.$$

Secondly, for $\gamma \in (1,2)$, we show the proof in two steps. On the one hand, a direct application of the inverse estimate (11.26) leads to

$$\|\varrho_h\|_{L^2 L^2} \lesssim h^{d(\frac{1}{2}-\frac{1}{\gamma})} \|\varrho_h\|_{L^\infty L^\gamma} \lesssim h^{\frac{\gamma-2}{2\gamma} d}.$$

On the other hand, recalling the algebraic inequality

$$a\gamma \left(\varrho_L^{\gamma/2} - \varrho_K^{\gamma/2}\right)^2 \leq \frac{\partial^2 P(\zeta)}{\partial \varrho^2} (\varrho_L - \varrho_K)^2, \ \forall \varrho_{h,\dagger} \in \text{co}\{\varrho_L, \varrho_K\}, \ \varrho_L, \varrho_K > 0$$

for $\gamma \in (1,2)$, we deduce from the estimate (11.17d) that

$$\begin{aligned} \left\|\nabla_{\mathcal{D}} \varrho_h^{\gamma/2}\right\|_{L^2L^2}^2 &= \int_0^T \int_\Omega |\nabla_{\mathcal{D}} \varrho_h^{\gamma/2}|^2 \,\mathrm{d}x\,\mathrm{d}t \lesssim \int_0^T \int_\Omega P''(\varrho_{h,\dagger}) |\nabla_{\mathcal{D}} \varrho_h|^2 \,\mathrm{d}x\,\mathrm{d}t \\ &\lesssim h^{-\varepsilon-1}. \end{aligned}$$

Using the above inequalities together with the Sobolev–Poincaré inequality (34), inverse estimate (11.26), and the second estimate of (11.17a), we derive

$$\begin{aligned} \|\varrho_h\|_{L^1 L^\infty} &= \int_0^T \left\|\varrho_h^{\gamma/2}\right\|_{L^\infty}^{2/\gamma} \mathrm{d}t \leq \int_0^T \left(h^{-d/6} \left\|\varrho_h^{\gamma/2}\right\|_{L^6}\right)^{2/\gamma} \mathrm{d}t \\ &\leq h^{-d/(3\gamma)} \int_0^T \left(\left\|\varrho_h^{\gamma/2}\right\|_{L^2}^2 + \left\|\nabla_{\mathcal{D}} \varrho_h^{\gamma/2}\right\|_{L^2}^2\right)^{1/\gamma} \mathrm{d}t \\ &\leq h^{-d/(3\gamma)} \left(\|\varrho_h\|_{L^1 L^\gamma} + \left\|\nabla_{\mathcal{D}} \varrho_h^{\gamma/2}\right\|_{L^{\gamma/2}L^2}^{2/\gamma}\right) \\ &\leq h^{-d/(3\gamma)} \left(\|\varrho_h\|_{L^\infty L^\gamma} + \left\|\nabla_{\mathcal{D}} \varrho_h^{\gamma/2}\right\|_{L^2L^2}^{2/\gamma}\right) \lesssim h^{-\frac{3\varepsilon+3+d}{3\gamma}}. \end{aligned}$$

Furthermore, the above inequality together with the Hölder inequality and the second estimate in (11.17a) immediately yield (11.18a), specifically

$$\|\varrho_h\|_{L^2L^2} = \left(\int_0^T \|\varrho_h\|_{L^2}^2 \, \mathrm{d}t\right)^{1/2} \leq \left(\int_0^T \|\varrho_h\|_{L^1} \|\varrho_h\|_{L^\infty} \, \mathrm{d}t\right)^{1/2}$$
$$\leq \|\varrho_h\|_{L^\infty L^1}^{1/2} \|\varrho_h\|_{L^1 L^\infty}^{1/2} \lesssim h^{-\frac{3\varepsilon+3+d}{6\gamma}}.$$

Collecting the above results we finish the proof of (11.18a).

Next, we show (11.18b). Firstly, it is obvious for $\gamma \geq \frac{6}{5}$ that

$$\|\varrho_h\|_{L^2L^{6/5}} \lesssim \|\varrho_h\|_{L^\infty L^\gamma} \lesssim 1.$$

Secondly, for $\gamma \in (1, 6/5)$ we show the proof of (11.18b) in two steps. On one hand, it is easy to observe for $h \in (0, 1)$ that

$$\|\varrho_h\|_{L^2L^{6/5}} \lesssim \|\varrho_h\|_{L^2 L^2} \lesssim h^{\beta_D} \leq h^{-\frac{3\varepsilon+3+d}{6\gamma}}.$$

On the other hand, due to the inverse estimates (11.26) and the density estimate in (11.17a) we have

$$\|\varrho_h\|_{L^2L^{6/5}} \lesssim \|\varrho_h\|_{L^\infty L^{6/5}} \lesssim h^{d\left(\frac{5}{6}-\frac{1}{\gamma}\right)} \|\varrho_h\|_{L^\infty L^\gamma} \lesssim h^{\frac{5\gamma-6}{6\gamma}d}.$$

Finally, we proceed to show the momentum estimate (11.18c). Firstly, for the case $\gamma \in (1, 2)$ we have

$$\|\varrho_h \boldsymbol{u}_h\|_{L^2L^2} \lesssim \left\|\sqrt{\varrho_h}\right\|_{L^2 L^\infty} \left\|\sqrt{\varrho_h}\boldsymbol{u}_h\right\|_{L^\infty L^2} = \|\varrho_h\|_{L^1 L^\infty}^{1/2} \left\|\varrho_h \boldsymbol{u}_h^2\right\|_{L^\infty L^1}^{1/2} \lesssim h^{-\frac{3\varepsilon+3+d}{6\gamma}}.$$

Secondly, for the case $\gamma \geq 3$ and $d = 3$ it follows by Hölder's inequality that

$$\|\varrho_h \boldsymbol{u}_h\|_{L^2L^2} \lesssim \|\varrho_h\|_{L^\infty L^3} \|\boldsymbol{u}_h\|_{L^2 L^6} \lesssim \|\varrho_h\|_{L^\infty L^\gamma} \|\boldsymbol{u}_h\|_{L^2 L^6} \lesssim 1.$$

Analogous estimates for $\gamma > 2$ and $d = 2$ yield

$$\|\varrho_h \boldsymbol{u}_h\|_{L^2L^2} \lesssim \|\varrho_h\|_{L^\infty L^\gamma} \|\boldsymbol{u}_h\|_{L^2 L^{\frac{2\gamma}{\gamma-2}}} \lesssim 1.$$

Thirdly, for $\gamma \in [2, 3)$ we have by inverse estimate (11.26) and Hölder's inequality that

$$\|\varrho_h \boldsymbol{u}_h\|_{L^2L^2} \lesssim \|\varrho_h\|_{L^\infty L^3} \|\boldsymbol{u}_h\|_{L^2 L^6} \lesssim h^{d(\frac{1}{3}-\frac{1}{\gamma})} \|\varrho_h\|_{L^\infty L^\gamma} \|\boldsymbol{u}_h\|_{L^2 L^6} \lesssim h^{\frac{\gamma-3}{3\gamma}d},$$

which completes the proof.

Further, we state a dissipation estimate for the density.

Lemma 11.5 (Density dissipation)
Let the pressure p satisfy (11.2) *with* $\gamma \geq 2$ *and let* $(\varrho_h, \boldsymbol{u}_h)$ *be a solution of the FV method (11.5) in the sense of (47). Then there holds*

$$\int_0^T \int_{\mathcal{E}} \frac{[\![\varrho_h]\!]^2}{\max\{\varrho_h^{\mathrm{in}}, \varrho_h^{\mathrm{out}}\}} |\langle \boldsymbol{u}_h \rangle_\sigma \cdot \boldsymbol{n}| \, \mathrm{dS}_x \, \mathrm{d}t \stackrel{<}{\sim} 1. \tag{11.19}$$

Proof First, recalling Lemma 11.2 we know that the FV method (11.5) preserves positivity of the density. Thus we may test (11.5a) with $\ln \varrho_h$ which leads to

$$\begin{aligned}
\int_{\mathbb{T}^d} D_t \varrho_h \ln \varrho_h \, \mathrm{d}x &= \int_{\mathcal{E}} F_h^{\mathrm{up}}[\varrho_h, \boldsymbol{u}_h] [\![\ln \varrho_h]\!] \, \mathrm{dS}_x \\
&= \int_{\mathcal{E}} \varrho_h^{\mathrm{up}} \langle \boldsymbol{u}_h \rangle_\sigma \cdot \boldsymbol{n} [\![\ln \varrho_h]\!] \, \mathrm{dS}_x - h^\varepsilon \int_{\mathcal{E}} \underbrace{[\![\varrho_h]\!] [\![\ln \varrho_h]\!]}_{\geq 0} \, \mathrm{dS}_x \\
&\leq \sum_{\sigma = K|L \in \mathcal{E}} |\sigma| \left(\varrho_K [\langle \boldsymbol{u}_h \rangle_\sigma \cdot \boldsymbol{n}_{\sigma K}]^+ + \varrho_L [\langle \boldsymbol{u}_h \rangle_\sigma \cdot \boldsymbol{n}_{\sigma K}]^- \right) (\ln \varrho_L - \ln \varrho_K) \\
&= \sum_{\sigma = K|L \in \mathcal{E}} |\sigma| \left(\varrho_K [\langle \boldsymbol{u}_h \rangle_\sigma \cdot \boldsymbol{n}_{\sigma K}]^+ - \varrho_L [\langle \boldsymbol{u}_h \rangle_\sigma \cdot \boldsymbol{n}_{\sigma L}]^+ \right) (\ln \varrho_L - \ln \varrho_K).
\end{aligned} \tag{11.20}$$

Next, noticing that $B(\varrho) = \varrho \ln \varrho - \varrho$ is a convex function of ϱ, with $B'(\varrho) = \ln \varrho$ for $\varrho > 0$, we find in view of the mass conservation (11.7) that

$$\int_{\mathbb{T}^d} D_t (\varrho_h \ln \varrho_h) \, \mathrm{d}x = \int_{\mathbb{T}^d} D_t B(\varrho_h) \, \mathrm{d}x \leq \int_{\mathbb{T}^d} \ln \varrho_h D_t \varrho_h \, \mathrm{d}x.$$

Further, substituting the above inequality into (11.20) we derive

$$\begin{aligned}
&\int_{\mathbb{T}^d} D_t (\varrho_h \ln \varrho_h) \, \mathrm{d}x \\
&\quad \leq \sum_{\sigma = K|L \in \mathcal{E}} |\sigma| \left(\varrho_K [\langle \boldsymbol{u}_h \rangle_\sigma \cdot \boldsymbol{n}_{\sigma K}]^+ - \varrho_L [\langle \boldsymbol{u}_h \rangle_\sigma \cdot \boldsymbol{n}_{\sigma L}]^+ \right) (\ln \varrho_L - \ln \varrho_K),
\end{aligned}$$

which implies

$$
\begin{aligned}
&\sum_{\sigma=K|L\in\mathcal{E}} |\sigma|[\langle \boldsymbol{u}_h\rangle_\sigma \cdot \boldsymbol{n}_{\sigma K}]^+ \left(\varrho_K(\ln\varrho_K - \ln\varrho_L) - (\varrho_K - \varrho_L)\right) \\
&\quad + \sum_{\sigma=K|L\in\mathcal{E}} |\sigma|[\langle \boldsymbol{u}_h\rangle_\sigma \cdot \boldsymbol{n}_{\sigma L}]^+ \left(\varrho_L(\ln\varrho_L - \ln\varrho_K) - (\varrho_L - \varrho_K)\right) \\
&\leq \sum_{\sigma=K|L\in\mathcal{E}} |\sigma| \left([\langle \boldsymbol{u}_h\rangle_\sigma \cdot \boldsymbol{n}_{\sigma K}]^+(\varrho_L - \varrho_K) + [\langle \boldsymbol{u}_h\rangle_\sigma \cdot \boldsymbol{n}_{\sigma L}]^+(\varrho_K - \varrho_L)\right) \\
&\quad - \int_{\mathbb{T}^d} D_t(\varrho_h \ln\varrho_h)\ \mathrm{d}x \\
&= \sum_{\sigma\in\mathcal{E}} |\sigma|\ \langle \boldsymbol{u}_h\rangle_\sigma \cdot \boldsymbol{n}\ [\![\varrho_h]\!] - \int_{\mathbb{T}^d} D_t(\varrho_h \ln\varrho_h)\ \mathrm{d}x \\
&= \sum_{K\in\mathcal{T}_h} \varrho_K \sum_{\sigma\in\mathcal{E}(K)} |\sigma|\ \langle \boldsymbol{u}_h\rangle_\sigma \cdot \boldsymbol{n} - \int_{\mathbb{T}^d} D_t(\varrho_h \ln\varrho_h)\ \mathrm{d}x \\
&= \int_{\mathbb{T}^d} \varrho_h \mathrm{div}_h \boldsymbol{u}_h\ \mathrm{d}x - \int_{\mathbb{T}^d} D_t(\varrho_h \ln\varrho_h)\ \mathrm{d}x.
\end{aligned}
$$

Moreover, we observe

$$
\begin{aligned}
&\frac{1}{2}\sum_{\sigma=K|L\in\mathcal{E}} |\sigma|\,|\langle \boldsymbol{u}_h\rangle_\sigma \cdot \boldsymbol{n}|\, \frac{[\![\varrho_h]\!]^2}{\max\{\varrho_K, \varrho_L\}} \\
&\leq \sum_{\sigma=K|L\in\mathcal{E}} |\sigma|[\langle \boldsymbol{u}_h\rangle_\sigma \cdot \boldsymbol{n}_{\sigma K}]^+ \left(\varrho_K(\ln\varrho_K - \ln\varrho_L) - (\varrho_K - \varrho_L)\right) \\
&\quad + \sum_{\sigma=K|L\in\mathcal{E}} |\sigma|[\langle \boldsymbol{u}_h\rangle_\sigma \cdot \boldsymbol{n}_{\sigma L}]^+ \left(\varrho_L(\ln\varrho_L - \ln\varrho_K) - (\varrho_L - \varrho_K)\right) \\
&\leq \int_{\mathbb{T}^d} \varrho_h \mathrm{div}_h \boldsymbol{u}_h\ \mathrm{d}x - \int_{\mathbb{T}^d} D_t(\varrho_h \ln\varrho_h)\ \mathrm{d}x,
\end{aligned}
$$

where we have used

$$
\begin{aligned}
b(\ln b - \ln a) - (b-a) &= b\int_a^b \frac{1}{y}\mathrm{d}y - \int_a^b \mathrm{d}y = \int_a^b \frac{1}{y}(b-y)\mathrm{d}y \\
&= c\int_a^b (b-y)dy = c\frac{(b-a)^2}{2} \geq \frac{1}{\max\{a,b\}}\frac{(b-a)^2}{2}
\end{aligned}
$$

with $c \in \mathrm{co}\{\frac{1}{a}, \frac{1}{b}\}$ for any $a, b > 0$.

Consequently, we derive (11.19)

$$\int_0^T \sum_{\sigma=K|L\in\mathcal{E}} |\sigma|\, |\langle \boldsymbol{u}_h \rangle_\sigma \cdot \boldsymbol{n}| \frac{[\![\varrho_h]\!]^2}{\max\{\varrho_K, \varrho_L\}} \,\mathrm{d}t$$
$$\leq 2 \int_0^T \int_{\mathbb{T}^d} \varrho_h \mathrm{div}_h \boldsymbol{u}_h - D_t(\varrho_h \ln \varrho_h) \,\mathrm{d}x \,\mathrm{d}t$$
$$\lesssim \|\varrho_h\|_{L^2L^2} \|\mathrm{div}_h \boldsymbol{u}_h\|_{L^2L^2} + \|\varrho_h \ln \varrho_h\|_{L^\infty L^2} \lesssim 1,$$

which concludes the proof.

Finally, the following lemma completes the list of estimates that will be useful for the derivation of the consistency formulation.

Lemma 11.6 *Let the pressure $p(\varrho)$ satisfy* (11.2) *and let $(\varrho_h, \boldsymbol{u}_h)$ be a solution of the FV method (11.5) in the sense of (47) with $h \in (0,1)$. Then there exists $c > 0$ independent of h and Δt such that*

$$\int_0^T \int_{\mathcal{E}} \left| [\![\varrho_h]\!] \, \boldsymbol{u}_h \cdot \boldsymbol{n} \right| \mathrm{dS}_x \,\mathrm{d}t \leq c h^\beta,$$

where

$$\beta = \begin{cases} -\frac{1}{2} & \text{if } \gamma \geq \frac{6}{5}, \\ -\frac{1}{2} + \frac{1}{2} d(\frac{5}{6} - \frac{1}{\gamma}) & \text{if } \gamma \in (1, \frac{6}{5}), \end{cases}$$

Proof For $\gamma \geq 2$ we recall Lemma 11.5 and get

$$\int_0^T \int_{\mathcal{E}} \left| [\![\varrho_h]\!] \, \langle \boldsymbol{u}_h \rangle_\sigma \cdot \boldsymbol{n} \right| \mathrm{dS}_x \,\mathrm{d}t$$
$$\leq \left(\int_0^T \int_{\mathcal{E}} \frac{[\![\varrho_h]\!]^2}{\max\{\varrho_h^{\mathrm{in}}, \varrho_h^{\mathrm{out}}\}} |\langle \boldsymbol{u}_h \rangle_\sigma \cdot \boldsymbol{n}| \,\mathrm{dS}_x \,\mathrm{d}t \right)^{1/2} \times$$
$$\times \left(\int_0^T \int_{\mathcal{E}} |\langle \boldsymbol{u}_h \rangle_\sigma \cdot \boldsymbol{n}| \max\{\varrho_h^{\mathrm{in}}, \varrho_h^{\mathrm{out}}\} \mathrm{dS}_x \,\mathrm{d}t \right)^{1/2}$$
$$\leq h^{-1/2} \|\varrho_h\|_{L^2L^{6/5}}^{1/2} \|\boldsymbol{u}_h\|_{L^2L^6}^{1/2}.$$

Further, for $\gamma \in (1,2)$ it is easy to check that $P''(\varrho_{h,\dagger})(\varrho_{h,\dagger} + 1) \geq 1$ for all $\varrho_{h,\dagger} > 0$, where $\varrho_{h,\dagger} \in \mathrm{co}\{\varrho_h^{\mathrm{in}}, \varrho_h^{\mathrm{out}}\}$. Note that $\varrho_h > 0$ owing to Lemma 11.2. Then

$$
\begin{aligned}
&\int_0^T \int_{\mathcal{E}} \left| [\![\varrho_h]\!] \langle \boldsymbol{u}_h \rangle_\sigma \cdot \boldsymbol{n} \right| \mathrm{dS}_x \, \mathrm{d}t \\
&\leq \int_0^T \int_{\mathcal{E}} \left| [\![\varrho_h]\!] \langle \boldsymbol{u}_h \rangle_\sigma \cdot \boldsymbol{n} \right| \underbrace{\sqrt{P''(\varrho_{h,\dagger})(\varrho_{h,\dagger}+1)}}_{\geq 1} \mathrm{dS}_x \, \mathrm{d}t \\
&\leq \left(\int_0^T \int_{\mathcal{E}} P''(\varrho_{h,\dagger}) [\![\varrho_h]\!]^2 \left| \langle \boldsymbol{u}_h \rangle_\sigma \cdot \boldsymbol{n} \right| \mathrm{dS}_x \, \mathrm{d}t \right)^{1/2} \times \\
&\quad \times \left(\int_0^T \int_{\mathcal{E}} (\varrho_{h,\dagger}+1) \left| \langle \boldsymbol{u}_h \rangle_\sigma \cdot \boldsymbol{n} \right| \mathrm{dS}_x \, \mathrm{d}t \right)^{1/2} \\
&\leq c h^{-1/2} \left(\|\varrho_h\|_{L^2 L^{6/5}} + 1 \right)^{1/2} \|\boldsymbol{u}_h\|_{L^2 L^6}^{1/2} .
\end{aligned}
$$

Consequently, collecting the above two estimates and recalling (11.18b) finish the proof.

11.3 Consistency

Having established stability of the scheme (11.5), we can prove consistency in the sense of Definition 8.6. In particular, we show that the numerical solutions satisfy the weak formulation of the continuous problem up to consistency error terms vanishing for $\Delta t, h \to 0$.

Theorem 11.2 (Consistency of the FV method)
Let the pressure p satisfy (11.2) with $\gamma > 1$ and let $(\varrho_h, \boldsymbol{u}_h)$ be a solution of the discrete problem (11.5) on the time interval $[0, T]$ in the sense of (47). Let

$$
\Delta t \approx h \in (0,1) \quad \text{and} \quad \varepsilon > -1.
$$

If $\gamma \in (1,2)$ we moreover assume that $\varepsilon < 2\gamma - 1 - d/3$.
Then there hold

$$
-\int_{\mathbb{T}^d} \varrho_h^0 \varphi(0,\cdot) \, \mathrm{d}x = \int_0^T \int_{\mathbb{T}^d} \left[\varrho_h \partial_t \varphi + \varrho_h \boldsymbol{u}_h \cdot \nabla_x \varphi \right] \mathrm{d}x \, \mathrm{d}t + \int_0^T e_{1,h}(t, \varphi) \, \mathrm{d}t, \tag{11.21}
$$

for any $\varphi \in C_c^2([0,T) \times \mathbb{T}^d)$;

$$-\int_{\mathbb{T}^d} \varrho_h^0 \boldsymbol{u}_h^0 \boldsymbol{\varphi}(0,\cdot)\,\mathrm{d}x = \int_0^T \int_{\mathbb{T}^d} \left[\varrho_h \boldsymbol{u}_h \cdot \partial_t \boldsymbol{\varphi} + (\varrho_h \boldsymbol{u}_h \otimes \boldsymbol{u}_h + p_h \mathbb{I}) : \nabla_x \boldsymbol{\varphi}\right]\,\mathrm{d}x\,\mathrm{d}t,$$

$$-\mu \int_0^T \int_{\mathbb{T}^d} \nabla_{\mathcal{D}} \boldsymbol{u}_h : \nabla_x \boldsymbol{\varphi}\,\mathrm{d}x\,\mathrm{d}t - \nu \int_0^T \int_{\mathbb{T}^d} \mathrm{div}_h \boldsymbol{u}_h\, \mathrm{div}_x \boldsymbol{\varphi}\,\mathrm{d}x\,\mathrm{d}t$$

$$+\int_0^T e_{2,h}(t,\boldsymbol{\varphi})\,\mathrm{d}t, \tag{11.22}$$

for any $\boldsymbol{\varphi} \in C_c^2([0,T) \times \mathbb{T}^d; R^d)$*;*

$$\|e_{1,h}(\cdot,\varphi)\|_{L^1(0,T)} \lesssim h^\beta \|\varphi\|_{C^2},\ \|e_{2,h}(\cdot,\boldsymbol{\varphi})\|_{L^1(0,T)} \lesssim h^\beta \|\boldsymbol{\varphi}\|_{C^2},\ \textit{for some}\ \beta > 0.$$

Proof Let $\varphi \in C_c^2([0,T) \times \mathbb{T}^d)$ and $\boldsymbol{\varphi} \in C_c^2([0,T) \times \mathbb{T}^d; R^d)$. We test the equations (11.5a) and (11.5b) with $\Pi_Q \varphi$ and $\Pi_Q \boldsymbol{\varphi}$, respectively, and deal with each term separately in four steps:

(1) Time derivative terms:

First we recall (48) and get

$$\int_0^T \int_{\mathbb{T}^d} D_t r_h \Pi_Q \varphi\,\mathrm{d}x\,\mathrm{d}t + \int_0^T \int_{\mathbb{T}^d} r_h \partial_t \varphi\,\mathrm{d}x\,\mathrm{d}t + \int_{\mathbb{T}^d} r_h^0 \varphi(0,\cdot)\,\mathrm{d}x$$
$$\leq \Delta t c(\|\varphi\|_{C^2}) \|r_h\|_{L^1L^1} + \Delta t c(\|\varphi\|_{C^1}) \left\|r_h^0\right\|_{L^1},$$

where r_h stands for ϱ_h or $\varrho_h u_{i,h}$, $i = 1, \ldots, d$. Then by using the estimate (11.17a) we know that

$$\|\varrho_h\|_{L^1L^1} \lesssim \|\varrho_h\|_{L^\infty L^\gamma} \lesssim 1 \quad \text{and} \quad \|\varrho_h \boldsymbol{u}_h\|_{L^1L^1} \lesssim \|\varrho_h \boldsymbol{u}_h\|_{L^\infty L^{\frac{2\gamma}{\gamma+1}}} \lesssim 1.$$

Thus, we have

$$\int_0^T \int_{\mathbb{T}^d} D_t \varrho_h \Pi_Q \varphi\,\mathrm{d}x\,\mathrm{d}t + \int_0^T \int_{\mathbb{T}^d} \varrho_h \partial_t \varphi\,\mathrm{d}x\,\mathrm{d}t + \int_{\mathbb{T}^d} \varrho_h^0 \varphi(0,\cdot)\,\mathrm{d}x \lesssim \Delta t, \tag{11.23a}$$

$$\int_0^T \int_{\mathbb{T}^d} \left(D_t(\varrho_h \boldsymbol{u}_h) \cdot \Pi_Q \boldsymbol{\varphi} + \varrho_h \boldsymbol{u}_h \cdot \partial_t \boldsymbol{\varphi}\right)\,\mathrm{d}x\,\mathrm{d}t + \int_{\mathbb{T}^d} \varrho_h^0 \boldsymbol{u}_h^0 \cdot \boldsymbol{\varphi}(0,\cdot)\,\mathrm{d}x \lesssim \Delta t, \tag{11.23b}$$

for the continuity and the momentum equations, respectively.

(2) Convective terms:

To deal with the convective terms, it is convenient to recall the following identity from Lemma 8.1:

$$\int_0^T \int_{\mathbb{T}^d} r_h \boldsymbol{u}_h \cdot \nabla_x \varphi \, \mathrm{d}x \, \mathrm{d}t - \int_0^T \int_{\mathcal{E}} F_h^{\mathrm{up}}[r_h, \boldsymbol{u}_h] \left[\!\left[\Pi_Q \varphi \right]\!\right] \mathrm{dS}_x \, \mathrm{d}t = \sum_{j=1}^4 E_j(r_h),$$

where

$$E_1(r_h) = \frac{1}{2} \int_0^T \int_{\mathcal{E}} | \{\!\{ \boldsymbol{u}_h \}\!\} \cdot \boldsymbol{n} | \, [\![r_h]\!] \left[\!\left[\Pi_Q \varphi \right]\!\right] \mathrm{dS}_x \, \mathrm{d}t,$$

$$E_2(r_h) = \frac{1}{4} \int_0^T \int_{\mathcal{E}} [\![\boldsymbol{u}_h]\!] \cdot \boldsymbol{n} \, [\![r_h]\!] \left[\!\left[\Pi_Q \varphi \right]\!\right] \mathrm{dS}_x \, \mathrm{d}t,$$

$$E_3(r_h) = \int_0^T \int_{\mathcal{E}} h^{\varepsilon} \, [\![r_h]\!] \left[\!\left[\Pi_Q \varphi \right]\!\right] \mathrm{dS}_x \, \mathrm{d}t,$$

$$E_4(r_h) = \int_0^T \int_{\mathbb{T}^d} r_h \boldsymbol{u}_h \cdot \Big(\nabla_x \varphi - \nabla_h \big(\Pi_Q \varphi \big) \Big) \, \mathrm{d}x \, \mathrm{d}t,$$

are the error terms to be estimated. Again, r_h is either ϱ_h or $\varrho_h u_{i,h}$, $i = 1, \ldots, d$.

- **Term** E_1: We can write

$$\begin{aligned}
E_1(r_h) =& \frac{1}{2} \int_0^T \int_{\mathcal{E}} | \{\!\{ \boldsymbol{u}_h \}\!\} \cdot \boldsymbol{n} | \left[\!\left[r_h \right]\!\right] \left[\!\left[\Pi_Q \varphi \right]\!\right] \mathrm{dS}_x \, \mathrm{d}t \\
=& \frac{1}{2} \int_0^T \sum_{i=1}^d \sum_{\sigma \in \mathcal{E}_i} \int_\sigma | \{\!\{ u_{i,h} \}\!\} | \, [\![r_h]\!] \, [\![\Pi_Q \varphi]\!] \, \mathrm{dS}_x \, \mathrm{d}t \\
=& \frac{1}{2} \int_0^T \sum_{i=1}^d \sum_{\sigma \in \mathcal{E}_i} \int_{D_\sigma} h_i \left| \Pi_W^{(i)} u_{i,h} \right| \eth_{\mathcal{D}_i} r_h \eth_{\mathcal{D}_i} \Pi_Q \varphi \, \mathrm{d}x \, \mathrm{d}t \\
=& -\frac{1}{2} \int_0^T \sum_{i=1}^d \sum_{K \in \mathcal{T}_h} \int_K h_i r_h \eth_{\mathcal{T}}^{(i)} \left(\left| \Pi_W^{(i)} u_{i,h} \right| \eth_{\mathcal{D}_i} \Pi_Q \varphi \right) \mathrm{d}x \, \mathrm{d}t \\
=& -\frac{1}{2} \int_0^T \sum_{i=1}^d \sum_{K \in \mathcal{T}_h} \int_K r_K h_i \Big(\Pi_Q \left| \Pi_W^{(i)} u_{i,h} \right| \eth_{\mathcal{T}}^{(i)} (\eth_{\mathcal{D}_i} \Pi_Q \varphi) + \\
& + \left(\eth_{\mathcal{T}}^{(i)} \left| \Pi_W^{(i)} u_{i,h} \right| \right) \Pi_Q (\eth_{\mathcal{D}_i} \Pi_Q \varphi) \Big) \, \mathrm{d}x \, \mathrm{d}t,
\end{aligned}$$

where we have used the integration by parts formula (44), and a product rule

$$r_2 q_2 - r_1 q_1 = \frac{r_1 + r_2}{2} (q_2 - q_1) + \frac{q_1 + q_2}{2} (r_2 - r_1).$$

Further, noticing the mesh is uniform and employing twice the inequality $\left(\frac{a+b}{2}\right)^2 \leq \frac{a^2+b^2}{2}$ yield

$$\left\| \Pi_Q \left| \Pi_W^{(i)} u_{i,h} \right| \right\|_{L^2} \lesssim \|u_{i,h}\|_{L^2} .$$

Similarly, identity $\left(\eth_T^{(i)} \Pi_W^{(i)} u_{i,h}\right)_K = \Pi_Q \left(\eth_{\mathcal{D}_i} u_{i,h}\right)_K$ implies

$$\left\|\eth_T^{(i)} \Pi_W^{(i)} u_{i,h}\right\|_{L^2} \lesssim \left\|\eth_{\mathcal{D}_i} u_{i,h}\right\|_{L^2} .$$

Then applying Hölder's inequality, interpolation error estimates (11.25), the velocity estimates (11.17b), the fact $|\partial_x u_i| \geq \partial_x |u_i|$, and noticing $\Delta_h^{(i)} r \equiv \eth_T^{(i)} \eth_{\mathcal{D}_i} r$, we derive

$$\begin{aligned}
E_1(r_h) = & -\frac{1}{2} \int_0^T \sum_{i=1}^d \sum_{K \in \mathcal{T}_h} \int_K r_K h_i \Big(\Pi_Q \left| \Pi_W^{(i)} u_{i,h} \right| \, \eth_T^{(i)} (\eth_{\mathcal{D}_i} \Pi_Q \varphi) + \\
& + \left(\eth_T^{(i)} \left| \Pi_W^{(i)} u_{i,h} \right| \right) \Pi_Q (\eth_{\mathcal{D}_i} \Pi_Q \varphi) \Big) \, \mathrm{d}x \, \mathrm{d}t, \\
\lesssim & \; h \sum_{i=1}^d \|r_h\|_{L^2 L^2} \Bigg(\left\| \Delta_h^{(i)} \Pi_Q \varphi \right\|_{L^\infty L^\infty} \|u_{i,h}\|_{L^2 L^2} + \\
& + \left\| \eth_{\mathcal{D}_i} \Pi_Q \varphi \right\|_{L^\infty L^\infty} \left\| \eth_{\mathcal{D}_i} u_{i,h} \right\|_{L^2 L^2} \Bigg) \\
\lesssim & \; h \|r_h\|_{L^2 L^2} \left(\left\| \Delta_h \Pi_Q \varphi \right\|_{L^\infty L^\infty} \|\boldsymbol{u}_h\|_{L^2 L^2} + \left\| \nabla_h \Pi_Q \varphi \right\|_{L^\infty L^\infty} \|\nabla_{\mathcal{D}} \boldsymbol{u}_h\|_{L^2 L^2} \right) \\
\lesssim & \; h \|r_h\|_{L^2 L^2} .
\end{aligned}$$

Consequently, employing the negative estimates (11.18a) and (11.18c) for $\gamma \in (1, 2)$ we derive for r_h being ϱ_h or $\varrho_h u_{i,h}$, $i = 1, \ldots, d$ that

$$E_1(r_h) \lesssim h^\beta, \quad \beta = 1 - \frac{3\varepsilon + 3 + d}{6\gamma}.$$

Obviously, $\beta > 0$ provided $\varepsilon < 2\gamma - 1 - d/3$.
If $\gamma \in [2, 3)$ the more restrictive estimates are due to $r_h = \varrho_h u_{i,h}$, cf. (11.18a), (11.18c) and yield

$$E_1(r_h) \lesssim h^\beta, \quad \beta = 1 + d \frac{\gamma - 3}{3\gamma}.$$

Clearly, $\beta > 0$ if $\gamma \in [2, 3)$. Furthermore, for $\gamma \geq 3$ we have due to (11.18a), (11.18c)

$$E_1(r_h) \lesssim h.$$

- **Term** E_2:
Thanks to Hölder's inequality, the velocity estimate (11.17b), interpolation error (11.19) and trace inequality (11.28) we derive

$$E_2(r_h) = \frac{1}{4}\int_0^T \int_{\mathcal{E}} [\![\boldsymbol{u}_h]\!] \cdot \boldsymbol{n} [\![r_h]\!] \left[\!\left[\Pi_Q \varphi \right]\!\right] \mathrm{dS}_x \, \mathrm{d}t$$

$$\lesssim h \left\| \varphi \right\|_{C^1} \left(\int_0^T \int_{\mathcal{E}} [\![\boldsymbol{u}_h]\!]^2 \mathrm{dS}_x \, \mathrm{d}t \right)^{1/2} \left(\int_0^T \int_{\mathcal{E}} [\![r_h]\!]^2 \mathrm{dS}_x \, \mathrm{d}t \right)^{1/2}$$

$$\lesssim h h^{1/2} \left\| \nabla_{\mathcal{D}} \boldsymbol{u}_h \right\|_{L^2 L^2} \left(\int_0^T \int_{\mathcal{E}} \{\!\{ r_h \}\!\}^2 \mathrm{dS}_x \, \mathrm{d}t \right)^{1/2}$$

$$\lesssim h \left\| r_h \right\|_{L^2 L^2},$$

Thus in view of the negative estimates (11.18a) and (11.18c) we derive for $\gamma \in (1, 2)$

$$E_2(r_h) \lesssim h^{\beta}, \quad \beta = 1 - \frac{3\varepsilon + 3 + d}{6\gamma}$$

for $r_h = \varrho_h$ and $\varrho_h u_{i,h}$, $i = 1, \dots, d$. Analogously to the above, we have

$$E_2(r_h) \lesssim h^{\beta}, \quad \beta = 1 + d\frac{\gamma - 3}{3\gamma} \text{ for } \gamma \in [2, 3) \text{ and } \beta = 1 \text{ for } \gamma \geq 3.$$

- **Term** E_3: Analogously to the above, the integration by parts formula (42), Hölder's inequality, and the interpolation error estimate (11.25) yield

$$E_3(r_h) = h^{\varepsilon} \int_0^T \int_{\mathcal{E}} [\![r_h]\!] \left[\!\left[\Pi_Q \varphi \right]\!\right] \mathrm{dS}_x \, \mathrm{d}t = h^{\varepsilon+1} \int_0^T \int_{\mathbb{T}^d} \nabla_{\mathcal{D}} r_h \cdot \nabla_{\mathcal{D}} \Pi_Q \varphi \, \mathrm{d}x \, \mathrm{d}t$$

$$= -h^{\varepsilon+1} \int_0^T \int_{\mathbb{T}^d} r_h \Delta_h \Pi_Q \varphi \, \mathrm{d}x \, \mathrm{d}t \leq h^{\varepsilon+1} c(\left\| \varphi \right\|_{C^2}) \left\| r_h \right\|_{L^1 L^1} \lesssim h^{\varepsilon+1} \left\| r_h \right\|_{L^1 L^1}.$$

Using the estimates (11.17a) we can conclude for r_h being ϱ_h or $\varrho_h u_{i,h}$, $i = 1, \dots, d$, that

$$E_3(r_h) \lesssim h^{\varepsilon+1}.$$

- **Term** E_4: Using the uniform bounds on the kinetic energy and momentum (see (11.17a)) together with the interpolation error (11.25) we obtain for r_h being ϱ_h

or $\varrho_h u_{i,h}$, $i = 1, \ldots, d$ that

$$E_4(r_h) = \int_0^T \int_{\mathbb{T}^d} r_h \boldsymbol{u}_h \cdot \big(\nabla_x \varphi - \nabla_h (\Pi_Q \varphi)\big) \, \mathrm{d}x \, \mathrm{d}t \lesssim h \, \|\varphi\|_{C^2} \, \|r_h \boldsymbol{u}_h\|_{L^1 L^1}$$
$$\lesssim h \, \|r_h \boldsymbol{u}_h\|_{L^\infty L^1} \lesssim h.$$

Consequently, we conclude the consistency formulation of the convective terms by collecting the above estimates of the four terms E_j, $j = 1, \ldots, 4$.

$$\int_0^T \int_{\mathbb{T}^d} \varrho_h \boldsymbol{u}_h \cdot \nabla_x \varphi \, \mathrm{d}x \, \mathrm{d}t - \int_0^T \int_{\mathcal{E}} F_h^{\mathrm{up}}[\varrho_h, \boldsymbol{u}_h] \, [\![\Pi_Q \varphi]\!] \, \mathrm{dS}_x \, \mathrm{d}t \lesssim h^{\beta_1},$$
$$\int_0^T \int_{\mathbb{T}^d} \varrho_h \boldsymbol{u}_h \otimes \boldsymbol{u}_h : \nabla_x \boldsymbol{\varphi} \, \mathrm{d}x \, \mathrm{d}t - \int_0^T \int_{\mathcal{E}} \boldsymbol{F}_h^{\mathrm{up}}[\varrho_h \boldsymbol{u}_h, \boldsymbol{u}_h] \, [\![\Pi_Q \boldsymbol{\varphi}]\!] \, \mathrm{dS}_x \, \mathrm{d}t \lesssim h^{\beta_2} \tag{11.24}$$

for some $\beta_1, \beta_2 > 0$. More precisely, in the case $\gamma \in (1, 2)$ we require $-1 < \varepsilon < 2\gamma - 1 - d/3$ to obtain $\beta_1, \beta_2 > 0$. If $\gamma \geq 2$ the above estimates hold for $\varepsilon > -1$.

(3) Viscosity terms:
In accordance with (11.25) and (11.17b) we can control the viscosity terms. Indeed, we have

$$\int_0^T \int_{\mathbb{T}^d} \nabla_{\mathcal{D}} \boldsymbol{u}_h : \nabla_x \boldsymbol{\varphi} \, \mathrm{d}x \, \mathrm{d}t - \int_0^T \int_{\mathbb{T}^d} \nabla_{\mathcal{D}} \boldsymbol{u}_h : \nabla_{\mathcal{D}} (\Pi_Q \boldsymbol{\varphi}) \, \mathrm{d}x \, \mathrm{d}t$$
$$= \int_0^T \int_{\mathbb{T}^d} \nabla_{\mathcal{D}} \boldsymbol{u}_h : (\nabla_x \boldsymbol{\varphi} - \nabla_{\mathcal{D}} \Pi_Q \boldsymbol{\varphi}) \, \mathrm{d}x \, \mathrm{d}t \lesssim \|\nabla_{\mathcal{D}} \boldsymbol{u}_h\|_{L^2 L^2} \, h \, \|\boldsymbol{\varphi}\|_{C^2} \lesssim h, \tag{11.25a}$$

and for the divergence term we get

$$\int_0^T \int_{\mathbb{T}^d} \mathrm{div}_h \boldsymbol{u}_h \, \mathrm{div}_h \big(\Pi_Q \boldsymbol{\varphi}\big) \, \mathrm{d}x \, \mathrm{d}t - \int_0^T \int_{\mathbb{T}^d} \mathrm{div}_h \boldsymbol{u}_h \, \mathrm{div}_x \boldsymbol{\varphi} \, \mathrm{d}x \, \mathrm{d}t$$
$$= \int_0^T \int_{\mathbb{T}^d} \mathrm{div}_h \boldsymbol{u}_h \Big(\mathrm{div}_h \big(\Pi_Q \boldsymbol{\varphi}\big) - \mathrm{div}_x \boldsymbol{\varphi} \Big) \, \mathrm{d}x \, \mathrm{d}t \lesssim \|\mathrm{div}_h \boldsymbol{u}_h\|_{L^2 L^2} \, h \, \|\boldsymbol{\varphi}\|_{C^2} \tag{11.25b}$$
$$\lesssim h,$$

by using (11.25) and (11.17b).

(4) Pressure term: The pressure term can be controlled by using the integration by parts formula (39), the interpolation error (11.25), and the estimate (11.17a) which yield

$$
\begin{aligned}
&\int_0^T \int_{\mathbb{T}^d} p_h \mathrm{div}_h(\Pi_Q \boldsymbol{\varphi}) \, \mathrm{d}x \, \mathrm{d}t - \int_0^T \int_{\mathbb{T}^d} p_h \mathrm{div}_x \boldsymbol{\varphi} \, \mathrm{d}x \, \mathrm{d}t \\
&= \int_0^T \int_{\mathbb{T}^d} p_h \left(\mathrm{div}_h \left(\Pi_Q \boldsymbol{\varphi} \right) - \mathrm{div}_x \boldsymbol{\varphi} \right) \, \mathrm{d}x \, \mathrm{d}t \lesssim \| p_h \|_{L^\infty L^1} \, h \, \| \boldsymbol{\varphi} \|_{C^2} \lesssim h.
\end{aligned}
\tag{11.26}
$$

Collecting the inequalities (11.23)–(11.26) we complete the proof of Theorem 11.2.

Remark 11.3 We would like to emphasize that the additional numerical diffusion of order $\mathcal{O}(h^{\varepsilon+1})$ plays a crucial role in order to obtain the consistency formulation of the convective terms. Indeed in the derivation of (11.24) we have used Lemma 11.2, which benefits from the uniform bounds (11.17d) of the h^ε-terms.

11.4 Convergence

Theorem 11.3 (Convergence of the FV method)

Let the pressure p satisfy (11.2) *with* $\gamma > 1$ *and* $(\varrho_h, \boldsymbol{u}_h)$ *be a solution of the discrete problem (11.5) on the time interval* $[0, T]$ *in the sense of (47). Let*

$$\Delta t \approx h \in (0, 1) \quad \text{and} \quad \varepsilon > -1.$$

If $\gamma \in (1, 2)$ *we moreover assume that* $\varepsilon < 2\gamma - 1 - d/3$.

Then the FV method (11.5) converges in the following way:

1. **Convergence to a DMV solution**. *There is a sequence* $h_n \to 0$ *such that* $\{\varrho_{h_n}^k, \boldsymbol{u}_{h_n}^k\}_{h \searrow 0}$ *generates a Young measure* $\{\mathcal{V}_{t,x}\}_{(t,x) \in (0,T) \times \mathbb{T}^d}$ – *a dissipative measure-valued solution of the Navier–Stokes system in the sense of Definition 5.10.*
2. **Convergence to a strong solution**. *Let the Navier–Stokes system endowed with the initial data* $[\varrho_0, \boldsymbol{u}_0]$ *admit a strong solution* $[\varrho, \boldsymbol{u}]$ *belonging to the class*

$$
\begin{aligned}
&\varrho > 0, \ \varrho \in C([0, T] \times \mathbb{T}^d), \ \nabla_x \varrho, \boldsymbol{u} \in C([0, T] \times \mathbb{T}^d; R^d), \\
&\nabla_x \boldsymbol{u} \in C([0, T] \times \mathbb{T}^d; R^{d \times d}), \ \partial_t \boldsymbol{u} \in L^2(0, T; C(\mathbb{T}^d; R^d)).
\end{aligned}
$$

Then

$$
\begin{aligned}
&\varrho_h \to \varrho \ \textit{(strongly) in} \ L^\gamma((0, T) \times \mathbb{T}^d), \\
&\boldsymbol{u}_h \to \boldsymbol{u} \ \textit{(strongly) in} \ L^2((0, T) \times \mathbb{T}^d; R^d).
\end{aligned}
$$

3. **Unconditional convergence with bounded density**. *Suppose that*

- *the initial data belong to the class* $\varrho_0 \in W^{3,2}(\mathbb{T}^d)$, $\varrho_0 > 0$, $\boldsymbol{u}_0 \in W^{3,2}(\mathbb{T}^d; R^d)$;
- *bulk viscosity vanishes, meaning* $\lambda = 0$;
- *uniform bound of density* $\varrho_{h_n} \lesssim 1$ *for some* $h_n \to 0$.

Then

$$\varrho_h \to \varrho \text{ (strongly) in } L^q((0,T)\times\mathbb{T}^d), \text{ for any } 1 \le q < \infty,$$
$$\boldsymbol{u}_h \to \boldsymbol{u} \text{ (strongly) in } L^2((0,T)\times\mathbb{T}^d; R^d),$$

where $[\varrho, \boldsymbol{u}]$ *is the strong solution to the Navier–Stokes system (11.1) with the initial data* $[\varrho_0, \boldsymbol{u}_0]$.

Proof We have proven in Theorem 11.1 and Theorem 11.2, respectively, the stability and consistency of the numerical solutions obtained by the FV method (11.5). Moreover, due to (11.25) the consistency of the discrete gradient and divergence operators acting on $\boldsymbol{Q}_h$, cf. (5.107), holds. Furthermore, we have due to the Sobolev–Poincaré inequality (11.15) and a priori estimates presented in Corollary 11.1 that

$$\varrho_h \to \varrho \text{ weakly-(*) in } L^\infty(0,T; L^\gamma(\mathbb{T}^d)),$$
$$\boldsymbol{m}_h = (\varrho_h \boldsymbol{u}_h) \to \boldsymbol{m} \text{ weakly-(*) in } L^\infty(0,T; L^{\frac{2\gamma}{\gamma+1}}(\mathbb{T}^d; R^d)),$$

and

$$\boldsymbol{u}_h \to \boldsymbol{u} \text{ weakly in } L^2((0,T)\times\mathbb{T}^d; R^d),$$
$$\nabla_{\mathcal{D}} \boldsymbol{u}_h \to \nabla_x \boldsymbol{u} \text{ weakly in } L^2((0,T)\times\mathbb{T}^d; R^{d\times d}),$$
$$\mathrm{div}_h \boldsymbol{u}_h \to \mathrm{div}_x \boldsymbol{u} \text{ weakly in } L^2((0,T)\times\mathbb{T}^d)$$

passing to a suitable subsequence as the case may be. Moreover,

$$\boldsymbol{u} \in L^2(0,T; W^{1,2}(\mathbb{T}^d; R^d)).$$

This already tells us that any solution to (11.5) is a consistent approximation of the Navier–Stokes system (11.1), cf. Definition 5.9. Thanks to Theorem 5.5 the limit of the consistent approximation for $h \to 0$ generates a dissipative measure-valued (DMV) solution of the Navier–Stokes system (11.1). Note that in Definition 5.10 of the DMV solution we also require that the Korn–Poincaré inequality (5.126) holds. Due to the consistency of the discrete differential operators div_h and $\nabla_{\mathcal{D}}$ and the discrete Sobolev–Poincaré inequality (11.15) we obtain that, cf. (5.121),

$$
\begin{aligned}
\int_0^\tau \int_{\mathbb{T}^d} \left\langle \mathcal{V}; |\widetilde{\boldsymbol{u}} - \boldsymbol{U}|^2 \right\rangle \, \mathrm{d}x \, \mathrm{d}t &\leq \liminf_{h \to 0} \int_0^\tau \int_{\mathbb{T}^d} |\boldsymbol{u}_h - \boldsymbol{U}|^2 \, \mathrm{d}x \, \mathrm{d}t \\
&\lesssim \int_0^\tau \int_{\mathbb{T}^d} \left\langle \mathcal{V}; \widetilde{\varrho} |\widetilde{\boldsymbol{u}} - \boldsymbol{U}|^2 \right\rangle \, \mathrm{d}x + \int_0^\tau \int_{\mathbb{T}^d} |\nabla_x (\boldsymbol{u} - \boldsymbol{U})|^2 \, \mathrm{d}x \, \mathrm{d}t \qquad (11.27) \\
&+ \int_0^\tau \left(\int_{\mathbb{T}^d} \mathrm{d}\mathfrak{E}(t) \right) \mathrm{d}t + \int_0^\tau \int_{\mathbb{T}^d} \mathrm{d}\mathfrak{D}.
\end{aligned}
$$

Consequently, (5.126) from the definition of DMV solution holds. This concludes the proof of the first item. The second and third convergence result follow from Theorem 7.12 and Theorem 7.13, respectively.

Remark 11.4 When assuming the upper bound for density, all results of Theorems 11.3 may be extended to an unstructured grid. Indeed the only difference of the proof would be showing the consistency of the convective terms in (11.24). The estimate of the error terms $E_1(\varrho_h)$ and $E_1(\varrho_h \boldsymbol{u}_h)$ could be done without the discrete integration by parts thanks to L^∞-bound on the density.

11.5 Numerical Experiment

In this section, we study a two-dimensional example to illustrate behavior of the FV method (11.5) and confirm theoretical convergence result proved in Theorem 11.3 via numerical experiments. In particular, we consider the following error norms

$$
\begin{aligned}
e_\varrho &= \|\varrho_h - \varrho_{\mathrm{ref}}\|_{L^\infty(0,T;L^\gamma(\mathbb{T}^d))}, & e_{\boldsymbol{u}} &= \|\boldsymbol{u}_h - \boldsymbol{u}_{\mathrm{ref}}\|_{L^2((0,T)\times\mathbb{T}^d)}, \\
e_{\nabla \boldsymbol{u}} &= \|\nabla_{\mathcal{D}}(\boldsymbol{u}_h - \boldsymbol{u}_{\mathrm{ref}})\|_{L^2((0,T)\times\mathbb{T}^d)}, & e_p &= \|p(\varrho_h) - p(\varrho_{\mathrm{ref}})\|_{L^\infty(0,T;L^1(\mathbb{T}^d))}
\end{aligned}
\qquad (11.28)
$$

between a numerical solution $(\varrho_h, \boldsymbol{u}_h)$ and a reference solution $(\varrho_{\mathrm{ref}}, \boldsymbol{u}_{\mathrm{ref}})$ computed on a fine grid with $h = 1/1024$.

The computational domain Ω is set to $\Omega = [0, 1]^2$, we apply periodic boundary conditions and the following initial conditions

$$
\varrho_0(x_1, x_2, t) = 2 + \cos(2\pi(x_1 + x_2)), \quad \boldsymbol{u}_0(x_1, x_2, t) = 0. \qquad (11.29)
$$

Further, we use the parameters $\mu = 0.1$, $\nu = 0$, $\gamma = 1.4$ and $\varepsilon = 0.6$ that satisfy the restriction stated in Theorem 11.3. Note that the FV method (11.5) yields a nonlinear algebraic system that we solve by fixed point iterations. For each sub-iteration a linear system is solved with the time step $\Delta t = 0.4h/(|\boldsymbol{u}_h| + c_h)_{\max}$, where $c_h = \sqrt{\gamma p_h/\varrho_h}$ is the sound speed.

Table 11.1 documents the convergence behavior of the FV method (11.5). Note that the experimental order of convergence (EOC) was defined in (10.76). As expected, we can observe the first order convergence rate for the density and slightly better convergence rate for the velocity. Recall that we use piecewise constant approximation in space (on a primary mesh) and the first order implicit approximation in time. Moreover, the discrete velocity gradients are approximated by means of a dual mesh, which actually yields piecewise linear representation of the velocity on a dual mesh.

In Fig. 11.1 we present time evolution of the kinetic energy $\int_{\mathbb{T}^d} \frac{1}{2}\varrho|\boldsymbol{u}|^2 \, \mathrm{d}x$, internal energy $\int_{\mathbb{T}^d} P(\varrho) \, \mathrm{d}x$ and the total energy. Figure 11.2 illustrates fluid flow evolution that is driven by the pressure gradient. We can see that the fluid velocity increases at the beginning and then decreases afterwards.

11.6 Conclusion, Bibliographical Remarks

In the literature we can find a variety of numerical schemes for viscous compressible flows, such as the Marker–and–Cell scheme Gallouët et al. [110, 113, 115], Hošek and She [129], the finite element schemes Ansanay–Alex et al. [5], Karper [140], Zienkiewicz et al. [201], the finite volume schemes Feistauer [99], Eymard et

Table 11.1 EOC of FV method for Navier–Stokes at $T = 0.1$

h	e_ϱ	EOC	$e_{\boldsymbol{u}}$	EOC	$e_{\nabla\boldsymbol{u}}$	EOC	e_p	EOC
				$\gamma = 1.4$				
1/32	5.51e-01	–	2.77e-02	–	2.62e-01	–	4.34e-01	–
1/64	3.30e−01	0.74	1.37e−02	1.02	1.30e−01	1.01	2.12e−01	1.03
1/128	1.92e−01	0.78	6.41e−03	1.09	6.11e−02	1.09	9.92e−02	1.10
1/256	1.05e−01	0.87	2.75e−03	1.22	2.62e−02	1.22	4.25e−02	1.22
1/512	4.78e−02	1.13	9.17e−04	1.58	8.74e−03	1.58	1.42e−02	1.58
				$\gamma = 5/3$				
1/32	7.51e−01	–	3.70e−02	–	3.55e−01	–	6.20e−01	–
1/64	4.88e−01	0.62	1.83e−02	1.02	1.77e−01	1.01	3.03e−01	1.03
1/128	3.10e−01	0.66	8.58e−03	1.09	8.32e−02	1.09	1.42e−01	1.10
1/256	1.86e−01	0.73	3.68e−03	1.22	3.57e−02	1.22	6.07e−02	1.22
1/512	9.63e−02	0.95	1.23e−03	1.58	1.19e−02	1.58	2.02e−02	1.58
				$\gamma = 2$				
1/32	9.69e−01	–	5.03e−02	–	5.01e−01	–	9.39e−01	–
1/64	6.77e−01	0.52	2.49e−02	1.01	2.52e−01	0.99	4.58e−01	1.03
1/128	4.63e−01	0.55	1.17e−02	1.09	1.19e−01	1.08	2.14e−01	1.10
1/256	3.03e−01	0.61	5.03e-03	1.22	5.13e-02	1.22	9.20e-02	1.22
1/512	1.75e-01	0.79	1.68e-03	1.58	1.71e-02	1.58	3.07e-02	1.58

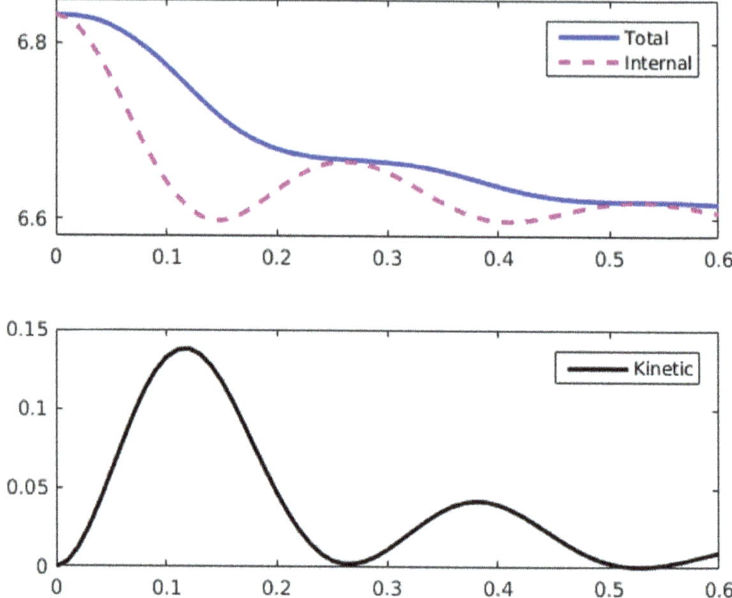

Fig. 11.1 FV method for the Navier–Stokes equations: time evolution of the energy

al. [78], Haack et al. [127], Meister and Sonar [163] and the discontinuous Galerkin schemes Dolejší and Feistauer [71], Gassner et al. [116], Ioriatti and Dumbser [130].

Although these methods are frequently used in practical simulations, their convergence for multidimensional viscous compressible flows remains open in general. For a mixed finite element–discontinuous Galerkin method, the convergence to a weak solution has been shown by Karper in his pioneering work [140] under the assumption that the adiabatic coefficient $\gamma > 3$. Note that the convergence in this case holds up to a subsequence as the weak solutions are not known to be unique. Moreover, any generalization of the proof of Karper [140] for other numerical schemes, in particular for cell-centered finite volume methods discussed in this chapter, is highly nontrivial. In [133] Jovanović obtained the error estimate for the Navier–Stokes equations for entropy dissipative finite volume – finite difference methods under some rather restrictive assumptions on the global smooth solution.

We should also mention the recent results on the analysis of the Marker–and–Cell schemes, cf. [113, 115, 129], which are based on the staggered grid approximation of the velocity and the primary grid approximation of the density. In [113] the convergence to a weak solution of stationary Navier–Stokes equations for $\gamma > 3$ has been proved. In [129] the consistency and the energy stability of the Marker–and–Cell scheme has been shown for instationary Navier–Stokes equations. The error estimates for $\gamma > 3/2$ have been presented in [115] using the relative entropy method.

The finite volume method studied in this chapter was proposed in [94]. Due to the use of the dissipative upwinding we get an additional artificial diffusion which

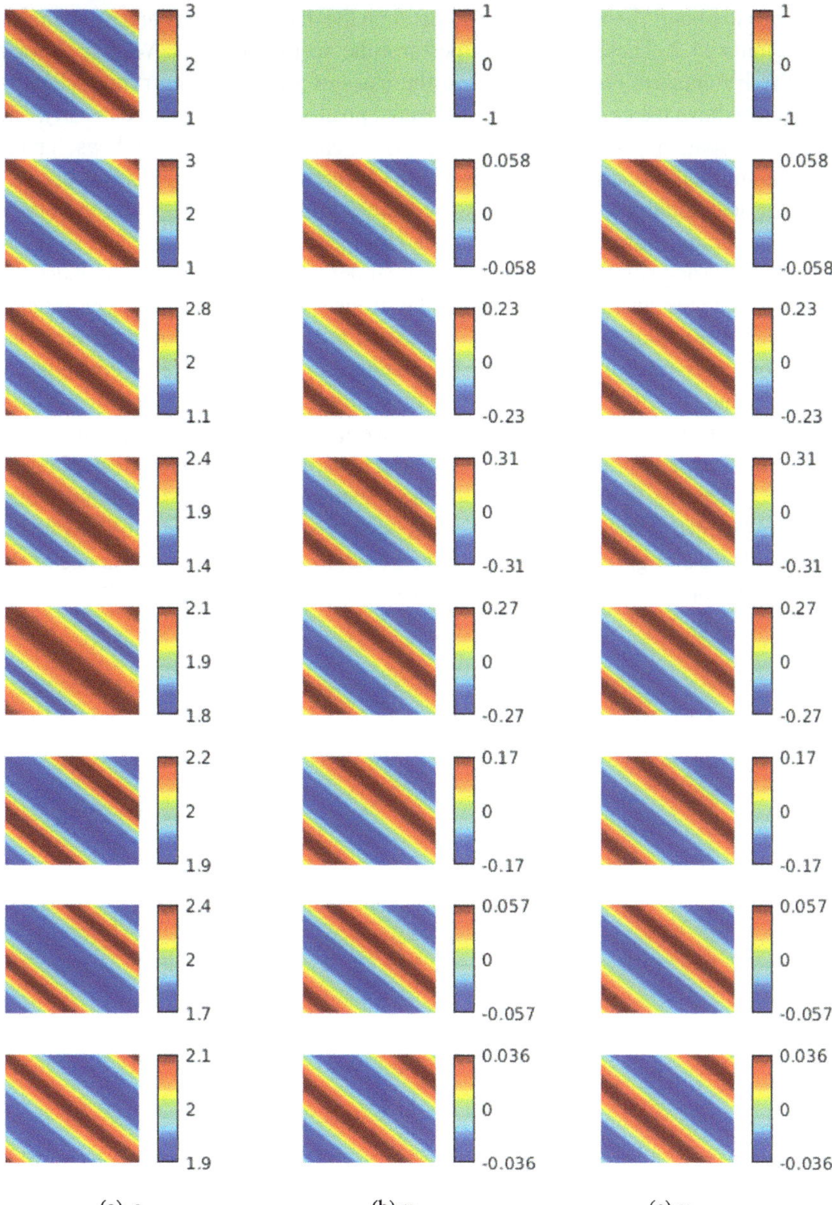

(a) ϱ (b) u_1 (c) u_2

Fig. 11.2 FV method for the Navier–Stokes equations: time evolution of the flow. From top to down are $t = 0, 0.01, 0.05, 0.1, 0.15, 0.2, 0.25, 0.5$. From left to right are ϱ, u_1, and u_2

allows us to obtain “better” a priori estimates on the discrete density and momentum, see Lemma 11.2. These allow us to perform the desired analysis. We have shown that the FV method (11.5) belongs to the class of structure preserving schemes. Indeed, we have the conservation of the mass, positivity of the discrete density, and energy dissipation. As illustrated in the previous chapters for inviscid flows, having obtained the stability and consistency of the FV method (11.5) implies convergence of numerical solutions. In general, we get weak convergence of the finite volume solutions (up to a subsequence) towards a DMV solution, cf. Theorem 11.3. In the case that a strong solution exists or numerical densities are uniformly bounded we have obtained the strong convergence to the strong solution. The results presented in this chapter are valid for any adiabatic constant $\gamma > 1$, which is more general than other convergence results available in the literature so far. For $\gamma \geq 2$ the convergence holds for any $\varepsilon > -1$ and thus only a very little amount of numerical diffusion is needed. On the other case, if $\gamma \in (1, 2)$ we have a condition on the numerical diffusion, which implies that we need enough numerical diffusion depending on γ and d to prove the convergence of the FV method (11.4). More importantly, the convergence analysis via the measure-valued solutions covers the physically reasonable case of gases $\gamma \in \left(1, \frac{5}{3}\right]$. For a generalization of the convergence of the FV method for the Navier–Stokes–Fourier system we refer a reader to [95].

Chapter 12
Finite Volume Method for the Barotropic Euler System – Revisited

On the one hand, we have seen in Chap. 10 that under the condition that the numerical solution remains in the gas nondegenerate region $0 < \underline{\varrho} \leq \varrho_h$ and $\vartheta_h \leq \overline{\vartheta}$, the finite volume method (10.10) is consistent, stable and consequently convergent. On the other hand, as shown in Chap. 11, the unconditional convergence of the finite volume method (11.4) for the Navier–Stokes system can be obtained. Being inspired by these results and realizing that the finite volume method (10.10) can be seen as a vanishing viscosity approximation for the corresponding inviscid system, we propose another approximation scheme for the barotropic Euler system that is based on the Brenner-type regularization (10.4), (10.5). In this approach, the numerical viscosity is enhanced by "natural" viscosity of Newtonian type, where both vanish in the asymptotic regime. As a result, the scheme is *unconditionally* convergent, in particular, the density remains strictly positive at any level of approximation without imposing any extra condition of CFL type.

Before formulating a new finite volume method we recall for convenience the continuous system, the Euler equations for barotropic fluids with the pressure-density EOS $p(\varrho) = a\varrho^\gamma$, $\gamma > 1$, cf. (2.71):

$$\begin{aligned} &\partial_t \varrho + \mathrm{div}_x \boldsymbol{m} = 0, \\ &\partial_t \boldsymbol{m} + \mathrm{div}_x \left(\frac{\boldsymbol{m} \otimes \boldsymbol{m}}{\varrho} \right) + \nabla_x p = 0. \end{aligned} \tag{12.1}$$

Recall that for the barotropic Euler equation, the total energy plays the role of a mathematical entropy η, cf. (2.73):

$$\eta \equiv E = \frac{1}{2} \frac{|\boldsymbol{m}|^2}{\varrho} + P(\varrho), \quad P(\varrho) = \frac{a}{\gamma - 1} \varrho^\gamma,$$

E. Feireisl et al., *Numerical Analysis of Compressible Fluid Flows*, MS&A 20,
https://doi.org/10.1007/978-3-030-73788-7_12

with the associated entropy flux $\boldsymbol{g} = \left(\frac{1}{2} \frac{|\boldsymbol{m}^2|}{\varrho} + P(\varrho) + p(\varrho) \right) \frac{\boldsymbol{m}}{\varrho}$. We accompany the Euler equations with the periodic boundary conditions ($\Omega = \mathbb{T}^d$), and the initial conditions $[\varrho_0, \boldsymbol{m}_0]$ belonging to the class

$$\varrho_0 > 0, \ \ \varrho_0 \in L^\infty(\mathbb{T}^d), \ \ \boldsymbol{m}_0 \in L^\infty(\mathbb{T}^d; R^d).$$

Since the proposed scheme can be seen as a hybrid mixing up the Navier–Stokes system in the zero viscosity regime with the inviscid Euler system, it is convenient to work with the fluid velocity $\boldsymbol{u} = \frac{\boldsymbol{m}}{\varrho}$ rather than the momentum $\boldsymbol{m}$. Accordingly, the initial data $[\varrho_0, \boldsymbol{u}_0]$ should satisfy

$$\varrho_0 > 0, \ \ \varrho_0 \in L^\infty(\mathbb{T}^d), \ \ \boldsymbol{u}_0 \in L^\infty(\mathbb{T}^d; R^d).$$

12.1 Numerical Method

We propose a fully discrete numerical method based on the implicit time discretization and diffusive upwind numerical flux (8.8) for the approximation of the system (12.1). In addition, analogously as in the finite volume method for the (complete) Euler system, we include a diffusion term of order h^α, $\alpha > 0$ in the momentum equation. Thus, the resulting viscous finite volume (VFV) method shares some similarities with the finite volume methods (11.4) and (10.10) proposed for the Navier–Stokes and (complete) Euler system, respectively.

Let the initial values $(\varrho_h^0, \boldsymbol{m}_h^0) = (\Pi_Q \varrho_0, \Pi_Q \boldsymbol{m}_0) \in Q_h \times \boldsymbol{Q}_h$ be given, $\boldsymbol{u}_h^0 = \frac{\boldsymbol{m}_h^0}{\varrho_h^0}$. Our aim is to find a piecewise constant approximation $(\varrho_h^k, \boldsymbol{u}_h^k) \in Q_h \times \boldsymbol{Q}_h$ satisfying for $k = 1, \dots, N_T$ the following equations:

$$D_t \varrho_K^k + \sum_{\sigma \in \mathcal{E}(K)} \frac{|\sigma|}{|K|} F_h^{\rm up}[\varrho_h^k, \boldsymbol{u}_h^k] = 0, \tag{12.2}$$

$$D_t (\varrho_h^k \boldsymbol{u}_h^k)_K + \sum_{\sigma \in \mathcal{E}(K)} \frac{|\sigma|}{|K|} \left(\boldsymbol{F}_h^{\rm up}[\varrho_h^k \boldsymbol{u}_h^k, \boldsymbol{u}_h^k] + \{\{ p_h^k \}\} \boldsymbol{n} \right) = h^{\alpha-1} \sum_{\sigma \in \mathcal{E}(K)} \frac{|\sigma|}{|K|} [\![\boldsymbol{u}_h^k]\!],$$

for any $K \in \mathcal{T}_h$, where $\mathcal{T}_h$ is a structured mesh approximation of $\mathbb{T}^d$ in the sense of Definition 3. Recall that $F_h^{\rm up}[r_h, \boldsymbol{u}_h]$ is the diffusive upwind flux given in (8.8). The term on the right hand side of the momentum equation represents the diffusion term $h^\alpha \Delta_h \boldsymbol{u}_h^k$ with $\alpha > 0$. Analogously as in Chap. 11 we have a fully discrete implicit scheme, where the value of the solution at the time level $k-1$ is incorporated through the discrete time derivative

$$D_t \varrho_K^k = \frac{\varrho_K^k - \varrho_K^{k-1}}{\Delta t}.$$

For numerical analysis it will be convenient to use the integral formulation that is analogous to (11.5)

$$\int_{\mathbb{T}^d} D_t \varrho_h^k \phi_h \, \mathrm{d}x - \sum_{\sigma \in \mathcal{E}} \int_\sigma F_h^{\mathrm{up}}[\varrho_h^k, \boldsymbol{u}_h^k] [\![\phi_h]\!] \, \mathrm{dS}_x = 0 \quad \text{for all } \phi_h \in Q_h, \tag{12.3a}$$

$$\int_{\mathbb{T}^d} D_t(\varrho_h^k \boldsymbol{u}_h^k) \cdot \boldsymbol{\phi}_h \, \mathrm{d}x - \sum_{\sigma \in \mathcal{E}} \int_\sigma \boldsymbol{F}_h^{\mathrm{up}}[\varrho_h^k \boldsymbol{u}_h^k, \boldsymbol{u}_h^k] \cdot [\![\boldsymbol{\phi}_h]\!] \, \mathrm{dS}_x - \int_{\mathbb{T}^d} p_h^k \mathrm{div}_h \boldsymbol{\phi}_h \, \mathrm{d}x$$

$$= -h^\alpha \int_{\mathbb{T}^d} \nabla_{\mathcal{D}} \boldsymbol{u}_h^k : \nabla_{\mathcal{D}} \boldsymbol{\phi}_h \, \mathrm{d}x \quad \text{for all } \boldsymbol{\phi}_h \in \boldsymbol{Q}_h. \tag{12.3b}$$

We recall that the discrete gradient $\nabla_{\mathcal{D}}$ is based on jumps over faces σ, cf. (15).

Direct comparison of our new finite volume method (12.3) and the finite volume method (11.5) for the Navier–Stokes equations shows that (12.3) can be viewed as the vanishing viscosity approximation of the barotropic Euler system. Indeed, setting $\mu = h^\alpha$ and $\nu = 0$ in (11.5) we obtain (12.3). Consequently, most of the results obtained in Chap. 11 can be directly applied to our new finite volume method (12.3) as we will see in the next sections.

12.2 Stability

In what follows we summarize the properties of discrete solutions $\{\varrho_h^k, \boldsymbol{u}_h^k\}_{k=1}^{N_T}$ and discuss the stability of (12.3).

Lemma 12.1 *For any $k = 1, \ldots, N_T$ there exists a solution $(\varrho_h^k, \boldsymbol{u}_h^k) \in Q_h \times \boldsymbol{Q}_h$ to the finite volume method (12.3). Moreover, its piecewise constant interpolant in time $(\varrho_h(t), \boldsymbol{u}_h(t))$, $t \in [0, T]$, enjoys the following properties:*

- **Discrete conservation of mass.**

$$\int_{\mathbb{T}^d} \varrho_h(t) \, \mathrm{d}x = \int_{\mathbb{T}^d} \varrho_0 \, \mathrm{d}x, \ t \in (0, T]. \tag{12.4}$$

- **Positivity of discrete density.**
  Let $\varrho_0 > 0$. Then it holds for all $t \in (0, T]$ that $\varrho_h(t) > 0$.

- **Discrete energy balance.**

$$
\begin{aligned}
& D_t \int_{\mathbb{T}^d} \left(\frac{1}{2} \varrho_h^k |\boldsymbol{u}_h^k|^2 + P(\varrho_h^k) \right) \mathrm{d}x + h^\alpha \left\| \nabla_{\mathcal{D}} \boldsymbol{u}_h^k \right\|_{L^2}^2 \\
& = -h^\varepsilon \sum_{\sigma \in \mathcal{E}} \int_\sigma \{\{\varrho_h^k\}\} \left[\!\left[\boldsymbol{u}_h^k \right]\!\right]^2 \mathrm{dS}_x - \frac{\Delta t}{2} \int_{\mathbb{T}^d} P''(\xi) |D_t \varrho_h^k|^2 \, \mathrm{d}x \\
& - \sum_{\sigma \in \mathcal{E}} \int_\sigma P''(\varrho_{h,\dagger}^k) \left[\!\left[\varrho_h^k \right]\!\right]^2 \left(h^\varepsilon + \frac{1}{2} \left| \{\{\boldsymbol{u}_h^k\}\} \cdot \boldsymbol{n} \right| \right) \mathrm{dS}_x \\
& - \frac{\Delta t}{2} \int_{\mathbb{T}^d} \varrho_h^{k-1} |D_t \boldsymbol{u}_h^k|^2 \, \mathrm{d}x - \frac{1}{2} \sum_{\sigma \in \mathcal{E}} \int_\sigma (\varrho_h^k)^{\mathrm{up}} \left| \{\{\boldsymbol{u}_h^k\}\} \cdot \boldsymbol{n} \right| \left[\!\left[\boldsymbol{u}_h^k \right]\!\right]^2 \mathrm{dS}_x,
\end{aligned}
\tag{12.5}
$$

where $\xi \in \mathrm{co}\{\varrho_h^{k-1}, \varrho_h^k\}$ *and* $\varrho_{h,\dagger}^k \in \mathrm{co}\{(\varrho_h^k)^{\mathrm{in}}, (\varrho_h^k)^{\mathrm{out}}\}$ *for any* $\sigma \in \mathcal{E}_{int}$.

Proof of the existence of a numerical solution follows from Lemma 11.2. The proof of the three properties, conservation of mass, positivity of density and energy balance can be done analogously as in (8.16), Lemma 8.3 and Theorem 11.1, respectively.

It will be helpful to summarize the uniform estimates that follow from the discrete energy inequality for the numerical solution $(\varrho_h, \boldsymbol{u}_h)$ analogously as in Corollary 11.1:

$$
\begin{aligned}
\left\| \varrho_h \, |\boldsymbol{u}_h|^2 \right\|_{L^\infty L^1} \lesssim 1, \quad \|\varrho_h\|_{L^\infty L^\gamma} \lesssim 1, \quad \|\varrho_h \boldsymbol{u}_h\|_{L^\infty L^{\frac{2\gamma}{\gamma+1}}} \lesssim 1, \\
h^{\alpha/2} \left\| \nabla_{\mathcal{D}} \boldsymbol{u}_h \right\|_{L^2 L^2} \lesssim 1, \quad h^{\alpha/2} \|\boldsymbol{u}_h\|_{L^2 L^p} \lesssim 1, \\
h^\varepsilon \int_0^T \sum_{\sigma \in \mathcal{E}} \int_\sigma \{\{\varrho_h\}\} \left[\!\left[\boldsymbol{u}_h \right]\!\right]^2 \mathrm{dS}_x \, \mathrm{d}t \lesssim 1, \\
\int_0^T \sum_{\sigma \in \mathcal{E}} \int_\sigma P''(\varrho_{h,\dagger}) \left[\!\left[\varrho_h \right]\!\right]^2 (h^\varepsilon + | \{\{\boldsymbol{u}_h\}\} \cdot \boldsymbol{n}|) \, \mathrm{dS}_x \, \mathrm{d}t \lesssim 1,
\end{aligned}
$$

where $\varrho_{h,\dagger} \in \mathrm{co}\{\varrho_h^{\mathrm{in}}, \varrho_h^{\mathrm{out}}\}$ for any $\sigma \in \mathcal{E}_{int}$, $p < \infty$ for $d = 2$ and $p \leq 6$ for $d = 3$.

Moreover, we have negative estimates for the density and momentum that can be proved analogously as in Lemma 11.4:

Lemma 12.2 (Negative estimates)
Let $(\varrho_h, \boldsymbol{u}_h)$ *be a solution of the FV method (12.3). Then*

$$
\|\varrho_h\|_{L^2 L^2} \lesssim h^{\beta_D}, \quad \beta_D = \begin{cases} -\frac{3\varepsilon+3+d}{6\gamma}, & \text{if } \gamma \in (1,2), \\ 0, & \text{if } \gamma \geq 2, \end{cases}
\tag{12.7}
$$

$$\|\varrho_h \boldsymbol{u}_h\|_{L^2L^2} \stackrel{<}{\sim} h^{\beta_M}, \quad \beta_M = \begin{cases} -\frac{3\varepsilon+3+d}{6\gamma}, & \textit{if } \gamma \in (1,2), \\ -\frac{d}{2\gamma}, & \textit{if } \gamma \geq 2, \\ -\frac{\alpha}{2}, & \textit{if } \gamma > 2 \textit{ for } d = 2 \textit{ or } \gamma \geq 3 \textit{ for } d = 3, \\ -\frac{\alpha}{2} + \frac{\gamma-3}{\gamma}, & \textit{if } \gamma \in [2,3) \textit{ for } d = 3. \end{cases} \tag{12.8}$$

Proof First, recalling Lemma 11.4 directly proves (12.7) and

$$\|\varrho_h \boldsymbol{u}_h\|_{L^2L^2} \stackrel{<}{\sim} h^{-\frac{3\varepsilon+3+d}{6\gamma}}, \text{ if } \gamma \in (1,2).$$

Next, for $\gamma \geq 2$ we apply the inverse estimate (26), and the first two estimates of uniform bounds (12.6) in order to get

$$\|\varrho_h \boldsymbol{u}_h\|_{L^2L^2} \leq \|\varrho_h\|_{L^1L^\infty}^{1/2} \left\|\varrho_h |\boldsymbol{u}_h|^2\right\|_{L^\infty L^1}^{1/2} \stackrel{<}{\sim} h^{-\frac{d}{2\gamma}}.$$

Further, for $d = 2$ and $\gamma > 2$ we have

$$\|\varrho_h \boldsymbol{u}_h\|_{L^2L^2} \leq \|\varrho_h\|_{L^\infty L^\gamma} \|\boldsymbol{u}_h\|_{L^2L^{\frac{2\gamma}{\gamma-2}}} \stackrel{<}{\sim} h^{-\frac{\alpha}{2}}.$$

For $d = 3$ we obtain

$$\|\varrho_h \boldsymbol{u}_h\|_{L^2L^2} \leq \|\varrho_h\|_{L^\infty L^3} \|\boldsymbol{u}_h\|_{L^2L^6} \stackrel{<}{\sim} h^{-\frac{\alpha}{2}} \|\varrho_h\|_{L^\infty L^\gamma} h^{\beta_\gamma} = h^{-\frac{\alpha}{2}+\beta_\gamma},$$

where $\beta_\gamma = 0$ if $\gamma \geq 3$ and $\beta_\gamma = d\frac{\gamma-3}{3\gamma} = \frac{\gamma-3}{\gamma}$ due to the inverse estimate (26), which completes the proof.

12.3 Consistency

Having shown the stability of the finite volume method (12.3) we can proceed with discussing the consistency. Thus, our aim is to show that consistency errors in (8.24) vanish as $h \to 0$. These errors have been already identified in Chap. 11 in the proof of Theorem 11.2.

We focus only on the integrals depending on the velocity that must be handled differently in the present setting. These are:

$$E_1(r_h) = \frac{1}{2} \int_0^T \sum_{\sigma \in \mathcal{E}} \int_\sigma |\{\!\{\boldsymbol{u}_h\}\!\} \cdot \boldsymbol{n}| \, [\![r_h]\!] \left[\!\left[\Pi_Q \varphi\right]\!\right] \mathrm{dS}_x \, \mathrm{d}t,$$

$$E_2(r_h) = \frac{1}{4} \int_0^T \sum_{\sigma \in \mathcal{E}} \int_\sigma [\![\boldsymbol{u}_h]\!] \cdot \boldsymbol{n} \, [\![r_h]\!] \left[\!\left[\Pi_Q \varphi\right]\!\right] \mathrm{dS}_x \, \mathrm{d}t$$

for r_h being ϱ_h or $\varrho_h u_{i,h}$, $i = 1, \ldots, d$.

Term E_1 can be estimated in the following way

$$E_1(r_h) \stackrel{<}{\sim} h||\varphi||_{C^2}||r_h||_{L^2L^2} (\|\boldsymbol{u}_h\|_{L^2L^2} + \|\nabla_{\mathcal{D}}\boldsymbol{u}_h\|_{L^2L^2}) \stackrel{<}{\sim} h^{1-\frac{\alpha}{2}} \|r_h\|_{L^2L^2},$$

where we have used the velocity bound in the second line of (12.6). Further, applying (12.7) and (12.8) we get

$$E_1(\varrho_h) \stackrel{<}{\sim} h^{\delta_1}, \; \delta_1 = 1 - \frac{\alpha}{2} + \beta_D,$$

and

$$E_1(\varrho_h \boldsymbol{u}_h) \stackrel{<}{\sim} h^{\delta_2}, \; \delta_2 = 1 - \frac{\alpha}{2} + \beta_M,$$

respectively, where β_D and β_M are given in Lemma 12.2.

Furthermore, we have

$$\begin{aligned}
E_2(\varrho_h) &\stackrel{<}{\sim} h||\varphi||_{C^2} \left(\int_0^T \sum_{\sigma\in\mathcal{E}} \int_\sigma [\![\boldsymbol{u}_h]\!]^2 \,\mathrm{dS}_x \,\mathrm{d}t \right)^{1/2} \left(\int_0^T \sum_{\sigma\in\mathcal{E}} \int_\sigma [\![\varrho_h]\!]^2 \,\mathrm{dS}_x \,\mathrm{d}t \right)^{1/2} \\
&\stackrel{<}{\sim} h||\varphi||_{C^2} h^{\frac{1}{2}} \|\nabla_{\mathcal{D}}\boldsymbol{u}_h\|_{L^2L^2} \left(\int_0^T \sum_{\sigma\in\mathcal{E}} \int_\sigma \{\!\{\varrho_h\}\!\}^2 \,\mathrm{dS}_x \,\mathrm{d}t \right)^{1/2} \\
&\stackrel{<}{\sim} h^{1-\frac{\alpha}{2}}||\varphi||_{C^2}||\varrho_h||_{L^2L^2} \stackrel{<}{\sim} h^{\delta_1}||\varphi||_{C^2}, \quad \text{for } \delta_1 = 1 - \frac{\alpha}{2} + \beta_D,
\end{aligned}$$

where β_D is given in Lemma 12.2, and, exactly in the same way as in the proof of Theorem 11.2,

$$E_2(\varrho_h \boldsymbol{u}_h) \stackrel{<}{\sim} h^{1-\frac{\alpha}{2}} \|\boldsymbol{\varphi}\|_{C^2} \|\varrho_h \boldsymbol{u}_h\|_{L^2L^2} \stackrel{<}{\sim} h^{\delta_2}, \qquad \delta_2 = 1 - \frac{\alpha}{2} + \beta_M,$$

where β_M is given in Lemma 12.2. The above estimates yield the following conditions on the diffusion parameters h^ε, h^α. In order to keep $\delta_1, \delta_2 > 0$ we require $\varepsilon < 2\gamma(1 - \frac{\alpha}{2}) - \frac{d}{3} - 1$ for $\gamma \in (1, 2)$ and

$$0 < \alpha < \alpha_0, \quad \alpha_0 = \begin{cases} 2 - \frac{d/3+1+\varepsilon}{\gamma} & \text{if } \gamma \in (1, 2), \\ 2 - \frac{d}{\gamma} & \text{if } \gamma \in [2, \infty). \end{cases}$$

Finally, diffusive term on the right hand side of the momentum equation

$$d(h, \boldsymbol{\varphi}) \equiv -h^\alpha \int_{\mathbb{T}^d} \nabla_{\mathcal{D}} \boldsymbol{u}_h^k : \nabla_{\mathcal{D}}(\Pi_{\mathcal{Q}} \boldsymbol{\varphi}) \,\mathrm{d}x$$

can be handled as follows

$$\left| h^\alpha \int_{\mathbb{T}^d} \nabla_{\mathcal{D}} \boldsymbol{u}_h^k : \nabla_{\mathcal{D}}(\Pi_Q \boldsymbol{\varphi}) \, \mathrm{d}x \right|$$
$$\leq h^{\frac{\alpha}{2}} \left(h^\alpha \int_{\mathbb{T}^d} (\nabla_{\mathcal{D}} \boldsymbol{u}_h^k)^2 \, \mathrm{d}x \right)^{1/2} \left(\int_{\mathbb{T}^d} (\nabla_{\mathcal{D}}(\Pi_Q \boldsymbol{\varphi}))^2 \, \mathrm{d}x \right)^{1/2}$$
$$\lesssim h^{\frac{\alpha}{2}} ||\boldsymbol{\varphi}||_{C^1} \to 0 \ \text{ as } h \to 0 \ \text{ for } \alpha > 0.$$

Combining the above estimates with those obtained in Theorem 11.2 we obtain the consistency of the finite volume method (12.3).

Theorem 12.1 (Consistency of the finite volume method (12.3)**)**
Let the parameters ε and α satisfy

$$\begin{aligned} & -1 < \varepsilon \text{ and } 0 < \alpha < 2 - \frac{d/3 + 1 + \varepsilon}{\gamma} \quad \text{for } \gamma \in (1, 2), \\ \text{or} \quad & -1 < \varepsilon \text{ and } 0 < \alpha < 2 - \frac{d}{\gamma} \quad \text{for } \gamma \geq 2. \end{aligned} \tag{12.9}$$

Then the finite volume method (12.3) *is consistent with the barotropic Euler equations* (12.1) *with the periodic boundary conditions and initial data* $[\varrho_0, \boldsymbol{m}_0 \equiv \varrho_0 \boldsymbol{u}_0]$. *Specifically, setting the discrete initial conditions as* $(\varrho_h^0, \boldsymbol{m}_h^0) = (\Pi_Q \varrho_0, \Pi_Q \boldsymbol{m}_0)$ *with* $\boldsymbol{u}_h^0 = \frac{\boldsymbol{m}_h^0}{\varrho_h^0}$, *the numerical solutions* $(\varrho_h, \varrho_h \boldsymbol{u}_h \equiv \boldsymbol{m}_h)$ *satisfy*

$$- \int_{\mathbb{T}^d} \varrho_h^0 \varphi(0, \cdot) \, \mathrm{d}x = \int_0^T \int_{\mathbb{T}^d} \left[\varrho_h \partial_t \varphi + \boldsymbol{m}_h \cdot \nabla_x \varphi \right] \, \mathrm{d}x \, \mathrm{d}t + \int_0^T e_{1,h}(t, \varphi) \, \mathrm{d}t \tag{12.10a}$$

for any $\varphi \in C_c^2([0, T) \times \mathbb{T}^d)$;

$$\begin{aligned} - \int_{\mathbb{T}^d} \boldsymbol{m}_h^0 \cdot \boldsymbol{\varphi}(0, \cdot) \, \mathrm{d}x = & \int_0^T \int_{\mathbb{T}^d} \left[\boldsymbol{m}_h \cdot \partial_t \boldsymbol{\varphi} + \left(\frac{\boldsymbol{m}_h \otimes \boldsymbol{m}_h}{\varrho_h} \right) : \nabla_x \boldsymbol{\varphi} + p_h \mathrm{div}_x \boldsymbol{\varphi} \right] \, \mathrm{d}x \, \mathrm{d}t \\ & + \int_0^T e_{2,h}(t, \boldsymbol{\varphi}) \, \mathrm{d}t, \quad p_h = p(\varrho_h) \end{aligned} \tag{12.10b}$$

for any $\boldsymbol{\varphi} \in C_c^2([0, T) \times \mathbb{T}^d; R^d)$;

$$\|e_{1,h}(\cdot, \varphi)\|_{L^1(0,T)} \lesssim h^\beta \|\varphi\|_{C^2}, \quad \|e_{2,h}(\cdot, \boldsymbol{\varphi})\|_{L^1(0,T)} \lesssim h^\beta \|\boldsymbol{\varphi}\|_{C^2} \text{ for some } \beta > 0.$$

12.4 Convergence

At this stage we are ready to apply the machinery developed in Chap. 5 and Sect. 7.5 to prove the convergence of the finite volume method (12.3) to a solution of the barotropic Euler system (12.1) with periodic boundary conditions and initial data $[\varrho_0, \boldsymbol{m}_0 \equiv \varrho_0 \boldsymbol{u}_0]$ belonging to the class

$$\varrho_0 > 0, \quad E_0 = \int_{\mathbb{T}^d} \left[\frac{1}{2} \frac{|\boldsymbol{m}_0|^2}{\varrho_0} + P(\varrho_0) \right] \mathrm{d}x < \infty.$$

We concentrate on the physically relevant case $1 < \gamma < 2$. According to Theorem 5.3 the sequence $\{\varrho_h, \boldsymbol{m}_h = \varrho_h \boldsymbol{u}_h\}_{h \searrow 0}$ of finite volume solutions of (12.3) generates a Young measure $\left\{ \mathcal{V}_{t,x} \right\}_{(t,x) \in (0,T) \times \mathbb{T}^d}$,

$$\mathcal{V}_{t,x} \in \mathcal{P}(R^{d+1}), \ R^{d+1} = \left\{ \widetilde{\varrho} \in R, \widetilde{\boldsymbol{m}} \in R^d \right\}$$

that is a DMV solution of the barotropic Euler equations, see Definition 5.5. In addition, the Young measure admits finite first moments and its barycenter

$$\varrho = \langle \mathcal{V}; \widetilde{\varrho} \rangle, \ \boldsymbol{m} = \langle \mathcal{V}, \widetilde{\boldsymbol{m}} \rangle$$

is a DW solution in the sense of Definition 5.6, cf. Theorem 5.4.

The following two theorems can be proved analogously as Theorems 9.3, 9.5, 9.6, 9.1. Note however that the convergence results in Chap. 9 are conditional, i.e. numerical density, and for the Lax–Friedrichs scheme the velocity as well, is supposed to be uniformly bounded. On the contrary, the convergence results presented below are unconditional. The main reason is that the present VFV method (12.5) mimics a vanishing viscosity approximation of the Euler system. Thus, the additional numerical diffusion in the momentum equation controls the blow up rates of discrete velocity gradients, a piece of information that is not available in standard finite volume schemes designed for general hyperbolic conservation laws.

Theorem 12.2 (Weak convergence of finite volume solutions)
Let $\{\varrho_h, \boldsymbol{m}_h \equiv \varrho_h \boldsymbol{u}_h\}_{h \searrow 0}$ be a family of numerical solutions obtained by the finite volume method (12.3) with the conditions (12.9) *for ε and α.*
Then there exists a subsequence $\{\varrho_{h_n}, \boldsymbol{m}_{h_n}\}_{h_n \searrow 0}$ such that

- ***weak convergence***

$$\varrho_{h_n} \to \varrho \text{ weakly-(*) in } L^\infty(0,T; L^\gamma(\mathbb{T}^d)),$$
$$\boldsymbol{m}_{h_n} \to \boldsymbol{m} \text{ weakly-(*) in } L^\infty(0,T; L^{\frac{2\gamma}{\gamma+1}}(\mathbb{T}^d; R^d)),$$

$$E(\varrho_{h_n}, \boldsymbol{m}_{h_n}) \equiv \frac{1}{2}\frac{|\boldsymbol{m}_{h_n}|^2}{\varrho_{h_n}} + P(\varrho_{h_n}) \to \overline{\frac{1}{2}\frac{|\boldsymbol{m}|^2}{\varrho} + P(\varrho)}$$

weakly-(*) in $L^\infty(0, T; \mathcal{M}(\mathbb{T}^d))$ *as* $h_n \to 0$.

- $\mathcal{K}$*–convergence*

$$\frac{1}{N}\sum_{n=1}^{N} \varrho_{h_n} \to \varrho \ \ \textit{as} \ N \to \infty \ \textit{in} \ L^q(0, T; L^\gamma(\mathbb{T}^d)), \ 1 \le q < \infty,$$

$$\frac{1}{N}\sum_{n=1}^{N} \boldsymbol{m}_{h_n} \to \boldsymbol{m} \ \ \textit{as} \ N \to \infty \ \textit{in} \ L^q(0, T; L^{\frac{2\gamma}{\gamma+1}}(\mathbb{T}^d; R^d)), \ 1 \le q < \infty,$$

$$\frac{1}{N}\sum_{n=1}^{N} \left(\frac{1}{2}\frac{|\boldsymbol{m}_{h_n}|^2}{\varrho_{h_n}} + P(\varrho_{h_n}) \right) \to E \ \textit{as} \ N \to \infty \ \textit{a.a. in} \ (0, T) \times \mathbb{T}^d,$$

$$E \ge \left\langle \mathcal{V}; \frac{1}{2}\frac{|\widetilde{\boldsymbol{m}}|^2}{\widetilde{\varrho}} + P(\widetilde{\varrho}) \right\rangle.$$

- **Strong convergence to the Young measure in the Wasserstein distance**

$$d_{W_r}\left[\frac{1}{N}\sum_{n=1}^{N} \delta_{[\varrho_{h_n}(t,x); \boldsymbol{m}_{h_n}(t,x)]}; \mathcal{V}_{t,x} \right] \to 0 \ \textit{as} \ N \to \infty$$

for a.a. $(t, x) \in (0, T) \times \mathbb{T}^d$,

and in $L^s((0, T) \times \mathbb{T}^d)$ *for any* $r = \frac{2\gamma}{\gamma+1}$ *and* $1 \le s < r$. *Here,* d_{W_r} *denotes the Wasserstein* r*-distance introduced in (3).*

- L^1 **convergence of the deviations**

$$\frac{1}{N}\sum_{n=1}^{N} \left| [\varrho_{h_n}, \boldsymbol{m}_{h_n}] - \frac{1}{N}\sum_{k=1}^{N} [\varrho_{h_n}, \boldsymbol{m}_{h_n}] \right|$$
$$\equiv \left\langle \frac{1}{N}\sum_{n=1}^{N} \delta_{[\varrho_{h_n}, \boldsymbol{m}_{h_n}]}; \left| [\widetilde{\varrho}, \widetilde{\boldsymbol{m}}] - \left\langle \frac{1}{N}\sum_{n=1}^{N} \delta_{[\varrho_{h_n}, \boldsymbol{m}_{h_n}]}; [\widetilde{\varrho}, \widetilde{\boldsymbol{m}}] \right\rangle \right| \right\rangle$$
$$\to \langle \mathcal{V}; |[\widetilde{\varrho}, \widetilde{\boldsymbol{m}}] - [\varrho, \boldsymbol{m}]| \rangle \ \textit{as} \ N \to \infty \ \textit{in} \ L^1((0, T) \times \mathbb{T}^d).$$

Theorem 12.3 (Strong convergence of finite volume solutions)
Suppose that $\{\varrho_h, \boldsymbol{m}_h\}_{h \searrow 0}$ *is a sequence of finite volume solutions obtained by the finite volume method (12.3) with the conditions* (12.9) *for* ε *and* α.

- **strong solution**
  Assume that the Euler system (12.1) *admits a strong solution* $[\varrho, \boldsymbol{m}]$, *such that*

$$\varrho \in W^{1,\infty}((0,T) \times \mathbb{T}^d), \ \boldsymbol{m} \in W^{1,\infty}((0,T) \times \mathbb{T}^d; R^d).$$

  Then for any $1 \leq q < \infty$

$$\begin{aligned} &\varrho_h \to \varrho \ \textit{in} \ L^q(0,T; L^\gamma(\mathbb{T}^d)), \\ &\boldsymbol{m}_h \to \boldsymbol{m} \ \textit{in} \ L^q(0,T; L^{\frac{2\gamma}{\gamma+1}}(\mathbb{T}^d; R^d)), \\ &E(\varrho_h, \boldsymbol{m}_h) \to E(\varrho, \boldsymbol{m}) \ \textit{in} \ L^q(0,T; L^1(\mathbb{T}^d)) \ \textit{as} \ h \to 0. \end{aligned}$$

- **classical solution**
  Suppose there is a sequence $h_n \to 0$ *such that*

$$\begin{aligned} &\varrho_{h_n} \to \varrho \ \textit{weakly-(*) as} \ h_n \to 0 \ \textit{in} \ L^\infty(0,T; L^\gamma(\mathbb{T}^d)) \\ &\boldsymbol{m}_{h_n} \to \boldsymbol{m} \ \textit{weakly-(*) as} \ h_n \to 0 \ \textit{in} \ L^\infty(0,T; L^{\frac{2\gamma}{\gamma+1}}(\mathbb{T}^d; R^d)), \end{aligned}$$

  where

$$\varrho \in C^1([0,T] \times \mathbb{T}^d), \ \boldsymbol{m} \in C^1([0,T] \times \mathbb{T}^d; R^d).$$

  Then $[\varrho, \boldsymbol{m}]$ *is a classical solution to the limit Euler system, and*

$$\varrho_h \to \varrho \ \textit{in} \ L^q(0,T; L^\gamma(\mathbb{T}^d)), \ \boldsymbol{m}_h \to \boldsymbol{m} \ \textit{in} \ L^q(0,T; L^{\frac{2\gamma}{\gamma+1}}(\mathbb{T}^d; R^d))$$

  as $h \to 0$ *for any* $1 \leq q < \infty$.

Proof In order to prove the strong convergence to the weak solution that belongs to the class (1.14) we apply the weak-strong uniqueness result. Indeed, if $\mathcal{V}$ is a DW solution generated by the finite volume method (12.3) and a weak solution $[\varrho, \boldsymbol{u}]$ belonging to the class (1.14) exists then

$$\mathcal{V} = \delta_{[\varrho, \boldsymbol{m}]} \ \text{a.a. in} \ (0,T) \times \Omega$$

and the strong convergence follows.

The strong convergence to the strong or classical solution can be shown in the same way as Theorems 9.1, 9.3.

Applying the results from Sect. 9.5 we can avoid the subsequence argument as soon as the approximate sequence $\{\varrho_{h_n}, \boldsymbol{m}_{h_n}\}_{n=1}^\infty$ is (S)–convergent.

Theorem 12.4 (Statistical convergence of finite volume solutions) *Let* $\{\varrho_{h_n}, \boldsymbol{m}_{h_n}\}_{n=1}^\infty$ *be a sequence of approximate solutions obtained by the finite volume method (12.3) with the conditions* (12.9) *for* ε *and* α.
In addition, suppose that $\{\varrho_{h_n}, \boldsymbol{m}_{h_n}\}_{n=1}^\infty$ *is (S)–convergent in the sense of Definition 9.2.*

Then

$$\frac{1}{N}\sum_{n=1}^{N}\varrho_{h_n} \to \varrho \text{ as } N \to \infty \text{ in } L^q((0,T)\times\mathbb{T}^d) \text{ for } 1 \le q < \gamma,$$

$$\frac{1}{N}\sum_{n=1}^{N}\boldsymbol{m}_{h_n} \to \boldsymbol{m} \text{ as } N \to \infty \text{ in } L^q((0,T)\times\mathbb{T}^d; R^d) \text{ for } 1 \le q < \frac{2\gamma}{\gamma+1}.$$

Moreover, the sequence of numerical solutions $\{\varrho_{h_n}, \boldsymbol{m}_{h_n}\}_{n=1}^{\infty}$ *generates a Young measure* $\{\mathcal{V}_{t,x}\}_{(t,x)\in(0,T)\times\mathbb{T}^d}$ *and*

$$d_{W_q}\left[\frac{1}{N}\sum_{n=1}^{N}\delta_{[\varrho_{h_n};\boldsymbol{m}_{h_n}]};\mathcal{V}\right] \to 0 \text{ as } N \to \infty \text{ in } L^q((0,T)\times\mathbb{T}^d)$$

for any $1 \le q < \frac{2\gamma}{\gamma+1}$.

The measure $\mathcal{V}$ *is a DMV solution of the Euler system in the sense of Definition 5.5, whereas its barycenter* $[\varrho, \boldsymbol{m}]$ *is a dissipative solution in the sense of Definition 5.6.*

12.5 Error Estimates

The relative energy inequality, derived in Sects. 4.1.3, 6.1.1, together with the stability-consistency formulation (12.5), (12.10), can be used to derive *error estimates* as soon as the Euler system admits a smooth solution. Indeed a short inspection of the proof of the relative energy inequality in Sect. 4.1.3 reveals that it is enough to add the consistency errors computed in (12.10) to the right-hand side of the relative energy inequality (6.5) evaluated for $\varrho = \varrho_h$, $\boldsymbol{m} = \boldsymbol{m}_h$, and $\widetilde{\varrho} = \varrho$, $\widetilde{\boldsymbol{u}} = \boldsymbol{u}$ – the smooth solution of the limit system. Realizing that the defect measures $\mathfrak{E}$ and $\mathfrak{R}$ vanish, we obtain

$$\begin{aligned}
&\left[\int_{\mathbb{T}^d} E\left(\varrho_h, \boldsymbol{m}_h \,\middle|\, \varrho, \boldsymbol{u}\right)\, \mathrm{d}x\right]_{t=0}^{t=\tau} \\
&\le -\int_0^\tau\int_{\mathbb{T}^d} \varrho_h \nabla_x \boldsymbol{u} \cdot \left(\frac{\boldsymbol{m}_h}{\varrho_h} - \boldsymbol{u}\right)\cdot\left(\frac{\boldsymbol{m}_h}{\varrho_h} - \boldsymbol{u}\right)\, \mathrm{d}x\, \mathrm{d}t \\
&\quad -\int_0^\tau\int_{\mathbb{T}^d} \Big[p(\varrho_h) - p'(\varrho)(\varrho_h - \varrho) - p(\varrho)\Big]\mathrm{div}_x \boldsymbol{u}\, \mathrm{d}x\, \mathrm{d}t \\
&\quad + \left\| e_{1,h}\left(\cdot, P'(\varrho) - \frac{1}{2}|\boldsymbol{u}|^2\right)\right\|_{L^1(0,T)} + \|e_{2,h}(\cdot, \boldsymbol{u})\|_{L^1(0,T)},
\end{aligned} \tag{12.11}$$

where $e_{1,h}, e_{2,h}$ are the consistency errors identified in (12.10), and $E\left(\varrho_h, \boldsymbol{m}_h \middle| \varrho, \boldsymbol{u}\right)$ is the relative energy introduced in (4.25).

Thus we may apply Gronwall's lemma to conclude that

$$\begin{aligned}
&\int_{\mathbb{T}^d} E\left(\varrho_h, \boldsymbol{m}_h \middle| \varrho, \boldsymbol{u}\right)(\tau, \cdot)\, \mathrm{d}x \\
&\leq c\left(T, \|\boldsymbol{u}\|_{C^1}\right) \left[\int_{\mathbb{T}^d} E\left(\varrho_h, \boldsymbol{m}_h \middle| \varrho, \boldsymbol{u}\right)(0, \cdot)\, \mathrm{d}x \right. \\
&\qquad \left. + \left\| e_{1,h}\left(\cdot, P'(\varrho) - \frac{1}{2}|\boldsymbol{u}|^2\right) \right\|_{L^1(0,T)} + \|e_{2,h}(\cdot, \boldsymbol{u})\|_{L^1(0,T)} \right]
\end{aligned} \tag{12.12}$$

for any $0 \leq \tau < T$. Thus we have an explicit estimate of the *rate of convergence* of the numerical method in terms of the error in the initial data approximation and the consistency errors. More precisely, choosing the initial data $\varrho_h^0 = \Pi_Q \varrho_0$, $\boldsymbol{m}_h^0 = \Pi_Q \boldsymbol{m}_0$ we have $E\left(\varrho_h^0, \boldsymbol{m}_h^0 \middle| \varrho_0, \boldsymbol{u}_0\right) \lesssim h$. Further,

$$\left\| e_{1,h}\left(\cdot, P'(\varrho) - \frac{1}{2}|\boldsymbol{u}|^2\right) \right\|_{L^1(0,T)} + \|e_{2,h}(\cdot, \boldsymbol{u})\|_{L^1(0,T)} \leq h^\delta$$

with $\delta = \min\left[\frac{\alpha}{2}, 1 - \left(\frac{3\varepsilon+3+d}{6\gamma} + \frac{\alpha}{2}\right)\right]$, which determines rate of the convergence.

12.6 Numerical Experiments

As in Chap. 10 we want to illustrate the concept of $\mathcal{K}$–convergence for barotropic Euler system. To this end we consider the Kelvin–Helmholtz problem as stated in Sect. 10.5.

We chose a computational domain $\Omega = [0, 1]^2$, apply periodic boundary conditions and the following initial data

$$(\varrho, u_1, u_2)(x) = \begin{cases} (2, -0.5, 0), & \text{if } I_1 < x_2 < I_2 \\ (1, 0.5, 0), & \text{otherwise.} \end{cases}$$

It is well-known that the Kelvin–Helmholtz problem describes the uniform velocity shear for fluids of different densities superposed one over the other. The interface profiles $I_j = I_j(x) \equiv J_j + \epsilon Y_j(x),\ j = 1, 2$, are chosen to be small perturbations around the lower $x_2 = J_1 = 0.25$ and the upper $x_2 = J_2 = 0.75$ interface, respectively. Further,

$$Y_j(x) = \sum_{k=1}^{m} a_j^k \cos(b_j^k + 2k\pi x_1), \quad j = 1, 2,$$

where $a_j^k \in [0, 1]$ and b_j^k, $j = 1, 2$, $k = 1, \ldots, m$ are arbitrary but fixed numbers. The coefficients a_j^k have been normalized such that $\sum_{k=1}^{m} a_j^k = 1$ to guarantee that $|I_j(x) - J_j| \leq \epsilon$ for $j = 1, 2$. Analogously as in the case of complete Euler system we set $m = 10$ and $\epsilon = 0.01$.

In Figs. 12.1 and 12.4 the results obtained by the VFV method (12.2) for $\gamma = 1.1$ and $\gamma = 1.4$ are presented. The coefficients of numerical viscosity are chosen to be $\alpha = 1.0 = \varepsilon$. Grid resolution is consecutively refined using 256^2, 512^2, 1024^2, and 2048^2 finite volume cells. As a comparison we also present in Figs. 12.2, 12.3, 12.5 and 12.6 the results obtained by more classical finite volume methods such as the Rusanov and the Lax–Friedrichs method, cf. (9.12) and (9.13). It is well-known that the Lax–Friedrichs finite volume method is quite diffusive which can be observed by results on a grid with 2048^2 cells. In order to obtain better resolution of shear effects numerical simulations on a finer grid would need to be performed. Another option is, of course, to apply a higher order approximation

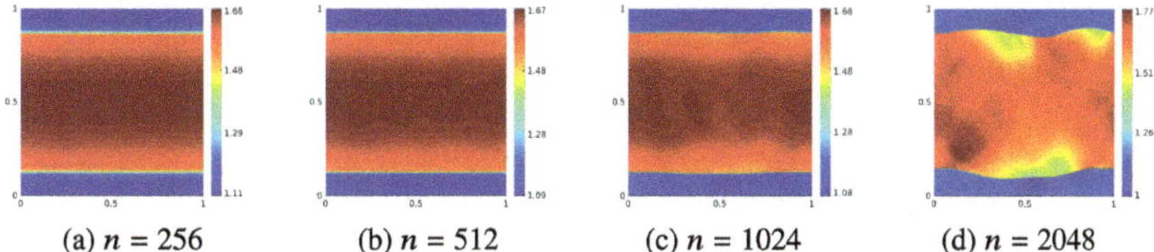

(a) $n = 256$ (b) $n = 512$ (c) $n = 1024$ (d) $n = 2048$

Fig. 12.1 VFV scheme with $\alpha = 1.0$, $\varepsilon = 1.0$, $\gamma = 1.1$

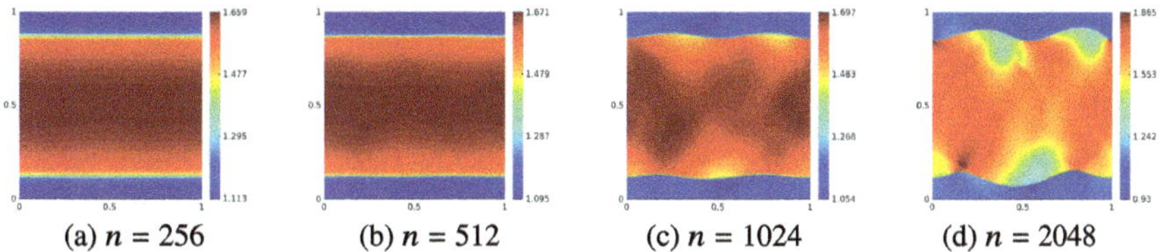

(a) $n = 256$ (b) $n = 512$ (c) $n = 1024$ (d) $n = 2048$

Fig. 12.2 Rusanov finite volume method, $\gamma = 1.1$

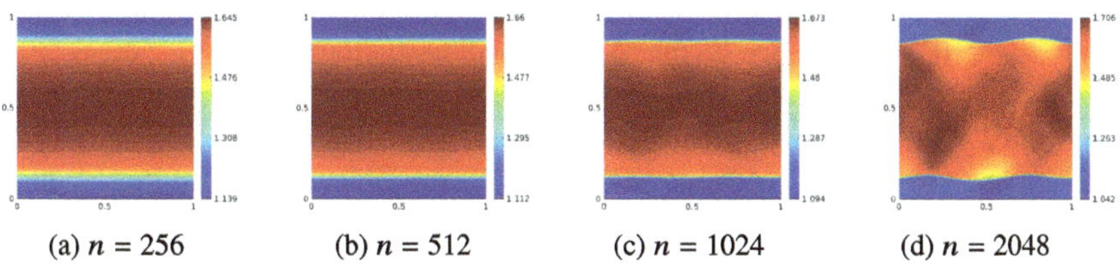

(a) $n = 256$ (b) $n = 512$ (c) $n = 1024$ (d) $n = 2048$

Fig. 12.3 Lax–Friedrichs finite volume method, $\gamma = 1.1$

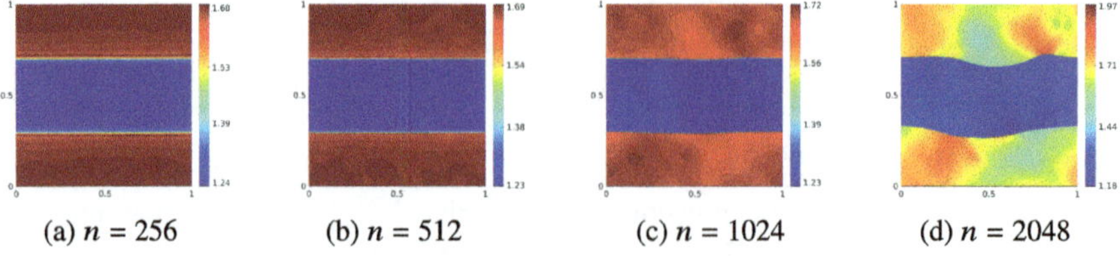

(a) $n = 256$ (b) $n = 512$ (c) $n = 1024$ (d) $n = 2048$

Fig. 12.4 VFV scheme with $\alpha = 1.0$, $\varepsilon = 1.0$, $\gamma = 1.4$

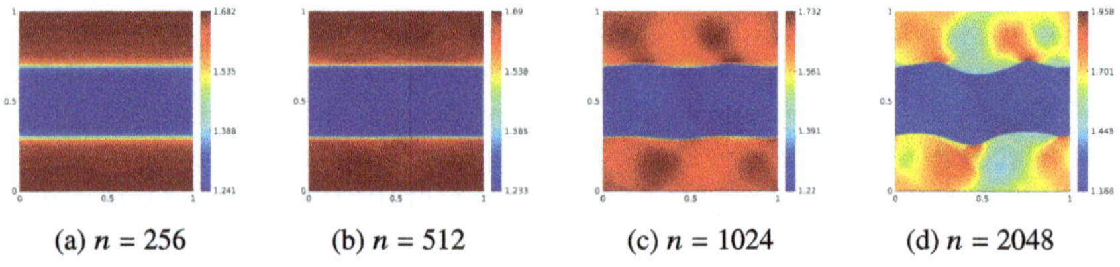

(a) $n = 256$ (b) $n = 512$ (c) $n = 1024$ (d) $n = 2048$

Fig. 12.5 Rusanov finite volume method, $\gamma = 1.4$

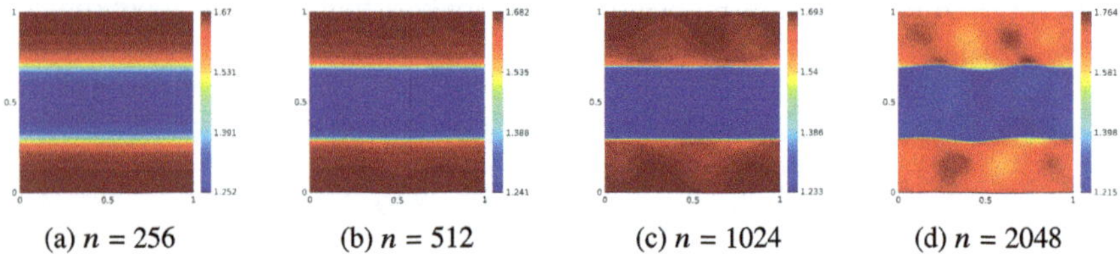

(a) $n = 256$ (b) $n = 512$ (c) $n = 1024$ (d) $n = 2048$

Fig. 12.6 Lax–Friedrichs finite volume method, $\gamma = 1.4$

using, e.g., a piecewise linear reconstruction. For simplicity, all simulations are done with the first order explicit time discretization using the CFL stability condition $\Delta t \max(|\boldsymbol{u}_h| + \sqrt{\gamma \varrho_h^{\gamma-1}})/h = 0.3$. Recall that $h = 1/n$, where n is the number of grid cells in one spatial direction.

Figure 12.7 documents the convergence behavior of the vanishing viscosity finite volume method (12.2) for $\gamma = 1.1$ and $\gamma = 1.4$. The coefficients of numerical diffusion are taken to be $\varepsilon = 1.0$ and $\alpha = 1.5$ or $\alpha = 1.0$. Let us recall that the error E_1 measures the L^1-error of the difference between the numerical solution and the reference solution computed on the finest grid, the L^1-error of the Cesàro averages, the L^1-error of the deviation, and the L^1-error of the Wasserstein distance for the Cesàro average of the Dirac measures concentrated on numerical solutions, respectively, see (10.75). We observe that due to small scale oscillations the convergence of individual numerical solutions is not achieved, see the first column. On the other hand the $\mathcal{K}$-convergence approach using the Cesàro averages yields better convergence rates, see the second column. This can be seen by considering the gradients of the error curves, in particular for the density and smaller numerical diffusion with $\alpha = 1.5$. The larger the gradient is, the better convergence rate is achieved. Similar effects are observed for the errors E_3 and E_4 of deviation and the Cesàro averages of the Dirac measures, respectively.

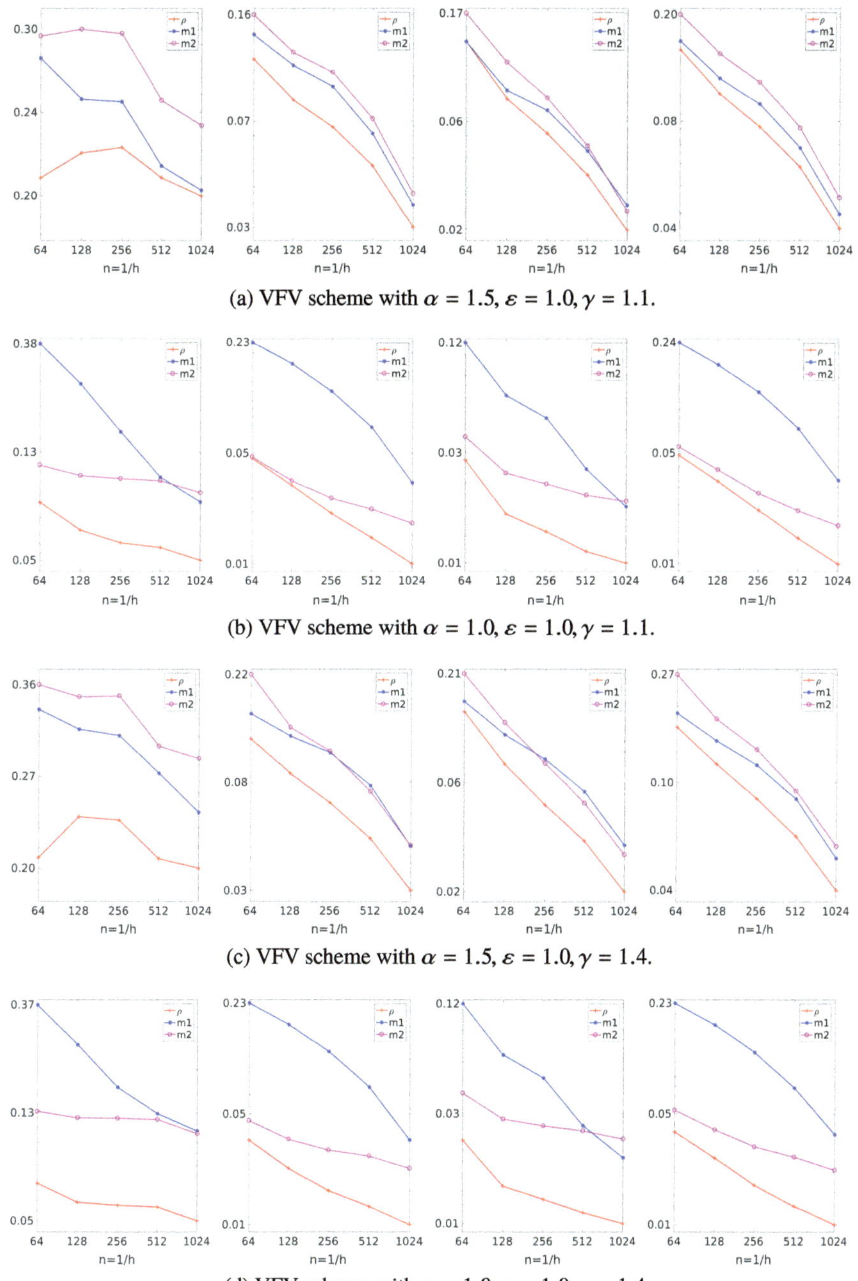

(a) VFV scheme with $\alpha = 1.5, \varepsilon = 1.0, \gamma = 1.1$.

(b) VFV scheme with $\alpha = 1.0, \varepsilon = 1.0, \gamma = 1.1$.

(c) VFV scheme with $\alpha = 1.5, \varepsilon = 1.0, \gamma = 1.4$.

(d) VFV scheme with $\alpha = 1.0, \varepsilon = 1.0, \gamma = 1.4$.

Fig. 12.7 Convergence study for the Kelvin–Helmholtz problem: E_1, E_2, E_3, and E_4 errors (left to right)

Chapter 13
Mixed Finite Volume – Finite Element Method for the Navier–Stokes System

In Chap. 11 we have studied the convergence of a finite volume method for the Navier–Stokes system using piecewise constant approximations for density and velocity. Due to the viscosity terms in the momentum equation another rather natural option would be to work with piecewise linear approximation for the velocity. Moreover, due to the parabolic character of the momentum equation it is suitable to approximate it by means of finite element method. This gives rise to a mixed finite volume – finite element method that combines the advantage of finite volume approximation of the convective terms in the continuity and momentum equations, with the finite element approximation of the dissipation terms. The goal of present chapter is to introduce such a mixed method for the Navier–Stokes system and analyze its convergence.

Let us consider the Navier–Stokes system in the time-space cylinder $(0, T) \times \Omega$ ($\Omega \subset R^d, d = 2, 3$):

$$\partial_t \varrho + \mathrm{div}_x(\varrho \boldsymbol{u}) = 0, \tag{13.1a}$$

$$\partial_t(\varrho \boldsymbol{u}) + \mathrm{div}_x(\varrho \boldsymbol{u} \otimes \boldsymbol{u}) + \nabla_x p(\varrho) = \mathrm{div}_x \mathbb{S}(\nabla_x \boldsymbol{u}), \tag{13.1b}$$

$$p = a\varrho^\gamma, \ a > 0, \ \gamma > 1, \tag{13.1c}$$

where $\mathbb{S}$ is the viscous stress tensor

$$\mathbb{S}(\nabla_x \boldsymbol{u}) = \mu \left(\nabla_x \boldsymbol{u} + \nabla_x^T \boldsymbol{u} - \frac{2}{d} \mathrm{div}_x \boldsymbol{u} \mathbb{I} \right) + \lambda \mathrm{div}_x \boldsymbol{u} \mathbb{I}$$

for $\mu > 0$ and $\lambda \geq -\frac{d-2}{d}\mu$. As the viscosity coefficients are constant, we may rewrite

$$\mathrm{div}_x \mathbb{S}(\nabla_x \boldsymbol{u}) = \mu \Delta_x \boldsymbol{u} + \nu \nabla_x \mathrm{div}_x \boldsymbol{u},$$

where $\nu = \frac{d-2}{d}\mu + \lambda \geq 0$.

E. Feireisl et al., *Numerical Analysis of Compressible Fluid Flows*, MS&A 20,
https://doi.org/10.1007/978-3-030-73788-7_13

The system is supplemented with the initial data

$$\varrho(0) = \varrho_0 \geq \underline{\varrho} > 0, \ \varrho_0 \in L^\infty(\Omega), \quad \boldsymbol{u}(0) = \boldsymbol{u}_0 \in W_0^{1,2}(\Omega; R^d) \cap L^\infty(\Omega; R^d) \tag{13.2}$$

and the no-slip boundary condition

$$\boldsymbol{u}|_{\partial\Omega} = 0. \tag{13.3}$$

Alternatively we may apply the periodic boundary conditions. In that case the domain can be identified with the flat torus ($\Omega = \mathbb{T}^d$), cf. Chap. 11.

The main program is to apply the theoretical results on the convergence of consistent approximations to a DMV solution (see Theorem 5.5) and the convergence to strong solutions (see Theorems 7.12 and 7.13). Thus the main goal is to show that the numerical solution is a consistent approximation of the Navier–Stokes system (13.1) in the sense of Definition 5.9, namely the numerical solution is stable in the sense of Definition 8.3 and consistent in the sense of Definition 8.6. Note, however, that the analysis performed in the theoretical part concerns exclusively the case of spatially periodic boundary conditions. Although there are no major obstacles to accommodate the no-slip condition (13.3), there is one issue that deserves attention. The present approach yields strong convergence to a strong solution as long as the latter *exists*. As computational domains are usually of polygonal type, meaning merely Lipschitz, the existence of a smooth solution for a problem involving second order in space differential operators is a delicate issue. To avoid this difficulty, we approximate a smooth physical domain Ω by a family of computational domains Ω_h,

$$\Omega_h \subset \Omega, \ K \subset \Omega_h \text{ for any compact } K \subset \Omega \text{ and any } 0 < h \leq h_0. \tag{13.4}$$

As this is the only part of the book, where variable computational domain is considered, we feel it is more convenient to explain the changes on this particular example rather than elaborate a new version of the abstract theory.

Our approach is based on the following steps:

- we derive the discrete energy inequality in the sense of Definition 8.3;
- we derive the consistency formulation of the problem in the sense of Definition 8.6 upon the uniform bounds obtained from the previous step of the energy stability;
- we combine the results of previous two steps to gain a consistent approximation of the Navier–Stokes system in the sense of Definition 5.9;
- we verify the compatibility of discrete differential operators in the sense of Definition 5.8;
- finally, we apply suitable modifications of Theorems 5.5 and 7.12 (or Theorem 7.13 for bounded density) to derive the convergence results.

We start by introducing the numerical method and continue with studying its stability, consistency, and convergence step by step.

13.1 Numerical Method

In this section we introduce a mixed finite volume – finite element method for the Navier–Stokes equations (13.1). Concerning the notation used in the scheme, we invite the reader to Preliminary material and Chap. 8 for the definitions. In accordance with (13.4), we assume that the physical domain Ω is approximated by a polygonal, resp. polyhedral approximation $\Omega_h \subset \Omega$ using an unstructured mesh $\mathcal{T}_h$, cf. Definition 1.

Definition 13.1 (FV–FE SCHEME FOR THE NAVIER–STOKES SYSTEM)
Given the initial data (13.2) we set

$$\varrho_h^0 = \Pi_Q \varrho_0, \quad \boldsymbol{u}_h^0 = \Pi_V \boldsymbol{u}_0$$

so that $\varrho_h^0 \in Q_h$ and $\boldsymbol{u}_h^0 \in \boldsymbol{V}_h$, where Q_h and $\boldsymbol{V}_h$ are the space of piecewise constant and piecewise linear Crouzeix–Raviart elements, respectively, see (13.10). A sequence $(\varrho_h^k, \boldsymbol{u}_h^k) \in Q_h \times \boldsymbol{V}_{0,h}$, $k = 1, \ldots, N_T$ is a numerical solution of the finite volume – finite element method if the following holds for any $\phi_h \in Q_h$ and $\boldsymbol{\phi}_h \in \boldsymbol{V}_{0,h}$

$$\int_{\Omega_h} D_t \varrho_h^k \phi_h \, \mathrm{d}x - \int_{\mathcal{E}_{int}} F_h^{\mathrm{up}}[\varrho_h^k, \boldsymbol{u}_h^k] \, [\![\phi_h]\!] \, \mathrm{dS}_x = 0, \tag{13.5a}$$

$$\begin{aligned} &\int_{\Omega_h} D_t \left(\varrho_h^k \overline{\boldsymbol{u}_h}^k\right) \cdot \boldsymbol{\phi}_h \, \mathrm{d}x - \int_{\mathcal{E}_{int}} \boldsymbol{F}_h^{\mathrm{up}}[\varrho_h^k \overline{\boldsymbol{u}_h}^k, \boldsymbol{u}_h^k] \cdot [\![\overline{\boldsymbol{\phi}_h}]\!] \, \mathrm{dS}_x \\ &\quad - \int_{\Omega_h} p(\varrho_h^k) \mathrm{div}_h \boldsymbol{\phi}_h \, \mathrm{d}x + \mu \int_{\Omega_h} \nabla_h \boldsymbol{u}_h^k : \nabla_h \boldsymbol{\phi}_h \, \mathrm{d}x + \nu \int_{\Omega_h} \mathrm{div}_h \boldsymbol{u}_h^k \mathrm{div}_h \boldsymbol{\phi}_h \, \mathrm{d}x = 0. \end{aligned} \tag{13.5b}$$

Here, the discrete operators div_h and ∇_h are given in (15), and the operator $[\![\cdot]\!]$ is given in (13.6), respectively. Recall that the operator $\overline{f} = \Pi_Q f$ represents the elementwise constant interpolation and the symbol $\int_{\mathcal{E}_{int}}$ stands for $\sum_{\sigma \in \mathcal{E}_{int}} \int_\sigma$. Moreover, the diffusive upwind numerical flux reads

$$F_h^{\mathrm{up}}[r, \boldsymbol{u}] = Up[r, \boldsymbol{u}] - \frac{h^\varepsilon}{2} [\![r]\!] \chi\left(\frac{\langle \boldsymbol{u} \rangle \cdot \boldsymbol{n}}{h^\varepsilon}\right), \quad \chi(z) = \begin{cases} 1 - |z| & \text{if } |z| \leq 1, \\ 0 & \text{otherwise,} \end{cases} \tag{13.6}$$

where $Up[r, \boldsymbol{u}]$ represents the standard upwind flux, see (8.6).

Remark 13.1 Note that the diffusive upwind flux F_h^{up} in this chapter is slightly different from those used previously. In (8.8) we have defined the following version

$$F_h^{\rm up}[r,\boldsymbol{u}] = Up[r,\boldsymbol{u}] - h^\varepsilon [\![r]\!] = \{\!\{r\}\!\}\, \langle \boldsymbol{u}\rangle \cdot \boldsymbol{n} - [\![r]\!]\,(h^\varepsilon + \frac{1}{2}|\langle \boldsymbol{u}\rangle \cdot \boldsymbol{n}|). \tag{13.7}$$

By an easy calculation we may reformulate the numerical flux (13.6) as

$$F_h^{\rm up}[r,\boldsymbol{u}] = \{\!\{r\}\!\}\, \langle \boldsymbol{u}\rangle \cdot \boldsymbol{n} - \frac{1}{2}[\![r]\!] \max\{h^\varepsilon, |\langle \boldsymbol{u}\rangle \cdot \boldsymbol{n}|\},$$

which implies a smaller viscosity coefficient $\frac{1}{2}\max\{h^\varepsilon, |\langle \boldsymbol{u}\rangle \cdot \boldsymbol{n}|\}$ compared to $(h^\varepsilon + |\langle \boldsymbol{u}\rangle \cdot \boldsymbol{n}|)$, which is the artificial viscosity coefficient of the flux function defined in (13.7).

The numerical flux function (13.6) satisfies the following identities.

Lemma 13.1 (Diffusive numerical flux)
Let $r_h, \phi_h \in Q_h$, $\boldsymbol{u}_h \in \boldsymbol{V}_{0,h}$, $\phi \in C^1(\Omega)$. Then we have

$$\int_{\mathcal{E}_{int}} \left(F_h^{\rm up}[r_h,\boldsymbol{u}_h] \left[\!\!\left[\frac{|\overline{\boldsymbol{u}_h}|^2}{2} \right]\!\!\right] - \boldsymbol{F}_h^{\rm up}[r_h\boldsymbol{u}_h,\boldsymbol{u}_h] \cdot [\![\overline{\boldsymbol{u}_h}]\!] \right) {\rm dS}_x = D_1 + D_2\,, \tag{13.8}$$

where

$$\begin{aligned} D_1 &= \frac{1}{2}\int_{\mathcal{E}_{int}} r_h^{\rm up}\, |\langle \boldsymbol{u}_h\rangle \cdot \boldsymbol{n}|\, |[\![\overline{\boldsymbol{u}_h}]\!]|^2 \,{\rm dS}_x \\ &= \frac{1}{2}\int_{\mathcal{E}_{int}} \left(r_h^{\rm in}[\langle \boldsymbol{u}_h\rangle \cdot \boldsymbol{n}]^+ - r_h^{\rm out}[\langle \boldsymbol{u}_h\rangle \cdot \boldsymbol{n}]^- \right) |[\![\overline{\boldsymbol{u}_h}]\!]|^2 \,{\rm dS}_x, \\ D_2 &= \frac{h^\varepsilon}{2}\int_{\mathcal{E}_{int}} \{\!\{r_h\}\!\}\, |[\![\overline{\boldsymbol{u}_h}]\!]|^2\, \chi\left(\frac{\langle \boldsymbol{u}_h\rangle \cdot \boldsymbol{n}}{h^\varepsilon} \right) {\rm dS}_x; \end{aligned}$$

and

$$\begin{aligned} &\int_{\Omega_h} r_h \boldsymbol{u}_h \cdot \nabla_x \phi \,{\rm d}x - \int_{\mathcal{E}_{int}} F_h^{\rm up}[r_h,\boldsymbol{u}_h]\, [\![\phi_h]\!] \,{\rm dS}_x \\ &= \frac{h^\varepsilon}{2}\int_{\mathcal{E}_{int}} [\![r_h]\!]\, [\![\phi_h]\!]\, \chi\left(\frac{\langle \boldsymbol{u}_h\rangle \cdot \boldsymbol{n}}{h^\varepsilon} \right) {\rm dS}_x \\ &+ \sum_{K\in\mathcal{T}_h} \sum_{\sigma\in\mathcal{E}(K)} \int_\sigma (\phi_h - \phi)\, [\![r_h]\!]\, [\langle \boldsymbol{u}_h\rangle \cdot \boldsymbol{n}]^- \,{\rm dS}_x \\ &+ \sum_{K\in\mathcal{T}_h} \sum_{\sigma\in\mathcal{E}(K)} \int_\sigma \phi r_h \big(\boldsymbol{u}_h - \langle \boldsymbol{u}_h\rangle \big) \cdot \boldsymbol{n} \,{\rm dS}_x \\ &+ \int_{\Omega_h} r_h(\phi_h - \phi) {\rm div}_h \boldsymbol{u}_h \,{\rm d}x. \end{aligned} \tag{13.9}$$

Proof The proof is analogous to the proof of Lemma 8.1 and we omit it here.

Next, replacing ϕ by $\varphi \in C_c^\infty([0,T)\times\Omega)$ and ϕ_h by $\varphi_h = \Pi_Q\varphi$ in (13.9), we deduce

$$\int_0^T\int_{\Omega_h} r_h \boldsymbol{u}_h \cdot \nabla_x\varphi \,\mathrm{d}x\,\mathrm{d}t - \int_0^T\int_{\mathcal{E}_{int}} F_h^{\mathrm{up}}[r_h, \boldsymbol{u}_h] [\![\varphi_h]\!] \,\mathrm{dS}_x\,\mathrm{d}t = \sum_{i=1}^4 e_i(r_h). \tag{13.10}$$

Using the interpolation estimate (13.9), Poincaré's inequality (13.2) and Hölder's inequality we get the following estimates

$$|e_1(r_h)| \lesssim h^{\varepsilon+1}\int_0^T\int_{\mathcal{E}_{int}} \left| [\![r_h]\!]\, \chi\left(\frac{\langle \boldsymbol{u}_h\rangle\cdot\boldsymbol{n}}{h^\varepsilon}\right)\right| \mathrm{dS}_x\,\mathrm{d}t,$$

$$|e_2(r_h)| \lesssim h\int_0^T\int_{\mathcal{E}_{int}} \left| [\![r_h]\!]\, [\langle \boldsymbol{u}_h\rangle\cdot\boldsymbol{n}]^- \right| \mathrm{dS}_x\,\mathrm{d}t,$$

$$\begin{aligned} |e_3(r_h)| &= \int_0^T \sum_{K\in\mathcal{T}_h}\sum_{\sigma\in\mathcal{E}(K)}\int_\sigma r_h(\varphi - \Pi_Q\varphi)\big(\boldsymbol{u}_h - \langle\boldsymbol{u}_h\rangle\big)\cdot\boldsymbol{n}\,\mathrm{dS}_x\,\mathrm{d}t \\ &\lesssim h\,\|r_h\|_{L^2L^2}\,\|\nabla_h\boldsymbol{u}_h\|_{L^2L^2}\,, \\ |e_4(r_h)| &\lesssim h\,\|r_h\|_{L^2L^2}\,\|\mathrm{div}_h\boldsymbol{u}_h\|_{L^2L^2}\,. \end{aligned}$$

Analogously, replacing $\boldsymbol{\phi}$ by $\boldsymbol{\varphi} \in C_c^\infty([0,T)\times\Omega)$ and $\boldsymbol{\phi}_h$ by $\boldsymbol{\varphi}_h = \overline{\Pi_V\boldsymbol{\varphi}}$ in (13.9), we deduce

$$\int_0^T\int_{\Omega_h} \boldsymbol{r}_h \otimes \boldsymbol{u}_h : \nabla_x\boldsymbol{\varphi}\,\mathrm{d}x\,\mathrm{d}t - \int_0^T\int_{\mathcal{E}_{int}} F_h^{\mathrm{up}}[\boldsymbol{r}_h, \boldsymbol{u}_h]\left[\!\left[\overline{\Pi_V\boldsymbol{\varphi}}\right]\!\right] \mathrm{dS}_x\,\mathrm{d}t = \sum_{i=1}^4 e_i(\boldsymbol{r}_h), \tag{13.11}$$

where $e_i(\boldsymbol{r}_h)$, $i = 1,2,3,4$, are bounded as in (13.10) by using the interpolation estimate (13.21), Poincaré's inequality (13.2) and Hölder's inequality .

Remark 13.2 In view of hypothesis (13.4), we have

$$C_c^\infty([0,T)\times\Omega) \subset \bigcup_{k>0}\bigcap_{h<k} C_c^\infty([0,T)\times\Omega_h).$$

In particular, the test functions belonging to this space are eligible in the numerical scheme defined only on the computational domain Ω_h.

Hereafter, we recall the piecewise constant approximation in time (47) to extend the discrete values $\{f_h^k\}_{k=1}^{N_T}$ to the whole time interval $(0, T)$, i.e.

$$f_h(t) = \sum_{k=0}^{N_T} 1_{[k\Delta t,(k+1)\Delta t)} f_h^k, \quad f_h \in \{\varrho_h, \boldsymbol{u}_h\}.$$

In what follows we list a few properties of the scheme (13.5).

Lemma 13.2 (Properties of the mixed FV–FE scheme)
Let the initial data $[\varrho_0, \boldsymbol{u}_0]$ satisfy (13.2). The mixed FV–FE scheme (13.5) enjoys the following properties:

1. **Existence of a numerical solution**. For any $h > 0$, given $(\varrho_h^{k-1}, \boldsymbol{u}_h^{k-1}) \in Q_h \times \boldsymbol{V}_h$, there exists at least one solution $(\varrho_h^k, \boldsymbol{u}_h^k) \in Q_h \times \boldsymbol{V}_{0,h}$ to the scheme (13.5) for all $k = 1, \ldots, N_T$.
2. **Positivity of discrete density**. Any solution to (13.5) satisfies $\varrho_h(t) > 0$ for all $t \in (0, T)$.
3. **Discrete mass conservation**. Any solution to (13.5) satisfies the conservation of mass
$$\int_{\Omega_h} \varrho_h(t)\,\mathrm{d}x = \int_{\Omega_h} \varrho_0\,\mathrm{d}x, \ t \in (0, T).$$
4. **Internal energy balance**. There exist $\xi \in \mathrm{co}\{\varrho_h^{k-1}, \varrho_h^k\}$, $\zeta \in \mathrm{co}\{(\varrho_h^k)^{\mathrm{in}}, (\varrho_h^k)^{\mathrm{out}}\}$ for any $\sigma \in \mathcal{E}_{int}$ such that
$$\begin{aligned}\int_{\Omega_h} D_t P(\varrho_h^k) + p(\varrho_h^k)\mathrm{div}_h \boldsymbol{u}_h^k \,\mathrm{d}x &= -\frac{\Delta t}{2}\int_{\Omega_h} P''(\xi)|D_t \varrho_h^k|^2\,\mathrm{d}x \\ &- \frac{1}{2}\int_{\mathcal{E}_{int}} P''(\zeta)\left[\!\left[\varrho_h^k\right]\!\right]^2 \max\{h^{\varepsilon}, |\langle \boldsymbol{u}_h^k\rangle \cdot \boldsymbol{n}|\}\,\mathrm{dS}_x.\end{aligned} \tag{13.12}$$

Proof We refer to Lemma 11.3 and Lemma 6 for the proof as it can be done exactly in the same way.

13.2 Stability

The first step in the convergence analysis of the mixed FV–FE scheme is showing its stability. Indeed, the energy stability in the sense of Definition 8.3 is crucial for the proof of convergence.

Theorem 13.1 (Energy stability)
Let $(\varrho_h, \boldsymbol{u}_h)$ be a numerical solution obtained by the mixed FV–FE method (13.5) in the sense of (47). Then, for any $k = 1, \ldots, N_T$, there exist $\xi \in \text{co}\{\varrho_h^{k-1}, \varrho_h^k\}$ and $\zeta \in \text{co}\{(\varrho_h^k)^{\text{in}}, (\varrho_h^k)^{\text{out}}\}$ for any $\sigma \in \mathcal{E}_{int}$, such that

$$\int_{\Omega_h} D_t \left[\frac{1}{2}\varrho_h^k \left|\overline{\boldsymbol{u}_h}^k\right|^2 + P(\varrho_h^k)\right] \, \mathrm{d}x + \int_{\Omega_h} \left[\mu|\nabla_h \boldsymbol{u}_h^k|^2 + \nu|\text{div}_h \boldsymbol{u}_h^k|^2\right] \, \mathrm{d}x = -D_{\Delta t,h},$$

where $P(\varrho) = \frac{a}{\gamma-1}\varrho^\gamma$ is the pressure potential and $D_{\Delta t,h} \geq 0$ represents the numerical diffusion

$$\begin{aligned} D_{\Delta t,h} = &\frac{1}{2}\int_{\Omega_h} P''(\xi)\frac{\left(\varrho_h^k - \varrho_h^{k-1}\right)^2}{\Delta t} \, \mathrm{d}x + \int_{\Omega_h} \frac{\Delta t}{2}\varrho_h^{k-1} \left|\frac{\overline{\boldsymbol{u}_h}^k - \overline{\boldsymbol{u}_h}^{k-1}}{\Delta t}\right|^2 \, \mathrm{d}x \\ &+ \frac{1}{2}\int_{\mathcal{E}_{int}} P''(\zeta) \left[\!\left[\varrho_h^k\right]\!\right]^2 \max\{h^\varepsilon, |\langle \boldsymbol{u}_h^k\rangle \cdot \boldsymbol{n}|\} \, \mathrm{dS}_x \\ &+ \frac{h^\varepsilon}{2}\int_{\mathcal{E}_{int}} \{\{\varrho_h^k\}\} \cdot \left[\!\left[\overline{\boldsymbol{u}_h}^k\right]\!\right]^2 \chi\left(\frac{\langle \boldsymbol{u}_h^k\rangle \cdot \boldsymbol{n}}{h^\varepsilon}\right) \, \mathrm{dS}_x \\ &+ \frac{1}{2}\int_{\mathcal{E}_{int}} \left((\varrho_h^k)^{\text{in}}[\langle \boldsymbol{u}_h^k\rangle \cdot \boldsymbol{n}]^+ - (\varrho_h^k)^{\text{out}}\left[\langle \boldsymbol{u}_h^k\rangle \cdot \boldsymbol{n}\right]^-\right) \left[\!\left[\overline{\boldsymbol{u}_h}^k\right]\!\right]^2 \, \mathrm{dS}_x. \end{aligned}$$

Proof First, we test the discrete continuity equation (13.5a) by $-\frac{|\overline{\boldsymbol{u}_h}^k|^2}{2}$ to get

$$-\frac{1}{2}\int_{\Omega_h} D_t\varrho_h^k \left|\overline{\boldsymbol{u}_h}^k\right|^2 \, \mathrm{d}x + \frac{1}{2}\int_{\mathcal{E}_{int}} F_h^{\text{up}}[\varrho_h^k, \boldsymbol{u}_h^k] \left[\!\left[\left|\overline{\boldsymbol{u}_h}^k\right|^2\right]\!\right] \mathrm{dS}_x = 0.$$

Next, we test the discrete momentum equation (13.5b) by $\boldsymbol{u}_h^k$ to obtain

$$\begin{aligned} &\int_{\Omega_h} D_t \left(\varrho_h^k \overline{\boldsymbol{u}_h}^k\right) \cdot \boldsymbol{u}_h^k \, \mathrm{d}x - \int_{\mathcal{E}_{int}} F_h^{\text{up}}[\varrho_h^k \overline{\boldsymbol{u}_h}^k, \boldsymbol{u}_h^k] \cdot \left[\!\left[\overline{\boldsymbol{u}_h}^k\right]\!\right] \, \mathrm{dS}_x \\ &\qquad - \int_{\Omega_h} p(\varrho_h^k)\text{div}_h \boldsymbol{u}_h^k \, \mathrm{d}x + \mu \left\|\nabla_h \boldsymbol{u}_h^k\right\|_{L^2}^2 + \nu \left\|\text{div}_h \boldsymbol{u}_h^k\right\|_{L^2}^2 = 0. \end{aligned}$$

Summing up the above two equations and recalling (13.8) we derive

$$\int_{\Omega_h} D_t\left(\frac{1}{2}\varrho_h^k\left|\overline{\boldsymbol{u}_h}^k\right|^2\right)\,\mathrm{d}x + \int_{\Omega_h}\frac{\Delta t}{2}\varrho_h^{k-1}\left|D_t\overline{\boldsymbol{u}_h}^k\right|^2\,\mathrm{d}x$$

$$+\mu\left\|\nabla_h\boldsymbol{u}_h^k\right\|_{L^2}^2+\nu\left\|\mathrm{div}_h\boldsymbol{u}_h^k\right\|_{L^2}^2+\frac{h^\varepsilon}{2}\int_{\mathcal{E}_{int}}\{\{\varrho_h^k\}\}\cdot\left[\!\left[\overline{\boldsymbol{u}_h}^k\right]\!\right]^2\chi\left(\frac{\langle\boldsymbol{u}_h^k\rangle\cdot\boldsymbol{n}}{h^\varepsilon}\right)\mathrm{dS}_x$$

$$+\frac{1}{2}\int_{\mathcal{E}_{int}}\left((\varrho_h^k)^{\mathrm{in}}[\langle\boldsymbol{u}_h^k\rangle\cdot\boldsymbol{n}]^+-(\varrho_h^k)^{\mathrm{out}}\left[\langle\boldsymbol{u}_h^k\rangle\cdot\boldsymbol{n}\right]^-\right)\left[\!\left[\overline{\boldsymbol{u}_h^k}\right]\!\right]^2\mathrm{dS}_x$$

$$=\int_{\Omega_h}p(\varrho_h^k)\mathrm{div}_h\boldsymbol{u}_h^k\,\mathrm{d}x.$$

Finally, summing up the above equality with the internal energy balance (13.12) completes the proof.

As a consequence of the energy balance, see Theorem 13.1, we have the following uniform bounds.

Corollary 13.1 (Uniform bounds)
Let $(\varrho_h,\boldsymbol{u}_h)$ be a solution of the mixed FV–FE method (13.5). Then we have the following uniform estimates:

$$\left\|\varrho_h|\overline{\boldsymbol{u}_h}|^2\right\|_{L^\infty L^1}\lesssim 1,\quad \|\varrho_h\|_{L^\infty L^\gamma}\lesssim 1,\quad \|\varrho_h\overline{\boldsymbol{u}_h}\|_{L^\infty L^{2\gamma/(\gamma+1)}}\lesssim 1, \tag{13.13a}$$

$$\|\nabla_h\boldsymbol{u}_h\|_{L^2L^2}\lesssim 1,\quad \|\mathrm{div}_h\boldsymbol{u}_h\|_{L^2L^2}\lesssim 1,\quad \|\boldsymbol{u}_h\|_{L^2L^q}\lesssim 1, \tag{13.13b}$$

$$\int_0^T\int_{\mathcal{E}_{int}}P''(\zeta)\left[\!\left[\varrho_h\right]\!\right]^2\max\{h^\varepsilon,|\langle\boldsymbol{u}_h\rangle\cdot\boldsymbol{n}|\}\,\mathrm{dS}_x\,\mathrm{d}t\lesssim 1, \tag{13.13c}$$

$$h^\varepsilon\int_0^T\int_{\mathcal{E}_{int}}\{\{\varrho_h\}\}\cdot|[\![\overline{\boldsymbol{u}_h}]\!]|^2\chi\left(\frac{\langle\boldsymbol{u}_h\rangle\cdot\boldsymbol{n}}{h^\varepsilon}\right)\mathrm{dS}_x\,\mathrm{d}t\lesssim 1, \tag{13.13d}$$

$$\int_0^T\int_{\mathcal{E}_{int}}\left((\varrho_h)^{\mathrm{in}}[\langle\boldsymbol{u}_h\rangle\cdot\boldsymbol{n}]^+-(\varrho_h)^{\mathrm{out}}[\langle\boldsymbol{u}_h\rangle\cdot\boldsymbol{n}]^-\right)|[\![\overline{\boldsymbol{u}_h}]\!]|^2\,\mathrm{dS}_x\,\mathrm{d}t\lesssim 1, \tag{13.13e}$$

where $q=6$ if $d=3$ and $1\le q<\infty$ if $d=2$.

Further application of the above estimates leads to the following corollary, for the proof we refer to the proofs of Lemmas 11.4, 11.5, and 11.6 as it can be done exactly in the same way.

Corollary 13.2 (Some useful estimates)
Let $(\varrho_h, \boldsymbol{u}_h)$ be a numerical solution of the mixed FV–FE method (13.5). Then the following estimates hold:

$$\|\varrho_h\|_{L^2L^2} \stackrel{<}{\sim} h^{\beta_D}, \quad \beta_D = \begin{cases} -\frac{3\varepsilon+3+d}{6\gamma} & \text{if } \gamma \in (1,2), \\ 0, & \text{if } \gamma \geq 2, \end{cases} \tag{13.14a}$$

$$\|\varrho_h\|_{L^2L^{6/5}} \stackrel{<}{\sim} h^{\beta_R}, \quad \beta_R = \begin{cases} \frac{5\gamma-6}{6\gamma}d & \text{if } \gamma \in (1, \frac{6}{5}), \\ 0, & \text{if } \gamma \geq \frac{6}{5}, \end{cases} \tag{13.14b}$$

$$\|\varrho_h\overline{\boldsymbol{u}_h}\|_{L^2L^2} \stackrel{<}{\sim} h^{\beta_M}, \quad \beta_M = \begin{cases} -\frac{3\varepsilon+3+d}{6\gamma}, & \text{if } \gamma \in (1,2), \\ \frac{\gamma-3}{3\gamma}d, & \text{if } \gamma \in [2,3), \\ 0, & \text{if } \gamma \geq 3 \text{ and } d = 3, \\ 0, & \text{if } \gamma \geq 2 \text{ and } d = 2, \end{cases} \tag{13.14c}$$

$$\int_0^T \int_{\mathcal{E}_{int}} \frac{[\![\varrho_h]\!]^2}{\max\{\varrho_h^{\text{in}}, \varrho_h^{\text{out}}\}} |\langle \boldsymbol{u}_h \rangle \cdot \boldsymbol{n}| \, \mathrm{dS}_x \, \mathrm{d}t \stackrel{<}{\sim} 1 \ \text{ for } \gamma \geq 2, \tag{13.14d}$$

$$\int_0^T \int_{\mathcal{E}_{int}} \left| [\![\varrho_h]\!] \, \langle \boldsymbol{u}_h \rangle \cdot \boldsymbol{n} \right| \mathrm{dS}_x \, \mathrm{d}t \stackrel{<}{\sim} h^{\beta_F}, \ \beta_F = \begin{cases} -\frac{1}{2} & \text{if } \gamma \geq \frac{6}{5}, \\ \frac{d}{2}(\frac{5}{6} - \frac{1}{\gamma}) - \frac{1}{2} & \text{if } \gamma \in (1, \frac{6}{5}). \end{cases} \tag{13.14e}$$

13.3 Consistency

The next step towards the convergence analysis is showing the consistency of numerical solutions in the sense of Definition 8.6, meaning that the numerical solutions satisfy the weak formulation of the Navier–Stokes system (13.1) with certain remainder terms that tend to zero as $h \to 0$.

Theorem 13.2 (Consistency of the mixed FV–FE scheme)
Let $(\varrho_h, \boldsymbol{u}_h)$ be a numerical solution of the mixed FV–FE scheme (13.5) with $\Delta t \approx h \in (0, 1)$. Further, we assume that

$$0 < \varepsilon \ \text{ if } \ \gamma \geq 2 \ \text{ and } \ \varepsilon \in (0, 2\gamma - 1 - d/3) \ \text{ if } \ \gamma \in (4d/(1+3d), 2).$$

Then for any test functions $\varphi \in C_c^\infty([0,T) \times \overline{\Omega})$ and $\boldsymbol{\varphi} \in C_c^\infty([0,T) \times \Omega; R^d)$ the following holds:

$$-\int_\Omega \varrho_h^0 \varphi(0,\cdot)\,\mathrm{d}x = \int_0^T\int_\Omega \left[\varrho_h \partial_t \varphi + \varrho_h \boldsymbol{u}_h \cdot \nabla_x \varphi\right]\,\mathrm{d}x\,\mathrm{d}t + \int_0^T e_{1,h}(t,\varphi)\,\mathrm{d}t; \tag{13.15a}$$

$$-\int_\Omega \varrho_h^0 \overline{\boldsymbol{u}_h}^0 \cdot \boldsymbol{\varphi}(0,\cdot)\,\mathrm{d}x$$

$$= \int_0^T\int_\Omega \left[\varrho_h \overline{\boldsymbol{u}_h} \cdot \partial_t \boldsymbol{\varphi} + \varrho_h \overline{\boldsymbol{u}_h} \otimes \boldsymbol{u}_h : \nabla_x \boldsymbol{\varphi} + p(\varrho_h)\mathrm{div}_x \boldsymbol{\varphi}\right]\mathrm{d}x\,\mathrm{d}t \tag{13.15b}$$

$$-\int_0^T\int_\Omega \left[\mu \nabla_h \boldsymbol{u}_h : \nabla_x \boldsymbol{\varphi} + \nu \mathrm{div}_h \boldsymbol{u}_h \mathrm{div}_x \boldsymbol{\varphi}\right]\mathrm{d}x\,\mathrm{d}t + \int_0^T e_{2,h}(t,\boldsymbol{\varphi})\,\mathrm{d}t;$$

$$\|e_{1,h}(\cdot,\varphi)\|_{L^1(0,T)} \lesssim h^\beta \|\varphi\|_{C^2},\ \|e_{2,h}(\cdot,\boldsymbol{\varphi})\|_{L^1(0,T)} \lesssim h^\beta \|\boldsymbol{\varphi}\|_{C^2}, \quad \text{for some } \beta > 0.$$

Proof Given $\varphi \in C_c^\infty([0,T) \times \overline{\Omega})$ and $\boldsymbol{\varphi} \in C_c^\infty([0,T) \times \Omega; R^d)$ we set $\varphi_h = \Pi_Q \varphi$ and $\boldsymbol{\varphi}_h = \Pi_V \boldsymbol{\varphi}$ to be the test functions in the discrete continuity and momentum equation, respectively. As pointed out in Remark 13.2, we have $\boldsymbol{\varphi} \in C_c^\infty([0,T) \times \Omega_h; R^d)$ for all h small enough so that we may replace Ω by Ω_h in the momentum equation (13.15b). Similarly, extending $\boldsymbol{u}_h$ to be zero outside Ω_h, we may adopt the same convention in the continuity equation (13.15a). Adopting this convention, we simply write Ω instead of Ω_h in the remaining part of the proof.

In what follows we deal with each term in the following four steps.

1. Time derivative terms:

We recall (48a) with $r_h = \varrho_h$ and apply the density bound in (13.13a) as well as the initial condition (13.2) to derive

$$\begin{aligned}\int_0^T\int_\Omega D_t \varrho_h \Pi_Q \varphi\,\mathrm{d}x\,\mathrm{d}t + \int_0^T\int_\Omega \varrho_h \partial_t \varphi\,\mathrm{d}x\,\mathrm{d}t + \int_\Omega \varrho_h^0 \varphi(0)\,\mathrm{d}x \\ \lesssim \Delta t c(\|\varphi\|_{C^2}) \|\varrho_h\|_{L^1L^1} + \Delta t c(\|\varphi\|_{C^1}) \left\|\varrho_h^0\right\|_{L^1} \\ \lesssim \Delta t \|\varrho_h\|_{L^\infty L^\gamma} + \Delta t \left\|\varrho_h^0\right\|_{L^1} \lesssim \Delta t.\end{aligned}$$

Analogously, by setting $r_h = \varrho_h \overline{\boldsymbol{u}_h}$ in (48b) and using the momentum bound in (13.13a) as well as the initial condition (13.2) we derive

$$\int_0^T\int_\Omega \left(D_t(\varrho_h \overline{\boldsymbol{u}_h}) \cdot \Pi_V \boldsymbol{\varphi} + \varrho_h \overline{\boldsymbol{u}_h} \cdot \partial_t \boldsymbol{\varphi}\right)\,\mathrm{d}x\,\mathrm{d}t + \int_\Omega \varrho_h^0 \overline{\boldsymbol{u}_h}^0 \cdot \boldsymbol{\varphi}(0)\,\mathrm{d}x \lesssim \Delta t + h.$$

2. Convective terms:

We first deal with the convective term of the continuity equation. To begin, we recall (13.10) for $r_h = \varrho_h$ to get

$$\int_0^T \int_\Omega \varrho_h \boldsymbol{u}_h \cdot \nabla_x \varphi \, \mathrm{d}x \, \mathrm{d}t - \int_0^T \int_{\mathcal{E}_{int}} F_h^{\mathrm{up}}[\varrho_h, \boldsymbol{u}_h] \llbracket \varphi_h \rrbracket \, \mathrm{dS}_x \, \mathrm{d}t = \sum_{j=1}^4 e_j(\varrho_h),$$

where

$$|e_1(\varrho_h)| \lesssim h^{\varepsilon+1} \int_0^T \int_{\mathcal{E}_{int}} \left| \llbracket \varrho_h \rrbracket \chi \left(\frac{\langle \boldsymbol{u}_h \rangle \cdot \boldsymbol{n}}{h^\varepsilon} \right) \right| \mathrm{dS}_x \, \mathrm{d}t,$$

$$|e_2(\varrho_h)| \lesssim h \int_0^T \int_{\mathcal{E}_{int}} \left| \llbracket \varrho_h \rrbracket \, [\langle \boldsymbol{u}_h \rangle \cdot \boldsymbol{n}]^- \right| \mathrm{dS}_x \, \mathrm{d}t,$$

$$|e_3(\varrho_h)| \lesssim h \, \|\varrho_h\|_{L^2L^2} \, \|\nabla_h \boldsymbol{u}_h\|_{L^2L^2},$$

$$|e_4(\varrho_h)| \lesssim h \, \|\varrho_h\|_{L^2L^2} \, \|\mathrm{div}_h \boldsymbol{u}_h\|_{L^2L^2}.$$

Now we deal with the above residual terms $e_j(\varrho_h)$ for $j = 1, 2, 3, 4$.

Term $e_1(\varrho_h)$. We apply the trace estimates (28), the fact $\left| \llbracket \varrho_h \rrbracket \right| \leq 2 \{\!\{ \varrho_h \}\!\}$ and the density estimate (13.13a) to get

$$|e_1(\varrho_h)| \lesssim h^{\varepsilon+1} \int_0^T \int_{\mathcal{E}_{int}} \left| \llbracket \varrho_h \rrbracket \right| \mathrm{dS}_x \, \mathrm{d}t \lesssim h^{\varepsilon+1} \int_0^T \int_{\mathcal{E}_{int}} \{\!\{ \varrho_h \}\!\} \, \mathrm{dS}_x \, \mathrm{d}t$$
$$\lesssim h^\varepsilon \, \|\varrho_h\|_{L^1L^1} \lesssim h^\varepsilon \, \|\varrho_h\|_{L^\infty L^\gamma} \lesssim h^\varepsilon.$$

Term $e_2(\varrho_h)$. Recalling the estimate (13.14e) we get a positive β such that

$$|e_2(\varrho_h)| \lesssim h^\beta, \quad \beta = 1 + \beta_F = \begin{cases} \frac{1}{2}, & \text{if } \gamma \geq \frac{6}{5}, \\ \frac{1}{2} + \frac{1}{2} d \left(\frac{5}{6} - \frac{1}{\gamma} \right) > \frac{1}{4}, & \text{if } \gamma \in (1, \frac{6}{5}), \end{cases}$$

where β_F is given in (13.14e).

Term $e_3(\varrho_h)$ and $e_4(\varrho_h)$. We use the negative estimate (13.14a) and the velocity bound (13.13b) to get

$$|e_3(\varrho_h)| + |e_4(\varrho_h)| \lesssim h\big(\|\nabla_h \boldsymbol{u}_h\|_{L^2L^2} + \|\mathrm{div}_h \boldsymbol{u}_h\|_{L^2L^2} \big) \, \|\varrho_h\|_{L^2L^2} \lesssim h^\beta.$$

Note that $\beta = 1 + \beta_D$, where β_D is given in (13.14a). Clearly, for $\gamma \geq 2$ we have $\beta > 0$. Further, for $\gamma < 2$ it holds $\beta > 0$ if $\varepsilon < 2\gamma - 1 - d/3$.

For the sum $\sum_{i=1}^{4} |e_i(\varrho_h)|$ we then have the estimate

$$\left| \int_0^T \int_\Omega \varrho_h \boldsymbol{u}_h \cdot \nabla_x \varphi \, \mathrm{d}x \, \mathrm{d}t - \int_0^T \int_{\mathcal{E}_{int}} F_h^{\mathrm{up}}[\varrho_h, \boldsymbol{u}_h] \, [\![\varphi_h]\!] \, \mathrm{dS}_x \, \mathrm{d}t \right| \lesssim h^\beta. \tag{13.16}$$

It holds $\beta > 0$ for all $\varepsilon \in (0, 2\gamma - 1 - d/3)$ if $\gamma \in (1, 2)$, and for all $\varepsilon > 0$ if $\gamma \geq 2$.

Next, to show the consistency of the convective term of the momentum equation, we set $\boldsymbol{r}_h = \varrho_h \overline{\boldsymbol{u}_h}$ in (13.11), which implies

$$\int_0^T \int_\Omega (\varrho_h \overline{\boldsymbol{u}_h} \otimes \boldsymbol{u}_h) : \nabla_x \boldsymbol{\varphi} \, \mathrm{d}x \, \mathrm{d}t - \int_0^T \int_{\mathcal{E}_{int}} F_h^{\mathrm{up}}[\varrho_h \overline{\boldsymbol{u}_h}, \boldsymbol{u}_h] \cdot \left[\!\left[\Pi_V \boldsymbol{\varphi} \right]\!\right] \, \mathrm{dS}_x \, \mathrm{d}t$$
$$= \sum_{j=1}^{4} e_j(\varrho_h \overline{\boldsymbol{u}_h}),$$

where the estimates of residual terms $e_j(\varrho_h \overline{\boldsymbol{u}_h}), j = 1, 2, 3, 4$, read

$$|e_1(\varrho_h \overline{\boldsymbol{u}_h})| \lesssim h^{\varepsilon+1} \int_0^T \int_{\mathcal{E}_{int}} \left| \left[\!\left[\varrho_h \overline{\boldsymbol{u}_h} \right]\!\right] \chi \left(\frac{\langle \boldsymbol{u}_h \rangle \cdot \boldsymbol{n}}{h^\varepsilon} \right) \right| \, \mathrm{dS}_x \, \mathrm{d}t,$$
$$|e_2(\varrho_h \overline{\boldsymbol{u}_h})| \lesssim h \int_0^T \int_{\mathcal{E}_{int}} \left| \left[\!\left[\varrho_h \overline{\boldsymbol{u}_h} \right]\!\right] [\langle \boldsymbol{u}_h \rangle \cdot \boldsymbol{n}]^- \right| \, \mathrm{dS}_x \, \mathrm{d}t,$$
$$|e_3(\varrho_h \overline{\boldsymbol{u}_h})| \lesssim h \, \|\varrho_h \overline{\boldsymbol{u}_h}\|_{L^2 L^2} \, \|\nabla_h \boldsymbol{u}_h\|_{L^2 L^2} \, ,$$
$$|e_4(\varrho_h \overline{\boldsymbol{u}_h})| \lesssim h \, \|\varrho_h \overline{\boldsymbol{u}_h}\|_{L^2 L^2} \, \|\mathrm{div}_h \boldsymbol{u}_h\|_{L^2 L^2} \, .$$

Term $e_1(\varrho_h \overline{\boldsymbol{u}_h})$. First we apply the chain rule (13.23) and the triangle inequality to get

$$|e_1(\varrho_h \overline{\boldsymbol{u}_h})| \lesssim h^{\varepsilon+1} \int_0^T \int_{\mathcal{E}_{int}} |\{\!\{\varrho_h\}\!\} \, [\![\overline{\boldsymbol{u}_h}]\!]| \, \chi \left(\frac{\langle \boldsymbol{u}_h \rangle \cdot \boldsymbol{n}}{h^\varepsilon} \right) \mathrm{dS}_x \, \mathrm{d}t$$
$$+ h^{\varepsilon+1} \int_0^T \int_{\mathcal{E}_{int}} \left| \left[\!\left[\varrho_h \right]\!\right] \{\!\{\overline{\boldsymbol{u}_h}\}\!\} \right| \, \chi \left(\frac{\langle \boldsymbol{u}_h \rangle \cdot \boldsymbol{n}}{h^\varepsilon} \right) \mathrm{dS}_x \, \mathrm{d}t$$
$$\equiv e_{11} + e_{12}.$$

Next, we apply Hölder's inequality, the density bound (13.13a), numerical dissipation (13.13d), and the trace inequality (28) to control the first term e_{11}, i.e.

$$
\begin{aligned}
e_{11} &= h^{1+\varepsilon} \int_0^T \int_{\mathcal{E}_{int}} |\{\!\{\varrho_h\}\!\} [\![\overline{\boldsymbol{u}_h}]\!]| \ \chi\left(\frac{\langle \boldsymbol{u}_h \rangle \cdot \boldsymbol{n}}{h^\varepsilon}\right) \mathrm{dS}_x \, \mathrm{d}t \\
&\lesssim h^{1+\varepsilon} \left(\int_0^T \int_{\mathcal{E}_{int}} \{\!\{\varrho_h\}\!\} \, |[\![\overline{\boldsymbol{u}_h}]\!]|^2 \, \chi\left(\frac{\langle \boldsymbol{u}_h \rangle \cdot \boldsymbol{n}}{h^\varepsilon}\right) \mathrm{dS}_x \, \mathrm{d}t \right)^{1/2} \times \\
&\quad \times \left(\int_0^T \int_{\mathcal{E}_{int}} \{\!\{\varrho_h\}\!\} \, \mathrm{dS}_x \, \mathrm{d}t \right)^{1/2} \\
&\lesssim h^{(1+\varepsilon)/2}.
\end{aligned}
$$

Further, the term e_{12} can be estimated in two steps. On one hand, for $\gamma \geq 6/5$ it is obvious by the trace inequality (28) that

$$
\begin{aligned}
e_{12} &= h^{\varepsilon+1} \int_0^T \int_{\mathcal{E}_{int}} \left| [\![\varrho_h]\!] \, \{\!\{\overline{\boldsymbol{u}_h}\}\!\} \right| \, \chi\left(\frac{\langle \boldsymbol{u}_h \rangle \cdot \boldsymbol{n}}{h^\varepsilon}\right) \mathrm{dS}_x \, \mathrm{d}t \\
&\lesssim h^\varepsilon \, \|\varrho_h\|_{L^2 L^{6/5}} \, \|\boldsymbol{u}_h\|_{L^2 L^6} \lesssim h^\varepsilon.
\end{aligned}
$$

On the other hand, for $\gamma \in (1, 6/5)$ we may apply Hölder's inequality, trace inequality (28) and a priori estimates (13.13b) and (13.13c) to estimate the second term e_{12},

$$
\begin{aligned}
e_{12} &= h^{\varepsilon+1} \int_0^T \int_{\mathcal{E}_{int}} \left| [\![\varrho_h]\!] \, \{\!\{\overline{\boldsymbol{u}_h}\}\!\} \right| \, \chi\left(\frac{\langle \boldsymbol{u}_h \rangle \cdot \boldsymbol{n}}{h^\varepsilon}\right) \mathrm{dS}_x \, \mathrm{d}t \\
&\lesssim h^{\varepsilon+1} \int_0^T \int_{\mathcal{E}_{int}} \left| \sqrt{P''(\varrho_{h,\dagger})} \, [\![\varrho_h]\!] \, \sqrt{\varrho_{h,\dagger}^{2-\gamma}} \, \{\!\{\overline{\boldsymbol{u}_h}\}\!\} \right| \mathrm{dS}_x \, \mathrm{d}t \\
&\lesssim h^{\varepsilon+1} \int_0^T \sum_{\sigma \in \mathcal{E}_{int}} \left(\int_\sigma P''(\varrho_{h,\dagger}) \, [\![\varrho_h]\!]^2 \, \mathrm{dS}_x \right)^{1/2} \|\overline{\boldsymbol{u}_h}\|_{L^6(\sigma)} \left\| \sqrt{\varrho_{h,\dagger}^{2-\gamma}} \right\|_{L^3(\sigma)} \mathrm{d}t \\
&\lesssim h^{\frac{1+\varepsilon}{2}} \left(h^\varepsilon \int_0^T \int_{\mathcal{E}_{int}} P''(\varrho_{h,\dagger}) \, [\![\varrho_h]\!]^2 \, \mathrm{dS}_x \, \mathrm{d}t \right)^{1/2} \|\boldsymbol{u}_h\|_{L^2 L^6} \left\| \sqrt{\varrho_{h,\dagger}^{2-\gamma}} \right\|_{L^\infty L^3} \\
&\lesssim h^{\frac{1+\varepsilon}{2}} \, \|\varrho_h\|_{L^\infty L^{3(2-\gamma)/2}}^{(2-\gamma)/2} \lesssim h^{\frac{\varepsilon}{2}}
\end{aligned}
$$

where $\varrho_{h,\dagger} \in \mathrm{co}\{\varrho_h^{\mathrm{in}}, \varrho_h^{\mathrm{out}}\}$ is given in Corollary 13.1 and we have used the inverse estimate (26) in the last inequality, i.e.

$$h^{\frac{1+\varepsilon}{2}} \|\varrho_h\|^{(2-\gamma)/2}_{L^\infty L^{3(2-\gamma)/2}} \lesssim h^{\frac{1+\varepsilon}{2}} \left(h^{d(\frac{2}{3(2-\gamma)} - \frac{1}{\gamma})} \|\varrho_h\|_{L^\infty L^\gamma} \right)^{\frac{2-\gamma}{2}} \lesssim h^{\frac{\varepsilon}{2} + \frac{(3+5d)\gamma - 6d}{6\gamma}} \lesssim h^{\frac{\varepsilon}{2}}$$

as $(5d+3)\gamma > 5d+3 \geq 6d$. Hence

$$|e_1(\varrho_h \overline{\boldsymbol{u}_h})| \lesssim e_{11} + e_{12} \lesssim h^{\frac{1+\varepsilon}{2}} + h^{\frac{\varepsilon}{2}}.$$

Term $e_2(\varrho_h \overline{\boldsymbol{u}_h})$. We apply the product rule (13.23) and triangular inequality to get

$$\begin{aligned}
|e_2(\varrho_h \overline{\boldsymbol{u}_h})| &\lesssim h \int_0^T \int_{\mathcal{E}_{int}} \left| \left((\varrho_h)^{\text{out}} [\![\overline{\boldsymbol{u}_h}]\!] + [\![\varrho_h]\!] \overline{\boldsymbol{u}_h} \right) [\langle \boldsymbol{u}_h \rangle \cdot \boldsymbol{n}]^- \right| \, \mathrm{dS}_x \, \mathrm{d}t \\
&\lesssim h \int_0^T \int_{\mathcal{E}_{int}} \left(\left| (\varrho_h)^{\text{out}} [\![\overline{\boldsymbol{u}_h}]\!] [\langle \boldsymbol{u}_h \rangle \cdot \boldsymbol{n}]^- \right| + \left| [\![\varrho_h]\!] \overline{\boldsymbol{u}_h} [\langle \boldsymbol{u}_h \rangle \cdot \boldsymbol{n}]^- \right| \right) \mathrm{dS}_x \, \mathrm{d}t \\
&\equiv e_{21} + e_{22}.
\end{aligned}$$

In order to estimate e_{21}, we use Hölder's inequality, the numerical diffusion (13.13e), trace inequality, velocity bound (13.13b) and negative density estimate (13.14b), i.e.

$$\begin{aligned}
e_{21} &= h \int_0^T \int_{\mathcal{E}_{int}} \left| (\varrho_h)^{\text{out}} [\![\overline{\boldsymbol{u}_h}]\!] [\langle \boldsymbol{u}_h \rangle \cdot \boldsymbol{n}]^- \right| \, \mathrm{dS}_x \, \mathrm{d}t \\
&\lesssim h \left(\int_0^T \int_{\mathcal{E}_{int}} -(\varrho_h)^{\text{out}} |[\![\overline{\boldsymbol{u}_h}]\!]|^2 [\langle \boldsymbol{u}_h \rangle \cdot \boldsymbol{n}]^- \, \mathrm{dS}_x \, \mathrm{d}t \right)^{1/2} \times \\
&\quad \times \left(\int_0^T \int_{\mathcal{E}_{int}} (\varrho_h)^{\text{out}} |\langle \boldsymbol{u}_h \rangle| \, \mathrm{dS}_x \, \mathrm{d}t \right)^{1/2} \\
&\lesssim h \left(h^{-1} \|\varrho_h\|_{L^2 L^{6/5}} \|\boldsymbol{u}_h\|_{L^2 L^6} \right)^{1/2} \lesssim h^\beta,
\end{aligned}$$

where $\beta > \frac{1}{4}$ depends on β_R given in (13.14b),

$$\beta = \frac{1}{2} + \frac{1}{2}\beta_R = \begin{cases} \frac{1}{2}, & \text{if } \gamma \geq \frac{6}{5}, \\ \frac{1}{2}\left(1 + \frac{(5\gamma - 6)d}{6\gamma}\right) > \frac{1}{4}, & \text{if } \gamma \in (1, \frac{6}{5}). \end{cases}$$

Further, we estimate the term e_{22} in two steps. On one hand, for $\gamma \geq 2$ we have

$$
\begin{aligned}
e_{22} &= h \int_0^T \int_{\mathcal{E}_{int}} \left| [\![\varrho_h]\!] \overline{\boldsymbol{u}_h} [\langle \boldsymbol{u}_h \rangle \cdot \boldsymbol{n}]^- \right| \mathrm{dS}_x \, \mathrm{d}t \\
&\lesssim h \left(\int_0^T \int_{\mathcal{E}_{int}} \frac{[\![\varrho_h]\!]^2}{\max\{\varrho_h^{\mathrm{in}}, \varrho_h^{\mathrm{out}}\}} \left| [\langle \boldsymbol{u}_h \rangle \cdot \boldsymbol{n}]^- \right| \mathrm{dS}_x \, \mathrm{d}t \right)^{1/2} \times \\
&\quad \times \left(\int_0^T \int_{\mathcal{E}_{int}} \max\{\varrho_h^{\mathrm{in}}, \varrho_h^{\mathrm{out}}\} \left| \overline{\boldsymbol{u}_h} \right|^2 \left| [\langle \boldsymbol{u}_h \rangle \cdot \boldsymbol{n}]^- \right| \mathrm{dS}_x \, \mathrm{d}t \right)^{1/2} \\
&\lesssim h \left(h^{-1} \| \varrho_h \|_{L^\infty L^2} \| \boldsymbol{u}_h \|_{L^2 L^6} \| \boldsymbol{u}_h \|_{L^\infty L^6} \right)^{1/2} \\
&\lesssim h^{1/2} (\Delta t)^{-1/4} \approx h^{1/4},
\end{aligned}
$$

where we have applied the uniform bounds (13.13a), (13.13b), (13.14d), inverse estimate (26) and the trace inequality (28).

On the other hand, for $\gamma \in (1, 2)$ we use $\Delta t \approx h$ and apply the uniform bounds (13.13a), (13.13b), (13.13c), inverse estimate (26) and trace inequality (28) to deduce

$$
\begin{aligned}
e_{22} &= h \int_0^T \int_{\mathcal{E}_{int}} \left| [\![\varrho_h]\!] \overline{\boldsymbol{u}_h} [\langle \boldsymbol{u}_h \rangle \cdot \boldsymbol{n}]^- \right| \mathrm{dS}_x \, \mathrm{d}t \\
&\lesssim h \int_0^T \int_{\mathcal{E}_{int}} \sqrt{P''(\varrho_{h,\dagger}) \varrho_{h,\dagger}^{2-\gamma}} \left| [\![\varrho_h]\!] \overline{\boldsymbol{u}_h} [\langle \boldsymbol{u}_h \rangle \cdot \boldsymbol{n}]^- \right| \mathrm{dS}_x \, \mathrm{d}t \\
&\lesssim h \left(\int_0^T \int_{\mathcal{E}_{int}} P''(\varrho_{h,\dagger}) [\![\varrho_h]\!]^2 \left| [\langle \boldsymbol{u}_h \rangle \cdot \boldsymbol{n}]^- \right| \mathrm{dS}_x \, \mathrm{d}t \right)^{1/2} \times \\
&\quad \times \left(\int_0^T \int_{\mathcal{E}_{int}} \{\!\{ 2\varrho_h \}\!\}^{2-\gamma} \left| \overline{\boldsymbol{u}_h} \right|^2 \left| [\langle \boldsymbol{u}_h \rangle \cdot \boldsymbol{n}]^- \right| \mathrm{dS}_x \, \mathrm{d}t \right)^{1/2} \\
&\lesssim h \left(h^{-1} \left\| \varrho_h^{2-\gamma} \right\|_{L^\infty L^2} \| \boldsymbol{u}_h \|_{L^2 L^6} \| \boldsymbol{u}_h \|_{L^\infty L^6} \right)^{1/2} \\
&\lesssim h^{1/2} (\Delta t)^{-1/4} \| \varrho_h \|_{L^\infty L^{4-2\gamma}}^{(2-\gamma)/2} \| \boldsymbol{u}_h \|_{L^2 L^6}^{3/2} \lesssim h^{1/4} \| \varrho_h \|_{L^\infty L^{4-2\gamma}}^{(2-\gamma)/2},
\end{aligned}
$$

where $\varrho_{h,\dagger} \in \mathrm{co}\{\varrho_h^{\mathrm{in}}, \varrho_h^{\mathrm{out}}\}$ for any $\sigma \in \mathcal{E}$. Clearly, for $4 - 2\gamma \leq \gamma$ (meaning $\gamma \geq 4/3$), we have

$$
h^{1/4} \| \varrho_h \|_{L^\infty L^{4-2\gamma}}^{(2-\gamma)/2} \lesssim h^{1/4} \| \varrho_h \|_{L^\infty L^\gamma}^{(2-\gamma)/2}.
$$

Concerning the case $4 - 2\gamma > \gamma$ (meaning $\gamma < 4/3$), thanks to the inverse estimate (26), we have

$$e_{22} \stackrel{<}{\sim} h^{1/4} \|\varrho_h\|_{L^\infty L^{4-2\gamma}}^{(2-\gamma)/2} \stackrel{<}{\sim} h^{\frac{1}{4}} \left(h^{d(\frac{1}{4-2\gamma} - \frac{1}{\gamma})} \|\varrho_h\|_{L^\infty L^\gamma} \right)^{(2-\gamma)/2} \stackrel{<}{\sim} h^{\frac{1}{4} + d\frac{3\gamma-4}{4\gamma}}.$$

Thus we get

$$e_{22} \stackrel{<}{\sim} h^\beta, \ \beta = \begin{cases} \frac{1}{4} & \text{if } \gamma \geq \frac{4}{3}, \\ \frac{(1+3d)\gamma - 4d}{4\gamma} & \text{if } \gamma \in (1, \frac{4}{3}). \end{cases}$$

Obviously $\beta > 0$ requires $\gamma > 4d/(1 + 3d)$. Collecting the estimates of the terms e_{21} and e_{22} we derive

$$|e_2(\varrho_h \overline{\boldsymbol{u}_h})| \stackrel{<}{\sim} h^\beta \text{ for a } \beta > 0 \text{ provided } \gamma > \frac{4d}{1+3d}.$$

Term $e_3(\varrho_h \overline{\boldsymbol{u}_h})$ and $e_4(\varrho_h \overline{\boldsymbol{u}_h})$. We use the negative estimate (13.14c) and the velocity bounds (13.13b) to get

$$|e_3(\varrho_h \overline{\boldsymbol{u}_h})| + |e_4(\varrho_h \overline{\boldsymbol{u}_h})| \stackrel{<}{\sim} h \|\varrho_h \overline{\boldsymbol{u}_h}\|_{L^2 L^2} \left(\|\nabla_h \boldsymbol{u}_h\|_{L^2 L^2} + \|\mathrm{div}_h \boldsymbol{u}_h\|_{L^2 L^2}\right) \stackrel{<}{\sim} h^\beta,$$

where $\beta = 1 + \beta_M$, see β_M in (13.14c). If $\gamma \in (1, 2)$ then $\beta > 0$ provided $\varepsilon < 2\gamma - 1 - d/3$, else if $\gamma \geq 2$ then $\beta > 0$.

Consequently, collecting the above terms leads to the consistency of the convective term of the momentum method, i.e.

$$\left| \int_0^T \int_\Omega \varrho_h \left(\overline{\boldsymbol{u}_h} \otimes \boldsymbol{u}_h \right) : \nabla_x \boldsymbol{\varphi} \, \mathrm{d}x \, \mathrm{d}t - \int_0^T \int_{\mathcal{E}_{int}} F_h^{\mathrm{up}}[\varrho_h \overline{\boldsymbol{u}_h}, \boldsymbol{u}_h] \cdot [\![\Pi_V \boldsymbol{\varphi}]\!] \, \mathrm{dS}_x \, \mathrm{d}t \right|$$
$$\stackrel{<}{\sim} h^\beta.$$

We have $\beta > 0$ provided $\varepsilon \in (0, 2\gamma - 1 - d/3)$ and $\gamma \in (4d/(1 + 3d), 2)$. Further, for $\gamma \geq 2$ we have $\beta > 0$ for all $\varepsilon > 0$.

3. Viscosity and pressure terms:

It is easy to derive, analogously as (18), that

$$\int_K \nabla_x \boldsymbol{\varphi} \, \mathrm{d}x = \int_K \nabla_h \Pi_V \boldsymbol{\varphi} \, \mathrm{d}x \ \text{ and } \int_K \mathrm{div}_x \boldsymbol{\varphi} \, \mathrm{d}x = \int_K \mathrm{div}_h \Pi_V \boldsymbol{\varphi} \, \mathrm{d}x.$$

Noticing that $\nabla_h \boldsymbol{u}_h$, $\mathrm{div}_h \boldsymbol{u}_h$ and $p(\varrho_h)$ are piecewise constant functions, we have

$$\int_\Omega \nabla_h \boldsymbol{u}_h : \nabla_h \Pi_V \boldsymbol{\varphi}\ \mathrm{d}x = \int_\Omega \nabla_h \boldsymbol{u}_h : \nabla_x \boldsymbol{\varphi}\ \mathrm{d}x,$$

$$\int_\Omega \mathrm{div}_h \boldsymbol{u}_h \mathrm{div}_h\ \Pi_V \boldsymbol{\varphi}\ \mathrm{d}x = \int_\Omega \mathrm{div}_h \boldsymbol{u}_h\ \mathrm{div}_x \boldsymbol{\varphi}\ \mathrm{d}x,$$

$$\int_\Omega p(\varrho_h)\ \mathrm{div}_h \Pi_V \boldsymbol{\varphi}\ \mathrm{d}x = \int_\Omega p(\varrho_h)\ \mathrm{div}_x \boldsymbol{\varphi}\ \mathrm{d}x.$$

Finally, summing up the previous observations, we obtain (13.15), i.e.

$$-\int_\Omega \varrho_h^0 \varphi(0,\cdot)\ \mathrm{d}x = \int_0^T \int_\Omega \left[\varrho_h \partial_t \varphi + \varrho_h \boldsymbol{u}_h \cdot \nabla_x \varphi\right]\ \mathrm{d}x\, \mathrm{d}t + \mathcal{O}(h^{\beta_1})$$

for $\beta_1 > 0$ provided $\varepsilon \in (0, 2\gamma - 1 - d/3)$ if $\gamma \in (1,2)$ and $\varepsilon > 0$ if $\gamma \geq 2$.

$$\begin{aligned}
&-\int_\Omega \varrho_h^0 \overline{\boldsymbol{u}_h}^0 \cdot \boldsymbol{\varphi}(0,\cdot)\ \mathrm{d}x \\
&= \int_0^T \int_\Omega \left[\varrho_h \overline{\boldsymbol{u}_h} \cdot \partial_t \boldsymbol{\varphi} + \varrho_h \overline{\boldsymbol{u}_h} \otimes \boldsymbol{u}_h : \nabla_x \boldsymbol{\varphi} + p(\varrho_h) \mathrm{div}_x \boldsymbol{\varphi}\right]\ \mathrm{d}x\ \mathrm{d}t \\
&- \int_0^T \int_\Omega \left[\mu \nabla_h \boldsymbol{u}_h : \nabla_x \boldsymbol{\varphi} + \nu \mathrm{div}_h \boldsymbol{u}_h \cdot \mathrm{div}_x \boldsymbol{\varphi}\right]\ \mathrm{d}x\, \mathrm{d}t + \mathcal{O}(h^{\beta_2})
\end{aligned}$$

for $\beta_2 > 0$ provided $\varepsilon \in (0, 2\gamma - 1 - d/3)$ if $\gamma \in (4d/(1+3d), 2)$ and $\varepsilon > 0$ if $\gamma \geq 2$.

Remark 13.3 A close inspection of the previous discussion shows that the same method can be used to handle a variable time step Δt_k adjusted for each step of iteration by means of a CFL-type condition, such as $||\boldsymbol{u}_h^{k-1} + c_h^{k-1}||_{L^\infty(\Omega)} \Delta t_k / h \leq CFL$. Here $CFL \in (0,1]$ and $c_h^{k-1} \equiv \sqrt{p'(\varrho_h^{k-1})}$ denotes the sound speed. Though this condition is necessary for stability of time-explicit numerical schemes, it still may be appropriate even for implicit schemes for areas of high-speed flows.

13.4 Convergence

Before stating our main result concerning convergence of the mixed FV–FE element scheme, we specify the necessary modification in Definition 5.10 that accommodates the no-slip boundary conditions (13.3). We recall the identity

$$\int_\Omega \left[\mu|\nabla_x \boldsymbol{u}|^2 + \nu|\mathrm{div}_x \boldsymbol{u}|^2\right] \mathrm{d}x = \int_\Omega \left[F(\mathbb{D}\boldsymbol{u}) + F^*(\mathbb{S})\right] \mathrm{d}x$$

that can be verified for the limit exactly as in Definition 5.10.

Definition 13.2 (**DMV** SOLUTION, NO SLIP BOUNDARY CONDITION) A parametrized probability measure $\{\mathcal{V}_{t,x}\}_{(t,x)\in(0,T)\times\Omega}$,

$$\mathcal{V} \in L^\infty((0,T)\times\Omega; \mathcal{P}(R^{d+1})),\ R^{d+1} = \left\{[\widetilde{\varrho}, \widetilde{\boldsymbol{u}}] \ \middle|\ \widetilde{\varrho} \in R,\ \widetilde{\boldsymbol{u}} \in R^d\right\},$$
$$\mathcal{V}_{t,x}\{\widetilde{\varrho} \geq 0\} = 1 \text{ for a.a. } (t,x),$$

is called *dissipative measure-valued (DMV) solution* of the Navier–Stokes system (13.1), (13.3), with the initial conditions $[\varrho_0, \boldsymbol{m}_0]$, if the following hold (10):

- (**energy inequality**)

$$\boldsymbol{u} \equiv \langle \mathcal{V}; \widetilde{\boldsymbol{u}}\rangle \in L^2(0,T; W_0^{1,2}(\Omega; R^d)),\ \mathbb{S} \in L^2(0,T; L^2(\Omega; R^{d\times d}_{\mathrm{sym}})),$$

and the integral inequality

$$\begin{aligned} &\int_\Omega \left\langle \mathcal{V}_{\tau,x}; \frac{1}{2}\widetilde{\varrho}|\widetilde{\boldsymbol{u}}|^2 + P(\widetilde{\varrho})\right\rangle \mathrm{d}x + \int_0^\tau \int_\Omega \left[F(\mathbb{D}_x \boldsymbol{u}) + F^*(\mathbb{S})\right] \mathrm{d}x\,\mathrm{d}t \\ &+ \int_{\overline{\Omega}} \mathrm{d}\mathfrak{E}(\tau) + \int_0^\tau \int_{\overline{\Omega}} \mathrm{d}\mathfrak{D} \leq \int_\Omega \left[\frac{1}{2}\frac{|\boldsymbol{m}_0|^2}{\varrho_0} + P(\varrho_0)\right] \mathrm{d}x \end{aligned} \tag{13.17}$$

holds for a.a. $0 \leq \tau \leq T$, with the energy concentration defect

$$\mathfrak{E} \in L^\infty(0,T; \mathcal{M}^+(\overline{\Omega})),$$

and the dissipation defect

$$\mathfrak{D} \in \mathcal{M}^+([0,T]\times\overline{\Omega});$$

- (**equation of continuity**)

$$\langle \mathcal{V}; \widetilde{\varrho}\rangle \in C_{\mathrm{weak}}([0,T]; L^\gamma(\Omega)),\ \left\langle \mathcal{V}_{0,x}; \widetilde{\varrho}\right\rangle = \varrho_0(x) \text{ for a.a. } x \in \Omega,$$

and the integral identity

$$\left[\int_\Omega \langle \mathcal{V}; \widetilde{\varrho}\rangle \varphi \,\mathrm{d}x\right]_{t=0}^{t=\tau} = \int_0^\tau \int_\Omega \left[\langle \mathcal{V}; \widetilde{\varrho}\rangle \partial_t \varphi + \langle \mathcal{V}; \widetilde{\varrho}\widetilde{\boldsymbol{u}}\rangle \cdot \nabla_x \varphi\right] \mathrm{d}x\,\mathrm{d}t \tag{13.18}$$

for any $0 \leq \tau \leq T$, and any $\varphi \in W^{1,\infty}((0,T) \times \Omega)$;

- **(momentum equation)**

$$\langle \mathcal{V}; \widetilde{\varrho \boldsymbol{u}} \rangle \in C_{\rm weak}([0,T]; L^{\frac{2\gamma}{\gamma+1}}(\Omega; R^d)),\ \left\langle \mathcal{V}_{0,x}; \widetilde{\varrho \boldsymbol{u}} \right\rangle = \boldsymbol{m}_0(x) \text{ for a.a. } x \in \Omega,$$

and the integral identity

$$\begin{aligned} &\left[\int_\Omega \langle \mathcal{V}; \widetilde{\varrho \boldsymbol{u}} \rangle \cdot \boldsymbol{\varphi} \, {\rm d}x \right]_{t=0}^{t=\tau} \\ &= \int_0^\tau \int_\Omega \left[\langle \mathcal{V}; \widetilde{\varrho \boldsymbol{u}} \rangle \cdot \partial_t \boldsymbol{\varphi} + \langle \mathcal{V}; \widetilde{\varrho \boldsymbol{u}} \otimes \widetilde{\boldsymbol{u}} \rangle : \nabla_x \boldsymbol{\varphi} + \langle \mathcal{V}; p(\widetilde{\varrho}) \rangle {\rm div}_x \boldsymbol{\varphi} \right] \, {\rm d}x \, {\rm d}t \\ &- \int_0^\tau \int_\Omega \mathbb{S} : \nabla_x \boldsymbol{\varphi} \, {\rm d}x \, {\rm d}t + \int_0^\tau \int_\Omega \nabla_x \boldsymbol{\varphi} : {\rm d}\mathfrak{R}(t) \, {\rm d}t \end{aligned} \tag{13.19}$$

holds for any $0 \leq \tau \leq T$, and any $\boldsymbol{\varphi} \in C^1_c([0,T] \times \Omega; R^d)$, with the Reynolds concentration defect

$$\mathfrak{R} \in L^\infty(0,T; \mathcal{M}^+(\Omega; R^{d \times d}_{\rm sym}))$$

satisfying

$$\underline{d}\ \mathfrak{E} \leq {\rm tr}[\mathfrak{R}] \leq \overline{d}\ \mathfrak{E} \text{ for some constants } 0 < \underline{d} \leq \overline{d}; \tag{13.20}$$

- **(Poincaré inequality)** the velocity field $\boldsymbol{u} = \langle \mathcal{V}; \widetilde{\boldsymbol{u}} \rangle \in L^2(0,T; W^{1,2}(\Omega; R^d))$ satisfies

$$\begin{aligned} &\int_0^\tau \int_\Omega \left\langle \mathcal{V}; |\widetilde{\boldsymbol{u}} - \boldsymbol{U}|^2 \right\rangle \, {\rm d}x \, {\rm d}t \\ &\lesssim \int_0^\tau \int_\Omega |\nabla_x (\boldsymbol{u} - \boldsymbol{U})|^2 \, {\rm d}x \, {\rm d}t + \int_0^\tau \left(\int_{\overline{\Omega}} {\rm d}\mathfrak{E}(t) \right) {\rm d}t + \int_0^\tau \int_{\overline{\Omega}} {\rm d}\mathfrak{D} \end{aligned} \tag{13.21}$$

for a.a. $0 \leq \tau \leq T$, and any $\boldsymbol{U} \in C^1_c([0,T] \times \Omega; R^d)$.

The only difference with respect to Definition 5.10 is the class of admissible test functions in the momentum equation (13.19), and the Poincaré inequality (13.21) that replaces the Korn–Poincaré inequality (5.126). It is easy to check that all fundamental results stated in Theorems 7.12, 7.13 remain valid in the present setting.

Indeed, it is enough to observe that the momentum balance (13.19) can be extended to the class of test functions including the strong solution, namely $\boldsymbol{\varphi} \in C^1([0,T] \times \overline{\Omega}; R^d)$, $\boldsymbol{\varphi}|_{\partial\Omega} = 0$. As $\partial\Omega$ is smooth, it is easy to construct a sequence of functions

$$\chi_n \in C^1_c(\Omega),\ 0 \leq \chi_n \leq 1,$$
$$\chi_n(x) = 0 \text{ if } \mathrm{dist}[x, \partial\Omega] < \frac{1}{2n},\ \chi_n(x) = 1 \text{ for } \mathrm{dist}[x, \partial\Omega] > \frac{1}{n},$$
$$|\nabla_x \chi_n(x)| \stackrel{<}{\sim} n \text{ for } x \in \Omega.$$

For $\boldsymbol{\varphi} \in C^1([0, T] \times \overline{\Omega}; R^d)$, we have

$$\boldsymbol{\varphi}_n(t, x) = \chi_n(x)\boldsymbol{\varphi}(t, x) \in C^1_c([0, T] \times \Omega; R^d),$$

and

$$\boldsymbol{\varphi}_n(t, x) \to \boldsymbol{\varphi}(t, x),\ \nabla_x \boldsymbol{\varphi}_n(t, x) \to \nabla_x \boldsymbol{\varphi}(t, x) \text{ for any } t \in [0, T],\ x \in \Omega,$$

$$|\nabla_x \boldsymbol{\varphi}_n(t, x)| \leq |\nabla_x \chi_n(x) \cdot \boldsymbol{\varphi}(t, x)| + |\chi_n(x)\nabla_x \boldsymbol{\varphi}(t, x)| \stackrel{<}{\sim} 1$$

as

$$|\nabla_x \chi_n(x) \cdot \boldsymbol{\varphi}(t, x)| \stackrel{<}{\sim} n \sup_{\mathrm{dist}[x,\partial\Omega] \leq \frac{1}{n}} |\boldsymbol{\varphi}(t, x)| \stackrel{<}{\sim} 1 \text{ for any } (t, x) \in [0, T] \times \Omega.$$

Using $\boldsymbol{\varphi}_n$ as a test function in the momentum balance (13.19) and passing to the limit for $n \to \infty$ we get the desired conclusion. Note that, by virtue of the Lebesgue theorem,

$$\int_0^T \int_\Omega \nabla_x \boldsymbol{\varphi}_n : \mathrm{d}\mathfrak{R}(t)\, \mathrm{d}t \to \int_0^T \int_\Omega \nabla_x \boldsymbol{\varphi} : \mathrm{d}\mathfrak{R}(t)\, \mathrm{d}t,$$

in particular, the limit integral is over the *open* set Ω.

Our ultimate goal is to establish the convergence of the numerical solutions obtained by the mixed FV–FE method (13.5) to a DMV solution. The principal difficulty is that certain nonlinear terms in the consistency formulation (13.15b) contain both $\boldsymbol{u}_h$ and its projection $\overline{\boldsymbol{u}_h}$.

Theorem 13.3 (**Convergence of the FV–FE method**)
Let $\Omega \subset R^d$, $d = 2, 3$, be a bounded domain of class $C^{2+\alpha}$, $\alpha > 0$. Let $(\varrho_h, \boldsymbol{u}_h)$ be a solution to the mixed FV–FE method (13.5) with the initial data (13.2) and $\Delta t \approx h$. Further, assume that $\varepsilon > 0$ if $\gamma \geq 2$ and $0 < \varepsilon < 2\gamma - 1 - d/3$ if $\gamma \in (4d/(1 + 3d), 2)$.

1. **Convergence to DMV solution**. Any Young measure $\{\mathcal{V}_{t,x}\}_{(t,x)\in(0,T)\times\Omega}$ generated by a sequence of FV–FE solutions $\{\varrho_h^k, \overline{\boldsymbol{u}_h}^k\}_{h\searrow 0}$ represents a DMV solution of the Navier–Stokes system in $(0, T) \times \Omega$ in the sense of Definition 13.2.
2. **Convergence to strong solution**. Suppose that the Navier–Stokes system (13.1), (13.3), endowed with the initial data (13.2), admits a regular solution $[\varrho, \boldsymbol{u}]$ belonging to the class

$$\varrho > 0,\ \varrho \in C([0,T] \times \overline{\Omega}),\ \nabla_x \varrho,\ \boldsymbol{u} \in C([0,T] \times \overline{\Omega}; R^d),$$
$$\nabla_x \boldsymbol{u} \in C([0,T] \times \overline{\Omega}; R^{d\times d}),\ \partial_t \boldsymbol{u} \in L^2(0,T; C(\overline{\Omega}; R^d)),\ \boldsymbol{u}|_{\partial\Omega} = 0.$$

Then

$$\varrho_h \to \varrho \text{ (strongly) in } L^\gamma((0,T) \times K),\ \boldsymbol{u}_h \to \boldsymbol{u} \text{ (strongly) in } L^2((0,T) \times K; R^d)$$

for any compact $K \subset \Omega$.

3. **Convergence to strong solution with bounded density**. Suppose that
   - the initial data belong to the class $\varrho_0 \in W^{3,2}(\Omega),\ \boldsymbol{u}_0 \in W^{3,2}(\Omega; R^d)$;
   - bulk viscosity vanishes, meaning $\lambda = 0$;
   - uniform bound of density $\|\varrho_h\|_{L^\infty((0,T)\times\Omega)} \lesssim 1$ for $h \to 0$.

   Then

$$\varrho_h \to \varrho \text{ (strongly) in } L^q((0,T) \times \Omega),\ q \geq 1,$$
$$\boldsymbol{u}_h \to \boldsymbol{u} \text{ (strongly) in } L^2((0,T) \times \Omega; R^d),$$

   $[\varrho, \boldsymbol{u}]$ is the strong solution to the Navier–Stokes system (11.1) with the initial data $[\varrho_0, \boldsymbol{u}_0]$.

Proof In order to prove the convergence to a DMV solution, we follow the arguments of Theorem 5.5. In view of Theorem 13.1, we may suppose, passing to a subsequence if necessary, that $\{\varrho_h, \overline{\boldsymbol{u}_h}\}_{h\searrow 0}$ generates a Young measure

$$\{\mathcal{V}_{t,x}\}_{(t,x)\in(0,T)\times\Omega},\ \mathcal{V} \in L^\infty_{\text{weak}-(*)}((0,T) \times \Omega; \mathcal{P}(R^{d+1})).$$

Moreover, we can deduce from Theorem 13.1 that for a suitable subsequence

$$\varrho_h \to \varrho \text{ weakly-(*) in } L^\infty(0,T; L^\gamma(\Omega)),\ \varrho \geq 0$$
$$\overline{\boldsymbol{u}_h},\ \boldsymbol{u}_h \to \boldsymbol{u} \text{ weakly in } L^2((0,T) \times \Omega; R^d),$$
$$\text{where } \boldsymbol{u} \in L^2(0,T; W_0^{1,2}(\Omega; R^d)),\ \nabla_h \boldsymbol{u}_h \to \nabla_x \boldsymbol{u} \text{ weakly in } L^2((0,T) \times \Omega; R^{d\times d}),$$
$$\varrho_h \overline{\boldsymbol{u}_h} \to \overline{\varrho \boldsymbol{u}} \text{ weakly-(*) in } L^\infty(0,T; L^{\frac{2\gamma}{\gamma+1}}(\Omega; R^d)).$$

Further, by virtue of the discrete Poincaré inequality (30) and the energy estimates, cf. Theorem 13.1, we also have

$$\|\boldsymbol{u}_h - \overline{\boldsymbol{u}_h}\|_{L^2(0,T;L^2(K;R^d))} \lesssim h \text{ for any compact } K \in \Omega.$$

Consequently, the Young measures generated by $\{\varrho_h, \overline{\boldsymbol{u}_h}\}_{h\searrow 0}$ and $\{\varrho_h, \boldsymbol{u}_h\}_{h\searrow 0}$ coincide for a.a. $(t,x) \in (0,T) \times \Omega$. In particular, their barycenter's second coordinate represented by the limits of $\boldsymbol{u}_h$ and $\overline{\boldsymbol{u}_h}$ coincide on $(0, T) \times \Omega$. Indeed this can be checked by considering a globally Lipschitz compactly supported function $b = b(\varrho, \boldsymbol{u})$ for which we can estimate

$$\int_0^T \int_K |b(\varrho_h, \boldsymbol{u}_h) - b(\varrho_h, \overline{\boldsymbol{u}_h})| \, \mathrm{d}x \, \mathrm{d}t \lesssim \int_0^T \int_K |\boldsymbol{u}_h - \overline{\boldsymbol{u}_h}| \, \mathrm{d}x \, \mathrm{d}t$$

$$\lesssim \|\boldsymbol{u}_h - \overline{\boldsymbol{u}_h}\|_{L^2(0,T;L^2(K;R^d))} \lesssim h.$$

Now, passing to the limit with $h \to 0$ in the consistency formulation of the continuity equation (13.15a), see Theorem 13.2, we obtain

$$\left[\int_\Omega \varrho \varphi(t, \cdot) \, \mathrm{d}x \right]_{t=0}^{t=\tau} = \int_0^\tau \int_\Omega \left[\varrho \partial_t \varphi + \overline{\varrho \boldsymbol{u}} \cdot \nabla_x \varphi \right] \, \mathrm{d}x \, \mathrm{d}t \tag{13.22}$$

for any $0 \leq \tau \leq T$ and any $\varphi \in C^\infty([0, T] \times \overline{\Omega})$. This can be rewritten as (13.18) from Definition 13.2 of the DMV solution, i.e.

$$\left[\int_\Omega \langle \mathcal{V}; \widetilde{\varrho} \rangle \, \varphi \, \mathrm{d}x \right]_{t=0}^{t=\tau} = \int_0^\tau \int_\Omega \left[\langle \mathcal{V}; \widetilde{\varrho} \rangle \, \partial_t \varphi + \langle \mathcal{V}; \widetilde{\varrho} \widetilde{\boldsymbol{u}} \rangle \cdot \nabla_x \varphi \right] \mathrm{d}x \, \mathrm{d}t$$

for any $0 \leq \tau \leq T$, and any $\varphi \in W^{1,\infty}((0, T) \times \Omega)$.

In order to apply a similar treatment to the momentum equation (13.15b), we have to replace the expression $\varrho_h \overline{\boldsymbol{u}_h} \otimes \boldsymbol{u}_h$ in the convective term by $\varrho_h \overline{\boldsymbol{u}_h} \otimes \overline{\boldsymbol{u}_h}$. This is possible as

$$\|\varrho_h \overline{\boldsymbol{u}_h} \otimes \boldsymbol{u}_h - \varrho_h \overline{\boldsymbol{u}_h} \otimes \overline{\boldsymbol{u}_h}\|_{L^1 L^1} = \|\varrho_h \overline{\boldsymbol{u}_h} \otimes (\boldsymbol{u}_h - \overline{\boldsymbol{u}_h})\|_{L^1 L^1}$$

$$\lesssim \|\varrho_h \overline{\boldsymbol{u}_h}\|_{L^2 L^2} \|\boldsymbol{u}_h - \overline{\boldsymbol{u}_h}\|_{L^2 L^2} \lesssim h^{1+\beta_M},$$

where, by virtue of (13.14c), $1 + \beta_M > 0$ as soon as $0 < \varepsilon < 2\gamma - 1 - d/3$ for $\gamma \in (1, 2)$. For $\gamma \geq 2$ it holds that $1 + \beta_M > 0$.

Moreover, we have

$$\varrho_h \overline{\boldsymbol{u}_h} \otimes \overline{\boldsymbol{u}_h} + p(\varrho_h)\mathbb{I} \to \overline{\varrho \boldsymbol{u} \otimes \boldsymbol{u} + p(\varrho)\mathbb{I}} \text{ weakly-(*) in } L^\infty(0, T; \mathcal{M}(\Omega; R^{d \times d}_{\mathrm{sym}}));$$

whence letting $h \to 0$ in (13.15b) gives rise to the integral equality for the momentum equation (13.19) from Definition 13.2 of the DMV solution.

Finally, in order to assure that the limit is a DMV solution in the sense of Definition 13.2, it remains to show that the discrete differential operators are compatible, cf. Definition 5.8. Indeed, due to (13.20), (13.2), the consistency of the discrete gradient and divergence operators holds. Thus, the limit $\mathcal{V}$ is a DMV solution in the sense of Definition 13.2. Consequently, we may apply Theorems 7.12, and 7.13 to obtain to the second and the third convergence statements.

Remark 13.4 Note that the Navier–Stokes system (13.1) admits local-in-time strong solutions for arbitrary smooth initial data, see Theorem 3.4, as soon as the physical domain Ω is sufficiently smooth.

13.5 Numerical Experiment

The aim of this section is to illustrate behavior of the mixed FV–FE method (13.5) and confirm theoretical convergence result proved in Theorem 13.3 via numerical experiments. To this end, we study the experimental order of convergence (EOC), cf. (10.76), with respect to the following norms

$$
\begin{aligned}
e_{\nabla \boldsymbol{u}} &= \|\nabla_{\mathcal{D}}(\boldsymbol{u}_h - \boldsymbol{u}_{\mathrm{ref}})\|_{L^2((0,T)\times\Omega)}, & e_{\boldsymbol{u}} &= \|\boldsymbol{u}_h - \boldsymbol{u}_{\mathrm{ref}}\|_{L^2((0,T)\times\Omega)}, \\
e_{\varrho} &= \|\varrho_h - \varrho_{\mathrm{ref}}\|_{L^1((0,T)\times\Omega)}, & e_p &= \|p(\varrho_h) - p(\varrho_{\mathrm{ref}})\|_{L^\infty(0,T;L^1(\Omega))}, \\
e_{\varrho}^{\dagger} &= \|\varrho_h - \varrho_{\mathrm{ref}}\|_{L^\infty(0,T;L^\gamma(\Omega))}. & &
\end{aligned}
\tag{13.23}
$$

Here $(\varrho_h, \boldsymbol{u}_h)$ denotes a numerical solution obtained by the mixed FV–FE method (13.5). Since the exact solution is not explicitly known we choose for the reference solution $(\varrho_{\mathrm{ref}}, \boldsymbol{u}_{\mathrm{ref}})$ the numerical solution computed on a fine grid with the mesh size $h = 1/512$.

Analogously as in Chap. 11 we consider the following initial data

$$
\varrho_0(x_1, x_2, t) = 2 + \cos(2\pi(x_1 + x_2)), \quad \boldsymbol{u}_0(x_1, x_2, t) = \mathbf{0}
$$

on a two-dimensional domain $\Omega = [0, 1]^2$, no-slip boundary conditions and the following parameters $\mu = 1.0$, $\lambda = 0$, $\gamma = 1.4$ and $\varepsilon = 0.6$. We apply the fixed point iteration method to solve the nonlinear system resulting from (13.5). For each subiteration we apply additionally the CFL condition $\Delta t = 0.6h/(|\boldsymbol{u}_h| + c_h)_{\max}$, where $c_h = \sqrt{\gamma p_h/\varrho_h}$ is the sound speed.

Table 13.1 illustrates the numerical convergence measured in the norms presented in (13.23). Clearly, we observe the first order convergence rate for the density and the gradient of velocity. As expected the convergence rate of the velocity is higher due to piecewise linear approximation.

Table 13.1 EOC of mixed FV–FE method for the Navier–Stokes equations at $T = 0.1$

h	$e_\varrho^\dagger$	EOC	$e_{\boldsymbol{u}}$	EOC	$e_{\nabla \boldsymbol{u}}$	EOC	e_p	EOC	e_ϱ	EOC
					$\gamma = 1.4$					
1/32	3.12e−02	–	3.08e−03	–	1.69e−01	–	7.78e−03	–	1.03e−03	–
1/64	1.91e−02	0.71	1.30e−03	1.24	8.60e−02	0.97	3.92e−03	0.99	4.99e−04	1.04
1/128	1.16e−02	0.72	7.00e−04	0.89	4.41e−02	0.96	1.94e−03	1.01	2.46e−04	1.02
1/256	5.87e−03	0.98	2.53e−04	1.47	2.23e−02	0.98	7.53e−04	1.37	1.02e−04	1.27
					$\gamma = 5/3$					
1/32	8.64e−02	–	5.04e−03	–	2.31e−01	–	1.69e−02	–	1.02e−03	–
1/64	5.73e−02	0.59	2.22e−03	1.18	1.18e−01	0.97	8.52e−03	0.99	4.92e−04	1.05
1/128	3.76e−02	0.61	1.00e−03	1.15	6.01e−02	0.97	4.21e−03	1.02	2.38e−04	1.05
1/256	2.13e−02	0.82	3.60e−04	1.48	3.03e−02	0.99	1.64e−03	1.36	9.83e−05	1.27
					$\gamma = 2$					
1/32	2.16e−01	–	6.58e−03	–	4.15e−01	–	4.65e−02	–	9.94e−04	–
1/64	1.30e−01	0.73	3.25e−03	1.02	1.69e−01	1.30	1.68e−02	1.47	4.90e−04	1.02
1/128	9.15e−02	0.50	1.49e−03	1.13	8.62e−02	0.97	8.37e−03	1.01	2.36e−04	1.05
1/256	5.71e−02	0.68	5.16e−04	1.53	4.33e−02	0.99	3.26e−03	1.36	9.61e−05	1.30

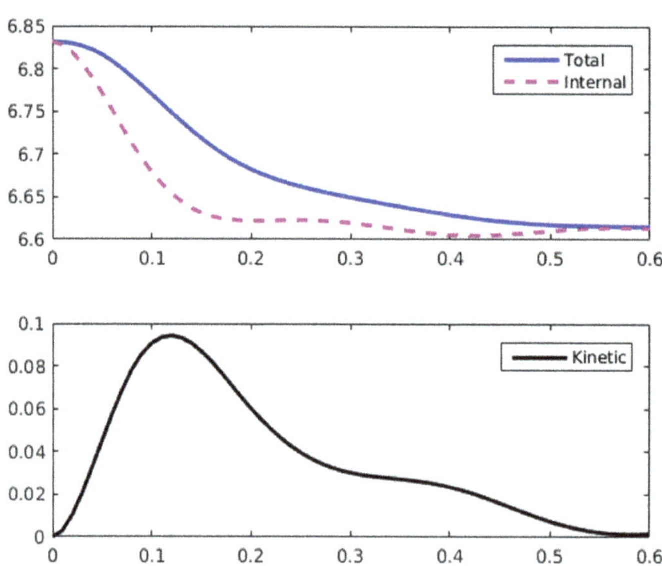

Fig. 13.1 Mixed FV–FE method for the Navier–Stokes equations: time evolution of the energy till $t = 0.6$

Figure 13.1 demonstrates energy dissipation of the flow. Finally, we show in Fig. 13.2 time evolution of the solution. Note that the results are now quite different to Fig. 11.2, which are computed by the VFV method (11.5) due to the different type of boundary conditions.

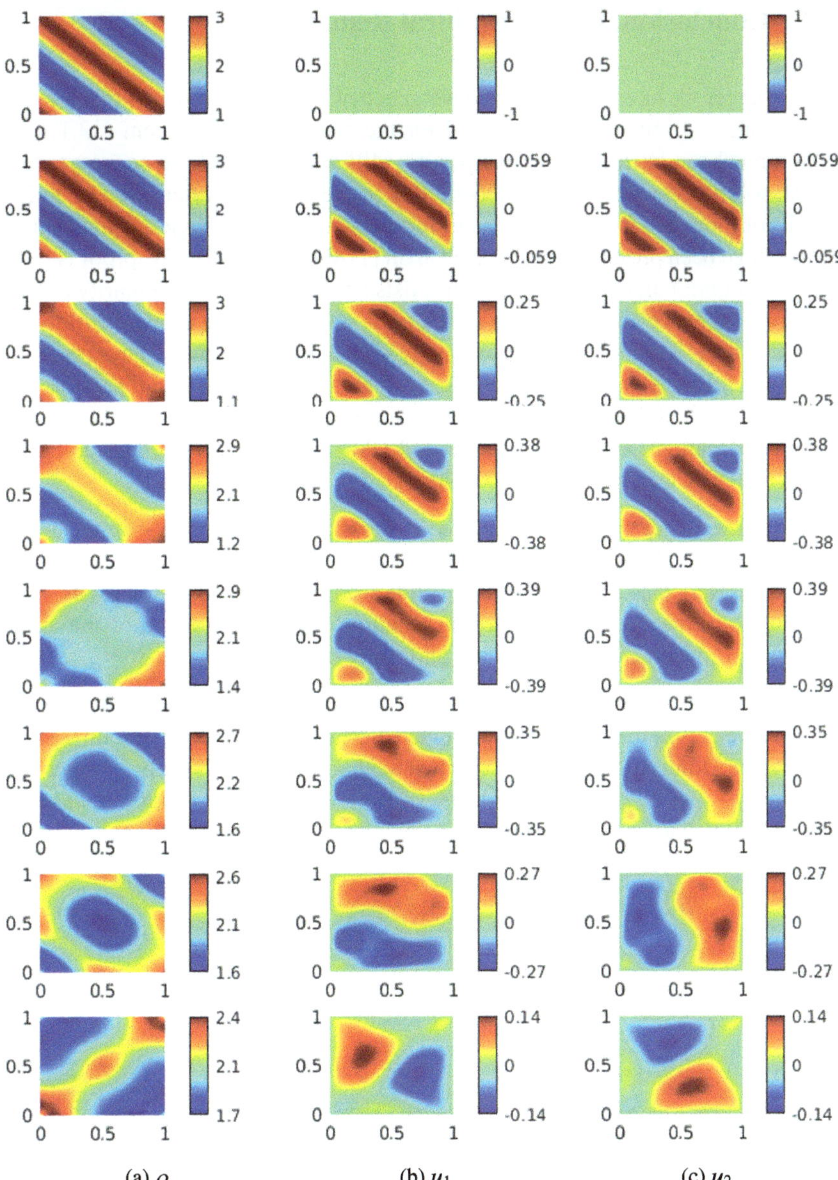

(a) ϱ (b) u_1 (c) u_2

Fig. 13.2 Mixed FV–FE method for the Navier–Stokes equations: time evolution of the flow. From top to bottom are $t = 0, 0.01, 0.05, 0.1, 0.15, 0.2, 0.25, 0.5$. From left to right are ϱ, u_1, and u_2

13.6 Conclusion, Bibliographical Remarks

In this chapter we have studied the convergence of numerical solutions to the mixed finite volume – finite element method for the Navier–Stokes system (13.1). The method has been introduced by Karper in [140] who obtained the convergence to a weak solution for the adiabatic coefficient $\gamma > 3$. We also refer a reader to [88, Part II] for more details. Error estimates for the mixed finite volume – finite element method have been proved by Gallouët et al. in [114] for $\gamma > 3/2$. The proof is based on the assumption that a strong solution exists and belongs to C^2 regularity class.

The convergence analysis of the mixed finite volume – finite element method for the case $\gamma \in (1, 2)$ and $d = 3$ was originally studied in [92]. Here we have presented a more careful analysis and extended the result also to $d = 2$ and $\gamma \geq 2$ for $d = 2, 3$.

In order to establish the convergence result we have used the concept of consistent approximations, see Definition 5.9 and DMV solutions for the Navier–Stokes system (5.10). These tools allow us to show

- weak convergence of a suitable subsequence of the finite volume – finite element solutions to a DMV solution of the Navier–Stokes equations
- strong convergence of the numerical solutions, if the strong solution exists
- strong convergence of the numerical solutions and the existence of the strong solution, if the discrete densities are uniformly bounded.

Comparing the convergence results for the finite volume method obtained in Chap. 11 we realize that the results presented there are slightly more general. Indeed, for the mixed finite volume – finite element method we have a slightly stronger condition on γ, $\gamma > 4d/(1 + 3d)$ and $\varepsilon > 0$. Both convergence results, obtained in Chap. 11 as well as here, hold for physically interesting cases including dry air with $\gamma = 1.4$.

Chapter 14
Finite Difference Method for the Navier-Stokes System

In the previous chapters we have rigorously proved convergence of a finite volume method and a mixed finite volume – finite element method for the Navier–Stokes system, see Theorems 11.3 and 13.3. As already discussed previously these convergence results follow from the consistency and stability of the corresponding numerical schemes and can be seen as a nonlinear analogue of the *Lax equivalence theorem*. In this chapter, our aim is to apply similar ideas to the convergence analysis of a finite difference method for the whole range of adiabatic coefficient $\gamma \in (1, \infty)$. In addition, we will discuss the convergence rate of the finite difference approximation towards a regular solution for $\gamma > d/2$.

To begin, let us recall the Navier–Stokes system, cf. (3.14)–(3.16):

$$\partial_t \varrho + \mathrm{div}_x(\varrho \boldsymbol{u}) = 0, \tag{14.1a}$$

$$\partial_t(\varrho \boldsymbol{u}) + \mathrm{div}_x(\varrho \boldsymbol{u} \otimes \boldsymbol{u}) + \nabla_x p(\varrho) = \mathrm{div}_x \mathbb{S}(\nabla_x \boldsymbol{u}) \tag{14.1b}$$

in the time-space cylinder $[0, T] \times \Omega$, with the space periodic boundary conditions (meaning $\Omega = \mathbb{T}^d, d = 2, 3$). Here ϱ is the density, $\boldsymbol{u}$ is the velocity field, and $\mathbb{S}$ is the viscous stress tensor given by

$$\mathbb{S}(\nabla_x \boldsymbol{u}) = \mu\left(\nabla_x \boldsymbol{u} + \nabla_x^T \boldsymbol{u} - \frac{2}{d}\mathrm{div}_x \boldsymbol{u}\mathbb{I}\right) + \lambda \mathrm{div}_x \boldsymbol{u}\mathbb{I}, \quad \mu > 0,\ \lambda \geq -\frac{d-2}{d}\mu.$$

The pressure is assumed to satisfy the *isentropic* law

$$p = a\varrho^\gamma,\ a > 0,\ \gamma > 1. \tag{14.2}$$

The system (14.1) is supplemented with the initial data

$$\varrho(0) = \varrho_0 \geq \underline{\varrho} > 0,\ \varrho_0 \in L^\infty(\mathbb{T}^d) \text{ and } \boldsymbol{u}(0) = \boldsymbol{u}_0 \in L^\infty(\mathbb{T}^d; R^d). \tag{14.3}$$

E. Feireisl et al., *Numerical Analysis of Compressible Fluid Flows*, MS&A 20,
https://doi.org/10.1007/978-3-030-73788-7_14

14.1 Numerical Method

In this section we define a finite difference Marker–and–Cell (MAC) scheme for the Navier–Stokes system and list useful notations, identities, and estimates.

Concerning the time discretization we recall the notation $f_h^k = f_h(t^k)$ for $t^k = k\Delta t$, $k = 1, 2, ..., N_T (\equiv T/\Delta t)$ and adopt analogously as in (47) a piecewise constant approximation in time

$$f_h(t,\cdot) = f_h^0 \text{ for } t < \Delta t,\ \ f_h(t,\cdot) = f_h^k \text{ for } t \in [t^k, t^{k+1}),\ k = 1, 2, \ldots, N_T,$$

where f_h represents a generic discrete function.

For the spatial discretization we firstly define a staggered grid $\mathcal{G} = \left(\mathcal{T}_h, \{\mathcal{D}_{i,h}\}_{i=1}^d\right)$, where $\mathcal{T}_h$ is a structured mesh given in Definition 3 and $\mathcal{D}_{i,h}$, $i = 1, \ldots, d$, are dual grids of $\mathcal{T}_h$ given in Definition 4. Recall that $\mathcal{E}_{int} = \mathcal{E}$ due to the periodicity of the domain.

We set the discrete density and pressure in the center of each element $K \in \mathcal{T}_h$ to be ϱ_K and $p_K = p(\varrho_K)$, respectively. We further approximate the ith velocity component in the center of each dual element $D_\sigma \in \mathcal{D}_{i,h}$ by $u_{i,\sigma}$ for all $i = 1, \ldots, d$. It is convenient to extend these quantities to functions defined on $\mathbb{T}^d$:

$$\varrho_h(x) = \sum_{K\in\mathcal{T}_h} \varrho_K 1_K(x), \quad p_h(x) = \sum_{K\in\mathcal{T}_h} p_K 1_K(x), \quad u_{i,h}(x) = \sum_{\sigma\in\mathcal{E}_i} u_{i,\sigma} 1_{D_\sigma}(x),$$

for all $x \in \mathbb{T}^d$, where 1_K and 1_{D_σ} are characteristic functions, see (12). Clearly, for any $t \in (0, T)$ we have $\varrho_h(t), p_h(t) \in Q_h$ and $\boldsymbol{u}_h(t) \in \boldsymbol{W}_h$, see the definition of the discrete function spaces (10).

We further introduce a bidual grid that shall be useful only for the definition of a discrete gradient operator for a piecewise constant velocity on the dual grid. This will allow us to integrate by parts at the discrete level in order to derive the weak formulation of the scheme. Note that the bidual grid is not necessary for the implementation of the scheme defined below.

Definition 14.1 **(Bidual grid)** Let $\widetilde{\mathcal{E}}(D_\sigma)$ be the set of all faces of the dual cell D_σ given in Definition 4 and $\widetilde{\mathcal{E}}_i$ be the set of all faces of the dual grid $\mathcal{D}_{i,h}$, $i = 1, \ldots, d$. A generic dual face and its barycenter are denoted by $\epsilon \in \widetilde{\mathcal{E}}(D_\sigma)$ and x_ϵ, respectively. We write $\epsilon = D_\sigma|D_{\sigma'}$ if $\epsilon \in \widetilde{\mathcal{E}}(D_\sigma) \cap \widetilde{\mathcal{E}}(D_{\sigma'})$ separates the dual cells D_σ and $D_{\sigma'}$. Further, we write $\epsilon = \overrightarrow{D_\sigma|D_{\sigma'}}$ if, moreover, $x_{\sigma'} - x_\sigma = h\mathbf{e}_i$ for some $i = 1, \ldots, d$. The bidual cell D_ϵ associated to $\epsilon = D_\sigma|D_{\sigma'}$ is defined as the union of adjacent halves of D_σ and $D_{\sigma'}$, see Fig. 14.1 for a graphic illustration in two dimensions.

We recall the definition of discrete operators from Preliminary material and moreover define the following discrete divergence operator $\text{div}_{\mathcal{T}}^{\text{up}}$ and discrete Laplace operator $\Delta_{\mathcal{E}}$:

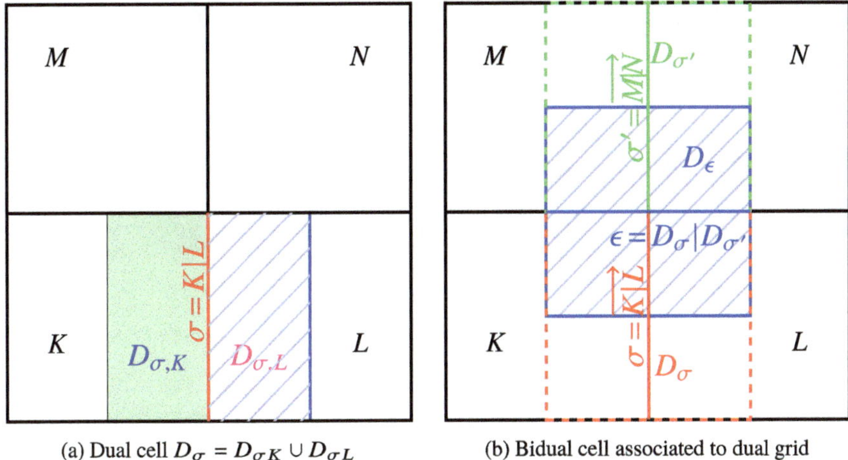

(a) Dual cell $D_\sigma = D_{\sigma K} \cup D_{\sigma L}$

(b) Bidual cell associated to dual grid

Fig. 14.1 MAC grid in two dimensions

$$\mathrm{div}_{\mathcal{T}}^{\mathrm{up}}(r, \boldsymbol{u})(x) = \sum_{K \in \mathcal{T}_h} 1_K \mathrm{div}_{\mathcal{T}}^{\mathrm{up}}(r, \boldsymbol{u})_K, \quad \mathrm{div}_{\mathcal{T}}^{\mathrm{up}}(r, \boldsymbol{u})_K = \sum_{\sigma \in \mathcal{E}(K)} \frac{|\sigma|}{|K|} Up[r, \boldsymbol{u}]_\sigma,$$

$$\Delta_{\mathcal{E}} u_i = \sum_{\sigma \in \mathcal{E}_i} 1_{D_\sigma} (\Delta_{\mathcal{E}} u_i)_\sigma, \quad (\Delta_{\mathcal{E}} u_i)_\sigma = \frac{1}{h^2} \sum_{\sigma' \in \mathcal{N}^\star(\sigma)} (u_{i,\sigma'} - u_{i,\sigma}),$$

where we have denoted $\mathcal{N}^\star(\sigma)$ the set of all faces whose associated dual elements are neighbors of D_σ, meaning,

$$\mathcal{N}^\star(\sigma) = \{\sigma' \mid D_{\sigma'} \text{ is a neighbor of } D_\sigma\}.$$

Based on the bidual grid, we define a new *discrete velocity gradient*, different from the gradient operators defined in (15) and (16). Specifically,

$$\nabla_\epsilon \boldsymbol{u}(x) = \big(\nabla_\epsilon u_1(x), \dots, \nabla_\epsilon u_d(x)\big) \text{ with } \nabla_\epsilon u_i(x) = \big(\eth_1 u_i(x), \dots, \eth_d u_i(x)\big),$$

where

$$\eth_j u_i(x) = \sum_{\epsilon \in \widetilde{\mathcal{E}}_j} (\eth_j u_i)_{D_\epsilon} 1_{D_\epsilon}(x)$$

and

$$(\eth_j u_i)_{D_\epsilon} = \frac{u_{\sigma'} - u_\sigma}{h} \text{ for } \epsilon = \overrightarrow{D_\sigma | D_{\sigma'}} \in \widetilde{\mathcal{E}}_j, \ \sigma, \sigma' \in \mathcal{E}_i.$$

For the proper implementation of the scheme we need to interpolate the functions defined on the primary grid to the dual grid and vice versa. First, we recall the definition of the average operator for any scalar function $r_h \in Q_h$,

$$\{\!\{r_h\}\!\}_\sigma = \frac{r_K + r_L}{2}, \ \sigma = K|L \in \mathcal{E}.$$

If in addition, $\sigma = K|L \in \mathcal{E}_i$ for an $i \in \{1, \dots, d\}$, we write

$$\{\!\{r_h\}\!\}_\sigma^{(i)} = \frac{r_K + r_L}{2}, \text{ and } \{\!\{r_h\}\!\}^{(i)} = \sum_{\sigma \in \mathcal{E}_i} 1_{D_\sigma} \{\!\{r_h\}\!\}_\sigma^{(i)}.$$

Further, for vector valued functions $\mathbf{r}_h = (r_{1,h}, \dots, r_{d,h}) \in \boldsymbol{Q}_h$ and $\boldsymbol{v}_h = (v_{1,h}, \dots, v_{d,h}) \in \boldsymbol{W}_h$, we define

$$\{\!\{\mathbf{r}_h\}\!\} = \left(\left\{\!\left\{r_{1,h}\right\}\!\right\}^{(1)}, \dots, \left\{\!\left\{r_{d,h}\right\}\!\right\}^{(d)}\right),$$

$$\overline{v_{i,h}} = \sum_{K \in \mathcal{T}_h} 1_K (\overline{v_{i,h}})_K, \quad (\overline{v_{i,h}})_K = \frac{v_{i,\sigma_{K,i+}} + v_{i,\sigma_{K,i-}}}{2}, \quad \text{and } (\overline{\boldsymbol{v}_h})_K = \sum_{i=1}^d (\overline{v_{i,h}})_K \mathbf{e}_i,$$

where $\sigma_{K,i-}$ and $\sigma_{K,i+}$ are the left and right edge of K in the ith-direction.

It is easy to check that for $\boldsymbol{u}, \boldsymbol{U} \in \boldsymbol{W}_h$

$$\int_{\mathbb{T}^d} \Delta_{\mathcal{E}} \boldsymbol{u} \cdot \boldsymbol{U} \, \mathrm{d}x = - \int_{\mathbb{T}^d} \nabla_\epsilon \boldsymbol{u} : \nabla_\epsilon \boldsymbol{U} \, \mathrm{d}x. \tag{14.4}$$

Moreover, it follows from (42) that

$$\int_{\mathbb{T}^d} \left\{\!\left\{\left(\eth_{\mathcal{T}}^{(j)} u_{j,h}\right)\right\}\!\right\}^{(i)} U_{i,h} \, \mathrm{d}x = \int_{\mathbb{T}^d} \left(\eth_{\mathcal{T}}^{(j)} u_{j,h}\right)\overline{U_{i,h}} \, \mathrm{d}x = - \int_{\mathbb{T}^d} u_{j,h}\left(\eth_{\mathcal{D}_j} \overline{U_{i,h}}\right) \mathrm{d}x. \tag{14.5}$$

Applying the mean value theorem and the Taylor expansion yield

$$\left\| \nabla_{\mathcal{D}} \Pi_{\mathcal{Q}} \phi \right\|_{L^\infty} \stackrel{<}{\sim} \|\phi\|_{C^1}, \quad \left\| \Pi_W^{(i)} \phi - \phi \right\|_{L^\infty} \stackrel{<}{\sim} h \, \|\phi\|_{C^1} \text{ for any } \phi \in C^1(\mathbb{T}^d). \tag{14.6a}$$

Further, for any $\phi \in C^2(\mathbb{T}^d)$ and $\boldsymbol{\phi} \in C^2(\mathbb{T}^d; R^d)$ we have

$$\left\| \nabla_{\mathcal{D}} \Pi_{\mathcal{Q}} \phi - \nabla_x \phi \right\|_{L^\infty} \stackrel{<}{\sim} h \, \|\phi\|_{C^2}, \quad \|\nabla_\epsilon \Pi_{\boldsymbol{W}} \boldsymbol{\phi} - \nabla_x \boldsymbol{\phi}\|_{L^\infty} \stackrel{<}{\sim} h \, \|\boldsymbol{\phi}\|_{C^2}, \tag{14.6b}$$

$$\left|\Delta_h \Pi_{\mathcal{Q}} \phi\right| \stackrel{<}{\sim} \|\phi\|_{C^2}, \ \left|\eth_{\mathcal{T}}^{(j)} \eth_{\mathcal{D}_j} \overline{\Pi_W^{(i)} \phi}\right| \stackrel{<}{\sim} \|\phi\|_{C^2}, \ \left\| D_t \overline{\Pi_{\boldsymbol{W}} \boldsymbol{\phi}} \right\|_{L^\infty L^\infty} \stackrel{<}{\sim} \|\partial_t \boldsymbol{\phi}\|_{L^\infty L^\infty}. \tag{14.6c}$$

Finally, we point out that

$$\|\boldsymbol{u}_h - \overline{\boldsymbol{u}_h}\|_{L^2(K)} = h \, \|\mathrm{div}_{\mathcal{T}} \boldsymbol{u}_h\|_{L^2(K)} \text{ for any } K \in \mathcal{T}_h. \tag{14.7}$$

Note that the equality (14.7) is different from the Poincaré inequality due to the staggered discretization of velocity components and the construction of discrete divergence operator $\mathrm{div}_{\mathcal{T}}$.

We are now ready to introduce an implicit in time MAC scheme for the Navier–Stokes system and its weak formulation.

Definition 14.2 (MAC scheme for the Navier–Stokes system) Given the initial data (14.3) we set $(\varrho_h^0, \boldsymbol{u}_h^0) = (\Pi_Q \varrho_0, \Pi_{\boldsymbol{W}} \boldsymbol{u}_0) \in Q_h \times \boldsymbol{W}_h$. The MAC approximation of the Navier–Stokes system (14.1a)–(14.1b) is a sequence $(\varrho_h^k, \boldsymbol{u}_h^k) \in Q_h \times \boldsymbol{W}_h$ which solves the following system of algebraic equations:

$$
\begin{aligned}
D_t \varrho_h^k + \mathrm{div}_{\mathcal{T}}^{\mathrm{up}}(\varrho_h^k, \boldsymbol{u}_h^k) - h^{\varepsilon+1} \Delta_h \varrho_h^k &= 0, \\
D_t \left\{\!\left\{ \varrho_h^k \overline{u_{i,h}}^k \right\}\!\right\}^{(i)} + \left\{\!\left\{ \mathrm{div}_{\mathcal{T}}^{\mathrm{up}}(\varrho_h^k \overline{u_{i,h}}^k, \boldsymbol{u}_h^k) \right\}\!\right\}^{(i)} + \eth_{\mathcal{D}_i} p(\varrho_h^k) - \mu \Delta_{\mathcal{E}} u_{i,h}^k & \\
- \nu \eth_{\mathcal{D}_i} \mathrm{div}_{\mathcal{T}} \boldsymbol{u}_h^k = h^{\varepsilon+1} \sum_{j=1}^{d} \left\{\!\left\{ \eth_{\mathcal{T}}^{(j)} \left(\left\{\!\left\{ \overline{u_{i,h}}^k \right\}\!\right\}^{(j)} (\eth_{\mathcal{D}_j} \varrho_h^k) \right) \right\}\!\right\}^{(i)}&,
\end{aligned}
\tag{14.8}
$$

for all $i = 1, \ldots, d$, and for all $k = 1, \ldots, N_T$. Recall $\boldsymbol{u}_h^k = (u_{1,h}^k, \ldots, u_{d,h}^k)$, $u_{i,h} \in W_h^{(i)}$. The parameter ε satisfies

$$
\varepsilon \in \left(0, 2\gamma - 1 - \frac{d}{3}\right) \quad \text{for } \gamma \in (1,2), \quad \text{and } \varepsilon > 0 \ \text{ for } \gamma \geq 2. \tag{14.9}
$$

Lemma 14.1 (MAC scheme for the Navier–Stokes system: weak formulation) *Let* $(\varrho_h^k, \boldsymbol{u}_h^k) \in Q_h \times \boldsymbol{W}_h$ *be a solution of the MAC scheme* (14.8) *with the initial data as in Definition 14.2.*

Then the sequence $(\varrho_h^k, \boldsymbol{u}_h^k)$, $k = 1, \ldots, N_T$, *fulfils the following weak formulation for any* $\phi_h \in Q_h$ *and* $\boldsymbol{\phi}_h = (\phi_{1,h}, \ldots, \phi_{d,h}) \in \boldsymbol{W}_h$,

$$
\int_{\mathbb{T}^d} D_t \varrho_h^k \phi_h \, \mathrm{d}x - \int_{\mathcal{E}} Up[\varrho_h^k, \boldsymbol{u}_h^k] [\![\phi_h]\!] \, \mathrm{dS}_x + h^{\varepsilon+1} \int_{\mathbb{T}^d} \nabla_{\mathcal{D}} \varrho_h^k \cdot \nabla_{\mathcal{D}} \phi_h \, \mathrm{d}x = 0, \tag{14.10a}
$$

$$
\begin{aligned}
&\int_{\mathbb{T}^d} D_t (\varrho_h^k \overline{\boldsymbol{u}_h}^k) \cdot \overline{\boldsymbol{\phi}_h} \, \mathrm{d}x - \int_{\mathcal{E}} Up[\varrho_h^k \overline{\boldsymbol{u}_h}^k, \boldsymbol{u}_h^k] \cdot [\![\overline{\boldsymbol{\phi}_h}]\!] \, \mathrm{dS}_x \\
&+ \mu \int_{\mathbb{T}^d} \nabla_{\epsilon} \boldsymbol{u}_h^k : \nabla_{\epsilon} \boldsymbol{\phi}_h \, \mathrm{d}x + \nu \int_{\mathbb{T}^d} \mathrm{div}_{\mathcal{T}} \boldsymbol{u}_h^k \, \mathrm{div}_{\mathcal{T}} \boldsymbol{\phi}_h \, \mathrm{d}x \\
&= \int_{\mathbb{T}^d} p_h^k \mathrm{div}_{\mathcal{T}} \boldsymbol{\phi}_h \, \mathrm{d}x - h^{\varepsilon+1} \sum_{i=1}^{d} \sum_{j=1}^{d} \int_{\mathbb{T}^d} \left\{\!\left\{ \overline{u_{i,h}}^k \right\}\!\right\}^{(j)} (\eth_{\mathcal{D}_j} \varrho_h) \eth_{\mathcal{D}_j} \overline{\phi_{i,h}} \, \mathrm{d}x,
\end{aligned}
\tag{14.10b}
$$

where the parameter ε *satisfies* (14.9) *and* $Up[\varrho_h \overline{\boldsymbol{u}_h}, \boldsymbol{u}_h] = \sum_{i=1}^d Up[\varrho_h \overline{u_{i,h}}, \boldsymbol{u}_h] \mathbf{e}_i$.

Analogously to Lemma 8.1, it is easy to derive the following relations for $r_h \in Q_h$, $\boldsymbol{u}_h \in \boldsymbol{W}_h$, and $\phi \in C^1(\mathbb{T}^d)$:

$$\begin{aligned}
&\int_{\mathcal{E}} Up[r_h, \boldsymbol{u}_h] \left[\!\!\left[\frac{|\overline{\boldsymbol{u}_h}|^2}{2} \right]\!\!\right] \mathrm{dS}_x - \int_{\mathcal{E}} Up[r_h \overline{\boldsymbol{u}_h}, \boldsymbol{u}_h] \cdot [\![\overline{\boldsymbol{u}_h}]\!] \, \mathrm{dS}_x \\
&= \frac{1}{2} \int_{\mathcal{E}} r_h^{\mathrm{up}} \, |\boldsymbol{u}_h \cdot \boldsymbol{n}| \, [\![\overline{\boldsymbol{u}_h}]\!]^2 \, \mathrm{dS}_x,
\end{aligned} \tag{14.11a}$$

$$\begin{aligned}
&\int_{\mathbb{T}^d} Up[r_h, \boldsymbol{u}_h] \left[\!\left[\Pi_Q \phi \right]\!\right] \, \mathrm{d}x = \int_{\mathbb{T}^d} r_h \boldsymbol{u}_h \cdot \nabla_x \phi \, \mathrm{d}x \\
&+ \int_{\mathbb{T}^d} r_h \boldsymbol{u}_h \cdot \left(\nabla_{\mathcal{D}} (\Pi_Q \phi) - \nabla_x \phi \right) \mathrm{d}x + \frac{h}{2} \sum_{i=1}^{d} \sum_{K \in \mathcal{T}_h} \int_K r_h \Delta_h^{(i)} (\Pi_Q \phi) \overline{|u_{i,h}|} \, \mathrm{d}x \\
&+ \frac{h}{2} \sum_{i=1}^{d} \sum_{K \in \mathcal{T}_h} \int_K r_h \overline{\eth_{\mathcal{D}_i} (\Pi_Q \phi) \eth_{\mathcal{T}}^{(i)} |u_{i,h}|} \, \mathrm{d}x,
\end{aligned} \tag{14.11b}$$

and

$$\begin{aligned}
&\int_{\mathbb{T}^d} Up[r_h, \boldsymbol{u}_h] \left[\!\!\left[\Pi_W^{(i)} \phi \right]\!\!\right] \, \mathrm{d}x = \int_{\mathbb{T}^d} r_h \boldsymbol{u}_h \cdot \nabla_x \phi \, \mathrm{d}x \\
&+ \int_{\mathbb{T}^d} r_h \boldsymbol{u}_h \cdot \left(\nabla_{\mathcal{D}} (\Pi_Q \Pi_W^{(i)} \phi) - \nabla_x \phi \right) \mathrm{d}x \\
&+ \frac{h}{2} \sum_{j=1}^{d} \sum_{K \in \mathcal{T}_h} \int_K r_h \Delta_h^{(j)} (\Pi_Q \Pi_W^{(i)} \phi) \overline{|u_{j,h}|} \, \mathrm{d}x \\
&+ \frac{h}{2} \sum_{j=1}^{d} \sum_{K \in \mathcal{T}_h} \int_K r_h \overline{\eth_{\mathcal{D}_j} (\Pi_Q \Pi_W^{(i)} \phi) \, \eth_{\mathcal{T}}^{(j)} |u_{j,h}|} \, \mathrm{d}x.
\end{aligned} \tag{14.11c}$$

The lemma below describes the main properties of the MAC scheme (14.10).

Lemma 14.2 (Properties of the MAC scheme) *Let $\varrho_0 > 0$. Then the MAC scheme (14.10) enjoys the following properties:*

1. **Existence of a numerical solution.**
   There exists a solution $(\varrho_h, \boldsymbol{u}_h) = \{\varrho_h^k, \boldsymbol{u}_h^k\}_{k=1}^{N_T} \in Q_h \times \boldsymbol{W}_h$ to (14.10).
2. **Positivity of discrete density.**
   Any solution to (14.10) satisfies $\varrho_h(t) > 0$ for $t \in (0, T)$.
3. **Discrete conservation of mass.**
   Any solution to (14.10) satisfies the conservation of mass

$$\int_{\mathbb{T}^d} \varrho_h(t)\, \mathrm{d}x = \int_{\mathbb{T}^d} \varrho_0\, \mathrm{d}x \equiv M_0,\ t \in (0, T).$$

4. **Internal energy balance.**
There exist $\zeta \in \mathrm{co}\{(\varrho_h^k)^{\mathrm{in}}, (\varrho_h^k)^{\mathrm{out}}\}$ *for any* $\sigma \in \mathcal{E}$ *and* $\xi \in \mathrm{co}\{\varrho_h^{k-1}, \varrho_h^k\}$ *such that*

$$\int_{\mathbb{T}^d} D_t P(\varrho_h^k)\, \mathrm{d}x + \int_{\mathbb{T}^d} p(\varrho_h^k) \mathrm{div}_h \boldsymbol{u}_h^k\, \mathrm{d}x = -\frac{\Delta t}{2} \int_{\mathbb{T}^d} P''(\xi) \left|D_t \varrho_h^k\right|^2\, \mathrm{d}x - \\ - \int_{\mathcal{E}} P''(\zeta) \left[\!\left[\varrho_h^k\right]\!\right]^2 \left(h^\varepsilon + \frac{1}{2}|\boldsymbol{u}_h^k \cdot \boldsymbol{n}|\right) \mathrm{dS}_x, \tag{14.12}$$

where $P(\varrho) = \frac{a}{\gamma-1}\varrho^\gamma$ *is the pressure potential.*

Proof We refer to Lemma 11.3 and 11.2 for the proof as it can be done exactly in the same way.

14.2 Stability

The finite difference scheme (14.10) is stable in the sense of Definition 8.3. Indeed, we have the following energy estimates.

Theorem 14.1 (Energy stability of the MAC scheme) *Let* $(\varrho_h, \boldsymbol{u}_h)$ *be a numerical solution obtained from the MAC scheme (14.10). Then there exist* $\xi \in \mathrm{co}\{\varrho_h^{k-1}, \varrho_h^k\}$ *and* $\zeta \in \mathrm{co}\{(\varrho_h^k)^{\mathrm{in}}, (\varrho_h^k)^{\mathrm{out}}\}$, *such that for any* $k = 1, \ldots, N_T$

$$D_t \int_{\mathbb{T}^d} \left(\frac{1}{2}\varrho_h^k \left|\overline{\boldsymbol{u}_h}^k\right|^2 + P(\varrho_h^k)\right) \mathrm{d}x + \mu \int_{\mathbb{T}^d} \left|\nabla_\epsilon \boldsymbol{u}_h^k\right|^2\, \mathrm{d}x + \nu \int_{\mathbb{T}^d} \left|\mathrm{div}_{\mathcal{T}} \boldsymbol{u}_h^k\right|^2\, \mathrm{d}x \\ = -\frac{\Delta t}{2} \int_{\mathbb{T}^d} P''(\xi) \left|D_t \varrho_h^k\right|^2\, \mathrm{d}x - \int_{\mathcal{E}} P''(\zeta) \left[\!\left[\varrho_h^k\right]\!\right]^2 \left(h^\varepsilon + \frac{1}{2}\left|\boldsymbol{u}_h^k \cdot \boldsymbol{n}\right|\right) \mathrm{dS}_x \\ - \frac{\Delta t}{2} \int_{\mathbb{T}^d} \varrho_h^{k-1} \left|D_t \overline{\boldsymbol{u}_h}^k\right|^2\, \mathrm{d}x - \frac{1}{2} \int_{\mathcal{E}} (\varrho_h^k)^{\mathrm{up}} \left|\boldsymbol{u}_h^k \cdot \boldsymbol{n}\right| \left[\!\left[\overline{\boldsymbol{u}_h}^k\right]\!\right]^2 \mathrm{dS}_x. \tag{14.13}$$

Proof First, taking $\phi_h = -\frac{\left|\overline{\boldsymbol{u}_h}^k\right|^2}{2}$ in (14.10a) we get

$$-\frac{1}{2}\int_{\mathbb{T}^d} D_t\varrho_h^k \left|\overline{\boldsymbol{u}_h}^k\right|^2 \,\mathrm{d}x + \frac{1}{2}\int_{\mathcal{E}} Up[\varrho_h^k, \boldsymbol{u}_h^k] \left[\!\left[\left|\overline{\boldsymbol{u}_h}^k\right|^2 \right]\!\right] \,\mathrm{dS}_x$$
$$= \frac{h^{\varepsilon+1}}{2}\int_{\mathbb{T}^d} \nabla_{\mathcal{D}}\varrho_h^k \cdot \nabla_{\mathcal{D}} \left|\overline{\boldsymbol{u}_h}^k\right|^2 \,\mathrm{d}x.$$

Next, setting $\boldsymbol{\phi}_h = \boldsymbol{u}_h^k$ in (14.10b) and noticing $\{\{\overline{u_{i,h}}\}\}^{(j)}\,\eth_{\mathcal{D}_j}\overline{u_{i,h}} = \frac{1}{2}\eth_{\mathcal{D}_j}|\overline{u_{i,h}}|^2$ we derive

$$\int_{\mathbb{T}^d} D_t(\varrho_h^k\overline{\boldsymbol{u}_h}^k)\cdot\overline{\boldsymbol{u}_h}^k \,\mathrm{d}x - \int_{\mathcal{E}} Up[\varrho_h^k\overline{\boldsymbol{u}_h}^k, \overline{\boldsymbol{u}_h}^k]\cdot[\![\boldsymbol{u}_h^k]\!] \,\mathrm{dS}_x$$
$$+\mu\int_{\mathbb{T}^d} \left|\nabla_\epsilon \boldsymbol{u}_h^k\right|^2 \,\mathrm{d}x + \nu\int_{\mathbb{T}^d} \left|\mathrm{div}_{\mathcal{T}}\boldsymbol{u}_h^k\right|^2 \,\mathrm{d}x$$
$$= \int_{\mathbb{T}^d} p(\varrho_h^k)\mathrm{div}_{\mathcal{T}}\boldsymbol{u}_h^k \,\mathrm{d}x - \frac{h^{\varepsilon+1}}{2}\sum_{i=1}^{d}\sum_{j=1}^{d}\int_{\mathbb{T}^d} (\eth_{\mathcal{D}_j}\varrho_h)\left(\eth_{\mathcal{D}_j}|\overline{u_{i,h}}^k|^2\right) \,\mathrm{d}x$$
$$= \int_{\mathbb{T}^d} p(\varrho_h^k)\mathrm{div}_{\mathcal{T}}\boldsymbol{u}_h^k \,\mathrm{d}x - \frac{h^{\varepsilon+1}}{2}\int_{\mathbb{T}^d} \nabla_{\mathcal{D}}\varrho_h^k\cdot\nabla_{\mathcal{D}}\left|\overline{\boldsymbol{u}_h}^k\right|^2 \,\mathrm{d}x.$$

Further, summing up the previous two equalities and recalling (14.11b) we infer that

$$D_t\int_{\mathbb{T}^d} \frac{1}{2}\varrho_h^k\left|\overline{\boldsymbol{u}_h}^k\right|^2 \,\mathrm{d}x + \mu\int_{\mathbb{T}^d}\left|\nabla_\epsilon\boldsymbol{u}_h^k\right|^2 \,\mathrm{d}x + \nu\int_{\mathbb{T}^d}\left|\mathrm{div}_{\mathcal{T}}\boldsymbol{u}_h^k\right|^2 \,\mathrm{d}x$$
$$= -\frac{\Delta t}{2}\int_{\mathbb{T}^d} \varrho_h^{k-1}\left|D_t\overline{\boldsymbol{u}_h}^k\right|^2 \,\mathrm{d}x - \int_{\mathbb{T}^d} p_h^k\mathrm{div}_{\mathcal{T}}\boldsymbol{u}_h^k \,\mathrm{d}x \tag{14.14}$$
$$-\frac{1}{2}\int_{\mathcal{E}} (\varrho_h^k)^{\mathrm{up}}\left|\boldsymbol{u}_h^k\cdot\boldsymbol{n}\right|[\![\overline{\boldsymbol{u}_h}^k]\!]^2 \,\mathrm{dS}_x.$$

Finally, combining (14.14) with (14.12) we get

$$D_t\int_{\mathbb{T}^d}\left(\frac{1}{2}\overline{\varrho_h}^k\left|\overline{\boldsymbol{u}_h}^k\right|^2 + P(\varrho_h^k)\right) \,\mathrm{d}x + \mu\int_{\mathbb{T}^d}\left|\nabla_\epsilon\boldsymbol{u}_h^k\right|^2 \,\mathrm{d}x + \nu\int_{\mathbb{T}^d}\left|\mathrm{div}_{\mathcal{T}}\boldsymbol{u}_h^k\right|^2 \,\mathrm{d}x$$
$$= -\frac{\Delta t}{2}\int_{\mathbb{T}^d}\varrho_h^{k-1}\left|D_t\overline{\boldsymbol{u}_h}^k\right|^2 \,\mathrm{d}x - \frac{\Delta t}{2}\int_{\mathbb{T}^d} P''(\xi)\left|D_t\varrho_h^k\right|^2 \,\mathrm{d}x$$
$$-\int_{\mathcal{E}} P''(\zeta)[\![\varrho_h^k]\!]^2\left(h^\varepsilon + \frac{1}{2}\left|\boldsymbol{u}_h^k\cdot\boldsymbol{n}\right|\right) \,\mathrm{dS}_x - \frac{1}{2}\int_{\mathcal{E}}(\varrho_h^k)^{\mathrm{up}}\left|\boldsymbol{u}_h^k\cdot\boldsymbol{n}\right|[\![\overline{\boldsymbol{u}_h}^k]\!]^2 \,\mathrm{dS}_x,$$

which completes the proof.

The energy inequality (14.13), the Sobolev–Poincaré inequality (37) and Remark 1 yield the following corollary.

Corollary 14.1 (Uniform bounds)

1. *Let $(\varrho_h, \boldsymbol{u}_h)$ be a solution to the MAC scheme (14.10) with the pressure satisfying* (14.2). *Then there exists $c > 0$ depending on the initial energy E_0 such that*

$$\left\| \varrho_h |\overline{\boldsymbol{u}_h}|^2 \right\|_{L^\infty L^1} \leq c, \quad \|\varrho_h\|_{L^\infty L^\gamma} \leq c, \quad \|\boldsymbol{m}_h\|_{L^\infty L^{\frac{2\gamma}{\gamma+1}}} \leq c, \tag{14.15a}$$

$$\|\mathrm{div}_{\mathcal{T}} \boldsymbol{u}_h\|_{L^2 L^2} \leq c, \quad \|\nabla_\epsilon \boldsymbol{u}_h\|_{L^2 L^2} \leq c, \quad \|\boldsymbol{u}_h\|_{L^2 L^6} \leq c, \tag{14.15b}$$

$$\Delta t \int_0^T \int_{\mathbb{T}^d} \varrho_h(t - \Delta t) \left| D_t \overline{\boldsymbol{u}_h} \right|^2 \, \mathrm{d}x \, \mathrm{d}t \leq c, \tag{14.15c}$$

$$h \int_0^T \sum_{\sigma \in \mathcal{E}} \int_{D_\sigma} \varrho_h^{\mathrm{up}} \left| \boldsymbol{u}_h \cdot \boldsymbol{n} \right| \, \left| \nabla_{\mathcal{D}} \overline{\boldsymbol{u}_h} \right|^2 \, \mathrm{d}x \, \mathrm{d}t \leq c, \tag{14.15d}$$

$$\int_0^T \int_{\mathbb{T}^d} \left((h^{\varepsilon+1} + h|u_\sigma|) P''(\varrho_{h,\dagger}) \left| \nabla_{\mathcal{D}} \varrho_h \right|^2 \right) \mathrm{d}x \, \mathrm{d}t \leq c, \tag{14.15e}$$

where $\boldsymbol{m}_h = \varrho_h \overline{\boldsymbol{u}_h}$, $\varrho_{h,\dagger} \in \mathrm{co}\{\varrho_h^{\mathrm{in}}, \varrho_h^{\mathrm{out}}\}$ for any $\sigma \in \mathcal{E}$.

2. *In addition, let $[r, \boldsymbol{U}]$ belong to the class*

$$0 < \underline{r} \leq r(t,x) \leq \overline{r}, \quad \boldsymbol{U} \in C^1([0,T] \times \mathbb{T}^d; R^d).$$

Then there exists $c = c(M_0, E_0, \underline{r}, \overline{r}, \|\boldsymbol{U}\|_{C^1}) > 0$ such that

$$\left\| \boldsymbol{u}_h^k - \boldsymbol{U}_h^k \right\|_{L^q} \stackrel{<}{\sim} \left\| \nabla_\epsilon (\boldsymbol{u}_h^k - \boldsymbol{U}_h^k) \right\|_{L^2} + c E(\varrho_h^k, \boldsymbol{u}_h^k | r_h^k, \boldsymbol{U}_h^k)^{1/2}, \tag{14.16}$$

$$\sup_{0 \leq k \leq N_T} E(\varrho_h^k, \boldsymbol{u}_h^k | r_h^k, \boldsymbol{U}_h^k) \leq c, \tag{14.17}$$

where $q \in [1,6]$ if $d = 3$ and $q \in [1, \infty)$ if $d = 2$, $(r_h, \boldsymbol{U}_h) = (\Pi_Q r, \Pi_W \boldsymbol{U})$, and E is the discrete version of the relative energy functional, cf. Section 6.3, given by

$$E\left(\varrho_h^k, \boldsymbol{u}_h^k \middle| r_h^k, \boldsymbol{U}_h^k\right) \equiv \int_{\mathbb{T}^d} \left[\frac{1}{2} \varrho_h^k \left| \overline{\boldsymbol{u}_h}^k - \overline{\boldsymbol{U}_h}^k \right|^2 + \mathbb{E}(\varrho_h^k | r_h^k) \right] \mathrm{d}x. \tag{14.18}$$

Here the relative energy part related to the potential energy is denoted by

$$\mathbb{E}(\varrho_h | r_h) = P(\varrho_h) - P'(r_h)(\varrho_h - r_h) - P(r_h).$$

Analogously to Lemmas 11.4–11.6, we claim the following estimates,

$$\|\varrho_h\|_{L^2L^2} \lesssim h^{\beta_D}, \quad \beta_D = \begin{cases} \max\left\{-\frac{3\varepsilon+d+3}{6\gamma}, \frac{\gamma-2}{2\gamma}d\right\}, & \text{if } \gamma \in (1,2), \\ 0, & \text{if } \gamma \geq 2, \end{cases} \tag{14.19a}$$

$$\|\varrho_h\|_{L^2L^{6/5}} \lesssim h^{\beta_R}, \quad \beta_R = \begin{cases} \max\left\{-\frac{3\varepsilon+d+3}{6\gamma}, \frac{5\gamma-6}{6\gamma}d\right\}, & \text{if } \gamma \in (1,\frac{6}{5}), \\ 0, & \text{if } \gamma \geq \frac{6}{5}, \end{cases} \tag{14.19b}$$

$$\|\varrho_h\overline{\boldsymbol{u}_h}\|_{L^2L^2} \lesssim h^{\beta_M}, \quad \beta_M = \begin{cases} -\frac{3\varepsilon+d+3}{6\gamma}, & \text{if } \gamma \in (1,2), \\ \frac{\gamma-3}{3\gamma}d, & \text{if } \gamma \in [2,3), \\ 0, & \text{if } \gamma \geq 3 \text{ for } d=3, \\ 0, & \text{if } \gamma > 2 \text{ for } d=2, \end{cases} \tag{14.19c}$$

$$\int_0^T \int_{\mathcal{E}} \frac{[\![\varrho_h]\!]^2}{\max\{\varrho_h^{\text{in}}, \varrho_h^{\text{out}}\}} |\boldsymbol{u}_h \cdot \boldsymbol{n}| \,\mathrm{dS}_x \,\mathrm{d}t \lesssim 1, \quad \text{for } \gamma \geq 2, \tag{14.19d}$$

$$\int_0^T \int_{\mathcal{E}} \left|[\![\varrho_h]\!]\, \boldsymbol{u}_h \cdot \boldsymbol{n}\right| \mathrm{dS}_x \,\mathrm{d}t \lesssim h^{\beta_F}, \beta_F = \begin{cases} -\frac{1}{2} & \text{if } \gamma \geq \frac{6}{5}, \\ \frac{d}{2}(\frac{5}{6}-\frac{1}{\gamma}) - \frac{1}{2} & \text{if } \gamma \in (1,\frac{6}{5}). \end{cases} \tag{14.19e}$$

14.3 Consistency

Another step towards convergence is the consistency in the sense of Definition 8.6.

Theorem 14.2 (Consistency of the MAC scheme) *Let the pressure p satisfy* (14.2) *with $\gamma > 1$. Let $(\varrho_h, \boldsymbol{u}_h)$ be a solution of the MAC scheme (14.10) with $\Delta t \approx h$.*

Then for any $\varphi \in C_c^2([0,T) \times \mathbb{T}^d)$ and $\boldsymbol{\varphi} \in C_c^2([0,T) \times \mathbb{T}^d; R^d)$ there holds

$$-\int_{\mathbb{T}^d} \varrho_h^0 \varphi(0,\cdot)\,\mathrm{d}x = \int_0^T \int_{\mathbb{T}^d} \left[\varrho_h \partial_t \varphi + \varrho_h \overline{\boldsymbol{u}_h} \cdot \nabla_x \varphi\right] \mathrm{d}x\,\mathrm{d}t \tag{14.20a}$$

$$+ \int_0^T e_{1,h}(t,\varphi)\,\mathrm{d}t, \ \beta_1 > 0; \tag{14.20b}$$

$$
\begin{aligned}
&- \int_{\mathbb{T}^d} \varrho_h^0 \overline{\boldsymbol{u}_h}^0 \cdot \boldsymbol{\varphi}(0,\cdot)\, \mathrm{d}x \\
&= \int_0^T \int_{\mathbb{T}^d} \left[\varrho_h \overline{\boldsymbol{u}_h} \cdot \partial_t \boldsymbol{\varphi} + \varrho_h \overline{\boldsymbol{u}_h} \otimes \overline{\boldsymbol{u}_h} : \nabla_x \boldsymbol{\varphi} + p_h \mathrm{div}_x \boldsymbol{\varphi} \right] \mathrm{d}x\, \mathrm{d}t \qquad (14.20c) \\
&- \mu \int_0^T \int_{\mathbb{T}^d} \nabla_\epsilon \boldsymbol{u}_h : \nabla_x \boldsymbol{\varphi}\, \mathrm{d}x\, \mathrm{d}t - \nu \int_0^T \int_{\mathbb{T}^d} \mathrm{div}_{\mathcal{T}} \boldsymbol{u}_h\, \mathrm{div}_x \boldsymbol{\varphi}\, \mathrm{d}x\, \mathrm{d}t \\
&+ \int_0^T e_{2,h}(t, \boldsymbol{\varphi})\, \mathrm{d}t,
\end{aligned}
$$

where

$$
\|e_{1,h}(\cdot,\varphi)\|_{L^1(0,T)} \lesssim h^\beta \|\varphi\|_{C^2},\ \|e_{2,h}(\cdot,\boldsymbol{\varphi})\|_{L^1(0,T)} \lesssim h^\beta \|\boldsymbol{\varphi}\|_{C^2},\ \textit{for some } \beta > 0.
$$

Proof To show the consistency formulation (14.20) we test the continuity Eq. (14.10a) and momentum equation (14.10b) with $\Pi_Q \varphi$ and $\Pi_W \boldsymbol{\varphi}$, respectively. In what follows we handle each term step by step.

(1) Time derivative terms:

First, we recall (48a) and apply the density estimate of (14.15a) to derive

$$
\begin{aligned}
&\int_0^T \int_{\mathbb{T}^d} D_t \varrho_h \Pi_Q \varphi\, \mathrm{d}x\, \mathrm{d}t + \int_0^T \int_{\mathbb{T}^d} \varrho_h \partial_t \varphi\, \mathrm{d}x\, \mathrm{d}t + \int_{\mathbb{T}^d} \varrho_h^0 \varphi(0)\, \mathrm{d}x \\
&\lesssim \Delta t\, \|\varphi\|_{C^2} \|\varrho_h\|_{L^1 L^1} + \Delta t\, \|\varphi\|_{C^1} \left\| \varrho_h^0 \right\|_{L^1} \lesssim \Delta t.
\end{aligned}
$$

Next, we recall (48c) for $\boldsymbol{m}_h = \varrho_h \overline{\boldsymbol{u}_h}$ and apply the momentum estimate of (14.15a) to obtain

$$
\begin{aligned}
&\int_0^T \int_{\mathbb{T}^d} D_t \boldsymbol{m}_h \cdot \overline{\Pi_W \boldsymbol{\varphi}}\, \mathrm{d}x\, \mathrm{d}t + \int_0^T \int_{\mathbb{T}^d} \boldsymbol{m}_h \partial_t \boldsymbol{\varphi}\, \mathrm{d}x\, \mathrm{d}t + \int_{\mathbb{T}^d} \boldsymbol{m}_h^0 \cdot \boldsymbol{\varphi}(0)\, \mathrm{d}x \\
&\lesssim \Delta t \left(\|\boldsymbol{\varphi}\|_{C^2} \|\boldsymbol{m}_h\|_{L^1 L^1} + \|\boldsymbol{\varphi}\|_{C^1} \left\| \boldsymbol{m}_h^0 \right\|_{L^1} \right) \lesssim \Delta t.
\end{aligned}
$$

(2) Convective terms:

To deal with the convective term of the continuity Eq. (14.10a) we recall (14.11b) with $(r_h, \phi) = (\varrho_h, \varphi)$ to get

$$\int_0^T \int_{\mathbb{T}^d} Up[\varrho_h, \boldsymbol{u}_h] [\![\Pi_Q \varphi]\!] \, dx \, dt - \int_0^T \int_{\mathbb{T}^d} \varrho_h \overline{\boldsymbol{u}_h} \cdot \nabla_x \varphi \, dx \, dt = \sum_{i=1}^4 e_i,$$

where the terms e_i, $i = 1, 2, 3, 4$, read

$$e_1 = \int_0^T \int_{\mathbb{T}^d} \varrho_h \boldsymbol{u}_h \cdot \left(\nabla_{\mathcal{D}}(\Pi_Q \varphi) - \nabla_x \varphi \right) dx \, dt,$$

$$e_2 = \frac{h}{2} \sum_{i=1}^d \int_0^T \sum_{K \in \mathcal{T}_h} \int_K \varrho_h \Delta_h^{(i)} (\Pi_Q \varphi) \overline{|u_{i,h}|} \, dx \, dt,$$

$$e_3 = \frac{h}{2} \sum_{i=1}^d \int_0^T \sum_{K \in \mathcal{T}_h} \int_K \varrho_h \overline{\eth_{\mathcal{D}_i} (\Pi_Q \varphi) \eth_{\mathcal{T}}^{(i)}} |u_{i,h}| \, dx \, dt,$$

$$e_4 = \int_0^T \int_{\mathbb{T}^d} \varrho_h (\boldsymbol{u}_h - \overline{\boldsymbol{u}_h}) \cdot \nabla_x \varphi \, dx \, dt.$$

Now, we use Hölder's inequality, the interpolation error (14.6), and the uniform bounds (14.15b) to obtain

$$|e_1| \lesssim h \, \|\varrho_h\|_{L^2 L^{6/5}} \, \|\boldsymbol{u}_h\|_{L^2 L^6} \, \|\varphi\|_{C^2} \lesssim h^{\beta_1},$$
$$|e_2| \lesssim h \, \|\varrho_h\|_{L^2 L^{6/5}} \, \|\varphi\|_{C^2} \, \|\boldsymbol{u}_h\|_{L^2 L^6} \lesssim h^{\beta_1},$$
$$|e_3| \lesssim h \, \|\varrho_h\|_{L^2 L^2} \, \|\varphi\|_{C^1} \, \|\mathrm{div}_h \boldsymbol{u}_h\|_{L^2 L^2} \lesssim h^{\beta_2},$$
$$|e_4| \lesssim h \, \|\varrho_h\|_{L^2 L^2} \, \|\varphi\|_{C^1} \, \|\nabla_\epsilon \boldsymbol{u}_h\|_{L^2 L^2} \lesssim h^{\beta_2},$$

where $\beta_1 = 1 + \beta_R > 0$ and $\beta_2 = 1 + \beta_D > 0$ provided ε satisfies (14.9), see the definition of β_R and β_D in (14.19).

Next, setting $(r_h, \phi) = (\varrho_h \overline{u_{i,h}}, \varphi_i)$, $i = 1, \dots, d$, in formula (14.11c) for the convective term, we get

$$\int_0^T \int_{\mathbb{T}^d} \left[Up[\boldsymbol{m}_h, \boldsymbol{u}_h] \cdot [\![\Pi_W \boldsymbol{\varphi}]\!] - (\varrho_h \overline{\boldsymbol{u}_h} \otimes \boldsymbol{u}_h) : \nabla_x \boldsymbol{\varphi} \right] dx \, dt = \sum_{i=1}^4 \widetilde{e}_i,$$

where the terms $\widetilde{e}_i$, $i = 1, 2, 3, 4$, read

$$\widetilde{e_1} = \sum_{i=1}^{d} \int_0^T \int_{\mathbb{T}^d} \varrho_h \overline{u_{i,h}} \boldsymbol{u}_h \cdot \left(\nabla_{\mathcal{D}} (\Pi_Q \Pi_W^{(i)} \varphi_i) - \nabla_x \varphi_i \right) \,\mathrm{d}x \,\mathrm{d}t,$$

$$\widetilde{e_2} = \sum_{i=1}^{d} \sum_{j=1}^{d} \int_0^T \frac{h}{2} \sum_{K \in \mathcal{T}_h} \int_K \varrho_h \overline{u_{i,h}} \Delta_h^{(j)} (\Pi_Q \Pi_W^{(i)} \varphi_i) \overline{|u_{j,h}|} \,\mathrm{d}x \,\mathrm{d}t,$$

$$\widetilde{e_3} = \sum_{i=1}^{d} \sum_{j=1}^{d} \int_0^T \frac{h}{2} \sum_{K \in \mathcal{T}_h} \int_K \varrho_h \overline{u_{i,h}} \overline{\eth_{\mathcal{D}_j} (\Pi_Q \Pi_W^{(i)} \varphi_i) \eth_{\mathcal{T}}^{(j)}} \left| u_{j,h} \right| \,\mathrm{d}x \,\mathrm{d}t,$$

$$\widetilde{e_4} = \int_0^T \int_{\mathbb{T}^d} \varrho_h \overline{\boldsymbol{u}_h} \otimes (\boldsymbol{u}_h - \overline{\boldsymbol{u}_h}) : \nabla_x \boldsymbol{\varphi} \,\mathrm{d}x \,\mathrm{d}t.$$

Then we apply Hölder's inequality and the uniform bounds (14.15) to control the terms $\widetilde{e_i}$, $i = 1, 2, 3, 4$ as follows

$$\begin{aligned}
|\widetilde{e_1}| &\lesssim h \left\| \varrho_h |\overline{\boldsymbol{u}_h}|^2 \right\|_{L^1 L^1} \|\boldsymbol{\varphi}\|_{C^2} \lesssim h, \\
|\widetilde{e_2}| &\lesssim h \left\| \varrho_h \overline{\boldsymbol{u}_h} \right\|_{L^2 L^2} \|\boldsymbol{u}_h\|_{L^2 L^6} \|\boldsymbol{\varphi}\|_{C^2} \lesssim h^\beta, \\
|\widetilde{e_3}| &\lesssim h \left\| \varrho_h \overline{\boldsymbol{u}_h} \right\|_{L^2 L^2} \|\mathrm{div}_{\mathcal{T}} \boldsymbol{u}_h\|_{L^2 L^2} \|\varphi\|_{C^2} \lesssim h^\beta, \\
|\widetilde{e_4}| &\lesssim h \left\| \varrho_h \overline{\boldsymbol{u}_h} \right\|_{L^2 L^2} \|\nabla_\epsilon \boldsymbol{u}_h\|_{L^2 L^2} \|\boldsymbol{\varphi}\|_{C^1} \lesssim h^\beta,
\end{aligned}$$

where in view of the negative estimate (14.19c) we have $\beta = 1 + \beta_M > 0$ provided ε satisfies (14.9) and $\beta_M > -1$ is given in (14.19c).

(3) Pressure and viscosity terms:

Due to the identity (18) we have

$$\int_0^T \int_{\mathbb{T}^d} p(\varrho_h) \mathrm{div}_{\mathcal{T}} \Pi_W \boldsymbol{\varphi} \,\mathrm{d}x \,\mathrm{d}t = \int_0^T \int_{\mathbb{T}^d} p(\varrho_h) \mathrm{div}_x \boldsymbol{\varphi} \,\mathrm{d}x \,\mathrm{d}t,$$

$$\int_0^T \int_{\mathbb{T}^d} \mathrm{div}_{\mathcal{T}} \boldsymbol{u}_h \,\mathrm{div}_{\mathcal{T}} \Pi_W \boldsymbol{\varphi} \,\mathrm{d}x \,\mathrm{d}t = \int_0^T \int_{\mathbb{T}^d} \mathrm{div}_{\mathcal{T}} \boldsymbol{u}_h \,\mathrm{div}_x \boldsymbol{\varphi} \,\mathrm{d}x \,\mathrm{d}t.$$

Next, we apply Hölder's inequality and the interpolation estimate (14.6b) to control

$$\begin{aligned}
&\int_0^T \int_{\mathbb{T}^d} \nabla_\epsilon \boldsymbol{u}_h : \nabla_\epsilon \Pi_W \boldsymbol{\varphi} \,\mathrm{d}x \,\mathrm{d}t - \int_0^T \int_{\mathbb{T}^d} \nabla_\epsilon \boldsymbol{u}_h : \nabla_x \boldsymbol{\varphi} \,\mathrm{d}x \,\mathrm{d}t \\
&\lesssim \|\nabla_\epsilon \boldsymbol{u}_h\|_{L^2 L^2} \|\nabla_\epsilon \Pi_W \boldsymbol{\varphi} - \nabla_x \boldsymbol{\varphi}\|_{L^2 L^2} \lesssim h.
\end{aligned}$$

(4) Artificial diffusion term:

We apply the integration by parts formula (44) and the estimate (14.6c) to control the artificial diffusion term in the continuity Eq. (14.10a),

$$h^{\varepsilon+1}\int_0^T\int_{\mathbb{T}^d} \nabla_{\mathcal{D}}\varrho_h\cdot\nabla_{\mathcal{D}}\varphi \,\mathrm{d}x\,\mathrm{d}t = -h^{\varepsilon+1}\int_0^T\int_{\mathbb{T}^d} \varrho_h\Delta_h\varphi\,\mathrm{d}x\,\mathrm{d}t$$
$$\lesssim h^{\varepsilon+1}\left\|\varrho_h\right\|_{L^\infty L^\gamma}\left\|\varphi\right\|_{C^2}\lesssim h^{\varepsilon+1}.$$

Similarly, using the integration by parts formulas (44) and (14.5), together with the product rule and the negative density estimate (14.19a) we obtain for the artificial diffusion term in the momentum equation (14.10b) the following estimate:

$$-h^{\varepsilon+1}\int_0^T\sum_{i=1}^d\sum_{j=1}^d\int_{\mathbb{T}^d}\{\{\overline{u_{i,h}}^k\}\}^{(j)}\left(\eth_{\mathcal{D}_j}\varrho_h\right)\eth_{\mathcal{D}_j}\overline{\Pi_W^{(i)}\varphi_i}\,\mathrm{d}x\,\mathrm{d}t$$
$$=h^{\varepsilon+1}\int_0^T\sum_{i=1}^d\sum_{j=1}^d\int_{\mathbb{T}^d}\varrho_h\eth_{\mathcal{T}}^{(j)}\left(\{\{\overline{u_{i,h}}^k\}\}^{(j)}\eth_{\mathcal{D}_j}\overline{\Pi_W^{(i)}\varphi_i}\right)\mathrm{d}x\,\mathrm{d}t$$
$$=h^{\varepsilon+1}\int_0^T\sum_{i=1}^d\sum_{j=1}^d\int_{\mathbb{T}^d}\varrho_h\eth_{\mathcal{T}}^{(j)}\left(\{\{\overline{u_{i,h}}^k\}\}^{(j)}\right)\overline{\eth_{\mathcal{D}_j}\overline{\Pi_W^{(i)}\varphi_i}}\,\mathrm{d}x\,\mathrm{d}t$$
$$+h^{\varepsilon+1}\int_0^T\sum_{i=1}^d\sum_{j=1}^d\int_{\mathbb{T}^d}\varrho_h\overline{\{\{\overline{u_{i,h}}^k\}\}^{(j)}}\eth_{\mathcal{T}}^{(j)}\left(\eth_{\mathcal{D}_j}\overline{\Pi_W^{(i)}\varphi_i}\right)\mathrm{d}x\,\mathrm{d}t$$
$$\lesssim h^{\varepsilon+1}\left\|\varrho_h\right\|_{L^2L^2}\left(\left\|\nabla_\epsilon\boldsymbol{u}_h\right\|_{L^2L^2}\left\|\boldsymbol{\varphi}\right\|_{C^1}+\left\|\boldsymbol{u}_h\right\|_{L^2L^2}\left\|\boldsymbol{\varphi}\right\|_{C^2}\right)$$
$$\lesssim h^{\zeta},$$

where $\zeta=\varepsilon+1+\beta_D>0$ as $\varepsilon>0$ and $\beta_D>-1$ is given in (14.19a) provided ε satisfies (14.9).

Consequently, collecting all the above estimates yields (14.20a) and (14.20c) and thus completes the proof of Theorem 14.2.

14.4 Convergence

We present the convergence result for the MAC scheme (14.10).

Theorem 14.3 (Convergence of the MAC scheme) *Let* $[\varrho_0,\boldsymbol{u}_0]$ *be the initial data given in* (14.3). *Let the pressure satisfy* (14.2) *with* $\gamma>1$. *Let* $\{\varrho_h,\boldsymbol{u}_h\}_{h\searrow 0}(=\{\varrho_h^k,\boldsymbol{u}_h^k\}_{k=1}^{N_T})$ *be a family of solutions obtained by the MAC scheme (14.10) with* $\Delta t\approx h$.

Then we have the following convergence results:

1. **Convergence to DMV solution.**
   Any Young measure $\{\mathcal{V}_{t,x}\}_{(t,x)\in(0,T)\times\mathbb{T}^d}$ *generated by* $(\varrho_{h_n}, \boldsymbol{u}_{h_n})$ *for a suitable sequence* $h_n \searrow 0$ *represents a dissipative measure-valued solution of the Navier–Stokes system (14.1) in the sense of Definition 5.10.*
2. **Convergence to a strong solution.** *Let the Navier–Stokes system endowed with the initial data* $[\varrho_0, \boldsymbol{u}_0]$ *admit a strong solution* $[\varrho, \boldsymbol{u}]$ *belonging to the class*

$$\varrho > 0,\ \varrho \in C([0,T]\times\mathbb{T}^d),\ \nabla_x\varrho, \boldsymbol{u} \in C([0,T]\times\mathbb{T}^d; R^d),$$
$$\nabla_x\boldsymbol{u} \in C([0,T]\times\mathbb{T}^d; R^{d\times d}),\ \partial_t\boldsymbol{u} \in L^2(0,T; C(\mathbb{T}^d; R^d)).$$

   Then

$$\varrho_h \to \varrho \text{ (strongly) in } L^\gamma((0,T)\times\mathbb{T}^d),$$
$$\boldsymbol{u}_h \to \boldsymbol{u} \text{ (strongly) in } L^2((0,T)\times\mathbb{T}^d; R^d).$$

3. **Convergence to a strong solution for bounded density.**
   Suppose that

   - *the initial data belong to the class* $\varrho_0 \in W^{3,2}(\mathbb{T}^d),\ \boldsymbol{u}_0 \in W^{3,2}(\mathbb{T}^d; R^d)$;
   - *bulk viscosity vanishes, meaning* $\lambda = 0$;
   - *discrete density is uniformly bounded, i.e.* $\|\varrho_h\|_{L^\infty((0,T)\times\mathbb{T}^d)} \lesssim 1$ *for* $h \to 0$.

   Then

$$\varrho_h \to \varrho \text{ (strongly) in } L^q((0,T)\times\mathbb{T}^d),\ q \geq 1,$$
$$\boldsymbol{u}_h \to \boldsymbol{u} \text{ (strongly) in } L^2((0,T)\times\mathbb{T}^d; R^d),$$

   where $[\varrho, \boldsymbol{u}]$ *is a strong solution to the Navier–Stokes system (14.1) with the initial data* $[\varrho_0, \boldsymbol{u}_0]$.

Proof Due to Theorems 7.12 and 7.13, it suffices to show that the numerical solution of the MAC scheme is a consistent approximation of the Navier–Stokes system in the sense of Definition 5.9. Note that the essential prerequisites to achieve a consistent approximation are the energy stability and consistency formulation of the numerical scheme in the sense of Definitions 8.3 and 8.6 which were shown in Sects. 14.2 and 14.3, respectively. Thus, the proof can be done analogously as for Theorem 13.3. For completeness, we repeat the main arguments here.

First, from the uniform bounds given in Corollary 14.1 we derive for a suitable subsequence (not relabeled) that

$$\overline{\boldsymbol{u}_h},\ \boldsymbol{u}_h \to \boldsymbol{u} \text{ weakly in } L^2((0,T)\times\mathbb{T}^d; R^d),$$
$$\nabla_\epsilon\boldsymbol{u}_h \to \nabla_x\boldsymbol{u} \text{ weakly in } L^2((0,T)\times\mathbb{T}^d; R^{d\times d}),$$
$$\mathrm{div}_{\mathcal{T}}\boldsymbol{u}_h \to \mathrm{div}_x\boldsymbol{u} \text{ weakly in } L^2((0,T)\times\mathbb{T}^d),$$

where $\boldsymbol{u} \in L^2(0, T; W^{1,2}(\mathbb{T}^d; R^d))$. Further, recall that the mass conservation, energy stability and consistency of the numerical solution have been proven in Lemma 14.2, Theorems 14.1 and 14.2, respectively. Thus, there is a subsequence of $\{\varrho_h, \overline{\boldsymbol{u}_h}\}_{h\searrow 0}$ generating a Young measure

$$\{\mathcal{V}_{t,x}\}_{(t,x)\in(0,T)\times\mathbb{T}^d}, \ \mathcal{V} \in L^\infty_{\text{weak}-(*)}((0,T)\times\mathbb{T}^d; \mathcal{P}(R^{d+1})).$$

Using the inequality (14.7) for each $K \in \mathbb{T}^d$ we can deduce that

$$\|\boldsymbol{u}_h - \overline{\boldsymbol{u}_h}\|_{L^2L^2} \stackrel{<}{\sim} h.$$

Consequently, the Young measures generated by $\{\varrho_h, \overline{\boldsymbol{u}_h}\}_{h\searrow 0}$ and $\{\varrho_h, \boldsymbol{u}_h\}_{h\searrow 0}$ coincide for a.a. $(t, x) \in (0, T) \times \mathbb{T}^d$. Particularly, their barycenter's second coordinate represented by the limits of $\boldsymbol{u}_h$ and $\overline{\boldsymbol{u}_h}$ coincide on $(0, T) \times \mathbb{T}^d$, see also the proof of Theorem 13.3. Finally, recalling the interpolation estimates (18) and (14.6b) we know that the discrete operators $\mathrm{div}_\mathcal{T}$ and ∇_ϵ are compatible with div_x and ∇_x, respectively. Consequently, the sequence $\{\varrho_h, \overline{\boldsymbol{u}_h}\}_{h\searrow 0}$ is a consistent approximation of the Navier–Stokes system in the sense of Definition 5.9. Applying Theorem 5.5 proves the first statement, which is the weak convergence to a DMV solution. Employing Theorems 7.12 and 7.13 finally leads to the second and the third convergence statement.

14.5 Error Estimates

Assuming a smooth solution to the Navier–Stokes system exists, we can use the discrete relative energy functional introduced in (14.18) to derive the error estimates.

The following algebraic inequality related to the relative energy functional shall be used at the end of this section in the final step of proving the convergence rate for the MAC scheme.

Lemma 14.3 *Let $\gamma > 1$, $0 < \underline{r} < \overline{r} < \infty$ and $r \in [\underline{r}, \overline{r}]$.*

Then there exists a $c = c(\underline{r}, \overline{r}) > 0$ such that

$$\mathbb{E}(\varrho|r) \geq c(\underline{r}, \overline{r}) \left((1 + \varrho^\gamma) 1_{R_+ \setminus [\underline{r}/2, 2\overline{r}]}(\varrho) + (\varrho - r)^2 1_{[\underline{r}/2, 2\overline{r}]}(\varrho)\right),$$

where $\mathbb{E}(\varrho|r) = P(\varrho) - P'(r)(\varrho - r) - P(r)$, $P(r) = ar^\gamma$, $a > 0$, $\gamma > 1$.

Proof As $\varrho \mapsto \mathbb{E}(\varrho|r)$ is a strictly convex function on $[0, \infty)$ attaining its minimum (zero) at $\varrho = r$, we get

$$\mathbb{E}(\varrho|r) \geq \min_{\xi\in[\underline{r}/2, 2\overline{r}]} P''(\xi)(\varrho - r)^2, \quad \varrho \in [\underline{r}/2, 2\overline{r}],$$

and

$$\mathbb{E}(\varrho|r) \geq \min\left\{\mathbb{E}(\underline{r}/2, \underline{r}); \mathbb{E}(2\overline{r}, \overline{r})\right\} > 0 \text{ for any } \varrho \in [0, \underline{r}/2] \cup [2\overline{r}, \infty).$$

In particular, we have

$$\mathbb{E}(\varrho|r) \geq \mathbb{E}(\underline{r}/2)\frac{\underline{r}^\gamma}{\underline{r}^\gamma} \geq \frac{\mathbb{E}(\underline{r}/2)}{\underline{r}^\gamma}\varrho^\gamma \text{ for all } \varrho \in [0, \underline{r}/2].$$

Finally,

$$\lim_{\varrho\to\infty} \frac{\mathbb{E}(\varrho|r)}{\varrho^\gamma} = \frac{a}{\gamma - 1},$$

which yields the desired conclusion for $\varrho \in [2\overline{r}, \infty)$.

The result on error estimates for the MAC scheme (14.10) follows.

Theorem 14.4 (Convergence rate of the MAC scheme) *Let $\gamma > \frac{d}{2}$ and the assumptions of Theorem 14.3 hold. Suppose further that the Navier–Stokes system (14.1) admits a strong solution $[\varrho = r, \boldsymbol{u} = \boldsymbol{U}]$ belonging to the class*

$$r \in C^2([0, T] \times \mathbb{T}^d),\ 0 < \underline{r} \leq r(t, x) \leq \overline{r}, \quad \boldsymbol{U} \in C^2([0, T] \times \mathbb{T}^d; R^d). \tag{14.21}$$

Then, for $\Delta t << 1$, there exists a positive number c depending on M_0, E_0, $\underline{r}$, $\overline{r}$, $\|(r, \boldsymbol{U})\|_{C^2}$, $\|p\|_{C^2([\underline{r},\overline{r}])}$, T, γ, such that, for

$$(r_h^k, \boldsymbol{U}_h^k) = (\Pi_Q r(t^k), \Pi_W \boldsymbol{U}(t^k))$$

there holds

$$\begin{aligned} \sup_{0\leq k\leq N_T} E(\varrho_h^k, \boldsymbol{u}_h^k | r_h^k, \boldsymbol{U}_h^k) + \Delta t \sum_{k=1}^{N_T} \frac{\mu}{2} \int_{\mathbb{T}^d} |\nabla_\epsilon(\boldsymbol{u}_h^k - \boldsymbol{U}_h^k)|^2 \,\mathrm{d}x \\ \leq c\left(h^A + \sqrt{\Delta t} + E(\varrho_h^0, \boldsymbol{u}_h^0 | r_h^0, \boldsymbol{U}_h^0)\right) \end{aligned} \tag{14.22}$$

with the exponent

$$A = \min\left\{\frac{2\gamma - d}{\gamma}, \frac{1}{2}\right\}. \tag{14.23}$$

Before showing the convergence rate of the MAC scheme (14.10) for the Navier–Stokes system (14.1a)–(14.1b) as stated above we prepare the main ingredients for doing so, namely

1. *Exact discrete relative energy equality:* the discrete counterpart of the continuous version of the relative energy inequality;

2. *Approximate discrete relative energy inequality:* the exact discrete relative energy with a particularly chosen discrete test functions and suitably transformed terms;
3. *Consistency error:* the identity satisfied by a strong solution tested with $\boldsymbol{u}_h - \boldsymbol{U}_h$, which shall be useful to compensate the terms in step 2;

1. Exact discrete relative energy equality

In the first step we derive the discrete relative energy equality satisfied by a solution $(\varrho_h, \boldsymbol{u}_h)$ of the MAC scheme (14.10) and a generic function pair $(r_h, \boldsymbol{U}_h)$.

Lemma 14.4 (Exact discrete relative energy) *Let $(\varrho_h, \boldsymbol{u}_h)$ be a solution to the MAC scheme (14.10). Then for any $\{r_h^k, \boldsymbol{U}_h^k\}_{k=1}^{N_T} \in Q_h \times \boldsymbol{W}_h$, $r_h > 0$, and for $m = 1, \dots N_T$ it holds*

$$
\begin{aligned}
&\int_{\mathbb{T}^d} \frac{1}{2}\left(\varrho_h^m \left|\overline{\boldsymbol{u}_h}^m - \overline{\boldsymbol{U}_h}^m\right|^2 - \varrho_h^0 |\overline{\boldsymbol{u}_h}^0 - \overline{\boldsymbol{U}_h}^0|^2\right) \, \mathrm{d}x \\
&+ \int_{\mathbb{T}^d} \left(\mathbb{E}(\varrho_h^m | r_h^m) - \mathbb{E}(\varrho_h^0 | r_h^0)\right) \, \mathrm{d}x \\
&+ \Delta t \sum_{k=1}^m \mu \int_{\mathbb{T}^d} |\nabla_\epsilon(\boldsymbol{u}_h^k - \boldsymbol{U}_h^k)|^2 \, \mathrm{d}x + \Delta t \sum_{k=1}^m \nu \int_{\mathbb{T}^d} |\mathrm{div}_{\mathcal{T}}(\boldsymbol{u}_h^k - \boldsymbol{U}_h^k)|^2 \, \mathrm{d}x \\
&+ \sum_{i=1}^8 T_i = -\sum_{i=1}^5 D_i,
\end{aligned}
\tag{14.24}
$$

where $T_i = \sum_{k=1}^m \Delta t \; T_i^k$ and $D_i = \sum_{k=1}^m \Delta t \; D_i^k \geq 0$ with

$$
\begin{aligned}
T_1^k &= \mu \int_{\mathbb{T}^d} \nabla_\epsilon \boldsymbol{U}_h^k : \nabla_\epsilon(\boldsymbol{u}_h^k - \boldsymbol{U}_h^k) \, \mathrm{d}x + \nu \int_{\mathbb{T}^d} \mathrm{div}_{\mathcal{T}} \boldsymbol{U}_h^k \, \mathrm{div}_{\mathcal{T}}(\boldsymbol{u}_h^k - \boldsymbol{U}_h^k) \, \mathrm{d}x \\
T_2^k &= \int_{\mathbb{T}^d} \varrho_h^{k-1} D_t \overline{\boldsymbol{U}_h}^k \left(\overline{\boldsymbol{u}_h}^{k-1} - \frac{\overline{\boldsymbol{U}_h}^{k-1} + \overline{\boldsymbol{U}_h}^k}{2}\right) \mathrm{d}x, \\
T_3^k &= \sum_{K \in \mathcal{T}_h} \sum_{\sigma \in \mathcal{E}(K)} \int_\sigma \varrho_h^{k,\mathrm{up}} \left(\left\{\!\!\left\{\overline{\boldsymbol{U}_h}^k\right\}\!\!\right\} - \overline{\boldsymbol{u}_h}^{k,\mathrm{up}}\right) \cdot \overline{\boldsymbol{U}_h}^k (\boldsymbol{u}_h^k \cdot \boldsymbol{n}) \mathrm{dS}_x, \\
T_4^k &= \int_{\mathbb{T}^d} p(\varrho_h^k) \mathrm{div}_{\mathcal{T}} \boldsymbol{U}_h^k \, \mathrm{d}x, \\
T_5^k &= \int_{\mathbb{T}^d} \frac{r_h^k - \varrho_h^k}{\Delta t} \left(P'(r_h^{k-1}) - P'(r_h^k)\right) \mathrm{d}x, \\
T_6^k &= -\sum_{K \in \mathcal{T}_h} \sum_{\sigma \in \mathcal{E}(K)} \int_\sigma \varrho_h^{k,\mathrm{up}} P'(r_h^{k-1}) (\boldsymbol{u}_h^k \cdot \boldsymbol{n}) \mathrm{dS}_x,
\end{aligned}
$$

$$T_7^k = h^{\varepsilon+1} \int_{\mathbb{T}^d} \Delta_h \varrho_h^k P'(r_h^{k-1}) \, dx,$$

$$T_8^k = h^{\varepsilon+1} \int_{\mathbb{T}^d} \nabla_{\mathcal{D}} \varrho_h^k \cdot \nabla_{\mathcal{D}} \overline{\boldsymbol{U}_h}^k \cdot \left\{\!\!\left\{ \overline{\boldsymbol{U}_h}^k - \overline{\boldsymbol{u}_h}^k \right\}\!\!\right\} \, dx,$$

$$D_1^k = \frac{1}{2\Delta t} \int_{\mathbb{T}^d} \varrho_h^{k-1} \left| \overline{\boldsymbol{u}_h}^{k-1} - \overline{\boldsymbol{u}_h}^k \right|^2 \, dx \geq 0,$$

$$D_2^k = \frac{1}{\Delta t} \int_{\mathbb{T}^d} \left(\mathbb{E}(\varrho_h^{k-1} | \varrho_h^k) + \mathbb{E}(r_h^k | r_h^{k-1}) \right) \, dx \geq 0,$$

$$D_3^k = \frac{1}{2} \int_{\mathcal{E}} \varrho_h^{k,\mathrm{up}} \left[\!\left[\overline{\boldsymbol{u}_h}^k \right]\!\right]^2 \left| \boldsymbol{u}_h^k \cdot \boldsymbol{n} \right| \mathrm{dS}_x \geq 0,$$

$$D_4^k = \int_{\mathcal{E}} \left| \boldsymbol{u}_h^k \cdot \boldsymbol{n} \right| \mathbb{E}\left(\varrho_h^{k,\mathrm{in}} | \varrho_h^{k,\mathrm{out}} \right) \mathrm{dS}_x \geq 0,$$

$$D_5^k = h^{\varepsilon+1} \sum_{K \in \mathcal{T}_h} \sum_{\sigma \in \mathcal{E}(K)} \int_{\sigma} \nabla_{\mathcal{D}} \varrho_h^k \cdot \nabla_{\mathcal{D}} P'(\varrho_h^k) \mathrm{dS}_x$$

$$= h^{\varepsilon+1} \sum_{K \in \mathcal{T}_h} \sum_{\sigma \in \mathcal{E}(K)} \int_{\sigma} |\nabla_{\mathcal{D}} \varrho_h^k|^2 P''(\varrho_{h,\dagger}^k) \mathrm{dS}_x \geq 0.$$

Proof We set $\phi_h = \frac{1}{2}\left(\left| \overline{\boldsymbol{U}_h}^k \right|^2 - \left| \overline{\boldsymbol{u}_h}^k \right|^2 \right)$ in (14.10a), $\phi_h = \left(P'(\varrho_h^k) - P'(r_h^{k-1}) \right)$ in (14.10a), and $\boldsymbol{\phi}_h = (\boldsymbol{u}_h^k - \boldsymbol{U}_h^k)$ in (14.10b), respectively, to deduce

$$\sum_{k=1}^{3} I_k = 0, \quad \sum_{k=4}^{6} I_k = 0, \quad \sum_{k=7}^{11} I_k = 0,$$

where

$$I_1 = \frac{1}{2} \int_{\mathbb{T}^d} D_t \varrho_h^k \left(\left| \overline{\boldsymbol{U}_h}^k \right|^2 - \left| \overline{\boldsymbol{u}_h}^k \right|^2 \right) \, dx,$$

$$I_2 = -\frac{1}{2} \int_{\mathcal{E}} \mathrm{Up}[\varrho_h^k, \boldsymbol{u}_h^k] \left[\!\!\left[\left| \overline{\boldsymbol{U}_h}^k \right|^2 - \left| \overline{\boldsymbol{u}_h}^k \right|^2 \right]\!\!\right] \mathrm{dS}_x,$$

$$I_3 = \frac{h^{\varepsilon+1}}{2} \int_{\mathbb{T}^d} \nabla_{\mathcal{D}} \varrho_h^k \cdot \nabla_{\mathcal{D}} \left(\left| \overline{\boldsymbol{U}_h}^k \right|^2 - \left| \overline{\boldsymbol{u}_h}^k \right|^2 \right) \, dx,$$

$$I_4 = \int_{\mathbb{T}^d} D_t \varrho_h^k \left(P'(\varrho_h^k) - P'(r_h^{k-1}) \right) \, dx,$$

$$I_5 = -\int_{\mathcal{E}} Up[\varrho_h^k, \boldsymbol{u}_h^k] \left[\!\left[P'(\varrho_h^k) - P'(r_h^{k-1}) \right]\!\right] \mathrm{dS}_x,$$

$$I_6 = h^{\varepsilon+1} \int_{\mathbb{T}^d} \nabla_{\mathcal{D}} \varrho_h^k \cdot \nabla_{\mathcal{D}} \left(P'(\varrho_h^k) - P'(r_h^{k-1}) \right) \, \mathrm{d}x,$$

$$I_7 = \int_{\mathbb{T}^d} D_t \varrho_h^k \overline{\boldsymbol{u}_h}^k \cdot \left(\overline{\boldsymbol{u}_h}^k - \overline{\boldsymbol{U}_h}^k \right) \mathrm{d}x,$$

$$I_8 = -\int_{\mathcal{E}} Up[\varrho_h^k \overline{\boldsymbol{u}_h}^k, \boldsymbol{u}_h^k] \cdot \left[\!\left[\overline{\boldsymbol{u}_h}^k - \overline{\boldsymbol{U}_h}^k \right]\!\right] \mathrm{dS}_x,$$

$$I_9 = -\int_{\mathbb{T}^d} p(\varrho_h^k) \cdot \mathrm{div}_{\mathcal{T}} \left(\boldsymbol{u}_h^k - \boldsymbol{U}_h^k \right) \mathrm{d}x,$$

$$I_{10} = \mu \int_{\mathbb{T}^d} \nabla_\epsilon \boldsymbol{u}_h^k \cdot \nabla_\epsilon \left(\boldsymbol{u}_h^k - \boldsymbol{U}_h^k \right) \mathrm{d}x + \nu \int_{\mathbb{T}^d} \mathrm{div}_{\mathcal{T}} \boldsymbol{u}_h^k \, \mathrm{div}_{\mathcal{T}} \left(\boldsymbol{u}_h^k - \boldsymbol{U}_h^k \right) \mathrm{d}x,$$

$$I_{11} = h^{\varepsilon+1} \sum_{i=1}^d \sum_{j=1}^d \int_{\mathbb{T}^d} \left\{\!\!\left\{ \overline{u_{i,h}}^k \right\}\!\!\right\}^{(j)} \eth_{\mathcal{D}_j} \varrho_h^k \, \eth_{\mathcal{D}_j} \left(\overline{u_{i,h}}^k - \overline{U_{i,h}}^k \right) \, \mathrm{d}x.$$

Now we sum up all I_k terms and derive the desired inequality in 7 steps:

- The term I_1 and I_7:

$$I_1 + I_7 = \frac{1}{2} \int_{\mathbb{T}^d} D_t \varrho_h^k \left(\left| \overline{\boldsymbol{U}_h}^k \right|^2 - \left| \overline{\boldsymbol{u}_h}^k \right|^2 \right) \, \mathrm{d}x + \int_{\mathbb{T}^d} D_t \left(\varrho_h^k \overline{\boldsymbol{u}_h}^k \right) \cdot \left(\overline{\boldsymbol{u}_h}^k - \overline{\boldsymbol{U}_h}^k \right) \mathrm{d}x$$

$$= \frac{1}{2\Delta t} \int_{\mathbb{T}^d} \varrho_h^k \left(\left| \overline{\boldsymbol{U}_h}^k \right|^2 - \left| \overline{\boldsymbol{u}_h}^k \right|^2 + 2 \overline{\boldsymbol{u}_h}^k \cdot \left(\overline{\boldsymbol{u}_h}^k - \overline{\boldsymbol{U}_h}^k \right) \right) \, \mathrm{d}x$$

$$- \frac{1}{2\Delta t} \int_{\mathbb{T}^d} \varrho_h^{k-1} \left(\left| \overline{\boldsymbol{U}_h}^k \right|^2 - \left| \overline{\boldsymbol{u}_h}^k \right|^2 + 2 \overline{\boldsymbol{u}_h}^{k-1} \cdot \left(\overline{\boldsymbol{u}_h}^k - \overline{\boldsymbol{U}_h}^k \right) \right) \, \mathrm{d}x$$

$$= \frac{1}{2\Delta t} \int_{\mathbb{T}^d} \left(\varrho_h^k \left| \overline{\boldsymbol{u}_h}^k - \overline{\boldsymbol{U}_h}^k \right|^2 - \varrho_h^{k-1} \left| \overline{\boldsymbol{u}_h}^{k-1} - \overline{\boldsymbol{U}_h}^{k-1} \right|^2 \right) \, \mathrm{d}x$$

$$+ \int_{\mathbb{T}^d} \varrho_h^{k-1} D_t \overline{\boldsymbol{U}}_h^k \left(\overline{\boldsymbol{u}_h}^{k-1} - \frac{\overline{\boldsymbol{U}_h}^{k-1} + \overline{\boldsymbol{U}_h}^k}{2} \right) \, \mathrm{d}x$$

$$+ \frac{1}{2\Delta t} \int_{\mathbb{T}^d} \varrho_h^{k-1} \left| \overline{\boldsymbol{u}_h}^{k-1} - \overline{\boldsymbol{u}_h}^k \right|^2 \, \mathrm{d}x$$

$$= \frac{1}{2\Delta t} \int_{\mathbb{T}^d} \left(\varrho_h^k \left| \overline{\boldsymbol{u}_h}^k - \overline{\boldsymbol{U}_h}^k \right|^2 - \varrho_h^{k-1} \left| \overline{\boldsymbol{u}_h}^{k-1} - \overline{\boldsymbol{U}_h}^{k-1} \right|^2 \right) \, \mathrm{d}x + T_2^k + D_1^k.$$

- The term I_4:

$$
\begin{aligned}
I_4 &= \int_{\mathbb{T}^d} D_t \varrho_h^k \left(P'(\varrho_h^k) - P'(r_h^{k-1})\right) \,\mathrm{d}x \\
&= \frac{1}{\Delta t} \int_{\mathbb{T}^d} \left(P(\varrho_h^{k-1}) - P(\varrho_h^k) - P'(\varrho_h^k)(\varrho_h^{k-1} - \varrho_h^k)\right) \,\mathrm{d}x \\
&+ \frac{1}{\Delta t} \int_{\mathbb{T}^d} \left(P(\varrho_h^k) - P(\varrho_h^{k-1}) + P'(r_h^{k-1})(\varrho_h^{k-1} - \varrho_h^k)\right) \,\mathrm{d}x \\
&= \frac{1}{\Delta t} \int_{\mathbb{T}^d} \mathbb{E}(\varrho_h^{k-1} | \varrho_h^k) \,\mathrm{d}x + \frac{1}{\Delta t} \int_{\mathbb{T}^d} \left(P(\varrho_h^k) - P(r_h^k) - P'(r_h^k)(\varrho_h^k - r_h^k)\right) \,\mathrm{d}x \\
&- \frac{1}{\Delta t} \int_{\mathbb{T}^d} \left(P(\varrho_h^{k-1}) - P(r_h^{k-1}) - P'(r_h^{k-1})(\varrho_h^{k-1} - r_h^{k-1})\right) \,\mathrm{d}x \\
&+ \frac{1}{\Delta t} \int_{\mathbb{T}^d} \left(P(r_h^k) - P(r_h^{k-1}) + P'(r_h^k)(\varrho_h^k - r_h^k) + P'(r_h^{k-1})(r_h^{k-1} - \varrho_h^k)\right) \,\mathrm{d}x \\
&= \frac{1}{\Delta t} \int_{\mathbb{T}^d} \left(\mathbb{E}(\varrho_h^{k-1} | \varrho_h^k) + \mathbb{E}(\varrho_h^k | r_h^k) - \mathbb{E}(\varrho_h^{k-1} | r_h^{k-1})\right) \,\mathrm{d}x \\
&+ \frac{1}{\Delta t} \int_{\mathbb{T}^d} \left(\mathbb{E}(r_h^k | r_h^{k-1}) + P'(r_h^k)(\varrho_h^k - r_h^k) + P'(r_h^{k-1})(r_h^k - \varrho_h^k)\right) \,\mathrm{d}x \\
&= \frac{1}{\Delta t} \int_{\mathbb{T}^d} \left(\mathbb{E}(\varrho_h^k | r_h^k) - \mathbb{E}(\varrho_h^{k-1} | r_h^{k-1})\right) \,\mathrm{d}x \\
&+ \frac{1}{\Delta t} \int_{\mathbb{T}^d} \left(\mathbb{E}(\varrho_h^{k-1} | \varrho_h^k) + \mathbb{E}(r_h^k | r_h^{k-1})\right) \,\mathrm{d}x \\
&+ \int_{\mathbb{T}^d} \frac{r_h^k - \varrho_h^k}{\Delta t} \left(P'(r_h^{k-1}) - P'(r_h^k)\right) \,\mathrm{d}x \\
&= \frac{1}{\Delta t} \int_{\mathbb{T}^d} \left(\mathbb{E}(\varrho_h^k | r_h^k) - \mathbb{E}(\varrho_h^{k-1} | r_h^{k-1})\right) \,\mathrm{d}x + D_2^k + T_5^k.
\end{aligned}
$$

- The term I_2 and I_8:

$$
\begin{aligned}
I_2 + I_8 = &-\frac{1}{2} \int_{\mathcal{E}} Up[\varrho_h^k, \boldsymbol{u}_h^k] \left[\!\left[\left|\overline{\boldsymbol{U}_h}^k\right|^2 - \left|\overline{\boldsymbol{u}_h}^k\right|^2 \right]\!\right] \,\mathrm{dS}_x \\
&- \int_{\mathcal{E}} Up[\varrho_h^k \overline{\boldsymbol{u}_h}^k, \boldsymbol{u}_h^k] \cdot \left[\!\left[\overline{\boldsymbol{u}_h}^k - \overline{\boldsymbol{U}_h}^k \right]\!\right] \,\mathrm{dS}_x
\end{aligned}
$$

$$= \int_{\mathcal{E}} \left(Up[\varrho_h^k \overline{\boldsymbol{u}_h}^k, \boldsymbol{u}_h^k] \cdot \left[\!\left[\overline{\boldsymbol{U}_h}^k \right]\!\right] - \frac{1}{2} Up[\varrho_h^k, \boldsymbol{u}_h^k] \left[\!\left[\left| \overline{\boldsymbol{U}_h}^k \right|^2 \right]\!\right] \right) \mathrm{dS}_x$$

$$+ \int_{\mathcal{E}} \left(\frac{1}{2} Up[\varrho_h^k, \boldsymbol{u}_h^k] \left[\!\left[|\overline{\boldsymbol{u}_h}^k|^2 \right]\!\right] - Up[\varrho_h^k \overline{\boldsymbol{u}_h}^k, \boldsymbol{u}_h^k] \cdot [\![\overline{\boldsymbol{u}_h}^k]\!] \right) \mathrm{dS}_x$$

$$= \int_{\mathcal{E}} \varrho_h^{k,\mathrm{up}} (\boldsymbol{u}_h^k \cdot \boldsymbol{n}) \left[\!\left[\overline{\boldsymbol{U}_h}^k \right]\!\right] \cdot \left(\overline{\boldsymbol{u}_h}^{k,\mathrm{up}} - \left\{\!\left\{ \overline{\boldsymbol{U}_h}^k \right\}\!\right\} \right) \mathrm{dS}_x$$

$$+ \frac{1}{2} \int_{\mathcal{E}} \varrho_h^{k,\mathrm{up}} [\![\overline{\boldsymbol{u}_h}^k]\!]^2 |\boldsymbol{u}_h^k \cdot \boldsymbol{n}| \, \mathrm{dS}_x$$

$$= T_3^k + D_3^k,$$

where we have used (14.11a).

- The sum of I_5 and I_9 yields both T_4^k and T_6^k:

$$I_5 + I_9 = - \int_{\mathcal{E}} Up[\varrho_h^k, \boldsymbol{u}_h^k] [\![P'(\varrho_h^k) - P'(r_h^{k-1})]\!] \, \mathrm{dS}_x$$

$$- \int_{\mathbb{T}^d} p(\varrho_h^k) \cdot \mathrm{div}_{\mathcal{T}} \big(\boldsymbol{u}_h^k - \boldsymbol{U}_h^k \big) \, \mathrm{d}x$$

$$= \int_{\mathbb{T}^d} p(\varrho_h^k) \mathrm{div}_{\mathcal{T}} \boldsymbol{U}_h^k \, \mathrm{d}x + \int_{\mathcal{E}} Up[\varrho_h^k, \boldsymbol{u}_h^k] [\![P'(r_h^{k-1})]\!] \, \mathrm{dS}_x$$

$$- \int_{\mathbb{T}^d} p(\varrho_h^k) \mathrm{div}_{\mathcal{T}} \boldsymbol{u}_h^k \, \mathrm{d}x - \int_{\mathcal{E}} Up[\varrho_h^k, \boldsymbol{u}_h^k] [\![P'(\varrho_h^k)]\!] \, \mathrm{dS}_x$$

$$= T_4^k + T_6^k + D_4^k,$$

where

$$D_4^k = - \int_{\mathbb{T}^d} p(\varrho_h^k) \mathrm{div}_{\mathcal{T}} \boldsymbol{u}_h^k \, \mathrm{d}x - \int_{\mathcal{E}} Up[\varrho_h^k, \boldsymbol{u}_h^k] [\![P'(\varrho_h^k)]\!] \, \mathrm{dS}_x$$

$$= \sum_{K \in \mathcal{T}_h} \sum_{\sigma \in \mathcal{E}(K)} \int_{\sigma} \big(\varrho_h^{k,\mathrm{up}} P'(\varrho_h^k) - p(\varrho_h^k) \big) \boldsymbol{u}_h^k \cdot \boldsymbol{n} \mathrm{dS}_x$$

$$= \sum_{\sigma = K|L \in \mathcal{E}} \int_{\sigma} [\boldsymbol{u}_h^k \cdot \boldsymbol{n}]^+ (P(\varrho_K^k) - P'(\varrho_L^k)(\varrho_K^k - \varrho_L^k) - P(\varrho_L^k)) \mathrm{dS}_x$$

$$+ \sum_{\sigma = K|L \in \mathcal{E}} \int_{\sigma} [\boldsymbol{u}_h^k \cdot \boldsymbol{n}_{\sigma,L}]^+ (P(\varrho_L^k) - P'(\varrho_K^k)(\varrho_L^k - \varrho_K^k) - P(\varrho_K^k)) \mathrm{dS}_x$$

$$= \sum_{\sigma=K|L\in\mathcal{E}} \int_\sigma [\boldsymbol{u}_h^k \cdot \boldsymbol{n}_{\sigma K}]^+ \mathbb{E}(\varrho_K^k|\varrho_L^k) + [\boldsymbol{u}_h^k \cdot \boldsymbol{n}_{\sigma L}]^+ \mathbb{E}(\varrho_L^k|\varrho_K^k) \mathrm{dS}_x$$

$$= \int_{\mathcal{E}} |\boldsymbol{u}_h^k \cdot \boldsymbol{n}| \, \mathbb{E}\left(\varrho_h^{k,\mathrm{in}}|\varrho_h^{k,\mathrm{out}}\right) \mathrm{dS}_x \geq 0.$$

- The term I_3 and I_{11}:

$$I_3 + I_{11} = \frac{h^{\varepsilon+1}}{2} \int_{\mathbb{T}^d} \nabla_{\mathcal{D}} \varrho_h^k \cdot \nabla_{\mathcal{D}} \left(\left|\overline{\boldsymbol{U}_h}^k\right|^2 - \left|\overline{\boldsymbol{u}_h}^k\right|^2 \right) \mathrm{d}x$$

$$+ h^{\varepsilon+1} \sum_{i=1}^d \sum_{j=1}^d \int_{\mathbb{T}^d} \{\{\overline{u_{i,h}}^k\}\}^{(j)} \eth_{\mathcal{D}_j} \varrho_h^k \eth_{\mathcal{D}_j} (\overline{u_{i,h}}^k - \overline{U_{i,h}}^k) \, \mathrm{d}x$$

$$= h^{\varepsilon+1} \sum_{i=1}^d \sum_{j=1}^d \int_{\mathbb{T}^d} \eth_{\mathcal{D}_j} \varrho_h^k \eth_{\mathcal{D}_j} \overline{U_{i,h}}^k \left\{\left\{\overline{U_{i,h}}^k\right\}\right\}^{(j)} \mathrm{d}x$$

$$- h^{\varepsilon+1} \sum_{i=1}^d \sum_{j=1}^d \int_{\mathbb{T}^d} \{\{\overline{u_{i,h}}^k\}\}^{(j)} \eth_{\mathcal{D}_j} \varrho_h^k \eth_{\mathcal{D}_j} \overline{U_{i,h}}^k \, \mathrm{d}x$$

$$= T_8^k,$$

where we have used the equality

$$\sum_{i=1}^d \sum_{j=1}^d \int_{\mathbb{T}^d} \{\{\overline{u_{i,h}}^k\}\}^{(j)} \eth_{\mathcal{D}_j} \varrho_h^k \eth_{\mathcal{D}_j} (\overline{u_{i,h}}^k) \, \mathrm{d}x = \int_{\mathbb{T}^d} \frac{1}{2} \nabla_{\mathcal{D}} \varrho_h^k \cdot \nabla_{\mathcal{D}} \left|\overline{\boldsymbol{u}_h}^k\right|^2 \, \mathrm{d}x.$$

- Term I_6:

$$I_6 = h^{\varepsilon+1} \int_{\mathbb{T}^d} \nabla_{\mathcal{D}} \varrho_h^k \cdot \nabla_{\mathcal{D}} \left(P'(\varrho_h^k) - P'(r_h^{k-1}) \right) \, \mathrm{d}x$$

$$= h^{\varepsilon+1} \int_{\mathbb{T}^d} \Delta_h \varrho_h^k \, P'(r_h^{k-1}) \, \mathrm{d}x + h^{\varepsilon+1} \int_{\mathbb{T}^d} |\nabla_{\mathcal{D}} \varrho_h^k|^2 \frac{P''(\varrho_{h,\dagger}^k)}{2} \, \mathrm{d}x$$

$$= T_7^k + D_5^k.$$

- Term I_{10}:

$$I_{10} = \mu \int_{\mathbb{T}^d} \nabla_\epsilon \boldsymbol{u}_h^k : \nabla_\epsilon (\boldsymbol{u}_h^k - \boldsymbol{U}_h^k) \, \mathrm{d}x + \nu \int_{\mathbb{T}^d} \mathrm{div}_{\mathcal{T}} \boldsymbol{u}_h^k \, \mathrm{div}_{\mathcal{T}} (\boldsymbol{u}_h^k - \boldsymbol{U}_h^k) \, \mathrm{d}x$$

$$= \mu \left\| \nabla_\epsilon (\boldsymbol{u}_h^k - \boldsymbol{U}_h^k) \right\|_{L^2}^2 + \nu \left\| \mathrm{div}_{\mathcal{T}} (\boldsymbol{u}_h^k - \boldsymbol{U}_h^k) \right\|_{L^2}^2 + T_1^k.$$

Finally, collecting all the above calculations and summing up from $k = 1$ to m finishes the proof.

2. Approximate relative energy inequality

In this step we set $(r_h, \boldsymbol{U}_h) = (\Pi_Q r, \Pi_W \boldsymbol{U})$ in Lemma 14.4 for $[r, \boldsymbol{U}]$ belonging to the class (14.21), and we further analyze the equality (14.24)—mainly the T_i terms given in Lemma 14.4.

Lemma 14.5 (Approximate discrete relative energy) *Let* $(\varrho_h, \boldsymbol{u}_h) \in Q_h \times \boldsymbol{W}_h$ *be a solution to the MAC scheme (14.10), and let* $(r_h, \boldsymbol{U}_h) = (\Pi_Q r, \Pi_W \boldsymbol{U})$ *for* $[r, \boldsymbol{U}]$ *belonging to the class* (14.21). *Then there exists a positive constant*

$$c = c\left(M_0, E_0, \underline{r}, \overline{r}, \|p\|_{C^2([\underline{r},\overline{r}])}, \|(r, \boldsymbol{U})\|_{C^2}\right)$$

such that for all $m = 1, \ldots, N_T$ *we have*

$$E(\varrho_h^m, \boldsymbol{u}_h^m | r_h^m, \boldsymbol{U}_h^m) - E(\varrho_h^0, \boldsymbol{u}_h^0 | r_h^0, \boldsymbol{U}_h^0) + \Delta t \sum_{k=1}^{m} \mu \int_{\mathbb{T}^d} |\nabla_\epsilon (\boldsymbol{u}_h^k - \boldsymbol{U}_h^k)|^2 \, \mathrm{d}x$$

$$+ \Delta t \sum_{k=1}^{m} \nu \int_{\mathbb{T}^d} |\mathrm{div}_{\mathcal{T}} (\boldsymbol{u}_h^k - \boldsymbol{U}_h^k)|^2 \, \mathrm{d}x \leq \sum_{i=1}^{6} Q_i + R_h^m + G^m, \tag{14.25}$$

where the terms $Q_i, i = 1, \ldots, 6$, *read*

$$Q_1 = -\Delta t \sum_{k=1}^{m} \int_{\mathbb{T}^d} \left(\mu \nabla_\epsilon \boldsymbol{U}_h^k : \nabla_\epsilon (\boldsymbol{u}_h^k - \boldsymbol{U}_h^k) + \nu \mathrm{div}_{\mathcal{T}} \boldsymbol{U}_h^k \, \mathrm{div}_{\mathcal{T}} (\boldsymbol{u}_h^k - \boldsymbol{U}_h^k)\right) \, \mathrm{d}x,$$

$$Q_2 = \Delta t \sum_{k=1}^{m} \int_{\mathbb{T}^d} \varrho_h^{k-1} D_t \overline{\boldsymbol{U}_h}^k \left(\overline{\boldsymbol{U}_h}^k - \overline{\boldsymbol{u}_h}^k\right) \, \mathrm{d}x,$$

$$Q_3 = \Delta t \sum_{k=1}^{m} \sum_{K \in \mathcal{T}_h} \sum_{\sigma \in \mathcal{E}(K)} \int_{\sigma} \varrho_h^{k,\mathrm{up}} \left(\overline{\boldsymbol{U}_h}^{k,\mathrm{up}} - \overline{\boldsymbol{u}_h}^{k,\mathrm{up}}\right) \cdot (\boldsymbol{U}_h^k - \overline{\boldsymbol{U}_h}^k) \overline{\boldsymbol{U}_h}^{k,\mathrm{up}} \cdot \boldsymbol{n} \mathrm{dS}_x,$$

$$Q_4 = -\Delta t \sum_{k=1}^{m} \int_{\mathbb{T}^d} p(\varrho_h^k) (\mathrm{div}_x \boldsymbol{U})^k \, \mathrm{d}x,$$

$$Q_5 = \Delta t \sum_{k=1}^{m} \int_{\mathbb{T}^d} (r_h^k - \varrho_h^k) \frac{p'(r_h^k)}{r_h^k} (\partial_t r)^k \, \mathrm{d}x,$$

$$Q_6 = -\Delta t \sum_{k=1}^{m} \int_{\mathbb{T}^d} \varrho_h^k \frac{p'(r_h^k)}{r_h^k} \overline{\boldsymbol{u}_h}^k \cdot (\nabla_x r)^k \, \mathrm{d}x,$$

and

$$\left|G^m\right| \le c\Delta t \sum_{k=1}^{m} E(\varrho_h^k, \boldsymbol{u}_h^k | r_h^k, \boldsymbol{U}_h^k),$$

$$\left|R_h^m\right| \le c(\sqrt{\Delta t} + h^A),\ A = \begin{cases} \frac{2\gamma-d}{\gamma} & \text{if } \gamma \in (\frac{d}{2}, 2), \\ \frac{1}{2} & \text{if } \gamma \ge 2. \end{cases}$$

Proof We start the proof from the inequality (14.24) derived in the previous Lemma 14.4. We only need to deal with the terms T_i, $i = 1, \ldots, 8$, as the other terms will remain the same.

- We keep the term T_1 unchanged and set $Q_1 = -T_1$.
- The second term T_2 can be rewritten as

$$- T_2 = \Delta t \sum_{k=1}^{m} \int_{\mathbb{T}^d} \varrho_h^{k-1} D_t \overline{\boldsymbol{U}}_h^k \left(\frac{\overline{\boldsymbol{U}_h}^{k-1} + \overline{\boldsymbol{U}_h}^k}{2} - \overline{\boldsymbol{u}_h}^{k-1} \pm \frac{1}{2}\overline{\boldsymbol{U}_h}^k \pm \overline{\boldsymbol{u}_h}^k \right) \mathrm{d}x$$

$$= Q_2 + R_1,$$

where

$$R_1 = \Delta t \sum_{k=1}^{m} \int_{\mathbb{T}^d} \frac{1}{2} \varrho_h^{k-1} D_t \overline{\boldsymbol{U}}_h^k (\overline{\boldsymbol{U}_h}^{k-1} - \overline{\boldsymbol{U}_h}^k) + \varrho_h^{k-1} D_t \overline{\boldsymbol{U}}_h^k (\overline{\boldsymbol{u}_h}^k - \overline{\boldsymbol{u}_h}^{k-1})\ \mathrm{d}x.$$

By the interpolation estimate (14.6c) and the uniform bounds (14.15) we have

$$|R_1| \stackrel{<}{\sim} \Delta t\, \|\varrho_h\|_{L^\infty L^1} \|\partial_t \boldsymbol{U}\|^2_{L^\infty W^{1,\infty}}$$
$$+ \|\varrho_h\|^{1/2}_{L^1 L^1} \|\partial_t \boldsymbol{U}\|_{L^\infty W^{1,\infty}} \left(\int_0^T \int_{\mathbb{T}^d} \varrho_h(t - \Delta t) |D_t \overline{\boldsymbol{u}_h}|^2 (\Delta t)^2\ \mathrm{d}x\,\mathrm{d}t \right)^{1/2}$$
$$\stackrel{<}{\sim} c(E_0, \|\boldsymbol{U}\|_{C^2})(\Delta t + \Delta t^{1/2}).$$

- From the third term T_3 we get

$$- T_3 = -\Delta t \sum_{k=1}^{m} \sum_{K \in \mathcal{T}_h} \sum_{\sigma \in \mathcal{E}(K)} \int_\sigma \varrho_h^{k,\mathrm{up}} \left(\left\{\!\!\left\{ \overline{\boldsymbol{U}_h}^k \right\}\!\!\right\} - \overline{\boldsymbol{u}_h}^{k,\mathrm{up}} \right) \cdot \overline{\boldsymbol{U}_h}^k (\boldsymbol{u}_h^k \cdot \boldsymbol{n}) \mathrm{dS}_x$$

$$= -\Delta t \sum_{k=1}^{m} \sum_{K \in \mathcal{T}_h} \sum_{\sigma \in \mathcal{E}(K)} \int_\sigma \varrho_h^{k,\mathrm{up}} \left(\overline{\boldsymbol{U}_h}^{k,\mathrm{up}} - \overline{\boldsymbol{u}_h}^{k,\mathrm{up}} \right) \cdot \overline{\boldsymbol{U}_h}^k (\boldsymbol{u}_h^k \cdot \boldsymbol{n}) \mathrm{dS}_x + R_{21},$$

where

$$R_{21} = \Delta t \sum_{k=1}^{m} \sum_{K \in \mathcal{T}_h} \sum_{\sigma \in \mathcal{E}(K)} \int_{\sigma} \varrho_h^{k,\mathrm{up}} \left(\overline{\boldsymbol{U}_h}^{k,\mathrm{up}} - \left\{\left\{ \overline{\boldsymbol{U}_h}^{k} \right\}\right\} \right) \cdot \overline{\boldsymbol{U}_h}^{k} (\boldsymbol{u}_h^k \cdot \boldsymbol{n}) \mathrm{dS}_x$$

$$= \frac{\Delta t}{2} \sum_{k=1}^{m} \sum_{\sigma = K|L \in \mathcal{E}} \int_{\sigma} \Big[\varrho_K^k |\overline{\boldsymbol{U}}_K^k - \overline{\boldsymbol{U}}_L^k|^2 [\boldsymbol{u}_h^k \cdot \boldsymbol{n}]^+$$

$$+ \varrho_L^k |\overline{\boldsymbol{U}}_K^k - \overline{\boldsymbol{U}}_L^k|^2 [\boldsymbol{u}_h^k \cdot \boldsymbol{n}_{\sigma,L}]^+ \Big] \mathrm{dS}_x.$$

Seeing the equality

$$\sum_{K \in \mathcal{T}_h} \sum_{\sigma \in \mathcal{E}(K)} \int_{\sigma} f_h^{k,\mathrm{up}} \cdot \boldsymbol{U}_h^k (\boldsymbol{u}_h^k \cdot \boldsymbol{n}) \mathrm{dS}_x = 0 \text{ for } f_h^{k,\mathrm{up}} = \varrho_h^{k,\mathrm{up}} \left(\overline{\boldsymbol{U}_h}^{k,\mathrm{up}} - \overline{\boldsymbol{u}_h}^{k,\mathrm{up}} \right)$$

we have

$$- T_3 = \Delta t \sum_{k=1}^{m} \sum_{K \in \mathcal{T}_h} \sum_{\sigma \in \mathcal{E}(K)} \int_{\sigma} \varrho_h^{k,\mathrm{up}} \left(\overline{\boldsymbol{U}_h}^{k,\mathrm{up}} - \overline{\boldsymbol{u}_h}^{k,\mathrm{up}} \right) \cdot (\boldsymbol{U}_h^k - \overline{\boldsymbol{U}_h}^{k}) (\boldsymbol{u}_h^k \cdot \boldsymbol{n}) \mathrm{dS}_x$$

$$+ R_{21}$$

$$= \Delta t \sum_{k=1}^{m} \sum_{K \in \mathcal{T}_h} \sum_{\sigma \in \mathcal{E}(K)} \int_{\sigma} \varrho_h^{k,\mathrm{up}} \left(\overline{\boldsymbol{U}_h}^{k,\mathrm{up}} - \overline{\boldsymbol{u}_h}^{k,\mathrm{up}} \right) \cdot (\boldsymbol{U}_h^k - \overline{\boldsymbol{U}_h}^{k}) \overline{\boldsymbol{U}_h}^{k,\mathrm{up}} \cdot \boldsymbol{n} \mathrm{dS}_x$$

$$+ R_{21} + R_{22} = Q_3 + R_{21} + R_{22},$$

where

$$R_{22} = \Delta t \sum_{k=1}^{m} \sum_{K \in \mathcal{T}_h} \sum_{\sigma \in \mathcal{E}(K)} \int_{\sigma} \varrho_h^{k,\mathrm{up}} \left(\overline{\boldsymbol{U}_h}^{k,\mathrm{up}} - \overline{\boldsymbol{u}_h}^{k,\mathrm{up}} \right)$$

$$\cdot (\overline{\boldsymbol{U}_h}^{k} - \boldsymbol{U}_h^k)(\overline{\boldsymbol{U}_h}^{k,\mathrm{up}} - \boldsymbol{u}_h^k) \cdot \boldsymbol{n} \, \mathrm{dS}_x$$

$$= \Delta t \sum_{k=1}^{m} \sum_{K \in \mathcal{T}_h} \sum_{\sigma \in \mathcal{E}(K)} \int_{\sigma} \varrho_h^{k,\mathrm{up}} \left(\overline{\boldsymbol{U}_h}^{k,\mathrm{up}} - \overline{\boldsymbol{u}_h}^{k,\mathrm{up}} \right)$$

$$\cdot (\overline{\boldsymbol{U}_h}^{k} - \boldsymbol{U}_h^k)(\overline{\boldsymbol{U}_h}^{k,\mathrm{up}} - \overline{\boldsymbol{u}_h}^{k,\mathrm{up}}) \cdot \boldsymbol{n} \, \mathrm{dS}_x$$

$$+ \Delta t \sum_{k=1}^{m} \sum_{K \in \mathcal{T}_h} \sum_{\sigma \in \mathcal{E}(K)} \int_{\sigma} \varrho_h^{k,\mathrm{up}} \left(\overline{\boldsymbol{U}_h}^{k,\mathrm{up}} - \overline{\boldsymbol{u}_h}^{k,\mathrm{up}} \right)$$

$$\cdot (\overline{\boldsymbol{U}_h}^{k} - \boldsymbol{U}_h^k)(\overline{\boldsymbol{u}_h}^{k,\mathrm{up}} - \boldsymbol{u}_h^k) \cdot \boldsymbol{n} \, \mathrm{dS}_x$$

$$\equiv R_{221} + R_{222}.$$

By Hölder's inequality, the uniform bounds (14.15), and the trace inequality (28) we have

$$|R_{21}| \lesssim h\, \|\varrho_h\|_{L^2L^{6/5}} \|\boldsymbol{u}_h\|_{L^2L^6} (\|\nabla_x \boldsymbol{U}\|_{L^\infty L^\infty})^2 \lesssim c(E_0, \|\boldsymbol{U}\|_{C^1})h,$$

where $\beta = 1 + \beta_R > \frac{2}{3}$ for $\gamma > \frac{d}{2}$ and β_R is given in (14.19b).
Further, by similar argument it holds

$$|R_{221}| \lesssim c(\|\boldsymbol{U}\|_{C^1})\Delta t \sum_{k=1}^{m} E(\varrho_h^k, \boldsymbol{u}_h^k | r_h^k, \boldsymbol{U}_h^k),$$

and

$$\begin{aligned}
|R_{222}| &\lesssim \|\boldsymbol{U}\|_{C^1} \left\| \sqrt{\varrho_h} \right\|_{L^\infty L^{2\gamma}} \Delta t \sum_{k=1}^{m} E(\varrho_h^k, \boldsymbol{u}_h^k | r_h^k, \boldsymbol{U}_h^k)^{1/2} \left\| \overline{\boldsymbol{u}_h}^{k,\mathrm{up}} - \boldsymbol{u}_h^k \right\|_{L^q} \\
&\lesssim \|\boldsymbol{U}\|_{C^1} \|\varrho_h\|_{L^\infty L^\gamma}^{1/2} \left(h^{2d\left(\frac{1}{q}-\frac{1}{2}\right)} h^2 \|\nabla_\epsilon \boldsymbol{u}_h\|_{L^2L^2}^2 + \Delta t \sum_{k=1}^{m} E(\varrho_h^k, \boldsymbol{u}_h^k | r_h^k, \boldsymbol{U}_h^k) \right) \\
&\lesssim c(E_0, \|\boldsymbol{U}\|_{C^1}) h^{\frac{2\gamma-d}{\gamma}} + c(E_0, \|\boldsymbol{U}\|_{C^1})\Delta t \sum_{k=1}^{m} E(\varrho_h^k, \boldsymbol{u}_h^k | r_h^k, \boldsymbol{U}_h^k),
\end{aligned}$$

where we have also used Young's inequality. Here $q = \frac{2\gamma}{\gamma-1} \in (2, 6)$ provided $\gamma > \frac{d}{2}$.

- The fourth term T_4 directly yields, due to (18),

$$-T_4 = -\Delta t \sum_{k=1}^{m} \int_{\mathbb{T}^d} p(\varrho_h^k) \mathrm{div}_{\mathcal{T}} \boldsymbol{U}_h^k \, \mathrm{d}x = -\Delta t \sum_{k=1}^{m} \int_{\mathbb{T}^d} p(\varrho_h^k) [\mathrm{div}_x \boldsymbol{U}]^k \, \mathrm{d}x = Q_4.$$

- We proceed with the fifth term T_5 and obtain

$$\begin{aligned}
-T_5 &= \Delta t \sum_{k=1}^{m} \int_{\mathbb{T}^d} \frac{r_h^k - \varrho_h^k}{\Delta t} \left(P'(r_h^k) - P'(r_h^{k-1}) \right) \mathrm{d}x \\
&= \Delta t \sum_{k=1}^{m} \int_{\mathbb{T}^d} \frac{r_h^k - \varrho_h^k}{\Delta t} \left(P''(r_h^k)(r_h^k - r_h^{k-1}) - \frac{P'''(r_h^{n,\star})}{2} (r_h^k - r_h^{k-1})^2 \right) \mathrm{d}x \\
&= \Delta t \sum_{k=1}^{m} \int_{\mathbb{T}^d} (r_h^k - \varrho_h^k) \frac{p'(r_h^k)}{r_h^k} [\partial_t r]^k \, \mathrm{d}x + R_3 \\
&= Q_5 + R_3,
\end{aligned}$$

where we have used the fact $P''(r) = \frac{p'(r)}{r}$. Here, the residual term R_3 reads

$$R_3 = -\Delta t \sum_{k=1}^{m} \int_{\mathbb{T}^d} \frac{r_h^k - \varrho_h^k}{\Delta t} \frac{P'''(r_h^{n,\star})}{2} (r_h^k - r_h^{k-1})^2 \, dx,$$

$$+ \Delta t \sum_{k=1}^{m} \int_{\mathbb{T}^d} (r_h^k - \varrho_h^k) \frac{p'(r_h^k)}{r_h^k} (D_t r_h^k - [\partial_t r]^k) \, dx.$$

It is easy to get

$$\begin{aligned} |R_3| &\lesssim \Delta t \, \|r_h - \varrho_h\|_{L^1 L^1} \, |p|_{C^2([\underline{r}, \overline{r}])} \, \|\partial_t r\|_{L^\infty L^\infty} \\ &\quad + \Delta t \, \|r_h - \varrho_h\|_{L^\infty L^1} \, |p|_{C^1([\underline{r}, \overline{r}])} \left(\left\| \partial_t^2 r \right\|_{L^\infty L^\infty} + \|\partial_t \nabla_x r\|_{L^\infty L^\infty} \right) \\ &\lesssim \Delta t. \end{aligned}$$

- Term T_6 yields Q_6 after a suitable manipulation and estimating three residual terms. Indeed,

$$\begin{aligned} - T_6 &= \Delta t \sum_{k=1}^{m} \sum_{K \in \mathcal{T}_h} \sum_{\sigma \in \mathcal{E}(K)} \int_{\sigma} \varrho_h^{k,\mathrm{up}} P'(r_h^{k-1}) (\boldsymbol{u}_h^k \cdot \boldsymbol{n}) \mathrm{dS}_x \\ &= \Delta t \sum_{k=1}^{m} \sum_{K \in \mathcal{T}_h} \sum_{\sigma \in \mathcal{E}(K)} \int_{\sigma} \varrho_h^k \big(P'(r_h^{k-1}) - P'(\Pi_W r^{k-1}) \big) (\boldsymbol{u}_h^k \cdot \boldsymbol{n}) \mathrm{dS}_x \\ &\quad + \underbrace{\Delta t \sum_{k=1}^{m} \sum_{K \in \mathcal{T}_h} \sum_{\sigma \in \mathcal{E}(K)} \int_{\sigma} (\varrho_h^{k,\mathrm{up}} - \varrho_h^k) \big(P'(r_h^{k-1}) - P'(\Pi_W r^{k-1}) \big) (\boldsymbol{u}_h^k \cdot \boldsymbol{n}) \mathrm{dS}_x}_{R_{41}} \\ &= \Delta t \sum_{k=1}^{m} \sum_{K \in \mathcal{T}_h} \sum_{\sigma \in \mathcal{E}(K)} \int_{\sigma} \varrho_h^k P''(r_h^{k-1}) (r_h^{k-1} - \Pi_W r^{k-1}) (\boldsymbol{u}_h^k \cdot \boldsymbol{n}) \mathrm{dS}_x + R_{41} \\ &\quad \underbrace{- \frac{\Delta t}{2} \sum_{k=1}^{m} \sum_{K \in \mathcal{T}_h} \sum_{\sigma \in \mathcal{E}(K)} \int_{\sigma} \varrho_h^k P'''(r_{h,\dagger}^k) (r_h^{k-1} - \Pi_W r^{k-1})^2 (\boldsymbol{u}_h^k \cdot \boldsymbol{n}) \mathrm{dS}_x}_{R_{42}} \\ &= -\Delta t \sum_{k=1}^{m} \int_{\mathbb{T}^d} \varrho_h^k P''(r_h^{k-1}) \overline{\boldsymbol{u}_h}^k \cdot (\nabla_x r)^{k-1} \, \mathrm{d}x + R_{41} + R_{42} \\ &\quad + \underbrace{\Delta t \sum_{k=1}^{m} \sum_{K \in \mathcal{T}_h} \sum_{\sigma \in \mathcal{E}(K)} \int_{\sigma} \varrho_h^k P''(r_h^{k-1}) (r_h^{k-1} - \Pi_W r^{k-1}) (\overline{\boldsymbol{u}_h}^k - \boldsymbol{u}_h^k) \cdot \boldsymbol{n}_{\sigma,K} \mathrm{dS}_x}_{R_{43}} \\ &= Q_6 + R_{41} + R_{42} + R_{43}, \end{aligned}$$

where $\Pi_W = \Pi_W^{(i)}$ for any $\sigma \in \mathcal{E}_i$. Here we have used the following equality in the last second line

$$\sum_{\sigma \in \mathcal{E}(K)} \int_{\sigma} (r_K - \Pi_W r)\overline{\boldsymbol{u}_h} \cdot \boldsymbol{n} \mathrm{dS}_x = -\int_K \overline{\boldsymbol{u}_h} \cdot \nabla_x r \,\mathrm{d}x.$$

Now we estimate the residual terms $R_{4i}, i = 1, 2, 3$. It holds

$$\begin{aligned}
&|R_{41}| \\
&= \left| \Delta t \sum_{k=1}^{m} \sum_{K \in \mathcal{T}_h} \sum_{\sigma \in \mathcal{E}(K)} \int_{\sigma} (\varrho_h^{k,\mathrm{up}} - \varrho_h^k)\big(P'(r_h^{k-1}) - P'(\Pi_W r^{k-1})\big)\boldsymbol{u}_h^k \cdot \boldsymbol{n} \mathrm{dS}_x \right| \\
&\lesssim \Delta t \sum_{k=1}^{m} \sum_{K \in \mathcal{T}_h} \sum_{\sigma \in \mathcal{E}(K)} \int_{\sigma} |(\varrho_h^{k,\mathrm{up}} - \varrho_h^k)\boldsymbol{u}_h^k \cdot \boldsymbol{n}|\; P''(r_{h,\dagger}^{k-1})|r_h^{k-1} - \Pi_W r^{k-1}| \mathrm{dS}_x \\
&\lesssim h|p|_{C^1([\underline{r},\overline{r}])} \|\nabla_x r\|_{L^\infty L^\infty} \int_0^T \sum_{K \in \mathcal{T}_h} \sum_{\sigma \in \mathcal{E}(K)} \int_{\sigma} \left| [\![\varrho_h]\!] (\boldsymbol{u}_h \cdot \boldsymbol{n})^- \right| \mathrm{dS}_x \,\mathrm{d}t \\
&\lesssim c\left(E_0, |p|_{C^1([\underline{r},\overline{r}])}, \|\nabla_x r\|_{L^\infty L^\infty}\right) h^{1/2},
\end{aligned} \tag{14.26}$$

where we have used (14.19e). Further, we derive

$$\begin{aligned}
|R_{42}| &= \frac{\Delta t}{2} \sum_{k=1}^{m} \sum_{K \in \mathcal{T}_h} \sum_{\sigma \in \mathcal{E}(K)} \int_{\sigma} \varrho_h^k P'''(r_{h,\dagger}^k)(r_h^{k-1} - \Pi_W r^{k-1})^2 (\boldsymbol{u}_h^k \cdot \boldsymbol{n}) \mathrm{dS}_x \\
&\lesssim h\,|p|_{C^2([\underline{r},\overline{r}])} \|\nabla_x r\|_{L^\infty L^\infty}^2 \|\varrho_h\|_{L^2 L^{6/5}} \|\boldsymbol{u}_h\|_{L^2 L^6} \\
&\leq c\left(E_0, |p|_{C^2([\underline{r},\overline{r}])}, \|\nabla_x r\|_{L^\infty L^\infty}^2\right) h^{\beta},
\end{aligned}$$

where $\beta = 1 + \beta_R > \frac{2}{3}$ for $\gamma > \frac{d}{2}$ and β_R is given in (14.19b).
Finally, using Hölder's inequality, the trace inequality (28), the velocity bounds (14.15b), and the negative estimate (14.19b) we get

$$\begin{aligned}
&|R_{43}| \\
&= \left| \Delta t \sum_{k=1}^{m} \sum_{K \in \mathcal{T}_h} \sum_{\sigma \in \mathcal{E}(K)} \int_{\sigma} \varrho_h^k P''(r_h^{k-1})(r_h^{k-1} - \Pi_W r^{k-1})(\overline{\boldsymbol{u}_h}^k - \boldsymbol{u}_h^k) \cdot \boldsymbol{n} \mathrm{dS}_x \right| \\
&\lesssim h|p|_{C^1([\underline{r},\overline{r}])} \|\nabla_x r\|_{L^\infty L^\infty} \|\varrho_h\|_{L^2 L^2} \|\nabla_\epsilon \boldsymbol{u}_h\|_{L^2 L^2} \\
&\lesssim c\left(E_0, |p|_{C^1([\underline{r},\overline{r}])} \|\nabla_x r\|_{L^\infty L^\infty}\right) h^{\beta},
\end{aligned}$$

where

$$\beta = \begin{cases} 1 + \frac{\gamma-2}{2\gamma} d > \frac{2\gamma-d}{\gamma} & \text{if } \gamma \in (\frac{d}{2}, 2), \\ 1 & \text{if } \gamma \geq 2. \end{cases} \tag{14.27}$$

- For the seventh term T_7 we get by integration by parts formula (44) and Hölder's inequality that

$$\begin{aligned} |T_7| &= \left| \Delta t h^{\varepsilon+1} \sum_{k=1}^{m} \int_{\mathbb{T}^d} \Delta_h \varrho_h^k P'(r_h^{k-1}) \, \mathrm{d}x \right| = \left| \Delta t h^{\varepsilon+1} \sum_{k=1}^{m} \int_{\mathbb{T}^d} \varrho_h^k \Delta_h P'(r_h^{k-1}) \, \mathrm{d}x \right| \\ &\lesssim h^{\varepsilon+1} |p|_{C^2([\underline{r},\overline{r}])} \, \|\nabla_x r\|_{L^\infty L^\infty} \, \|\varrho_h\|_{L^1 L^1} \\ &\lesssim h^{\varepsilon+1} |p|_{C^2([\underline{r},\overline{r}])} \, \|\nabla_x r\|_{L^\infty L^\infty} \, \|\varrho_h\|_{L^\infty L^\gamma} \\ &\leq c(E_0, |p|_{C^2([\underline{r},\overline{r}])}, \|\nabla_x r\|_{L^\infty L^\infty}) \, h^{\varepsilon+1}. \end{aligned}$$

- The last term T_8 admits the bound

$$\begin{aligned} |T_8| &= \Delta t h^{\varepsilon+1} \left| \sum_{k=1}^{m} \sum_{i=1}^{d} \sum_{j=1}^{d} \int_{\mathbb{T}^d} \eth_{\mathcal{D}_j} \varrho_h^k \eth_{\mathcal{D}_j} \overline{U_{i,h}}^k \left\{\!\!\left\{ \overline{U_{i,h}^k - u_{i,h}^k} \right\}\!\!\right\}^{(j)} \, \mathrm{d}x \right| \\ &= \Delta t h^{\varepsilon+1} \left| \sum_{k=1}^{m} \sum_{i=1}^{d} \sum_{j=1}^{d} \int_{\mathbb{T}^d} \varrho_h^k \eth_T^{(j)} \left(\eth_{\mathcal{D}_j} \overline{U_{i,h}}^k \left\{\!\!\left\{ \overline{U_{i,h}}^k - \overline{u_{i,h}}^k \right\}\!\!\right\}^{(j)} \right) \, \mathrm{d}x \right| \\ &\lesssim \Delta t h^{\varepsilon+1} \left| \sum_{k=1}^{m} \int_{\mathbb{T}^d} \varrho_h^k \left(\Delta_h \overline{\boldsymbol{U}_h}^k \cdot \overline{\left\{\!\!\left\{ \overline{\boldsymbol{U}_h}^k - \overline{\boldsymbol{u}_h}^k \right\}\!\!\right\}} \right) \, \mathrm{d}x \right| \\ &\quad + \Delta t h^{\varepsilon+1} \left| \sum_{k=1}^{m} \sum_{i=1}^{d} \sum_{j=1}^{d} \int_{\mathbb{T}^d} \varrho_h^k \overline{\eth_{\mathcal{D}_j} \overline{U_{i,h}}^k} \cdot \eth_T^{(j)} \left\{\!\!\left\{ \overline{U_{i,h}}^k - \overline{u_{i,h}}^k \right\}\!\!\right\}^{(j)} \, \mathrm{d}x \right| \\ &\lesssim h^{\varepsilon+1} \, \|\varrho_h\|_{L^2 L^{6/5}} \, \|\boldsymbol{U}\|_{C^2} \, (\|\boldsymbol{U}_h\|_{C^0} + \|\boldsymbol{u}_h\|_{L^2 L^6}) \\ &\quad + h^{\varepsilon+1} \, \|\varrho_h\|_{L^1 L^1} \, \|\boldsymbol{U}\|_{C^1}^2 + h^{\varepsilon+1} \, \|\boldsymbol{U}\|_{C^1} \, \|\varrho_h\|_{L^2 L^2} \, \|\nabla_\epsilon \boldsymbol{u}_h\|_{L^2 L^2} \\ &\lesssim h^{\varepsilon+1} + h^\zeta, \end{aligned}$$

where we have used the inequality

$$\left\| \eth_T^{(j)} \left\{\!\!\left\{ \overline{u_{i,h}}^k \right\}\!\!\right\}^{(j)} \right\|_{L^2} \lesssim \left\| \eth_j u_{i,h} \right\|_{L^2},$$

and $\zeta = \varepsilon + \beta > A$ with β being given in (14.27). Here, we have used the same trick as in the estimate of the term R_{43}.

3. Consistency error

We proceed by assuming now that the "test functions" in the relative energy inequality solve the Navier–Stokes system.

Lemma 14.6 (Consistency error) *Let* $(\varrho_h, \boldsymbol{u}_h) = \{\varrho_h^k, \boldsymbol{u}_h^k\}_{k=1}^{N_T}$ *be a family of solutions obtained by the MAC scheme (14.10). Let* $[r, \boldsymbol{U}]$ *be a solution of the Navier–Stokes system (14.1) that belongs to the class* (14.21). *Then for any* $m = 1, \ldots, N_T$ *and* $(r_h, \boldsymbol{U}_h) = (\Pi_Q r, \Pi_W \boldsymbol{U})$ *the following identity holds true:*

$$\sum_{i=1}^{5} J_i + \mathcal{R}_h = 0, \tag{14.28}$$

where $|\mathcal{R}_h| \lesssim h + \Delta t$ *and*

$$J_1 = \Delta t \sum_{k=1}^{m} \int_{\mathbb{T}^d} r_h^{k-1} D_t \overline{\boldsymbol{U}_h}^k \left(\overline{\boldsymbol{u}_h}^k - \overline{\boldsymbol{U}_h}^k\right) \, \mathrm{d}x,$$

$$J_2 = \Delta t \sum_{k=1}^{m} \sum_{K \in \mathcal{T}_h} \sum_{\sigma \in \mathcal{E}(K)} \int_{\sigma} r_h^{k,\mathrm{up}} \left(\overline{\boldsymbol{u}_h}^{k,\mathrm{up}} - \overline{\boldsymbol{U}_h}^{k,\mathrm{up}}\right) \cdot (\boldsymbol{U}_h^k - \overline{\boldsymbol{U}_h}^k) \overline{\boldsymbol{U}_h}^{k,\mathrm{up}} \cdot \boldsymbol{n} \mathrm{dS}_x,$$

$$J_3 = \Delta t \sum_{k=1}^{m} \int_{\mathbb{T}^d} p(r_h^k) \mathrm{div}_x \boldsymbol{U}^k + p'(r_h^k) \overline{\boldsymbol{u}_h}^k \cdot \nabla_x r^k \, \mathrm{d}x,$$

$$J_4 = \mu \Delta t \sum_{k=1}^{m} \int_{\mathbb{T}^d} \nabla_\epsilon \boldsymbol{U}_h^k : \nabla_\epsilon (\boldsymbol{u}_h^k - \boldsymbol{U}_h^k) \, \mathrm{d}x,$$

$$J_5 = \nu \Delta t \sum_{k=1}^{m} \int_{\mathbb{T}^d} \mathrm{div}_{\mathcal{T}} \boldsymbol{U}_h^k \, \mathrm{div}_{\mathcal{T}} (\boldsymbol{u}_h^k - \boldsymbol{U}_h^k) \, \mathrm{d}x. \tag{14.29}$$

Proof (of Lemma 14.6) Noticing $[r, \boldsymbol{U}]$ is a solution of (14.1), we subtract (14.1a) from (14.1b) to derive

$$r \partial_t \boldsymbol{U} + r \boldsymbol{U} \cdot \nabla_x \boldsymbol{U} + \nabla_x p(r) = \mu \Delta_x \boldsymbol{U} + \nu \nabla_x \mathrm{div}_x \boldsymbol{U}.$$

Further, taking the above equation at time $t = t^k$, multiplying with $\Delta t (\boldsymbol{u}_h^k - \boldsymbol{U}_h^k)$, integrating over $\mathbb{T}^d$, and summing up from $k = 1$ to m $(\leq N_T)$ we derive

$$\sum_{i=1}^{5} G_i = 0 \ \text{ with } G_1 = \Delta t \sum_{k=1}^{m} \int_{\mathbb{T}^d} r^k \partial_t \boldsymbol{U}^k \cdot (\boldsymbol{u}_h^k - \boldsymbol{U}_h^k) \, \mathrm{d}x,$$

$$G_2 = \Delta t \sum_{k=1}^{m} \int_{\mathbb{T}^d} r^k \boldsymbol{U}^k \cdot \nabla_x \boldsymbol{U}^k \cdot (\boldsymbol{u}_h^k - \boldsymbol{U}_h^k) \, \mathrm{d}x,$$

$$G_3 = \Delta t \sum_{k=1}^{m} \int_{\mathbb{T}^d} \nabla_x p(r^k) \cdot (\boldsymbol{u}_h^k - \boldsymbol{U}_h^k) \, \mathrm{d}x,$$

$$G_4 = -\mu \Delta t \sum_{k=1}^{m} \int_{\mathbb{T}^d} \Delta_x \boldsymbol{U}^k \cdot (\boldsymbol{u}_h^k - \boldsymbol{U}_h^k) \, \mathrm{d}x,$$

$$G_5 = -\nu \Delta t \sum_{k=1}^{m} \int_{\mathbb{T}^d} \nabla_x \mathrm{div}_x \boldsymbol{U}^k \cdot (\boldsymbol{u}_h^k - \boldsymbol{U}_h^k) \, \mathrm{d}x.$$

In what follows we analyze the above G_i terms.

- Term G_1 yields J_1:

$$G_1 = \Delta t \sum_{k=1}^{m} \int_{\mathbb{T}^d} r^k \partial_t \boldsymbol{U}^k \cdot (\boldsymbol{u}_h^k - \boldsymbol{U}_h^k) \, \mathrm{d}x = J_1 + R_1,$$

where

$$\begin{aligned}
R_1 = G_1 - J_1 &= \Delta t \sum_{k=1}^{m} \int_{\mathbb{T}^d} r^k \partial_t \boldsymbol{U}^k \cdot (\boldsymbol{u}_h^k - \boldsymbol{U}_h^k) - r_h^{k-1} D_t \overline{\boldsymbol{U}_h}^k \left(\overline{\boldsymbol{u}_h}^k - \overline{\boldsymbol{U}_h}^k \right) \mathrm{d}x \\
&= \Delta t \sum_{k=1}^{m} \int_{\mathbb{T}^d} r^k \partial_t \boldsymbol{U}^k \cdot \left((\boldsymbol{u}_h^k - \boldsymbol{U}_h^k) - (\overline{\boldsymbol{u}_h}^k - \overline{\boldsymbol{U}_h}^k) \right) \mathrm{d}x \\
&\quad + \Delta t \sum_{k=1}^{m} \int_{\mathbb{T}^d} \left(r^k - r_h^{k-1} \right) \partial_t \boldsymbol{U}^k \cdot (\overline{\boldsymbol{u}_h}^k - \overline{\boldsymbol{U}_h}^k) \, \mathrm{d}x \\
&\quad + \Delta t \sum_{k=1}^{m} \int_{\mathbb{T}^d} r_h^{k-1} \left(\partial_t \boldsymbol{U}^k - D_t \overline{\boldsymbol{U}_h}^k \right) \cdot (\overline{\boldsymbol{u}_h}^k - \overline{\boldsymbol{U}_h}^k) \, \mathrm{d}x \\
&\equiv R_{11} + R_{12} + R_{13}.
\end{aligned}$$

Due to the regularity (14.21) and the velocity bound (14.15b), the residual terms R_{11}, R_{12} and R_{13} can be estimated in the following steps.

$$\begin{aligned}
R_{11} &= \Delta t \sum_{k=1}^{m} \int_{\mathbb{T}^d} r^k \partial_t \boldsymbol{U}^k \cdot \left((\boldsymbol{u}_h^k - \boldsymbol{U}_h^k) - (\overline{\boldsymbol{u}_h}^k - \overline{\boldsymbol{U}_h}^k) \right) \mathrm{d}x \\
&= \Delta t \sum_{i=1}^{d} \sum_{k=1}^{m} \sum_{K \in \mathcal{T}_h} \int_K r^k \partial_t U_i^k \left((u_{i,h}^k - U_{i,h}^k) - (\overline{u_{i,h}}^k - \overline{U_{i,h}}^k) \right) \mathrm{d}x \\
&\lesssim h \, \|r\|_{L^\infty L^\infty} \, \|\partial_t \boldsymbol{U}\|_{L^\infty L^\infty} \, \|\nabla_\epsilon (\boldsymbol{u}_h - \boldsymbol{U}_h)\|_{L^2 L^2} \\
&\lesssim h \, \|r\|_{L^\infty L^\infty} \, \|\partial_t \boldsymbol{U}\|_{L^\infty L^\infty} \, (\|\nabla_\epsilon \boldsymbol{u}_h\|_{L^2 L^2} + \|\nabla_x \boldsymbol{U}\|_{L^\infty L^\infty}) \\
&\lesssim h,
\end{aligned}$$

$$
\begin{aligned}
R_{12} &= \Delta t \sum_{k=1}^{m} \int_{\mathbb{T}^d} \left(r^k - r_h^k + r_h^k - r_h^{k-1}\right) \partial_t \boldsymbol{U}^k \cdot (\overline{\boldsymbol{u}_h}^k - \overline{\boldsymbol{U}_h}^k) \,\mathrm{d}x \\
&\lesssim (h \left\|\nabla_x r\right\|_{L^\infty L^\infty} + \Delta t \left\|\partial_t r\right\|_{L^\infty L^\infty}) \left\|\partial_t \boldsymbol{U}^k\right\|_{L^\infty L^\infty} \left\|\overline{\boldsymbol{u}_h}^k - \overline{\boldsymbol{U}_h}^k\right\|_{L^2 L^6} \\
&\lesssim \Delta t + h,
\end{aligned}
$$

$$
\begin{aligned}
R_{13} &= \Delta t \sum_{k=1}^{m} \int_{\mathbb{T}^d} r_h^{k-1} \left(\partial_t \boldsymbol{U}^k - D_t \boldsymbol{U}_h^k + D_t \boldsymbol{U}_h^k - D_t \overline{\boldsymbol{U}_h}^k\right) \cdot (\overline{\boldsymbol{u}_h}^k - \overline{\boldsymbol{U}_h}^k) \,\mathrm{d}x \\
&\lesssim \|r\|_{L^\infty L^\infty} (\Delta t \left\|\partial_{tt} \boldsymbol{U}\right\|_{L^\infty L^\infty} + h \left\|\partial_t \nabla_x \boldsymbol{U}\right\|_{L^\infty L^\infty}) \left\|\boldsymbol{u}_h - \boldsymbol{U}_h\right\|_{L^2 L^6} \\
&\lesssim \Delta t + h.
\end{aligned}
$$

Collecting the above three estimates, we have

$$
G_1 = J_1 + R_1, \quad R_1 = R_{11} + R_{12} + R_{13} \lesssim \Delta t + h.
$$

- Term G_2 yields J_2:
  We first reformulate the term G_2 as

$$
\begin{aligned}
G_2 &= \Delta t \sum_{k=1}^{m} \int_{\mathbb{T}^d} r^k \boldsymbol{U}^k \cdot \nabla_x \boldsymbol{U}^k \cdot (\boldsymbol{u}_h^k - \boldsymbol{U}_h^k) \,\mathrm{d}x \\
&= \Delta t \sum_{k=1}^{m} \int_{\mathbb{T}^d} r_h^k \overline{\boldsymbol{U}_h}^k \cdot \nabla_x \boldsymbol{U}^k \cdot (\overline{\boldsymbol{u}_h}^k - \overline{\boldsymbol{U}_h}^k) \,\mathrm{d}x + R_{21} \equiv G_{21} + R_{21},
\end{aligned}
$$

where $G_{21} = \Delta t \sum_{k=1}^{m} \int_{\mathbb{T}^d} r_h^k \overline{\boldsymbol{U}_h}^k \cdot \nabla_x \boldsymbol{U}^k \cdot (\overline{\boldsymbol{u}_h}^k - \overline{\boldsymbol{U}_h}^k) \,\mathrm{d}x$ and

$$
\begin{aligned}
R_{21} &= \Delta t \sum_{k=1}^{m} \int_{\mathbb{T}^d} r^k \boldsymbol{U}^k \cdot \nabla_x \boldsymbol{U}^k \cdot \left((\boldsymbol{u}_h^k - \boldsymbol{U}_h^k) - (\overline{\boldsymbol{u}_h}^k - \overline{\boldsymbol{U}_h}^k)\right) \,\mathrm{d}x \\
&\quad + \Delta t \sum_{k=1}^{m} \int_{\mathbb{T}^d} r^k (\boldsymbol{U}^k - \overline{\boldsymbol{U}_h}^k) \cdot \nabla_x \boldsymbol{U}^k \cdot (\overline{\boldsymbol{u}_h}^k - \overline{\boldsymbol{U}_h}^k) \,\mathrm{d}x \\
&\quad + \Delta t \sum_{k=1}^{m} \int_{\mathbb{T}^d} (r^k - r_h^k) \overline{\boldsymbol{U}_h}^k \cdot \nabla_x \boldsymbol{U}^k \cdot (\overline{\boldsymbol{u}_h}^k - \overline{\boldsymbol{U}_h}^k) \,\mathrm{d}x \\
&\lesssim c\,(\overline{r}, \left\|\nabla_x r, \boldsymbol{U}, \nabla_x \boldsymbol{U}\right\|_{L^\infty L^\infty}, \left\|\nabla_\epsilon \boldsymbol{u}_h\right\|_{L^2 L^2}, \left\|\boldsymbol{u}_h\right\|_{L^2 L^6})\, h.
\end{aligned}
$$

Further, by Stokes' formula and using the fact that r_h^k and $\overline{\boldsymbol{U}_h}^k$ are piecewise constants, we get

$$
\begin{aligned}
G_{21} &= \Delta t \sum_{k=1}^{m} \int_{\mathbb{T}^d} r_h^k \overline{\boldsymbol{U}_h}^k \cdot \nabla_x \boldsymbol{U}^k \cdot (\overline{\boldsymbol{u}_h}^k - \overline{\boldsymbol{U}_h}^k)\, \mathrm{d}x \\
&= \Delta t \sum_{k=1}^{m} \sum_{K \in \mathcal{T}_h} \sum_{\sigma \in \mathcal{E}(K)} |\sigma| r_K^k \overline{\boldsymbol{U}}_K^k \cdot \boldsymbol{n} \boldsymbol{U}_h^k \cdot (\overline{\boldsymbol{u}_h}^k - \overline{\boldsymbol{U}_h}^k) \\
&= \Delta t \sum_{k=1}^{m} \sum_{K \in \mathcal{T}_h} \sum_{\sigma \in \mathcal{E}(K)} |\sigma| r_K^k \overline{\boldsymbol{U}}_K^k \cdot \boldsymbol{n} (\boldsymbol{U}_h^k - \overline{\boldsymbol{U}_h}^k) \cdot (\overline{\boldsymbol{u}_h}^k - \overline{\boldsymbol{U}_h}^k) \\
&= \Delta t \sum_{k=1}^{m} \sum_{K \in \mathcal{T}_h} \sum_{\sigma \in \mathcal{E}(K)} |\sigma| (r_h^k \overline{\boldsymbol{U}}_K^k)^{\mathrm{up}} \cdot \boldsymbol{n} (\boldsymbol{U}_h^k - \overline{\boldsymbol{U}_h}^k) \cdot (\overline{\boldsymbol{u}_h}^k - \overline{\boldsymbol{U}_h}^k)^{\mathrm{up}} + R_{22} \\
&= J_2 + R_{22},
\end{aligned}
$$

where we have used the equality $\sum_{\sigma \in (K)} \int_\sigma n \,\mathrm{dS}_x = 0$ and

$$
\begin{aligned}
R_{22} &= \Delta t \sum_{k=1}^{m} \sum_{K \in \mathcal{T}_h} \sum_{\sigma \in \mathcal{E}(K)} |\sigma| \big(r_K^k - (r_h^k)^{\mathrm{up}}\big) \overline{\boldsymbol{U}}_K^k \cdot \boldsymbol{n} (\boldsymbol{U}_h^k - \overline{\boldsymbol{U}_h}^k) \cdot (\overline{\boldsymbol{u}_h}^k - \overline{\boldsymbol{U}_h}^k) \\
&+ \Delta t \sum_{k=1}^{m} \sum_{K \in \mathcal{T}_h} \sum_{\sigma \in \mathcal{E}(K)} |\sigma| (r_h^k)^{\mathrm{up}} \big(\overline{\boldsymbol{U}}_K^k - (\overline{\boldsymbol{U}}_K^k)^{\mathrm{up}}\big) \cdot \boldsymbol{n} (\boldsymbol{U}_h^k - \overline{\boldsymbol{U}_h}^k) \cdot (\overline{\boldsymbol{u}_h}^k - \overline{\boldsymbol{U}_h}^k) \\
&+ \Delta t \sum_{k=1}^{m} \sum_{K \in \mathcal{T}_h} \sum_{\sigma \in \mathcal{E}(K)} |\sigma| (r_h^k \overline{\boldsymbol{U}}_K^k)^{\mathrm{up}} \cdot \boldsymbol{n} (\boldsymbol{U}_h^k - \overline{\boldsymbol{U}_h}^k) \\
&\qquad \cdot \Big((\overline{\boldsymbol{u}_h}^k - \overline{\boldsymbol{U}_h}^k) - (\overline{\boldsymbol{u}_h}^k - \overline{\boldsymbol{U}_h}^k)^{\mathrm{up}} \Big) \\
&\lesssim h \, \|\nabla_x r\|_{L^\infty L^\infty} \|\boldsymbol{U}\|_{L^\infty L^\infty} \|\nabla_x \boldsymbol{U}\|_{L^\infty L^\infty} \left(\|\boldsymbol{u}_h\|_{L^2 L^6} + \|\boldsymbol{U}\|_{L^\infty L^\infty} \right) \\
&+ h \, \|r\|_{L^\infty L^\infty} \|\nabla_x \boldsymbol{U}\|_{L^\infty L^\infty}^2 \left(\|\boldsymbol{u}_h\|_{L^2 L^6} + \|\boldsymbol{U}\|_{L^\infty L^\infty} \right) \\
&+ h \, \|r\|_{L^\infty L^\infty} \|\boldsymbol{U}\|_{L^\infty L^\infty} \|\nabla_x \boldsymbol{U}\|_{L^\infty L^\infty} \left(\|\nabla_\epsilon \boldsymbol{u}_h\|_{L^2 L^2} + \|\nabla_x \boldsymbol{U}\|_{L^\infty L^\infty} \right) \\
&\lesssim h.
\end{aligned}
$$

Collecting the above two estimates we have

$$
G_2 = J_2 + R_2, \quad R_2 = R_{21} + R_{22} \lesssim h.
$$

- Term G_3 yields J_3:

$$
\begin{aligned}
G_3 &= \Delta t \sum_{k=1}^{m} \int_{\mathbb{T}^d} \nabla_x p(r^k) \cdot (\boldsymbol{u}_h^k - \boldsymbol{U}_h^k) \,\mathrm{d}x \\
&= \Delta t \sum_{k=1}^{m} \int_{\mathbb{T}^d} \big(p'(r^k)\nabla_x r^k \cdot \boldsymbol{u}_h^k - \nabla_x p(r^k) \cdot \boldsymbol{U}^k \\
&\qquad\qquad\qquad + \nabla_x p(r^k) \cdot (\boldsymbol{U}^k - \boldsymbol{U}_h^k)\big) \,\mathrm{d}x \\
&= \Delta t \sum_{k=1}^{m} \int_{\mathbb{T}^d} \Big(p'(r_h^k)\overline{\boldsymbol{u}_h}^k \cdot \nabla_x r^k + p(r_h^k)\mathrm{div}_x \boldsymbol{U}^k\Big) \,\mathrm{d}x + R_3 \\
&= J_3 + R_3,
\end{aligned}
$$

where R_3 reads

$$
\begin{aligned}
R_3 &= \Delta t \sum_{k=1}^{m} \int_{\mathbb{T}^d} \big(p'(r^k)\boldsymbol{u}_h^k - p'(r_h^k)\overline{\boldsymbol{u}_h}^k\big) \cdot \nabla_x r^k \,\mathrm{d}x \\
&\quad + \Delta t \sum_{k=1}^{m} \int_{\mathbb{T}^d} \big(p(r^k) - p(r_h^k)\big) \,\mathrm{div}_x \boldsymbol{U}^k \,\mathrm{d}x \\
&\quad + \Delta t \sum_{k=1}^{m} \int_{\mathbb{T}^d} \nabla_x p(r^k) \cdot (\boldsymbol{U}^k - \boldsymbol{U}_h^k) \,\mathrm{d}x \\
&\equiv R_{31} + R_{32} + R_{33}.
\end{aligned}
$$

First, the term R_{31} can be estimated by

$$
\begin{aligned}
R_{31} &= \Delta t \sum_{k=1}^{m} \int_{\mathbb{T}^d} \big(p'(r^k)\boldsymbol{u}_h^k - p'(r_h^k)\overline{\boldsymbol{u}_h}^k\big) \cdot \nabla_x r^k \,\mathrm{d}x \\
&= \Delta t \sum_{k=1}^{m} \int_{\mathbb{T}^d} \big((p'(r^k) - p'(r_h^k))\boldsymbol{u}_h^k - p'(r_h^k)(\boldsymbol{u}_h^k - \overline{\boldsymbol{u}_h}^k)\big) \cdot \nabla_x r^k \,\mathrm{d}x \\
&\lesssim h \left(|p|_{C^2(\underline{r},\overline{r})} \|\nabla_x r\|_{L^\infty L^\infty} \|\boldsymbol{u}_h\|_{L^2 L^6} + |p|_{C^1(\underline{r},\overline{r})} \|\nabla_\epsilon \boldsymbol{u}_h\|_{L^2 L^2}\right) \|\nabla_x r\|_{L^\infty L^\infty} \\
&\lesssim h.
\end{aligned}
$$

Next, it is easy to estimate the term R_{32} by

$$R_{32} = \Delta t \sum_{k=1}^{m} \int_{\mathbb{T}^d} \left(p(r^k) - p(r_h^k)\right) \mathrm{div}_x \boldsymbol{U}^k \, \mathrm{d}x$$
$$\lesssim h|p|_{C^1(\underline{r},\overline{r})} \|\nabla_x r\|_{L^\infty L^\infty} \|\boldsymbol{U}_h\|_{C^1} \lesssim h.$$

Finally, we have

$$R_{33} = \Delta t \sum_{k=1}^{m} \int_{\mathbb{T}^d} \nabla_x p(r^k) \cdot (\boldsymbol{U}^k - \boldsymbol{U}_h^k) \, \mathrm{d}x$$
$$= \Delta t \sum_{k=1}^{m} \int_{\mathbb{T}^d} p'(r^k) \nabla_x r^k \cdot (\boldsymbol{U}^k - \boldsymbol{U}_h^k) \, \mathrm{d}x$$
$$\lesssim h|p|_{C^1(\underline{r},\overline{r})} \|\nabla_x r\|_{L^\infty L^\infty} \|\nabla_x \boldsymbol{U}_h\|_{L^\infty L^\infty} \lesssim h.$$

Collecting the above three estimates, we have

$$G_3 = J_3 + R_3, \quad R_3 = R_{31} + R_{32} + R_{33} \lesssim h.$$

- Term G_4 yields J_4:

$$G_4 = -\mu \Delta t \sum_{k=1}^{m} \int_{\mathbb{T}^d} \Delta_x \boldsymbol{U}^k \cdot (\boldsymbol{u}_h^k - \boldsymbol{U}_h^k) \, \mathrm{d}x$$
$$= -\mu \Delta t \sum_{k=1}^{m} \sum_{i=1}^{d} \sum_{\sigma \in \mathcal{E}_i} (u_{i,h}^k - U_{i,h}^k) \int_{D_\sigma} \sum_{j=1}^{d} \frac{\partial}{\partial x_j} \frac{\partial U_i^k}{\partial x_j} \, \mathrm{d}x$$
$$= -\mu \Delta t \sum_{k=1}^{m} \sum_{i=1}^{d} \sum_{j=1}^{d} \sum_{\sigma \in \mathcal{E}_i} (u_{i,\sigma}^k - U_{i,\sigma}^k) \sum_{\epsilon \in \widetilde{\mathcal{E}}(D_\sigma)} \int_{\epsilon} \frac{\partial U_i^k}{\partial x_j} \boldsymbol{n}_{\epsilon, D_\sigma} \mathrm{dS}_x$$
$$= \mu \Delta t \sum_{k=1}^{m} \sum_{i=1}^{d} \sum_{j=1}^{d} \sum_{\epsilon \in \widetilde{\mathcal{E}}_i} |D_\epsilon| \eth_j (u_{i,h}^k - U_{i,h}^k) \frac{1}{|\epsilon|} \int_{\epsilon} \frac{\partial U_i^k}{\partial x_j} \mathrm{dS}_x$$
$$= \mu \Delta t \sum_{k=1}^{m} \int_{\mathbb{T}^d} \nabla_\epsilon (\boldsymbol{u}_h^k - \boldsymbol{U}_h^k) : \nabla_\epsilon \boldsymbol{U}_h^k \, \mathrm{d}x + R_4 = J_4 + R_4,$$

where $\boldsymbol{n}_{\epsilon, D_\sigma}$ is the unit normal vector on ϵ pointing outwards D_σ and

$$R_4 = \mu \Delta t \sum_{k=1}^{m} \sum_{i=1}^{d} \sum_{j=1}^{d} \sum_{\epsilon \in \widetilde{\mathcal{E}}_i} |D_\epsilon| \eth_j (u_{i,h}^k - U_{i,h}^k) \left(\frac{1}{|\epsilon|} \int_\epsilon \frac{\partial U_i^k}{\partial x_j} \mathrm{dS}_x - \eth_j U_{i,h}^k \right)$$
$$\lesssim h \left(\|\nabla_\epsilon \boldsymbol{u}_h\|_{L^2 L^2} + \|\boldsymbol{U}\|_{C^1} \right) \|\boldsymbol{U}\|_{C^2} \lesssim h.$$

- Term G_5 yields J_5:

$$\begin{aligned} G_5 &= -\nu \Delta t \sum_{k=1}^{m} \int_{\mathbb{T}^d} \nabla_x \mathrm{div}_x \boldsymbol{U}^k \cdot (\boldsymbol{u}_h^k - \boldsymbol{U}_h^k) \,\mathrm{d}x \\ &= -\nu \Delta t \sum_{k=1}^{m} \sum_{i=1}^{d} \sum_{\sigma \in \mathcal{E}_i} (u_{i,h}^k - U_{i,h}^k) \mathbf{e}_i \cdot \sum_{\substack{\epsilon \in \widetilde{\mathcal{E}}(D_\sigma) \\ \epsilon \perp \mathbf{e}_i}} \int_\epsilon \mathrm{div}_x \boldsymbol{U}^k \boldsymbol{n}_{\epsilon, D_\sigma} \mathrm{dS}_x \\ &= \nu \Delta t \sum_{k=1}^{m} \sum_{i=1}^{d} \sum_{K \in \mathcal{T}_h} |K| \eth_{\mathcal{T}}^{(i)} (u_{i,h}^k - U_{i,h}^k) \frac{1}{|\epsilon|} \int_\epsilon \mathrm{div}_x \boldsymbol{U}^k \mathrm{dS}_x \\ &= \nu \Delta t \sum_{k=1}^{m} \int_{\mathbb{T}^d} \mathrm{div}_{\mathcal{T}} (\boldsymbol{u}_h^k - \boldsymbol{U}_h^k) \mathrm{div}_{\mathcal{T}} \boldsymbol{U}_h^k \,\mathrm{d}x + R_5 = J_5 + R_5, \end{aligned}$$

where $\epsilon = \{x \in K | x_i = (x_K)_i\}$ is the plane (or the line if $d = 2$) crossing the point x_K (center of element K) that separates K into two halves and

$$R_5 = \nu \Delta t \sum_{k=1}^{m} \sum_{i=1}^{d} \sum_{K \in \mathcal{T}_h} |K| \eth_{\mathcal{T}}^{(i)} (u_{i,h}^k - U_{i,h}^k) \left(\frac{1}{|\epsilon|} \int_\epsilon \mathrm{div}_x \boldsymbol{U}^k \mathrm{dS}_x - \mathrm{div}_{\mathcal{T}} \boldsymbol{U}_h^k \right)$$
$$\lesssim h \left(\|\mathrm{div}_{\mathcal{T}} \boldsymbol{u}_h\|_{L^2 L^2} + \|\boldsymbol{U}\|_{C^1} \right) \|\boldsymbol{U}\|_{C^2} \lesssim h.$$

Consequently, collecting the above implies

$$\sum_{i=1}^{5} J_i + \mathcal{R}_h = 0 \text{ with } |\mathcal{R}_h| = \left| \sum_{i=1}^{5} R_i \right| \lesssim h + \Delta t,$$

which completes the proof.

Now we are ready to prove Theorem 14.4.

Proof (of Theorem 14.4) In order to deduce the desired convergence rate we apply the discrete Gronwall lemma to the sum of the two estimates derived in the second and the third step above.

More precisely, we sum up (14.25) and (14.28) to get

$$E(\varrho_h^m, \boldsymbol{u}_h^m | r_h^m, \boldsymbol{U}_h^m) - E(\varrho_h^0, \boldsymbol{u}_h^0 | r_h^0, \boldsymbol{U}_h^0) + \Delta t \sum_{k=1}^{m} \mu \int_{\mathbb{T}^d} |\nabla_\epsilon (\boldsymbol{u}_h^k - \boldsymbol{U}_h^k)|^2 \, \mathrm{d}x$$
$$\leq \mathcal{P}_1 + \mathcal{P}_2 + \mathcal{P}_3 + R_h^m + G^m,$$

where $\left|G^m\right| \leq c\Delta t \sum_{k=1}^{m} E(\varrho_h^k, \boldsymbol{u}_h^k | r_h^k, \boldsymbol{U}_h^k)$, $\left|R_h^m\right| \leq c(\sqrt{\Delta t} + h^A)$ are given in Lemma 14.5, and $\mathcal{P}_i = \Delta t \sum_{k=1}^{m} \mathcal{P}_i^k$, $i = 1, 2, 3$, read

$$\mathcal{P}_1^k = \int_{\mathbb{T}^d} (\varrho_h^{k-1} - r_h^{k-1}) D_t \overline{\boldsymbol{U}_h}^k \left(\overline{\boldsymbol{U}_h}^k - \overline{\boldsymbol{u}_h}^k \right) \, \mathrm{d}x,$$

$$\mathcal{P}_2^k = \sum_{K \in \mathcal{T}_h} \sum_{\sigma \in \mathcal{E}(K)} \int_{\sigma} (\varrho_h^{k,\mathrm{up}} - r_h^{k,\mathrm{up}}) \left(\overline{\boldsymbol{U}_h}^{k,\mathrm{up}} - \overline{\boldsymbol{u}_h}^{k,\mathrm{up}} \right) \cdot (\boldsymbol{U}_h^k - \overline{\boldsymbol{U}_h}^k) \overline{\boldsymbol{U}_h}^{k,\mathrm{up}} \cdot \boldsymbol{n} \mathrm{d}S_x,$$

$$\mathcal{P}_3^k = \int_{\mathbb{T}^d} \frac{p'(r_h^k)}{r_h^k} (r_h^k - \varrho_h^k) \left([\partial_t r]^k + \overline{\boldsymbol{u}_h}^k \cdot \nabla_x r^k \right) \, \mathrm{d}x$$
$$+ \int_{\mathbb{T}^d} \left(p(r_h^k) - p(\varrho_h^k) \right) \mathrm{div}_x \boldsymbol{U}^k \, \mathrm{d}x.$$

Recalling that $[r, \boldsymbol{U}]$ satisfies (14.1a) we have

$$\partial_t r^k = -\mathrm{div}_x (r^k \boldsymbol{U}^k) = -\boldsymbol{U}^k \cdot \nabla_x r^k - r^k \mathrm{div}_x \boldsymbol{U}^k.$$

Then we may rewrite $\mathcal{P}_3^k$ as

$$\mathcal{P}_3^k = \int_{\mathbb{T}^d} \frac{p'(r_h^k)}{r_h^k} (r_h^k - \varrho_h^k) \left(-\boldsymbol{U}^k \cdot \nabla_x r^k - r^k \mathrm{div}_x \boldsymbol{U}^k + \overline{\boldsymbol{u}_h}^k \cdot \nabla_x r^k \right) \, \mathrm{d}x$$
$$+ \int_{\mathbb{T}^d} \left(p(r_h^k) - p(\varrho_h^k) \right) \mathrm{div}_x \boldsymbol{U}^k \, \mathrm{d}x$$
$$= \int_{\mathbb{T}^d} \left(\frac{p'(r_h^k)}{r_h^k} (r_h^k - \varrho_h^k) (\overline{\boldsymbol{u}_h}^k - \boldsymbol{U}^k) \cdot \nabla_x r^k + \mathbb{E}(r_h^k | \varrho_h^k) \mathrm{div}_x \boldsymbol{U}^k \right) \, \mathrm{d}x.$$

Next, we estimate the terms $\mathcal{P}_1$, $\mathcal{P}_2$, and $\mathcal{P}_3$.

• **Term** $\mathcal{P}_1$:
Thanks to Hölder's inequality, the estimate (14.16), Lemma 14.3, Young's inequality, and the bound (14.17) we can estimate $\mathcal{P}_1$ in the following way

$$
\begin{aligned}
\mathcal{P}_1^k &= \int_{\mathbb{T}^d} (\varrho_h^{k-1} - r_h^{k-1}) D_t \overline{\boldsymbol{U}_h}^k \left(\overline{\boldsymbol{U}_h}^k - \overline{\boldsymbol{u}_h}^k \right) \, \mathrm{d}x \\
&\le \left\| \partial_t \boldsymbol{U}^k \right\|_{L^\infty} \left\| \overline{\boldsymbol{U}_h}^k - \overline{\boldsymbol{u}_h}^k \right\|_{L^6} \left\| \varrho_h^{k-1} - r_h^{k-1} \right\|_{L^{6/5}} \\
&\le \left\| \partial_t \boldsymbol{U}^k \right\|_{L^\infty} \left\| \overline{\boldsymbol{U}_h}^k - \overline{\boldsymbol{u}_h}^k \right\|_{L^6} \times \\
&\quad \times \left(\left\| (\varrho_h^{k-1} - r_h^{k-1}) 1_{R_+ \setminus [\underline{r}/2, 2\overline{r}]}(\varrho_h^{k-1}) \right\|_{L^{6/5}} + \left\| (\varrho_h^{k-1} - r_h^{k-1}) 1_{[\underline{r}/2, 2\overline{r}]}(\varrho_h^{k-1}) \right\|_{L^2} \right) \\
&\le c(\underline{r}, \overline{r}) \left\| \partial_t \boldsymbol{U}^k \right\|_{L^\infty} \left(\left\| \nabla_\epsilon (\boldsymbol{U}_h^k - \boldsymbol{u}_h^k) \right\|_{L^2} + E(\varrho_h^k, \boldsymbol{u}_h^k | r_h^k, \boldsymbol{U}_h^k)^{1/2} \right) \times \\
&\quad \times \left(E(\varrho_h^{k-1}, \boldsymbol{u}_h^{k-1} | r_h^{k-1}, \boldsymbol{U}_h^{k-1})^{5/6} + E(\varrho_h^{k-1}, \boldsymbol{u}_h^{k-1} | r_h^{k-1}, \boldsymbol{U}_h^{k-1})^{1/2} \right) \\
&\le \frac{c}{\delta} E(\varrho_h^{k-1}, \boldsymbol{u}_h^{k-1} | r_h^{k-1}, \boldsymbol{U}_h^{k-1}) + \delta \left\| \nabla_\epsilon (\boldsymbol{U}_h^k - \boldsymbol{u}_h^k) \right\|_{L^2}^2 \\
&\quad + c E(\varrho_h^k, \boldsymbol{u}_h^k | r_h^k, \boldsymbol{U}_h^k) + \frac{c}{\delta} E(\varrho_h^{k-1}, \boldsymbol{u}_h^{k-1} | r_h^{k-1}, \boldsymbol{U}_h^{k-1})^{5/3} \\
&\le c E(\varrho_h^{k-1}, \boldsymbol{u}_h^{k-1} | r_h^{k-1}, \boldsymbol{U}_h^{k-1}) + \delta \left\| \nabla_\epsilon (\boldsymbol{U}_h^k - \boldsymbol{u}_h^k) \right\|_{L^2}^2 \\
&\quad + c E(\varrho_h^k, \boldsymbol{u}_h^k | r_h^k, \boldsymbol{U}_h^k),
\end{aligned}
$$

where c depends on $\underline{r}$, $\overline{r}$, $\|\boldsymbol{U}\|_{C^1}$, and the initial data of the problem (mass and total energy). Consequently, we have derived

$$
\mathcal{P}_1 \le c \Delta t E(\varrho_h^0, \boldsymbol{u}_h^0 | r_h^0, \boldsymbol{U}_h^0) + c \Delta t \sum_{k=1}^m E(\varrho_h^k, \boldsymbol{u}_h^k | r_h^k, \boldsymbol{U}_h^k) + \delta \Delta t \sum_{k=1}^m \left\| \nabla_\epsilon (\boldsymbol{U}_h^k - \boldsymbol{u}_h^k) \right\|_{L^2}^2 .
$$

• **Term** $\mathcal{P}_2$:

$$
\begin{aligned}
\mathcal{P}_2^k &= \sum_{K \in \mathcal{T}_h} \sum_{\sigma \in \mathcal{E}(K)} \int_\sigma (\varrho_h^{k,\mathrm{up}} - r_h^{k,\mathrm{up}}) \left(\overline{\boldsymbol{U}_h}^{k,\mathrm{up}} - \overline{\boldsymbol{u}_h}^{k,\mathrm{up}} \right) \cdot (\boldsymbol{U}_h^k - \overline{\boldsymbol{U}_h}^k) \overline{\boldsymbol{U}_h}^{k,\mathrm{up}} \cdot \boldsymbol{n} \mathrm{dS}_x \\
&\lesssim \left\| \partial_t \boldsymbol{U}^k \right\|_{L^\infty} \left\| \nabla_x \boldsymbol{U}^k \right\|_{L^\infty} \left\| \overline{\boldsymbol{U}_h}^k - \overline{\boldsymbol{u}_h}^k \right\|_{L^6} \left\| \varrho_h^{k-1} - r_h^{k-1} \right\|_{L^{6/5}} .
\end{aligned}
$$

By the same reasoning, we have

$$
\begin{aligned}
\mathcal{P}_2 \le & c \Delta t E(\varrho_h^0, \boldsymbol{u}_h^0 | r_h^0, \boldsymbol{U}_h^0) + c \Delta t \sum_{k=1}^m E(\varrho_h^k, \boldsymbol{u}_h^k | r_h^k, \boldsymbol{U}_h^k) \\
& + \delta \Delta t \sum_{k=1}^m \left\| \nabla_\epsilon (\boldsymbol{U}_h^k - \boldsymbol{u}_h^k) \right\|_{L^2}^2 ,
\end{aligned}
$$

where c depends on δ, $\underline{r}$, $\overline{r}$, $\|\boldsymbol{U}\|_{C^1}$, and the initial data.

- **Term** $\mathcal{P}_3$:

$$\mathcal{P}_3^k = \int_0^T \int_{\mathbb{T}^d} \left(\frac{p'(r_h^k)}{r_h^k} (r_h^k - \varrho_h^k)(\overline{\boldsymbol{u}_h}^k - \boldsymbol{U}^k) \cdot \nabla_x r^k + \mathbb{E}(r_h^k | \varrho_h^k) \mathrm{div}_x \boldsymbol{U}^k \right) \, \mathrm{d}x \, \mathrm{d}t$$
$$\leq \left\| \mathrm{div}_x \boldsymbol{U}^k \right\|_{L^\infty} E(\varrho_h^k, \boldsymbol{u}_h^k | r_h^k, \boldsymbol{U}_h^k)$$
$$+ \left| \frac{p'(r)}{r} \right|_{C[\underline{r},\overline{r}]} \left\| \nabla_x r^k \right\|_{L^\infty} \left\| \overline{\boldsymbol{U}_h}^k - \overline{\boldsymbol{u}_h}^k \right\|_{L^6} \left\| \varrho_h^{k-1} - r_h^{k-1} \right\|_{L^{6/5}}.$$

Analogously, we have

$$\mathcal{P}_3 \leq c\Delta t \sum_{k=1}^m E(\varrho_h^k, \boldsymbol{u}_h^k | r_h^k, \boldsymbol{U}_h^k) + c\Delta t E(\varrho_h^0, \boldsymbol{u}_h^0 | r_h^0, \boldsymbol{U}_h^0)$$
$$+ \delta \Delta t \sum_{k=1}^m \left\| \nabla_\epsilon (\boldsymbol{U}_h^k - \boldsymbol{u}_h^k) \right\|_{L^2}^2,$$

where c depends on δ, $\underline{r}$, $\overline{r}$, $\|\boldsymbol{U}\|_{C^1}$, $\|\nabla_x r\|_{L^\infty L^\infty}$, and the initial data.

Collecting the above estimates we conclude

$$\begin{aligned} &E(\varrho_h^m, \boldsymbol{u}_h^m | r_h^m, \boldsymbol{U}_h^m) + \Delta t \sum_{k=1}^m \frac{\mu}{2} \int_{\mathbb{T}^d} |\nabla_\epsilon (\boldsymbol{u}_h^k - \boldsymbol{U}_h^k)|^2 \, \mathrm{d}x \\ &\leq c \left[h^A + \sqrt{\Delta t} + E(\varrho_h^0, \boldsymbol{u}_h^0 | r_h^0, \boldsymbol{U}_h^0) \right] + c\Delta t \sum_{k=1}^m E(\varrho_h^k, \boldsymbol{u}_h^k | r_h^k, \boldsymbol{U}_h^k), \end{aligned} \tag{14.30}$$

for all $m = 1, \ldots, N_T$. Here, the parameter A is given in (14.23) and

$$c = c \left(M_0, E_0, \underline{r}, \overline{r}, \gamma, \|p\|_{C^2([\underline{r},\overline{r}])}, \|[r, \boldsymbol{U}]\|_{C^2([0,T] \times \mathbb{T}^d)} \right)$$

is a positive constant.

Finally, applying Gronwall's inequality, see Lemma 9, to (14.30) we derive the desired estimate of the relative energy

$$E(\varrho_h^m, \boldsymbol{u}_h^m | r_h^m, \boldsymbol{U}_h^m) + \Delta t \sum_{k=1}^m \frac{\mu}{2} \int_{\mathbb{T}^d} |\nabla_\epsilon (\boldsymbol{u}_h^k - \boldsymbol{U}_h^k)|^2 \, \mathrm{d}x$$
$$\leq c \left[h^A + \sqrt{\Delta t} + E(\varrho_h^0, \boldsymbol{u}_h^0 | r_h^0, \boldsymbol{U}_h^0) \right] \exp \left(\frac{m \, c \, \Delta t}{1 - c \, \Delta t} \right)$$

for $0 < \Delta t < \frac{1}{c}$. This concludes the proof of Theorem 14.4.

14.6 Numerical Experiments

In this section we illustrate theoretical results stated in Theorem 14.4, in particular, the convergence rate derived in terms of the relative energy. Hence, we focus on the following errors

$$\begin{aligned} e_E &= \sup_{0 \le n \le N_T} E(\varrho_h^k, \boldsymbol{u}_h^k | \varrho_{\rm ref}^k, \boldsymbol{u}_{\rm ref}^k), & e_{\nabla \boldsymbol{u}} &= \| \nabla_\epsilon (\boldsymbol{u}_h - \boldsymbol{u}_{\rm ref}) \|_{L^2((0,T)\times \mathbb{T}^d)}, \\ e_\varrho &= \| \varrho_h - \varrho_{\rm ref} \|_{L^1((0,T)\times \mathbb{T}^d)}, & e_\varrho^\dagger &= \| \varrho_h - \varrho_{\rm ref} \|_{L^\infty(0,T;L^\gamma(\mathbb{T}^d))}, \\ e_{\boldsymbol{u}} &= \| \boldsymbol{u}_h - \boldsymbol{u}_{\rm ref} \|_{L^2((0,T)\times \mathbb{T}^d)}, & e_p &= \| p(\varrho_h) - p(\varrho_{\rm ref}) \|_{L^\infty(0,T;L^1(\mathbb{T}^d))} \end{aligned} \tag{14.31}$$

between the numerical solution $(\varrho_h, \boldsymbol{u}_h)$ and the reference solution $(\varrho_{\rm ref}, \boldsymbol{u}_{\rm ref})$. The latter is a numerical solution obtained by the MAC scheme (14.10) on a fine mesh with $h = 1/1024$. We present the results of two numerical experiments where numerical solutions to the Navier–Stokes system are computed on the domain $\Omega = [0, 1]^2$ with periodic boundary conditions. In our numerical simulations we use the following parameters $\mu = 0.1$, $\nu = 0$, $\varepsilon = 0.6$, and $\Delta t = \mathrm{CFL} \frac{h}{|\boldsymbol{u}_h|_{\max} + c_h}$, where $\mathrm{CFL} = 0.4$, $c_h = \sqrt{a\gamma \varrho_h^{\gamma-1}}$.

Experiment 1

In the first experiment we take the initial data

$$\varrho_0(x_1, x_2, t) = 2 + \cos(2\pi(x_1 + x_2)), \quad \boldsymbol{u}_0(x_1, x_2, t) = \boldsymbol{0}. \tag{14.32}$$

The fluid dynamics emanates from the initial pressure gradient. Table 14.1 presents the errors measured in the norms defined in (14.31) and the experimental order of convergence (EOC), cf. (10.76), for different γ. The final time was set to $T = 0.1$. We can observe the second order convergence rate for the relative entropy and the first order convergence rate for the density, velocity and the gradient of velocity. Our numerical simulations indicate that theoretical convergence rate obtained in Theorem 14.4 may be suboptimal.

Further, Fig. 14.2 depicts time evolution of the kinetic energy $\int_{\mathbb{T}^d} \frac{1}{2}\varrho |\boldsymbol{u}|^2$, internal energy $\int_{\mathbb{T}^d} P(\varrho)\,\mathrm{d}x$ and total energy. Comparing Fig. 13.1 with 11.1 we observe a higher peak in the kinetic energy. Finally, we present in Fig. 14.3 time evolution of the flow that is driven by the pressure gradient. We can see that the fluid velocity increases at the beginning and decreases afterwards.

Experiment 2

In this experiment we consider vortex flow and take an initial vortex centered at $x_c = (0.5, 0.5)$ with the radius $r_0 = 0.2$:

$$\varrho_0 = 1, \quad \boldsymbol{u}_0 = \frac{u_r(r)}{r} \begin{pmatrix} x_2 - 0.5 \\ 0.5 - x_1 \end{pmatrix},$$

Table 14.1 Experiment 1: EOC of the MAC scheme for the Navier–Stokes equations

h	e_E	EOC	$e_{\nabla u}$	EOC	$e_\varrho^\dagger$	EOC	e_u	EOC	e_p	EOC	e_ϱ	EOC
						$\gamma = 1.4$						
1/32	2.09e-03	–	8.28e-02	–	1.60e-01	–	8.01e-03	–	7.66e-02	–	3.93e-03	–
1/64	5.29e-04	1.99	4.18e-02	0.98	9.75e-02	0.71	4.11e-03	0.96	3.84e-02	1.00	1.94e-03	1.02
1/128	1.25e-04	2.08	2.05e-02	1.03	5.89e-02	0.73	2.02e-03	1.02	1.90e-02	1.02	9.40e-04	1.04
1/256	2.69e-05	2.22	9.66e-03	1.08	3.49e-02	0.75	9.57e-04	1.08	9.13e-03	1.05	4.44e-04	1.08
1/512	4.63e-06	2.54	4.07e-03	1.25	1.84e-02	0.92	4.03e-04	1.25	3.73e-03	1.29	1.81e-04	1.30
						$\gamma = 1.67$						
1/32	3.40e-03	–	1.22e-01	–	2.66e-01	–	1.14e-02	–	1.10e-01	–	3.67e-03	–
1/64	8.25e-04	2.04	6.01e-02	1.03	1.75e-01	0.60	5.66e-03	1.01	5.47e-02	1.01	1.79e-03	1.04
1/128	1.96e-04	2.07	2.94e-02	1.03	1.15e-01	0.60	2.79e-03	1.02	2.72e-02	1.01	8.68e-04	1.04
1/256	4.19e-05	2.22	1.38e-02	1.10	7.40e-02	0.64	1.31e-03	1.09	1.30e-02	1.06	4.08e-04	1.09
1/512	6.75e-06	2.63	5.71e-03	1.27	4.32e-02	0.78	5.47e-04	1.26	5.32e-03	1.29	1.66e-04	1.29
						$\gamma = 2$						
1/32	6.19e-03	–	2.01e-01	–	4.10e-01	–	1.69e-02	–	1.68e-01	–	3.15e-03	–
1/64	1.53e-03	2.02	1.00e-01	1.01	2.89e-01	0.50	8.31e-03	1.02	8.37e-02	1.01	1.55e-03	1.02
1/128	3.55e-04	2.11	4.84e-02	1.05	2.03e-01	0.51	4.01e-03	1.05	4.14e-02	1.01	7.54e-04	1.04
1/256	7.47e-05	2.25	2.23e-02	1.12	1.41e-01	0.53	1.87e-03	1.10	1.99e-02	1.06	3.55e-04	1.09
1/512	1.16e-05	2.69	8.92e-03	1.32	8.99e-02	0.65	7.68e-04	1.28	8.08e-03	1.30	1.44e-04	1.30

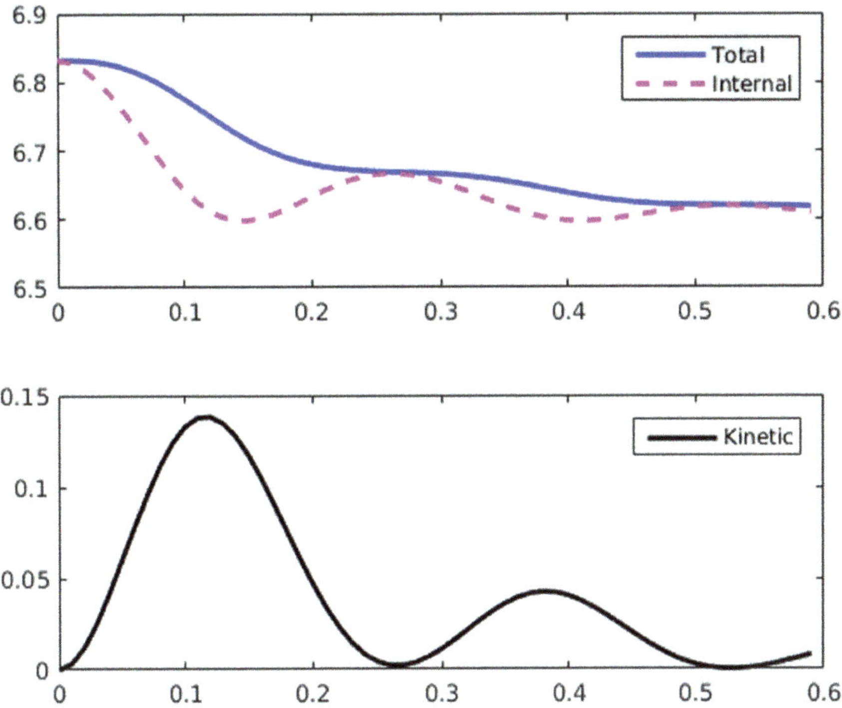

Fig. 14.2 Experiment 1: MAC scheme for the Navier–Stokes equations, time evolution of the energy

where $r = \sqrt{(x_1 - 0.5)^2 + (x_2 - 0.5)^2}$ and the radial velocity of the vortex u_r is

$$u_r(r) = \sqrt{\gamma} \begin{cases} 2r/r_0 & \text{if } 0 \le r < r_0/2, \\ 2(1 - r/r_0) & \text{if } r_0/2 \le r < r_0, \\ 0 & \text{if } R \ge r_0. \end{cases}$$

Numerical solutions are computed on different meshes up to a final time $T = 0.1$. In Table 14.2 we present the errors measured in the norms defined in (14.31), and the experimental order of convergence (EOC), cf. (10.76), for different γ. Convergence rates obtained in the previous experiment are confirmed again. Similarly as before we can see the second order convergence rate for the relative entropy and the first order convergence rate for the density, velocity and the gradient of velocity.

Further, we also present in Fig. 14.4 time evolution of the fluid flow emanating from the initial vortex. We can see that the vortex dissipates slowly during time evolution. This is also confirmed by Fig. 14.5 that presents the energy dissipation.

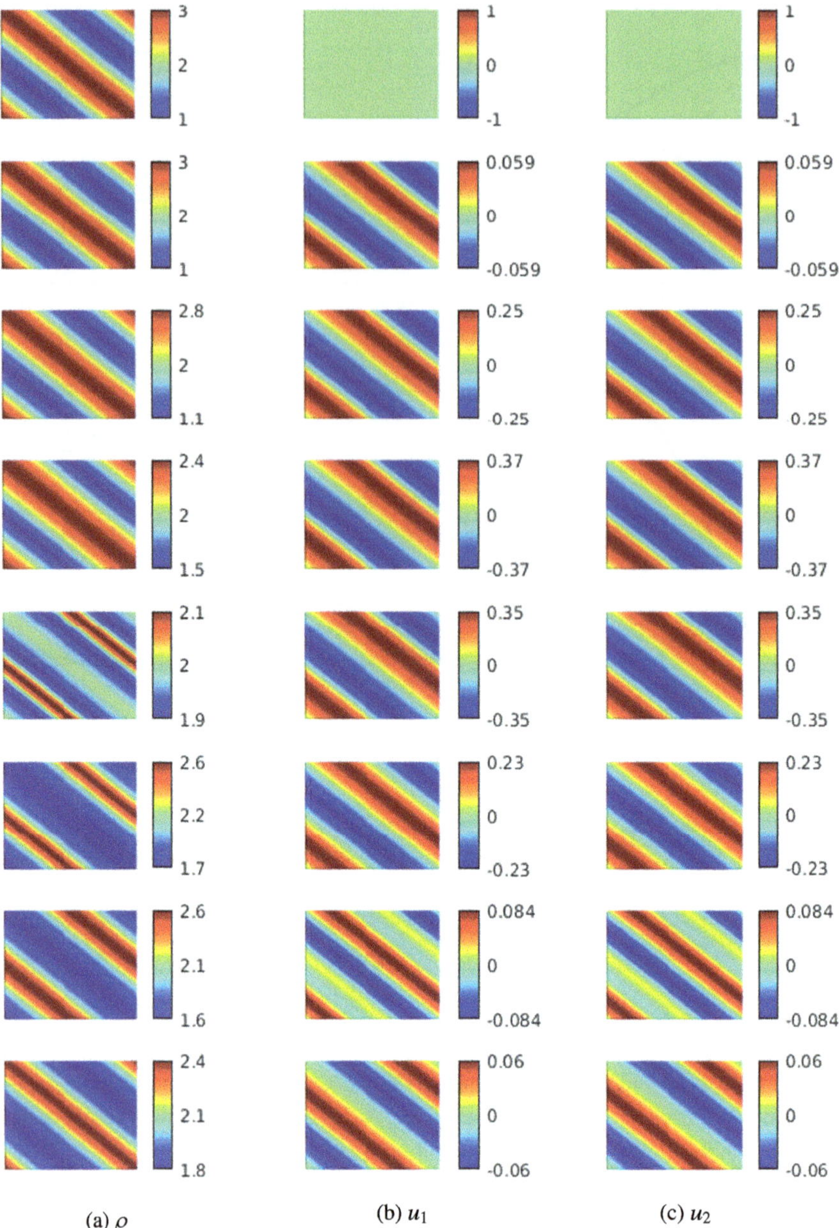

(a) ϱ (b) u_1 (c) u_2

Fig. 14.3 Experiment 1: MAC method for the Navier–Stokes equations, time evolution of the flow. From top to down are $t = 0, 0.01, 0.05, 0.1, 0.15, 0.2, 0.25, 0.5$. From left to right are ϱ, u_1, and u_2

Table 14.2 Experiment 2: EOC of the MAC scheme for the Navier–Stokes equations

h	e_E	EOC	$e_{\nabla \boldsymbol{u}}$	EOC	$e_\varrho^\dagger$	EOC	$e_{\boldsymbol{u}}$	EOC	e_p	EOC	e_ϱ	EOC
						$\gamma = 1.4$						
1/32	5.39e–04	–	1.30e–01	–	1.06e–02	–	4.57e–03	–	1.71e–03	–	9.61e–05	–
1/64	1.35e–04	2.00	6.53e–02	0.99	5.95e–03	0.83	2.30e–03	0.99	7.67e–04	1.16	4.17e–05	1.20
1/128	3.34e–05	2.02	3.17e–02	1.04	3.47e–03	0.78	1.15e–03	1.00	3.60e–04	1.09	1.92e–05	1.12
1/256	7.95e–06	2.07	1.45e–02	1.13	1.95e–03	0.83	5.58e–04	1.05	1.61e–04	1.17	8.50e–06	1.18
1/512	1.56e–06	2.35	5.65e–03	1.36	9.75e–04	1.00	2.43e–04	1.20	6.09e–05	1.40	3.25e–06	1.39
						$\gamma = 1.67$						
1/32	6.42e–04	–	1.42e–01	–	2.58e–02	–	4.99e–03	–	2.25e—03	–	1.07e–04	–
1/64	1.61e–04	2.00	7.10e–02	1.00	1.60e–02	0.69	2.49e–03	1.00	1.01e–03	1.15	4.72e–05	1.18
1/128	3.96e–05	2.02	3.44e–02	1.05	1.02e–02	0.65	1.24e–03	1.00	4.80e–04	1.08	2.20e–05	1.10
1/256	9.37e–06	2.08	1.56e–02	1.14	6.30e–03	0.70	6.02e–04	1.05	2.15e–04	1.16	9.72e–06	1.18
1/512	1.85e–06	2.34	6.12e–03	1.36	3.52e–03	0.84	2.64e–04	1.19	8.14e–05	1.40	3.72e–06	1.39
						$\gamma = 2$						
1/32	7.64e–04	–	1.55e–01	–	5.45e–02	–	5.45e–03	–	2.98e–03	–	1.20e–04	–
1/64	1.91e–04	2.00	7.73e–02	1.00	3.68e–02	0.57	2.71e–03	1.01	1.36e–03	1.13	5.37e–05	1.16
1/128	4.72e–05	2.02	3.74e–02	1.05	2.55e–02	0.53	1.35e–03	1.01	6.49e–04	1.06	2.52e–05	1.09
1/256	1.12e–05	2.08	1.70e–02	1.14	1.70e–02	0.58	6.54e–04	1.05	2.90e–04	1.16	1.12e–05	1.18
1/512	2.22e–06	2.34	6.66e–03	1.35	1.05e–02	0.70	2.87e–04	1.19	1.10e–04	1.40	4.27e–06	1.39

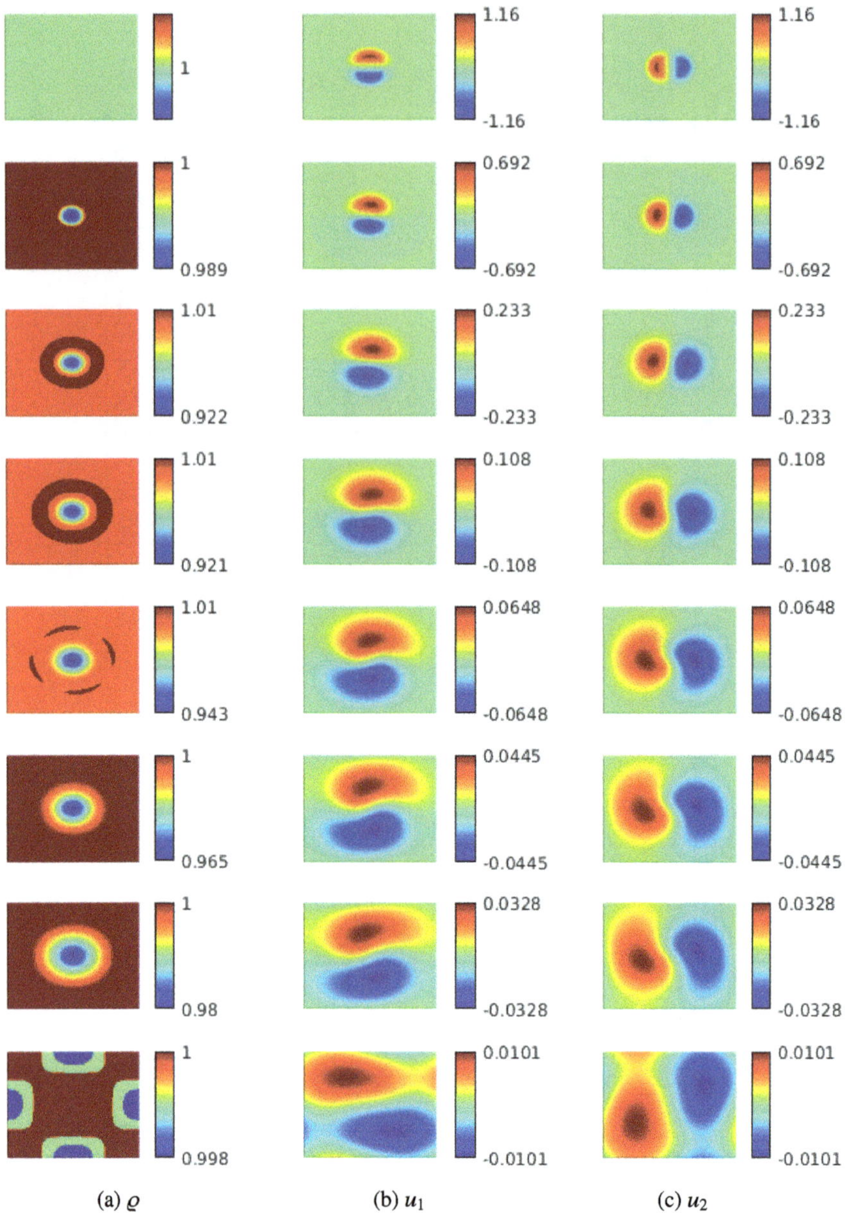

Fig. 14.4 Experiment 2: MAC scheme for the Navier–Stokes equations, time evolution of the flow. From top to down are $t = 0, 0.01, 0.05, 0.1, 0.15, 0.2, 0.25, 0.5$. From left to right are ϱ, u_1, and u_2

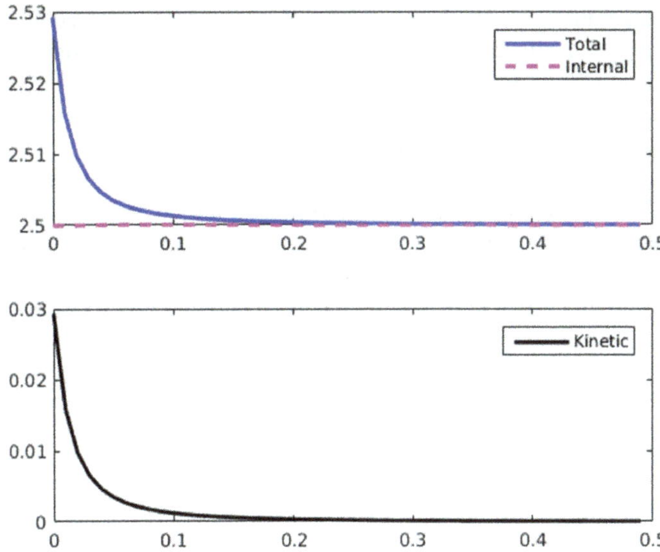

Fig. 14.5 Experiment 2: MAC scheme for the Navier–Stokes equations, time evolution of the energy

14.7 Conclusion, Bibliographical Remarks

Staggered mesh approach for the finite difference approximation is very popular in computational fluid dynamics due to its computational efficiency. A typical example is the well-known MAC (Marker–and–Cell) method frequently used in engineering, physics or geophysics in order to approximate viscous flows. The MAC method was firstly introduced for incompressible viscous flows by the Los Alamos group: B. J. Daly, F. H. Harlow, J. P. Shannon and J. E. Welch in their 1965 report [61], but an idea of this discretization has already existed in the work of V. I. Lebedev in 1964, cf. [150]. Afterwards the method has been quickly generalized for compressible viscous fluids and it is nowadays a basis of several academic or industrial codes.

Since the MAC scheme has been originally introduced for the incompressible Navier–Stokes equations it is quite instructive to look on the history of the error analysis of the MAC scheme for the incompressible Navier–Stokes equations. Although the method has been used successfully since 1965, its theoretical numerical analysis was not carried out until 1992 by Nicolaides and his collaborators [172, 173]. They reinterpreted the MAC scheme as a finite volume approximation of the velocity-vorticity equations on the dual meshes and proved the first order error estimates for pressure and velocity in the standard L^2 norms. Afterwards there has been quite a number of papers where the convergence order of the MAC scheme was analyzed from the theoretical point of view. For example, in 1996 Girault and Lopez [118] showed that the finite difference equations of the MAC method can be derived by combining a velocity-vorticity mixed finite element method of degree one with an adequate quadrature formula. Later, in 2008 Kanschat [136] showed that the MAC scheme is algebraically equivalent to the first order local discontinuous Galerkin method with a proper quadrature and also obtained first order convergence of the scheme. Further, in 2018 Gallouët, Herbin et al. [112] showed the convergence of the MAC scheme on nonuniform grids.

In 2015 Li and Sun [152] succeeded to show the superconvergence of the MAC scheme and proved the second order convergence of velocity and pressure measured in the L^2 norms even on irregular rectangular grids. The authors do not reinterpret the MAC method as finite volume or finite element scheme, but work directly in the finite difference framework. Careful and elegant analysis of various sources of errors shows that although the local truncation errors are only first order, a suitable cancellation of local errors yields after summation the second order global errors for both the velocity and pressure.

Analogous superconvergence results for compressible Navier–Stokes equations are not yet available. Indeed, there are only relatively few results for the convergence and error analysis of the MAC method for compressible Navier–Stokes equations. We refer to the convergence analysis for stationary Navier–Stokes equations by Gallouët, Herbin, Latché and Maltese [113], stability and consistency analysis by Hošek and She [129] and the error estimates results by Gallouët et al. [115], and Mizerová, She [165]. The latter are obtained using the relative energy as a tool to quantify the error between the numerical and a strong exact solution.

The convergence analysis discussed in the present chapter follows the same lines as in the previous chapters of the book. First, we have shown the stability and consistency of the MAC scheme, see Theorems 14.1 and 14.2, respectively. Consequently, applying Theorems 5.5, 7.12 and 7.13 we have proven the convergence of numerical solutions, see Theorem 14.3. Despite the fact that the methodology of convergence analysis for the Navier–Stokes equations had been already used in the previous chapters, the proof for the MAC scheme remains nontrivial. The analysis required an elaborate treatment and technical estimates linked to the staggered grid and suitable discrete differential operators applied to staggered and non-staggered discrete functions.

Further, assuming the existence of a regular solution to the Navier–Stokes system, we have derived error estimates for the MAC scheme (14.10) in terms of the relative energy, see Theorem 14.4. Theoretical convergence rate is 1/2 for the relative energy functional. On the other hand, in numerical experiments we can observe the second order convergence rate, cf. Tables 14.1 and 14.2. Apparently, theoretical convergence rate is suboptimal, but we point out that the result is unconditional, meaning there is no assumption on the regularity or boundedness of numerical solutions. Assuming that numerical solutions are uniformly bounded would allow us to obtain the second order convergence rate for the relative energy, too.

Relative energy approach has been used recently also in the context of the error analysis of a mixed finite volume – finite element method, see [84, 85, 96, 114].

References

1. Abbatiello, A., Feireisl, E.: On strong continuity of the weak solutions to the compressible Euler system. J. Nonlinear Sci. **31**(2), No. 33, 16pp, (2021)
2. Abbatiello, A., Feireisl, E., Novotný, A.: Generalized solutions to models of compressible viscous fluids. Discrete Cont. Dyn.-A **41**(1), 1–28 (2021)
3. Adams, R.: Sobolev spaces. Academic Press, New York (1975)
4. Alibert, J.J., Bouchitté, G.: Non-uniform integrability and generalized Young measures. J. Convex Anal. **4**(1), 129–147 (1997)
5. Ansanay-Alex, G., Babik, F., Latché, J.C., Vola, D.: An L^2-stable approximation of the Navier-Stokes convection operator for low-order non-conforming finite elements. Internat. J. Numer. Methods Fluids **66**(5), 555–580 (2011)
6. Antontsev, S.N., Kazhikhov, A.V., Monakhov, V.N.: Krajevyje zadaci mechaniki neodnorodnych zidkostej. Novosibirsk (1983)
7. Baire, R.: Leçons sur les fonctions discontinues. Les Grands Classiques Gauthier-Villars. [Gauthier-Villars Great Classics]. Éditions Jacques Gabay, Sceaux (1995). Reprint of the 1905 original
8. Ball, J.: A version of the fundamental theorem for Young measures. In Lect. Notes in Physics 344, Springer-Verlag pp. 207–215 (1989)
9. Ball, J., Murat, F.: Remarks on Chacons biting lemma. Proc. Amer. Math. Soc. **107**, 655–663 (1989)
10. Banach, S.: Theorie des operations lineaires. Monografie Matematyczne I, Warszawa (1932)
11. Banach, S., Saks, S.: Sur la convergence forte dans les champs L^p. Studia Math. **2**, 51–57 (1930)
12. Bardow, A., Öttinger, H.: Consequences of the Brenner modification to the Navier–Stokes equations for dynamic light scattering. Phys. A **373**, 88–96 (2007)
13. Batchelor, G.K.: An introduction to fluid dynamics. Cambridge University Press, Cambridge (1967)
14. Bechtel, S.E., Rooney, F., Forest, M.: Connection between stability, convexity of internal energy, and the second law for compressible Newtonian fuids. J. Appl. Mech. **72**, 299–300 (2005)
15. Ben-Artzi, M., Falcovitz, J.: A second-order Godunov-type scheme for compressible fluid dynamics. J. Comput. Phys. **55**(1), 1–32 (1984)
16. Ben-Artzi, M., Falcovitz, J.: Generalized Riemann problems in computational fluid dynamics, *Cambridge Monographs on Applied and Computational Mathematics*, vol. 11. Cambridge University Press, Cambridge (2003)

E. Feireisl et al., *Numerical Analysis of Compressible Fluid Flows*, MS&A 20,
https://doi.org/10.1007/978-3-030-73788-7

17. Ben-Artzi, M., Li, J.: Hyperbolic balance laws: Riemann invariants and the generalized Riemann problem. Numer. Math. **106**(3), 369–425 (2007)
18. Benzoni-Gavage, S., Serre, D.: Multidimensional hyperbolic partial differential equations, First order systems and applications. Oxford Mathematical Monographs. The Clarendon Press Oxford University Press, Oxford (2007)
19. Berkes, I.: An extension of the Komlós subsequence theorem. Acta Math. Hungar. **55**(1-2), 103–110 (1990)
20. Berthelin, F.: Convergence of flux vector splitting schemes with single entropy inequality for hyperbolic systems of conservation laws. Numer. Math. **99**(4), 585–604 (2005)
21. Berthelin, F., Bouchut, F.: Relaxation to isentropic gas dynamics for a BGK system with single kinetic entropy. Methods Appl. Anal. **9**(2), 313–327 (2002)
22. Bessemoulin-Chatard, M., Chainais-Hillairet, C., Filbet, F.: On discrete functional inequalities for some finite volume schemes. IMA J. Numer. Anal. **35**(3), 1125–1149 (2015)
23. Boffi, D., Brezzi, F., Fortin, M.: Mixed finite element methods and applications, *Springer Series in Computational Mathematics*, vol. 44. Springer, Heidelberg (2013)
24. Bouchut, F.: Entropy satisfying flux vector splittings and kinetic BGK models. Numer. Math. **94**(4), 623–672 (2003)
25. Breit, D., Feireisl, E., Hofmanová, M.: Generalized solutions to models of inviscid fluids. Discrete Cont. Dyn.-B **25**(10), 3831–3842 (2020)
26. Brenier, Y., De Lellis, C., Székelyhidi Jr., L.: Weak-strong uniqueness for measure-valued solutions. Commun. Math. Phys. **305**(2), 351–361 (2011)
27. Brenner, H.: Kinematics of volume transport. Phys. A **349**, 11–59 (2005)
28. Brenner, H.: Navier-Stokes revisited. Phys. A **349**(1-2), 60–132 (2005)
29. Brenner, H.: Fluid mechanics revisited. Phys. A **370**, 190–224 (2006)
30. Bresch, D., Desjardins, B.: Stabilité de solutions faibles globales pour les équations de Navier-Stokes compressibles avec température. C.R. Acad. Sci. Paris **343**, 219–224 (2006)
31. Bresch, D., Desjardins, B.: On the existence of global weak solutions to the Navier-Stokes equations for viscous compressible and heat conducting fluids. J. Math. Pures Appl. **87**, 57–90 (2007)
32. Bresch, D., Jabin, P.E.: Global existence of weak solutions for compressible Navier-Stokes equations: thermodynamically unstable pressure and anisotropic viscous stress tensor. Ann. of Math. **188**(2), 577–684 (2018)
33. Brezis, H.: Operateurs maximaux monotones et semi-groupes de contractions dans les espaces de Hilbert. North-Holland, Amsterdam (1973)
34. Buckmaster, T., Colombo, M., Vicol, V.: Wild solutions of the Navier–Stokes equations whose singular sets in time have Hausdorff dimension strictly less than 1. J. Eur. Math. Soc. (2021)
35. Buckmaster, T., de Lellis, C., Székelyhidi Jr., L., Vicol, V.: Onsager's conjecture for admissible weak solutions. Commun. Pure Appl. Math. **72**(2), 229–274 (2019)
36. Buckmaster, T., Shkoller, S., Vicol, V.: Formation of shocks for 2D isentropic compressible Euler. Commun. Pure Appl. Math. (2021)
37. Buckmaster, T., Vicol, V.: Convex integration and phenomenologies in turbulence. EMS Surv. Math. Sci. **6**(1), 173–263 (2019)
38. Buckmaster, T., Vicol, V.: Nonuniqueness of weak solutions to the Navier-Stokes equation. Ann. of Math. **189**(1), 101–144 (2019)
39. Buffa, A., Ortner, C.: Compact embeddings of broken Sobolev spaces and applications. IMA J. Numer. Anal. **29**(4), 827–855 (2009)
40. Březina, J., Feireisl, E.: Measure-valued solutions to the complete Euler system. J. Math. Soc. Japan **70**(4), 1227–1245 (2018)
41. Carroll, R.W.: Abstract methods in partial differential equations. Harper & Row Publishers, New York (1969). Harper's Series in Modern Mathematics
42. Chainais-Hillairet, C., Droniou, J.: Finite-volume schemes for noncoercive elliptic problems with Neumann boundary conditions. IMA J. Numer. Anal. **31**(1), 61–85 (2011)
43. Chen, G.Q., Glimm, J.: Kolmogorov-type theory of compressible turbulence and inviscid limit of the Navier-Stokes equations in R^3. Phys. D **400**, 132138, 10 (2019)

44. Chen, G.Q., Torres, M.: Divergence-measure fields, sets of finite perimeter, and conservation laws. Arch. Ration. Mech. Anal. **175**(2), 245–267 (2005)
45. Chen, G.Q., Torres, M., Ziemer, W.P.: Gauss-Green theorem for weakly differentiable vector fields, sets of finite perimeter, and balance laws. Commun. Pure Appl. Math. **62**(2), 242–304 (2009)
46. Chiodaroli, E.: A counterexample to well-posedness of entropy solutions to the compressible Euler system. J. Hyperbolic Differ. Equ. **11**(3), 493–519 (2014)
47. Chiodaroli, E., Feireisl, E., Flandoli, F.: Ill posedness for the full Euler system driven by multiplicative white noise. Indiana Univ. Math. J. **70**(4), 1267–1282 (2021)
48. Chiodaroli, E., Feireisl, E., Kreml, O., Wiedemann, E.: $\mathcal{A}$-free rigidity and applications to the compressible Euler system. Ann. Mat. Pura Appl. (4) **196**(4), 1557–1572 (2017)
49. Chiodaroli, E., Kreml, O.: On the energy dissipation rate of solutions to the compressible isentropic Euler system. Arch. Ration. Mech. Anal. **214**(3), 1019–1049 (2014)
50. Chiodaroli, E., Kreml, O., Mácha, V., Schwarzacher, S.: Non–uniqueness of admissible weak solutions to the compressible Euler equations with smooth initial data. Trans. Amer. Math. Soc. 374(4), 2269–2295 (2021)
51. Chiodaroli, E., De Lellis, C., Kreml, O.: Global ill-posedness of the isentropic system of gas dynamics. Commun. Pure Appl. Math. **68**(7), 1157–1190 (2015)
52. Cho, Y., Choe, H., Kim, H.: Unique solvability of the initial boundary value problems for compressible viscous fluids. J. Math. Pures. Appl. **83**, 243–275 (2004)
53. Chorin, A.J., Marsden, J.E.: A mathematical introduction to fluid mechanics. Springer-Verlag, New York (1979)
54. Christodoulou, D., Miao, S.: Compressible flow and Euler's equations, *Surveys of Modern Mathematics*, vol. 9. International Press, Somerville, MA; Higher Education Press, Beijing (2014)
55. Ciarlet, P.G.: The finite element method for elliptic problems. North-Holland Publishing Co., Amsterdam-New York-Oxford (1978). Studies in Mathematics and its Applications, Vol. 4
56. Ciarlet, P.G.: Mathematical elasticity: I. Three-dimensional elasticity. North-Holland, Amsterdam (1988)
57. Cobzaş, Ş., Miculescu, R., Nicolae, A.: Lipschitz functions, *Lecture Notes in Mathematics*, vol. 2241. Springer, Cham (2019)
58. Crouzeix, M., Raviart, P.: Conforming and nonconforming finite element methods for solving the stationary Stokes equations. Rev. Française Automat. Informat. Recherche Opérationnelle Sér. Rouge **7**, 33–75 (1973)
59. Dafermos, C.M.: The second law of thermodynamics and stability. Arch. Rational Mech. Anal. **70**, 167–179 (1979)
60. Dafermos, C.M.: Hyperbolic conservation laws in continuum physics. Springer-Verlag, Berlin (2000)
61. Daly, B.J., Harlow, F.R., Shannon, J.P., Welch, J.E.: The MAC method. Technical report LA-3425, Los Alamos Scientific Laboratory, University of California (1965)
62. De Lellis, C., Székelyhidi Jr., L.: On admissibility criteria for weak solutions of the Euler equations. Arch. Ration. Mech. Anal. **195**(1), 225–260 (2010)
63. Di Pietro, D.A., Ern, A.: Mathematical aspects of discontinuous Galerkin methods, *Mathématiques & Applications (Berlin) [Mathematics & Applications]*, vol. 69. Springer, Heidelberg (2012)
64. DiPerna, R.: Convergence of the viscosity method for isentropic gas dynamics. Commun. Math. Phys. **91**, 1–30 (1983)
65. DiPerna, R.: Measure-valued solutions to conservation laws. Arch. Rat. Mech. Anal. **88**, 223–270 (1985)
66. DiPerna, R., Majda, A.: Reduced Hausdorff dimension and concentration cancellation for two-dimensional incompressible flow. J. Amer. Math. Soc. **1**, 59–95 (1988)
67. DiPerna, R.J.: Uniqueness of solutions to hyperbolic conservation laws. Indiana Univ. Math. J. **28**(1), 137–188 (1979)

68. DiPerna, R.J., Lions, P.L.: Ordinary differential equations, transport theory and Sobolev spaces. Invent. Math. **98**, 511–547 (1989)
69. DiPerna, R.J., Majda, A.J.: Concentrations in regularizations for 2-d incompressible flow. Commun. Pure Appl. Math. **40**(3), 301–345 (1987)
70. DiPerna, R.J., Majda, A.J.: Oscillations and concentrations in weak solutions of the incompressible fluid equations. Commun. Math. Phys. **108**(4), 667–689 (1987)
71. Dolejší, V., Feistauer, M.: Discontinuous Galerkin method, *Springer Series in Computational Mathematics*, vol. 48. Springer, Cham (2015). Analysis and applications to compressible flow
72. Edwards, R.E.: Functional Analysis. Holt-Rinehart-Winston, New York (1965)
73. Ekeland, I., Temam, R.: Convex analysis and variational problems. North-Holland, Amsterdam (1976)
74. Eliezer, S., Ghatak, A., Hora, H.: An introduction to equations of states, theory and applications. Cambridge University Press, Cambridge (1986)
75. Erdös, P., Magidor, M.: A note on regular methods of summability and the Banach-Saks property. Proc. Amer. Math. Soc. **59**(2), 232–234 (1976)
76. Ern, A., Guermond, J.L.: Theory and practice of finite elements, *Applied Mathematical Sciences*, vol. 159. Springer-Verlag, New York (2004)
77. Evans, L., Gariepy, R.: Measure theory and fine properties of functions. CRC Press, Boca Raton (1992)
78. Eymard, R., Gallouët, T., Herbin, R.: Finite volume methods. In: Handbook of numerical analysis, Vol. VII, pp. 713–1020. North-Holland, Amsterdam (2000)
79. Fefferman, C.L.: Existence and smoothness of the Navier-Stokes equation. In: The millennium prize problems, pp. 57–67. Clay Math. Inst., Cambridge, MA (2006)
80. Feireisl, E.: Dynamics of viscous compressible fluids. Oxford University Press, Oxford (2004)
81. Feireisl, E.: Weak solutions to problems involving inviscid fluids. In: Mathematical Fluid Dynamics, Present and Future, *Springer Proceedings in Mathematics and Statistics*, vol. 183, pp. 377–399. Springer, New York (2016)
82. Feireisl, E., Gwiazda, P., Świerczewska-Gwiazda, A., Wiedemann, E.: Dissipative measure-valued solutions to the compressible Navier–Stokes system. Calc. Var. Partial Differ. Equ. **55**(6), 55–141 (2016)
83. Feireisl, E., Hofmanová, M.: On convergence of approximate solutions to the compressible Euler system. Ann. PDE **6**(2), 11 (2020)
84. Feireisl, E., Hošek, R., Maltese, D., Novotný, A.: Error estimates for a numerical method for the compressible Navier-Stokes system on sufficiently smooth domains. ESAIM: M2AN **51**(1), 279–319 (2017)
85. Feireisl, E., Hošek, R., Maltese, D., Novotný, A.: Unconditional convergence and error estimates for bounded numerical solutions of the barotropic Navier-Stokes system. Numer. Meth. Part. D. E. **33**(4), 1208–1223 (2017)
86. Feireisl, E., Karper, T., Michálek, M.: Convergence of a numerical method for the compressible Navier–Stokes system on general domains. Numer. Math. **134**(4), 667–704 (2016)
87. Feireisl, E., Karper, T., Novotný, A.: A convergent numerical method for the Navier–Stokes–Fourier system. IMA J. Numer. Anal. **36**(4), 1477–1535 (2016)
88. Feireisl, E., Karper, T., Pokorný, M.: Mathematical theory of compressible viscous fluids: Analysis and numerics. Birkhäuser–Verlag, Basel (2017)
89. Feireisl, E., Klingenberg, C., Kreml, O., Markfelder, S.: On oscillatory solutions to the complete Euler system. J. Differ. Equ. **269**(2), 1521–1543 (2020)
90. Feireisl, E., Lukáčová-Medvid'ová, M., Mizerová, H.: $\mathcal{K}-$convergence as a new tool in numerical analysis. IMA J. Numer. Anal. **40**(4), 2227–2255 (2020)
91. Feireisl, E., Lukáčová-Medvid'ová, M., She, B., Wang, Y.: Computing oscillatory solutions of the Euler system via $\mathcal{K}$-convergence. Math. Models Methods Appl. Sci. **31**, 537–576 (2021)
92. Feireisl, E., Lukáčová-Medvid'ová, M.: Convergence of a mixed finite element–finite volume scheme for the isentropic Navier-Stokes system via dissipative measure-valued solutions. Found. Comput. Math. **18**(3), 703–730 (2018)

93. Feireisl, E., Lukáčová-Medvid'ová, M., Mizerová, H.: A finite volume scheme for the Euler system inspired by the two velocities approach. Numer. Math. **144**(1), 89–132 (2020)
94. Feireisl, E., Lukáčová-Medvid'ová, M., Mizerová, H., She, B.: Convergence of a finite volume scheme for the compressible Navier-Stokes system. ESAIM: M2AN **53**(6), 1957–1979 (2019)
95. Feireisl, E., Lukáčová-Medvid'ová, M., Mizerová, H., She, B.: On the convergence of a finite volume scheme for the Navier–Stokes–Fourier system. IMA J. Numer. Anal. (2020). https://doi.org/10.1093/imanum/draa060
96. Feireisl, E., Lukáčová-Medvid'ová, M., Nečasová, Š., Novotný, A., She, B.: Asymptotic preserving error estimates for numerical solutions of compressible Navier-Stokes equations in the low Mach number regime. Multiscale Model. Simul. **16**(1), 150–183 (2018)
97. Feireisl, E., Novotný, A.: Singular limits in thermodynamics of viscous fluids. Birkhäuser-Verlag, Basel (2009)
98. Feireisl, E., Novotný, A.: Singular limits in thermodynamics of viscous fluids. Advances in Mathematical Fluid Mechanics. Birkhäuser/Springer, Cham (2017). Second edition
99. Feistauer, M.: Mathematical methods in fluid dynamics, *Pitman Monographs and Surveys in Pure and Applied Mathematics*, vol. 67. Longman Scientific & Technical, Harlow; copublished in the United States with John Wiley & Sons, Inc., New York (1993)
100. Feistauer, M., Felcman, J., Straškraba, I.: Mathematical and computational methods for compressible flow. Numerical Mathematics and Scientific Computation. The Clarendon Press, Oxford University Press, Oxford (2003)
101. Fjordholm, U.S.: High-order accurate entropy stable numerical schemes for hyperbolic conservation laws. Zürich dissertation Nr. 21025 (2013)
102. Fjordholm, U.S., Käppeli, R., Mishra, S., Tadmor, E.: Construction of approximate entropy measure valued solutions for hyperbolic systems of conservation laws. Found. Comp. Math. **17**, 763–827 (2017)
103. Fjordholm, U.S., Lye, K., Mishra, S.: Numerical approximation of statistical solutions of scalar conservation laws. SIAM J. Numer. Anal. **56**(5), 2989–3009 (2018)
104. Fjordholm, U.S., Lye, K., Mishra, S., Weber, F.: Statistical solutions of hyperbolic systems of conservation laws: numerical approximation. Math. Models Methods Appl. Sci. **30**(3), 539–609 (2020)
105. Fjordholm, U.S., Mishra, S., Tadmor, E.: Arbitrarily high-order accurate entropy stable essentially nonoscillatory schemes for systems of conservation laws. SIAM J. Numer. Anal. **50**(2), 544–573 (2012)
106. Fjordholm, U.S., Mishra, S., Tadmor, E.: On the computation of measure-valued solutions. Acta Numer. **25**, 567–679 (2016)
107. Fjordholm, U.S., Wiedemann, E.: Statistical solutions and Onsager's conjecture. Phys. D **376/377**, 259–265 (2018)
108. Fleischer, I., Porter, J.E.: Convergence of metric space-valued BV functions. Real Anal. Exchange **27**(1), 315–319 (2001/02)
109. Gajewski, H., Gröger, K., Zacharias, K.: Nichtlineare Operatorgleichungen und Operatordifferentialgleichungen. Akademie Verlag, Berlin (1974)
110. Gallouët, T., Gastaldo, L., Herbin, R., Latché, J.C.: An unconditionally stable pressure correction scheme for the compressible barotropic Navier-Stokes equations. ESAIM: M2AN **42**(2), 303–331 (2008)
111. Gallouët, T., Herbin, R., Latché, J.C.: A convergent finite element-finite volume scheme for the compressible Stokes problem. I. The isothermal case. Math. Comput. **78**(267), 1333–1352 (2009)
112. Gallouët, T., Herbin, R., Latché, J.C., Mallem, K.: Convergence of the marker-and-cell scheme for the incompressible Navier-Stokes equations on non-uniform grids. Found. Comput. Math. **18**(1), 249–289 (2018)
113. Gallouët, T., Herbin, R., Latché, J.C., Maltese, D.: Convergence of the MAC scheme for the compressible stationary Navier-Stokes equations. Math. Comp. **87**(311), 1127–1163 (2018)
114. Gallouët, T., Herbin, R., Maltese, D., Novotný, A.: Error estimates for a numerical approximation to the compressible barotropic Navier-Stokes equations. IMA J. Numer. Anal. **36**(2), 543–592 (2016)

115. Gallouët, T., Maltese, D., Novotný, A.: Error estimates for the implicit MAC scheme for the compressible Navier-Stokes equations. Numer. Math. **141**(2), 495–567 (2019)
116. Gassner, G.J., Winters, A.R., Hindenlang, F.J., Kopriva, D.A.: The BR1 scheme is stable for the compressible Navier-Stokes equations. J. Sci. Comput. **77**(1), 154–200 (2018)
117. Germain, P.: Weak-strong uniqueness for the isentropic compressible Navier-Stokes system. J. Math. Fluid Mech. **13**(1), 137–146 (2011)
118. Girault, V., Lopez, H.: Finite-element error estimates for the MAC scheme. IMA J. Numer. Anal. **16**(3), 247–379 (1996)
119. Godlewski, E., Raviart, P.A.: Numerical approximation of hyperbolic systems of conservation laws, *Applied Mathematical Sciences*, vol. 118. Springer-Verlag, New York (1996)
120. Gottlieb, S., Shu, C.W., Tadmor, E.: Strong stability-preserving high-order time discretization methods. SIAM Rev. **43**(1), 89–112 (2001)
121. Greengard, C., Thomann, E.: On DiPerna-Majda concentration sets for two-dimensional incompressible flow. Commun. Pure Appl. Math. **41**(3), 295–303 (1988)
122. Greenshields, C.J., Reese, J.M.: The structure of shock waves as a test of Brenner's modifications to the Navier-Stokes equations. J. Fluid Mech. **580**, 407–429 (2007)
123. Guermond, J.L., Popov, B.: Viscous regularization of the Euler equations and entropy principles. SIAM J. Appl. Math. **74**(2), 284–305 (2014)
124. Guermond, J.L., Popov, B.: Invariant domains and first-order continuous finite element approximation for hyperbolic systems. SIAM J. Numer. Anal. **54**(4), 2466–2489 (2016)
125. Guo, Z., Xu, K.: Numerical validation of Brenner's hydrodynamic model by force driven Poiseuille flow. Adv. Appl. Math. Mech. **1**(3), 391–401 (2009)
126. Gwiazda, P., Świerczewska-Gwiazda, A., Wiedemann, E.: Weak-strong uniqueness for measure-valued solutions of some compressible fluid models. Nonlinearity **28**(11), 3873–3890 (2015)
127. Haack, J., Jin, S., Liu, J.G.: An all-speed asymptotic-preserving method for the isentropic Euler and Navier-Stokes equations. Commun. Comput. Phys. **12**(4), 955–980 (2012)
128. Heywood, J.G., Rannacher, R.: Finite-element approximation of the nonstationary Navier-Stokes problem part iv: Error analysis for second-order time discretization. SIAM J. Numer. Anal. **27**(2), 353–384 (1990)
129. Hošek, R., She, B.: Stability and consistency of a finite difference scheme for compressible viscous isentropic flow in multi-dimension. J. Numer. Math. **26**(3), 111–140 (2018)
130. Ioriatti, M., Dumbser, M.: Semi-implicit staggered discontinuous Galerkin schemes for axially symmetric viscous compressible flows in elastic tubes. Comput. & Fluids **167**, 166–179 (2018)
131. Isett, P.: A proof of Onsager's conjecture. Ann. of Math. **188**(3), 871–963 (2018)
132. John, V.: Finite element methods for incompressible flow problems, *Springer Series in Computational Mathematics*, vol. 51. Springer, Cham (2016)
133. Jovanović, V.: An error estimate for a numerical scheme for the compressible Navier-Stokes system. Kragujevac J. Math. **30**, 263–275 (2007)
134. Jovanović, V., Rohde, C.: Error estimates for finite volume approximations of classical solutions for nonlinear systems of hyperbolic balance laws. SIAM J. Numer. Anal. **43**(6), 2423–2449 (2006)
135. Kakutani, T.: Weak convergence in uniformly convex spaces. Tôhoku Math. J. **45**, 188–193 (1938)
136. Kanschat, G.: Divergence-free discontinuous Galerkin schemes for the Stokes equations and the MAC scheme. Internat. J. Numer. Methods Fluids **56**(7), 941–950 (2008)
137. Karlsen, K.H., Karper, T.K.: A convergent nonconforming finite element method for compressible Stokes flow. SIAM J. Numer. Anal. **48**(5), 1846–1876 (2010)
138. Karlsen, K.H., Karper, T.K.: Convergence of a mixed method for a semi-stationary compressible Stokes system. Math. Comp. **80**(275), 1459–1498 (2011)
139. Karlsen, K.H., Karper, T.K.: A convergent mixed method for the Stokes approximation of viscous compressible flow. IMA J. Numer. Anal. **32(3)**, 725–764 (2012)
140. Karper, T.K.: A convergent FEM-DG method for the compressible Navier-Stokes equations. Numer. Math. **125**(3), 441–510 (2013)

141. Kelley, J.L.: General topology. Van Nostrand, Inc., Princeton (1957)
142. Komlós, J.: A generalization of a problem of Steinhaus. Acta Math. Acad. Sci. Hungar. **18**, 217–229 (1967)
143. Kröner, D.: Numerical schemes for conservation laws. Wiley-Teubner Series Advances in Numerical Mathematics. John Wiley & Sons, Ltd., Chichester; B. G. Teubner, Stuttgart (1997)
144. Kröner, D., Zajaczkowski, W.M.: Measure-valued solutions of the Euler equations for ideal compressible polytropic fluids. Math. Methods Appl. Sci. **19**(3), 235–252 (1996)
145. Kufner, A., John, O., Fučík, S.: Function spaces. Noordhoff International Publishing, Leyden (1977). Monographs and Textbooks on Mechanics of Solids and Fluids; Mechanics: Analysis
146. Kukučka, P.: On the existence of finite energy weak solutions to the Navier-Stokes equations in irregular domains. Math. Methods Appl. Sci. **32**(11), 1428–1451 (2009)
147. Lamb, H.: Hydrodynamics. Cambridge University Press, Cambridge (1932)
148. Lasis, A., Süli, E.: Poincaré-type inequalities for broken Sobolev spaces. Technical Report **03/10** (2003). Oxford University Computing Laboratory
149. Lax, P.D., Richtmyer, R.D.: Survey of the stability of linear finite difference equations. Commun. Pure Appl. Math. **9**, 267–293 (1956)
150. Lebedev, V.: Difference analogoues of orthogonal decompositions, fundamental differential operators and certain boundary-value problems of mathematical physics. Z. Vycisl. Mat. i Mat. Fiz. (in Russian) **4**, 449–465 (1964)
151. LeVeque, R.J.: Finite volume methods for hyperbolic problems. Cambridge Texts in Applied Mathematics. Cambridge University Press, Cambridge (2002)
152. Li, J., Sun, S.: The superconvergence phenomenon and proof of the MAC scheme for the Stokes equations on non-uniform rectangular meshes. J. Sci. Comput **65**, 341–362 (2015)
153. Li, J., Sun, Z.: Remark on the generalized Riemann problem method for compressible fluid flows. J. Comput. Phys. **222**(2), 796–808 (2007)
154. Lions, P.L.: Mathematical topics in fluid dynamics, Vol.2, Compressible models. Oxford Science Publication, Oxford (1998)
155. Lions, P.L., Perthame, B., Souganidis, E.: Existence and stability of entropy solutions for the hyperbolic systems of isentropic gas dynamics in Eulerian and Lagrangian coordinates. Commun. Pure Appl. Math. **49**, 599–638 (1996)
156. Majda, A.: Compressible fluid flow and systems of conservation laws in several space variables, *Applied Mathematical Sciences*, vol. 53. Springer-Verlag, New York (1984)
157. Málek, J., Nečas, J., Rokyta, M., Ružička, M.: Weak and measure-valued solutions to evolutionary PDE's. Chapman and Hall, London (1996)
158. Markfelder, S., Klingenberg, C.: The Riemann problem for the multidimensional isentropic system of gas dynamics is ill-posed if it contains a shock. Arch. Ration. Mech. Anal. **227**(3), 967–994 (2018)
159. Matsumura, A., Nishida, T.: The initial value problem for the equations of motion of viscous and heat-conductive gases. J. Math. Kyoto Univ. **20**, 67–104 (1980)
160. Matsumura, A., Nishida, T.: The initial value problem for the equations of motion of compressible and heat conductive fluids. Commun. Math. Phys. **89**, 445–464 (1983)
161. Matušů-Nečasová, Š., Novotný, A.: Measure-valued solution for non-Newtonian compressible isothermal monopolar fluid. Acta Appl. Math. **37**, 109–128 (1994)
162. Maz'ya, V.: Sobolev spaces. Springer, Berlin (1985)
163. Meister, A., Sonar, T.: Finite-volume schemes for compressible fluid flow. Surveys Math. Indust. **8**(1), 1–36 (1998)
164. Mellet, A., Vasseur, A.: On the barotropic compressible Navier-Stokes equations. Commun. Part. Diff. Eq. **32**(1-3), 431–452 (2007)
165. Mizerová, H., She, B.: Convergence and error estimates for a finite difference scheme for the multi-dimensional compressible Navier-Stokes system. J. Sci. Comput. **84**, 25 (2020)
166. Murat, F.: Compacité par compensation. Ann. Sc. Norm. Sup. Pisa, Cl. Sci. Ser. 5 **IV**, 489–507 (1978)
167. Nečas, J.: Les méthodes directes en théorie des équations elliptiques. Academia, Praha (1967)

168. Nečas, J., Novotný, A., Šilhavý, M.: Global solution to the ideal compressible heat-conductive multipolar fluid. Comment. Math. Univ. Carolinae **30**, 551–564 (1989)
169. Nečas, J., Novotný, A., Šilhavý, M.: Global solutions to the compressible isothermal multipolar fluid. J. Math. Anal. Appl. **162**, 223–241 (1991)
170. Nečas, J., Šilhavý, M.: Viscous multipolar fluids. Quart. Appl. Math. **49**, 247–266 (1991)
171. Neustupa, J.: Measure-valued solutions of the Euler and Navier-Stokes equations for compressible barotropic fluids. Math. Nachr. **163**, 217–227 (1993)
172. Nicolaides, R.A.: Analysis and convergence of the MAC scheme. I. The linear problem. SIAM J. Numer. Anal. **29**(6), 1579–1591 (1992)
173. Nicolaides, R.A., Wu, X.: Analysis and convergence of the MAC scheme. II. Navier-Stokes equations. Math. Comp. **65**(213), 29–44 (1996)
174. Novotný, A., Straškraba, I.: Convergence to equilibria for compressible Navier-Stokes equations with large data. Annali Mat. Pura Appl. **169**, 263–287 (2001)
175. Novotný, A., Straškraba, I.: Introduction to the mathematical theory of compressible flow. Oxford University Press, Oxford (2004)
176. Payne, L.E., Weinberger, H.F.: An optimal Poincaré inequality for convex domains. Arch. Rational Mech. Anal. **5**, 286–292 (1960)
177. Pedregal, P.: Parametrized measures and variational principles. Birkhäuser, Basel (1997)
178. Pedregal, P.: Optimization, relaxation and Young measures. Bull. Amer. Math. Soc. **36**, 27–58 (1999)
179. Plotnikov, P.I., Weigant, W.: Isothermal Navier-Stokes equations and Radon transform. SIAM J. Math. Anal. **47**(1), 626–653 (2015)
180. Prokhorov, Y.V.: Convergence of random processes and limit theorems in probability theory. Teor. Veroyatnost. i Primenen. **1**, 177–238 (1956)
181. Rudin, W.: Real and complex analysis. McGraw-Hill, Singapore (1987)
182. Schochet, S.: The compressible Euler equations in a bounded domain: Existence of solutions and the incompressible limit. Commun. Math. Phys. **104**, 49–75 (1986)
183. Sideris, T.C., Thomases, B., Wang, D.: Long time behavior of solutions to the 3D compressible Euler equations with damping. Commun. Part. Diff. Eq. **28**(3-4), 795–816 (2003)
184. Smoller, J.: Shock waves and reaction-diffusion equations. Springer-Verlag, New York (1967)
185. Sprung, B.: Upper and lower bounds for the Bregman divergence. J. Inequal. Appl. 2019(1): 4 (2019)
186. Sun, Y., Wang, C., Zhang, Z.: A Beale-Kato-Majda criterion for the 3-D compressible Navier-Stokes equations. J. Math. Pures Appl. **95**(1), 36–47 (2011)
187. Sun, Y., Wang, C., Zhang, Z.: A Beale-Kato-Majda criterion for three dimensional compressible viscous heat-conductive flows. Arch. Ration. Mech. Anal. **201**(2), 727–742 (2011)
188. Székelyhidi Jr., L., Wiedemann, E.: Young measures generated by ideal incompressible fluid flows. Arch. Rational Mech. Anal. **206**, 333–366 (2012)
189. Szlenk, W.: Sur les suites faiblement convergentes dans l'espace L. Studia Math. **25**, 337–341 (1965)
190. Tadmor, E.: The numerical viscosity of entropy stable schemes for systems of conservation laws. I. Math. Comp. **49**(179), 91–103 (1987)
191. Tadmor, E.: Entropy stability theory for difference approximations of nonlinear conservation laws and related time-dependent problems. Acta Numer. **12**, 451–512 (2003)
192. Tadmor, E.: Perfect derivatives, conservative differences and entropy stable computation of hyperbolic conservation laws. Discrete Contin. Dyn. Syst. **36**(8), 4579–4598 (2016)
193. Tartar, L.: Compensated compactness and applications to partial differential equations. Nonlinear Anal. and Mech., Heriot-Watt Sympos., L.J. Knopps editor, Research Notes in Math 39, Pitman, Boston pp. 136–211 (1975)
194. Toro, E.F.: Riemann solvers and numerical methods for fluid dynamics, third edn. Springer-Verlag, Berlin (2009). A practical introduction
195. Vaigant, V.A., Kazhikhov, A.V.: On the existence of global solutions to two-dimensional Navier-Stokes equations of a compressible viscous fluid (in Russian). Sibirskij Mat. Z. **36**(6), 1283–1316 (1995)

196. Valli, A.: An existence theorem for compressible viscous fluids. Ann. Mat. Pura Appl. **130**(4), 197–213 (1982)
197. Valli, A., Zajaczkowski, M.: Navier-Stokes equations for compressible fluids: Global existence and qualitative properties of the solutions in the general case. Commun. Math. Phys. **103**, 259–296 (1986)
198. Villani, C.: Optimal transport, *Grundlehren der Mathematischen Wissenschaften [Fundamental Principles of Mathematical Sciences]*, vol. 338. Springer-Verlag, Berlin (2009). Old and new
199. Wiedemann, E.: Weak-strong uniqueness in fluid dynamics. In: Partial differential equations in fluid mechanics, *London Math. Soc. Lecture Note Ser.*, vol. 452, pp. 289–326. Cambridge Univ. Press, Cambridge (2018)
200. Ziemer, W.: Weakly differentiable functions. Springer-Verlag, New York (1989)
201. Zienkiewicz, O.C., Taylor, R.L., Nithiarasu, P.: The finite element method for fluid dynamics, seventh edn. Elsevier/Butterworth Heinemann, Amsterdam (2014)

Index

E. Feireisl et al., *Numerical Analysis of Compressible Fluid Flows*, MS&A 20,
https://doi.org/10.1007/978-3-030-73788-7

N

Y

www.ingramcontent.com/pod-product-compliance
Ingram Content Group UK Ltd.
Pitfield, Milton Keynes, MK11 3LW, UK
UKHW020140300726
14059UKWH00007B/18

* 9 7 8 3 0 3 0 7 3 7 8 7 0 *